Saas-Fee Advanced Course 33

P. Schneider C. Kochanek J. Wambsganss

Gravitational Lensing: Strong, Weak and Micro

Saas-Fee Advanced Course 33

Swiss Society for Astrophysics and Astronomy
Edited by G. Meylan, P. Jetzer and P. North

With 196 Illustrations, 36 in Color

Springer

Peter Schneider

Institut für Astrophysik und
Extraterrestrische Forschung
Universität Bonn
Auf dem Hügel 71
D-53121 Bonn, Germany
peter@astro.uni-bonn.de

Joachim Wambsganss

Zentrum für Astronomie
Universität Heidelberg (ZAH)
Mönchhofstr. 12-14
D-69120 Heidelberg, Germany
jkw@ari.uni-heidelberg.de

Christopher S. Kochanek

Department of Astronomy
The Ohio State University
4055 McPherson Lab
140 West 18th Avenue
Columbus, OH 43210 USA
ckochanek@astonomy.ohio-state.edu

Volume Editors:

Georges Meylan
Pierre North

Laboratoire d'Astrophysique
Ecole Polytechnique Fédérale de
Lausanne (EPFL)
Observatoire
CH-1290 Sauverny, Switzerland

Philippe Jetzer

Institute of Theoretical Physics
Universität Zürich
Winterthurerstrasse 190
CH-8057 Zürich, Switzerland

This series is edited on behalf of the Swiss Society for Astrophysics and Astronomy:
Société Suisse d'Astrophysique et d'Astronomie
Observatoire de Genève, ch. des Maillettes 51, 1290 Sauverny, Switzerland

---

Cover picture: (Left) Matterhorn, Zermatt, Switzerland, as seen in all its usual beauty (Kurt Müller, http://photo.zermatt.ch). *(Right)* Another vision of the same mountain, as observed on 1 April 2003, while suffering from the transiant phenomenon of a passing-by black hole of one Jupiter mass (with the help of B. McLeod, CfA, Castle, and F. Summers, STScI)

Library of Congress Control Number: 2006920099

ISBN-10 3-540-30309-X Springer Berlin Heidelberg New York
ISBN-13 978-3-540-30309-1 Springer Berlin Heidelberg New York

Springer is a part of Springer Science+Business Media
springer.com

Printed in The Netherlands

Typesetting by the authors and SPI Publisher Services using a Springer LaTeX macro package

Cover design: *design & production* GmbH, Heidelberg

Printed on acid-free paper SPIN: 11568278 55/3100/SPI - 5 4 3 2 1 0

To the memory of Dennis Walsh (12 June 1933–1 June 2005) who, with his two colleagues Bob Carswell and Ray Weymann, discovered in 1979 the first extragalactic gravitational lens, the quasar QSO 0957+0561

Preface

The observation, in 1919 by A.S. Eddington and collaborators, of the gravitational deflection of light by the Sun proved one of the many predictions of Einstein's Theory of General Relativity: The Sun was the first example of a gravitational lens.

In 1936, Albert Einstein published an article in which he suggested using stars as gravitational lenses. A year later, Fritz Zwicky pointed out that galaxies would act as lenses much more likely than stars, and also gave a list of possible applications, as a means to determine the dark matter content of galaxies and clusters of galaxies.

It was only in 1979 that the first example of an extragalactic gravitational lens was provided by the observation of the distant quasar QSO 0957+0561, by D. Walsh, R.F. Carswell, and R.J. Weymann. A few years later, the first lens showing images in the form of arcs was detected.

The theory, observations, and applications of gravitational lensing constitute one of the most rapidly growing branches of astrophysics. The gravitational deflection of light generated by mass concentrations along a light path produces magnification, multiplicity, and distortion of images, and delays photon propagation from one line of sight relative to another. The huge amount of scientific work produced over the last decade on gravitational lensing has clearly revealed its already substantial and wide impact, and its potential for future astrophysical applications.

The 33rd Saas-Fee Advanced Courses of the Swiss Society for Astronomy and Astrophysics, entitled *Gravitational Lensing: Strong, Weak, and Micro*, took place from 8–12 April, 2003, in Les Diablerets, a pleasant mountain resort of the Swiss Alps. The three lecturers were Peter Schneider, Christopher S. Kochanek, and Joachim Wambsganss.

These proceedings are provided in four complementary parts of a book on gravitational lensing. P. Schneider wrote Part 1, *Introduction to Gravitational Lensing and Cosmology*, the first draft of which was made available to all registered participants a week before the course. C.S. Kochanek wrote Part 2 about *Strong Gravitational Lensing*, while P. Schneider in Part 3 dealt with

Weak Gravitational Lensing, and J. Wambsganss in Part 4 about *Gravitational Microlensing*.

We are thankful to Nicole Tharin, the secretary of the Laboratoire d'Astrophysique de l'Ecole Polytechnique Fédérale de Lausanne (EPFL), for her continuous presence and efficient help, and to Yves Debernardi for his efficient logistic support during the course. We are equally thankful to Frédéric Courbin, Dominique Sluse, Christel Vuissoz, and Alexander Eigenbrod for help in the editorial process of this book.

The meeting was also sponsored by the Université de Lausanne, the Ecole Polytechnique Fédérale de Lausanne (EPFL), the Swiss Society for Astronomy and Astrophysics, the Académie Suisse des Sciences Naturelles, the Fonds National Suisse de la Recherche Scientifique, the Space Telescope Science Institute, the Universität Zürich, and the Observatoire de Genève.

Lausanne,
July 2005

Georges Meylan
Philippe Jetzer
Pierre North

Contents

List of Previous Saas-Fee Advanced Courses

!! 2004 The Sun, Solar Analogs and the Climate
J.D. Haigh, M. Lockwood, M.S. Giampapa

!! 2003 GravitationalLensing: Strong,Weak and Micro
P. Schneider, C. Kochanek, J. Wambsganss

!! 2002 The Cold Universe
A.W. Blain, F. Combes, B.T. Draine

!! 2001 Extrasolar Planets
T. Guillot, P. Cassen, A. Quirrenbach

!! 2000 High-Energy Spectroscopic Astrophysics
S.M. Kahn, P. von Ballmoos, R.A. Sunyaev

!! 1999 Physics of Star Formation in Galaxies
F. Palla, H. Zinnecker

!! 1998 Star Clusters
B.W. Carney, W.E. Harris

!! 1997 Computational Methods for Astrophysical Fluid Flow
R.J. LeVeque, D. Mihalas, E.A. Dorfi, E. Müller

!! 1996 Galaxies Interactions and Induced Star Formation
R.C. Kennicutt, F. Schweizer, J.E. Barnes

!! 1995 Stellar Remnants
S.D. Kawaler, I. Novikov, G. Srinivasan

* 1994 Plasma Astrophysics
J.G. Kirk, D.B. Melrose, E.R. Priest

* 1993 The Deep Universe
A.R. Sandage, R.G. Kron, M.S. Longair

* 1992 Interacting Binaries
S.N. Shore, M. Livio, E.J.P. van den Heuvel

* 1991 The Galactic Interstellar Medium
W.B. Burton, B.G. Elmegreen, R. Genzel

* 1990 Active Galactic Nuclei
R. Blandford, H. Netzer, L. Woltjer

! 1989 The Milky Way as a Galaxy
G. Gilmore, I. King, P. van der Kruit

! 1988 Radiation in Moving Gaseous Media
H. Frisch, R.P. Kudritzki, H.W. Yorke

! 1987 Large Scale Structures in the Universe
A.C. Fabian, M. Geller, A. Szalay

! 1986 Nucleosynthesis and Chemical Evolution
J. Audouze, C. Chiosi, S.E. Woosley

! 1985 High Resolution in Astronomy
R.S. Booth, J.W. Brault, A. Labeyrie

! 1984 Planets, Their Origin, Interior and Atmosphere
D. Gautier, W.B. Hubbard, H. Reeves

! 1983 Astrophysical Processes in Upper Main Sequence Stars
A.N. Cox, S. Vauclair, J.P. Zahn

* 1982 Morphology and Dynamics of Galaxies
J. Binney, J. Kormendy, S.D.M. White

! 1981 Activity and Outer Atmospheres of the Sun and Stars
F. Praderie, D.S. Spicer, G.L. Withbroe

* 1980 Star Formation
J. Appenzeller, J. Lequeux, J. Silk

* 1979 Extragalactic High Energy Physics
F. Pacini, C. Ryter, P.A. Strittmatter

* 1978 Observational Cosmology
J.E. Gunn, M.S. Longair, M.J. Rees

* 1977 Advanced Stages in Stellar Evolution
I. Iben Jr., A. Renzini, D.N. Schramm

* 1976 Galaxies
K. Freeman, R.C. Larson, B. Tinsley

* 1975 Atomic and Molecular Processes in Astrophysics
A. Dalgarno, F. Masnou-Seeuws, R.V.P. McWhirter

* 1974 Magnetohydrodynamics
L. Mestel, N.O. Weiss

* 1973 Dynamical Structure and Evolution of Stellar Systems
G. Contopoulos, M. Hénon, D. Lynden-Bell

* 1972 Interstellar Matter
N.C. Wickramasinghe, F.D. Kahn, P.G. Metzger

* 1971 Theory of the Stellar Atmospheres
D. Mihalas, B. Pagel, P. Souffrin

---

* Out of print
! May be ordered from Geneva Observatory
Saas-Fee Courses
Geneva Observatory
CH-1290 Sauverny
Switzerland
!! May be ordered from Springer-Verlag

Part 1: Introduction to Gravitational Lensing and Cosmology

P. Schneider

1 Introduction

Light rays are deflected when they propagate through a gravitational field. Long suspected before General Relativity – the theory which we believe provides the correct description of gravity – it was only after Einstein's final formulation of this theory that the effect was described quantitatively. The rich phenomena which are caused by this gravitational light deflection has led to the development of the rather recent active research field of gravitational lensing, and the fact that the 2003 Saas-Fee course is entirely devoted to this subject is just but one of the indications of the prominence this topic has achieved. In fact, the activities in this area have become quite diverse and are reflected by the three main lectures of this course. The phenomena of light propagation in strong gravitational fields, as it occurs near the surface of neutron stars or black holes, are usually not incorporated into gravitational lensing – although the physics is the same, these strong-field effects require a rather different mathematical description than the weak deflection phenomena.

In this introductory first part (PART 1) we shall provide an outline of the basics of gravitational lensing, covering aspects that are at the base of it and which will be used extensively in the three main lectures. We start in Sect. 1.1 with a brief historical account; the study of the influence of a gravitational field on the propagation of light started long before the proper theory of gravity – Einstein's General Relativity – was formulated. Illustrations of the most common phenomena of gravitational lensing will be given next, before we will introduce in Sect. 2 the basic equations of gravitational lensing theory. A few simple lens models will be considered in Sect. 3, in particular the point-mass lens and the singular isothermal sphere model. Since the sources and deflectors in gravitational lensing are often located at distances comparable to the radius of the observable Universe, the large-scale geometry of space-time needs to be accounted for. Thus, in Sect. 4 we give a brief introduction to the standard model of cosmology. We then proceed in Sect. 5 with some basic

Schneider P (2006), Introduction to gravitational lensing and cosmology. In: Meylan G, Jetzer Ph and North P (eds) Gravitational lensing: Strong, weak, and micro. Saas-Fee Adv Courses vol 33, pp 1–89

DOI 10.1007/3-540-30309-X_1

considerations about lensing statistics, i.e., the question of how probable it is that observations of a source at large distance are significantly affected by a lensing effect, and conclude with a description of the large-scale matter distribution in the Universe. The material covered in this introductory part will be used extensively in the later parts of this book; those will be abbreviated as SL (Strong Lensing, Kochanek, 2005, Part 2 of this book), WL (Weak Lensing, Schneider, 2005, Part 3 of this book), and ML (MicroLensing, Wambsganss, 2005 Part 4 of this book).

Gravitational lensing as a whole, and several particular aspects of it, has been reviewed previously. Two extensive monographs (Schneider et al. 1992, hereafter SEF; Petters, Levine and Wambsganss 2001, hereafter PLW) describe lensing in great detail, in particular providing a derivation of the gravitational lensing equations from General Relativity (see also Seitz et al. 1994). Blandford and Narayan (1992) review the cosmological applications of gravitational lensing, Refsdall and Surdej (1994) and Courbin et al. (2002) discuss quasar lensing by galaxies and provide an intuitive geometrical optics approach to lensing, Fort and Mellier (1994) describe the giant luminous arcs and arclets in clusters of galaxies, Paczyński (1996) reviews the effects of gravitational microlensing in the local group, the review by Narayan and Bartelmann (1999) provides a concise account of gravitational lensing theory and observations, and Mellier (1999), Bartelmann and Schneider (2001), Wittman (2002) and van Waerbeke and Mellier (2003) review the relatively young field of weak gravitational lensing.

1.1 History of Gravitational Light Deflection

We start with a (very) brief account on the history of gravitational lensing; the reader is referred to SEF and PLW for a more detailed presentation.

The Early Years, Before General Relativity

The Newtonian theory of gravitation predicts that the gravitational force $\mathbf{F}$ on a particle of mass m is proportional to m, so that the gravitational acceleration $\mathbf{a} = \mathbf{F}/m$ is independent of m. Therefore, the trajectory of a test particle in a gravitational field is independent of its mass but depends, for a given initial position and direction, only on the velocity of the test particle. About 200 years ago, several physicists and astronomers speculated that, if light could be treated like a particle, light rays may be influenced in a gravitational field as well. John Mitchell in 1784, in a letter to Henry Cavendish, and later Johann von Soldner in 1804, mentioned the possibility that light propagating in the field of a spherical mass M (like a star) would be deflected by an angle $\hat{\alpha}_{\rm N} = 2GM/(c^2\xi)$, where G and c are Newton constant of gravity and the velocity of light, respectively, and ξ is the impact parameter of the incoming light ray. At roughly the same time, Pierre-Simon Laplace in 1795 noted "that the gravitational force of a heavenly body could be so large, that light could not flow out of it" (Laplace 1975), i.e., that the escape velocity $v_e = \sqrt{2GM/R}$ from the surface of a spherical mass M of radius R becomes the velocity of light, which happens if $R = R_s \equiv 2GM/c^2$, nowadays called the *Schwarzschild radius* of a mass M.

Gravitational Light Deflection in GR

All these results were derived under the assumption that light somehow can be considered like a massive test particle; this was of course well before the concept of photons was introduced. Only after the formulation of General Relativity by Albert Einstein in 1915 could the behavior of light in a gravitational field be studied on a firm physical ground. Before the final formulation of GR, Einstein published a paper in 1911 where he recalculated the results of Mitchell and Soldner (of whose work he was unaware) for the deflection angle. Only after the completion of GR did it become clear that the 'Newtonian' value of the deflection angle was too small by a factor of 2. In the general theory of relativity, the deflection is

$$\hat{\alpha} = \frac{4GM}{c^2 \xi} = 1''.75 \left(\frac{M}{M_\odot}\right) \left(\frac{\xi}{R_\odot}\right)^{-1} . \tag{1}$$

The deflection of light by the Sun can be measured during a total solar eclipse when it is possible to observe stars projected near the Solar surface; light deflection then slightly changes their positions. A measurement of the deflection in 1919, with a sufficient accuracy to distinguish between the 'Newtonian' and the GR value, provided a tremendous success for Einstein's new theory of gravity.

Soon thereafter, Lodge (1919) used the term 'lens' in the context of gravitational light deflection, but noted that 'it has no focal length'. Chwolson (1924) considered a source perfectly coaligned with a foreground mass, concluding that the source should be imaged as a ring around the lens – in fact, only fairly recently did it become known that Einstein made some unpublished notes on this effect in 1912 (Renn et al. 1997) – hence, calling them 'Einstein rings' is indeed appropriate. If the alignment is not perfect, two images of the background source would be visible, one on either side of the foreground star. Einstein, in 1936, after being approached by the Czech engineer Rudi Mandl, wrote a paper where he considered this lensing effect by a star, including both the image positions, their separation, and their magnifications. He concluded that the angular separation between the two images would be far too small (of order milli-arcseconds) to be resolvable, so that "there is no great chance of observing this phenomenon" (Einstein 1936).

Zwicky's Visions

This pessimistic view was not shared by Fritz Zwicky , who in 1937 published two truly visionary papers. Instead of looking at lensing by stars in our Galaxy, he considered "extragalactic nebulae" (nowadays called galaxies) as lenses – with his mass estimates of these nebulae, he estimated typical image separation of a background source to be of order $10''$ – about one order of magnitude too high – and such pairs of images *can* be separated with telescopes. Observing such an effect, he noted, would furnish an additional test of GR, allow one to see galaxies at larger distances (due to the magnification effect), and would determine the masses of these nebulae acting as lenses (Zwicky 1937a). He then went on to estimate the probability that a distant source would be lensed to produce multiple images, concluded that about 1 out of 400 distant sources should be affected by lensing (this is about the fractional area covered by the bright parts of nebulae on photographic plates), and hence predicted that "the probability that nebulae which act as gravitational lenses will be found becomes practically a certainty" (Zwicky 1937b). As we shall see in due course, basically all of Zwicky's predictions became true.[1]

The Revival of Lensing

Until the beginning of the 1960's the subject rested, but in 1963/4, three authors independently reopened the field: Klimov (1963), Liebes (1964) and Refsdal (1964a,b). Klimov considered lensing of galaxies by galaxies, whereas Liebes and Refsdal mainly studied lensing by point-mass lenses. Their papers have been milestones in lensing research; for example, Liebes considered the possibility that stars in the Milky Way can act as lenses for stars in M31 – we shall see in ML, this is a truly modern idea. Refsdal calculated the difference of the light travel times between the two images of a source – since light propagates along different paths from the source to the observer, there will in general be a time delay which can be observed provided the source is variable, such like a supernova. Refsdal pointed out that the time delay depends on the mass of the lens and the distances to the lens and the source, and concluded that, if the image separation and the time delay could be measured, the lens mass and the Hubble constant could be determined. We shall see in SL (Part 2) how these predictions have been realized in the meantime.

In 1963, the first quasars were detected: luminous, compact ('quasi-stellar') and very distant sources – hence, a source population had been discovered which lies behind Zwicky's nebulae, and finding lens systems amongst them should be a certainty. Nevertheless, it took another 15 years until the first lens system was observed and identified as such.

[1] Zwicky thought he had found a gravitational lens system and said so at a conference in the 1950s. Munch, one of his Caltech colleagues, said that if it were a lens, he'd "eat his hat". Sargent (from whom this story was communicated) found the photographic plate after Zwicky's death, hoping to improve Munch's diet, but concluded it was a plate defect.

1.2 Discoveries

First Detections of Multiple Imaging (1979)

In their program to optically identify radio sources, Walsh et al. in 1979 discovered a pair of quasars separated by about 6 arcseconds, having identical colors, redshifts ($z_s = 1.41$) and spectra (see Walsh 1989 for the history of this discovery). The year 1979 also marked two important technical developments in astronomy: the first CCD detectors replaced photographic plates, thus providing much higher sensitivity, dynamic range and linearity, and the very large array (VLA), a radio interferometer providing radio images of subarcsecond image quality, went into operation. With the VLA it was soon demonstrated that both quasar images are compact radio sources, with similar radio spectra. Soon thereafter, a galaxy situated between the two quasar images was detected (Stockton 1980; Young et al. 1980). The galaxy has a redshift of $z_d = 0.36$ and it is the brightest galaxy in a small cluster. We now know that the cluster contributes its share to the large image separation in this system. Furthermore, the first very long baseline interferometry (VLBI) data of this system, known as QSO 0957+561, showed that both components have a core-jet structure with the symmetry expected for lensed images of a common source (see Fig. 1). The great similarities of the two optical spectra (Fig. 2) is another proof of the lensing nature of this system.

One year later, the so-called triple quasar PG 1115+080 was discovered (Weymann et al. 1980). It apparently consisted of three images, one of which was much brighter than the other two (see Fig. 3). Soon thereafter it was shown that the bright image was in fact a blend of two images separated by $\sim 0\farcs5$, and thus very difficult to resolve with optical telescopes from the ground. The fact that the close pair is much brighter than the other two images is a generic prediction of lens theory, as will be shown below.

Until 1990, a few more lens systems or lens candidate systems have been discovered, some of them from a systematic search for lenses amongst radio sources (e.g., Burke et al. 1992), but most of them serendipitously (such as the one shown in Fig. 4). The 1990s then have witnessed several systematic searches for lens systems, including programs carried out with the Hubble Space Telescope (HST; Maoz et al. 1993), lens searches amongst 15,000 radio sources (JVAS and CLASS; see King et al. 1999; Browne et al. 2003), and those amongst very bright high-redshift quasars – these surveys will be detailed in SL (Part 2). By now, more than 80 multiple-image lens systems with a galaxy acting as the (main) lens are known.

Giant Luminous Arcs (1986)

In 1986, two groups (Lynds and Petrosian 1986; Soucail et al. 1987) independently pointed out the existence of strongly elongated, curved features around

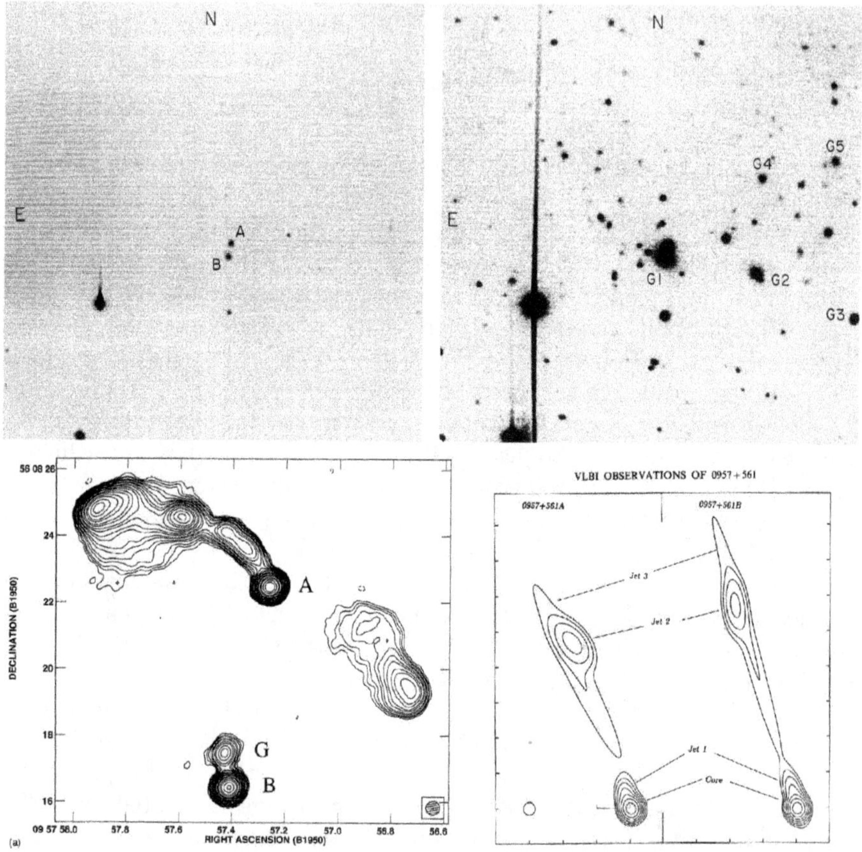

Fig. 1. The two upper panels show a short (*left*) and longer (*right*) optical exposure of the field of the double QSO 0957+561 (Young et al. 1981). In the short exposure, the two QSO images are clearly visible as a pair of point sources, separated by $\sim 6''$. The longer exposure reveals the presence of an extended source, the lens galaxy, between the two point sources, as well as a small cluster of galaxies of which the lens galaxy G1 is the brightest member. The lower left panel shows a 6 cm VLA map of the system (Harvanek et al. 1997), where besides the two QSO sources A and B, and the extended radio structure seen for image A, radio emission from the lens galaxy G is also visible. The milli-arcsecond structure of the two compact components A, B is shown in the lower right panel (Gorenstein et al. 1988a), where it is clearly seen that one VLBI jet is a linearly transformed version of the other, and they are mirror symmetric; this is predicted by any generic lens model which assigns opposite parity to the two images

two clusters of galaxies (see Figs. 5 and 6). Their tangential extent relative to the cluster center was at least ten times their radial extent, although the exact value was difficult to determine as they were not well resolved in width from the ground (HST has shown that this ratio is substantially larger than

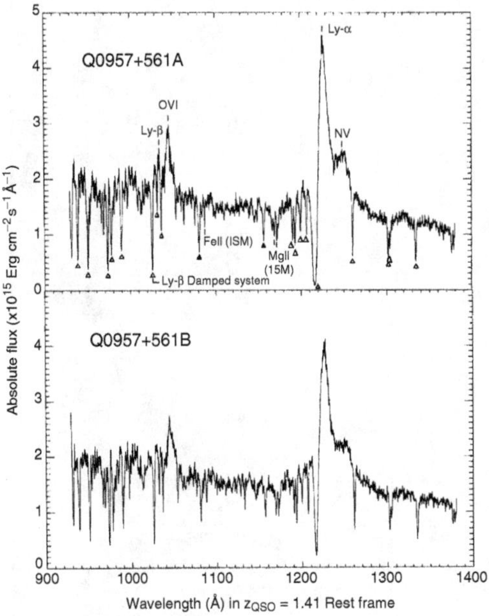

Fig. 2. Spectra of the two images of the lens system QSO 0957+561, obtained with the Faint Object Spectrograph on board HST (Michalitsianos et al. 1997). The strong similarities of the spectra, in particular the same line ratios and the identical redshift, verifies this system as a definite gravitational lens system

10:1 in many cases). These *giant luminous arcs* were seen displaced from the cluster center, and curving around it. Various hypotheses were put forward as to their nature, and all proven wrong, except for one (Paczyński 1987), when the redshift of the giant arc in A370 was measured (Soucail et al. 1988) and shown to be much larger than the redshift of the cluster. The arc was thus proven to be a highly distorted and magnified image of an otherwise normal, higher-redshift galaxy. By now, many clusters with giant arcs are known and have been investigated in detail. As with most optical studies of lenses, the high-resolution of the HST was essential to study the detailed brightness distribution of arcs and to identify multiple images by their morphology and colors. Less distorted images of background galaxies have been named *arclets* (Fort et al. 1988); they can be identified in many clusters, and they are generally stretched tangentially with respect to the cluster center. In addition, clusters can act as strong lenses also to produce multiple images of background galaxies. Some of these aspects will be covered in Sect. 4 of WL (Part 3).

Rings, After All (1988)

Whereas Einstein ring images were predicted in the case of a perfectly coaligned source with a spherically symmetric lens, the first multiple images lens systems have taught us that lenses are far from spherical – thus, the discovery of a radio ring in the source MG 1131+0456 (Hewitt et al. 1988) came as a big surprise. Unfortunately, owing to its faint optical counterpart, the lensing nature of this first system could not be proven easily, but the relative

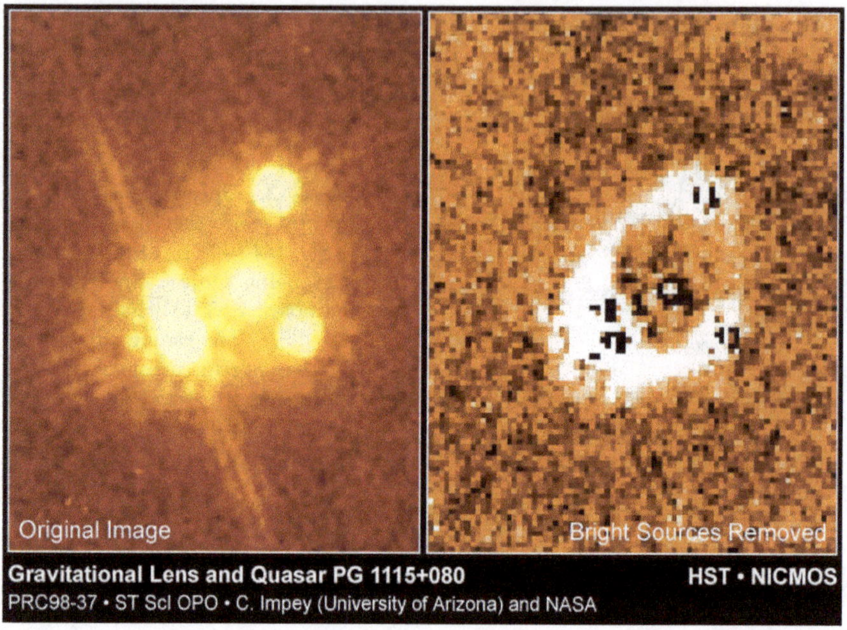

Fig. 3. In the *left panel*, a NIR image of the gravitational lens system PG 1115+080 is shown, taken with the NICMOS instrument on board HST. The QSO has a redshift of $z_s = 1.72$. The double nature of the brightest component is clearly recognized, as well as the lens galaxy with redshift $z_d = 0.31$, situated in the 'middle' of the four QSO images. When the QSO images and the lens galaxy are subtracted from the picture, the remaining image of the system (*right panel*) shows a nearly complete ring, which is the lensed image of the host galaxy of the QSO, mapped onto a nearly complete Einstein ring. In near-IR observations of lens systems, such rings occur frequently (source: C. Impey and NASA, see Impey et al. 1998)

ease by which the radio source morphology, at several frequencies, could be modeled by a simple gravitational lens (Kochanek et al. 1989) made a very strong case for its lensing nature. The second radio ring discovered (Langston et al. 1989) made a much cleaner case: of the two radio lobes of a redshift 1.72 quasar, one of them is imaged into a ring (see Fig. 7). At the center of this ring lies a bright, redshift $z_d = 0.25$ galaxy, responsible for the light deflection. High-resolution imaging with HST in optical and near-infrared filters revealed the presence of Einstein rings in many multiply imaged quasars (Fig. 8), where the host galaxy of the active nucleus is the corresponding (extended) source. We now know a lens needs not be exactly spherical; it is a combination of the asymmetry (ellipticity) of the mass distribution and the source size that determines whether we will see an Einstein ring (see SL Part 2, Sect. 10).

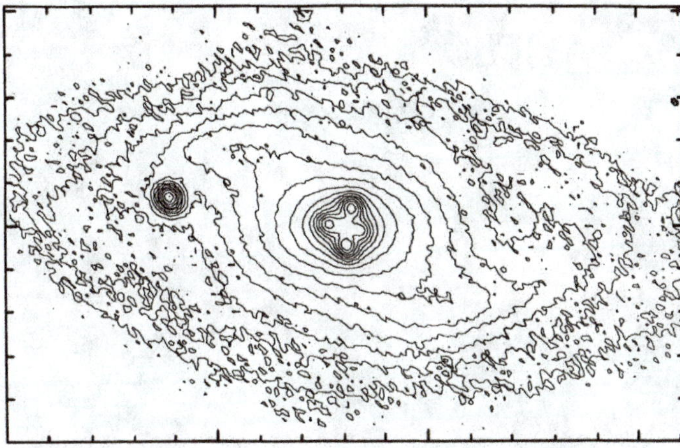

Fig. 4. Around the center of this nearby spiral galaxy ($z_d = 0.04$), four point-like sources are seen is a fairly symmetric geometry (Yee 1988). Their spectra identify them as four images of a background QSO with $z_s = 1.7$. This system, QSO 2237+0305, is the closest gravitational lens and one of the few systems where the lens is a spiral; it has been found in a spectroscopic redshift survey of nearby galaxies

Fig. 5. The giant arc in the cluster of galaxies Cl 2244−02, taken with the ISAAC instrument at the VLT (source: ESO Press Photo 46d/98). The arc has a redshift of $z_s = 2.24$, and was at the time of discovery the highest redshift normal galaxy. The high magnification caused by the gravitational lens renders this still (one of) the brightest galaxies with $z \geq 2$

Quasar Microlensing (1989)

The mass of galaxies is not distributed smoothly, since at least a fraction of it is in stars. These stars will split the (macro)images of a quasar into many microimages whose typical separations of few micro-arcseconds are unresolvable. However, these perturbations of the gravitational field change the magnification of the macroimages, provided the source is sufficiently compact. Since

Fig. 6. The cluster A2218 at $z = 0.175$ contains one of the most impressive systems of arcs, as can be seen in the multi-color images taken with the WFPC2 instrument on board HST (source: NASA/STScI). This cluster contains several multiple image systems of background galaxies which, together with the morphology of arcs, allows the construction of very detailed mass models for this cluster. Also remarkable is the thinness of several of the arcs, so that they are not resolved in width even with the HST; this implies very large length-to-width ratios of these arcs and, correspondingly, very high magnifications

the source, the lens and the observer are not stationary, and the stars in the galaxies move, this magnification will also change in time; the characteristic time-scales are of order a decade or less, and in one case (QSO 2237+0305, see Fig. 4) where the lens is very close to us ($z_{\rm d} = 0.0395$), even smaller. Hence, as predicted by Chang and Refsdal (1979, 1984), Paczyński (1986a), Kayser et al. (1986) and Schneider and Weiss (1987), this *microlensing* effect should yield flux variations of the images which are uncorrelated between the different images – an intrinsic variation of the source would affect the flux of all images in the same way, though with a time delay. In 1989, this microlensing effect was detected in the four image quasar lens QSO 2237+0305 as uncorrelated brightness variations in the four images (Irwin et al. 1989).

Weak Lensing (1990)

As mentioned before, arclets are images of background galaxies stretched by the lensing effect of a cluster. In order to identify an arclet as such, the image distortion must be significant; otherwise, owing to the intrinsic ellipticity distribution of galaxies, the stretching could not be distinguished from the intrinsic shape. However, if the distortion field varies slowly with position, then galaxy images lying close to each other should be distorted by a similar degree. Since we live in a Universe where the sky is densely covered with faint and small galaxies (e.g., Tyson 1988; Williams et al. 1996), an average over

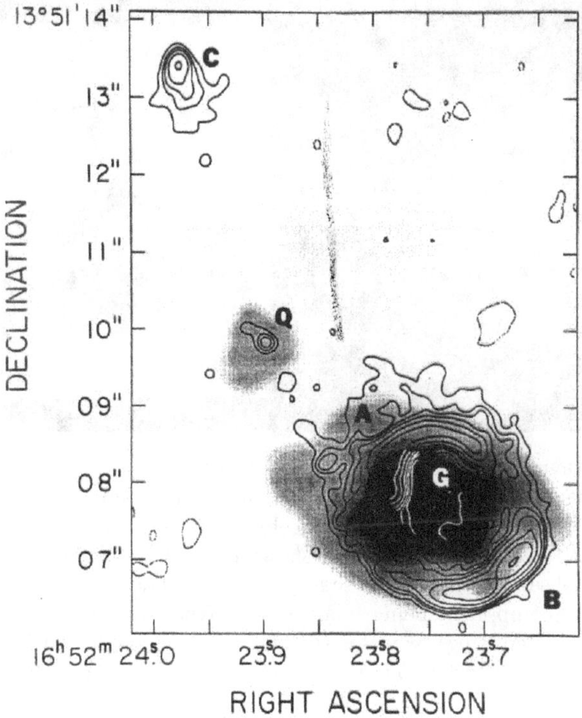

Fig. 7. The quasar MG 1654+13 at redshift $z_s = 1.72$ is shown, both as an optical image (gray scale) and in the radio (contours). The optical QSO is denoted as Q, and is the central component (or core) of a triple radio source. The Northern radio lobe is denoted by C, whereas the Southern radio lobe is mapped onto an Einstein ring. At the center of this ring, one sees a bright galaxy with spectroscopic redshift of $z_d = 0.25$. This galaxy lenses the second radio lobe into a complete Einstein ring. Within this ring, brightness peaks can be identified, and the components denoted A and B are similar to, but not multiple images of, the brightness peak in the Northern lobe C (source: G. Langston)

local ensembles of galaxies can be taken; the mean distortion of this ensemble is then a measure for the lens stretching. This *weak gravitational lensing effect* was first detected in two clusters in 1990 (Tyson et al. 1990). The advances in optical imaging cameras, in particular the availability of large mosaic CCD cameras which enable the mapping of nearly degree-sized fields in a single pointing, and the development of specific image analysis tools, have permitted the detection and quantitative analysis of weak lensing in many clusters. Even weaker lensing effects, those by an ensemble of galaxies and of the large-scale matter distribution in the Universe were discovered in 1996 (Brainerd et al. 1996) and 2000 (Bacon et al. 2000; Kaiser et al. 2000; van Waerbeke et al. 2000; Wittman et al. 2000); we shall report on this in WL (Part 3).

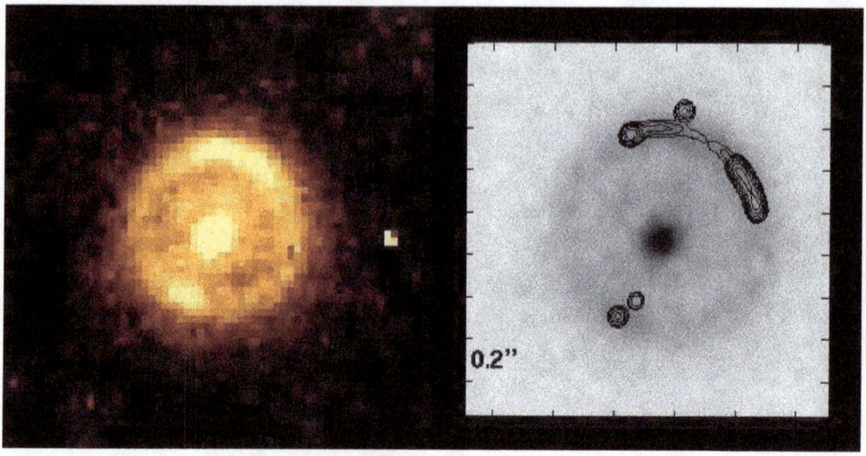

Fig. 8. The gravitational lens system B 1938+666. The *left panel* shows a NICMOS@HST image of the system, clearly showing a complete Einstein ring into which the Active Galaxy is mapped, together with the lens galaxy situated near the center of the ring. The *right panel* shows the NICMOS image as gray-scales, with the radio observations superposed as contours. The radio source is indeed a double, with one component being imaged twice (the two images just outside and just inside the Einstein ring), whereas the other source component has four images along the Einstein ring, with two of them close together (source: L.J. King, see King et al. 1998)

Time Delays ($\geq$ 1992)

Following Refsdal's idea to determine the Hubble constant from lensing by combining a good mass model for the lens with the time delay, the light curves of the first double QSO 095+561 were monitored by several groups in the optical and radio waveband (e.g., Vanderriest et al. 1989; Schild 1990; Lehár et al. 1992). From these light curves, estimates of the time delay were derived by a number of groups, and significantly different results were obtained. Difficulties include seasonal gaps in the optical light curves and the possibility of uncorrelated variability in the images due to microlensing by the lensing galaxy. To account for these effects, different methods were developed, yielding different results; broadly speaking, either delays of 410 days or 540 days were obtained. The issue was put to rest when a relatively sharp variation of the flux of the leading image was detected in December 1994 (Kundić et al. 1995; Fig. 9). Each of the two estimates for the time delay predicted a different epoch for the occurrence of the corresponding feature in the other image. With the observation of the feature in the trailing image in February 1996 (Kundić et al. 1997), the controversy was resolved in favor of the short delay, yielding 417 ± 3 days. Time delays have now been measured in 10 lens

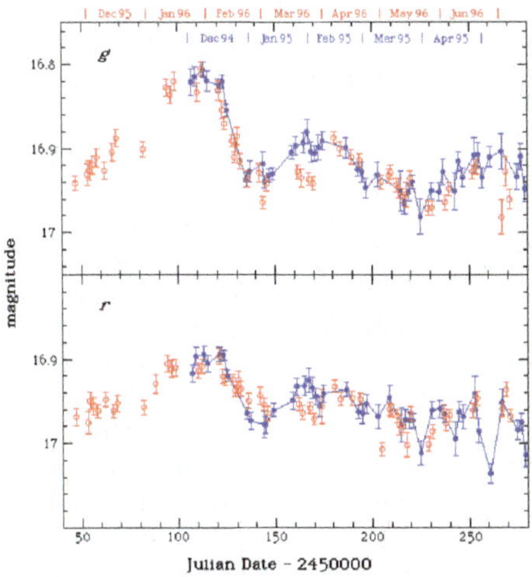

Fig. 9. Light curves of the two images of the QSO 0957+561A,B in two different filters. The two light curves have been shifted in time relative to each other by the measured time delay of 417 days, and in flux according to the flux ratio. The sharp drop measured in image A in Dec. 1994 and subsequently in image B in Feb. 1996 provides an accurate measurement of the time delay (data from Kundić et al. 1997)

systems, although the resulting estimates for the Hubble constant are still problematic – see SL (Part 2).

Galactic Microlensing (1993)

Stars in our Galaxy can act as lenses for other stars or extragalactic sources; however, the probability for this to occur is extremely small, as already noted by Liebes (1964). However, if one considers a sufficient number of background sources, even very small probabilities can be beaten. Such a lensing effect would be noted as a magnification of the background star; owing to transverse motion of source, lens and observer, the magnification changes in time and leads to a characteristic light curve. Paczyński (1986b) proposed in 1986 to monitor the brightness of stars in dense stellar fields of the Large Magellanic Cloud to search for such characteristic variability. The main idea behind this suggestion was to test whether the dark matter in the halo of our Galaxy, necessary to explain the flat rotation curve of the Milky Way (and other spiral galaxies) is made up of compact objects – brown dwarfs, neutron stars, 'Jupiters', black holes. The 'only' problem was that about 1 out of 10^7 stars in the LMC is expected to be lensed at any given time – the number of stars needed to be monitored is indeed large. Nevertheless, two groups started this adventure in the early 1990s, and reported in 1993 the first microlensing events toward the LMC (Alcock et al. 1993; Aubourg et al. 1993) (Fig. 10). Shortly thereafter, a third group announced the discovery of microlensing events toward the Galactic bulge (Udalski et al. 1993). Since then, this field has flourished, and will be covered in depth in ML. In addition to the discovery of a large number of microlensing events, these surveys provide unique data sets which are also useful for other branches of astronomy, most notably studies of stellar statistics and variability.

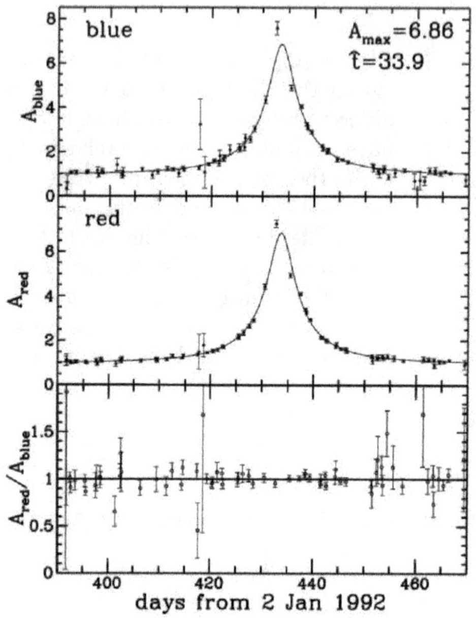

Fig. 10. Blue and red light curve of the first Galactic microlensing event MACHO-LMC-1 (Alcock et al. 1993). Data points with error bars show the measured brightness of a star in the LMC as a function of time, and the curve in both upper panels show the best fitting 'standard' microlensing lightcurve. Overall, the quality of the fit is impressive, and the lack of chromatic effects, demonstrated by the constancy of the flux ratio shown in the lowest panel, strongly argues for this being a microlensing event. However, some points (in particular one close to the maximum flux) deviate very significantly from the simple model lightcurve, indicating that this may be a binary microlens

1.3 What is Lensing Good for?

Hopefully, by the end of these lectures we will have provided convincing answers to this question, but for the impatient, we shall summarize some of the highlights of lensing applications.

Measure Mass and Mass Distributions

Gravitational light deflection is determined by the gravitational field through which light propagates. This in turn is related to the mass distribution via the Poisson equation (or its GR generalization). It is essential to realize that this simple fact implies that *gravitational light deflection is independent of the nature of the matter and of its state* – lensing is equally sensitive to dark and luminous matter, and to matter in equilibrium or far out of it. On the negative side, this implies that lensing alone cannot distinguish between these forms of matter, but on the positive side, it also cannot miss one of these matter forms. Hence, lensing is an ideal tool for measuring the total mass of astronomical bodies, dark and luminous.

From the Einstein deflection law (1), it is obvious that characteristic image separations scale with the lens mass like $M^{1/2}$; hence, the observation of multiple images and rings immediately allows an estimate of the mass of the lensing galaxy – or more precisely, the mass within a cylinder with a diameter of the image separation or the ring diameter, centered on the lens.[2] More detailed modeling, and additional observables, such as flux ratios, can yield very precise mass estimates. Indeed, as will be discussed in SL (Part 2), accurate mass estimates within galaxies, with an uncertainty of a few percent, have been achieved – by far the most precise mass determinations in (extragalactic) astronomy. Similarly, from the locations of giant arcs in clusters, the masses of the central parts of clusters can be determined (Sect. 4 of WL Part 3). With the advent of HST imaging and the discovery of multiple image systems in some strong lensing clusters, detailed mass models have been obtained, which led to very precise mass estimates in those clusters (needless to say, they confirm the dominance of dark matter in clusters).

Weak lensing studies of clusters estimate the mass distribution to much larger radii than the strong lensing regime, and, like strong lensing effects, probe for asymmetries and substructures in the cluster mass. For example, already the strong lensing properties of the cluster A2218 (Fig. 6) reveals the bimodal nature of the mass distribution. In fact, substructure in the mass distribution of lens galaxies has been detected, thereby confirming one of the robust predictions of the Cold Dark Matter model for our Universe (SL Part 2, Sect. 8). In addition, the mass distribution of galaxies at large radii, where one runs out of local dynamical tracers, can be studied statistically using an effect called galaxy–galaxy lensing (WL Part 3, Sect. 8).

Constraining the Number Density of Mass Concentrations

The probability for a lensing event to occur (e.g., the fraction of high-redshift sources that are multiply imaged, or the fraction of stars undergoing microlensing) depends on the projected number density of potential lenses. Hence, by investigating statistically well-defined samples of sources and their lensed fraction, we can infer the number density of lenses. Examples of such studies are estimates of the number density of compact objects in the dark halo of our Galaxy, the redshift evolution of the number density of galaxies acting as strong lenses, and the number density of clusters producing strong and weak lensing signals. Upper limits on the number of lensing events can also be translated into upper bounds on the number density of putative lenses: e.g., the fact that nearly all multiply-imaged sources have a visible lens galaxy puts strong upper bounds on the number density of dark lenses (they can at most provide a few percent of the galaxy-mass objects), and the non-detection of lens systems with image separations of tens of milli-arcseconds provides bounds on the number density of compact galaxies with masses $\sim 10^9 M_\odot$. In fact, by now lensing has put stringent constraints on the population of compact massive objects in the Universe over an extremely broad range of mass scales, from $\sim 10^{-3} M_\odot$ (from upper limits on the variability of distant

[2] Whereas this ‘cylinder’ contains all the mass inhomogeneities of the cosmic matter distribution between the source and the observer, it is dominated by the mass of the lensing galaxy.

quasars) to $\sim 10^{16} M_{\odot}$ (from the absence of very wide pairs of quasars), with only a few mass gaps within this range. Even lower-mass objects ($\sim 10^{-6} M_{\odot}$) can be ruled out as significant contributors to the dark matter in our Milky Way (see ML).

Providing Estimates of Cosmological Parameters

Following Refsdal's idea, the Hubble constant can be obtained from the time delay in multiple image systems. This method has the advantage of being independent of the usual distance ladder used in determinations of H_0, and it also measures the Hubble constant on a truly cosmic scale, in contrast to the quite local measurements based on Cepheid distances. Despite the determination of time delays in a number of systems, values for H_0 by lensing are burdened with the uncertainties of the lens models; however, there is a trend toward slightly lower values of the Hubble constant than obtained from Cepheids (see SL Part 2, Sect. 5). Other cosmological parameters can also be obtained from lensing. For example, the fraction of lensed high-redshift quasars when combined with the distribution of image separations can be used to estimate the cosmological model (SL Part 2, Sect. 6). Weak lensing by the large-scale structure is sensitive to the matter density parameter and the normalization of the density fluctuations, and significant constraints on these parameters have been obtained (WL Part 3, Sect. 7). In particular in combination with results from the anisotropy of the cosmic microwave background, future cosmic shear studies will provide an invaluable probe of the equation of state of the dark energy. Weak lensing has also successfully been used to determine the bias parameter, which describes the relation between the statistical distribution of galaxies and the underlying dark matter, and for which only few alternative methods are available (WL Part 3, Sect. 8).

Lenses as Natural Telescopes

Since a lens can magnify background sources, these appear brighter than they would without a lens. This makes it easier to investigate these sources in detail, e.g. through spectroscopic observations. In some cases, this magnification is even essential to detect the sources in the first place, provided their lensed brightness just exceeds the detection threshold of a survey or of the current instrumental sensitivity. This magnification effect has in fact yielded spectacular results, such as very detailed spectra of very distant galaxies, the detection of some of the highest redshift galaxies behind cluster lenses, and the detection of very faint sub-millimeter sources in cluster fields.[3] In fact,

[3] A magnification by a factor of, say, 5 implies that a spectrum of the source can be taken in 1/25-th of the time it would take to get the same signal-to-noise spectrum of the unlensed source. Needless to say that such a factor can make the difference between an observation being made and one that cannot be done.

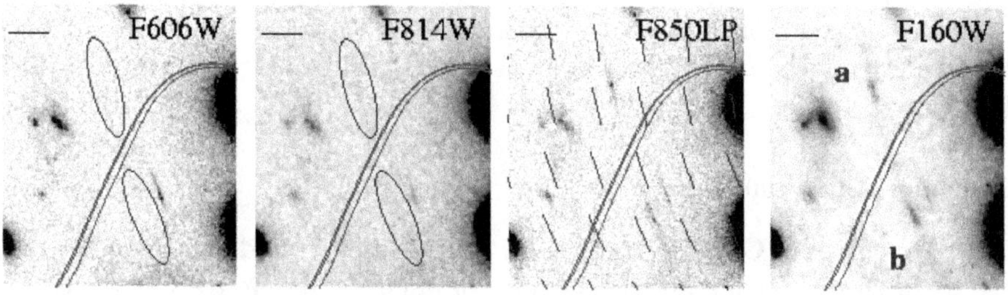

Fig. 11. Example for the use of a gravitational lens as a natural telescope. In a search for very high redshift objects, deep multi-band HST images are taken near the critical curves of clusters, where large magnifications are expected. Shown here are images of a field in the cluster A2218 (see Fig. 6) in four filters, ranging from 0.6 µm to the near-IR at 1.6 µm. In the two larger wavelength images, a double source is seen, which is absent at shorter wavelength. The two components are situated at opposite sides of the critical curve, which is drawn for three source redshifts of $z_s = 6$, 6.5 and 7; due to the large number of strong lensing constraints for this cluster, its mass distribution in the central part is very well determined. The sticks indicate the shear field of the cluster, and the elongation of the double images is parallel to this shear, as expected if they were gravitationally lensed images. From the location of the images with respect to the critical curve, and the drop-out of their flux at wavelengths shorter than $\sim 0.8\,\mu$m, the redshift of the source is estimated to be between $z_s = 6.5$ to 7 (from Kneib et al. 2004)

the lens magnification can be very large in some rare cases, but these rare cases truly stick out: some of the most extreme sources, with regards to their apparent luminosity, are strongly magnified – such as the spectacular IRAS galaxy F10214 (e.g., Broadhurst and Lehár 1995), the by-far brightest redshift ~ 3 galaxy cB58 (Seitz et al. 1998), or the extremely luminous $z = 3.87$ quasar QSO APM 08279+5255 (Irwin et al. 1998).[4] A good fraction of known galaxies with redshift larger than ~ 4 have been detected behind cluster lenses, including the redshift record holder candidate ($z = 10.0$) at present (Pelló et al. 2004); see Fig. 11 for an example. During high-magnification Galactic microlensing events, detailed spectra of stars at large distances (e.g. the Galactic bulge) have been taken. As the high magnification region crosses a distant star, observations have mapped the surface brightness distribution of the stars to test stellar atmosphere models.

With the lenses as magnifiers, larger effective angular resolution of the sources is obtained. Galaxies acting as sources for giant arcs can therefore be resolved in unprecedented detail, at least in one dimension. The host galaxy of quasars, which is difficult to study in unlensed objects owing to the large brightness contrast between the active nucleus and the surrounding host, can

[4] Such extremely bright quasars are of great importance for the study of the intergalactic absorption, e.g., the Ly-α forest; no surprise then that such objects, like the highly magnified $z = 3.62$ QSO 1422+231, are preferred targets for investigating absorption lines.

be studied much more easily when lensing allows the spatial resolution of the host – in many cases, the host galaxy is in fact mapped into an Einstein ring (see Figs. 3 and 8).

Searches for Planets

The light curves of Galactic microlensing events are affected by companions of the main lens. For example, light curves of binary stars are readily identified as such, provided their separation falls into a favorable range determined by the geometry of the lens system. Because of that, even planets will leave an observable trace in the microlensing light curves if they are situated at the right radius from the star and at the right orbital phase. Although these traces can be quite subtle, and last for a short time only, current observing campaigns aimed at the search for planets have the sensitivity for their detection, and several candidate events for the detection of planetary signals in microlensing light curves have been reported. Indeed, microlensing is considered to be the simplest (and cheapest) possibility to detect the presence of low-mass planets around distant stars (ML).

These few examples should suffice to illustrate the broad range of applications of gravitational lensing; the ever increased publication rate of articles investigating and applying gravitational lensing underlines the timeliness of the subject.

2 Gravitational Lens Theory

Assuming the validity of General Relativity, light propagates along the null geodesics of the space–time metric. However, most astrophysically relevant situations permit a much simpler approximate description of light rays, which is called gravitational lens theory. In this section, we summarize the basic equations for the description of light deflection in a gravitational field. The reader is referred to SEF and PLW for a more detailed account and further references.

2.1 The Deflection Angle

Consider first the deflection of a light ray by the exterior of a spherically symmetric mass M. Provided that the ray impact parameter ξ is much larger than the Schwarzschild radius of the mass, $\xi \gg R_\mathrm{s} \equiv 2GM\,c^{-2}$, then General Relativity predicts that the deflection angle $\hat\alpha$ is

$$\hat\alpha = \frac{4GM}{c^2\,\xi}\,. \tag{2}$$

This is just twice the value obtained in Newtonian gravity (see Sect. 1.1). According to the condition $\xi \gg R_\mathrm{s}$, the deflection angle is small, $\hat\alpha \ll 1$. This condition also implies that the Newtonian gravitational field strength is small, $\phi_N/c^2 \ll 1$.

The field equations of General Relativity can be linearized if the gravitational field is weak. The deflection angle of an ensemble of mass points is then the (vectorial) sum of the deflections due to the individual mass components. A three-dimensional mass distribution with volume density $\rho(\boldsymbol{r})$ can be divided into cells of size $\mathrm{d}V$ and mass $\mathrm{d}m = \rho(\boldsymbol{r})\,\mathrm{d}V$. Let a light ray pass this mass distribution, and describe its spatial trajectory by $(\xi_1(\lambda), \xi_2(\lambda), r_3(\lambda))$, where the coordinates are chosen such that the incoming light ray (i.e., far from the deflecting mass distribution) propagates along r_3. The actual light ray is deflected, but if the deflection angle is small, the ray can be approximated as a straight line in the neighborhood of the deflecting mass (note that this corresponds to the Born approximation in atomic and nuclear physics). A mass distribution for which this condition is satisfied is called a *geometrically-thin lens*. Then, $\boldsymbol{\xi}(\lambda) \approx \boldsymbol{\xi}$, independent of the affine parameter λ. Note that $\boldsymbol{\xi} = (\xi_1, \xi_2)$ is a two-dimensional vector. The impact vector of the light ray relative to the mass element $\mathrm{d}m$ at $\boldsymbol{r}' = (\xi'_1, \xi'_2, r'_3)$ is then $\boldsymbol{\xi} - \boldsymbol{\xi}'$, independent of r'_3, and the total deflection angle is

$$\hat{\boldsymbol{\alpha}}(\boldsymbol{\xi}) = \frac{4G}{c^2} \sum \mathrm{d}m(\xi'_1, \xi'_2, r'_3)\, \frac{\boldsymbol{\xi} - \boldsymbol{\xi}'}{|\boldsymbol{\xi} - \boldsymbol{\xi}'|^2} = \frac{4G}{c^2} \int \mathrm{d}^2\xi' \int \mathrm{d}r'_3\, \rho(\xi'_1, \xi'_2, r'_3)\, \frac{\boldsymbol{\xi} - \boldsymbol{\xi}'}{|\boldsymbol{\xi} - \boldsymbol{\xi}'|^2}\,, \tag{3}$$

which is also a two-dimensional vector. Since the last factor in (3) is independent of r'_3, the r'_3-integration can be carried out by defining the *surface mass density*

$$\Sigma(\boldsymbol{\xi}) \equiv \int \mathrm{d}r_3\, \rho(\xi_1, \xi_2, r_3)\,, \tag{4}$$

which is the mass density projected onto a plane perpendicular to the incoming light ray. Thus, the deflection angle produced by an arbitrary density distribution is

$$\hat{\boldsymbol{\alpha}}(\boldsymbol{\xi}) = \frac{4G}{c^2} \int \mathrm{d}^2\xi'\, \Sigma(\boldsymbol{\xi}')\, \frac{\boldsymbol{\xi} - \boldsymbol{\xi}'}{|\boldsymbol{\xi} - \boldsymbol{\xi}'|^2}\,, \tag{5}$$

provided that the deviation of the actual light ray from a straight (undeflected) line within the mass distribution is small compared to the scale on which the mass distribution changes significantly. This condition is satisfied in virtually all astrophysically relevant situations (i.e., lensing by galaxies and clusters of galaxies), unless the deflecting mass extends all the way from the source to the observer (a case which will be dealt with in WL Part 3). It should also be noted that in a lensing situation such as that displayed in Fig. 12, the incoming light rays are not mutually parallel, but fall within a beam with opening angle approximately equal to the angle which the mass distribution subtends on the sky. This angle, however, is typically *very* small (in the case of cluster lensing, the relevant angular scales are of order 1 arc min $\approx 3 \times 10^{-4}$ radians).

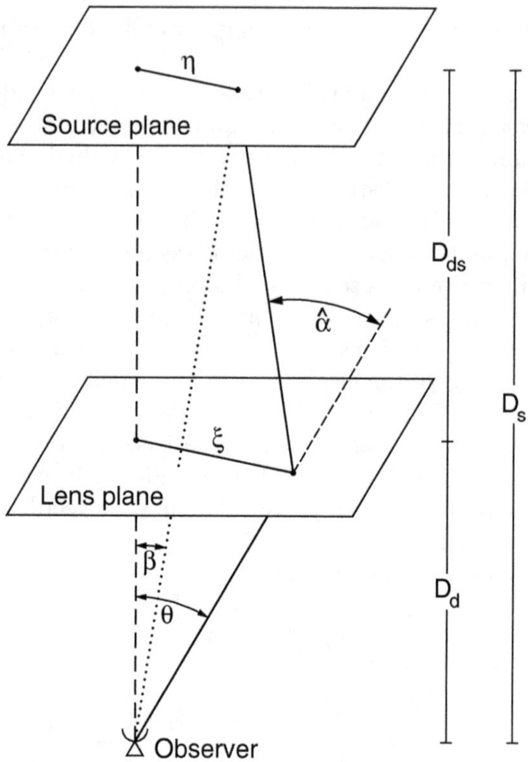

Fig. 12. Sketch of a typical gravitational lens system

2.2 The Lens Equation

A typical situation considered in gravitational lensing is sketched in Fig. 12, where a mass concentration at redshift $z_{\rm d}$ (or distance $D_{\rm d}$) deflects the light rays from a source at redshift $z_{\rm s}$ (or distance $D_{\rm s}$). If there are no other deflectors close to the line-of-sight, and if the extent of the deflecting mass along the line-of-sight is very much smaller than both $D_{\rm d}$ and the distance $D_{\rm ds}$ from the deflector to the source,[5] the actual light rays which are smoothly curved in the neighborhood of the deflector can be replaced by two straight rays with a kink near the deflector. The magnitude and direction of this kink is described by the deflection angle $\hat{\boldsymbol{\alpha}}$, which depends on the mass distribution of the deflector and the impact vector of the light ray. The lens equation relates the true position of the source to its observed position on the sky. As sketched in Fig. 12, the source and lens planes are defined as planes perpendicular to a straight line (the optical axis) from the observer to the lens at the distance

[5] This condition is very well satisfied in most astrophysical situations. A cluster of galaxies, for instance, has a typical size of a few Mpc, whereas the distances $D_{\rm d}$, $D_{\rm s}$, and $D_{\rm ds}$ are fair fractions of the Hubble length $cH_0^{-1} = 3\,h^{-1} \times 10^3$ Mpc.

of the source and of the lens, respectively. The exact definition of the optical axis does not matter because of the smallness of the angles involved in a typical lens system. Let $\boldsymbol{\eta}$ denote the two-dimensional position of the source on the source plane, measured with respect to the intersection point of the optical axis with the source plane. From Fig. 12 we can read off the geometric condition that (again making use of the smallness of angles occurring, so that $\sin \hat{\alpha} \approx \hat{\alpha} \approx \tan \hat{\alpha}$)

$$\boldsymbol{\eta} = \frac{D_{\rm s}}{D_{\rm d}} \boldsymbol{\xi} - D_{\rm ds} \hat{\boldsymbol{\alpha}}(\boldsymbol{\xi}) \ . \tag{6}$$

Introducing angular coordinates by

$$\boldsymbol{\eta} = D_{\rm s} \boldsymbol{\beta} \ \text{ and } \ \boldsymbol{\xi} = D_{\rm d} \boldsymbol{\theta} \ , \tag{7}$$

we can transform (6) to

$$\boldsymbol{\beta} = \boldsymbol{\theta} - \frac{D_{\rm ds}}{D_{\rm s}} \hat{\boldsymbol{\alpha}}(D_{\rm d}\boldsymbol{\theta}) \equiv \boldsymbol{\theta} - \boldsymbol{\alpha}(\boldsymbol{\theta}) \ , \tag{8}$$

where we defined the *scaled deflection angle* $\boldsymbol{\alpha}(\boldsymbol{\theta})$ in the last step. The interpretation of the lens equation (8) is that a source with true position $\boldsymbol{\beta}$ can be seen by an observer to be located at angular positions $\boldsymbol{\theta}$ satisfying (8). If (8) has more than one solution for fixed $\boldsymbol{\beta}$, a source at $\boldsymbol{\beta}$ has images at several positions on the sky, i.e., the lens produces multiple images. For this to happen, the lens must be 'strong'. We can express the scaled deflection angle in terms of the surface mass density as

$$\boldsymbol{\alpha}(\boldsymbol{\theta}) = \frac{1}{\pi} \int_{\mathbb{R}^2} {\rm d}^2\theta' \, \kappa(\boldsymbol{\theta}') \, \frac{\boldsymbol{\theta} - \boldsymbol{\theta}'}{|\boldsymbol{\theta} - \boldsymbol{\theta}'|^2} \ , \tag{9}$$

where we have defined the *dimensionless surface mass density* or *convergence*

$$\kappa(\boldsymbol{\theta}) := \frac{\Sigma(D_{\rm d}\boldsymbol{\theta})}{\Sigma_{\rm cr}} \quad \text{with} \quad \Sigma_{\rm cr} = \frac{c^2}{4\pi G} \frac{D_{\rm s}}{D_{\rm d} \, D_{\rm ds}} \ , \tag{10}$$

where the *critical surface mass density* $\Sigma_{\rm cr}$ depends on the distances to the source and the lens. As will be discussed later (Sect. 2.4), a mass distribution which has $\kappa \geq 1$ somewhere, i.e., $\Sigma \geq \Sigma_{\rm cr}$, produces multiple images for some source positions. Hence, $\Sigma_{\rm cr}$ is a characteristic value for the surface mass density which is the dividing line between 'weak' and 'strong' lenses.[6]

The lens equation (8) describes a mapping $\boldsymbol{\theta} \to \boldsymbol{\beta}$ from the lens plane to the source plane; for any mass distribution $\Sigma(\boldsymbol{\xi})$, this mapping can (in principle) be easily calculated. One problem of gravitational lens theory is the inversion of (8), i.e., to find all the image positions $\boldsymbol{\theta}$ for a given source

---

[6] In order to derive the foregoing equations, we have used Euclidean geometry to relate angles to length scales. We shall discuss in Sect. 4 that the equations still hold in an expanding universe, provided the distances D's are interpreted as angular diameter distances – hence, in the notation of Sect. 4, $D \equiv D^{\rm ang}$.

position $\boldsymbol{\beta}$. Since the mapping $\boldsymbol{\theta} \to \boldsymbol{\beta}$ is non-linear, the inversion of the lens equation can be carried out analytically only for very simple mass models of the lens. As the number of images $\boldsymbol{\theta}$ for a given source $\boldsymbol{\beta}$ is not known a priori, a numerical inversion is non-trivial in general; however, we shall see below that there are methods to determine the image multiplicity as a function of the source position.

The identity $\nabla \ln|\boldsymbol{\theta}| = \boldsymbol{\theta}/|\boldsymbol{\theta}|^2$, valid for any two-dimensional vector $\boldsymbol{\theta}$, shows that the (scaled) deflection angle can be written as a gradient of the *deflection potential*,

$$\psi(\boldsymbol{\theta}) = \frac{1}{\pi} \int_{\mathbb{R}^2} \mathrm{d}^2\theta' \, \kappa(\boldsymbol{\theta}') \ln|\boldsymbol{\theta} - \boldsymbol{\theta}'| \tag{11}$$

as

$$\boldsymbol{\alpha} = \nabla\psi \; , \tag{12}$$

so that the mapping $\boldsymbol{\theta} \to \boldsymbol{\beta}$ is a gradient mapping. Furthermore, using the identity $\nabla^2 \ln|\boldsymbol{\theta}| = 2\pi\delta_{\mathrm{D}}(\boldsymbol{\theta})$, where δ_{D} is the (two-dimensional) Dirac delta 'function', one obtains from (11) that

$$\nabla^2\psi = 2\kappa \; , \tag{13}$$

which is the *Poisson equation* in two dimensions. The similarity between these lensing relations and standard three-dimensional gravity (ψ corresponds to the gravitational potential ϕ_{N}, $\boldsymbol{\alpha}$ corresponds to the acceleration vector, κ corresponds to the volume mass density ρ) shall be noted.

For later purposes, we shall find it useful to define a further scalar function

$$\tau(\boldsymbol{\theta};\boldsymbol{\beta}) = \frac{1}{2}\left(\boldsymbol{\theta} - \boldsymbol{\beta}\right)^2 - \psi(\boldsymbol{\theta}) \; , \tag{14}$$

called the *Fermat potential*; this is a function of the lens plane coordinate $\boldsymbol{\theta}$, with the source position $\boldsymbol{\beta}$ entering as a parameter. It should be noted that

$$\nabla\tau(\boldsymbol{\theta};\boldsymbol{\beta}) = \mathbf{0} \tag{15}$$

is equivalent to the lens equation (8). As has been shown in Schneider (1985); see also SEF), the function $\tau(\boldsymbol{\theta};\boldsymbol{\beta})$ is, up to an affine transformation, the light travel time along a ray starting at position $\boldsymbol{\beta}$, traversing the lens plane at position $\boldsymbol{\theta}$ and arriving at the observer. Thus, (15) expresses the fact that physical light rays are those for which the light travel time is stationary – which thus expresses Fermat principle in the context of lensing by a geometrically thin matter distribution. We shall see that the Fermat potential – or time-delay function – is very useful for a classification of the multiple images in a gravitational lens system. Displaying lens properties in terms of the Fermat potential (Blandford and Narayan 1986) provides useful insight in the behavior of the lens mapping.

2.3 Magnification and Distortion

The solutions $\boldsymbol{\theta}$ of the lens equation yield the angular positions of the images of a source at $\boldsymbol{\beta}$. The shapes of the images will differ from the shape of the source because light bundles are deflected differentially, as we saw from the images of giant arcs in Fig. 6. In general, the shape of the images must be determined by solving the lens equation for all points within an extended source. Liouville theorem and the absence of emission and absorption of photons in gravitational light deflection imply that lensing conserves surface brightness (or specific intensity). Hence, if $I^{(\mathrm{s})}(\boldsymbol{\beta})$ is the surface brightness distribution in the source plane, the observed surface brightness distribution in the lens plane is

$$I(\boldsymbol{\theta}) = I^{(\mathrm{s})}[\boldsymbol{\beta}(\boldsymbol{\theta})] . \tag{16}$$

If a source is much smaller than the angular scale on which the lens properties change, the lens mapping can be linearized locally. The distortion of images is then described by the Jacobian matrix

$$\mathcal{A}(\boldsymbol{\theta}) = \frac{\partial \boldsymbol{\beta}}{\partial \boldsymbol{\theta}} = \left(\delta_{ij} - \frac{\partial^2 \psi(\boldsymbol{\theta})}{\partial \theta_i \partial \theta_j}\right) = \begin{pmatrix} 1-\kappa-\gamma_1 & -\gamma_2 \\ -\gamma_2 & 1-\kappa+\gamma_1 \end{pmatrix} , \tag{17}$$

where we have introduced the components of the *shear* $\gamma \equiv \gamma_1 + \mathrm{i}\gamma_2 = |\gamma| \mathrm{e}^{2\mathrm{i}\varphi}$,

$$\gamma_1 = \frac{1}{2}(\psi_{,11} - \psi_{,22}) , \quad \gamma_2 = \psi_{,12} , \tag{18}$$

and κ is related to ψ through Poisson equation (13). Hence, if $\boldsymbol{\theta}_0$ is a point within an image, corresponding to the point $\boldsymbol{\beta}_0 = \boldsymbol{\beta}(\boldsymbol{\theta}_0)$ within the source, we find from (16), using the locally linearized lens equation,

$$I(\boldsymbol{\theta}) = I^{(\mathrm{s})} \left[\boldsymbol{\beta}_0 + \mathcal{A}(\boldsymbol{\theta}_0) \cdot (\boldsymbol{\theta} - \boldsymbol{\theta}_0)\right] . \tag{19}$$

According to this equation, the images of a source with circular isophotes are ellipses. The ratios of the semi-axes of such an ellipse to the radius of the source are given by the inverse of the eigenvalues of $\mathcal{A}(\boldsymbol{\theta}_0)$, which are $1 - \kappa \pm |\gamma|$, and the ratio of the solid angles subtended by an image and the unlensed source is the inverse of the (absolute value of the) determinant of $\mathcal{A}$. The inverse of the Jacobian is called the magnification tensor,

$$M(\boldsymbol{\theta}) = \mathcal{A}^{-1} , \tag{20}$$

and yields the local mapping from the source to the image plane. The fluxes observed from the image and from the unlensed source are given as integrals over the brightness distributions $I(\boldsymbol{\theta})$ and $I^{(\mathrm{s})}(\boldsymbol{\beta})$, respectively, and their ratio is the *magnification* $|\mu(\boldsymbol{\theta}_0)|$. From (19), we find for the magnification of a 'small' source

$$\mu = \det M = \frac{1}{\det \mathcal{A}} = \frac{1}{(1-\kappa)^2 - |\gamma|^2} . \tag{21}$$

The images are thus distorted in shape and size. The shape distortion is due to the tidal gravitational field, described by the shear γ, whereas the magnification is caused by both isotropic focusing due to the local matter density κ and anisotropic focusing due to shear. The magnification as defined in (21) can have either sign; the sign of μ is called the *parity* of an image. Negative-parity images are mirror-symmetric images of the source. Of course, the observed fluxes of images are determined by the absolute value of μ. Since the intrinsic luminosity of sources is unknown, the magnification in a lens system is not an observable. However, the flux ratio of different images provides a direct measurement of the (absolute value of the) corresponding magnification ratio. In general, if two extended images of a source are observed, then their shapes depend on the shape of the source through $\mathcal{A}$. As the shape of the source is unknown, what can be determined from the shape of extended images is the relative magnification matrix $\mathcal{A}_{ij} = \mathcal{A}(\boldsymbol{\theta}_i)\mathcal{A}^{-1}(\boldsymbol{\theta}_j)$, which provides the linearized mapping of one image onto the other. Note that $\mathcal{A}_{ij}$ is in general not symmetric and thus has four independent components. For a pair of images with opposite parity, $\det \mathcal{A}_{ij} < 0$, and so these two images are mirror symmetric; an example of this can be seen in the VLBI images of QSO 0957+561 (see Fig. 1).

To consider the distortion of the shape of images in somewhat more detail, we shall rewrite the Jacobian in a slightly different form,

$$\mathcal{A}(\boldsymbol{\theta}) = (1-\kappa)\begin{pmatrix} 1-g_1 & -g_2 \\ -g_2 & 1+g_1 \end{pmatrix} , \tag{22}$$

where we have defined the *reduced shear*

$$g \equiv \frac{\gamma}{1-\kappa} = \frac{|\gamma|}{1-\kappa}\,\mathrm{e}^{2\mathrm{i}\varphi} . \tag{23}$$

As can be easily seen from (3), the factor $(1-\kappa)$ only yields an isotropic stretching of the image, but does not affect its shape. The reduced shear g – like γ – is considered to be a complex number, $g = g_1 + \mathrm{i}g_2$ and its components determine the change of shape between the source and the image. In particular, a circular source of unit radius is mapped onto an ellipse with axes $|(1-\kappa)(1+|g|)|^{-1}$ and $|(1-\kappa)(1-|g|)|^{-1}$, and the orientation of the ellipse is determined by the phase φ of g. As will be seen in WL (Part 3), the reduced shear is the central quantity in weak gravitational lensing.

The images of a small source (what that means depends on the context; see below) are therefore magnified by $|\mu(\boldsymbol{\theta}_i)|$, and the total magnification of a small source at position $\boldsymbol{\beta}$ is given by the sum of the magnifications over all its images,

$$\mu_{\mathrm{p}}(\boldsymbol{\beta}) = \sum_i |\mu(\boldsymbol{\theta}_i)| , \tag{24}$$

where the index ‘p’ indicates that this equation applies to the point-source limit. The magnification of real sources with finite extent is given by the weighted mean of μ_{p} over the source area,

$$\mu = \left[\int \mathrm{d}^2\beta \, I^{(\mathrm{s})}(\boldsymbol{\beta}) \right]^{-1} \int \mathrm{d}^2\beta \, I^{(\mathrm{s})}(\boldsymbol{\beta}) \, \mu_\mathrm{p}(\boldsymbol{\beta}) \,, \tag{25}$$

where $I^{(\mathrm{s})}(\boldsymbol{\beta})$ is the surface brightness profile of the source. Whereas gravitational lensing is achromatic, because the deflection of photons does not depend on their frequency, the finite resolution of observations can lead to color terms in practice, since the surface brightness distribution $I^{(\mathrm{s})}(\boldsymbol{\beta})$ can be different at different frequencies. Then, if the magnification $\mu_\mathrm{p}(\boldsymbol{\beta})$ varies on scales comparable to the source size, the magnification of an extended but unresolved source *can* depend on the frequency.

Since the shear is defined by the trace-free part of the symmetric Jacobian matrix $\mathcal{A}$, it has two independent components. There exists a one-to-one mapping from symmetric, trace-free 2×2 matrices onto complex numbers, and we shall extensively use complex notation. Note that the shear (and the reduced shear) transforms as $\mathrm{e}^{2\mathrm{i}\varphi}$ under rotations of the coordinate frame, and is therefore *not* a vector (but a polar, i.e., it has the same transformation properties as the linear polarization of electromagnetic waves). Equations (11) and (18) imply that the complex shear can be written as

$$\begin{aligned} \gamma(\boldsymbol{\theta}) &= \frac{1}{\pi} \int_{\mathbb{R}^2} \mathrm{d}^2\theta' \, \mathcal{D}(\boldsymbol{\theta} - \boldsymbol{\theta}') \, \kappa(\boldsymbol{\theta}') \,, \quad \text{with} \\ \mathcal{D}(\boldsymbol{\theta}) &\equiv \frac{\theta_2^2 - \theta_1^2 - 2\mathrm{i}\theta_1\theta_2}{|\boldsymbol{\theta}|^4} = \frac{-1}{(\theta_1 - \mathrm{i}\theta_2)^2} \,. \end{aligned} \tag{26}$$

2.4 Critical Curves and Caustics, and General Properties of Lenses

In any lens there can be closed, smooth curves, known as *critical curves*, on which the Jacobian vanishes, $\det \mathcal{A}(\boldsymbol{\theta}) = 0$. The curves in the source plane which are obtained by mapping the critical curves with the lens equation are called *caustics*, which are not necessarily smooth, but can develop cusps. Critical curves and caustics are of great importance for a qualitative understanding of the lens mapping, owing to their following properties:

1. The magnification $\mu = 1/\det \mathcal{A}$ formally diverges for an image on a critical curve. Infinite magnifications are of course unphysical. All astronomical sources have a finite size that keeps their observed magnification (25) finite. For a hypothetical source of vanishing extent, the magnification would be finite because the geometrical optics approximation then breaks down and we must use wave optics. The resulting diffraction patterns predict finite, though potentially very high magnifications (see e.g. Ohanian 1983 or Chap. 7 of SEF). Nevertheless, a source located near a caustic can produce very highly magnified images close to the corresponding critical curve in the lens plane.
2. The number of images a source produces depends on its location relative to the caustic curves. Assuming a mass profile of a lens for which the

deflection angle tends to zero for large $|\boldsymbol{\theta}|$ – as is true for all real lenses – and an upper bound to the deflection angle (i.e., excluding point-mass lenses for the moment), a source at large $|\boldsymbol{\beta}|$ will have only one image, at $\boldsymbol{\theta} \approx \boldsymbol{\beta}$, whereas it can have multiple images for small impact vectors. The lens mapping (8) is locally invertible at all locations for which $\det \mathcal{A} \neq 0$. This immediately implies that a change of the source position does not lead to the change of the number of images *unless* the source moves across a caustic – since caustics are obtained by mapping the critical curves (where the lens mapping in not invertible) onto the source plane. When a source position crosses a caustic, a pair of images near the corresponding critical curve is either created or destroyed, depending on the direction of crossing. The side of the caustic where the number of images is larger by two is often called the 'inner side'. A source close to, and on the inner side of a caustic possesses a pair of images with very high and nearly equal magnification on either side of the critical curve, in addition to any other images. The bright pair must have opposite parities because the magnification changes sign at the critical curve.

Whereas the critical curves are smooth, this does not need to be the case for caustics. To see that, let $\boldsymbol{\theta}(\lambda)$ be a parameterization of a critical curve; the caustic then is $\boldsymbol{\beta}(\boldsymbol{\theta}(\lambda))$. The tangent vector to the critical curve is the derivative $\dot{\boldsymbol{\theta}}(\lambda) \equiv \mathrm{d}\boldsymbol{\theta}(\lambda)/\mathrm{d}\lambda$, and the tangent vector to the caustic is

$$\frac{\mathrm{d}\boldsymbol{\beta}(\boldsymbol{\theta}(\lambda))}{\mathrm{d}\lambda} = \frac{\partial \boldsymbol{\beta}}{\partial \boldsymbol{\theta}} \frac{\mathrm{d}\boldsymbol{\theta}}{\mathrm{d}\lambda} = \mathcal{A}(\boldsymbol{\theta}(\lambda))\, \dot{\boldsymbol{\theta}}(\lambda) \; .$$

This vector, however, can vanish if the tangent vector to the critical curve $\dot{\boldsymbol{\theta}}(\lambda)$ is parallel to the eigenvector of $\mathcal{A}$ whose eigenvalue is 0 (remember that we are analyzing a critical curve, along which one eigenvalue of $\mathcal{A}$ is always zero). Hence, if the direction of the tangent vector to the critical curve is the singular direction of $\mathcal{A}$, the caustic curve need not be smooth; in fact, it has a *cusp*. Apart from any cusps the caustic curves are smooth curves called *fold caustics.* These names are taken from singularity theory, a mathematical discipline that studies the critical points of general mappings. We shall see the occurrence of cusps later in several specific examples of lens mappings. A source close to and inside a cusp has three highly magnified images near the corresponding point on the critical curve; one can show (see e.g. Schneider and Weiss 1992; Mao 1992) that the sum of the absolute values of the magnification of the two outer images equals the absolute value of the magnification of the central of these three images. A source just outside the cusp has one highly magnified image near the corresponding critical curve.

We thus obtain a qualitative understanding of the geometry of a lens mapping from the critical curves and caustics. The critical curves divide the lens plane into regions of positive (i.e., $\mu > 0$) and negative ($\mu < 0$) parity. The corresponding caustics divide the source plane into regions of different image multiplicity: whenever a source position changes across a caustic, the number of images changes by ± 2. Since for mass distributions without singularities (e.g., point masses) the number of images is 1 if the angular source position is sufficiently distant from the mass concentration, the number of images can be easily assigned to each of the regions in the source plane, once the caustics are known.

If an extended source is located on the caustic, either fold or cusp, the corresponding two or three images merge; only that part of the source which lies inside the inner region of the caustic is (locally) multiply imaged. Since $\det \mathcal{A} = 0$ implies that (at least) one of the two eigenvalues of $\mathcal{A}$ vanishes, the image(s) are highly distorted in the direction of the corresponding eigenvector; therefore, the image of a circular source can be very strongly elongated. This is the origin of the giant luminous arcs in clusters. From what has been said above, for a cusp the singular direction is tangent to the critical curve; hence, if an arc is produced by a source on a cusp, the direction of elongation displays approximately the local direction of the critical curve.

Types of Images

The Fermat potential $\tau(\boldsymbol{\theta}; \boldsymbol{\beta})$ introduced in Sect. 2.2 yields a convenient classification of images, according to whether an image $\boldsymbol{\theta}$ is located at a minimum, maximum, or saddle point of τ – remember that images of a source occur at points $\boldsymbol{\theta}$ for which τ is stationary. Since the Jacobian matrix is the Hessian of τ, $\mathcal{A}_{ij} = \partial^2\tau/(\partial\theta_i\partial\theta_j)$, these three types of images are distinguished by the signs of the two eigenvalues a_i of $\mathcal{A}$: At a minimum of τ, both are positive, implying that $\det \mathcal{A} > 0$, $\operatorname{tr} \mathcal{A} > 0$, whereas at a maximum, both are negative so that $\det \mathcal{A} > 0$, $\operatorname{tr} \mathcal{A} < 0$. At a saddle point, the signs of the a_i are different, so that $\det \mathcal{A} < 0$. Given that

$$\operatorname{tr} \mathcal{A} = 2(1 - \kappa) \ , \tag{27}$$

one sees that minima (maxima) occur at positions where $\kappa < 1$ ($\kappa > 1$), whereas nothing can be said about κ at saddles.

Odd-Number and Magnification Theorems

In a remarkable, one-page paper, Burke (1981) proved a theorem on the number of images a gravitational lens can produce: For a gravitational lens with a smooth surface mass density which decreases faster than $|\boldsymbol{\theta}|^{-1}$ as $|\boldsymbol{\theta}| \to \infty$, the number of images corresponding to extrema of τ, and thus to positive parity images, equals the number of saddle points plus 1, provided the source is not located on a caustic. Hence, the total number of images is odd. In addition, at least one of the images corresponds to a minimum of τ.

The proof of this theorem is obtained using the Poincaré–Hopf index theorem and can also be found in Sect. 5.4 of SEF. The fact that any source has at least one image corresponding to a minimum of τ is easily seen: the Fermat potential $\tau(\boldsymbol{\theta}; \boldsymbol{\beta})$ behaves like $|\boldsymbol{\theta}|^2/2$ for $|\boldsymbol{\theta}| \to \infty$, i.e., increases for large impact vectors, and since it is a smooth function, it must attain a minimum somewhere. In particular this implies that smooth lenses cannot make sources disappear. A simple way to see the validity of this theorem follows from what has been said above about the regions of different image multiplicity: A very misaligned source has one image, corresponding to a minimum of τ, and the number of images changes by ± 2 (one of either parity) whenever the source crosses a caustic, and thus is always odd.

As shown by Schneider (1984), a minimum image is magnified, provided $\kappa \geq 0$. This follows directly from the properties of minima,

$$0 < \det \mathcal{A} = (1 - \kappa)^2 - |\gamma|^2 < 1 ,$$

where the final inequality follows from $\operatorname{tr} \mathcal{A} > 0$. Since each source is mapped onto at least one minimum image, the positive density constraint implies that the total magnification of all sources is larger than unity; in other words, the flux of a source behind a lens is larger than the unlensed source. What may sound as a contradiction on first sight – 'all sources are magnified' (which has triggered a rich and often confusing literature on the 'flux conservation' issue) is due to the assumed positivity of the surface density κ which is certainly the case near to strong lenses. However, most lines-of-sight in the inhomogeneous Universe pass through regions which are slightly underdense relative to the homogeneous Universe, resulting in negative κ – since κ is defined as the projected mass overdensity relative to the smooth Universe. The mean magnification over the sphere of sources at given redshift indeed is unity (Weinberg 1976) if the magnification is defined relative to the flux the same source would have in a homogeneous universe of the same mean density.

These two theorems can also be generalized to the case that the deflecting matter distribution is not a geometrically-thin lens, both using heuristic arguments (SEF) or a rigorous proof (Seitz and Schneider 1992).

Necessary and Sufficient Conditions for Multiple Imaging

A matter distribution described by its dimensionless surface mass density κ may or may not be sufficiently strong to cause multiple images of sources. Two general criteria for the occurrence of multiple images can be obtained:

1. An isolated transparent lens can produce multiple images if, and only if, there is a point $\boldsymbol{\theta}$ with $\det \mathcal{A}(\boldsymbol{\theta}) < 0$. This can be shown as follows: if $\det \mathcal{A}(\boldsymbol{\theta}) > 0$ for all $\boldsymbol{\theta}$, then the lens equation is globally invertible, and so no multiple images can occur. On the other hand, if $\det \mathcal{A}(\boldsymbol{\theta}_0) < 0$ at some point $\boldsymbol{\theta}_0$, then a source at $\boldsymbol{\beta}_0 \equiv \boldsymbol{\beta}(\boldsymbol{\theta}_0)$ has an image (at $\boldsymbol{\theta}_0$) which corresponds to a saddle point; according to the odd-number theorem, there must be at least two additional images corresponding to extrema of τ.
2. A sufficient (but not necessary) condition for possible multiple images is that there exists a point $\boldsymbol{\theta}$ such that $\kappa(\boldsymbol{\theta}) > 1$. The argument is similar to the one above: if $\kappa(\boldsymbol{\theta}_0) > 1$, then the source at $\boldsymbol{\beta}_0 \equiv \boldsymbol{\beta}(\boldsymbol{\theta}_0)$ has an

image which cannot correspond to a minimum of τ, as for them $\kappa < 1$; hence, the source must have at least one additional image corresponding to a minimum.

The second criterion shows why lenses with $\kappa > 1$ are called 'strong': whereas $\kappa > 1$ is not a necessary condition for the possible occurrence of multiple images, the critical surface mass density $\Sigma_{\rm cr}$ is nevertheless the characteristic scale for the occurrence of strong lensing features likes arcs and multiple images. It should be noted that the critical surface mass density depends on the redshift (or distance) of the source; for a given physical surface mass density Σ, the lens strength increases with increasing source redshift since $\Sigma_{\rm cr}$ decreases. This also implies that the critical curves are different for sources at different redshifts; this effect is clearly seen in several clusters of galaxies where strong lensing phenomena occur at different separations from the cluster center for sources of different redshifts.

2.5 The Mass-Sheet Degeneracy

Suppose you observe a multiply-imaged source for which the image positions, their fluxes and perhaps their shapes (in the case of resolved images) can be measured. One then wants to find a mass model for the lens which can reproduce the observational constraints in order to obtain information about the mass distribution in the lens. Whereas this topic will be treated in SL (Part 2), Sect. 5, and in a somewhat different context in WL (Part 3), we can already here consider the question of how unique such models can be, even if one assumes a great number of observational constraints. A partial answer to the question is provided by the existence of the *mass-sheet degeneracy* (Falco et al. 1985; Gorenstein et al. 1988b; for the weak lensing case, see Schneider and Seitz 1995).

Let $\kappa(\boldsymbol{\theta})$ be a mass distribution which provides a good fit to the observables (i.e., image positions, flux ratios, relative image shapes in the case of extended images, etc.); then the whole family of lens models with mass distribution

$$\kappa_\lambda(\boldsymbol{\theta}) = (1-\lambda) + \lambda\,\kappa(\boldsymbol{\theta}), \tag{28}$$

provides an equally good fit to the data. The first term corresponds to adding a homogeneous surface mass density $\kappa_{\rm c} = 1-\lambda$ to the mass distribution, whereas the second term describes a rescaling of the 'original' mass distribution $\kappa(\boldsymbol{\theta})$. We shall now prove the statement made above.

The lens equation corresponding to κ_λ reads

$$\boldsymbol{\beta} = \boldsymbol{\theta} - \boldsymbol{\alpha}_\lambda(\boldsymbol{\theta}) \quad \text{with} \quad \boldsymbol{\alpha}_\lambda(\boldsymbol{\theta}) = (1-\lambda)\boldsymbol{\theta} + \lambda\boldsymbol{\alpha}(\boldsymbol{\theta})\;, \tag{29}$$

where quantities without index 'λ' correspond to the unscaled mass distribution $\kappa(\boldsymbol{\theta})$. Indeed,

$$\boldsymbol{\alpha}_\lambda(\boldsymbol{\theta}) = \nabla\psi_\lambda(\boldsymbol{\theta}) \quad \text{where} \quad \psi_\lambda(\boldsymbol{\theta}) = \frac{1-\lambda}{2}|\boldsymbol{\theta}|^2 + \lambda\psi(\boldsymbol{\theta})\;, \tag{30}$$

so that the Poisson equation (13) is satisfied, $\nabla^2\psi_\lambda = 2\kappa_\lambda$. By combining the two equations (29), one finds

$$\frac{\boldsymbol{\beta}}{\lambda} = \boldsymbol{\theta} - \boldsymbol{\alpha}(\boldsymbol{\theta}) , \tag{31}$$

so that the lens equation for the transformed mass distribution κ_λ has the same form as for the untransformed mass distribution, except that the coordinates in the source plane is multiplied by $1/\lambda$. However, this rescaling is not directly observable. As a consequence, the Jacobi matrix and the magnification behave as

$$\mathcal{A}_\lambda = \lambda\mathcal{A} \quad ; \quad \mu_\lambda = \frac{\mu}{\lambda^2} \, ; \tag{32}$$

the first of these relations then implies with (17) that $\gamma_\lambda(\boldsymbol{\theta}) = \lambda\gamma(\boldsymbol{\theta})$ and $(1-\kappa_\lambda) = \lambda(1-\kappa)$, in agreement with (28). However, the reduced shear g (23) is unchanged under the transformation, which means that the axis ratios of the elliptical images of a round source are unaffected by the transformation. In general, if nothing sets an absolute scale for the source (size or luminosity) or an absolute mass scale for the lens (e.g., from observations of its stellar dynamics), then one cannot distinguish the model described by κ from one described by κ_λ. In particular, the critical curves and the curves with $\kappa = 1$ are unaffected by the transformation (28). However, the Fermat potential transforms as

$$\tau_\lambda(\boldsymbol{\theta};\boldsymbol{\beta}) = \frac{1}{2}(\boldsymbol{\theta}-\boldsymbol{\beta})^2 - \psi_\lambda(\boldsymbol{\theta}) = \lambda\tau(\boldsymbol{\theta};\boldsymbol{\beta}/\lambda) + \text{const.} \, , \tag{33}$$

where the const. only depends on $\boldsymbol{\beta}$. As noted before, the Fermat potential is, up to an affine transformation, the light travel time from the source to the observer when passing through the lens plane at $\boldsymbol{\theta}$. Therefore, since the difference in the Fermat potential calculated at two image positions is proportional to the differences in light travel time, the mass-sheet degeneracy changes this observable time delay. If we know the value of H_0 from other cosmological observations, we can break the degeneracy and determine the absolute surface mass density of a lens. The implications of the mass-sheet degeneracy for lens determinations of the Hubble constant will be described in SL (Part 2). Furthermore, since the transformation (28) leaves the image shapes of extended sources unchanged, the weak lensing techniques to be described in WL (Part 3) are unable to break the mass-sheet degeneracy, unless magnification information can be used – see (32). In addition, the mass-sheet degeneracy can be broken if sources with different distances D_s are lensed, since for a given physical mass density Σ, the resulting convergence κ will be different for different source distances, owing to the dependence of Σ_cr on the source redshift.

Up to now we have not constrained the value of λ in (28); however, not all values are physically meaningful. For example, for some values of λ the resulting mass distribution κ_λ may attain negative values. Depending on κ, the non-negativity of the surface mass density will restrict the possible value of λ.

3 Simple Lens Models

For a general mass distribution, the deflection angle has to be obtained through numerical integration; however, for some relatively simple mass distributions, analytical expressions can be obtained. We shall introduce here a few simple mass models for lenses which turn out to be useful for understanding many of the lensing phenomena. The simplest lens models are obtained if the mass distribution is assumed to be spherically symmetric or, of relevance for lensing, if the projected mass distribution is axially symmetric, as then the lens equation reduces essentially to a one-dimensional equation. We shall consider the general properties of such lenses before specializing to two highly relevant cases, the point-mass lens, or more generally, the light deflection exterior to a spherically-symmetric mass distribution, and the isothermal sphere lens. The former one is of utmost relevance for Galactic microlensing, as will be demonstrated in ML, whereas the latter is often used as a simple prescription for the (dark) matter distribution of galaxies and clusters. Clusters and galaxies are not expected to have axisymmetric gravitational potentials; we shall consider the next simple lens models – those which have two axes of symmetry, like elliptical mass distributions – and their generic behavior next.

3.1 Axially Symmetric Lenses

An axisymmetric matter distribution is characterized by $\Sigma(\boldsymbol{\xi}) = \Sigma(|\boldsymbol{\xi}|)$, if the origin is chosen at the center of symmetry, implying $\kappa(\boldsymbol{\theta}) = \kappa(|\boldsymbol{\theta}|)$. The scaled deflection angle $\boldsymbol{\alpha}(\boldsymbol{\theta})$ is then collinear to $\boldsymbol{\theta}$, as follows from the symmetry of the situation; indeed, from (9) one obtains that

$$\begin{aligned} \boldsymbol{\alpha}(\boldsymbol{\theta}) &= \frac{\boldsymbol{\theta}}{|\boldsymbol{\theta}|^2} 2 \int_0^{|\boldsymbol{\theta}|} \mathrm{d}\theta' \, \theta' \, \kappa(\theta') \quad \text{or} \\ \hat{\boldsymbol{\alpha}}(\boldsymbol{\xi}) &= \frac{\boldsymbol{\xi}}{|\boldsymbol{\xi}|^2} \frac{4G}{c^2} \, 2\pi \int_0^{\xi} \mathrm{d}\xi' \, \xi' \, \Sigma(\xi') \equiv \frac{4GM(|\boldsymbol{\xi}|)}{c^2 \, |\boldsymbol{\xi}|^2} \boldsymbol{\xi} \, , \end{aligned} \tag{34}$$

where $M(\xi)$ is the projected mass enclosed by the circle of radius $\xi = |\boldsymbol{\xi}|$. The deflection due to a geometrically-thin axisymmetric mass distribution at a point $\boldsymbol{\xi}$ is thus the point-mass deflection angle (2) for the mass $M(|\boldsymbol{\xi}|)$ enclosed by the circle with radius $|\boldsymbol{\xi}|$. This fact is analogous to Birkhoff theorem in three-dimensional gravity which states that the gravitational force caused by a spherically-symmetric mass shell vanishes inside of it; here, the axisymmetric mass in rings causes no deflection at points within them.

Since $\boldsymbol{\alpha}$ is collinear with $\boldsymbol{\theta}$, so is $\boldsymbol{\beta}$, as seen from (8). Hence, if the source position is described by $\boldsymbol{\beta} = \beta \mathbf{e}$, where $\mathbf{e}$ is a unit vector, then $\boldsymbol{\theta} = \theta \mathbf{e}$ as well, and the lens equation becomes one-dimensional,

$$\beta = \theta - \alpha(\theta) \, , \tag{35}$$

where the deflection angle has the properties

$$\alpha(\theta) = -\alpha(-\theta) = \frac{m(\theta)}{\theta} = \bar{\kappa}(\theta)\theta \; , \tag{36}$$

where $m(\theta)$ is the dimensionless mass inside a circle of angular radius θ and $\bar{\kappa}(\theta)$ is the mean surface mass density inside of θ,

$$m(\theta) = 2\int_0^\theta \mathrm{d}\theta' \, \theta' \, \kappa(\theta') \; , \quad \bar{\kappa}(\theta) = \frac{m(\theta)}{\theta^2} \; . \tag{37}$$

For calculating the Jacobian matrix, it is useful to write the lens equation in the form

$$\boldsymbol{\beta} = \left[1 - \bar{\kappa}(|\boldsymbol{\theta}|)\right] \boldsymbol{\theta} \; ; \tag{38}$$

then, according to (17) one finds from differentiation that

$$\mathcal{A}(\boldsymbol{\theta}) = \left[1 - \bar{\kappa}(|\boldsymbol{\theta}|)\right] \mathcal{I} - \frac{\bar{\kappa}'}{|\boldsymbol{\theta}|} \begin{pmatrix} \theta_1^2 & \theta_1\theta_2 \\ \theta_1\theta_2 & \theta_2^2 \end{pmatrix} \; , \tag{39}$$

where $\mathcal{I}$ is the two-dimensional identity matrix, and $\bar{\kappa}'(\theta) \equiv \mathrm{d}\bar{\kappa}/\mathrm{d}\theta = 2\left[\kappa(\theta) - \bar{\kappa}(\theta)\right]/\theta$. Comparing (39) with the final form of (17), one sees that indeed $\mathrm{tr}\mathcal{A} = 2(1-\kappa)$, and the shear is

$$\gamma(\boldsymbol{\theta}) = \left[\kappa(\theta) - \bar{\kappa}(\theta)\right] \mathrm{e}^{2\mathrm{i}\varphi} \; , \tag{40}$$

where we set $\boldsymbol{\theta} = \theta(\cos\varphi, \sin\varphi)$; hence, the phase of the shear is the same as the polar angle of $\boldsymbol{\theta}$, as expected from symmetry. The determinant of the Jacobian matrix can be calculated either from (21) as

$$\det \mathcal{A} = (1-\kappa)^2 - |\gamma|^2 = (1-\kappa)^2 - (\bar{\kappa}-\kappa)^2 = (1-\bar{\kappa})\,(1+\bar{\kappa}-2\kappa) \; , \tag{41}$$

or, using the original definition (17) of $\mathcal{A}$ as

$$\det \mathcal{A} = \frac{\beta}{\theta}\frac{\mathrm{d}\beta}{\mathrm{d}\theta} = (1-\bar{\kappa})\,(1-\bar{\kappa}-\theta\bar{\kappa}') \; , \tag{42}$$

which can be seen, by inserting the derivative of $\bar{\kappa}$, to yield the same expression.

The fact that $\det \mathcal{A}$ factorizes allows a very simple characterization of the critical curves of these axisymmetric lenses: Critical curves, which of course are circles in this case, occur either when $1-\bar{\kappa}(\theta) = 0$, or when $1+\bar{\kappa}(\theta)-2\kappa(\theta) = 0$. The former ones are called *tangential critical curves*, the latter ones *radial critical curves*. The reason for naming them this way is found by considering the distortion of images close to these critical curves. Consider an image position on the θ_1-axis; according to (39), the Jacobian matrix is diagonal there, $\mathcal{A} = \mathrm{diag}(1+\bar{\kappa}-2\kappa, 1-\bar{\kappa})$. Near a tangential critical curve, the second eigenvalue becomes very small. If the image was a circle, the corresponding source in the source plane would be a highly flattened ellipse, with the minor axis in the β_2-direction being much smaller than the major axis. This implies that if the source is a circle, then the corresponding image near the tangential critical curve will be a highly elongated ellipse, with the highly stretched axis in the θ_2 direction, that is, tangent to the direction toward the center of the lens. Analogous reasoning shows that the image of a circular source near a radial critical curve will be strongly stretched in the radial direction. Recalling the shape and orientation of giant luminous arcs, this consideration suggests that arcs are images of (probably relatively round) sources occurring close to the tangential critical curves of the cluster lenses.

Tangential critical curves are thus characterized by the condition $\bar{\kappa} = 1$. The simplicity of this relation implies that from the location of the tangential critical curve, one can immediately determine the mass inside of it, using (37), namely $\pi\theta_{\rm E}^2\, D_{\rm d}^2\, \Sigma_{\rm cr}$, where $\theta_{\rm E}$ is the angular radius of the critical curve. The relation between $\theta_{\rm E}$ and the mass enclosed within $\theta_{\rm E}$ is

$$\theta_{\rm E} = \left(\frac{4GM}{c^2}\,\frac{D_{\rm ds}}{D_{\rm D} D_{\rm s}}\right)^{1/2} \approx 0\overset{\prime\prime}{.}9 \left(\frac{M(\le \theta_{\rm E})}{10^{12} M_\odot}\right)^{1/2} \left(\frac{D_{\rm ds}\, 1{\rm Gpc}}{D_{\rm D} D_{\rm s}}\right)^{1/2} , \tag{43}$$

where we used the definition (10) of the critical surface mass density. Thus, if giant arcs indeed trace the location of the tangential critical curve, their observation can be used to obtain a (at least approximate) mass estimate for the corresponding cluster mass inside of it (we shall come back to this issue in much mode detail in Sect. 4 of WL Part 3). The caustic corresponding to a tangential critical curve is a very special one: according to (38), the whole circle $\theta = \theta_{\rm E}$ is mapped onto the origin $\boldsymbol{\beta} = \mathbf{0}$ in the source plane: the caustic degenerates into a point. This degeneracy occurs solely due to the highly symmetric situation of the lens model; as we shall see later, any slight perturbation of the mass distribution will 'unfold' this caustic point into a curve of finite extent. This symmetric situation then leads to the following result: if a source is placed onto the caustic point, it will be imaged by the lens into a ring with radius $\theta_{\rm E}$, plus an additional image at the center of the lens with $\boldsymbol{\theta} = \mathbf{0}$. Such rings were predicted by Chwolson (1924), but already in 1911, Einstein has discussed their possible occurrence in his notebook, as shown in Renn et al. (1997). Whereas real lenses are not expected to be perfectly axisymmetric, and therefore one would also not expect to find such Einstein rings, they have indeed been detected, as shown in Fig. 7; as will be explained in Sect. 10 of SL (Part 2), the occurrence of rings depends on a combination of the mass asymmetry in a lens and the extent of the source. If the source in an axisymmetric lens is moved away from the caustic point, the ring will break up into two images, located near the Einstein radius, on opposite side of the lens center; their image separation will be $\Delta\theta \approx 2\theta_{\rm E}$.

Radial critical curves are circles where $1+\bar{\kappa}(\theta)-2\kappa(\theta) = 0$, or, equivalently, $\mathrm{d}\alpha/\mathrm{d}\theta = 1$. Their corresponding caustics are circles in the source plane. In clusters of galaxies, these radial critical curves give rise to radial arcs seen close to the cluster center, whereas they are not seen in galaxy lenses.

General Properties of Axisymmetric Lenses

One can derive several general properties of axisymmetric lenses; again we shall assume that $\alpha(\theta) \to 0$ as $|\theta| \to \infty$, that the deflection angle is bounded, $|\alpha| \le \alpha_{\rm max}$, and that $\alpha(\theta)$ is a differentiable function. Then one can show that a source with sufficiently large β has only a single image at $\theta \approx \beta$ (the validity of this property is intuitively clear, but can be proven rigorously; see SEF).

Further, a lens can produce multiple images if, and only if, there exists at least one value of θ where $\mathrm{d}\beta/\mathrm{d}\theta = 1 + \bar\kappa(\theta) - 2\kappa(\theta) < 0$. The necessity is obvious, since if $\mathrm{d}\beta/\mathrm{d}\theta \ge 0$ throughout, $\beta(\theta)$ is a monotonic function, which can be globally inverted, and no multiple solutions can occur. Sufficiency is seen as follows: if $\mathrm{d}\beta/\mathrm{d}\theta < 0$ at one point, then there must exist a pair of points such that $\mathrm{d}\beta/\mathrm{d}\theta = 0$ (note that these points lie on radial critical curves), since asymptotically for large $|\theta|$, $\mathrm{d}\beta/\mathrm{d}\theta \to 1$. Hence, $\beta(\theta)$ then has a local maximum (say at θ_1) and a local minimum (at $\theta_2 > \theta_1$), and between these two values the function $\beta(\theta)$ decreases. A source located at β_0 with $\beta(\theta_2) \le \beta_0 \le \beta(\theta_1)$ then has at least three images, one with $\theta < \theta_1$, one with $\theta > \theta_2$, and one with $\theta_1 < \theta < \theta_2$. The points $\beta(\theta_1)$ and $\beta(\theta_2)$ lie on radial caustics (see Sect. 3 of SL Part 2, for graphical illustrations of this point).

The conditions for the possible occurrence of multiple images can also be phrased in terms of the surface mass density: A necessary condition for the occurrence of multiple images is that $\kappa > 1/2$ at least at one point. This can be seen by noting that $\mathrm{d}\beta/\mathrm{d}\theta < 0$ implies $\kappa > (1+\bar\kappa)/2 > 1/2$. A sufficient condition for the possible occurrence of multiple images is $\kappa > 1$ at least at one point; this property has been shown already for a general mass distribution, and in this special situation can be seen as follows: if the maximum of κ occurs at $\theta_{\rm m}$, then $\kappa(\theta_{\rm m}) > 1$ and $\kappa(\theta_{\rm m}) \ge \bar\kappa(\theta_{\rm m})$, which implies $\mathrm{d}\beta/\mathrm{d}\theta < 0$ at $\theta_{\rm m}$, which according to the property shown before is a sufficient condition for possible multiple images.

The most useful statement on multiple imaging applies to centrally condensed lenses; those are mass distributions where $\kappa(\theta)$ does not increase with θ, or $\kappa'(\theta) \le 0$ for $\theta \ge 0$. These mass profiles are the only relevant ones in astrophysics. Centrally condensed lenses are capable of producing multiple images if, and only if, $\kappa(0) > 1$. Sufficiency was shown already. Necessity follows from this: if $\kappa(0) \le 1$, then $\bar\kappa \le 1$ for all θ; then, one finds that $\mathrm{d}\beta/\mathrm{d}\theta = (1-\bar\kappa) - \theta\bar\kappa' > 0$, since $\bar\kappa$ is also a non-increasing function of θ. Another way to phrase the multiple image condition for centrally condensed lenses is $\mathrm{d}\alpha/\mathrm{d}\theta > 1$ at the origin.

3.2 The Point-Mass Lens

Consider a point mass M or, equivalently, the outside region of a spherical mass distribution of total mass M; let the mass be located at the origin of the lens plane. Then the surface mass density is $\Sigma(\boldsymbol{\xi}) = M\delta_{\rm D}(\boldsymbol{\xi})$, and from (5) one finds for the deflection angle

$$\hat{\boldsymbol{\alpha}}(\boldsymbol{\xi}) = \frac{4GM}{c^2}\frac{\boldsymbol{\xi}}{|\boldsymbol{\xi}|^2}\,, \tag{44}$$

hence the amplitude of the deflection angle agrees with (2), and its direction is the same as that of $\boldsymbol{\xi}$, as expected from symmetry. Specializing (8) to the current lens model yields

$$\boldsymbol{\beta} = \boldsymbol{\theta} - \frac{4GMD_{\rm ds}}{c^2\,D_{\rm D}D_{\rm s}}\frac{\boldsymbol{\theta}}{|\boldsymbol{\theta}|^2} = \boldsymbol{\theta} - \theta_{\rm E}^2\frac{\boldsymbol{\theta}}{|\boldsymbol{\theta}|^2}\,, \tag{45}$$

where in the second step we have used the definition (43) of the Einstein angle which depends on the lens mass M and the distances to lens and source. If we choose without loss of generality the source position $\boldsymbol{\beta}$ to be on the positive β_1-axis, then $\boldsymbol{\theta}$ will also be on the θ_1 axis, and the lens equation becomes one-dimensional. Scaling the angles in terms of the Einstein angle as $y := \beta/\theta_{\rm E}$, $x := \theta/\theta_{\rm E}$, (45) becomes $y = x - 1/x$, with the two solutions

$$x_\pm = \frac{1}{2}\left(y \pm \sqrt{y^2+4}\right)\,, \tag{46}$$

i.e., one image on each side of the lens. Note that $x_+ \geq |x_-|$, hence the image on the same side of the lens as the source is further away from the lens than the other image. In the language of the previous section, $m(\theta) = \theta_{\rm E}^2$, $\bar{\kappa}(\theta) = (\theta_{\rm E}/\theta)^2 = x^{-2}$, so that we find from (41) the image magnification to be

$$\mu = \frac{1}{\det\mathcal{A}} = \frac{1}{1-\bar{\kappa}^2} = \left(1 - \frac{1}{x^4}\right)^{-1}\,. \tag{47}$$

As seen from (46), $x_+ \geq 1$, and so $\mu(x_+) \equiv \mu_+ \geq 1$. On the other hand, the magnification of the second image can be rather small if x_- becomes small. The magnification of the two images and the total magnification of the source is

$$\mu_\pm = \pm\frac{1}{4}\left[\frac{y}{\sqrt{y^2+4}} + \frac{\sqrt{y^2+4}}{y} \pm 2\right]\,, \quad \mu_{\rm p} = \mu_+ + |\mu_-| = \frac{y^2+2}{y\sqrt{y^2+4}}\,; \tag{48}$$

hence, unless $y \lesssim 1$, the secondary image will be strongly demagnified. The image separation

$$\Delta\theta = 2\theta_{\rm E}\sqrt{1+y^2/4} \gtrsim 2\theta_{\rm E} \tag{49}$$

is therefore only slightly larger than $2\theta_{\rm E}$ in relevant cases, since for values of $y \gtrsim 1$ the secondary image will be demagnified below the detection threshold. The sum of the two magnifications is $\mu_{\rm p} \approx 1.34$ for $y = 1$.

The magnification formally diverges for $x = 1$, or $\theta = \theta_{\rm E}$, which justifies using the same name as for the tangential critical curve in Sect. 3.1. But what about the odd-number theorem (see Sect. 2.4)? Remember, for its validity the smoothness of the mass distribution was assumed, but a point-mass lens is not smooth; in particular, the deflection potential ψ has a logarithmic spike at the origin. In fact, one can easily picture where the third image has been 'lost': Assume one would smear out the mass M over a small but finite region (say in the shape of a Gaussian), the central surface mass density κ_0 would be very high but finite; in that case, there would be a maximum of the Fermat potential close to the center (the exact position depending on the location of the source), hence the third image would appear there. Its magnification $\mu_3 \approx (\kappa_0 - 1)^{-2} \ll 1$ would then be very small.

3.3 The Singular Isothermal Sphere

A simple lens model which applies, at least to first order, to the lensing properties of galaxies and clusters is the so-called *singular isothermal sphere* (SIS). This spherical mass distribution yields flat rotation curves, such as are observed for spiral galaxies. Their density distribution is described by

$$\rho(r) = \frac{\sigma_v^2}{2\pi G r^2} \, . \tag{50}$$

Physically this model corresponds to a distribution of self-gravitating particles where the velocity distribution at all radii is a Maxwellian with one-dimensional velocity dispersion σ_v (hence, the term 'isothermal'). The three-dimensional velocity dispersion is $\sqrt{3}\sigma_v$, and the Keplerian rotation velocity (i.e., the velocity of particles on a circular orbit) is $v_{\rm c} = \sqrt{2}\sigma_v$.

The mass distribution (50) has two pathological properties: the central density diverges as $\rho \propto r^{-2}$ (hence the name 'singular'), and the total mass of this distribution diverges as $r \to \infty$. The former feature can be cured by introducing a finite core radius, whereas the distribution for large r does not affect the lensing properties at smaller radii. In SL (Part 2) and WL (Part 3) we shall discuss the constraints lensing provides on the core radius of galaxy and cluster lenses.

The SIS Lens Model

For the reasons just mentioned, the singular isothermal sphere is often used as a mass model for gravitational lenses; its surface mass density $\Sigma(\xi)$ follows from projection of (50) along the line-of-sight,

$$\Sigma(\xi) = \int_{-\infty}^{\infty} {\rm d}r_3 \, \rho\left(\sqrt{\xi^2 + r_3^2}\right) = \frac{\sigma_v^2}{2G} \, \xi^{-1} \, . \tag{51}$$

As will be shown immediately, the Einstein radius of this lens model is

$$\theta_{\rm E} = 4\pi \left(\frac{\sigma_v}{c}\right)^2 \frac{D_{\rm ds}}{D_{\rm s}} \, , \tag{52}$$

in terms of which one obtains

$$\kappa(\theta) = \frac{\theta_{\rm E}}{2|\theta|} \; ; \quad \bar\kappa(\theta) = \frac{\theta_{\rm E}}{|\theta|} \; ; \quad |\gamma|(\theta) = \frac{\theta_{\rm E}}{2|\theta|} \; ; \quad \alpha(\theta) = \theta_{\rm E} \, \frac{\theta}{|\theta|} \; ; \tag{53}$$

note that the magnitude of α is constant. Here we use the same notation as introduced before (34). The fact that $\bar\kappa(\theta_{\rm E}) = 1$ shows that $\theta_{\rm E}$ is the tangential critical curve of the SIS. The lens equation then reads

$$\beta = \theta - \theta_{\rm E} \, \frac{\theta}{|\theta|} \; , \quad \text{or} \quad y = x - \frac{x}{|x|} \; , \tag{54}$$

where the second form employs the scaled angles $x = \theta/\theta_{\rm E}$, $y = \beta/\theta_{\rm E}$. As before, we set $y \ge 0$; then, for $y < 1$, there are two images, at $x_+ = y + 1$ and $x_- = y - 1$, i.e., on opposite sides of the lens center, with image separation $\Delta\theta = 2\theta_{\rm E}$. For $y > 1$, only one image occurs, at $x_+ = y + 1$. x_+ corresponds to a minimum of the Fermat potential, whereas x_- to a saddle point, so that the subscripts denote the parity of the two images. The magnification can be calculated from (41), noting that $\bar\kappa = 2\kappa$, so that

$$\mu = \frac{1}{\det \mathcal{A}} = \frac{1}{1 - \bar\kappa} = \frac{|x|}{|x| - 1} \; ; \tag{55}$$

hence, since $x_+ > 1$, $\mu_+ > 1$, whereas the secondary image, with $|x_-| < 1$, can be strongly demagnified as $x_- \to 0$, or $y \to 1$. From (40) we find that $|\gamma(x)| = \kappa(x) = 1/(2x)$; thus, images are stretched in the tangential direction by a factor $|\mu|$, whereas the distortion factor in the radial direction is unity. The total magnification of a point source is $\mu_{\rm p} = 2/y$ for $y \le 1$, and $(1 + y)/y$ for $y \ge 1$.

Again, what about the odd-number theorem? As was true for the point-mass lens, the mass distribution of the SIS is not smooth, so the theorem does not apply. Another 'strange' property of the SIS is that the number of images changes by ± 1 when the source position crosses the circle $y = 1$ – this is in apparent conflict to what we said in Sect. 2.4. Both of these effects are due to the singular mass distribution as $\theta \to 0$, which causes $|\boldsymbol{\alpha}|$ to be constant. If we smoothed out the central mass singularity, by introducing a small but finite core, then the deflection angle would be constant, except very close to the center where it would make a smooth transition from $-\theta_{\rm E}$ for $\theta < 0$ to $+\theta_{\rm E}$ for $\theta > 0$. In this transition region, there will be two points (at $\theta = \pm\theta_{\rm r}$) where $\mathrm{d}\alpha/\mathrm{d}\theta = 1$, corresponding to a radial critical curve. The corresponding caustic circle will have radius $\beta_{\rm r} \lesssim \theta_{\rm E}$. A source with $|\beta| < \beta_{\rm r}$ will have three images, one at $x \approx y + 1$, one with $x \approx y - 1$ and one in the inner core region, whereas a source with $|\beta| > \beta_{\rm r}$ has just one. The lens equation maps the small circle with radius $\theta_{\rm r}$ onto the circle $\beta_{\rm r} \approx \theta_{\rm E}$. When we now let the core radius go to zero, $\beta_{\rm r} \to \theta_{\rm E}$, the magnification of the central image $\mu_3 \to 0$, and the central region of the lens that is mapped onto $\beta_{\rm r} \approx \theta_{\rm E}$ decreases to zero area. Hence, this limit process suggests that one can consider the third image to be present, located at $\theta = 0$, and having zero magnification.

3.4 Non-Symmetric Lenses

To describe the mass profile of real lenses, more complicated (and realistic) radial mass profiles can be used; even though the lens equation may no longer be analytically solvable, the fact that it is one-dimensional renders numerical investigations simple. The qualitative features of (centrally-condensed) axisymmetric lenses do not depend strongly on the details of the radial profile and can basically be read-off from the corresponding Young diagram (see Sect. 3 of SL Part 2).

Breaking the symmetry leads to qualitatively new properties of the lens. Most obvious of them, the central caustic point gets unfolded into a curve of finite size; a source situated inside this curve can then have five images. The fact that many of the observed lens systems have four images (i.e., five minus the one being invisible probably due to very strong demagnification at the center) shows that the axisymmetric models are definitely not sufficient to explain them.

The next more complicated gravitational potential is then one with two lines of symmetry, such as an ellipse has. Hence, one would be tempted to consider mass distributions where κ is constant on (confocal) ellipses. In fact, Bourassa et al. (1973), Bourassa and Kantowski (1975), and later Schramm (1990) have considered the lensing properties of such elliptical lenses – they turn out to be fairly complicated analytically in general; nevertheless, for some of the most relevant radial density profiles, explicit expressions for the deflection angle can be derived (e.g., Kormann et al. 1994; Keeton and Kochanek 1998; see SL Part 2), and such elliptical mass models are generally used for fitting observed lens systems. Here we consider a simpler class of lens models with similar symmetry, namely axisymmetric matter distributions with an external perturbation, henceforth called 'quadrupole lenses'.

Quadrupole Lenses

Even if the mass distribution of a lens is axisymmetric (like that of a star), the corresponding gravitational potential is not expected to share this symmetry, because lenses are typically not isolated: a galaxy is often situated inside or near a group of galaxies, and the other member galaxies, and the dark-matter halo of the group, will perturb the symmetry of the potential. In many cases of astrophysical interest, like the one just mentioned, the perturbing gravitational field changes very little over the relevant length scale of the main lens. As an example, consider a lens galaxy in a cluster of galaxies. The relevant length scale of the galaxy is about the region where the multiple images occur, i.e., a region with radius of the Einstein radius, or typically $1''$. In contrast, the relevant length scale of the cluster perturbation is either the separation of the galaxy from the cluster center, or the Einstein radius of the cluster, whatever is larger, and thus typically much larger than $1''$. It is thus natural to expand the deflection potential of the perturber about the center of the main deflector; the lowest-order, non-trivial term in the expansion is the quadratic term (tidal field). The analogous situation occurs for a star in a distant galaxy, where the symmetry of the point-mass lens is broken by the tidal field of its host galaxy; this is the situation considered by Chang and Refsdal (1984); see ML).

Here we study the lens action of an axisymmetric matter distribution which is perturbed by a larger-scale gravitational field, and we assume that the latter is locally (that is, over the region where we want to study the lensing properties of the main deflector) well described by its second-order Taylor approximation. Choosing the origin to be the center of the main lens, and the orientation of the coordinate system such that the Hessian of the deflection potential of the perturber (or tidal matrix) is diagonal at the origin, then the deflection caused by the perturber can be written as

$$\boldsymbol{\alpha}_{\rm p}(\boldsymbol{\theta}) = \boldsymbol{\alpha}_{\rm p}(\mathbf{0}) + \begin{pmatrix} \kappa_{\rm p} + \gamma_{\rm p} & 0 \\ 0 & \kappa_{\rm p} - \gamma_{\rm p} \end{pmatrix} \boldsymbol{\theta} \,, \tag{56}$$

where the surface mass density and shear of the perturber are labeled with subscript 'p'. Note that the strength of the perturbation is not assumed to be small. The lens equation then reads

$$\boldsymbol{\beta} = \left[1 - \bar{\kappa}(|\boldsymbol{\theta}|)\right] \boldsymbol{\theta} - \begin{pmatrix} \kappa_{\rm p} + \gamma_{\rm p} & 0 \\ 0 & \kappa_{\rm p} - \gamma_{\rm p} \end{pmatrix} \boldsymbol{\theta} \,, \tag{57}$$

where we have translated the origin in the source plane by the vector $\boldsymbol{\alpha}_{\rm p}(\mathbf{0})$. The perturber thus adds a uniform sheet of matter plus an *external shear*. The uniform sheet can be transformed away, recalling our discussion of the mass-sheet degeneracy in Sect. 2.5; indeed, (57) can be rewritten as

$$\hat{\boldsymbol{\beta}} := \frac{\boldsymbol{\beta}}{1 - \kappa_{\rm p}} = \begin{pmatrix} 1 - g_{\rm p} & 0 \\ 0 & 1 + g_{\rm p} \end{pmatrix} \boldsymbol{\theta} - \hat{\bar{\kappa}}(|\boldsymbol{\theta}|)\boldsymbol{\theta} \,, \tag{58}$$

where $g_{\rm p} = \gamma_{\rm p}/(1 - \kappa_{\rm p})$ is the reduced shear of the perturber, $\hat{\boldsymbol{\beta}}$ the rescaled source coordinate, and $\hat{\bar{\kappa}}(|\boldsymbol{\theta}|) = \bar{\kappa}(|\boldsymbol{\theta}|)/(1 - \kappa_{\rm p})$ the rescaled surface mass density. We shall in the following discard the hats on the variables in (58).

Although an axisymmetric lens with an external shear is too simple to represent real lenses, the resulting lens equation is sufficiently simple to allow some analytical progress; for didactic purposes, we shall discuss this lens model in somewhat more detail. The lens equation now is two-dimensional, and therefore more complicated to invert (i.e., to find all image positions for a given source position) than in the axisymmetric case. However, the lens equation can be recast into a one-dimensional equation, by introducing polar coordinates $\boldsymbol{\theta} = \theta(\cos\varphi, \sin\varphi)$ in the lens plane; then, (58) can be written as

$$\cos\varphi = \frac{\beta_1}{\theta\left[1 - \bar{\kappa}(\theta) - g_{\rm p}\right]} \quad , \quad \sin\varphi = \frac{\beta_2}{\theta\left[1 - \bar{\kappa}(\theta) + g_{\rm p}\right]} \,, \tag{59}$$

and by adding the squares of these two equations,

$$\theta^2 \left[(1-\bar\kappa)^2 - g_\mathrm{p}^2 \right]^2 - \beta_1^2 (1-\bar\kappa+g_\mathrm{p})^2 - \beta_2^2 (1-\bar\kappa-g_\mathrm{p})^2 = 0 \,, \tag{60}$$

the polar angle φ has been eliminated: (60) is an equation for θ only and can be solved numerically. For each solution θ, the polar angle can be calculated from (59). Not all solutions will have $|\cos\varphi| \le 1$ and $|\sin\varphi| \le 1$; those solutions θ have been generated by a number of algebraic manipulations needed to arrive at (60) and thus shall be discarded then. We just saw a nice example of reducing the effective dimension of a problem to make it more tractable.

The Jacobian for the quadrupole lens can be obtained from its definition (17), and its determinant reads

$$\det\mathcal{A} = (1-\bar\kappa)^2 - g_\mathrm{p}^2 - \theta\bar\kappa' \left(1-\bar\kappa+g_\mathrm{p}\cos 2\varphi\right) \,, \tag{61}$$

so that the critical curves can be easily calculated: for each value of θ, the condition $\det\mathcal{A} = 0$ yields a value for $\cos 2\varphi$; if this lies between ± 1, one has found a pair (θ, φ) of coordinates on the critical curve; in fact, one has obtained four different critical points, one in each quadrant of the lens plane, due to the symmetry of our lens model with respect to both reflections $(\theta_1, \theta_2) \mapsto \pm(\pm\theta_1, \pm\theta_2)$. Hence, the structure of critical curves and caustics for quadrupole lenses can be easily investigated, at least numerically.

The Non-Singular Isothermal Sphere with External Shear

We now consider a specific example of a quadrupole lens which has frequently been used in lens modeling: the perturbed non-singular isothermal sphere, for which

$$\kappa(\theta) = \frac{\theta_\mathrm{e}}{\theta_\mathrm{c}} \left(1 + \frac{\theta^2}{2\theta_\mathrm{c}^2}\right) \left(1 + \frac{\theta^2}{\theta_\mathrm{c}^2}\right)^{-3/2} \,, \quad \bar\kappa(\theta) = \frac{\theta_\mathrm{e}}{\sqrt{\theta^2+\theta_\mathrm{c}^2}} \,, \tag{62}$$

so that $\kappa(0) = \theta_\mathrm{e}/\theta_\mathrm{c}$ is finite. The complex form of $\kappa(\theta)$ is chosen so that the deflection profile is simple. We note that for $\theta \gg \theta_\mathrm{c}$, the mass distribution approaches that of an SIS with Einstein angle θ_e, but for a finite θ_c, θ_e is *not* the location of the critical curve, but in general, $\theta_\mathrm{E} = \left(\theta_\mathrm{e}^2 - \theta_\mathrm{c}^2\right)^{1/2}$ for $\theta_\mathrm{c} < \theta_\mathrm{e}$; otherwise, the lens is not critical. For this lens model, the one-dimensional form (60) of the lens equation can be even further simplified, by noting that (62) implies $\theta^2 = \theta_\mathrm{e}^2/\bar\kappa^2 - \theta_\mathrm{c}^2$. Inserting this expression into (60), one obtains after multiplying by $\bar\kappa^2$ an equation which is a sixth-order polynomial in $\bar\kappa$. Given that standard methods are known (e.g., Press et al. 1992) to find all solutions of polynomials, this latter form is much more useful; the roots of this polynomial are potential solution if they are real, and have $0 < \bar\kappa \le \theta_\mathrm{e}/\theta_\mathrm{c}$; those solutions can then be inserted into the original lens equation to check whether they are actual solutions.

The critical curves are found from (61), where now the relation $\theta\bar{\kappa}' = -\theta^2\bar{\kappa}^3/\theta_e^2$ can be used to replace $\bar{\kappa}'$ there. The equation $\det\mathcal{A} = 0$ can then be written in the form $\cos 2\varphi = f(\bar{\kappa})$, where f is a function of $\bar{\kappa}$ only. Thus for all radial coordinates $\theta > 0$, or equivalently, for all $0 < \bar{\kappa} \le \theta_e/\theta_c$ one can determine $f(\bar{\kappa})$; a value with $|f(\bar{\kappa})| \le 1$ yields four critical points at the radius corresponding to this mean surface mass density.

The critical curves of this special lens ("NIS plus external shear") can be studied analytically; because of its importance for understanding lens geometry, we shall provide a detailed description of the essential features in the following and illustrate the results in Fig. 13. First to note is that if the core radius is too large, or the central surface mass density too low, there is no critical curve. One finds that $\det\mathcal{A}(\mathbf{0}) = (1-\theta_e/\theta_c)^2 - g_p^2$ at the center, and can show that for $x_c \equiv \theta_c/\theta_e > (1-g_p)^{-1}$ no critical curves exist. In other words, for $\kappa_0 = 1/x_c < 1 - g_p$ the lens is not critical. Compare this condition with the one for an unperturbed lens ($g_p = 0$); there, in order for the lens to become critical, κ_0 must be larger than unity. If the core radius satisfies $(1+g_p)^{-1} < x_c < (1-g_p)^{-1}$, there is a single closed critical curve (see upper left panel in Fig. 13), and the corresponding caustic has two cusps. Owing to the shape of the caustic curve, one often calls it a lips caustic. A source located inside the caustic has three images, whereas one outside has a single image.

At $x_c = (1+g_p)^{-1}$, the Jacobian vanishes again at the origin, and for smaller values of the core radius, $x_c < (1+g_p)^{-1}$, there are two critical curves and caustics, as seen in the upper right panel of Fig. 13. A second lips caustic is located inside the first one, oriented perpendicular to it. Sources inside both caustics now have five images, and those inside the outer one but outside the inner one have three. When the core radius is further decreased, the two critical curves approach each other at two points, and correspondingly, the cusps of the inner lips caustic approach the outer caustic. At $x_c = (1-g_p)^{1/2}(1+g_p)^{-3/2}$, the critical curves and cusps merge, and for smaller values of x_c, there are again two separate critical curves and two caustics, but now, as shown in the lower left panel of Fig. 13, one of the caustics has four cusps, the other has none.[7] Two of these cusps lie inside the other caustic, the other two fall outside of it; these are called 'naked cusps'.[8]

[7] We have discussed folds and cusps before; one can show from singularity theory that these are the only two 'generic' singularities that occur in a lens mapping. However, if one considers a family of lens models, such as done here by varying x_c, higher-order singularities can occur. At the corresponding values of the lens model parameter – here x_c – the topology of critical curves can change. Examples are the creation of lips singularities, or the 'exchange of cusps' just mentioned, which technically speaking corresponds to a hyperbolic umbilic. Chapter of SEF provides a general description of singularities and their metamorphoses in lens mappings.

[8] Lensing geometries where a source is located inside a naked cusp, producing three bright images, are probably seen in clusters, indicating a relatively large core radius for them.

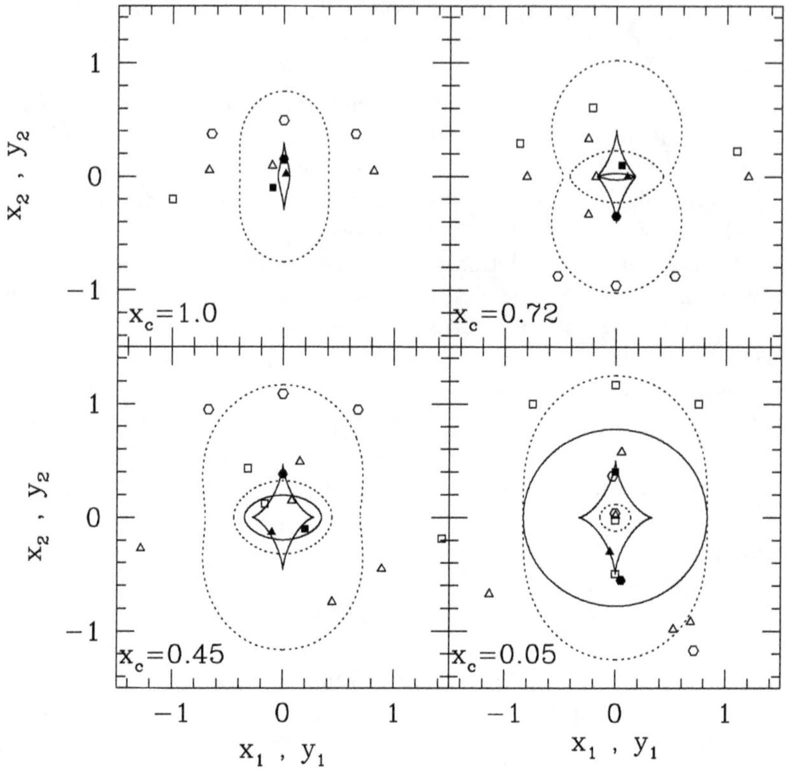

Fig. 13. Critical curves (*dashed*) and caustics (*solid curves*) for a non-singular isothermal sphere model with external shear. Angles in the source and lens plane have been scaled by $\theta_{\rm e}$, i.e., $\boldsymbol{\theta} = \theta_{\rm e}\mathbf{x}$, $\boldsymbol{\beta} = \theta_{\rm e}\mathbf{y}$. In all panels, the reduced shear is $g_{\rm p} = 0.2$. The four possible configurations are shown: for $x_{\rm c} > (1-g_{\rm p})^{-1}$, no critical curve exists; panel (**a**) shows a case with $(1+g_{\rm p})^{-1} < x_{\rm c} < (1-g_{\rm p})^{-1}$, for which a single critical curve exists, created from the previous case through a lips catastrophe. In panel (**b**), the case $(1-g_{\rm p})^{1/2}(1+g_{\rm p})^{-3/2} < x_{\rm c} < (1+g_{\rm p})^{-1}$ is shown, for which two critical curves exist, the second one created from the previous case by another lips catastrophe. The two corresponding caustics have two cusps each. In panel (**c**), $x_{\rm c} < (1-g_{\rm p})^{1/2}(1+g_{\rm p})^{-3/2}$; there, one caustic with four cusps, and one caustic without cusps occur. This case is obtained from the previous one through two hyperbolic umbilics where the two cusps of the inner caustic in (**b**) were transferred to the outer caustic; correspondingly, at that point the two critical curves intersect on the x_1-axis in this transition. Finally, panel (**d**) shows the same case as panel (**c**), except that now the two 'naked cusps' – the cusps outside the other caustic – are inside the other caustic. In addition, in all panels we have positioned three sources, indicated by the *filled triangle*, the *filled square* and the *filled hexagon*, together with their corresponding images, shown with the corresponding *open symbols*

The caustic with the cusps, also called the astroid or tangential caustic, corresponds to the outer critical curve, the one without cusp to the inner or radial critical curve. Decreasing $x_{\rm c}$ further, the inner critical curve decreases in size, whereas the corresponding caustic increases and finally completely encompasses the astroid caustic (Fig. 13, lower right panel).

In the figure, we have also illustrated the image locations (open symbols) for several source positions (filled symbols). As will be discussed in SL (Part 2), many of these image configurations have actually been observed. For galaxy lensing, configurations of the type shown in the lower right panel are most relevant since galaxies seem to have a small core radius. In that case, one of the images is located very close to the center of the lens where κ is much larger than unity, and therefore the magnification is very small – which is the canonical explanation for the absence of an observed odd image. Furthermore we see that in this case, the characteristic maximum image separation is $\Delta\theta \sim 2\theta_{\rm e}$, as expected. A source close to and inside a cusp produces three (highly magnified) images lying close together near the corresponding critical curve; in the case of a naked cusp, these are the only images of the source, whereas if the cusp lies inside the other caustic, two additional images are formed (one of which may be highly demagnified).

If the core radius is decreased to zero, the inner critical curve shrinks to zero size, the corresponding caustic becomes a circle with radius $\theta_{\rm e}$, and the number of images changes by ± 1 when a source crosses this curve – the behavior is identical to the one already encountered in the discussion of the SIS model. The other critical curve attains a simple parametric form,

$$\cos(2\varphi) = \frac{1-\left(1-g_{\rm p}^2\right)x}{g_{\rm p}} \qquad \Longleftrightarrow \qquad x = \frac{1-g_{\rm p}\cos(2\varphi)}{1-g_{\rm p}^2} \;, \tag{63}$$

describing a single closed curve around the origin, which is in fact an ellipse. By inserting this parameterized form into the lens equation and eliminating the parameter φ results in the equation describing the astroid caustic,

$$\left[2y_1^2\left(\frac{1+g_{\rm p}}{g_{\rm p}}\right)^2\right]^{1/3} + \left[2y_2^2\left(\frac{1-g_{\rm p}}{g_{\rm p}}\right)^2\right]^{1/3} = 2 \;, \tag{64}$$

from which the locations of the cusps can be read off. In particular, (64) shows that the size of the astroid caustic increases with increasing $g_{\rm p}$ and, to first order, its linear size is $\propto g_{\rm p}$.

In the limit $g_{\rm p} \to 0$, the two critical curves become circles, with the outer (inner) one being the tangential (radial) critical curves. Because of that, one often uses the same names for the critical curves also in the perturbed case $g_{\rm p} \neq 0$. The radial caustic then separates the three-image region from the single-image region in the source plane, and the tangential caustic degenerates into a single point. The fact that this point unfolds in the presence of a perturbation is nicely illustrated by (64).

General Discussion of 'Elliptical' Lenses

Mass distributions with elliptical isodensity contours are needed to realistically model gravitational lens systems. Although such models are considerably more difficult to handle analytically, their qualitative properties are similar to the NIS with external shear that was discussed above. In particular, the evolution of the critical curves and the caustics as a function of 'lens strength' or 'core size' for these models is the same as that shown in Fig. 13; the same is true for the properties regarding multiple imaging. Of course that does not mean that the choice of the lens model is arbitrary: for systems with sufficiently detailed observational constraints, a quantitative modeling technique can distinguish between the various classes of models; as we shall see in SL (Part 2), the NIS with external shear is often too simple; many lens systems require an elliptical mass distribution plus some external shear in addition.

4 The Cosmological Standard Model I: The Homogeneous Universe

We assume that the reader is familiar with the basic concepts of standard cosmology, such as the hot Big Bang occurring some 13.7 billion years ago, after which the Universes expanded and cooled down. During this expansion, the simplest atomic nuclei, predominantly helium, were formed about a minute after the Big Bang, and some 370,000 years later, the Universe became neutral and released a thermal radiation that is still visible today, the Cosmic Microwave Background radiation with a temperature of 2.73 K. In addition, it is assumed that you are aware of the existence of dark matter, material that reveals itself only through gravity, like in governing the rotation curves of the Milky Way and other spiral galaxies or in providing the deep potential wells of clusters of galaxies which can keep very hot X-ray emitting plasma and fastly moving galaxies gravitationally bound. Therefore, we shall only briefly summarize those relations which will be used later in this course. Excellent textbooks on cosmology are available, among them are Kolb and Turner (1990), Peacock (1999), Padmanabhan (1993), Peebles (1993) and Liddle and Lyth (2000).

4.1 The Cosmic Expansion

Metric and Coordinates

Observational evidence suggests that the Universe around us, when averaged over large angles, is isotropic (the Cosmic Microwave Background, or CMB; the faint galaxy distribution, etc.). Furthermore, if we assume that our location in the Universe is not special, the same property also holds for other observers: also for them the Universe should appear isotropic when averaged over large scales. Together, this implies that the Universe is spatially homogeneous and isotropic around every point. It has been shown by Robertson and Walker that for such a spacetime, the metric can be written in the form

$$\mathrm{d}s^2 = c^2\,\mathrm{d}t^2 - a^2(t)\left[\mathrm{d}w^2 + f_K^2(w)\left(\mathrm{d}\theta^2 + \sin^2\theta\,\mathrm{d}\varphi^2\right)\right] , \tag{65}$$

where t is the cosmic time [which agrees with the time measured by comoving observers, i.e., those with constant (w, θ, φ)], $a(t)$ the cosmic scale factor, normalized so that today, $a(t_0) = 1$, w the comoving radial coordinate, θ and φ are the angular coordinates on a unit sphere, and $f_K(w)$ the comoving angular diameter distance, which depends on the curvature parameter K in the following way:

$$\begin{aligned} f_K(w) &= |K|^{-1/2}\sinh\left(|K|^{1/2}w\right) \\ &\equiv \begin{cases} K^{-1/2}\sin(K^{1/2}w) & (K>0) \\ w & (K=0) \\ (-K)^{-1/2}\sinh[(-K)^{1/2}w] & (K<0) \end{cases} . \end{aligned} \tag{66}$$

Hence, (w, θ, φ) are spherical coordinates in a three-dimensional space of constant curvature K. Radiation from a comoving source emitted at time t_2 and received by a comoving observer at time $t_1 > t_2$ is *redshifted* by a factor $1 + z_{12} = a(t_1)/a(t_2)$.

Expansion Equation

Inserting the metric (65) into Einstein field equation of General Relativity shows that the matter contents must be that of a (homogeneous) perfect fluid with density $\rho(t)$ and pressure $p(t)$. The components of the field equation reduce to two independent dynamical equations for the scale factor $a(t)$,

$$\left(\frac{\dot a}{a}\right)^2 = \frac{8\pi G}{3}\rho - \frac{Kc^2}{a^2} + \frac{\Lambda}{3} \tag{67}$$

and

$$\frac{\ddot a}{a} = -\frac{4}{3}\pi G\left(\rho + \frac{3p}{c^2}\right) + \frac{\Lambda}{3} . \tag{68}$$

Equation (67) is called *Friedmann equation* (Friedmann 1922). The two equations (67) and (68) can be combined to yield the *adiabatic equation*

$$\frac{\mathrm{d}}{\mathrm{d}t}\left[a^3(t)\rho(t)c^2\right] + p(t)\frac{\mathrm{d}a^3(t)}{\mathrm{d}t} = 0 , \tag{69}$$

which has the following intuitive interpretation: the first term $a^3\rho$ is proportional to the energy contained in a fixed comoving volume, and hence the equation states that the change in 'internal' energy equals the pressure times the change in proper volume. Hence, (69) expresses the first law of thermodynamics in the cosmological context. The parameter Λ in (67) and (68) is the cosmological constant; Einstein introduced it into his field equation (in 1916) because without it, no static model of the Universe would be predicted by General Relativity – the Hubble expansion of the Universe was discovered only a decade later, after which Einstein dismissed this term. In recent years, the cosmological constant has regained great popularity – because, as will be discussed later, there is strong evidence in favor of $\Lambda \neq 0$. On the other hand, the interpretation of Λ has also changed, as we shall see.

World models for which the metric is given by (65) and where the scale factor $a(t)$ obeys Friedmann equation (67) and the adiabatic equation (69) are called Friedmann–Lemaître models. It should be noted that (68) can also be derived from Newtonian gravity except for the pressure term and the cosmological constant. Unlike in Newtonian theory, pressure acts as a source of gravity in General Relativity.

Matter Models

By themselves, these equations do not specify the expansion history $a(t)$; for this we have to add an equation of state (EOS). In general, matter components cannot be described by a simple equation of the form $p = p(\rho)$; however, for some limiting cases an equation of this form does exist. Fortunately, the matter contents in our Universe seems to be such that over most of its history it can be described by a few components, each of which having such a simple EOS.

If the constituents of matter have random (thermal) velocities much smaller than c, $p \ll \rho c^2$, then the pressure of this component can be neglected in the expansion equation; this kind of matter is approximated by $p = 0$ and called 'dust' (or simply 'matter'). For $p = 0$, (69) yields that

$$\rho_{\rm m} \propto a^{-3} , \tag{70}$$

a result that is intuitively clear: as the physical (or proper) volume of a fixed comoving volume behaves like $V \propto a^3$, and the number of matter particles is conserved, their number density, and thus mass density must decrease as $\rho \propto V^{-1} \propto a^{-3}$.

In the other limiting case, where the constituents of matter have a random velocity close to c (or even c, as must be the case for massless particles, like photons), one has $p = \rho c^2/3$. For obvious reasons, matter with this EOS is called 'radiation'. From (69) one then finds that the energy density of radiation evolves as

$$\rho_{\rm r} \propto a^{-4} , \tag{71}$$

a result that can also be easily understood. Whereas the number density of photons (assuming that they constitute the 'radiation') decreases as a^{-3}, again due to number conservation, their individual energy decreases as a^{-1}, owing to the redshift of their energy (or adiabatic decompression).

Finally, there may be a mass component which can be interpreted as the energy density of the vacuum, assumed to be a constant in time. If the density is independent of a, then (69) predicts that $p_{\rm v} = -\rho_{\rm v} c^2$ for this matter component.

The matter density and pressure of the Universe is then given by the sum of these three components,

$$\rho = \rho_{\rm m} + \rho_{\rm r} + \rho_{\rm v} = \frac{\rho_{\rm m0}}{a^3} + \frac{\rho_{\rm r0}}{a^4} + \rho_{\rm v} \; , \quad p = \frac{\rho_{\rm r} c^2}{3} - \rho_{\rm v} c^2 = \frac{\rho_{\rm r0} c^2}{3a^4} - \rho_{\rm v} c^2 \; , \tag{72}$$

where the additional index '0' indicates that these are the values at present time. Inserting these expressions into (67) and (68) (setting $\Lambda = 0$ in these equations) shows that a term of the same form as the Λ-term appears; the cosmological constant can therefore be interpreted as a vacuum energy density.

Cosmological Parameters

The ratio

$$H(t) = \dot{a}\, a^{-1} \tag{73}$$

is the expansion rate of the Universe, and its current value H_0 is called *Hubble constant.* This is the ratio of recession velocity to the distance of objects in the nearby Universe, and has the value

$$H_0 \approx 3.2 \times 10^{-18}\, h\, {\rm s}^{-1} \approx 1.0 \times 10^{-10}\, h\, {\rm yr}^{-1} \; , \tag{74}$$

where h parameterizes our lack of knowledge on the exact value of H_0; the currently best estimates yield $h \approx 0.72$ (see Sect. 6.3). If $\Lambda = 0$ and the spatial curvature vanishes, $K = 0$, then the current density of the Universe is directly related to H_0, as seen from (67); this density is called *critical density,*

$$\rho_{\rm cr} := \frac{3H_0^2}{8\pi G} \approx 1.9 \times 10^{-29}\, h^2\, {\rm g\, cm^{-3}} \; . \tag{75}$$

This characteristic density is used to scale the matter densities by defining the *density parameters*

$$\Omega_{\rm m} := \frac{\rho_{\rm m0}}{\rho_{\rm cr}} \; ; \quad \Omega_{\rm r} := \frac{\rho_{\rm r0}}{\rho_{\rm cr}} \; ; \quad \Omega_\Lambda := \frac{\rho_{\rm v}}{\rho_{\rm cr}} = \frac{\Lambda}{3H_0^2} \; . \tag{76}$$

The radiation density in the Universe is fairly well known: it is dominated by the energy density of the cosmic microwave background (CMB) which has a Planck spectrum with temperature of $T_{\rm CMB} \approx 2.73\,{\rm K}$ and whose energy density can be calculated from the Stefan–Boltzmann law to be

$$\rho_{\rm CMB} = \frac{1}{c^2} \frac{\pi^2}{15} \frac{(kT_{\rm CMB})^4}{(\hbar c)^3} \approx 4.5 \times 10^{-34}\, {\rm g\, cm^{-3}} \;\to\; \Omega_{\rm CMB} = 2.4 \times 10^{-5}\, h^{-2} \; . \tag{77}$$

In addition, from the era just before the primordial nucleosynthesis took place, there is a relic background of neutrinos, at a temperature corresponding to $T_\nu = (4/11)^{1/3} T_{\rm CMB} \approx 1.95\,{\rm K}$; their number density today is $113\,{\rm cm}^{-3}$ per species. If neutrinos were massless, they would contribute to the radiation density, and with three neutrino families one would have $\Omega_{\rm r} \approx 1.68 \Omega_{\rm CMB}$. If neutrinos have a small, but finite rest mass above $\sim 10^{-4}\,{\rm eV}$ they would be non-relativistic today, and contribute to the matter density instead. However, at earlier epochs the neutrinos were relativistic and therefore contributed to the radiation energy density. In any case, for the present epoch we can completely neglect the influence of the radiation on the expansion rate. This, of course, was not always the case; since the radiation density drops as a^{-4}, whereas the matter density only as a^{-3}, there was an epoch (or a scale factor) when both were equal, namely at

$$a_{\rm eq} = \frac{\Omega_{\rm r}}{\Omega_{\rm m}} = 3.2 \times 10^{-5} \Omega_{\rm m}^{-1}\, h^{-2} \,, \tag{78}$$

and we used $\Omega_{\rm r} \approx 1.68 \Omega_{\rm CMB}$ here, since at $a_{\rm eq}$, the neutrinos were relativistic. For scale factors $a \lesssim a_{\rm eq}$, radiation was the dominant component in the Universe.

Making use of (72) and the definitions (76) of the density parameters, the expansion equation (67) becomes

$$H^2 = H_0^2 \left[\frac{\Omega_{\rm r}}{a^4} + \frac{\Omega_{\rm m}}{a^3} - \frac{Kc^2}{a^2\, H_0^2} + \Omega_\Lambda \right] \,. \tag{79}$$

Specializing this to the current epoch, $a = 1$, yields an expression for the curvature, $K = (\Omega_{\rm m} + \Omega_\Lambda - 1) H_0^2/c^2$ (where we used $\Omega_{\rm r} \ll \Omega_{\rm m}$), which can be inserted into (80) to yield

$$H^2 = H_0^2 \left[\Omega_{\rm r} a^{-4} + \Omega_{\rm m} a^{-3} + (1 - \Omega_0) a^{-2} + \Omega_\Lambda \right] \,, \tag{80}$$

where we defined

$$\Omega_0 = \Omega_{\rm m} + \Omega_\Lambda + \Omega_{\rm r} \tag{81}$$

as the total density parameter of the present-day Universe. One sees that the sign of $\Omega_0 - 1$ agrees with that of K, so that the total matter density determines the spatial curvature of the Universe. Note that (80) is a first-order differential equation for $a(t)$, which can be integrated (numerically, if necessary) with the boundary condition $a(t_0) = 1$. The general discussion of the qualitative behavior of $a(t)$ (see, e.g., Peacock 1999, Sect. 3.2; also Fig. 19 below) yields the following results: The scale factor $a(t)$ is a monotonically increasing function for $a < 1$; hence, a decreases monotonically as we go backwards in time. Whereas in principle it is possible that a does not decrease below a finite positive value (so-called bouncing Universes), we happen not to live in one – such models predict a minimum a, and therefore a maximum redshift; the fact that we have discovered sources at redshift $z \gtrsim 6$, coupled with a matter density that certainly exceeds $\Omega_{\rm m} > 0.05$ excludes the possibility that our Universe is of that kind. Hence, formally $a \to 0$ as we go into the past, at a finite instant. This instant is called the Big Bang, an event when the Universe was extremely dense and hot. The behavior of $a(t)$ in the future depends on the values of the density parameters. If $\Omega_\Lambda = 0$, then $a(t)$ will continue to grow provided $\Omega_{\rm m} \le 1$, otherwise it will reach a maximum value of a and then recollapse. If $\Omega_\Lambda > 0$, the threshold value of $\Omega_{\rm m}$ for recollapse is slightly changed. Flat models, i.e., those with $\Omega_{\rm m} + \Omega_\Lambda = 1$ expand forever provided $\Omega_{\rm m} \le 1$. Defining $t = 0$ to be the instant of the Big Bang when $a = 0$, the cosmic time as a function of scale factor can be calculated from (80), since $\mathrm{d}t = \mathrm{d}a\,\dot a^{-1} = \mathrm{d}a(aH)^{-1}$; ignoring $\Omega_{\rm r}$ (which is important only over a very brief period at the beginning of the expansion), one has

$$t(a) = \frac{1}{H_0} \int_0^a \mathrm{d}a' \left[a'^{-1}\Omega_{\rm m} + (1 - \Omega_{\rm m} - \Omega_\Lambda) + a'^2 \Omega_\Lambda\right]^{-1/2} ; \qquad (82)$$

in particular, t_0 is obtained by setting $a = 1$. Apart from a numerical factor which depends on the density parameters, this yields $t_0 \sim H_0^{-1}$. Equation (82) can be inverted to yield $a(t)$.

Light propagates along null geodesics; in the coordinate system used to define the metric (65), it is easy to show from symmetry arguments that radial null curves (i.e., those with $\theta = $ const., $\varphi = $ const.) are geodesics; for them $c\,\mathrm{d}t = -a\,\mathrm{d}w$, if we choose our location at $w = 0$. The minus sign occurs since photons propagating to us have $\mathrm{d}t > 0$ but $\mathrm{d}w < 0$. Light from a source that we observe today was emitted at a time obtained from integrating $c\,\mathrm{d}t = -a\,\mathrm{d}w$; every observation of the distant Universe is inevitably a look into the past.

We therefore have a number of variables which can be used to describe the location of a source: its comoving distance w, the time t at which the light was emitted which we observe today from that source, the scale factor a at this time or, equivalently, the redshift $z = a^{-1} - 1$, and the temperature of the Universe (which is defined as the temperature of the microwave background radiation – note that cosmic expansion evolves a blackbody into a blackbody, with temperature $T \propto a^{-1}$). These variables are related to each other by

$$\mathrm{d}t = \frac{\mathrm{d}a}{\dot a} = \frac{\mathrm{d}a}{a\,H} ; \quad -\mathrm{d}w = \frac{c\,\mathrm{d}t}{a} = \frac{c\,\mathrm{d}a}{a\,\dot a} = \frac{c\,\mathrm{d}a}{a^2 H} . \qquad (83)$$

4.2 Distances and Volumes

The Meaning of Distance

Which of these descriptions of the location of a source is the 'correct distance'? Well, wrong question. This question is based on the Euclidean preconception that there is a uniquely defined correct distance, and that this is the outcome of all (correct) methods to measure the distance. However, in a general space-time, two complications occur. The harmless one is that space may be curved.

The more important one is that any observation measures distances not at a given instant of time, but along the backward light cone, and distances change in time as the Universe expands. There is not a unique meaning of distance. Nevertheless, one can construct methods on how to measure distance, and define distances according to these measurement procedures. The two most important definitions of distance are described next.

Distance Measures

Suppose one knows the physical diameter $2R$ of a source at redshift z (or scale factor a) which is observed to have an angular diameter of δ. In Euclidean space one would then measure the distance to this source to be $D = 2R/\delta$; accordingly, one defines the angular diameter distance as exactly this ratio,

$$D_{\rm ang}(z) = 2R/\delta = a(z)\, f_K(w)\;, \tag{84}$$

where the final expression follows from the metric by setting $\delta = {\rm d}\theta$ and ${\rm d}s = 2R$. In the foregoing expression, w is to be understood as a function of redshift; the corresponding relation can be obtained from (83) to be

$$\begin{aligned} w(z_1, z_2) &= \frac{c}{H_0} \int_{a(z_2)}^{a(z_1)} \left[a\Omega_{\rm m} + a^2(1 - \Omega_{\rm m} - \Omega_\Lambda) + a^4\Omega_\Lambda\right]^{-1/2} {\rm d}a \\ &= w(z_2) - w(z_1)\;, \end{aligned} \tag{85}$$

which is the *comoving distance* between two sources that we see to have redshifts $z_1 < z_2$, and we set $w(z) \equiv w(0, z)$. The comoving distance can be interpreted as the spatial distance between the intersections of the world-line of these two comoving sources with the spatial hypersurface $t = t_0$ (cf. the definition of comoving coordinates). Generalizing (84), we can define the angular-diameter distance $D_{\rm ang}(z_1, z_2)$ of a source at redshift z_2 seen by an observer at redshift $z_1 < z_2$ as

$$D_{\rm ang}(z_1, z_2) = a(z_2)\, f_K[w(z_1, z_2)]\;. \tag{86}$$

Note that in general, $D_{\rm ang}(z_1, z_2) \neq D_{\rm ang}(z_2) - D_{\rm ang}(z_1)$; on the other hand, such an additive relation *is* valid for the comoving angular diameter distances for a Universe with vanishing curvature $K = 0$, as seen from (85) and (66). Thus, it is often useful to employ the comoving angular diameter distance, i.e., the ratio between the comoving diameter of an object and its angular diameter.

Another method to measure distances is to relate the observed flux S of a source to its luminosity L; if we know the luminosity, then the distance to the source can be determined by

$$D_{\rm lum}(z) \equiv \sqrt{\frac{L}{4\pi S}}\;, \tag{87}$$

which is called the luminosity distance. In Euclidean space, this measurement would yield the same result as that from comparing diameters and angular sizes; in curved spacetimes, this is no longer true. In fact, one can show (Etherington 1933) that in general,

$$D_{\rm lum}(z) = (1+z)^2\, D_{\rm ang}(z) = (1+z)\, f_K(w)\ . \tag{88}$$

In this equation, flux and luminosity have to be interpreted as bolometric quantities, i.e., integrated over all frequencies. The flux at a given frequency ν is related to the specific luminosity of the source at a different frequency $\nu_{\rm e} = (1+z)\nu$, owing to redshift. This frequency shift is taken into account by the so-called K-correction in the relation between specific flux and luminosity.

We still need another distance concept, the proper distance. Suppose we measure the redshifts z and $z+\Delta z$ of two comoving sources, being very similar, and which also have small angular separation $\Delta\theta$ on the sky. What is the separation between these two sources that an observer would measure who lives somewhere near them? This separation can be measured by this fiducial observer in the same way as we can measure the distance to Virgo-cluster galaxies, without caring about the values of the cosmological parameters – locally space can be approximated as being Euclidean where distances have a unique meaning. The proper separation transverse to the line-of-sight is $D_{\rm ang}(z)\, \Delta\theta$, and that along the line-of-sight is

$$\begin{aligned} \Delta r_{\rm prop} &= a(z)\, \Delta w = a(z) \frac{{\rm d}w}{{\rm d}a} \frac{{\rm d}a}{{\rm d}z} \Delta z = \frac{c\, a(z)}{H(z)} \Delta z \\ &= \frac{c}{H_0} \frac{\Delta z}{\sqrt{\Omega_{\rm m} a^{-1} + (1-\Omega_{\rm m}-\Omega_\Lambda) + \Omega_\Lambda a^2}}\ . \end{aligned} \tag{89}$$

Volume Elements

We can now also calculate volume elements: suppose in a solid angle ω one measures ${\rm d}N$ sources with redshift between z and $z+{\rm d}z$, the proper number density of these sources is $n = {\rm d}N/{\rm d}V$, where the volume is given by the physical thickness of the redshift slice times the area transverse to the line-of-sight, which is $D^2_{\rm ang}(z)\, \omega$, so

$${\rm d}V_{\rm prop} = D^2_{\rm ang}(z)\, \omega\, \frac{{\rm d}r_{\rm prop}}{{\rm d}z}\, {\rm d}z\ , \tag{90}$$

where we indicated that this is the proper volume element. The corresponding comoving volume element is then

$${\rm d}V_{\rm com} = a^{-3}\, {\rm d}V_{\rm prop} = f^2_K[w(z)]\, \omega\, \frac{{\rm d}w}{{\rm d}z}\, {\rm d}z\ . \tag{91}$$

Finite volumes can be obtained from the foregoing equations by integration.

Special Cases

Whereas the foregoing expansion equations are easily evaluated through numerical integration, there are some cases where explicit expressions can be obtained. The simplest model is the *Einstein–de Sitter* (EdS) Universe, characterized by $\Omega_{\rm m} = 1$, $\Omega_\Lambda = 0$; this model has zero curvature. The expansion equation (neglecting radiation) reduces to $H = H_0 a^{-3/2}$, yielding $t = 2/(3H_0) a^{3/2}$; in particular, the current age of the Universe in this model is $t_0 = 2/(3H_0) \sim 6.7 \times 10^9\, h^{-1}$ years. The comoving and angular diameter distance for an EdS model are easily obtained as

$$\begin{aligned} D_{\rm ang}(z_1, z_2) &= \frac{2c}{H_0} \frac{1}{1+z_2} \left[(1+z_1)^{-1/2} - (1+z_2)^{-1/2} \right] , \\ D_{\rm com}(z_1, z_2) &= \frac{2c}{H_0} \left[(1+z_1)^{-1/2} - (1+z_2)^{-1/2} \right] . \end{aligned} \tag{92}$$

In particular, as for all flat models, the comoving angular diameter distance is the same as the comoving distance. Unfortunately, we seem to not be living in an EdS Universe (see Sect. 6.3).

For models without a cosmological constant, the angular-diameter distance can be written in closed from, using the famous Mattig (1958) relation,

$$\begin{aligned} D_{\rm ang}(z_1, z_2) &= \frac{2}{\Omega_{\rm m}^2 (1+z_1)(1+z_2)^2} \\ &\times \left[(\Omega_{\rm m} z_2 - \Omega_{\rm m} + 2)\sqrt{1+\Omega_{\rm m} z_1} - (\Omega_{\rm m} z_1 - \Omega_{\rm m} + 2)\sqrt{1+\Omega_{\rm m} z_2} \right] . \end{aligned} \tag{93}$$

Next we consider the expansion equation (80) qualitatively. The different dependencies of the four terms in (80) on the scale factor shows that for very small a, the expansion was dominated by radiation, for $a \gtrsim a_{\rm eq}$ it was dominated by matter; the effects of curvature (if different from zero) and the cosmological constant play a role only at later stages of the cosmic expansion. For small $a \ll 1$, (80) can be approximated as $H = H_0 \Omega_{\rm m}^{1/2} a^{-3/2} \sqrt{1 + a_{\rm eq}/a}$ which can be integrated to yield

$$t(a) = \frac{2}{3H_0}\, \Omega_{\rm m}^{-1/2} \left[a^{3/2} \left(1 - 2\frac{a_{\rm eq}}{a}\right) \left(1 + \frac{a_{\rm eq}}{a}\right)^{1/2} + 2\, a_{\rm eq}^{3/2} \right] , \tag{94}$$

and so $t = a^2 \left(2H_0 \sqrt{\Omega_{\rm m} a_{\rm eq}}\right)^{-1}$ for $a \ll a_{\rm eq}$, and $t = 2a^{3/2} \left(3H_0 \sqrt{\Omega_{\rm m}}\right)^{-1}$ for $a_{\rm eq} \ll a \ll 1$. For EdS, (94) describes the expansion for all a through the radiation and matter dominated phases.

4.3 Gravitational Lensing in Cosmology

The Meaning of Distance in Lensing

When we used distances to write the gravitational lens equations in Sect. 2 we have not discussed what 'distance' means there. Now we learned that the concept of distance in curved spacetimes, even if they are as simple as the Friedmann–Lemaître spacetimes, are more complicated than in the Euclidean case. Therefore, which of the many distances defined above is the one to be used in the gravitational lens equations?

The answer is quite obvious: recall that the basic lens equation (6) relates images and source positions by a geometrical consideration; for that one needs to relate angles with transverse distances. This is exactly the way the angular-diameter distance was defined; hence, all equations in Sect. 2 are also valid in a Friedmann–Lemaître spacetime if the distances D are taken to be the angular-diameter distances $D_{\rm ang}$.

In many cases, the equations of gravitational lensing become simpler if the comoving angular diameter distances $f_K(w)$ are used; one example is the expression (96) for the time delay. In particular this is true for flat cosmological models, for which $f_K(w) = w$. Furthermore, most equations of gravitational lensing contain distances only in form of the ratio $D_{\rm ds}/D_{\rm s}$, for which it is irrelevant whether $D = D_{\rm ang}$ or $D = f_K$ is used. *In order not to confuse the reader, we shall consistently use the following convention throughout the rest of this book (recalling that in all equations in Sects. 2 and 3, $D \equiv D_{\rm ang}$ is implied): the angular diameter distance is denoted by $D^{\rm ang}$, and the comoving distance is denoted by D or f_K.* For example, in this notation the critical surface mass density and the Einstein radius of a point mass read

$$\begin{aligned} \Sigma_{\rm cr} &= \frac{c^2}{4\pi G} \frac{D_{\rm s}^{\rm ang}}{D_{\rm d}^{\rm ang} D_{\rm ds}^{\rm ang}} = \frac{c^2(1+z_{\rm d})}{4\pi G} \frac{D_{\rm s}}{D_{\rm D} D_{\rm ds}} \\ \theta_{\rm E} &= \left(\frac{4GM(1+z_{\rm d})}{c^2} \frac{D_{\rm ds}}{D_{\rm D} D_{\rm s}} \right)^{1/2} . \end{aligned} \tag{95}$$

The Time Delay

We mentioned in Sect. 2.2 that the light travel times along the light rays that form the multiple images in a lens system are not the same, but have not given an expression for it. Now that we are armed with the necessary cosmological relations we can do so. There are two ways to derive an expression for the time delay, both of which shall briefly be described here. Cooke and Kantowski (1975) argued that the time delay must have two different components: first, a light ray that is bent is longer, and thus light needs more time to propagate along it, than for a straight ray. Since the individual light rays are bent by different angles, their geometrical lengths are different, giving rise to a 'geometrical time delay' between them. Second, light rays propagate through a gravitational potential which retards them; this is the well-known 'Shapiro effect', which has been amply tested by radar echo delay experiments towards the inner satellites in the Solar System. This is the 'gravitational time delay'. The total time delay is then simply the sum of the two. For this derivation, it is important to note that the gravitational time delay 'occurs' at the redshift of the lens, and hence gets redshifted by a factor $1 + z_{\rm d}$ owing to the cosmic expansion.

An alternative derivation of the time delay was given by Refsdal and his collaborators (see e.g. Kayser and Refsdal 1983 and references therein); they considered the wavefronts emitted from a source. Wavefronts are surfaces of equal light travel time (image for example a set of photons all emitted in a single flash from a bursting source; their location at fixed instant form a wavefront), and Fermat principle states that light rays are perpendicular to the wavefronts. Close to the source, the wavefronts are spheres; owing to perturbations in the gravitational potential, they get distorted. If propagating near a mass concentration, the wavefronts can become strongly distorted, and after passing it, they can actually intersect themselves. An observer located in that region would then be passed by the same folded wavefront more than once; since the different sheets of the wavefronts have different orientations, the observer will then see multiple images of the corresponding source, in the direction perpendicular to the individual wavefront sheets. The time delay between two images is then obtained as the time between the passing of the two corresponding wavefront sheets. From a purely geometrical consideration, the time delay can then be derived, yielding the same expression as adding together the geometrical and potential time delays of the first method.[9] The time delay can be written most conveniently in terms of the Fermat potential as (Schneider 1985)

$$\begin{aligned}\Delta t &= \frac{D_{\rm D}^{\rm ang} D_{\rm s}^{\rm ang}}{c\, D_{\rm ds}^{\rm ang}} (1+z_{\rm d}) \left[\tau(\boldsymbol{\theta}^{(1)};\boldsymbol{\beta}) - \tau(\boldsymbol{\theta}^{(2)};\boldsymbol{\beta})\right] \\ &= \frac{D_{\rm D} D_{\rm s}}{c\, D_{\rm ds}} \left[\tau(\boldsymbol{\theta}^{(1)};\boldsymbol{\beta}) - \tau(\boldsymbol{\theta}^{(2)};\boldsymbol{\beta})\right] \;, \end{aligned} \tag{96}$$

where in the second expression the comoving angular diameter distances were used, and $\tau(\boldsymbol{\theta};\boldsymbol{\beta})$ is the Fermat potential defined in (14). This result then confirms the statement made in Sect. 2.2: $\tau(\boldsymbol{\theta};\boldsymbol{\beta})$ is, up to an affine transformation, the light travel time along a ray originating at $\boldsymbol{\beta}$ in the source plane, traversing the lens plane at $\boldsymbol{\theta}$ and then propagating to the observer. The additive constant of this affine transformation is irrelevant, as only differences are observable; the factor in the linear term is given in (96). The potential time delay is described by the deflection potential $\psi(\boldsymbol{\theta})$ in τ, the geometric time delay by the $|\boldsymbol{\beta}-\boldsymbol{\theta}|^2/2 = |\boldsymbol{\alpha}|^2/2$-term.

5 Basics of Lensing Statistics

One is frequently interested in the probability that a specific gravitational lensing event occurs. For example, Zwicky (1937b) estimated the probability that a distant source is multiply imaged by "extragalactic nebulae" using the

[9] As Sjur Refsdal reports, the first referee of his paper on the wavefront method rejected it wholeheartedly, claiming that the resulting expression can contain only the geometrical contribution to the time delay. It remains unknown how this referee imagined the geometry of distorted and overlapping wavefronts without the effect of retardation provided by the gravitational field of the deflector.

surface density of these objects on the sky. This basic problem of statistical gravitational lensing has since been studied in considerable detail, as will be discussed in SL (Part 2). The results of such an investigation depend on the assumed distribution of lens masses and their individual density profiles. A comparison of these results with a statistically well defined sample of observed lens cases can in principle allow one to constrain the lens contents and/or the geometry of the Universe. The probability for microlensing events to occur depends on the density of compact objects along the sightline to the population of sources which are monitored, as will be detailed in ML.

Another typical problem of statistical lensing is the so-called magnification bias. Let us consider a sample which should include all sources of a certain kind in a region of the sky brighter than a given threshold (i.e., a flux-limited sample). From the observed fluxes of the sources and their distances (e.g., determined from their redshifts) we can derive the intrinsic luminosities of the sources. If a source is magnified by a gravitational lens, its derived luminosity will not be the true one, but will be higher in general. Moreover, there may be sources in the sample which do not belong there because they are intrinsically too faint to be included, but have been magnified above the threshold of the sample. Since flux-limited samples of extragalactic sources are used to derive information about the evolution of the sources and about the structure of the universe, the magnification can mislead astronomers. Statistical lensing investigations are used to estimate the importance of this effect and its consequences. In this section we shall provide the basics of lensing probability investigations, with details left to later sections when specific applications are discussed.

5.1 Cross-Sections

The lensing probabilities depend on the number density of lenses, as well as on their mass profile. The latter is used to define lensing cross-sections. We shall start with two specific examples which should motivate this concept.

Cross-Sections for a Point-Mass Lens

First, consider a (point) source at distance $D_{\rm S}$, and a point mass at distance $D_{\rm D}$ from Earth. The separation of the two images and their magnifications depend on the relative alignment of source, observer, and lens. There is a one-to-one relationship between the source position $\beta = y\theta_{\rm E}$ and the corresponding total magnification $\mu_{\rm p}$, see (48), where total magnification means the summed magnifications of the individual images. Thus, for any $\mu_{\rm p} > 1$ there is a value of y such that, if the distance β of the source is less than $y\theta_{\rm E}$ from the optical axis, the latter is magnified by more than $\mu_{\rm p}$:

$$y^2 = 2\left(\frac{\mu_{\rm p}}{\sqrt{\mu_{\rm p}^2 - 1}} - 1\right) \quad . \tag{97}$$

Hence, we can define the cross-section

$$\sigma(\mu_\mathrm{p}) = \pi\theta_\mathrm{E}^2\, y^2(\mu_\mathrm{p}) \tag{98}$$

for a point source magnification larger than μ_p. In other words, centered on the caustic point there is a solid angle $\sigma(\mu_\mathrm{p})$ within which the source must lie in order to be magnified by more than μ_p. One could also consider σ to be the solid angle in which a lens must be located such that a fixed source is magnified by more than μ_p; these two points of view are basically equivalent, though there are some subtleties involved (Ehlers and Schneider 1986) which shall not be discussed here.

As a second example, we consider the ratio $r = \mu_+/|\mu_-|$ of the absolute values of the magnifications (brightness ratio) for the two images produced by a Schwarzschild lens – see (48). In order for this ratio to be less than r, the impact parameter y needs to be less than $r^{1/4} - r^{-1/4}$, and so the cross-section for magnification ratio less than $r > 1$ is

$$\sigma(r) = \pi\theta_\mathrm{E}^2 \left(r^{1/2} + r^{-1/2} - 2\right) . \tag{99}$$

General Definition of a Lensing Cross-Section

After these two examples we now discuss the general definition of a cross-section. Consider a source and a lens, both at fixed distances from Earth. The lens may be described by a set of parameters, and the source is characterized, say, by its size and its brightness profile (if the source extent is relevant). If one is interested in a certain property Q of this gravitational lens system, one can ask where the source must be located such that the images have the property Q. Two examples for Q were given above, namely, that the total magnification is larger than μ and that the brightness ratio of the images is smaller than r. More complicated examples of Q will be considered in due course. The question can be answered through an analysis of the gravitational lens model, as demonstrated above for the point mass lens. One usually finds that the source must be in a certain region of the source plane. The solid angle of that region is then the Q-cross-section σ_Q for this lens–source system.

Lensing Cross-Section for a Singular Isothermal Sphere

To illustrate the concept further, we shall consider the lensing cross-sections of an SIS. From Sect. 3.3 we know that this lens produces two images if $\beta < \theta_\mathrm{E}$, and that the image separation is $2\theta_\mathrm{E}$. Hence, the cross-section of an SIS to produce two images with separation larger than $\Delta\theta$ is

$$\sigma(\Delta\theta) = \pi\theta_\mathrm{E}^2 \mathrm{H}(2\theta_\mathrm{E} - \Delta\theta) , \tag{100}$$

where $\mathrm{H}(x)$ is the Heaviside step function. Next the (total) magnification and the flux ratio r can be included; both are functions of $y = \beta/\theta_\mathrm{E}$, $\mu = 2/y$ and

$r = (1+y)/(1-y)$. Therefore, the cross-section for an SIS to produce two images with separation larger than $\Delta\theta$, total magnification larger than μ and flux ratio of the images smaller than r is

$$\sigma(\Delta\theta, r, \mu) = \pi\theta_{\rm E}^2 \left[\min\left(\frac{r-1}{r+1}, \frac{2}{\mu}\right)\right]^2 {\rm H}(2\theta_{\rm E} - \Delta\theta)\,{\rm H}(\mu - 2)\,. \qquad (101)$$

The Mass-Sheet Degeneracy and Scaling of Cross-Sections

In Sect. 2.5 we have seen that the imaging properties, such as angular separation and magnification ratios – and thus flux ratios, are unchanged if the surface mass density is transformed according to (28). As can be seen from (31), this transformation merely leads to a scaling of the lens plane, thereby affecting the magnifications. The scaling of the lens plane implies that the cross-section for an extended source of size ρ will, after the mass-sheet transformation, be related to the cross-section for a source of size $\lambda\rho$. Hence, the cross-section for flux ratio smaller than r, image separation larger than $\Delta\theta$, magnification larger than μ for a source of size ρ transforms like

$$\sigma_\lambda(r, \Delta\theta, \mu, \rho) = \lambda^2 \sigma\left(r, \Delta\theta, \lambda^2\mu, \frac{\rho}{\lambda}\right)\,. \qquad (102)$$

5.2 Lensing Probabilities; Optical Depth

The probability that a lensing event with specified properties Q occurs is given by the product of the number density of lenses and their cross-sections. Consider a solid angle ω toward sources at distance $D_{\rm s}$. To the distance interval ${\rm d}x$ around x (note that we use a different notation for distance along the line-of-sight here, for reasons which soon will become clear) corresponds a proper volume element ${\rm d}V = D_{\rm ang}^2(x)\,({\rm d}r_{\rm prop}/{\rm d}x)\,\omega$ within this solid angle. We consider lenses which are described by a set of parameters, summarized as χ; such parameters could be lens mass, core radius, ellipticity, etc. If $n(x,\chi)\,{\rm d}\chi$ is the (proper) number density of lenses at distance x with properties within ${\rm d}\chi$ of χ, the total cross-section of all lenses within the tube of solid angle ω is then

$$\sigma_{\rm tot}(Q) = \int_0^{x_{\rm s}} {\rm d}x\,\omega\,D_{\rm ang}^2(x)\,\frac{{\rm d}r_{\rm prop}}{{\rm d}x} \int {\rm d}\chi\,n(x,\chi)\,\sigma(Q; x, x_{\rm s}, \chi)\,, \qquad (103)$$

where the Q-cross-section of a lens depends on the lens parameters χ, the distance parameter x along the line-of-sight, and the source distance $x_{\rm s}$. The picture underlying this equation is that the cross-sections of the individual lenses simply add up linearly. This picture is justified as long as the projected separation between lenses is much larger than the linear size of the cross-sections, or in other words, the cross-sections of individual lenses do not overlap. The probability for a lensing event with property Q – also frequently called *optical depth* for lensing – is then given by the ratio of the cross-section $\sigma_{\rm tot}(Q)$ and the solid angle ω, i.e., the fraction of solid angle covered by the cross-sections,

$$P(Q) = \int_0^{x_{\rm s}} {\rm d}x\; D_{\rm ang}^2(x)\,\frac{{\rm d}r_{\rm prop}}{{\rm d}x}\int {\rm d}\chi\; n(x,\chi)\,\sigma(Q;x,x_{\rm s},\chi)\;. \qquad (104)$$

We shall now consider the case where the distance to the source is small, so that cosmological distances play no role. In this case, we can take $x = D_{\rm D}$, and (104) becomes

$$P(Q) = \int_0^{D_{\rm s}} {\rm d}D_{\rm D}\; D_{\rm D}^2 \int {\rm d}\chi\; n(D_{\rm D},\chi)\,\sigma(Q;D_{\rm D},D_{\rm s},\chi)\;. \qquad (105)$$

In the other case, where the sources of interest are at cosmological distances, redshift is a convenient distance variable, and (104) reads

$$P(Q) = \int_0^{z_{\rm s}} {\rm d}z_{\rm d}\; D_{\rm ang}^2(z_{\rm d})\,\frac{{\rm d}r_{\rm prop}}{{\rm d}z}\int {\rm d}\chi\; n(z_{\rm d},\chi)\,\sigma(Q;z_{\rm d},z_{\rm s},\chi)\;, \qquad (106)$$

where the proper distance interval along the line-of-sight is given in (89). In this cosmological context, it is often useful to specify the comoving number density $n_{\rm com}$ of lenses, instead of the proper density n; both are related by $n(z) = (1+z)^3\, n_{\rm com}$; furthermore, we can use the comoving distance w as integration variable, and work in terms of the comoving angular diameter distance $f_K(w)$; then, (106) becomes

$$P(Q) = \int_0^{w_{\rm s}} {\rm d}w\; f_K^2(w) \int {\rm d}\chi\; n_{\rm com}(w,\chi)\,\sigma(Q;w,w_{\rm s},\chi)\;, \qquad (107)$$

which is particularly convenient in the case of flat models ($f_K(w) = w$).

5.3 Magnification Bias

Besides the optical depth for a source to be multiply imaged (with image separation larger than an angular resolution limit of a survey), the magnification probability distribution has received great attention in the literature. Questions that have been studied include:

- Can all bright quasars be merely highly magnified images of much less luminous Seyfert galaxies (Barnothy 1965)? No, the lensing probabilities are far too small (e.g., Tyson 1981; Peacock 1982), even if the dark matter in the Universe consists of compact objects (Canizares 1982; Schneider 1987).
- Can magnification by (compact objects in the) halos of galaxies explain the apparent angular correlation (e.g., Arp 1987, and references therein) between nearby bright galaxies and high-redshift quasars? Again no, magnification probabilities are far too small (e.g., Canizares 1981; Vietri and Ostriker 1983; Schneider 1992).

- Does magnification affect the expected number of multiply-imaged QSOs in a gravitational lens survey? Yes (e.g., Wallington and Narayan 1993): for a lens survey with a bright flux threshold, magnification boosts the fraction of lensed sources by a large factor (see SL Part 2).

Magnification can cause sources to be included in a flux-limited sample which without lensing would be too faint to be included. Furthermore, sources with a very large magnification factor can attain apparent luminosities which exceed the maximum luminosities of the corresponding class of sources. In fact, several of the most luminosity-extreme sources are strongly magnified: The apparently most luminous IRAS galaxy F10214+4724 is magnified by a galaxy by about a factor of $\mu \sim 50$ (Broadhurst and Lehár 1995), the most luminous 'normal' Lyman-break galaxy cB58 (Yee et al. 1996) at redshift $z = 2.72$ is magnified by the cluster MS1512+36 with redshift $z \sim 0.3$ by a factor $\mu \sim 30$ (Seitz et al. 1998), the very bright $z = 3.87$ QSO APM 08279+5255 (Irwin et al. 1998) is gravitationally lensed and highly magnified by a foreground galaxy (Ibata et al. 1999), and several of the highest redshift galaxies have been found behind lensing clusters (e.g., Hu et al. 2002; Kneib et al. 2004; Pelló et al. 2004).

Consider a class of sources in a narrow redshift interval, and denote by $p(\mu)\,\mathrm{d}\mu$ the probability that one of these sources is magnified by a factor within $\mathrm{d}\mu$ of μ. Let $N_0(>S)$ be the number of these sources per unit solid angle that without lensing would be observed to have flux greater than S. If these sources get magnified by a factor μ, two things happen: first, a source with unlensed flux S will attain an observed flux μS. Second, since magnification enlarges the solid angle, sources that without lensing would be contained in a solid angle ω on the sky, will now be spread over the solid angle $\mu\omega$, i.e., the number density of sources decreases by a factor $1/\mu$. Together, if the magnification would be (locally) a constant μ, the observed source counts are

$$N(>S) = \frac{1}{\mu} N_0 \left(> \frac{S}{\mu}\right) \; . \tag{108}$$

Considering a probability distribution $p(\mu)$ in magnifications, this result generalizes to

$$N(>S) = \frac{1}{\langle\mu\rangle} \int \mathrm{d}\mu \; p(\mu)\, N_0 \left(> \frac{S}{\mu}\right) \; , \tag{109}$$

where $\langle\mu\rangle$ is the mean magnification within the region of the sky considered. If source counts are taken over random regions on the sky, then $\langle\mu\rangle = 1$, but if the magnification bias is considered around foreground galaxies, then in these regions, $\langle\mu\rangle > 1$. The probability $p(\mu)$ satisfies

$$\int \mathrm{d}\mu \; p(\mu) = 1 \; ; \quad \int \mathrm{d}\mu \; \mu \, p(\mu) = \langle\mu\rangle \; ; \tag{110}$$

the first relation expressing normalization, the second the definition of the mean magnification. Of course, $p(\mu)$ depends on the source redshift and the density of lenses, as mentioned in the previous section.

At this point, we need to enter briefly in the discussion on 'flux conservation'. As you recall, we have shown that any lens produces at least one image which is not demagnified, $\mu \geq 1$, provided $\kappa(\boldsymbol{\theta})$ is non-negative. If we consider source counts averaged over the whole sky, then of course $\langle\mu\rangle = 1$, and then (110) implies that if there are lines-of-sight where $\mu > 1$, then there must also exist those with $\mu < 1$. Hence, we get an apparent contradiction, which has led to much confusion in the literature. The resolution of this contradiction is seen when we consider the full matter distribution between us and the sphere of sources at a given redshift. The unmagnified flux of a source is defined as that flux which would be observed in a homogeneous Universe, since it is this model which underlies the definition of the luminosity distance. Now, the true mass distribution is inhomogeneous, consisting of large overdensities like galaxies and clusters, and underdensities like voids. A light bundle propagating through an underdense region of the Universe is less focused than one propagating through the mean density of the Universe, so that the effective $\kappa < 0$ for the former, and $\mu < 1$. Conversely, light bundles propagating through overdense regions get more focused, resulting in $\mu > 1$. We shall discuss these relations in more detail in WL (Part 3).

Note that there is a minimum magnification for each source redshift. Since the cosmic density $\rho \geq 0$, a light bundle cannot be more defocused than propagating through empty space; hence, μ is bound from below, and this bound depends on the source redshift and the density parameters $\Omega_{\rm m}$ and Ω_Λ.

We shall now consider the simple example of source counts which behave like a power law, $N_0(>S) = A\,S^{-\beta}$. Inserting this into (109) yields

$$N(>S) = \frac{1}{\langle\mu\rangle}\int \mathrm{d}\mu\; p(\mu)\,A\left(\frac{\mu}{S}\right)^{\beta} = N_0(>S)\,\frac{1}{\langle\mu\rangle}\int \mathrm{d}\mu\; p(\mu)\,\mu^{\beta}\;. \tag{111}$$

Thus, if the unlensed source counts behave like a power law, so do the lensed ones, with the same slope. The ratio between lensed and unlensed counts depends on the magnification probability distribution $p(\mu)$, as well as on the slope β of the counts. The first remarkable result is that, if $\beta = 1$, then (111) together with the second of (110) imply that $N(>S) = N_0(>S)$, i.e., the counts are unchanged in this case, independent of $p(\mu)$. Hence, in this case the enlargement of the solid angle over which sources are distributed just compensates the brightening of the sources. For $\beta < 1$, the number counts are depleted, whereas they are increased for $\beta > 1$. The larger the slope β, the larger is the ratio $N(>S)/N_0(>S)$, i.e., the stronger is the magnification bias.

If one considers point sources, or more generally, sources whose angular sizes are much smaller than the characteristic angular scale of the lenses, then one can show (Blandford and Narayan 1986) that for very high magnification, $p(\mu) \propto \mu^{-3}$, up to an upper limit for μ at which the finite size of the source limits the magnification. This functional dependence is due to the universal behavior of the lens equation near fold caustics, as was discussed in Sect. 2.4.

This functional behavior implies that the integral in (111) formally diverges as the slope β approaches 2. Hence, for a population of sources with steep number counts, the magnification bias can become very large. In fact, the formal divergence is due only to the assumption of a pure power law for $N_0(> S)$; whereas such a functional form is a good description, e.g., for the QSO counts over a limited range of fluxes (or luminosities), it cannot continue with a steep slope for arbitrarily faint sources, in order for the source population not to produce infinite total flux. Nevertheless, if the counts are steep, and one considers a value of S much larger than a break flux (where the steep counts turn into flatter ones towards lower fluxes), the ratio $N(> S)/N_0(> S)$ can be very high indeed. This is the reason why we see extreme QSOs like the one mentioned above, APM 08279+5255. Furthermore, if the source population is better described by a Schechter luminosity function, which implies an exponential decrease in the counts for high luminosities, the bias can be even larger: the probability with a Schechter function to find a single source far out in the exponential tail is very small, and if such an apparently luminous source is observed, it is most likely a lensed one, as is the case for F10214+4724 and cB58.

6 The Cosmological Standard Model II: The Inhomogeneous Universe

Whereas the Universe appears to be nearly homogeneous on large scales, it certainly is strongly inhomogeneous on smaller scales. Small fluctuations are imprinted onto the CMB, leading to tiny but measurable anisotropies in its temperature; in fact, these anisotropy measurements provide the strongest constraints on cosmological parameters currently available. Furthermore, the distribution of brighter (thus nearer) galaxies in the sky is highly anisotropic; galaxies tend to be strongly correlated, they tend to appear in groups or clusters of galaxies. Thus, on small scales the approximation of a homogeneous Universe must break down.

6.1 Structure Formation

Whereas the CMB fluctuations indicate very small inhomogeneities at the time of recombination (corresponding to a redshift $z \sim 1,100$) the inhomogeneities observed today in our neighborhood are much larger. A cluster of galaxies, for example, is a massive perturbation with a mean density more than hundred times larger than the mean density in the Universe. It is believed that the density inhomogeneities that we see today have evolved from much smaller fluctuations in the very early Universe. This evolution happens naturally through *gravitational instability*. A slightly overdense region has a somewhat higher self-gravity than the average region of the Universe, so its expansion rate will be slightly smaller than that of the Universe as a whole.

As a result of slower expansion, the density contrast of this region increases further, retarding expansion more, and so on. If the initial density contrast is sufficiently large, this instability can actually bring the expansion to a halt locally, after which the region recollapses under its own gravity to form galaxies and clusters.

In this picture of gravitational instability, one can study the evolution of structure in great detail. Since the major mass component in the Universe is dark matter, which by definition only interacts gravitationally, the dominant process is gravity. However, the laws of gravity are well understood. Furthermore, additional simplifications arise in certain regimes; e.g., at early stages in the evolution, density fluctuations are very small. One can therefore linearize the equations of gravity around the homogeneous Universe. If the length scale of the perturbations are much smaller than the characteristic scale of the Universe [a size given by $c/H(a)$], gravity can be approximated by the Newtonian equations. In addition, numerical simulations can follow the evolution of the density field under the influence of gravity, and great progress has been made in the level of detail these studies have achieved (e.g., Frenk et al. 1999; Springel et al. 2001).

We shall outline here a number of results which will be used in later sections; again, the reader is referred to the excellent textbooks mentioned at the beginning of this section for a much more detailed treatment.

Horizons

No signal propagates at speeds larger than c; at a given cosmic time t, this implies that the region within which matter has been in causal contact is finite, essentially given by $ct \sim c/H(t)$, where we used that $tH(t) \sim 1$. The size of this region is called the horizon size at time t. The comoving horizon size is

$$d_{\rm H} = \frac{c}{a\,H(a)} = \frac{c}{H_0}\,\Omega_{\rm m}^{-1/2}\,a^{1/2}\left(1+\frac{a_{\rm eq}}{a}\right)^{-1/2} , \qquad (112)$$

where in the second step we used the approximation for $a \ll 1$ when curvature and vacuum energy play hardly any role; cf. (94). As we shall see, the comoving horizon size at the epoch of matter and radiation equality is of particular importance and is

$$d_{\rm H}(a_{\rm eq}) = 2^{-1/2} c H_0^{-1} \Omega_{\rm m}^{-1/2} a_{\rm eq}^{1/2} \approx 12(\Omega_{\rm m} h^2)^{-1}\,{\rm Mpc}\,. \qquad (113)$$

Linear Density Evolution

If the density fluctuations are small, linear perturbation theory can be used to describe them. Specifically, one defines the *density contrast*

$$\delta(\mathbf{x}, t) := \frac{\rho(\mathbf{x}, t) - \bar\rho(t)}{\bar\rho(t)} , \qquad (114)$$

where $\bar{\rho}(t)$ is the mean cosmic matter density at time t, and $\mathbf{x}$ is the comoving spatial coordinate. The matter equations are linearized about the homogeneous model, and only terms of order δ are considered. If several matter components are relevant (e.g., non-relativistic matter and radiation, in early phases of the evolution), one defines a density contrast for each of them, and considers the (coupled) set of linear evolution equations. As soon as $|\delta| \sim 1$, this perturbation theory breaks down, and the full set of non-linear evolution equations needs to be treated (numerically).

If one considers fluctuations on a scale much smaller than the horizon scale, one can use Newtonian gravity. The relevant equations in this case are the Vlasov (or collisionless Boltzmann) equation and the Poisson equation; the former is usually approximated by the fluid equations, e.g. the continuity equation and the Euler equation. A homogeneous, expanding Universe is a solution to these equations, and the expansion factor follows the Friedmann equation (67). Setting $\rho(\mathbf{x}, t) = \bar{\rho}(t)[1 + \delta(\mathbf{x}, t)]$, these equations are then transformed into comoving coordinates, e.g. for the Poisson equation one finds

$$\nabla_x^2 \Phi = 4\pi G a^2 \bar{\rho} \delta = \frac{3H_0^2}{2a} \Omega_{\mathrm{m}} \delta \,. \tag{115}$$

Then, writing the velocity field as a sum of the Hubble expansion plus a small perturbation to it, one finds that the terms linear in δ lead to a single linear second-order differential equation in time whose coefficients do not depend on the spatial coordinates. Hence, there are two linearly independent solutions of the form $\delta(\mathbf{x}, t) = D_\pm(t)\, \Delta_\pm(\mathbf{x})$. One of the two functions, $D_-(t)$ say, decreases quickly with time and is therefore unimportant for structure growth; the other one grows with time, so that

$$\delta(\mathbf{x}, t) = D_+(t)\, \delta_0(\mathbf{x}) \tag{116}$$

is the relevant mode for structure growth. The function $D_+(t)$ is called the *linear growth factor*, which can be obtained from solving the aforementioned differential equation,

$$D_+ \propto H(a) \int_0^a \frac{\mathrm{d}a'}{[a'\, H(a')]^3} \,, \tag{117}$$

with the constant of proportionality chosen such that $D_+(t_0) = 1$. Because of this choice, one has $\delta(\mathbf{x}, t_0) = \delta_0(\mathbf{x})$; hence, $\delta_0(\mathbf{x})$ is the current density contrast provided the evolution of δ follows linear perturbation theory. Even if it does not, defining the field δ_0 is meaningful, since (116) still describes the evolution of the density contrast for epochs where δ was much smaller than unity. For obvious reasons, δ_0 is called the linearly-extrapolated density contrast. For an EdS Universe, the growth factor is $D_+(t) = a(t)$, for lower-density models, $D_+(t) \geq a(t)$ (see Fig. 14); Carroll et al. (1992) provide a fitting formula for D_+ for general cosmologies. Note that (116) predicts that the shape of fluctuations are time-independent in comoving coordinates, with only their amplitude being a function of time.

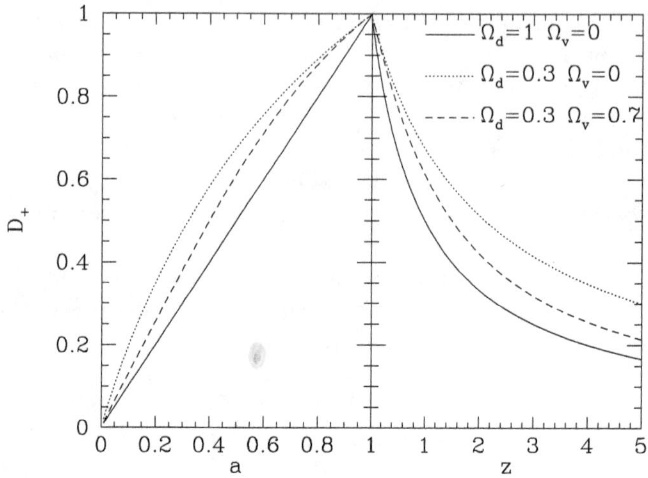

Fig. 14. The growth factor D_+ as a function of the scale factor a (*left*) and as a function of redshift (*right*), for three cosmological model: an EdS model ($\Omega_\mathrm{m} = 1$, $\Omega_\Lambda = 0$), a low density open model ($\Omega_\mathrm{m} = 0.3$, $\Omega_\Lambda = 0$), and a low-density flat Universe ($\Omega_\mathrm{m} = 0.3$, $\Omega_\Lambda = 0.7$). Notation: in this figure, $\Omega_\mathrm{m} = \Omega_\mathrm{D}$, $\Omega_\Lambda = \Omega_v$

Random Fields, Correlation Functions and Power Spectra

Cosmology will never be able to describe the specific density field of our Universe, since in order to do so, we would need to know the density fluctuation field $\delta(\mathbf{x}, t_\mathrm{i})$ at some initial time t_i. Instead, what a theory of structure formation should explain are the statistical properties of the density field as a function of time: how many clusters of galaxies per unit volume form as a function of redshift, how does matter cluster together, etc. This is analogous to statistical physics, where the behavior of a physical system is described by its macroscopic statistical properties, not by the trajectories of all molecules.

The density fluctuations $\delta(\boldsymbol{x})$ at some fixed time are considered to be a random field. A *random field* is characterized by the probability that a specific realization $\delta(\boldsymbol{x})$ of the density fluctuations occurs. This probability therefore is a mapping from the set of functions $\delta : \mathbb{R}^3 \to \mathbb{R}$ to $\mathbb{R}_0^+$. A conceptually simpler way to look at this is to assume that all probable realizations of the density field are 'smooth', so that $\delta(\boldsymbol{x})$ can be described, with sufficient accuracy, by its values on a regular grid in $\boldsymbol{x}$. Let $\boldsymbol{x}_i$ be a set of appropriately numbered gridpoints, and let $\delta_i = \delta(\boldsymbol{x}_i)$ be the density contrast at $\boldsymbol{x}_i$. The realization of the random field is then described by the (possibly infinite) set of the δ_i, and the random field is characterized by the joint probability distribution $p(\delta_1, \delta_2, \ldots)\mathrm{d}\delta_1\,\mathrm{d}\delta_2\,\ldots$ that $\delta(\boldsymbol{x}_i)$ lies within $\mathrm{d}\delta_i$ of δ_i. Hence, we have reduced the description of the random field to a joint probability distribution of (possibly infinite) discrete random variables. Since the Universe is assumed to be statistically homogeneous and isotropic, the density field should share these properties; this is formulated by the requirement that if all grid points are translated and rotated the same way, $\boldsymbol{x}_i \to \mathcal{R}(\boldsymbol{x}_i + \boldsymbol{y})$, where $\mathcal{R}$ is a rotation matrix and $\boldsymbol{y}$ a translation vector, the probability density p must remain unchanged.

More generally, let $g(\boldsymbol{x})$ be a real or complex homogeneous and isotropic random field in n dimensions. It is characterized by the probability distribution that a particular realization $g(\boldsymbol{x})$ can occur – note that we do not distinguish notationally between the random field and a particular realization, though these are two very different concepts. Hence, let $p(g(\boldsymbol{x}))\,\mathrm{d}g$ be the probability for the occurrence of the realization g within $\mathrm{d}g$, where $\mathrm{d}g$ is the volume element in function space. Let $\langle X \rangle$ denote the ensemble average of a quantity X, that is, we imagine to have many realizations of this random field with the same statistical properties, and we average X over all these realizations. Formally,

$$\langle X \rangle = \int \mathrm{d}g \; p(g(\boldsymbol{x}))\, X \; . \tag{118}$$

We shall assume that

$$\langle g(\boldsymbol{x}) \rangle = 0 \; , \tag{119}$$

so that the expectation value of g at every position $\boldsymbol{x}$ vanishes. Consider the (two-point) *correlation function*

$$\langle g(\boldsymbol{x}) g^*(\boldsymbol{y}) \rangle = C_{gg}(|\boldsymbol{x} - \boldsymbol{y}|) \; . \tag{120}$$

The correlation function can only depend on the separation $\boldsymbol{x} - \boldsymbol{y}$ of the two points because the homogeneity of the field g means that the correlator cannot depend on $\boldsymbol{x}$ and $\boldsymbol{y}$ individually. Furthermore, it depends only on $|\boldsymbol{x} - \boldsymbol{y}|$ because g is an isotropic random field. Note that C_{gg} is a real function, even if g is complex, as can be seen by taking the complex conjugate of (120), which is equivalent to interchanging $\boldsymbol{x}$ and $\boldsymbol{y}$, thus leaving the right-hand side unaffected. We define the Fourier transform of $g(\boldsymbol{x})$ as

$$\hat{g}(\boldsymbol{k}) = \int_{\mathbb{R}^n} \mathrm{d}^n x \; g(\boldsymbol{x})\, \mathrm{e}^{\mathrm{i}\boldsymbol{x}\cdot\boldsymbol{k}} \quad ; \quad g(\boldsymbol{x}) = \int_{\mathbb{R}^n} \frac{\mathrm{d}^n k}{(2\pi)^n}\, \hat{g}(\boldsymbol{k})\, \mathrm{e}^{-\mathrm{i}\boldsymbol{x}\cdot\boldsymbol{k}} \; . \tag{121}$$

We shall now calculate the ensemble average of $\langle \hat{g}(\boldsymbol{k})\, \hat{g}^*(\boldsymbol{k}') \rangle$, by inserting the Fourier representation,

$$\langle \hat{g}(\boldsymbol{k})\, \hat{g}^*(\boldsymbol{k}') \rangle = \int_{\mathbb{R}^n} \mathrm{d}^n x \; \mathrm{e}^{\mathrm{i}\boldsymbol{x}\cdot\boldsymbol{k}} \int_{\mathbb{R}^n} \mathrm{d}^n x' \; \mathrm{e}^{-\mathrm{i}\boldsymbol{x}'\cdot\boldsymbol{k}'} \langle g(\boldsymbol{x}) g^*(\boldsymbol{x}') \rangle \; . \tag{122}$$

Using (120) and substituting $\boldsymbol{x}' = \boldsymbol{x} + \boldsymbol{y}$, this becomes

$$\begin{aligned} \langle \hat{g}(\boldsymbol{k})\, \hat{g}^*(\boldsymbol{k}') \rangle &= \int_{\mathbb{R}^n} \mathrm{d}^n x \; \mathrm{e}^{\mathrm{i}\boldsymbol{x}\cdot\boldsymbol{k}} \int_{\mathbb{R}^n} \mathrm{d}^n y \; \mathrm{e}^{-\mathrm{i}(\boldsymbol{x}+\boldsymbol{y})\cdot\boldsymbol{k}'} \, C_{gg}(|\boldsymbol{y}|) \\ &= (2\pi)^n \delta_{\mathrm{D}}(\boldsymbol{k} - \boldsymbol{k}') \int_{\mathbb{R}^n} \mathrm{d}^n y \; \mathrm{e}^{\mathrm{i}\boldsymbol{y}\cdot\boldsymbol{k}}\, C_{gg}(|\boldsymbol{y}|) \\ &=: (2\pi)^n \delta_{\mathrm{D}}(\boldsymbol{k} - \boldsymbol{k}') P_g(|\boldsymbol{k}|) \; , \end{aligned} \tag{123}$$

where in the second step the x-integration was performed, and the final equality defines the *power spectrum of the quantity* g which obviously depends only on the modulus of $\boldsymbol{k}$. Hence, the power spectrum and the correlation function are Fourier transform pairs,

$$P_g(|\boldsymbol{k}|) = \int_{\mathbb{R}^n} \mathrm{d}^n y \, \mathrm{e}^{\mathrm{i}\boldsymbol{y}\cdot\boldsymbol{k}} \, C_{gg}(|\boldsymbol{y}|) \; . \tag{124}$$

Since the Fourier transform $\hat{g}(\boldsymbol{k})$ describes the same random field as $g(\boldsymbol{x})$, one can characterize the properties of the random field by the probability for the occurrence of a realization with Fourier transform $\hat{g}(\boldsymbol{k})$. As was done above for the real-space distribution, one can also discretize $\hat{g}$ on a grid in $\boldsymbol{k}$-space, denoted by $g_{\boldsymbol{k}}$.

A *Gaussian random field* is characterized by the properties that (1) the Fourier components $g_{\boldsymbol{k}}$ are mutually statistically independent, and that (2) the probability density for $g_{\boldsymbol{k}}$ is described by a Gaussian. The second property follows in many cases from the first, due to the central limit theorem. The first property implies that the phases of different Fourier components are mutually independent. A Gaussian random field is fully described by its power spectrum; a particular realization of such a random field can be obtained by drawing Gaussian deviates with dispersion $\sigma(\boldsymbol{k}) = \sqrt{P(|\boldsymbol{k}|)}$ and Fourier transforming the resulting $\hat{g}(\boldsymbol{k})$.

Gaussian random fields are almost universally used to describe the properties of the density perturbations in the early Universe. This is partly due to the argument given above, that the central limit theorem suggests that if the primordial perturbations were generated in a stochastic way (such as predicted from the inflationary theories), the resulting density field should be Gaussian. Another reason is that Gaussian random fields have very simple and convenient properties, which can be derived from the preceding results: The probability distribution of any linear combination of the random variable $g(\boldsymbol{x}_i)$ is a Gaussian, and more general, the joint probability distribution of a number M of linear combinations of the random variables $g(\boldsymbol{x}_i)$ is a multivariate Gaussian. In fact, this property can be used to define a Gaussian random field.

The Power Spectrum

Defining $\hat{\delta}(\boldsymbol{k}, t)$ to be the Fourier transform of the density fluctuation field, then (116) immediately yields that $\hat{\delta}(\boldsymbol{k}, t) = D_+(t)\hat{\delta}(\boldsymbol{k}, t_0)$, which implies for the corresponding power spectrum

$$P_\delta(k, t) = D_+^2(t) P_0(k) \; , \tag{125}$$

where $P_0(k)$ is the linearly extrapolated power spectrum which would be the true present-day power spectrum if the fluctuations follows the linear evolution characterized by (116). Furthermore, the factorization of $\hat{\delta}(\boldsymbol{k},t)$ implies that each Fourier mode evolves independently in time.

There are several situations where (116) is invalid. The obvious one is when the density contrast approaches unity, where the linearization of the evolution equations breaks down. We shall discuss this case further below. Second, since the comoving horizon scale grows in time, the characteristic comoving length scale $\lambda = 2\pi/k$ of each Fourier mode was larger than the horizon size some time in the past. For such superhorizon fluctuations, Newtonian theory of gravity necessarily breaks down, and one needs to use linear perturbation theory of the Einstein equations. Third, for $a \lesssim a_{\rm eq}$ radiation dominated the matter contents of the Universe, which affect the growth of structure. Fourth, particle populations with an appreciable intrinsic velocity dispersion will not simply fall into the potential wells, but can stream away from them; this certainly applies to all relativistic species. One distinguishes between cold dark matter (CDM), where the characteristic particle velocities are highly non-relativistic, $\sigma_v \ll c$, at the time when $a = a_{\rm eq}$, and hot dark matter (HDM), when the matter particles are relativistic at this epoch. If the Universe is dominated by HDM, small-scale fluctuations would be smeared out due to free-streaming, and the first objects to form would be clusters or superclusters of galaxies. Since the large-scale matter distribution obtained for such models are very different in several respects from the observed one, it is concluded that HDM (such as massive neutrinos) can only contribute a very small fraction to $\Omega_{\rm m}$. The favored model is one which is dominated by CDM.

The effects of radiation domination in the early Universe and the initial superhorizon scale of density modes can be summarized as follows: Suppose at some very early time $t_{\rm i}$ when all Fourier modes of interest had scales much larger than the horizon scale then, the power spectrum of the density fluctuations was $P_{\rm i}(k)$. The power spectrum at some later time when all scales of interest are much smaller than the horizon is then

$$P_\delta(k,t) = T^2(k)\,\frac{D_+^2(t)}{D_+^2(t_{\rm i})}P_{\rm i}(k)\;, \tag{126}$$

where the *transfer function* $T(k)$ accounts for the aforementioned effects. It can be calculated, and accurate approximation formulae for it are available (e.g., Bardeen et al. 1986; Eisenstein and Hu 1998). For large scales, i.e., small k, $T \approx 1$. For large k, $T(k) \propto k^{-2}$ in a CDM Universe [$T(k)$ decreases exponentially with k for a HDM model]. The transition between the two regimes depends on the scale of the comoving horizon at the time of matter-radiation equality, i.e., on $d_{\rm H}(a_{\rm eq})$. As cosmological length scales are measured in $h^{-1}\,$Mpc, this length is $\propto (\Omega_{\rm m}h)^{-1}$. Thus, the shape of the transfer function is determined by the *shape parameter*

$$\Gamma_{\rm spect} = h\,\Omega_{\rm m}\,\exp\left[-\Omega_{\rm b}\left(1+\sqrt{2h}\Omega_{\rm m}^{-1}\right)\right]\;, \tag{127}$$

and the final factor yields a small correction which accounts for the baryonic contribution (with density parameter $\Omega_{\rm b}$) to the density of the Universe. As the shape of the power spectrum, which is measurable, has imprinted on it the comoving horizon scale $d_{\rm H}(a_{\rm eq})$, this scale is actually observable.

Knowing the transfer function, one needs a prescription for the power spectrum $P_{\rm i}(k)$ at some very early times to predict P_δ for later stages of the evolution. Since at $t_{\rm i}$ all modes of relevance are much larger than the horizon scale, there is no characteristic length scale available; therefore, one assumes that the primordial power spectrum was a power law, $P_{\rm i}(k) \propto k^n$. Furthermore, if it is assumed that the total power of the fluctuations at the time when their scale equals the horizon size is independent of k, and for that matter, independent of time, then $n = 1$. Such a primordial power spectrum is called Harrison–Zeldovich power spectrum, and is also the favored value in theories which explain the origin of primordial fluctuations as initial quantum fluctuations blown up in a period of exponential expansion, the inflation period.

The linear power spectrum is thus determined in terms of n and the shape parameter, except for the overall normalization. Several methods exist to fix this normalization, three of which are mentioned here.

1. Normalization by density fluctuations in a sphere. The relative fluctuations of the galaxy number density in the local Universe, $\delta n/n$, is of order unity if one considers spheres of radius $R = 8h^{-1}\,$Mpc. If one assumes that the galaxies accurately trace the underlying dark matter field, this observation would imply that the fluctuation field $\delta(\boldsymbol{x}, t_0)$, averaged over a scale of $R = 8h^{-1}\,$Mpc, has a dispersion of 1. However, there is no guarantee that the galaxy number density field closely follows the dark matter distribution (see Fig. 15). Nevertheless, one might suspect that the

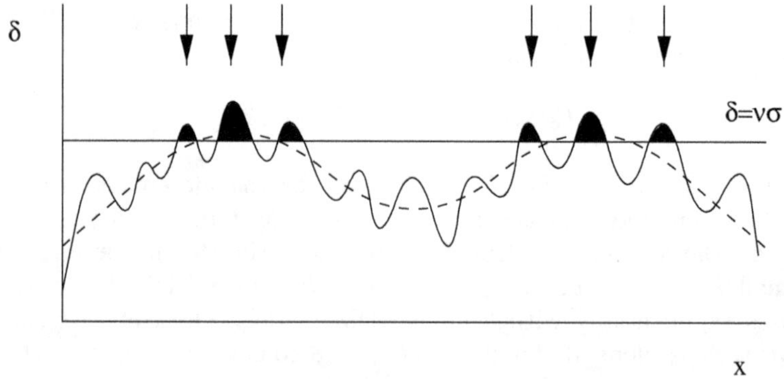

Fig. 15. This figure depicts a one-dimensional cut through a density field which contains power on small and large scales, the latter being shown by the *dashed curve*. If it is assumed that galaxies form only at locations where the density exceeds a critical threshold, here indicated by the *horizontal line*, then it is easily seen that they are more strongly clustered than the matter field itself, as this threshold is exceeded predominantly at the peaks of the long wavelength perturbations; therefore, this would lead to a bias of the galaxy distribution relative to that of the underlying matter (from Peacock 2003)

galaxy density is large at those locations where the dark matter density is also large. In particular, galaxies might find it easier to form in peaks of the dark matter distribution, and therefore galaxies can be clustered more than the dark matter. One often summarizes our ignorance about the relative distribution of galaxies and dark matter into a *linear bias factor* b, defined such that the fluctuations of the galaxy number density are a factor b larger than the fluctuations of the underlying dark matter distribution (for a detailed discussion on biasing, see, e.g., Bardeen et al. 1986; Kauffmann, Nusser and Steinmetz 1997). We define the density field smoothed on a scale R by

$$\delta_R(\boldsymbol{x}) = \int_{\mathbb{R}^3} \mathrm{d}^3 y\, \delta_0(\boldsymbol{y})\, W_R(|\boldsymbol{x}-\boldsymbol{y}|) \;, \tag{128}$$

where $W_R(x)$ is a filter function, normalized such that $\int \mathrm{d}^3 x W_R(x) = 1$. From the convolution theorem for Fourier transforms, one finds $\hat{\delta}_R(\boldsymbol{k}) = \hat{\delta}_0(\boldsymbol{k})\, \hat{W}_R(\boldsymbol{k})$, and the power spectrum of the smoothed field is $P_R(k) = |\hat{W}_R(k)|^2\, P_0(k)$. The dispersion of the smoothed density field is then

$$\sigma^2(R) = \langle \delta_R^2(\boldsymbol{x}) \rangle = \int \frac{\mathrm{d}^3 k}{(2\pi)^3}\, P_R(k) = \int \frac{\mathrm{d}^3 k}{(2\pi)^3} \left| \hat{W}_R(k) \right|^2 P_0(k) \;, \tag{129}$$

where we made use of (124). Note that $\sigma(R)$ describes the dispersion of the smoothed version of the linearly extrapolated density field today. If we take a so-called top-hat filter, which is constant for $x \le R$ and zero otherwise, one has

$$W_R(x) = \frac{3}{4\pi R^3} \mathrm{H}(R-x) \quad \longrightarrow \quad \hat{W}_R(k) = 3\frac{\sin kR - kR\cos kR}{(kR)^3} \;. \tag{130}$$

Coming back to the normalization, the dispersion in the galaxy number counts then implies that

$$\sigma(8h^{-1}\mathrm{Mpc}) \equiv \sigma_8 \approx \frac{1}{b} \;. \tag{131}$$

2. Normalization through the CMB. The DMR experiment on the COBE satellite mission (see, e.g., Bond 1996; Smoot 1997 for reviews) has detected the anisotropy of the microwave background on scales above $\sim$5 degrees. The degree of anisotropy is proportional to the fluctuation spectrum, and thus can be used directly to normalize the spectrum. The normalization of the power spectrum is hampered by the uncertainty whether the CMB anisotropy on large angular scales is caused solely by scalar perturbations (i.e., density and associated adiabatic temperature fluctuations), or whether tensor perturbations (gravitational waves) have been present at the recombination epoch. On smaller scales, the expected contribution from tensor modes becomes very small. The first

year data of the WMAP satellite (Bennett et al. 2003) have provided a much more accurate determination of the power spectrum normalization, yielding $\sigma_8 = 0.9 \pm 0.1$ (Spergel et al. 2003). Since the CMB anisotropies measured by WMAP probe inhomogeneities on considerably larger scales than $8h^{-1}$ Mpc, translating their amplitudes into a value of σ_8 depends on the shape of the power spectrum. In Sect. 6.3, we shall give the best current estimates of the whole set of cosmological parameters.

3. Normalization by the local abundance of clusters. The number density of clusters as a function of their mass can be estimated analytically in terms of the power spectrum $P_\delta(k)$, as will be shown later. By comparing the observed number density of clusters with these prediction, the normalization of the power spectrum is determined (e.g., Eke et al. 1996). This comparison yields normalizations which, when expressed in terms of σ_8, are of the form

$$\sigma_8 \approx 0.52 \Omega_{\mathrm{m}}^{-0.52+0.13\Omega_{\mathrm{m}}} \quad \text{for} \quad \Omega_{\mathrm{m}} + \Omega_\Lambda = 1 \; . \tag{132}$$

These estimates are relatively insensitive to the shape of the power spectrum (i.e., of Γ_{spect}), because the mass contained in a massive cluster is about the mass contained in a comoving sphere of radius $8h^{-1}$ Mpc, so that the cluster abundance directly measures σ_8. However, there has been some recent claims that (132) may overestimate σ_8 (e.g., Viana et al. 2002, and references therein). The main problem of the cluster normalization is to obtain a well-selected sample of clusters (e.g., from X-ray surveys) and to determine their masses reliably.

As we shall discuss in WL (Part 3), lensing by the large-scale structure (LSS) provides a powerful tool to determine the normalization of the power spectrum. As for clusters, this method yields, to lowest order, a degeneracy between σ_8 and the density parameter Ω_{m}.

To summarize, a CDM-dominated Universe has a (linear) power spectrum given by (126), which is determined by a few parameters. Together with the assumption that the primordial density field was Gaussian, then the evolved field will also be Gaussian as long as it stays in the linear regime. Thus, in a statistical sense, the density field is fully specified, so that this cosmological model is predictive.

Non-Linear Evolution

When $|\delta|$ is no longer much smaller than unity, the linear perturbation theory breaks down. The first idea, to consider higher-order perturbations, does not really solve the problem: the perturbation series is not converging, and only slightly larger perturbations of δ can be described satisfactorily. In addition, the fluid equations cease to be valid, since due to the converging velocity field, streams of matter start to intersect, and thus the Vlasov equation needs to be employed. A different approach, Lagrangian perturbation theory (Zeldovich

1970; Buchert and Ehlers 1993) is substantially more successful. However, with the advent of high-speed computers with large memory, the need for (semi)analytic approximations decreases, as the evolution of the density field can be obtained from N-body simulations tracing the dark matter particles (e.g. White et al. 1987; Pearce et al. 2001; Navarro et al. 2004). Each such simulation yields an evolved realization from an initially Gaussian density field with power spectrum according to (126), starting at a high redshift. Such simulations can be either used directly to study the properties of the matter distribution, or can be used to derive fit formulae for various quantities of interest, some of which will be discussed below.

Using a scaling argument, Hamilton et al. (1991) derived an approximate equation relating the linearly evolved power spectrum to the fully non-linear power spectrum P_δ; this equation contains a single function, whose parameters can be fitted to the results of N-body simulations. This approximation, later generalized and refined by Jain, Mo and White (1995) and Peacock and Dodds (1996), is truly remarkable as it yields an accurate description of the fully non-linear power spectrum for all values of k and t; example power spectra are displayed in Fig. 16. More recently, an even more accurate expression has been obtained by Smith et al. (2003).

6.2 Halo Abundance and Profile

Gravitationally bound objects like galaxies and clusters are of course highly non-linear structures in the Universe – their average density contrast is much larger that unity. Nevertheless, there are analytical approaches to determine their number density as a function of mass and redshift. The best known of these is the Press–Schechter approach, and more refined ones are deviates of it.

The Mass Function of Halos

The Press and Schechter (1974) approach is based on two considerations: the time evolution of a spherical overdensity and its collapse, and the statistical (Gaussian) property of the initial density field.

The spherical collapse model considers an overdensity with spherically symmetric density distribution. According to Birkhoff theorem, the evolution of a mass shell M is independent of the radial density profile at larger radii, as long as shells of matter do not cross each other. The radius of the mass shell as a function of time then follows an equation of motion. At early time, when the average density contrast δ inside the mass shell is small, the expansion closely follows the Hubble expansion, but being slightly slower, the density contrast grows. This leads to an increased deviation from the Hubble expansion, and thus further increased density contrast. Provided the latter is large enough, the expansion of the mass shell can come to a hold (at time $t_{\rm max}$, say), and the radius will decrease from there on: the shell is collapsing.

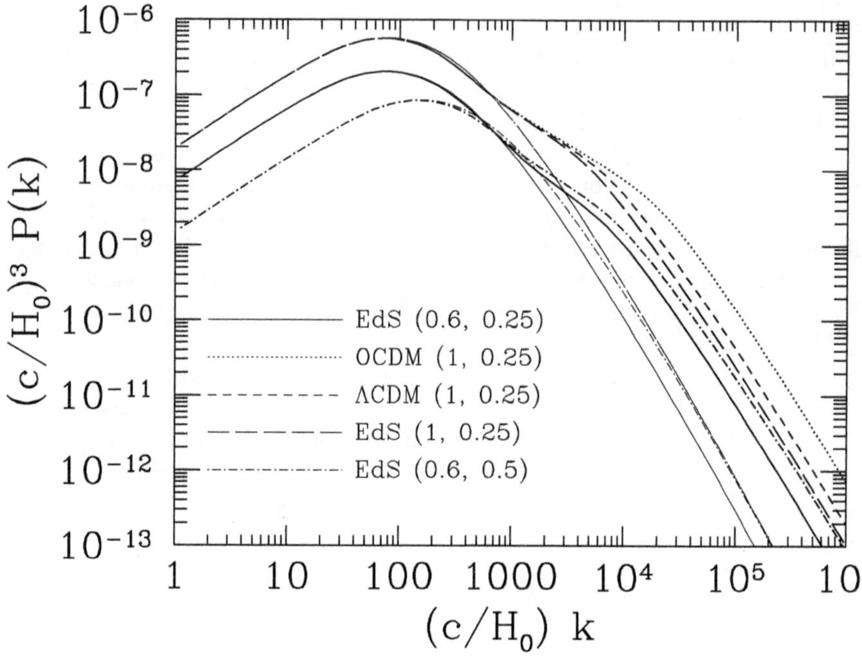

Fig. 16. The power spectrum of the cosmic density fluctuations at current epoch, for various cosmological models. $(c/H_0)^3 P(k)$ is plotted as a function of $(c/H_0)k$. The cosmological models are distinguished by line types [with values in parenthesis denoting $(\sigma_8, \Gamma_{\rm spect})$], with *thin lines* displaying the linearly extrapolated power spectrum, and *thick lines* the fully non-linear power spectrum following the prescription of Peacock and Dodds (1996). The *solid lines* correspond to an Einstein-de Sitter Universe, the *dotted lines* to an open Universe with $\Omega_{\rm m} = 0.3$, $\Omega_\Lambda = 0$, the *short-dashed lines* to a flat low-density Universe with $\Omega_{\rm m} = 0.3$, $\Omega_\Lambda = 0.7$; those three models are approximately normalized by the present-day cluster abundance, as discussed at the end of Sect. 6.1, and have the shape parameter $\Gamma_{\rm spect} = 0.25$. The remaining two models are EdS cosmologies, with different shape parameter or different normalization. The linear power spectrum depends only on σ_8 and $\Gamma_{\rm spect}$ (once the primordial slope $n = 1$ is fixed), so that those three models are degenerate. This degeneracy is broken in the non-linear spectrum. The non-linear spectrum deviates from the linear prediction at $(c/H_0)k \geq 1,000$, corresponding to length scales of $L = 2\pi/k \leq 20h^{-1}$ Mpc

The symmetry of the equation of motion with respect to $t \to -t$ then predicts that the collapse of the mass shell to very small radii takes the same time as the expansion, so that the shell collapses at time $t_{\rm coll} = 2t_{\rm max}$. If the mass distribution was exactly symmetric, the collapse would indeed proceed to basically a single point; however, small-scale inhomogeneities of the matter distribution will deflect the matter particles from their radial orbit, thereby enhancing the density fluctuations, and very quickly the orbits of particles become randomized. During this process the mass overdensity will virialize, and this process takes place with a time scale comparable to $t_{\rm coll}$ (violent relaxation; Lynden-Bell 1967; Binney and Tremaine 1987). The final state is then a spherical halo of (dark) matter in nearly virial equilibrium.

In an EdS model, the various parameters of this model can be calculated analytically: in order for the collapse to occur before the present time, $t_{\rm coll} \le t_0$, the linearly extrapolated mean density contrast δ_0 of the mass shell must satisfy $\delta_0 \ge \delta_{\rm c} = 3(12\pi)^{2/3}/20 \simeq 1.69$, and the condition that the collapse happened before redshift z is that $\delta_0 \ge \delta_{\rm c}(1+z)$. The mean density of the virialized halo is $\langle\rho\rangle = 18\pi^2\rho_{\rm cr}(1+z_{\rm coll})^3 \simeq 178(1+z_{\rm coll})^3\rho_{\rm cr}$. Note that $(1+z_{\rm coll})^3\rho_{\rm cr}$ is the critical density of the EdS Universe at redshift $z_{\rm coll}$. Hence, the mean overdensity of a virialized halo is of order 200 times the critical density of the Universe at the time of formation. For other cosmological parameters, these numbers change, but have been calculated (e.g., Eke et al. 1996). Given the idealization of the spherical model, one often defines the virial radius $r_{\rm vir}$ of a dark matter halo to be the radius within which the mean density is 200 times the critical density of the Universe.

Next, consider the linear density field at some early time being smoothed with a top-hat filter of comoving radius R [see (128)]; this corresponds to a mass inside the filter scale of $M = 4\pi R^3\rho_{\rm m0}/3$. A peak with density contrast $\delta_R \ge \delta_{\rm c}(1+z)$ in this smoothed density field will then collapse before redshift z to a virialized halo of mass M. Given that the linear density fluctuations are assumed to be Gaussian, one can calculate the abundance of peaks exceeding a certain threshold, and therefore the abundance of halos of a given mass (determined by the filter scale R) that form before redshift z. This then yields the Press–Schechter mass formula for the comoving density $n(M,z)\,{\rm d}M$ of halos of mass within ${\rm d}M$ of M at redshift z,

$$n(M,z) = -\frac{2\Omega_{\rm m}\rho_{\rm cr}}{\sqrt{2\pi}M}\,\frac{\delta_{\rm c}(z)}{\sigma^2(R)}\,\frac{{\rm d}\sigma(R)}{{\rm d}M}\,\exp\left(-\frac{\delta_{\rm c}(z)}{2\sigma^2(R)}\right), \tag{133}$$

where the radius R is related to the mass by the equation given above, $\sigma(R)$ is given by (129), and $\delta_{\rm c}(z)$ is the linearly extrapolated density contrast needed for a mass shell to collapse before redshift z. The mass spectrum of halos behaves approximately as a power law for masses $M \lesssim M_*(z)$, where $M_*(z)$ is the mass scale at redshift z at which the density field becomes non-linear; it is defined implicitly through $\sigma^2(R_*)D_+^2(z) = 1$ [cf. (129), with the linear power spectrum at redshift z being $D_+^2(z)$ times the one today]. For masses above $M_*(z)$, the mass function decreases exponentially. The redshift evolution of the mass function depends on the cosmological parameters: in low-density Universes, the evolution with redshift is slower than in an EdS Universe. Thus, for a given abundance of cluster-mass halos today, the expected number of massive clusters at high redshift is much smaller for an EdS model than for a low-density Universe (see Fig. 14). The normalization of the matter power spectrum through the local cluster abundance, as discussed in Sect. 6.1, is based on the prediction of the Press–Schechter function (133) or variants thereof. Comparison of the mass function with those obtained from N-body simulations (see Fig. 17), in which halos can be identified using a variety of techniques, leads to the conclusion that, although the Press–Schechter formula provides a very useful approximation of the halo abundance, it slightly over-predicts the number of halos with mass $\lesssim M_*(z)$ and underpredicts those with $M \gtrsim M_*(z)$. Various refinements to the original Press–Schechter approach have been conducted, including the collapse of ellipsoidal mass overdensities (Sheth and Tormen 1999). Jenkins et al. (2001) provided an accurate fit to the halo abundance obtained from their numerical models; it is very similar to the one obtained by Sheth and Tormen, and shares the simplicity of the Press–Schechter formula.

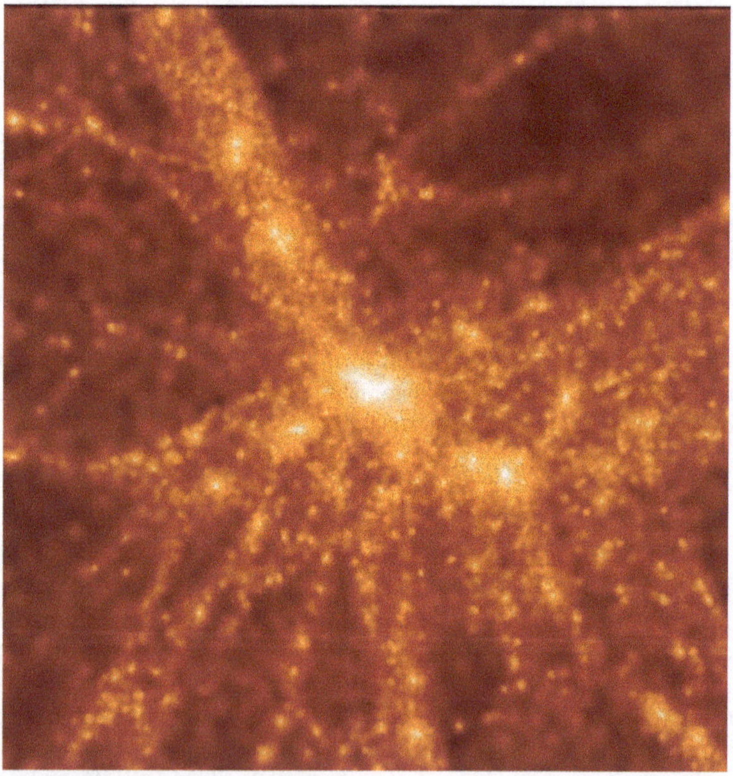

Fig. 17. One of the clusters obtained from N-body simulations. Shown is the dark matter distribution at redshift $z = 0$ in a region of $21 \times 21 \times 8 \left(h^{-1}\text{Mpc}\right)^3$. The strongly structured mass in and around the cluster is clearly visible; it has been formed through successive mergers of subclusters and groups (taken from the GIF collaboration; Kauffmann et al. 1999a,b)

The Press and Schechter approach can also be generalized to include statistical information about the merger history of dark matter halos (e.g., Bond et al. 1991; Lacey & Coles 1993). These merger histories form the starting point for semi-analytic models of galaxy formation and evolution (e.g., Kauffmann et al. 1993, 1994); see Figs. 17 and 18.

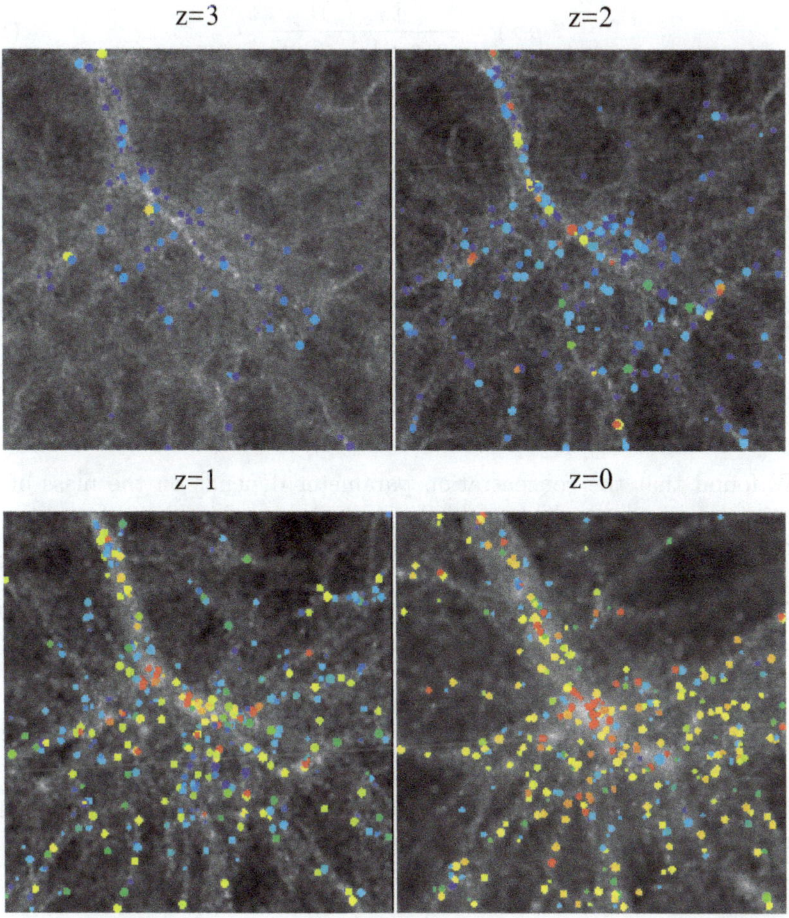

Fig. 18. The redshift evolution of the galaxy distribution, obtained by semi-analytic modeling based on the dark matter distribution as obtained from N-body simulation. The matter distribution is shown as gray scales and for $z = 0$ is the same as that shown in Fig. 17. Colors indicate the mean age of the stellar population in these galaxies, with blue (red) indicating a young (old) population; red galaxies are seen to be preferentially found in clusters. At high redshifts, there are of course no old stellar populations (taken from the GIF collaboration; Kauffmann et al. 1999a,b)

The 'Universal' Density Profile

From the numerical simulations, one can investigate the density profile of typical dark matter halos. Navarro, Frenk and White (1997; hereafter NFW) found that the density, averaged over spheres, of dark matter halos is described by a 'universal' profile given by

$$\rho(r) = \frac{\delta_c \rho_{cr}(z)}{(r/r_s)(1+r/r_s)^2} , \tag{134}$$

which is shallower than isothermal (r^{-2}) near the halo center and steeper than isothermal for $r \gtrsim r_s$. The *virial radius* is denoted by r_{200} and is the radius inside which the mean mass density of the halo equals $200\rho_{cr}(z)$, where $\rho_{cr}(z) = \frac{3H^2(z)}{8\pi G}$ is the critical density of the Universe at the redshift of the halo. Hence, r_{200} immediately yields the mass of the halos, $M = 200\rho_{cr}(z)\, 4\pi r_{200}^3/3$. The ratio of the virial radius r_{200} and the scale radius r_s is called the *concentration parameter* $c = r_{200}/r_s$. From the definition of r_{200}, the parameter δ_c can be related to the concentration parameter through

$$\delta_c = \frac{200}{3} \frac{c^3}{\ln(1+c) - c/(1+c)} . \tag{135}$$

NFW found that the concentration parameter depends on the mass of the halo; it is smaller for higher-mass halos. Takada and Jain (2002) found for the concentration parameter the dependence $c = c_0(1+z)^{-1}[M/M_*(z=0)]^{-\beta}$, with $c_0 \sim 10$ and $\beta \sim 0.2$.

There is no general agreement on the true 'universality' of the NFW profile; different groups obtain slightly different profile slopes for $r \to 0$, whereas the behavior $\rho \propto r^{-3}$ for large r is reproduced by other teams as well.

The NFW Gravitational Lens

The gravitational lensing properties of the NFW profile has been discussed by Bartelmann (1996), Golse and Kneib (2002) and others. By projecting (134) along the line-of-sight, the surface mass density can be written as

$$\kappa(\theta) = \kappa_k f(\theta/\theta_s) , \tag{136}$$

where $\theta_s = r_s/D_D^{ang}$ is the angular scale radius,

$$\kappa_k = \frac{2 r_s \delta_c \rho_{cr}(z)}{\Sigma_{cr}} \tag{137}$$

the characteristic surface mass density, and

$$f(x) = \frac{1}{x^2 - 1} [1 - \mathcal{F}(x)] , \tag{138}$$

where

$$\mathcal{F}(x) = \frac{\operatorname{acosh}(1/x)}{\sqrt{1-x^2}} \; ; \quad \mathcal{F}(x) = \frac{\operatorname{acos}(1/x)}{\sqrt{x^2-1}} \tag{139}$$

for $x < 1$ and $x > 1$, respectively; taking the limit $x \to 1$ in both cases yields $f(1) = 1/3$. Similarly, the mean surface mass density $\bar{\kappa}(\theta)$ inside θ is given by $\bar{\kappa}(\theta) = \kappa_k h(\theta/\theta_s)$, with

$$h(x) = \frac{2}{x^2}\left[\mathcal{F}(x) + \ln\left(\frac{x}{2}\right)\right] , \tag{140}$$

with $h(1) = 2[1 - \ln(2)]$. The absolute value of the shear is then, as usual, $\gamma = \bar{\kappa} - \kappa$. The surface mass density diverges logarithmically as $\theta \to 0$; therefore, the NWF lens is critical and thus has a tangential and radial critical curve. Since the deflection angle $\alpha = \theta\bar{\kappa}$ is a smooth function also at $\theta = 0$, the NFW lens produces either one or three images, i.e., the peculiarities of the SIS model do not occur here.

6.3 The Concordance Model

The past few years have seen great advances in the determination of the cosmological parameters, and the progress is continuing. At present, a set of cosmological parameters can be defined which seems to be in accord with all cosmological observations. Particularly notable is the fact that for each of the parameters there are at least two very different methods for its determination. Here we briefly mention the major results which led to the current concordance model, excluding the results from gravitational lensing, that will be described in the later sections.

The main observational results which led to the current set of cosmological parameters came from the following sources:

- Anisotropies in the CMB. The CMB is nearly isotropic, but there are temperature fluctuations of order $\Delta T/T \sim 10^{-5}$ superimposed on the isotropic field (plus, there is the dipole anisotropy reflecting our peculiar motion). The *primary anisotropies* are due to density, temperature, and potential inhomogeneities at the time of recombination, together with corresponding peculiar velocities of the matter at this epoch (see Hu and Dodelson 2002 for a recent review on the physics of the CMB anisotropies). Furthermore, anisotropies can be generated and modified during the propagation of the light from $z \sim 1,100$ to today, causing *secondary anisotropies*. The angular power spectrum of these anisotropies depends on basically all cosmological parameters; therefore, their measurements in recent years have yielded a wealth of cosmological constraints. Measurements before the release of the first WMAP data were summarized and analyzed by Wang et al. (2003). The breathtaking results obtained by WMAP (Bennett et al. 2003; Spergel et al. 2003) have confirmed earlier measurements, but with substantially smaller uncertainties.

- Light element abundances. During the first three minutes after the Big Bang, the Universe was hot and dense enough to form the lightest chemical elements. Their primordial abundances depend on the baryon density Ω_b of the Universe, as well as on the number of neutrino species which determine the expansion rate in the radiation-dominated phase. Whereas the primordial abundance of helium-4 (about 25% by mass) is fairly insensitive to Ω_b, the deuterium abundance is a very strong function of Ω_b. In the past few years, observations of intergalactic clouds in the form of the Lyman-α absorption lines in high-redshift quasars have yielded determinations of the deuterium-to-hydrogen abundance (e.g., Tytler et al. 2000). These measurements are extremely valuable, since this intergalactic material is thought to be fairly unprocessed chemically, and thus still reflects the primordial abundance ratios.
- Type Ia supernovae. This type of supernova explosions is believed to origin from the white dwarfs which just exceed their maximum possible (Chandasekhar) mass; hence, they all would have essentially the same explosion energy, which makes them excellent candidates for standard candles. In fact, although their maximum luminosity shows an intrinsic spread, this variation is correlated with the characteristic width of the light curve, which has been used for an empirical correction of the maximum luminosity; after this correction, the remaining spread in their peak luminosities is very small. No redshift evolution in their intrinsic properties (such as rest-frame colors or spectra) has been found. Hence, by measuring the flux at maximum light of SN Ia at different redshifts, one can measure the luminosity distance as a function of z; on the other hand, the luminosity distance depends on the cosmological parameters Ω_m and Ω_Λ. Two teams have systematically searched for high-redshift supernovae (Schmidt et al. 1998; Perlmutter et al. 1999), and constructed the redshift–distance relation from their events, extending up to $z \sim 1$. By now, many SN Ia have been found even with redshifts $\gtrsim 1$ (e.g. Riess et al. 2004); the analysis of their brightness shows the expected behavior for a Universe which is currently accelerating, but has been decelerating before $z \sim 0.7$, as expected in a model with $\Omega_\Lambda \sim 0.7$, $\Omega_\mathrm{m} \sim 0.3$, where the transition to vacuum domination has occurred rather recently.
- Large scale structure. The relation between the distribution of galaxies and the underlying dark matter distribution is not known; on the other hand, it is at least plausible that they follow each other closely. In particular on large spatial scales, say where the density field is still in the linear regime ($L \gtrsim 10h^{-1}\,\mathrm{Mpc}$), one assumes that the number density fluctuations of galaxies is proportional to that of the dark matter, with the proportionality factor being called bias factor b. Then, the power spectrum of the galaxy distribution is b^2 times the matter power spectrum. Large two-dimensional and three-dimensional (i.e., redshift) galaxy surveys have recently been performed (in particular, the 2dF galaxy redshift survey; see e.g. Hawkins et al. 2003; the Sloan Digital Sky Survey,

e.g, Tegmark et al. 2004) to construct the galaxy power spectrum. From that, the shape of the matter power spectrum, i.e., the shape parameter $\Gamma_{\rm spect}$, can be determined, among other parameters (e.g., Peacock 2003).

- Statistics of the Ly-α forest. The spectra of all QSOs show a dense ensemble of absorption lines shortward of the Ly-α emission line. These absorption lines are due to the inhomogeneous distribution of intergalactic hydrogen. At high redshift ($z \sim 3$) when these lines are observable in the optical spectrum of QSOs, the corresponding density field of the gas is still in the linear regime. One therefore expects that the gas follows the underlying dark matter closely. The gas is in photoionization equilibrium with the UV radiation field, and obeys a simple temperature-density relation. Most of the unknown physical parameters can be put into a mean absorption which can be measured from the flux decrement across the Ly-α emission line. The statistics of the Ly-α absorbers therefore probes the corresponding matter density fluctuation spectrum. Recently, large samples of QSO spectra became available for this kind of analysis; see Kim et al. (2004) and Seljak et al. (2005) for recent results.
- Cosmology from galaxy clusters. Clusters provide a wealth of cosmological information: their abundance depends strongly on the normalization σ_8 of the power spectrum, as mentioned in Sect. 6.1, the evolution of their abundance with redshift probes the rate of growth of structure, which in turn depends on the density parameters, and their correlation function probes the shape of the power spectrum on large scales. In addition, clusters are so large that one can assume their baryon-to-mass ratio $f_{\rm b}$ being very similar to the cosmological mean of this ratio. Since the baryon contents of clusters can be measured from their X-rays, and their mass can be determined by X-rays, dynamics of their member galaxies and by gravitational lensing, this baryon fraction can be determined and yields $f_{\rm b} \approx 0.17$, with rather little scatter between clusters.

From these and several other methods, the set of cosmological parameters can be determined. It must be realized that the various parameters are correlated in a given data set, and sometime rather degenerate [such as seen in (132)]. Estimates of one parameter need to be obtained by marginalizing over the remaining ones, and the estimated error bars will depend on how many parameters were considered in the analysis. This is not the place to discuss these issues; we therefore present the currently 'best' estimates and approximate 1-σ error bars of the relevant parameters. In Fig. 19, some of the constraints on the density parameters are summarized, and Fig. 20 illustrates the concordance in the determination of the power spectrum from a large variety of different methods.

- The Hubble constant, as determined from the Cepheid distances within the Hubble Key Project (Freedman et al. 2001) and from combining CMB data with galaxy redshift surveys, is

$$H_0 \approx 71\,{\rm km\,s^{-1}\,Mpc^{-1}}\;, \quad {\rm or} \quad h \approx 0.71 \pm 0.04, \tag{141}$$

where the error from the CMB plus LSS is formally smaller than from the Hubble Key Project, by about a factor of two.

- From the deuterium abundance in QSO absorption lines, as well as from WMAP combined with LSS and Ly-α statistics, the baryonic density parameter is estimated to be

$$h^2 \Omega_{\rm b} \approx 0.023 \pm 0.002 \; , \tag{142}$$

where we give a slightly larger error than quoted in some recent papers. Again, consistent results are obtained from totally different methods.

- The CMB anisotropies constrain the total density of the Universe to be very close to unity,

$$\Omega_{\rm m} + \Omega_\Lambda \approx 1.02 \pm 0.02 \; ; \tag{143}$$

combining the results from SN Ia studies with the evolution of the cluster abundance, a similar conclusion is obtained, though with a larger error estimate.

- The supernovae projects yield a joint constraint on the density parameters as is shown in Fig. 19, and hence by themselves require, a non-zero cosmological constant.

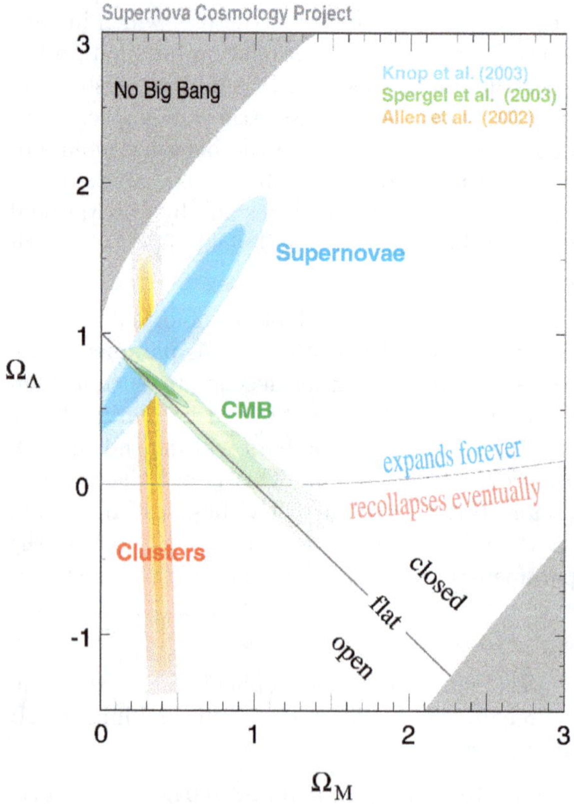

Fig. 19. This figure summarizes several constraints in the $\Omega_{\rm m}$–Ω_Λ-plane, from the WMAP measurement of the CMB, the abundance of clusters, and the high-redshift SN Ia. Each of these cosmological test probe very different physics; nevertheless, they yield consistent results – this form the basis of the concordance model (taken from Knop et al. 2003)

- The 2dF and SDSS galaxy redshift surveys determine the shape parameter; in particular, from the 2dFGRS one finds

$$\Gamma_{\rm spect} \approx \Omega_{\rm m} h \approx 0.18 \pm 0.02 \;, \tag{144}$$

while the SDSS yield a slightly larger value.
- CMB anisotropies, together with the LSS, yield a value for the matter density parameter of

$$\Omega_{\rm m} = 0.29 \pm 0.04 \;, \tag{145}$$

a value that is in excellent agreement from cluster abundance evolution, the determination of the shape parameter $\Gamma_{\rm spect}$ from the LSS, and the baryon fraction $f_{\rm b}$ in clusters, when combined with the baryon density $\Omega_{\rm b}$ and the value for the Hubble constant.
- The CMB anisotropy also determines the slope of the primordial density fluctuation spectrum, which turns out to be close to the Harrison–Zeldovich value, $n \approx 0.98 \pm 0.02$.
- Perhaps the parameter with the largest discrepancies between different methods is the normalization of the power spectrum; we quote here the value from Seljak et al. (2005), obtained by combining WMAP with the SDSS galaxy redshift survey and the Ly-α forest analysis,

$$\sigma_8 = 0.89 \pm 0.04 \;. \tag{146}$$

- The shape of the power spectrum, as shown in Fig. 20, is sufficiently well determined to rule out any significant contribution of Hot Dark Matter to the energy budget of the Universe. Translated into an upper bound on the sum of neutrino masses, this constraint reads

$$\sum m_\nu \lesssim 0.5\,{\rm eV} \;, \tag{147}$$

an upper limit that is better by a factor of about 10 for the electron neutrino, and tremendously much better for the other two neutrino species, than those obtained from laboratory measurements.

6.4 Challenges

One cannot finish a section on cosmology without pointing out the impressive developments that we are currently witnessing, and some of their implications. The concordance model that we have described in the previous section is indeed a remarkable achievement, if one considers the huge variety of methods and processes that have entered the determination of its parameters. There was no a priori guarantee that all of this would fit together. Constraints obtained from nuclear physics about one minute after the Big Bang are in agreement with those from the distribution of galaxies in the local Universe!

This concordance model has a large impact on other branches of physics, most noticeably particle physics. The tight constraints on neutrino masses

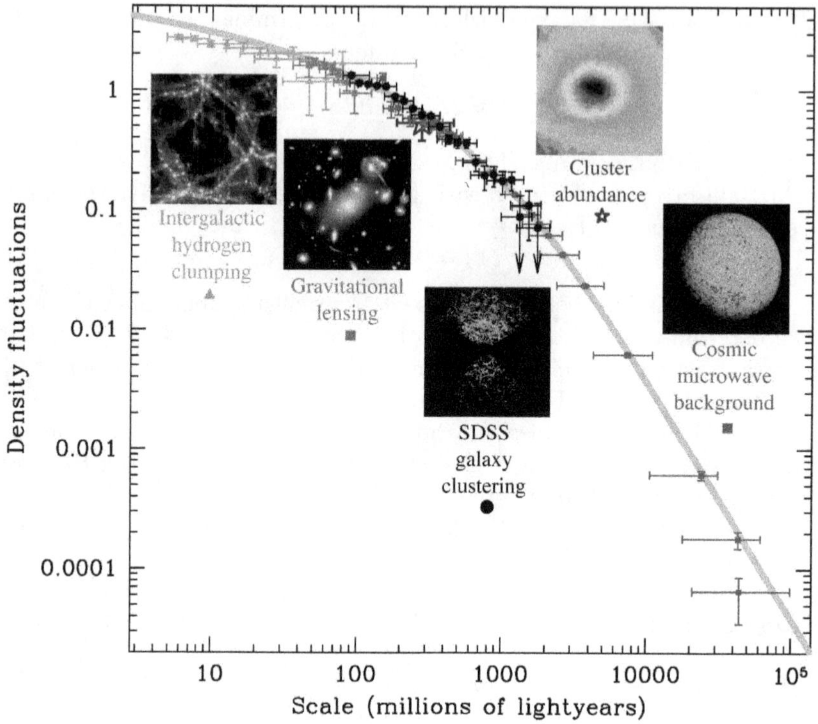

Fig. 20. The power spectrum of density fluctuations, as obtained from a variety of measurement methods. The CMB anisotropies measure the fluctuations on the largest spatial scale. Next come measurements from the clustering properties of galaxies, as obtained in galaxy redshift surveys. The cluster abundance provides a measurement of the fluctuation power near $\sim 10h^{-1}$ Mpc, i.e., close to the scale where σ_8 is defined. Cosmic shear and the statistical properties of the Ly-α forest provide reliable measurements on small spatial scales (taken from M. Tegmark's homepage, based on Tegmark et al. 2004)

obtained from cosmology is the most obvious example of a strong connection between these two branches of fundamental science. Even more important are the clear astronomical and cosmological evidences on the existence of non-baryonic dark matter which most likely is in the form of some as yet unknown species of elementary particles. This challenge has triggered a large number of underground experiments for a direct search for these dark matter particles. On the other hand, the next generation of particle accelerators may cross the energy threshold where new physics will be discovered. Many elementary particle physicist would put their eggs into the basket labeled Super-Symmetry, a theory which would provide a 'natural' candidate for the Cold Dark Matter particle, the neutralino.

Whereas one sees a possible solution to the Dark Matter problem, the situation is quite different with respect to vacuum energy, or the cosmological constant, or dark energy – this variety of names already tells a lot about how well this dominant component of the Universe is understood. Simple estimates of the density of vacuum energy from quantum field theory are barely 120 orders of magnitudes off, and we do not know why. It is not understood why the vacuum energy density is essentially zero (compared to the simple estimates), but in addition, why it is then not *exactly* zero. A constant vacuum energy density is not the only 'model' discussed for the dark energy; different equations of state $p = w\,\rho c^2$ cannot be ruled out by the current data, except that $w \lesssim -0.7$ at the present epoch. Obtaining tighter constraints on the equation of state of the dark energy from astronomical observations is probably the only way to investigate it empirically. The existence of this component to the cosmic energy budget arguably provides the largest challenge to fundamental physics, and its solution will almost certainly involve a unification of the laws of gravity and quantum mechanics – the long-sought theory of quantum gravity.

The concordance model also has made inflation a part of the standard model. Invented some 20 years ago, inflation provides a solution to the flatness problem (why the Universe has a total density parameter that is within an order of magnitude around unity), the horizon problem (why the CMB temperature on two opposite sides of the sky is the same within $\sim 10^{-5}$), and the apparent absence of magnetic monopoles and other topological defects. The model implies that the Universe underwent an early phase of exponential expansion, some 10^{-32} s after the Big Bang, before a phase transition ('reheating') brought it back on track for normal Friedmann expansion. In this model, the initial density fluctuations in the Universe are quantum fluctuations, inflated to macroscopic scales in the exponential expansion phase. The predictions of inflation, including that the Universe is nearly perfectly flat ($|-\Omega_{\rm m} + \Omega_\Lambda - 1| \ll 1$) and that the primordial fluctuation spectrum is very close to the Harrison–Zeldovich form, $1 - n \ll 1$, have been impressively verified with the recent cosmological observations. Probing the physics of inflation, e.g. through the presence of primordial gravitational waves which would leave an observable imprint on the polarization of the CMB, is one of the challenges of future cosmological studies.

7 Final Remarks

These notes are an extended version of two introductory lectures given at the beginning of the Saas-Fee course; they were intended to bring the participants up to speed on topics on which much of the rest of the course rested. Overlap with some of the later material was unavoidable, but given the different character and temperament of the three lecturers, maybe even not undesired. Some of this overlap will surely be present in these write-ups; hopefully, our readers do not mind.

Acknowledgments

I enjoyed the week of lecturing at Les Diablerets a lot. I thank Georges Meylan and his colleagues Philippe Jetzer and Pierre North for organizing this school so efficiently and smoothly. Furthermore, I thank my fellow lecturers Chris Kochanek and Joe Wambsganss for helpful comments on this manuscript, discussions, great company and good spirits. Thanks to Marusa Bradac for detailed comments on this manuscript. This work was supported by the German Ministry for Science and Education (BMBF) through the DLR under the project 50 OR 0106, by the German Ministry for Science and Education (BMBF) through DESY under the project 05AE2PDA/8, and by the Deutsche Forschungsgemeinschaft under the project SCHN 342/3–1.

References

Alcock, C., Akerloff, C.W., Allsman, R.A. et al. 1993, Nature 365, 621
Arp, H. 1987, *Quasars, redshifts, and controversies*, Interstellar Media, Berkeley
Aubourg, E., Bareyre, P., Brehin, S. et al. 1993, Nature 365, 623
Bacon, D.J., Refregier, A.R. & Ellis, R.S. 2000, MNRAS 318, 625
Bardeen, J.M., Bond, J.R., Kaiser, N. & Szalay, A.S. 1986, ApJS 304, 15
Barnothy, J.M. 1965, AJ 70, 666
Bartelmann, M. 1996, A&A 313, 697
Bartelmann, M. & Schneider, P. 2001, Phys. Rep. 340, 291
Bennett, C.L., Halpern, M., Hinshaw, G. et al. 2003, ApJS 148, 1
Binney, J. & Tremaine, S. 1987, Galactic Dynamics, Princeton University Press
Blandford, R.D. & Narayan, R. 1986, ApJ 310, 568
Blandford, R.D. & Narayan, R. 1992, ARA&A 30, 311
Bond, J.R. 1996, in Cosmology and large-scale structure, Schaeffer, R., Silk, J., Spiro, M. & Zinn-Justin, J. (eds.), Les Houches, Session LX, (Elsevier: North Holland), p. 469
Bond, J.R., Cole, S., Efstathiou, G. & Kaiser, N. 1991, ApJ 379, 440
Bourassa, R.R. & Kantowski, R. 1975, ApJ 195, 13
Bourassa, R.R., Kantowski, R. & Norton, T.D. 1973, ApJ 185, 747
Brainerd, T.G., Blandford, R.D., & Smail, I. 1996, ApJ 466, 623
Broadhurst, T. & Lehár, J. 1995, ApJ 450, L41
Browne, I.W.A., Wilkinson, P.N., Jackson, N.J.F. et al. 2003, MNRAS 341, 13
Buchert, T. & Ehlers, J. 1993, MNRAS 264, 375
Burke, B.F., Lehár, J. & Conner, S.R. 1992, in Gravitational Lenses, R. Kayser, T. Schramm & L. Nieser (eds.), (Springer-Verlag: Berlin), Lecture Notes in Physics, 406, p. 237
Burke, W.L. 1981, ApJ 244, L1
Canizares, C.R. 1982, ApJ 263, 508

Canizares, C.R. 1981, Nature 291, 620

Carroll, S.M., Press, W.H. & Turner, E.L. 1992, ARA&A 30, 499

Chang, K. & Refsdal, S. 1979, Nature 282, 561

Chang, K. & Refsdal, S. 1984, A&A 132, 168

Chwolson, O. 1924, AN 221, 329

Cooke, J.H. & Kantowski, R. 1975, ApJ 195, L11

Courbin, F., Saha, P. & Schechter, P.L. 2002, in Gravitational Lensing: An Astrophysical Tool, F. Courbin & D. Minniti (eds.), (Springer-Verlag: Berlin), Lecture Notes in Physics 608, p. 1

Ehlers, J., Schneider, P. 1986, A&A 168, 57

Einstein, A. 1936, Sci 84, 506

Eisenstein, D.J. & Hu, W. 1998, ApJ 496, 605

Eke, V.R., Cole, S. & Frenk, C.S. 1996, MNRAS 282, 263

Etherington, I.M.H. 1933, Philos. Mag. 15, 761

Falco, E.E., Gorenstein, M.V. & Shapiro, I.I. 1985, ApJ 289, L1

Fort, B. & Mellier, Y. 1994, A&AR 5, 239

Fort, B., Prieur, J.L., Mathez, G., Mellier, Y. & Soucail, G. 1988, A&A 200, L17

Freedman, W.L., Madore, B.F., Gibson, B.K. et al. 2001, ApJ 553, 47

Frenk, C.S., White, S.D.M., Bode, P. et al. 1999, ApJ 525, 554

Friedmann, A. 1922, Z. Phys. 10, 377

Golse, G. & Kneib, J.-P. 2002, A&A 390, 821

Gorenstein, M.V., Cohen, N.L., Shapiro, I.I. et al. 1988a, ApJ 334, 42

Gorenstein, M.V., Falco, E.E. & Shapiro, I.I. 1988b, ApJ 327, 693

Hamilton, A.J.S., Kumar, P., Lu, E. & Matthews, A. 1991, ApJ 374, L1

Harvanek, M., Stocke, J.T., Morse, J.A. & Rhee, G. 1997, AJ 114, 2240

Hawkins, E., Maddox, S., Cole, S. et al. 2003, MNRAS 346, 78

Hewitt, J.N., Turner, E.L., Schneider, D.P., Burke, B.F., Langston, G.I. & Lawrence, C.R. 1988, Nature 333, 537

Hu, W. & Dodelson, S. 2002, ARA&A 40, 171

Hu, E.M., Cowie, L.L., McMahon, R.G. et al. 2002, ApJ 568, L75

Ibata, R.A., Lewis, G.F., Irwin, M.J., Lehár, J. & Totten, E.J. 1999, AJ 118, 1922

Impey, C.D., Falco, E.E., Kochanek, C.S. et al. 1998, ApJ 509, 551

Irwin, M.J., Ibata, R.A., Lewis, G.F. & Totten, E.J. 1998, ApJ 505, 529

Irwin, M.J., Webster, R.L., Hewett, P.C., Corrigan, R.T. & Jedrzejewski, R.I. 1989, AJ 98, 1989

Jain, B., Mo, H.J. & White, S.D.M. 1995, MNRAS 276, L25

Jenkins, A., Frenk, C., White, S. et al. 2001, MNRAS 321, 372

Kaiser, N., Wilson, G. & Luppino, G. 2000, astro-ph/0003338

Kauffmann, G., Guiderdoni, B. & White, S.D.M. 1994, MNRAS 267, 981

Kauffmann, G., White, S.D.M. & Guiderdoni, B. 1993, MNRAS 264, 201

Kauffmann, G., Nusser, A. & Steinmetz, M. 1997, MNRAS 286, 795

Kauffmann, G, Colberg, J.M, Diaferio, A. & White, S.D.M. 1999a, MNRAS 303, 188

Kauffmann, G, Colberg, J.M, Diaferio, A. & White, S.D.M. 1999b, MNRAS 307, 529
Kayser, R. & Refsdal, S. 1983, A&A 128, 156
Kayser, R., Refsdal, S. & Stabell, R. 1986, A&A 166, 36
Keeton, C.R. & Kochanek, C.S. 1998, ApJ 495, 157
Kim, T.-S., Viel, M., Haehnelt, M.G., Carswell, R.F. & Cristiani, S. 2004, MNRAS 347, 355
King, L.J., Jackson, N., Blandford, R.D. et al. 1998, MNRAS 295, L41
King, L.J., Browne, I.W.A., Marlow, D.R., Patnaik, A.R. & Wilkinson, P.N. 1999, MNRAS 307, 225
Klimov, Yu.G. 1963, Soviet Phys. Doklady 8, 119
Kneib, J.-P., Ellis, R.S., Santos, M.R. & Richard, J. 2004, ApJ 607, 697
Knop, R.A., Aldering, G., Amanullah, R. et al. 2003, ApJ 598, 102
Kochanek, C.S. 2005, in Schneider, P., Kochanek, C.S., Wambsganss, J., Gravitational Lensing: Strong, Weak & Micro, Proceedings of the 33rd Saas-Fee Advanced Course, G. Meylan, P. Jetzer & P. North, eds. (Springer-Verlag: Berlin) (SL Part 2), also astro-ph/0407232
Kochanek, C.S., Blandford, R.D., Lawrence, C.R. & Narayan, R. 1989, MNRAS 238, 43
Kolb, E.W. & Turner, M.G. 1990, The Early Universe, Addison-Wesley Publ. Comp.: Redwood)
Kormann, R., Schneider, P. & Bartelmann, M. 1994, A&A 284, 285
Kundić, T., Colley, W.N., Gott, J.R. et al. 1995, ApJ 455, L5
Kundić, T., Turner, E.L., Colley, W.N. et al. 1997, ApJ 482, 75
Lacey, C. & Cole, S. 1993, MNRAS 262, 627
Langston, G.I., Schneider, D.P., Conner, S. et al. 1989, AJ 97, 1283
Laplace, P.S. 1795, Exposition du système du monde.
Lehár, J., Hewitt, J.N., Roberts, D.H. & Burke, B.F. 1992, ApJ 384, 453
Liddle, A.R. & Lyth, D.H. 2000, Cosmological Inflation and Large-Scale Structure, (Cambridge University Press: Cambridge)
Liebes Jr., S. 1964, Phys. Rev. 133, B835
Lodge, O.J. 1919, Nature 104, 354
Lynden-Bell, D. 1967, MNRAS 136, 101
Lynds, R. & Petrosian, V. 1986, BAAS 18, 1014
Mao, S. 1992, ApJ 389, 63
Maoz, D., Bahcall, J.N., Schneider, D.P. et al. 1993, ApJ 409, 28
Mattig, W. 1958, AN 284, 109
Mellier, Y. 1999, ARA&A 37, 127
Michalitsianos, A.G., Dolan, J.F., Kazanas, D. et al. 1997, ApJ 474, 598
Narayan, R. & Bartelmann, M. 1999, in Formation of Structure in the Universe, A. Dekel & J.P. Ostriker (eds.), (Cambridge University Press: Cambridge), p. 360
Navarro, J., Frenk, C. & White, S. 1997, ApJ 490, 493
Navarro, J.F., Hayashi, E., Power, C. et al. 2004, MNRAS 349, 1039
Ohanian, H.C. 1983, ApJ 271, 551

Paczyński, B. 1986a, ApJ 301, 503
Paczyński, B. 1986b, ApJ 304, 1
Paczyński, B. 1987, Nature 325, 572
Paczyński, B. 1996, ARA&A 34, 419
Padmanabhan, T. 1993, Structure Formation in the Universe, (Cambridge University Press: Cambridge)
Peacock, J.A. 1982, MNRAS 199, 987
Peacock, J.A. 1999, Cosmological Physics, (Cambridge University Press: Cambridge)
Peacock, J.A. 2003, astro-ph/0309240
Peacock, J.A. & Dodds, S.J. 1996, MNRAS 280, L19
Pearce, F.R., Jenkins, A., Frenk, C.S. 2001, MNRAS 326, 649
Peebles, P.J.E. 1993, Principles of Physical Cosmology, Princeton University Press
Pelló, R., Schaerer, D., Richard, J., Le Borgne, J.-F. & Kneib, J.-P. 2004, A&A 416, L35
Perlmutter, S., Aldering, G., Goldhaber, G. et al. 1999, ApJ 517, 565.
Petters, A.O., Levine, H. & Wambsganss, J. 2001, Singularity Theory and Gravitational Lensing (Birkhäuser: Boston) (PLW)
Press, W.H., Flannery, B.P., Teukolsky, S.A. & Vetterling, W.T. 1992, Numerical Recipes, (Cambridge University Press: Cambridge)
Press, W.H. & Schechter, P. 1974, ApJ 187, 425
Refsdal, S. 1964a, MNRAS 128, 295
Refsdal, S. 1964b, MNRAS 128, 307
Refsdal, S. & Surdej, J. 1994, Rep. Prog. Phys. 56, 117
Renn, J., Sauer, T. & Stachel, J. 1997, Sci 275, 184
Riess, A.G., Strolger, L.-G., Tonry, J. et al. 2004, ApJ 607, 665
Schild, R.E. 1990, AJ 100, 1771
Schmidt, B.P., Suntzeff, N.B.; Phillips, M.M. et al. 1998, ApJ 507, 46.
Schneider, P.1984, A&A 140, 119
Schneider, P. 1985, A&A 143, 413
Schneider, P. 1987, A&A 183, 189
Schneider, P. 1992, A&A 254, 14
Schneider, P. 2005, in Schneider, P., Kochanek, C.S., Wambsganss, J., Gravitational Lensing: Strong, Weak & Micro, Proceedings of the 33rd Saas-Fee Advanced Course, G. Meylan, P. Jetzer & P. North, eds. (Springer-Verlag: Berlin) (WL Part 3)
Schneider, P., Ehlers, J. & Falco, E.E. 1992, Gravitational Lenses (Springer-Verlag: Heidelberg), 2nd printing 1999 (SEF)
Schneider, P. & Seitz, C. 1995, A&A 294, 411
Schneider, P. & Weiss, A. 1987, A&A 171, 49
Schneider, P. & Weiss, A. 1992, A&A 260, 1
Schramm, T. 1990, A&A 231, 19
Seitz, S., Saglia, R.P., Bender, R., Hopp, U., Belloni, P. & Ziegler, B. 1998, MNRAS 298, 945

Seitz, S. & Schneider, P. 1992, A&A 265, 1
Seitz, S., Schneider, P. & Ehlers, J. 1994, Class. Quantum Gravity 11, 2345
Seljak, U., Makarov, A., McDonald, P. et al. 2005, PhRvD 71, 103515, also astro-ph/0407372
Sheth, R. & Tormen, G. 1999, MNRAS 308, 119
Smith, R.E., Peacock, J.A., Jenkins, A. et al. 2003, MNRAS 341, 1311
Smoot, G.F. 1997, in The Cosmic Microwave Background, C.H. Lineweaver, J.G. Bartlett, A. Blanchard, M. Signore & J. Silk (eds.), (Kluwer Academic Publishers: Dordrecht), p. 185
Soldner, J. 1804, Berliner Astron. Jahrb. 1804, p. 161
Soucail, G., Fort, B., Mellier, Y. & Picat, J.P. 1987, A&A 172, L14
Soucail, G., Mellier, Y., Fort, B., Mathez, G. & Cailloux, M. 1988, A&A 191, L19
Spergel, D.N., Verde, L., Peiris, H.V. et al. 2003, ApJS 148, 175
Springel, V., White, S.D.M., Tormen, G. & Kauffmann, G. 2001, MNRAS 328, 726
Stockton, A. 1980, ApJ 242, L141
Takada, M. & Jain, B. 2002, MNRAS 337, 875
Tegmark, M., Blanton, M.R., Strauss, M.A. et al. 2004, ApJ 606, 702
Tyson, J.A. 1981, ApJ 248, L89
Tyson, J.A. 1988, AJ 96, 1
Tyson, J.A., Valdes, F. & Wenk, R.A. 1990, ApJ 349, L1
Tytler, D., O'Meara, J.M., Suzuki, N. & Lubin, D. 2000, Phys. Rep. 333, 409
Udalski, A., Szymanski, M., Kaluzny, J. et al. 1993, Acta Astron. 43, 289
van Waerbeke, L. & Mellier, Y. 2003, astro-ph/0305089
van Waerbeke, L., Mellier, Y., Erben, T. et al. 2000, A&A, 358, 30
Vanderriest, C., Schneider, J., Herpe, G., Chevreton, M., Moles, M. & Wlérick, G. 1989, A&A 215, 1
Viana, P.T.P., Nichol, R.C. & Liddle, A.R. 2002, ApJ 569, L75
Vietri, M. & Ostriker, J.P. 1983, ApJ 267, 488
Wallington, S. & Narayan, R. 1993, ApJ 403, 517
Walsh, D. 1989, in Gravitational lenses, Moran, J.M., Hewitt, J.N. & Lo, K. Y. (eds.), Lecture Notes in Physics 330 (Springer-Verlag: New York), p. 11
Walsh, D., Carswell, R.F. & Weymann, R.J. 1979, Nature 279, 381
Wambsganss, J. 2005, in Schneider, P., Kochanek, C.S., Wambsganss, J., Gravitational Lensing: Strong, Weak & Micro, Proceedings of the 33rd Saas-Fee Advanced Course, G. Meylan, P. Jetzer & P. North (eds.), (Springer-Verlag: Berlin) (ML Part 4)
Wang, X., Tegmark, M., Jain, B. & Zaldarriaga, M. 2003, PhRvD 68, 123001
Weinberg, S. 1976, ApJ 208, L1
Weymann, R.J., Latham, D., Angel, J.R.P. et al. 1980, Nature 285, 641
White, S.D.M., Davis, M., Efstathiou, G. & Frenk, C.S. 1987, Nature 330, 451
Williams, R.E., Blacker, B., Dickinson, M. et al. 1996, AJ 112, 1335

Wittman, D. 2002, in Gravitational Lensing: An Astrophysical Tool, F. Courbin & D. Minniti (eds.), (Springer-Verlag: Berlin), Lecture Notes in Physics 608, p. 55
Wittman, D.M., Tyson, J.A., Kirkman, D., Dell'Antonio, I. & Bernstein, G. 2000, Nature 405, 143 ,
Yee, H.K.C. 1988, AJ 95, 1331
Yee, H.K.C., Ellingson, E., Bechtold, J., Carlberg, R.G. & Cuillandre, J.-C. 1996, AJ 111, 1783
Young, P., Gunn, J.E., Kristian, J., Oke, J.B. & Westphal, J.A. 1980, ApJ 241, 507
Young, P., Gunn, J.E., Kristian, J., Oke, J.B. & Westphal, J.A. 1981, ApJ 244, 736
Zeldovich, Ya.B. 1970, A&A 5, 84
Zwicky, F. 1937a, Phys. Rev. 51, 290
Zwicky, F. 1937b, Phys. Rev. 51, 679

Part 2: Strong Gravitational Lensing

C. S. Kochanek

1 Introduction

The objective of this lecture is to provide a practical introduction to strong gravitational lensing including the data, the theory, and the application of strong lensing to other areas of astrophysics. This is Part 2 of the complete Saas Fee lectures on gravitational lensing. Part 1 (Schneider, this book) provides a basic introduction, Part 2 (Kochanek, this book) examines strong gravitational lenses, Part 3 (Schneider, this book) explores cluster lensing and weak lensing, and Part 4 (Wambsganss, this book) examines microlensing. It is not my objective in this lecture to provide a historical review, carefully outlining the genealogy of every development in gravitational lensing, but to focus on current research topics. Part 1 of these lectures summarizes the history of lensing and introduces most of the basic equations of lensing. The discussion is divided into 9 sections. We start in Sect. 2 with an introduction to the observational data. In Sect. 3 we outline the basic principles of strong lenses, building on the general theory of lensing from Part 1. In Sect. 4 we discuss modeling gravitational lenses and the determination of the mass distribution of lens galaxies. In Sect. 5 we discuss time delays and the Hubble constant. In Sect. 6 we discuss gravitational lens statistics and the cosmological model. In Sect. 7 we discuss the differences between galaxies and clusters as lenses. In Sect. 8 we discuss the effects of substructure or satellites on gravitational lenses. In Sect. 9 we discuss the optical properties of lens galaxies and in Sect. 10 we discuss extended sources and quasar host galaxies. Finally in Sect. 11 we discuss the future of strong gravitational lensing.

It will also be clear to the reader that these are my lectures on strong lensing rather than an attempt to achieve some quasi-mythical consensus view. I have tried to make clear what matters (and what does not), what lensing can do (and cannot do) for astrophysics, where the field is serving the community well (and poorly), and where non-experts have understood the consequences (or have failed to do so). Doing so requires having definite opinions with which other researchers may well disagree. We will operate on the assumption that

Kochanek CS (2006), Strong gravitational lensing. In: Meylan G, Jetzer Ph and North P (eds) Gravitational lensing: Strong, weak, and micro. Saas-Fee Adv Courses vol 33, pp 91–268
DOI 10.1007/3-540-30309-X_2

anyone who disagrees sufficiently violently will have an opportunity to wreak a horrible revenge at a later date by spending six months doing their own set of lectures. I have tried to make the references to recent work complete – no doubt I will have failed in this endeavor. There are many earlier reviews of lensing (e.g. Blandford and Kochanek 1987a; Blandford and Narayan 1992; Refsdall and Surdej 1994; Wambsganss 1998; Narayan and Bartelmann 1999; Courbin, Saha and Schechter 2002b; Claeskens and Surdej 2002) as well as the book by Schneider, Ehlers and Falco 1992.

2 An Introduction to the Data

There are now 82 candidates for multiple image lenses besides those found in rich clusters. Of these candidates, there is little doubt about 74 of them, with the ambiguities resting in candidates consisting of faint galaxies with nearby arcs and no spectroscopic data. Indeed, the absence of complete spectroscopic information is the bane of most astrophysical applications of lenses. Less than half (38) of the good candidates have both source and lens redshifts – 43 have lens redshifts, 64 have source redshifts, and 5 have neither redshift. Much of this problem could be eliminated in about 5 clear nights of 8m time, but no TAC seems willing to devote the effort even though lens redshifts probably provide more cosmological information per redshift than any other sparsely distributed source. Of these 74 lenses, 11 have had their central velocity dispersions measured and 10 have measured time delays. A reasonably complete summary of the lens data is available at the CASTLES WWW site http://cfa-www.harvard.edu/castles/, although lack of manpower means that it is updated only episodically.

Figure 1 shows the distribution of the lenses in image separation and source redshift. The separations of the images range from $0\overset{''}{.}35$ to $15\overset{''}{.}9$ (using either half the image separations or the mean distance of the images from the lens). The observed distribution combines both the true separation distribution and selection effects. For example, in simple statistical models using standard models for galaxy properties (see Sect. 6) we would expect to find that the logarithmic separation distribution $dN/d\ln\Delta\theta$ is nearly constant at small separations (i.e. $dN/d\Delta\theta \propto \Delta\theta$), while the raw, observed distribution shows a cutoff due to the finite resolution of lens surveys (typically $0\overset{''}{.}25$ to $1\overset{''}{.}0$ depending on the survey). The cutoff at larger separations is real, and it is a consequence of the vastly higher lensing efficiency of galaxies relative to clusters created by the cooling of the baryons in galaxies (see Sect. 7).

Figure 2 shows the distribution in image separation and lens galaxy redshift. There is no obvious trend in the typical separation with redshift, as might be expected if there were rapid evolution in the typical masses of galaxies. Unfortunately, there is also an observational bias to measure the redshifts of large separation lenses, where the lens galaxies tend to be brighter and less confused with the images, which makes quantitative interpretation of

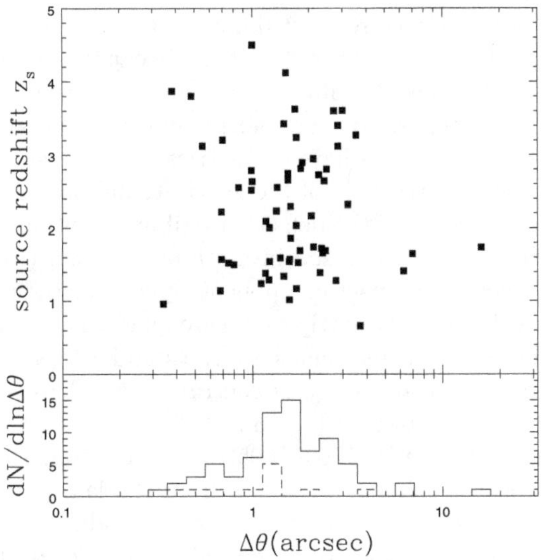

Fig. 1. The distribution of lens galaxies in separation $\Delta\theta$ and source redshift z_s. The *solid* histogram shows the distribution in separation for all the lenses while the *dashed* histogram shows the distribution of those with unmeasured source redshifts

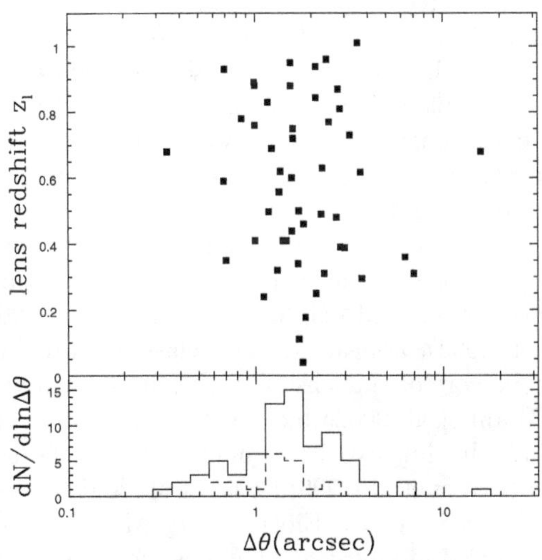

Fig. 2. The distribution of lens galaxies in separation $\Delta\theta$ and lens redshift z_l. The *solid* histogram shows the distribution in separation for all the lenses while the *dashed* histogram shows the distribution of those with unmeasured lens redshifts. There are no obvious correlations between lens redshift z_l and separation $\Delta\theta$, but the strong selection bias that small separation lenses are less likely to have measured redshifts makes this difficult to interpret. There may also be a deficit of low redshift, large separation lenses, which may be a selection bias created by the difficulty of finding quasar lenses embedded in bright galaxies

any trends in separation with redshift difficult. There is probably also a bias against finding large, low lens redshift systems because the flux from the lens galaxy will more easily mask the flux from the source. We examine the correlations between image separations and lens magnitudes in Sect. 9.

In almost all cases the lenses have geometries that are "standard" for models in which the angular structure of the gravitational potential is dominated by the quadrupole moments of the density distribution, either because the lens is ellipsoidal or because the lens sits in a strong external (tidal) shear field. Of the 60 lenses where a compact component (quasar or radio core) is clearly identifiable, 36 are doubles, 2 are triples, 20 are quads, 1 has five images and 1 has six images. The doubles and quads are the standard geometries produced by standard lenses with nearly singular central surface densities. Examples of these basic patterns are shown in Figs. 3 and 4.

In a two-image lens like HE1104–1805 (Wisotzki et al. 1993), the images usually lie at markedly different distances from the lens galaxy because the source must be offset from the lens center to avoid producing four images. The quads show three generic patterns depending on the location of the source relative to the lens center and the caustics. There are cruciform quads like HE0435–1223 (Wisotzki et al. 2002), where the images form a cross pattern bracketing the lens. These are created when the source lies almost directly behind the lens. There are fold-dominated quads like PG1115+080 (Weymann et al. 1980), where the source is close to a fold caustic and we observe a close pair of highly magnified images. Finally, there are cusp-dominated quads like RXJ1131–1231 (Sluse et al. 2003), where the source is close to a cusp caustic and we observe a close triple of highly magnified images. These are all generic patterns expected from caustic theory, as we discuss in Part 1 and Sect. 3. We will discuss the relative numbers of doubles and quads in Sect. 6.

The lenses with non-standard geometries all have differing origins. One triple, APM08279+5255 (Irwin et al. 1998; Ibata et al. 1999; Muñoz, Kochanek and Keeton 2001), is probably an example of a disk or exposed cusp lens (see Sect. 3), while the other, PMNJ1632–0033 (Winn et al. 2002a,b,c; Winn, Rusin and Kochanek 2004), appears to be a classical three image lens with the third image in the core of the lens (Fig. 5). The system with five images, PMNJ0134–0931 (Winn et al. 2002a,b,c; Keeton and Winn 2003; Winn et al. 2003a,b,c), is due to having two lens galaxies, while the system with six images, B1359+154 (Myers et al. 1999; Rusin et al. 2001), is a consequence of having three lens galaxies inside the Einstein ring. Many lenses have luminous satellites that are required in any successful lens model, such as the satellites known as "Object X" in MG0414+0534 (Hewitt et al. 1992; Schechter and Moore 1993) and object D in MG2016+112 (Lawrence et al. 1984) shown in Fig. 7. These satellite galaxies can be crucial parts of lens models, although there has been no systematic study of their properties in the lens sample.

If the structure of the source is more complicated, then the resulting image geometries become more complicated. For example, the source of the radio lens B1933+503 (Sykes et al. 1998) consists of a radio core and two radio lobes,

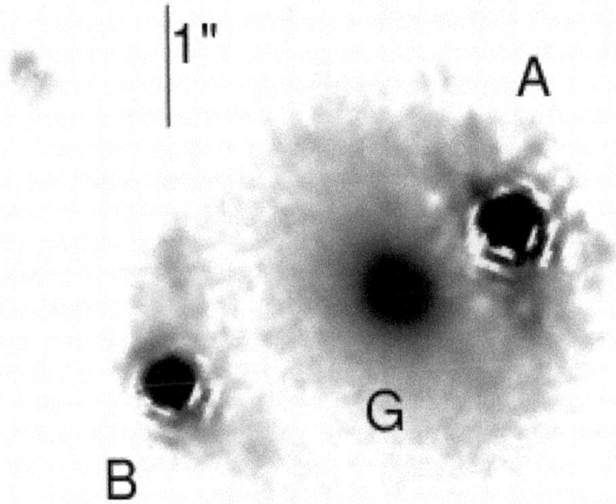

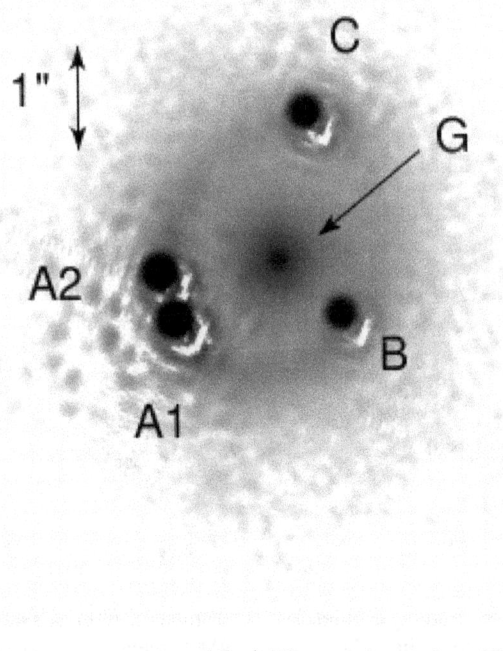

Fig. 3. Standard image geometries. (*Top*) The two-image lens HE1104–1805. G is the lens galaxy and A and B are the quasar images. We also see arc images of the quasar host galaxy underneath the quasar images. (*Bottom*) The four-image lens PG1115+080 showing the bright A_1 and A_2 images created by a fold caustic

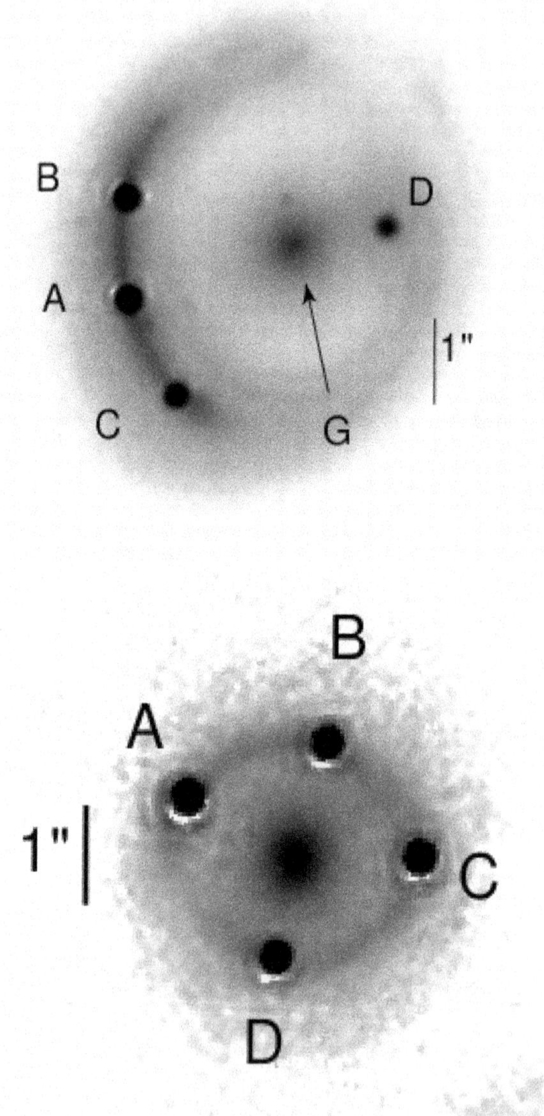

Fig. 4. Standard image geometries (continued). (*Top*) The four-image lens RXJ1131–1231 showing the bright A, B and C images created by a cusp caustic. (*Bottom*) The four-image lens HE0435–1223, showing the cruciform geometry created by a source near the center of the lens. For each lens in Figs. 3 and 4, we took the CASTLES H-band image, subtracted the bright quasars and then added them back as Gaussians with roughly the same FWHM as the real PSF. This removes the complex diffraction pattern of the HST PSF and makes it easier to see low surface brightness features

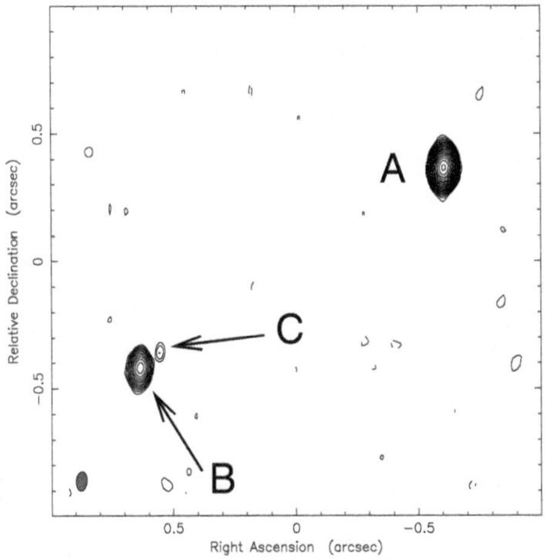

Fig. 5. PMN1632–0033 is the only known lens with a "classical" third image in the core of the lens galaxy. The center of the lens galaxy is close to the faint C image. Images A, B and C have identical radio spectra except for the longest wavelength flux of C, which can be explained by absorption in the core of the lens galaxy. Time delay measurements would be required to make the case absolutely secure. A central black hole in the lens galaxy might produce an additional image with a flux about 10% that of image C. (Winn et al. 2004)

leading to 10 observable images because the core and one lobe are quadruply imaged and the other lobe is doubly imaged (Fig. 6). If instead of discrete emission peaks there is a continuous surface brightness distribution, then we observe arcs or rings surrounding the lens galaxy usually of the host galaxy of the quasar or radio source. Figure 8 shows examples of Einstein rings for the case of MG1131+0456 in both the radio (Chen and Hewitt 1993) and the infrared (Kochanek et al. 2000a,b). The radio ring is formed from an extended radio jet, while the infrared ring is formed from the host galaxy of the radio source. We also chose most of the examples in Figs. 3 and 4 to show prominent arcs and rings formed by lensing the host galaxy of the source quasar. We discuss arcs and rings in Sect. 10.

3 Basic Principles

Most gravitational lenses have the standard configurations we illustrated in Sect. 2. These configurations are easily understood in terms of the caustic structures generic to simple lens models. In this section we illustrate the origin of these basic geometries using simple mathematical examples. We build on the general outline of lensing theory from Part 1.

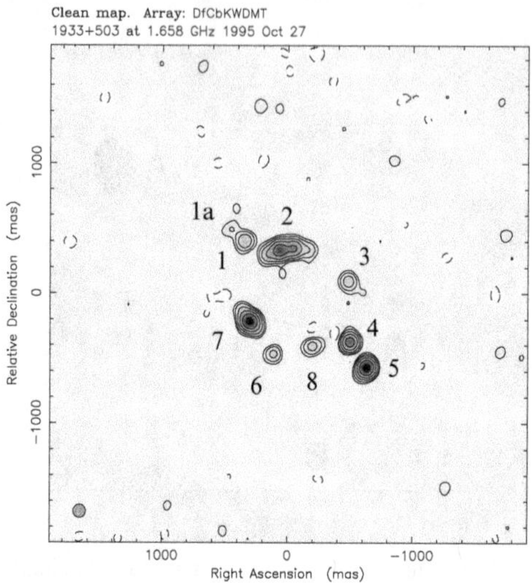

Fig. 6. A Merlin map of B1933+503 showing the 10 observed images of the three component source (Marlow et al. 1999). The flat radio spectrum core is lensed into images 1, 3, 4 and 6. One radio lobe is lensed into images 1a and 8, while the other is lensed into images 2, 7 and 5. Image 2 is really two images merging on a fold

3.1 Some Nomenclature

Throughout this lecture we use comoving angular diameter distances (also known as proper motion distances) rather than the more familiar angular diameter distances because almost every equation in gravitational lensing becomes simpler. The distance between two redshifts i and j is

$$D_{ij} = \frac{r_H}{|\Omega_k|^{1/2}} \text{sinn} \left[\int_i^j \frac{|\Omega_k|^{1/2} dz}{\left((1+z)^2(1+\Omega_M z) - z(2+z)\Omega_\Lambda\right)^{1/2}} \right], \quad (1)$$

where Ω_M, Ω_Λ and $\Omega_k = 1 - \Omega_M - \Omega_\Lambda$ are the present day matter density, cosmological constant and "curvature" density respectively, $r_H = c/H_0$ is the Hubble radius, and the function sinn(x) becomes sinh(x), x or sin(x) for open ($\Omega_k > 0$), flat ($\Omega_k = 0$) and closed ($\Omega_k < 0$) models (Carroll, Press and Turner 1992). We use D_d, D_s and D_{ds} for the distances from the observer to the lens, from the observer to the source and from the lens to the source. These distances are trivially related to the angular diameter distances, $D_{ij}^{ang} = D_{ij}/(1+z_j)$, and luminosity distances, $D_{ij}^{lum} = D_{ij}(1+z_j)$. In a flat universe, one can simply add comoving angular diameter distances ($D_s = D_d + D_{ds}$), which is not true of angular diameter distances. The comoving volume element also simplifies to

$$dV = \frac{D_d^2 dD_d d\omega}{\left(1 + \Omega_k r_H^{-2} D_d^2\right)^{1/2}} \rightarrow D_d^2 D_d d\omega \quad (2)$$

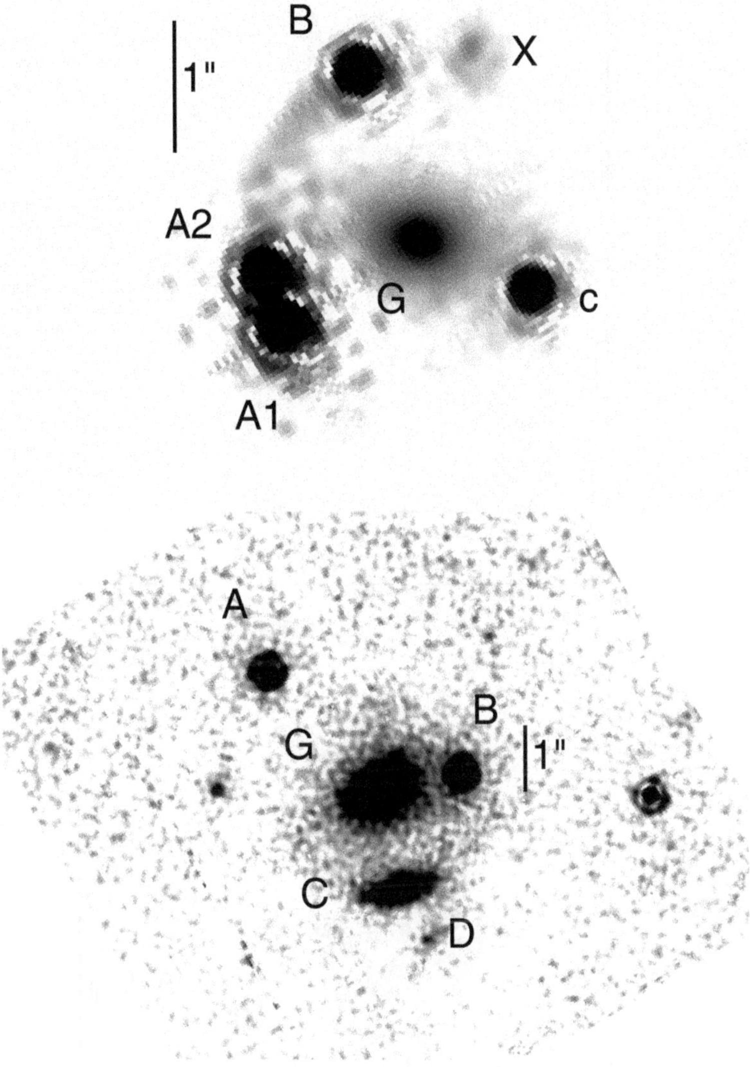

Fig. 7. H-band images of two lenses with small companions that are crucial for successful models. The upper image shows "Object X" in MG0414+0534, and the lower image shows component D of MG2016+112. MG2016+112 has the additional confusion that only A and B are images of the quasar (Koopmans et al. 2002). Image C is some combination of emission from the quasar jet (it is an extended X-ray source, Chartas et al. 2001) and the quasar host galaxy. Object D is known to be at the same redshift as the primary lens galaxy G (Koopmans and Treu 2002)

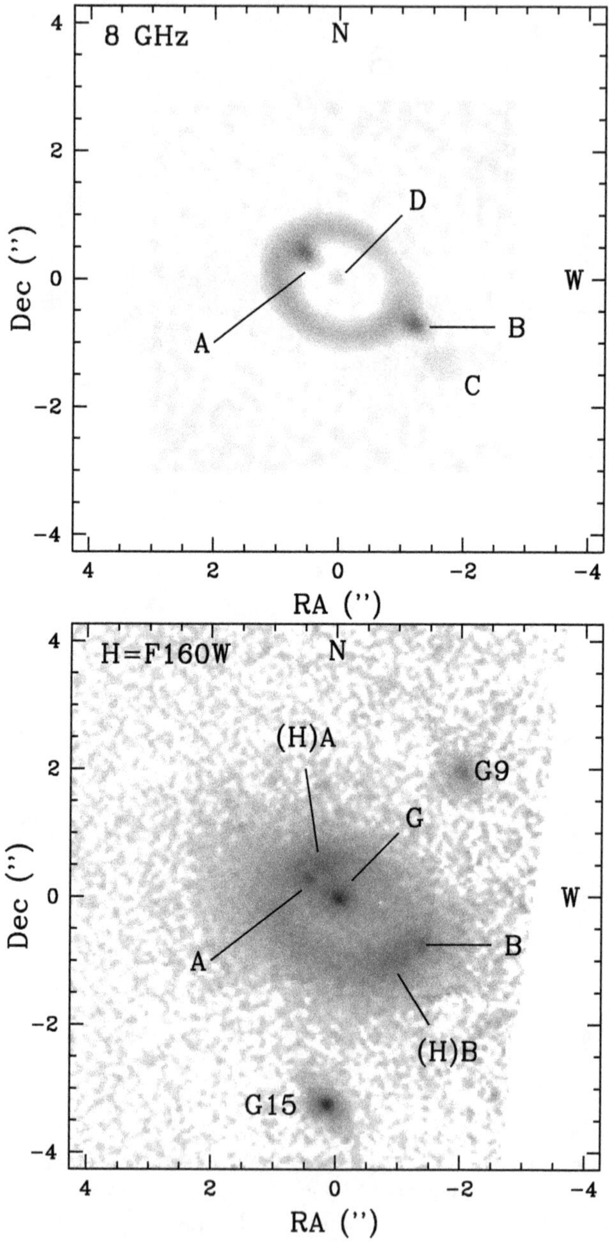

Fig. 8. The radio (*top*) and H-band (*bottom*) rings in MG1131+0456. The radio map was made at 8 GHz by Chen and Hewitt (1993), while the H-band image is from Kochanek et al. (2000a,b). The radio source D is probably another example of a central odd image, but the evidence is not as firm as that for PMN1632–0033. Note the perturbing satellite galaxies (G9 and G15) in the H-band image

for flat universes. We denote angles on the lens plane by $\boldsymbol{\theta} = \theta(\cos\chi, \sin\chi)$ and angles on the source plane by $\boldsymbol{\beta}$. Physical lengths on the lens plane are $\boldsymbol{\xi} = D_d^{ang}\boldsymbol{\theta}$. The lensing potential, denoted by $\Psi(\boldsymbol{\theta})$, satisfies the Poisson equation $\nabla^2\Psi = 2\kappa$ where $\kappa = \Sigma/\Sigma_c$ is the surface density Σ in units of the critical surface density $\Sigma_c = c^2 D_s^{ang}/(4\pi G D_d^{ang} D_{ds}^{ang})$. For a more detailed review of the basic physics, see Part 1.

3.2 Circular Lenses

While one of the most important lessons about modeling gravitational lenses in the real world is that you can never (EVER !)[1] safely neglect the angular structure of the gravitational potential, it is still worth starting with circular lens models. They provide a basic introduction to all the elements which are essential to realistic models without the need for numerical calculation. In a circular lens, the effective lens potential is a function only of the distance from the lens center $\theta = |\boldsymbol{\theta}|$. Rays are radially deflected by the angle

$$\alpha(\theta) = \frac{2}{\theta}\int_0^\theta \theta d\theta \kappa(\theta) = \frac{4GM(<\xi)}{c^2\xi}\frac{D_{ds}}{D_s}, \tag{3}$$

where we recall from Part 1 that $\kappa(\theta) = \Sigma(\theta)/\Sigma_c$ is the surface density in units of the critical surface density, D_{ds} and D_s are the lens-source and observer-source comoving distances and $\xi = D_d^{ang}\theta$ is the proper distance from the lens. The bend angle is simply twice the Schwarzschild radius of the enclosed mass, $4GM(<\xi)/c^2$, divided by the impact parameter ξ and scaled by the distance ratio D_{ds}/D_s.

The lens equation (see Part 1) becomes

$$\boldsymbol{\beta} = \boldsymbol{\theta}\left[1 - \alpha(\theta)/\theta\right] = \boldsymbol{\theta}\left[1 - \langle\kappa(\theta)\rangle\right], \tag{4}$$

where

$$\langle\kappa(\theta)\rangle = \frac{2}{\theta^2}\int_0^\theta \theta d\theta \kappa(\theta) = \alpha(\theta)/\theta \tag{5}$$

is the average surface density interior to θ in units of the critical density. Note that there must be a region with $\langle\kappa\rangle > 1$ to have solutions on both sides of the lens center. Because of the circular symmetry, all images will lie on a line passing through the source and the lens center.

The inverse magnification tensor (or Hessian, see Part 1) also has a simple form, with

$$M^{-1} = \frac{d\boldsymbol{\beta}}{d\boldsymbol{\theta}} = (1-\kappa)\begin{pmatrix} 1 & 0 \\ 0 & 1 \end{pmatrix} + \gamma\begin{pmatrix} \cos 2\chi & \sin 2\chi \\ \sin 2\chi & -\cos 2\chi \end{pmatrix}, \tag{6}$$

[1] AND I MEAN EVER ! DON'T EVEN THINK OF IT !

where $\boldsymbol{\theta} = \theta(\cos\chi, \sin\chi)$. The convergence (surface density) is

$$\kappa = \frac{1}{2}\left(\frac{\alpha}{\theta} + \frac{d\alpha}{d\theta}\right), \tag{7}$$

and the shear is

$$\gamma = \frac{1}{2}\left(\frac{\alpha}{\theta} - \frac{d\alpha}{d\theta}\right) = \langle\kappa\rangle - \kappa. \tag{8}$$

The eigenvectors of M^{-1} point in the radial and tangential directions, with a radial eigenvalue of $\lambda_+ = 1-\kappa+\gamma = 1-d\alpha/d\theta$ and a tangential eigenvalue of $\lambda_- = 1-\kappa-\gamma = 1-\alpha/\theta = 1-\langle\kappa\rangle$. If either one of these eigenvalues is zero, the magnification diverges and we are on either the radial or tangential critical curve. If we can resolve the images, we will see the images radially magnified near the radial critical curve and tangentially magnified near the tangential critical curve. For example, all the quasar host galaxies seen in Figs. 3 and 4 lie close to the tangential critical line and are stretched tangentially to form partial or complete Einstein rings. The signs of the eigenvalues $\lambda_\pm$ give the parities of the images and the type of time delay extremum associated with the images. If both eigenvalues are positive, the image is a minimum. If both are negative, the image is a maximum. If one is positive and the other negative, the image is a saddle point. The inverse of the total magnification $\mu^{-1} = |M^{-1}|$ is the product of the eigenvectors, so it is positive for minima and maxima and negative for saddle points. The signs of the eigenvalues are referred to as the partial parities of the images, while the sign of the total magnification is referred to as the total parity.

It is useful to use simple examples to illustrate the behavior of circular lenses for different density profiles. In most previous lensing reviews, the examples are based on lenses with finite core radii. However, most currently popular models of galaxies and clusters have central density cusps rather than core radii, so we will depart from historical practice and focus on the *power-law lens* (e.g. Evans and Wilkinson 1998). Suppose, in three dimensions, that the lens has a density distribution $\rho \propto r^{-n}$. Such a lens will produce deflections of

$$\alpha(\theta) = b\left(\frac{\theta}{b}\right)^{2-n} \tag{9}$$

as shown in Fig. 9, with convergence and shear profiles

$$\kappa(\theta) = \frac{3-n}{2}\left(\frac{\theta}{b}\right)^{1-n} \quad \text{and} \quad \gamma(\theta) = \frac{n-1}{2}\left(\frac{\theta}{b}\right)^{1-n}. \tag{10}$$

The power law lenses cover most of the simple, physically interesting models. The *point-mass* lens is the limit $n \to 3$, with deflection $\alpha = b^2/\theta$, convergence $\kappa = 0$ (with a central singularity) and shear $\gamma = b^2/\theta^2$. The *singular isothermal sphere (SIS)* is the case with $n = 2$. It has a constant deflection $\alpha = b$, and equal convergence and shear $\kappa = \gamma = b/2\theta$. A *uniform critical sheet* is the limit $n \to 1$ with $\alpha = \theta$, $\kappa = 1$ and $\gamma = 0$. Models with $n \to 3/2$ have the cusp

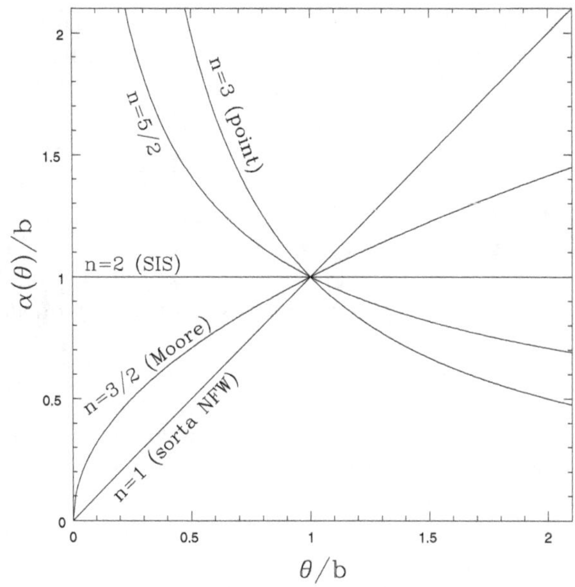

Fig. 9. The bending angles of the power law lens models. Profiles more centrally concentrated ($n > 2$) than the SIS ($n = 2$), have divergent central deflections, while profiles more extended ($n < 2$) than SIS have deflection profiles that become zero at the center of the lens. The $n = 1$ model is not quite an NFW model because the surface density is constant rather than logarithmic

exponent of the Moore, Quinn et al. (1999) halo model. The popular $\rho \propto 1/r$ NFW (Navarro, Frenk and White 1996, see Sect. 4.1) density cusps are not quite the same as the $n \to 1$ case because the projected surface density of a $\rho \propto 1/r$ cusp has $\kappa \propto \ln\theta$ rather than a constant. Nonetheless, the behavior of the power law models as $n \to 1$ will be very similar to the NFW model if the lens is dominated by the central cusp. The central regions of galaxies probably act like cusps with $1 \lesssim n \lesssim 2$.

The tangential magnification eigenvalue of these models is

$$1 - \kappa - \gamma = 1 - \frac{\alpha}{\theta} = 1 - \langle\kappa\rangle = 1 - (\theta/b)^{1-n}, \tag{11}$$

which is always equal to zero at $\theta = b \equiv \theta_E$. This circle defines the tangential critical curve or Einstein (ring) radius of the lens. We normalized the models in this fashion because the Einstein radius is usually the best-determined parameter of any lens model, in the sense that all successful models will find nearly the same Einstein radius (e.g. Kochanek 1991a; Wambsganss and Paczyński 1994). The source position corresponding to the tangential critical curve is the origin ($\beta = 0$), and the reason the magnification diverges is that a point source at the origin is converted into a ring on the tangential critical curve leading to a divergent ratio between the "areas" of the source and the image. The other important point to notice is that the mean surface density inside the tangential critical radius is $\langle\kappa\rangle \equiv 1$ independent of the model. This is true of any circular lens. With the addition of angular structure it is not strictly true, but it is a very good approximation unless the mass distribution is very flattened. The definition of b in terms of the properties of the lens galaxy will depend on the particular profile. For example, in a point mass lens ($n \rightarrow 3$), $b^2 = (4GM/c^2 D_d^{ang})(D_{ds}/D_s)$ where M is the mass, while in an SIS lens ($n = 2$), $b = 4\pi(\sigma_v/c)^2 D_{ds}/D_s$ where σ_v is the (1D) velocity dispersion of the lens. For the other profiles, b can be defined in terms of some velocity dispersion or mass estimate for the lens, as we will discuss later in Sect. 4.9 and Sect. 6. The radial magnification eigenvalue of these models is

$$1 - \kappa + \gamma = 1 - \frac{d\alpha}{d\theta} = 1 - (2-n)(\theta/b)^{1-n}, \tag{12}$$

which can be zero only if $n < 2$. If $n < 2$ the deflection goes to zero at the origin and the lens has a radial critical curve at $\theta = b(2-n)^{1/(n-1)} < b$ interior to the tangential critical curve. Models with $n \geq 2$ have constant ($n = 2$) or rising deflection profiles as we approach the lens center and have negative derivatives $d\alpha/d\theta$ at all radii.

A nice property of circular lenses is that they allow simple graphical solutions of the lens equation for arbitrary deflection profiles. There are two parts to the graphical solution – the first is to determine the radial positions θ_i of the images given a source position β, and the second is to determine the magnification by comparing the area of the images to the area of the source. Recall first, that by symmetry, all the images must lie on a line passing through the source and the lens. Let θ now be a signed radius that is positive along this line on one side of the lens and negative on the other. The lens (4) along the line is simply

$$\frac{\theta}{|\theta|}\alpha(|\theta|) = \theta - \beta\,, \tag{13}$$

where we have rearranged the terms to put the deflection on one side and the image and source positions on the other. One side of the equation is the bend angle (Fig. 9), while the other side of the equation, $\theta - \beta$, is simply a line of unit slope passing through the source position β. The solutions to the lens equation for any source position β are the radii θ_i where the line crosses the curve.

For understanding any observed lens, it is always useful to first sketch where the critical lines must lie. Recall from the discussion of caustics in Part 1, that images are always created and destroyed on critical lines as the source crosses a caustic, so the critical lines and caustics define the general structure of the lens. All our power-law models have a tangential critical line at $\theta = b$, which is the solution $\alpha(b) = b$ and corresponds to the source position $\beta = 0$. The origin, as the projection of the critical curve onto the source plane, is the tangential caustic (strictly speaking a degenerate pseudo-caustic) corresponding to the critical line. A point source at the origin is transformed into an Einstein ring of radius $\theta_E = b$.

The second step of the graphical construction is to determine the angular structure of the image. For simplicity, suppose the source is an arc with radial width $\Delta\beta$ and angular width $\Delta\chi$. By symmetry, the angle subtended by an image relative to the lens center must be the same as that subtended by the source. For an image at θ_i and a source at β, the tangential extent of the image is $|\theta_i|\Delta\chi$ while that of the source is $\beta\Delta\chi$. The tangential magnification of the image is simply $|\theta_i|/\beta = (1 - |\alpha(\theta_i)/\theta_i|)^{-1}$ after making use of the lens equation (13), and this is identical to the tangential magnification eigenvalue (11). The thickness of the arc requires finding the image radii for the inner and outer edges of the source, $\theta_i(\beta)$ and $\theta_i(\beta+\Delta\beta)$. The ratio of the thickness of the two arcs is the radial magnification,

$$\frac{\theta_i(\beta + \Delta\beta) - \theta_i(\beta)}{\Delta\beta} \simeq \frac{d\theta}{d\beta} = \left(1 - \frac{d\alpha}{d\theta}\right)^{-1}, \tag{14}$$

and this is simply the inverse of the radial eigenvalue of the magnification matrix (12) where we have taken the derivative of the lens equation (13) with respect to the source position to obtain the final result. Thus, the tangential magnification simply reflects the fact that the angle subtended by the source is the angle subtended by the image, while the radial magnification depends on the slope of the deflection profile with declining deflection profiles ($d\alpha/d\theta < 0$) demagnifying the source and rising profiles magnifying the source.

In Fig. 10 we illustrate this for the point mass lens ($n \to 3$). From the shape of the deflection profile, it is immediately obvious that there will be only two images, one on each side of the lens. If we assume $\beta > 0$, the first image is a minimum located at

$$\theta_1 = \frac{1}{2}\left(\beta + \sqrt{\beta^2 + 4b^2}\right) \tag{15}$$

with $\theta_1 > \theta_E$ and positive magnification

$$\mu_1 = \frac{1}{4}\left(\frac{\beta}{\sqrt{\beta^2 + 4b^2}} + \frac{\sqrt{\beta^2 + 4b^2}}{\beta} + 2\right) > 0, \tag{16}$$

while the second image is a saddle point located at

$$\theta_2 = \frac{1}{2}\left(\beta - \sqrt{\beta^2 + 4b^2}\right) \tag{17}$$

with $-\theta_E < \theta_2 < 0$ and negative magnification

$$\mu_2 = -\frac{1}{4}\left(\frac{\beta}{\sqrt{\beta^2 + 4b^2}} + \frac{\sqrt{\beta^2 + 4b^2}}{\beta} - 2\right) < 0. \tag{18}$$

As the source approaches the tangential caustic ($\beta \to 0$) the magnifications of both images diverge as β^{-1} and the image radii converge to θ_E. As the source

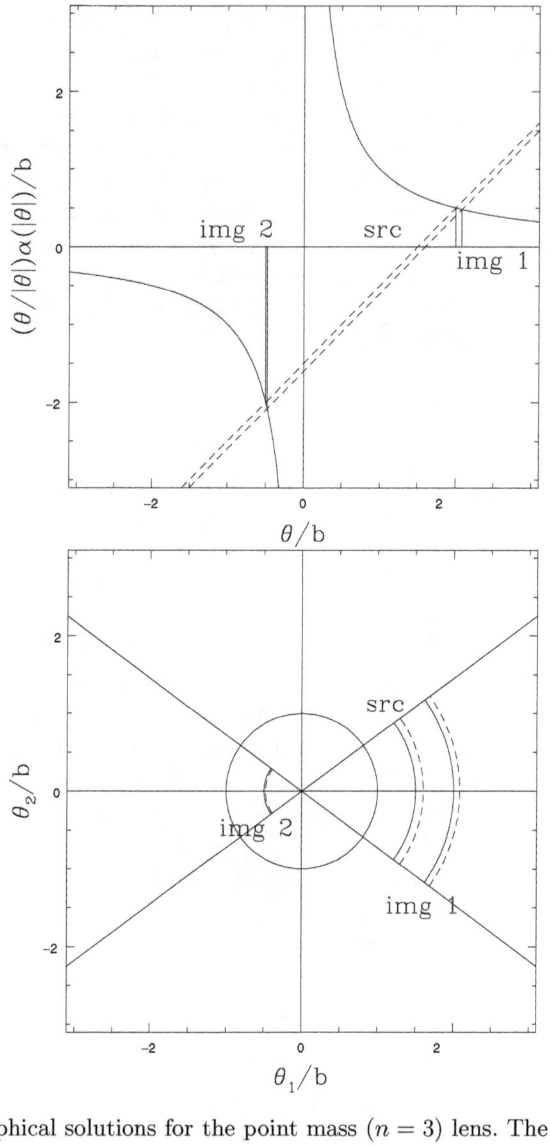

Fig. 10. Graphical solutions for the point mass ($n = 3$) lens. The *top* panel shows the graphical solution for the radial positions of the images, and the *bottom* panel shows the graphical solution for the image structure

moves to infinity, the magnification of the first image approaches unity and its position approaches that of the source, while the second image is demagnified by the factor $(1/2)(b/\beta)$ and converges to the position of the lens. The image separation

$$|\theta_1 - \theta_2| = 2b\sqrt{1 + \beta^2/4b^2} \geq 2b \tag{19}$$

is always larger than the diameter of the Einstein ring and the total magnification

$$|\mu_1| + |\mu_2| = \frac{2b^2 + \beta^2}{\beta\sqrt{\beta^2 + 4b^2}} \geq 1 \tag{20}$$

is the characteristic light curve expected for isolated Galactic microlensing events (see Part 4). The point mass lens has one peculiarity that makes it different from extended density distributions like galaxies in that it has two images independent of the impact parameter of the source and no radial caustic. This is a characteristic of any density distribution with a divergent central deflection ($n > 2$).

The SIS ($n = 2$) model is the "standard" lens model for galaxies. Figures 11 and 12 show the geometric constructions for the images of an SIS lens. If $0 < \beta < b$, then the SIS lens also produces two images (Fig. 11). The first image is a minimum located at

$$\theta_1 = \beta + b \quad \text{with } \theta_1 > b \text{ and positive magnification} \quad \mu_1 = 1 + b/\beta \tag{21}$$

and the second image is a saddle point located at

$$\theta_2 = \beta - b \quad \text{with } -b < \theta_2 < 0 \text{ and negative magnification} \quad \mu_2 = 1 - b/\beta. \tag{22}$$

The image separation $|\theta_1 - \theta_2| = 2b$ is constant, and the total magnification $|\mu_1| + |\mu_2| = 2b/\beta$ is a simple power law. The magnification produced by an SIS lens is purely tangential since the radial magnification is unity. If, however, $\beta > b$, then there is only one image, corresponding to the minimum located on the same side of the lens as the source (see Fig. 12). This boundary on the source plane at $\beta = b$ between having two images at smaller radii and only one image at larger radii is a radial (pseudo)-caustic that can be thought of as being associated with a radial critical curve at the origin. It is a pseudo-caustic because there are neither images nor a divergent magnification associated with it.

Historically the next step is to introduce a core radius to have a model with a true radial critical line and caustic (see Part 1, Blandford and Kochanek 1987a,b; Kochanek and Blandford 1987; Kovner 1987a; Hinshaw and Krauss 1987; Krauss and White 1992; Wallington and Narayan 1993; Kochanek 1996a,b). Instead we will consider the still softer power law model with $n = 3/2$, which would correspond to the central exponent of the "Moore" profile proposed for CDM halos (Moore et al. 1998). As Fig. 13 shows, there is only one solution for $|\beta| > b/4$, a minimum located at

$$\theta_1 = \frac{1}{2}\left(b + 2\beta + \sqrt{b + 4\beta}\right), \tag{23}$$

and with $\theta_1 > b$ assuming β is positive. The magnification expressions are too complex to be of much use, but the magnification μ_1 diverges at $\theta = b$ when the source is on the tangential pseudo-caustic at $\beta = 0$. As Fig. 14 shows,

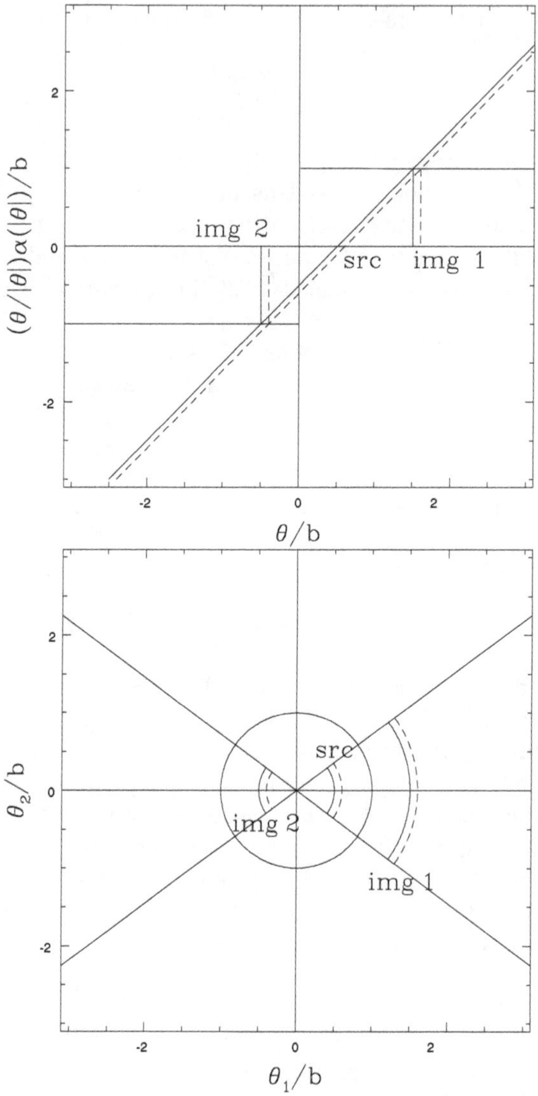

Fig. 11. Graphical solutions for the SIS ($n = 2$) lens when $\beta < b$ and there are two images

we find two additional images once $|\beta| < b/4$. The first additional image is a saddle point located at

$$\theta_2 = \frac{1}{2}\left(-b + 2\beta - \sqrt{b + 4\beta}\right) \tag{24}$$

with $-b < \theta_2 < -b/4$, which has a negative magnification that diverges at both $\theta_2 = -b$ (the tangential critical curve) and $\theta_2 = -b/4$. This latter radius

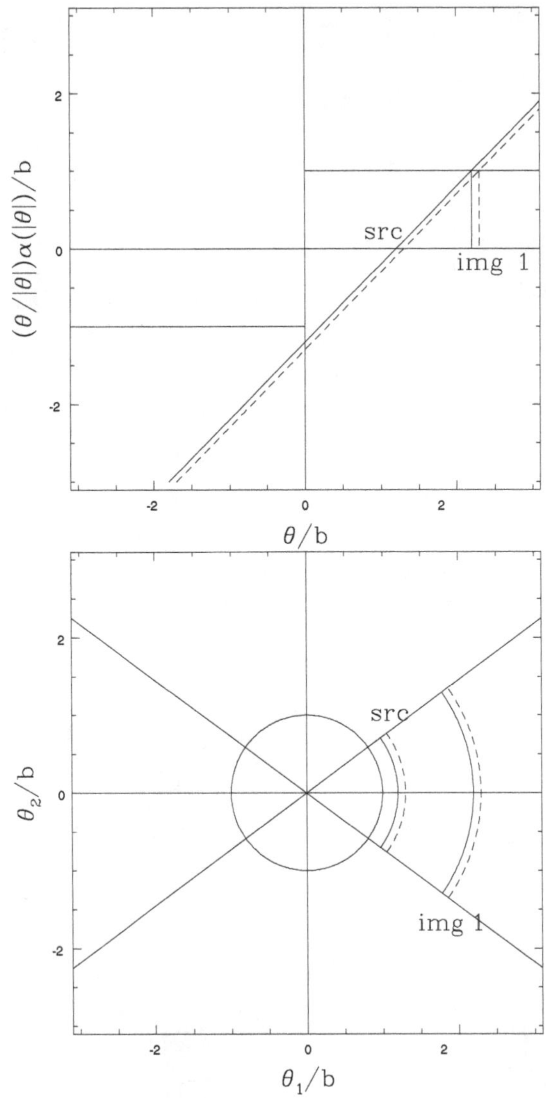

Fig. 12. Graphical solutions for the SIS ($n = 2$) lens when $\beta > b$ and there is only one image

defines the radial critical curve where the magnification diverges because the radial magnification eigenvalue $1 - \kappa + \gamma = 1 - d\alpha/d\theta = 0$ at radius $\theta = b/4$. The third image is a maximum located at

$$\theta_3 = \frac{1}{2}\left(-b + 2\beta + \sqrt{b + 4\beta}\right) \tag{25}$$

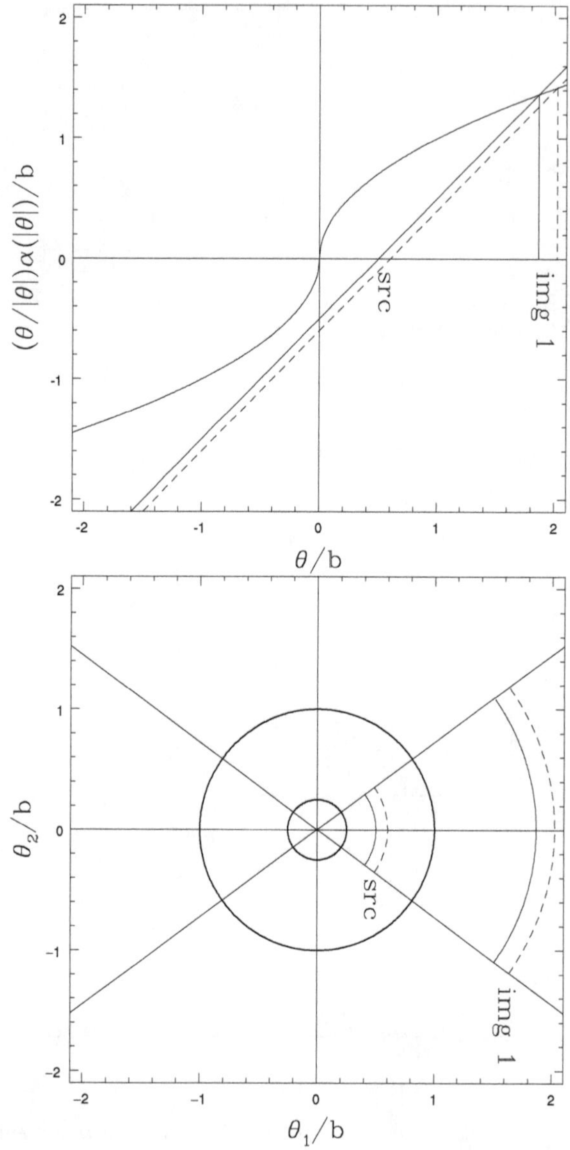

Fig. 13. Graphical solutions for the Moore profile cusp ($n = 3/2$) lens when $\beta > b/4$ and there is only one image

with $-b/4 < \theta_2 < 0$ and a positive magnification that diverges on the radial critical curve. As we move the source outward from the center we would see images 2 and 3 approach each other, merging on the radial critical line where they would have divergent magnifications, and then vanishing to leave only image 1. We would see the same pattern if instead of softening the exponent

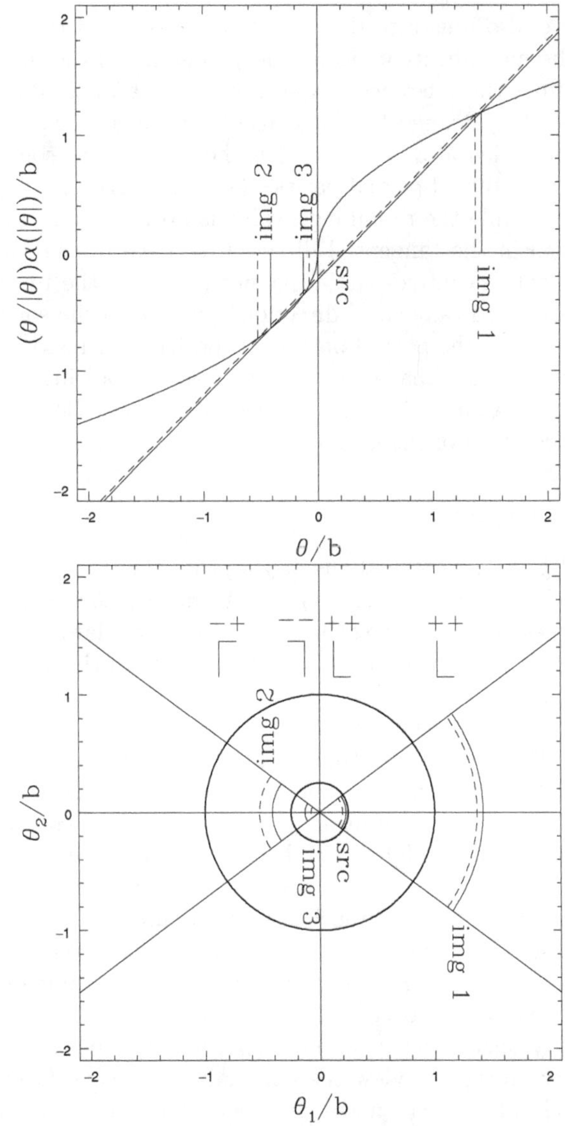

Fig. 14. Graphical solutions for the Moore profile cusp ($n = 3/2$) lens when $\beta < b/4$ and there are three images. At the top of the figure we illustrate the geometric meaning of the image partial parities defined by the signs of the magnification tensor eigenvalues (see text)

we had followed the traditional path and added a core radius to the SIS model. With a finite core radius the central deflection profile would pass through zero, and this would introduce a radial critical curve and a third image which would be a maximum of the time delay surface.

In Fig. 14 we also illustrate the geometric meaning of the partial parities (the signs of the magnification eigenvalues). A source structure (the L above the source) defines the reference shape. Image 1 is a minimum with positive partial parities (++) defined by the signs of the tangential and radial eigenvalues. The basic orientation of image 1 is the same as the source. Image 2 is a saddle point with mixed partial parities (−+) because the tangential eigenvalue is negative while the radial eigenvalue is positive. This means that the image is inverted in the tangential direction relative to the source. Image 3 is a maximum with negative partial parties (−−), so the image is inverted in both the radial and tangential directions relative to the source. The total parity, the product of the partial parities, is positive for maxima and minima so the orientation of the image can be produced by rotating the source. The total parity of the saddle point image is negative, so its orientation cannot be produced by a rotation of the source.

3.3 Non-Circular Lenses

The tangential pseudo-caustic at the origin producing Einstein ring images is unstable to the introduction of any angular structure into the gravitational potential of the lens. There are two generic sources of angular perturbations. The first source of angular perturbations is the ellipticity of the lens galaxy. What counts here is the ellipticity of the gravitational potential rather than of the surface density. For a lens with axis ratio q, ellipticity $\epsilon = 1 - q$, or eccentricity $e = (1-q^2)^{1/2}$, the ellipticity of the potential is usually $\epsilon_\Psi \sim \epsilon/3$ – potentials are always rounder than densities. The second source of angular perturbations is tidal perturbations from any nearby objects. This is frequently called the "external shear" or the "tidal shear" because it can be modeled as a linear shearing of the deflections. In all known lenses, quadrupole perturbations (i.e. $\Psi \propto \cos(2\chi)$ where χ is the azimuthal angle) dominate – higher order multipoles are certainly present and they can be quantitatively important, but they are smaller. For example, in an ellipsoid the amplitude of the $\cos 2m\chi$ multipole scales as ϵ_Ψ^m (see Sect. 4.4 and Sect. 8).

Unfortunately, there is no example of a non-circular lens that can be solved in full generality unless you view the nominally analytic solutions to quartic equations as helpful. We can make the greatest progress for the case of an SIS in an external (tidal) shear field. Tidal shear is due to perturbations from nearby objects and its amplitude can be determined by Taylor expanding its potential near the lens (see Part 1 and Sect. 4). Consider a lens with Einstein radius θ_E perturbed by an object with effective lens potential Ψ a distance θ_p away. For $\theta_E \ll \theta_p$ we can Taylor expand the potential of the nearby object about the center of the primary lens, dropping the leading two terms.[2] This

[2] The first term, a constant, gives an equal contribution to the time delays of all the images, so it is unobservable when all we can measure is relative delays. The second term is a constant deflection, which is unobservable when all we can measure is relative deflections.

leaves, as the first term with observable consequences,

$$\Psi(\theta) \simeq \frac{1}{2}\boldsymbol{\theta} \cdot \boldsymbol{\nabla\nabla}\Psi \cdot \boldsymbol{\theta} = \frac{1}{2}\kappa_p\theta^2 - \frac{1}{2}\gamma_p\theta^2\cos 2(\chi - \chi_p), \tag{26}$$

where κ_p is the surface density of the perturber at the center of the lens galaxy and $\gamma_p > 0$ is the tidal shear from the perturber. If the perturber is an SIS with critical radius b_p and distance θ_p from the primary lens, then $\kappa_p = \gamma_p = b_p/2\theta_p$. With this normalization, the angle χ_p points toward the perturber. For a circular lens, the shear $\gamma_p = \langle\kappa\rangle - \kappa$ can be expressed in terms of the surface density of the perturber, and it is larger (smaller) than the convergence if the density profile is steeper (shallower) than isothermal.

The effects of κ_p are observable only if we measure a time delay or have an independent estimate of the mass of the lens galaxy, while the effects of the shear are easily detected from the relative positions of the lensed images (see Part 1). Consider, for example, one component of the lens equation including an extra convergence,

$$\beta_1 = \theta_1(1 - \kappa_p) - d\Psi/d\theta_1, \tag{27}$$

and then simply divide by $1 - \kappa_p$ to get

$$\beta_1/(1 - \kappa_p) = \theta_1 - (d\Psi/d\theta_1)/(1 - \kappa_p). \tag{28}$$

The rescaling of the source position $\beta_1/(1-\kappa_p)$ has no consequences since the source position is not an observable quantity, while the rescaling of the deflection is simply a change in the mass of the lens. This is known as the "mass sheet degeneracy" because it corresponds to adding a constant surface density sheet to the lens model (Falco, Gorenstein and Shapiro 1985), and it is an important systematic problem for both strong lenses and cluster lenses (see Part 3).

Thus, while the extra convergence can be important for the quantitative understanding of time delays or lens galaxy masses, it is only the shear that introduces qualitatively new behavior to the lens equations. The effective potential of an SIS lens in an external shear is $\Psi = b\theta + (\gamma/2)\theta^2\cos 2\chi$ leading to the lens equations

$$\begin{aligned}\beta_1 &= \theta_1(1 - \gamma) - b\theta_1/|\boldsymbol{\theta}| \\ \beta_2 &= \theta_2(1 + \gamma) - b\theta_2/|\boldsymbol{\theta}|\end{aligned}, \tag{29}$$

where for $\gamma > 0$ the perturber is due North (or South) of the lens. The inverse magnification is

$$\mu^{-1} = 1 - \gamma^2 - \frac{b}{\theta}\left(1 - \gamma\cos 2\chi\right), \tag{30}$$

where $\boldsymbol{\theta} = (\theta_1, \theta_2) = \theta(\cos\chi, \sin\chi)$.

The first step in any general analysis of a new lens potential is to locate the critical lines and caustics. In this case we can easily solve $\mu^{-1} = 0$ to find that the tangential critical line

$$\theta = b\frac{1 - \gamma\cos 2\chi}{1 - \gamma^2} \tag{31}$$

is an ellipse whose axis ratio is determined by the amplitude of the shear γ and whose major axis points toward the perturber. We call it the tangential critical line because the associated magnifications are nearly tangential to the direction to the lens galaxy and because it is a perturbation to the Einstein ring of a circular lens. The tangential caustic, the image of the critical line on the source plane, is a curve called an astroid (Fig. 15, it is not a "diamond" despite repeated use of the term in the literature). The parametric expression for the astroid curve is

$$\beta_1 = -\frac{2b\gamma}{1+\gamma}\cos^3\chi = -\beta_+\cos^3\chi \qquad \beta_2 = +\frac{2b\gamma}{1-\gamma}\sin^3\chi = \beta_-\sin^3\chi, \quad (32)$$

where the parameter χ is the same as the angle appearing in the critical curve (31) and we have defined $\beta_\pm = 2b\gamma/(1\pm\gamma)$ for the locations of the cusp tips on the axes. The astroid consists of 4 cusp caustics on the symmetry axes of the lens connected by fold caustics with a major axis pointing toward the perturber. Like the SIS model without any shear, the origin plays the role of the radial critical line and there is a circular radial pseudo-caustic at $\beta = b$.

As mentioned earlier, there is no useful general solution for the image positions and magnifications. We can, however, solve the equations for a source on one of the symmetry axes of the lens. For example, consider a solution on the minor axis of the lens ($\beta_2 = 0$ for $\gamma > 0$). There are two ways of solving the lens equation to satisfy the criterion. One is to put the images on the same axis ($\theta_2 = 0$) and the other is to place them on the arc defined by $0 = 1 + \gamma - b/\theta$. The images with $\theta_2 = 0$ are simply the SIS solutions corrected for the effects of the shear. Image 1 is defined by

$$\theta_1 = \frac{\beta_1 + b}{1-\gamma} \quad \text{with} \quad \mu^{-1} = \left(1-\gamma^2\right)\frac{\beta_+ + \beta_1}{b + \beta_1}, \quad (33)$$

and image 2 is defined by

$$\theta_1 = \frac{\beta_1 - b}{1-\gamma} \quad \text{with} \quad \mu^{-1} = \left(1-\gamma^2\right)\frac{\beta_+ - \beta_1}{b - \beta_1}. \quad (34)$$

Image 1 exists if $\beta_1 > -b$, it is a saddle point for $-b < \beta_1 < -\beta_+$ and it is a minimum for $\beta_1 > -\beta_+$. Image 2 has the reverse ordering. It exists for $\beta_1 < b$, it is a saddle point for $\beta_+ < \beta_1 < b$ and it is a minimum for $\beta_1 < \beta_+$. The magnifications of both images diverge when they are on the tangential critical line ($\beta_1 = -\beta_+$ for image 1 and $\beta_1 = +\beta_+$ for image 2) and approach zero as they move into the core of the lens ($\beta_1 \to -b$ for image 1 and $\beta_1 \to +b$ for image 2). These two images shift roles as the source moves through the origin. The other two solutions are both saddle points, and they exist only if the source lies inside the astroid ($|\beta_1| < \beta_+$ along the axis). The positions of images 3 (+) and 4 (−) are

$$\theta_1 = -\frac{\beta_1}{2\gamma} \qquad \theta_2 = \pm\frac{b}{1+\gamma}\left[1 - \left(\frac{\beta_1}{\beta_+}\right)^2\right]^{1/2}, \quad (35)$$

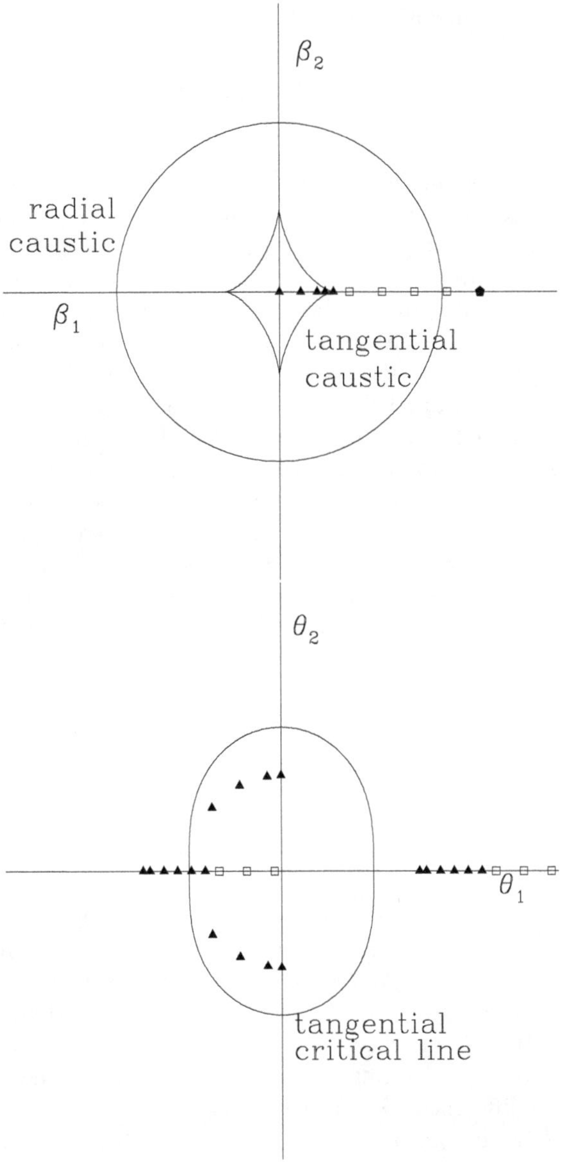

Fig. 15. Example of a minor axis cusp on the source (*top*) and image (*bottom*) planes. When the source is inside both the radial and tangential caustics (*triangles*) there are four images. As the source moves toward the cusp, three of the images head towards a merger on the critical line and become highly magnified to leave only one image once the source crosses the cusp and lies between the two caustics (*open squares*). In a minor axis cusp, the image surviving the cusp merger is a saddle point interior to the critical line. As the source approaches the radial caustic, one image approaches the center of the lens and then vanishes as it crosses the caustic to leave only one image (*pentagons*)

and they have equal magnifications

$$\mu^{-1} = -2\gamma(1+\gamma)\left[1 - \left(\frac{\beta_1}{\beta_+}\right)^2\right] . \tag{36}$$

The magnifications of the images diverge when the source reaches the cusp tip ($|\beta_1| = \beta_+$) and the image lies on the tangential critical curve.

Thus, if we start with a source at the origin we can follow the changes in the image structure (see Figs. 15 and 16). With the source at the origin we see 4 images on the symmetry axes with reasonably high magnifications, $\sum |\mu_i| = (2/\gamma)/(1-\gamma^2) \sim 10$. It is a generic result that the least magnified four-image system is found for an on-axis source, and this configuration has a total magnification of order the inverse of the ellipticity of the gravitational potential. As we move the source toward the tip of the cusp ($\beta \to \beta_+$, Fig. 15), image 1 simply moves out along the symmetry axis with slowly dropping magnification, while images 2, 3 and 4 move toward a merger on the tangential critical curve at $\boldsymbol{\theta} = (-\beta_+, 0)$. Their magnifications steadily rise and then diverge when the source reaches the cusp. If we move the source further outward we find only images 1 and 2 with 1 moving outward and 2 moving inward toward the origin. As it approaches the origin, image 2 becomes demagnified and vanishes when $\beta \to b$. Had we done the same calculation on the major axis (Fig. 16), there is a qualitative difference. As we moved image 1 outward along the β_2 axis, image 3 and 4 would merge with image 1 when the source reaches the tip of the cusp at $\beta_2 = \beta_-$ rather than with image 2.

Unfortunately once we move the source off a symmetry axis, there is no simple solution. It is possible to find the locations of the remaining images given that two images have merged on the critical line, and this is useful for determining the mean magnifications of the lensed images, a point we will return to when we discuss lens statistics in Sect. 6. Here we simply illustrate (Fig. 17) the behavior of the images when we move the source radially outward from the origin away from the symmetry axes. Rather than three images merging on the tangential critical line as the source approaches the tip of a cusp, we see two images merging as the source approaches the fold caustic of the astroid. This difference, two images merging versus three images merging, is a generic difference between folds and cusps as discussed in Part 1. All images in these four-image configurations are restricted to an annulus of width $\sim \gamma b$ around the critical line, so the mean magnification of all four image configurations is also of order γ^{-1} (see Finch et al. 2002).

There is one more possibility for the caustic structure of the lens if the external shear is large enough. For $1/3 < |\gamma| < 1$, the tip of the astroid caustic extends outside the radial caustic, as shown if Fig. 18. This allows a new image geometry, known as the cusp or disk geometry, where we see three images straddling the major axis of a very flattened potential. It is associated with the caustic region inside the astroid caustic associated with

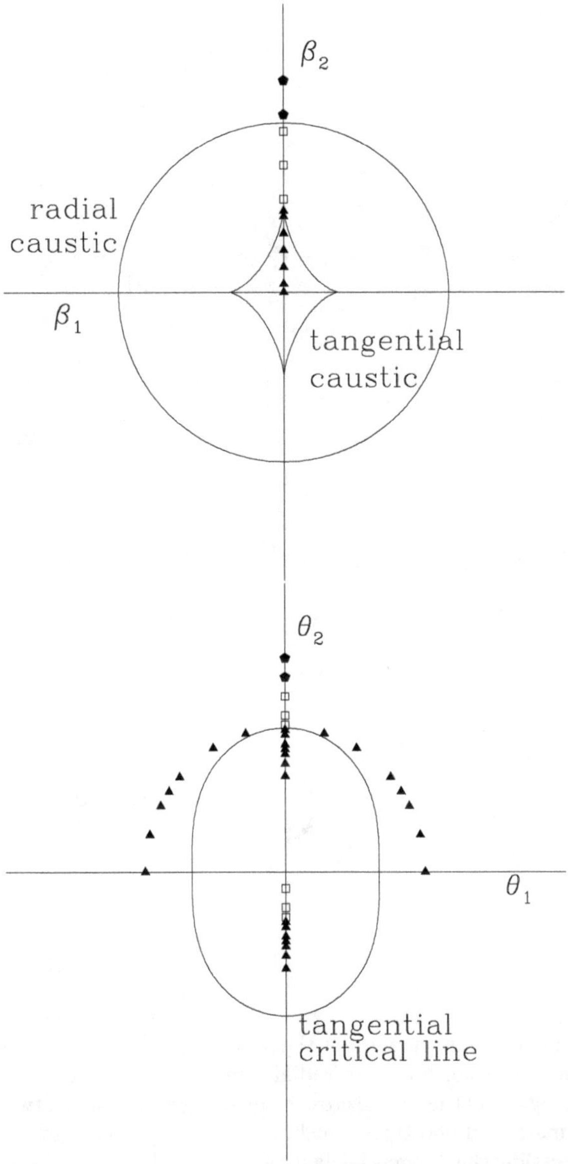

Fig. 16. Example of a major axis cusp on the source (*top*) and image (*bottom*) planes. When the source is inside both the radial and tangential caustics (*triangles*) there are four images. As the source moves toward the cusp, three of the images head towards a merger on the critical line and become highly magnified to leave only one image once the source crosses the cusp and lies between the two caustics (*open squares*). In a major axis cusp, the image surviving the cusp merger is the minimum corresponding to the image we would see in the absence of a lens. As the source approaches the radial caustic, one image approaches the center of the lens and then vanishes as the source crosses the caustic to leave only one image (*pentagons*)

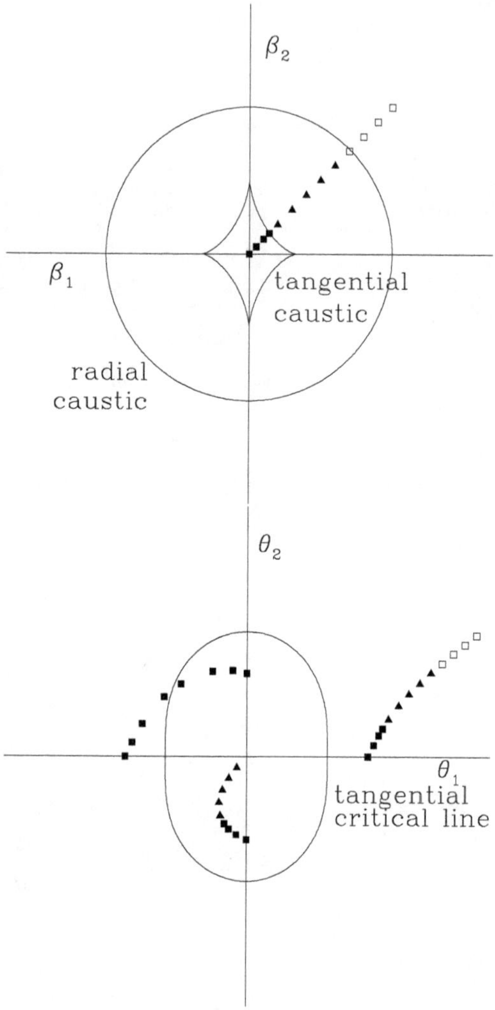

Fig. 17. Example of a fold merger on the source (*top*) and image (*bottom*) planes. When the source is inside both the radial and tangential caustics (*filled squares*) there are four images. As the source crosses the tangential caustic, two images merge, become highly magnified and then vanish, leaving only two images (*triangles*) when the source is outside the tangential caustic but inside the radial caustic. As the source approaches the radial caustic, one image moves into the center of the lens and then vanishes when the source crosses the radial caustic to leave only one image when the source is outside both caustics (*open squares*)

the tangential critical line but outside the radial caustic. This configuration appears to be rare for lenses produced by galaxies, with APM08279+5255 as the only likely candidate, but relatively more common in clusters. The difference is that clusters tend to have shallower density profiles than galaxies,

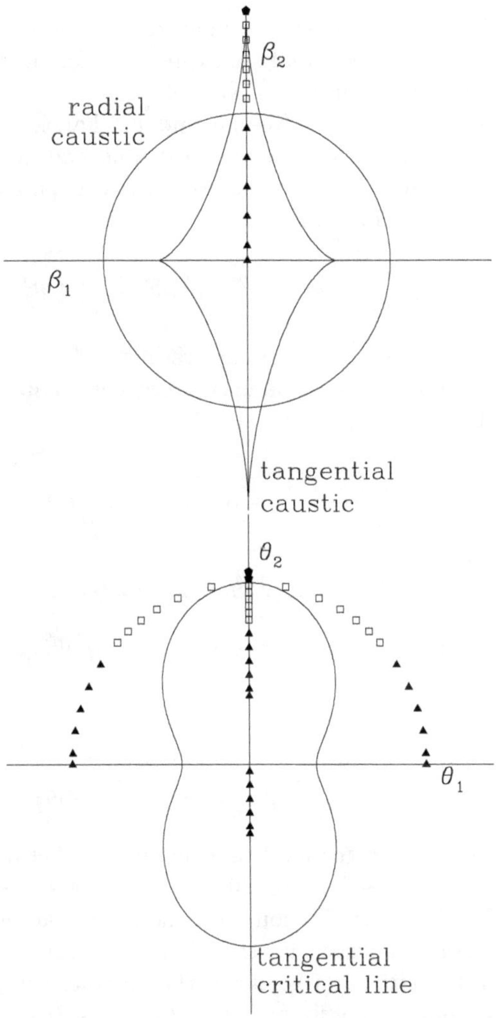

Fig. 18. Example of a cusp or disk image geometry on the source (*top*) and image (*bottom*) planes. The shear is high enough to make the tangential caustic extend outside the radial caustic. For a source inside both caustics (*triangles*) we see a standard four-image geometry as in Fig. 16. However, for a source outside the radial caustic but inside the tangential caustic (*squares*) we have three images all on one side of the lens. This is known as the cusp geometry because it is always associated with cusps, and the disk geometry because flattened disks are the only natural way to produce them. Once the source is outside the cusp tip (*pentagon*), a single image remains

which shrinks the radial caustics relative to the tangential caustics to allow more cross section for this image configuration and lower ellipticity thresholds before it becomes possible (Oguri and Keeton 2004 most recently, but also see Kochanek and Blandford 1987; Kovner 1987a; Wallington and Narayan 1993).

In general, it is far more difficult to analyze ellipsoidal lenses, in part because few ellipsoidal lenses have analytic expressions for their deflections. The exception is the isothermal ellipsoid (Kassiola and Kovner 1993; Kormann, Schneider and Bartelmann 1994; Keeton and Kochanek 1998), including a core radius s, which is both analytically tractable and generally viewed as the most likely average mass distribution for gravitational lenses. The surface density of the isothermal ellipsoid

$$\kappa = \frac{1}{2}\frac{b}{\omega} \quad \text{where} \quad \omega^2 = q^2(\theta_1^2 + s^2) + \theta_2^2 \tag{37}$$

depends on the axis ratio q and the core radius s. For $q = 1 - \epsilon < 1$ the major axis is the θ_1 axis and s is the major axis core radius. The deflections produced by this lens are remarkably simple,

$$\alpha_1 = \frac{b}{\sqrt{1-q^2}} \tan^{-1}\left[\frac{\theta_1\sqrt{1-q^2}}{\omega + s}\right], \quad \alpha_2 = \frac{b}{\sqrt{1-q^2}} \tanh^{-1}\left[\frac{\theta_2\sqrt{1-q^2}}{\omega + q^2 s}\right]. \tag{38}$$

The effective lens potential is cumbersome but analytic,

$$\Psi = \boldsymbol{\theta} \cdot \boldsymbol{\alpha} - bs \ln\left[(\omega + s)^2 + (1 - q^2)\theta_1^2\right], \tag{39}$$

the magnification is simple

$$\mu^{-1} = 1 - \frac{b}{\omega} - \frac{b^2 s}{\omega\left[(\omega+s)^2 + (1-q^2)\theta_1^2\right]} \tag{40}$$

and becomes even simpler in the limit of a singular isothermal ellipsoid (SIE) with $s = 0$ where $\mu^{-1} \to 1 - b/\omega$. In this case, contours of surface density κ are also contours of the magnification, and the tangential critical line is the $\kappa = 1/2$ isodensity contour just as for the SIS model. The critical radius scale b can be related to the circular velocity in the plane of the galaxy relatively easily. For an isothermal sphere we have that $b_{SIS} = 4\pi(\sigma_v/c)^2 D_{ds}/D_s$ where the circular velocity is $v_c = \sqrt{2}\sigma_v$. For the projection of a three-dimensional (3D) oblate ellipsoid of axis ratio q_3 and inclination i, so that $q^2 = q_3^2 \cos^2 i + \sin^2 i$, the deflection scale is $b = b_{SIS}(e_3/\sin^{-1} e_3)$ where $e_3 = \sqrt{1 - q_3^2}$ is the eccentricity of 3D mass distribution. In the limit that $q_3 \to 0$ the model becomes a Mestel (1963) disk, the infinitely thin disk producing a flat rotation curve, and $b = 2b_{SIS}/\pi$ (see Sect. 4.9 and Keeton, Kochanek and Seljak 1997; Keeton and Kochanek 1998; Chae 2003). At least for the case of a face-on disk, at fixed circular velocity you get a smaller Einstein radius as you make the 3D distribution flatter because a thin disk requires less mass to produce the same circular velocity.

We can generate several other useful models from the isothermal ellipsoids. For example, steeper ellipsoidal density distributions can be derived by differentiating with respect to s^2. The most useful of these is the first derivative with $\kappa \propto \omega^{-3/2}$ which is related to the Kuzmin (1956) disk (see Kassiola and Kovner 1993; Keeton and Kochanek 1998). It is also easy to generate models with flat inner rotation curves and truncated halos by taking the difference of two isothermal ellipsoids. In particular if $\kappa(s)$ is an isothermal ellipsoid with core radius s, the model

$$\kappa = \kappa(s) - \kappa(a) \tag{41}$$

with $a > s$ has a central core region with a rising rotation curve for $\theta \lesssim s$, a flat rotation curve for $s \lesssim \theta \lesssim a$ and a dropping rotation curve for $\theta \gtrsim a$. In the singular limit, it becomes the "pseudo-Jaffe model" corresponding to a 3D density distribution $\rho \propto (r^2 + s^2)^{-1}(r^2 + a^2)^{-1}$ whose name derives from the fact that it is very similar the Jaffe model with $\rho \propto r^{-2}(r + a)^{-2}$ (Kneib et al. 1996; Keeton and Kochanek 1998). We will discuss other common lens models in Sect. 4.1.

The last simple analytic models we mention are the generalized singular isothermal potentials of the form $\Psi = \theta F(\chi)$ with surface density $\kappa(\theta, \chi) = (1/2)(F(\chi) + F''(\chi))/\theta$. Both the SIS and SIE are examples of this model. The generalized isothermal sphere has a number of useful analytic properties. For example, the magnification contours are isodensity contours

$$\mu^{-1} = 1 - \frac{1}{\theta}\left[F(\chi) + F''(\chi)\right] = 1 - 2\kappa(\theta, \chi) \tag{42}$$

with the tangential critical line being the contour with $\kappa = 1/2$, and the time delays between images depend only on the distances from the images to the lens center (see Witt, Mao and Keeton 2000; Kochanek, Keeton and McLeod 2001a; Wucknitz 2002; Evans and Witt 2003).

4 The Mass Distributions of Galaxies

Contrary to popular belief, the modeling of gravitational lenses to determine the mass distribution of a lens is not a "black art." It is, however, an area in which the lensing community has communicated results badly. There are two main problems. First, many modeling results seem almost deliberately obfuscatory as to what models were actually used, what data were fit and what was actually constrained. Not only do many lens papers insist on taking well known density distributions from the dynamical literature and assigning them new names simply because they have been projected into two dimensions, but they then assign them a plethora of bizarre acronyms. Sometimes the model used is not actually the one named, for example using tidally truncated halos but calling them isothermal models. Second, there is a steady confusion between the parameters of models and the aspects of the mass distribution that have actually been constrained. Models with apparently very different parameters may be in perfect accord as to the properties of the mass distribution that are actually relevant to what is observed. Discussions of non-parametric mass models then confuse the issue further by conflating differences in parameters with differences in what is actually constrained to argue for non-parametric models when in fact they also are simply matching the same basic properties with lots of extra noise from the additional and uninteresting degrees of freedom. In short, the problem with lens modeling is not that it is a "black art" but that the practitioners try to make it seem to be a "black art" presumably so that people will believe they need wizards. The most important point to take from this section is that any idiot can model a lens and interpret it properly with a little thinking about what it is that lenses constrain.

There are two issues to think about in estimating the mass distributions of gravitational lenses. The first issue is how to model the mass distribution with a basic choice between parametric and non-parametric models. In Sect. 4.1 we summarize the most commonly used radial mass distributions for lens models. Ellipsoidal versions of these profiles combined with an external (tidal) shear are usually used to describe the angular structure, but there has been recent interest in deviations from ellipsoidal distributions which we discuss in Sect. 4.4 and Sect. 8. In Sect. 4.7 we summarize the most common approaches for non-parametric models of the mass distribution. Since this is my review, I will argue that the parametric models are all that is needed to model lenses and that they provide a better basis for understanding the results than non-parametric models (but the reader should be warned that if Prasenjit Saha was writing this you would probably get a different opinion).

The second issue is to determine the aspects of the lens data that actually constrain the mass distribution. Among the things that can be measured for a lens are the relative positions of the components (the astrometric constraints), the relative fluxes of the images, the time delays between the images, the dynamical properties of the lens galaxy, and the microlensing of the images. Of these, the most important constraints are the positions. We can usually measure the relative positions of the lensed components very accurately (5 mas or better) compared to the arc second scales of the component separations. Obviously the accuracy diminishes when components are faint, and the usual worst case is having very bright lensed quasars that make it difficult to detect the lens galaxy. As we discuss in Sect. 8, substructure and/or satellites of the lens galaxy set a lower limit of order 1–5 mas with which it is safe to impose astrometric constraints independent of the measurement accuracy. When the source is extended, the resulting arcs and rings discussed in Sect. 10 provide additional constraints. These are essentially astrometric in nature, but are considerably more difficult to use than multiply imaged point sources. Our general discussion of how lenses constrain the radial (Sect. 4.3) and angular structure (Sect. 4.4) focus on the use of astrometric constraints, and in Sect. 4.6 we discuss the practical details of fitting image positions in some detail.

The flux ratios of the images are one of the most easily measured constraints, but are presently unusable. Flux ratios measured at a single epoch are affected by time variability in the source (Sect. 5), microlensing by the stars in the lens galaxy in the optical continuum (see Part 4), magnification perturbations from substructure at all wavelengths (see Sect. 8), absorption by the interstellar medium (ISM) of the lens (dust in the optical, free-free in the radio) and scatter broadening in the radio (see Sect. 8 and Sect. 9). Most applications of flux ratios have focused on using them to probe these perturbing effects rather than for studying the mean mass distribution of the lens. Where radio sources have small scale VLBI structures, the changes in the relative astrometry of the components can constrain the components of the relative magnification tensors without needing to use any flux information (e.g. Garrett et al. 1994; Rusin et al. 2002).

Two types of measurements, time delays (Sect. 5) and microlensing by the stars or other compact objects in the lens galaxy (Part 4) constrain the surface density near the lensed images. Microlensing also constrains the fraction of that surface density that can be in the form of stars. To date, time delays have primarily been used to estimate the Hubble constant rather than the surface density, but if we view the Hubble constant as a known quantity, consider only time delay ratios, or simply want to compare surface densities between lenses, then time delays can be used to constrain the mass distribution. We discuss time delays separately because of their close association with attempts to measure the Hubble constant. Using microlensing variability to constrain the mass distribution is presently more theory than practice due to a lack of microlensing light curves for almost all lenses. However, the light curves of the one well monitored lens, Q2237+0305, appear to require a surface density composed mainly of stars as we would expect for a lens where we see the images deep in the bulge of a nearby spiral galaxy (Kochanek 2004). We will not discuss this approach further in Part 2.

Any independent measurement of the mass of a component will also help to constrain the structure of the lenses. At present this primarily means making stellar dynamical measurements of the lens galaxy and comparing the dynamical mass estimates to those from the lens geometry. We discuss this in detail in Sect. 4.9. For lenses associated with clusters, X-ray, weak lensing or cluster velocity dispersion measurements can provide estimates of the cluster mass. While this has been done in a few systems (e.g. X-rays, Morgan et al. 2001; Chartas et al. 2002; weak lensing, Fischer et al. 1997; velocity dispersions, Angonin-Willaime, Soucail and Vanderriest 1994), the precision of these mass estimates is not high enough to give strong constraints on lens models. X-ray observations are probably more important for locating the positions of groups and clusters relative to the lens than for estimating their masses.

The most useful way of thinking about lensing constraints on mass distributions is in terms of multipole expansions (e.g. Kochanek 1991a; Trotter, Winn and Hewitt 2000; Evans and Witt 2003; Kochanek and Dalal 2004). An arbitrary surface density $\kappa(\boldsymbol{\theta})$ can be decomposed into multipole components,

$$\kappa(\boldsymbol{\theta}) = \kappa_0(\theta) + \sum_{m=1}^{\infty} \left[\kappa_{cm}(\theta)\cos(m\chi) + \kappa_{sm}(\theta)\sin(m\chi)\right], \tag{43}$$

where the individual components are angular averages over the surface density

$$\kappa_0(\theta) = \frac{1}{2\pi}\int_0^{2\pi} d\chi \kappa(\boldsymbol{\theta}), \quad \text{and} \quad \begin{pmatrix} \kappa_{cm}(\theta) \\ \kappa_{sm}(\theta) \end{pmatrix} = \frac{1}{\pi}\int_0^{2\pi} d\chi \begin{pmatrix} \kappa(\boldsymbol{\theta}) cos(m\chi) \\ \kappa(\boldsymbol{\theta}) \sin(m\chi) \end{pmatrix}. \tag{44}$$

The first three terms are the monopole (κ_0), the dipole ($m = 1$) and the quadrupole ($m = 2$) of the lens. The Poisson equation $\nabla^2 \Psi = 2\kappa$ is separable in polar coordinates, so a multipole decomposition of the effective potential

$$\Psi(\boldsymbol{\theta}) = \Psi_0(\theta) + \sum_{m=1}^{\infty} \left[\Psi_{cm}(\theta)\cos(m\chi) + \Psi_{sm}(\theta)\sin(m\chi)\right] \tag{45}$$

will have terms that depend only on the corresponding multipole of the surface density, $\nabla^2 \Psi_{cm}(\theta)\cos(m\chi) = 2\kappa_{cm}(\theta)\cos(m\chi)$. The monopole of the potential is simply

$$\Psi_0(\theta) = 2\log(\theta)\int_0^{\theta} udu\kappa_0(u) + 2\int_{\theta}^{\infty} udu \log(u)\kappa(u) \tag{46}$$

and its derivative is the bend angle for a circular lens,

$$\alpha_0(\theta) = \frac{d\Psi_0}{d\theta} = \frac{2}{\theta}\int_0^{\theta} udu\kappa_0(u), \tag{47}$$

just as we derived earlier (3). The higher order multipoles are no more complicated, with

$$\begin{pmatrix} \Psi_{cm}(\theta) \\ \Psi_{sm}(\theta) \end{pmatrix} = -\frac{1}{m\theta^m}\int_0^{\theta} u^{1+m} du \begin{pmatrix} \kappa_{cm}(u) \\ \kappa_{sm}(u) \end{pmatrix} - \frac{\theta^m}{m}\int_{\theta}^{\infty} u^{1-m} du \begin{pmatrix} \kappa_{cm}(u) \\ \kappa_{sm}(u) \end{pmatrix}. \tag{48}$$

The angular multipoles are always composed of two parts. There is an interior pole $\Psi_{cm,int}(\theta)$ due to the multipole surface density interior to θ (the integral from $0 < u < \theta$) and an exterior pole $\Psi_{cm,ext}(\theta)$ due to the multipole surface density exterior to θ (the integral from $\theta < u < \infty$). The higher order multipoles produce deflections in both the radial

$$\alpha_{cm,rad} = \frac{d}{d\theta}\left[\Psi_{cm}\cos(m\chi)\right] = \frac{d\Psi_{cm}}{d\theta}\cos(m\chi), \tag{49}$$

and tangential

$$\alpha_{cm,tan} = \frac{1}{\theta}\frac{d}{d\chi}\left[\Psi_{cm}\cos(m\chi)\right] = -\frac{m}{\theta}\Psi_{cm}\sin(m\chi) \tag{50}$$

directions, where the radial deflection depends on the derivative of Ψ_{cm} and the tangential deflection depends only on Ψ_{cm}. This may seem rather formal, but the multipole expansion provides the basis for understanding which aspects of mass distributions will matter for lens models. Obviously it is the lowest order angular multipoles which are most important. The most common angular term added to lens models is the external shear

$$\Psi_{2,ext} = \frac{1}{2}\gamma_c\theta^2\cos 2(\chi - \chi_\gamma) + \frac{1}{2}\gamma_s\theta^2\sin 2(\chi - \chi_\gamma) \tag{51}$$

with dimensionless amplitudes γ_c and γ_s and axis χ_γ. The external (tidal) shear and any accompanying mean convergence are the lowest order perturbations from any object near the lens that have measurable effects on a gravitational lens (see (26)). While models usually consider only external (tidal) shears where these coefficients are constants, in reality γ_c, γ_s and χ_γ are functions of radius (i.e. (48)). Along with the external shear, there is an internal shear

$$\Psi_{2,int} = \frac{1}{2}\Gamma_1\frac{\langle\theta\rangle^4}{\theta^2}\cos 2(\chi - \chi_\Gamma) + \frac{1}{2}\Gamma_2\frac{\langle\theta\rangle^4}{\theta^2}\sin 2(\chi - \chi_\Gamma). \tag{52}$$

due to the quadrupole moment of the mass interior to a given radius. We introduce the mean radius of the lensed images $\langle\theta\rangle$ to make Γ_1 and Γ_2 dimensionless with magnitudes that can be easily compared to the external shear amplitudes γ_1 and γ_2. Arguably the critical radius of the lens is a better physical choice, but the mean image radius will be close to the critical radius and using it avoids any trivial covariances between the internal shear strength and the monopole mass. Usually the internal quadrupole is added as part of an ellipsoidal model for the central lens galaxy, but it is useful in analytic studies to consider it separately.

4.1 Common Models for the Monopole

Most attention in modeling lenses focuses on the monopole or radial mass distribution of the lenses. Unfortunately, much of the lensing literature uses an almost impenetrable array of ghastly non-standard acronyms to describe the mass models even though many of them are identical to well-known families of density distributions used in stellar dynamics. Here we summarize the radial mass distributions which are most commonly used and will keep reappearing in the remainder of Part 2.

The simplest possible choice for the mass distribution is to simply trace the light. The standard model for early-type galaxies or the bulges of spiral galaxies is the de Vaucouleurs (1948) profile with surface density

$$\Sigma(R) = I_e \exp\left[-7.67\left[(R/R_e)^{1/4} - 1\right]\right], \tag{53}$$

where the effective radius R_e encompasses half the total mass (or light) of the profile. Although the central density of a de Vaucouleurs model is finite, it actually acts like a rather cuspy density distribution and will generally fit the early-type lens data with no risk of producing a detectable central image (e.g. Lehár et al. 2000; Keeton 2003a). The simplest model for a disk galaxy is an exponential disk,

$$\Sigma(R) = I_0 \exp\left[-R/R_d\right], \tag{54}$$

where R_d is the disk scale length. An exponential disk by itself is rarely a viable lens model because it has so little density contrast between the center and the typical radii of images that detectable central images are almost always predicted but not observed. Some additional component, either a de Vaucouleurs bulge or a cuspy dark matter halo, is always required. This makes spiral galaxy lens models difficult because they generically require two stellar components (a bulge and a disk) and a dark matter halo, while the photometric data are rarely good enough to constrain the two stellar components (e.g. Maller, Flores and Primack 1997; Koopmans et al. 1998; Maller et al. 2000; Trott and Webster 2002; Winn, Hall and Schechter 2003c). Since spiral lenses are already relatively rare, and spiral lens galaxies with good photometry are rarer still, less attention has been given to these systems. The de Vaucouleurs and exponential disk models are examples of Sersic (1968) profiles

$$\Sigma(R) = I_0 \exp\left[-b_n \left[\left(R/R_e(n)\right)^{1/n}\right]\right], \tag{55}$$

where the effective radius $R_e(n)$ is defined to encompass half the light and $n = 4$ is a de Vaucouleurs model and $n = 1$ is an exponential disk. These profiles have not been used as yet for the study of lenses except for some quasar host galaxy models (Sect. 10). The de Vaucouleurs model can be approximated (or the reverse) by the Hernquist (1990)model with the 3D density distribution

$$\rho(r) = \frac{M}{\pi r} \frac{a}{(a+r)^3} \tag{56}$$

and $a \simeq 0.55 R_e$ if matched to a de Vaucouleurs model. For lensing purposes, the Hernquist model has one major problem. Its $\rho \propto 1/r$ central density cusp is shallower than the effective cusp of a de Vaucouleurs model, so Hernquist models tend to predict detectable central images even when the matching de Vaucouleurs model would not. As a result, the Hernquist model is more often used as a surrogate for dynamical normalization of the de Vaucouleurs model than as an actual lens model (see below).

Theoretical models for lenses started with simple, softened power laws of the form

$$\kappa(R) \propto \left(R^2 + s^2\right)^{-(n-1)/2} \rightarrow R^{1-n} \tag{57}$$

in the limit where there is no core radius. We are using these simple power law lenses in all our examples (see Sect. 3). These models include many well known stellar dynamical models such as the singular isothermal sphere (SIS, $n = 2$, $s = 0$), the modified Hubble profile ($n = 3$) and the Plummer model ($n = 5$). Since we only see the projected mass, these power laws are also related to common models for infinitely thin disks. The Mestel (1963) disk ($n = 2$, $s = 0$) is the disk that produces a flat rotation curve, and the Kuzmin (1956) disk ($n = 3$) can be used to mimic the rising and then falling rotation curve of an exponential disk. The softened power-law models have generally fallen out of favor other than as simple models for some of the visible components of lenses because the strong evidence for stellar and dark matter cusps makes models with core radii physically unrealistic. While ellipsoidal versions of these models are not available in useful form, there are fast series expansion methods for numerical models (Chae, Khersonsky and Turnshek 1998b; Barkana 1998).

Most "modern" discussions of galaxy density distributions are based on sub-cases of the density distribution

$$\rho(r) \propto \frac{1}{r^n} \frac{1}{(a^\alpha + r^\alpha)^{(m-n)/\alpha}}, \tag{58}$$

which has a central density cusp with $\rho \propto r^{-n}$, asymptotically declines as $\rho \propto r^{-m}$ and has a break in the profile near $r \simeq a$ whose shape depends on α (e.g. Zhao 1997). The most common cases are the Hernquist model ($n = 1$, $m = 4$, $\alpha = 1$) mentioned above, the Jaffe (1983) model ($n = 2$, $m = 4$, $\alpha = 1$), the NFW (Navarro, Frenk and White 1996) model ($n = 1$, $m = 3$, $\alpha = 1$) and the Moore et al. (1998) model ($n = 3/2$, $m = 3$, $\alpha = 1$). We can view the power-law models either as the limit $n \to 0$ and $\alpha = 2$, or we could generalize the r^{-n} term to $(r^2 + s^2)^{-n/2}$ and consider only regions with r and $s \ll a$. Projections of these models are similar to surface density distributions of the form

$$\kappa(R) \propto \frac{1}{R^{n-1}} \frac{1}{(a^\alpha + R^\alpha)^{(m-n)/\alpha}} \tag{59}$$

(although the definition of the break radius a may change) with the exception of the limit $n \to 1$ where the projection of a 3D density cusp $\rho \propto 1/r$ produces surface density terms $\kappa \propto \ln R$ that cannot be reproduced by the broken surface density power law. This surface density model is sometimes called the Nuker law (e.g. Byun et al. 1996). A particularly useful case for lensing is the pseudo-Jaffe model with $n = 2$, $m = 4$ and $\alpha = 2$ (where the normal Jaffe model has $\alpha = 1$) as the only example of a broken power law with simple analytic deflections even when ellipsoidal because the density distribution is the difference between two isothermal ellipsoids (see (41)). These cuspy models also allow fast approximate solutions for their ellipsoidal counterparts (see Chae 2002).

The most theoretically important of these cusped profiles is probably the NFW profile (Navarro et al. 1996) because it is the standard model for dark matter halos. Since it is such a common model, it is worth discussing it in a little more detail, particularly its peculiar normalization. The NFW profile is normalized by the mass M_{vir} inside the virial radius r_{vir}, with

$$\rho_{NFW}(r) = \frac{M_{vir}}{4\pi f(c)} \frac{1}{r(r+a)^2} \quad \text{and} \quad M_{NFW}(<r) = \frac{f(r/r_{vir})}{f(c)}, \tag{60}$$

where $f(c) = \ln(1+c) - c/(1+c)$ and the concentration $c = r_{vir}/a \sim 5$ for clusters and $c \sim 10$ for galaxies. The concentration is a function of mass whose scaling is determined from N-body simulations. A typical scaling for a halo at redshift z in an $\Omega_M = 0.3$ flat cosmological models is (Bullock et al. 2001a,b)

$$c(M) = \frac{9}{1+z}\left(\frac{M_{vir}}{8\times 10^{12} h M_\odot}\right)^{-0.14} \tag{61}$$

with a dispersion in $\log c$ of $\sigma_{\log(c)} \simeq 0.18$ dex. Because gravitational lensing is very sensitive to the central density of the lens, including the scatter in the concentration is quantitatively important for lensing by NFW halos (Keeton 2001a,b). The virial mass and radius are related and determined by the overdensity $\Delta_{vir}(z)$ required for a halo to collapse given the cosmological model and the redshift. This can be approximated by

$$M_{vir} = \frac{4\pi}{3}\Delta_{vir}(z)\rho_u(z) r_{vir}^3 \simeq 0.23\times 10^{12} h \left(\frac{(1+z) r_{vir}}{100 h^{-1}\text{kpc}}\right)^3 \left(\frac{\Omega_M \Delta_{vir}}{200}\right) M_\odot, \tag{62}$$

where $\rho_u(z) = 3H_0^2\Omega_M(1+z)^3/8\pi G$ is the mean matter density when the halo forms and $\Delta_{vir} \simeq (18\pi^2 + 82x - 39x^2)/\Omega(z)$ with $x = \Omega - 1$ is the overdensity needed for a halo to collapse. There are differences in normalizations between authors and with changes in the central cusp exponent γ, but models of this type are what we presently expect for the structure of dark matter halos around galaxies.

For most lenses, HST imaging allows us to measure the spatial distribution of the stars, thereby providing us with a model for the distribution of stellar mass with only the stellar mass-to-light ratio as a parameter. For present purposes, gradients in the stellar mass-to-light ratio are unimportant compared to the uncertainties arising from the dark matter. Unless we are prepared to abandon the entire paradigm for modern cosmology, the luminous galaxy is embedded in a dark matter halo and we must decide how to model the overall mass distribution. The most common approach, as suggested by the rich variety of mass profiles we introduced in Sect. 4.1, is to assume a parametric form for the total mass distribution rather than attempting to decompose it into luminous and dark components. The alternative is to try to embed the stellar component in a dark matter halo. Operationally, doing so is trivial – the lens is simply modeled as the sum of two mass components. However, there are theoretical models for how CDM halos should be combined with the stellar component.

Most non-gravitational lensing applications focus on embedding disk galaxies in halos because angular momentum conservation provides a means of estimating a baryonic scale length (e.g. Mo, Mao and White 1998). The spin parameter of the halo sets the angular momentum of the baryons, and the final disk galaxy is defined by the exponential disk with the same angular momentum. As the baryons become more centrally concentrated, they pull the dark matter inwards as well through a process known as adiabatic contraction

(Blumenthal et al. 1986). The advantage of this approach, which in lensing has been used only by Kochanek and White (2001), is that it allows a full *ab initio* calculation of lens statistical properties when combined with a model for the cooling of the baryons (see Sect. 7). It has the major disadvantage that most lens galaxies are early-type galaxies rather than spirals, and that there is no analog of the spin parameter and angular momentum conservation to set the scale length of the stellar component in a model for an early-type galaxy.

Models of early-type galaxies embedded in CDM halos have to start with an empirical estimate of the stellar effective radius. In models of individual lenses this is a measured property of the lens galaxy (e.g. Rusin et al. 2003a,b; Rusin and Kochanek 2005 or Koopmans and Treu 2002; Kochanek 2003a,b,c). Statistical models must use a model for the scaling of the effective radius with luminosity or other observable parameters of early-type galaxies (e.g. Keeton 2001a,b). From the luminosity, a mass-to-light ratio is used to estimate the stellar mass. If all baryons have cooled and been turned into stars, then the stellar mass provides the total baryonic mass of the halo, otherwise the stellar mass sets a lower bound on the baryonic mass. Combining the baryonic mass with an estimate of the baryonic mass fraction yields the total halo mass to be fed into the model for the CDM halo.

In general, there is no convincing evidence favoring either approach – for the regions over which the mass distributions are constrained by the data, both approaches will agree on the overall mass distribution. However, there can be broad degeneracies in how the total mass distribution is decomposed into luminous and dark components (see Sect. 4.6).

4.2 The Effective Single Screen Lens

Throughout these notes we will treat lenses as if all the lens components lay at a single redshift ("the single screen approximation"). The lens equations for handling multiple deflection screens (e.g. Blandford and Narayan 1986; Kovner 1987b; Barkana 1996) are known but little used except for numerical studies (e.g. Kochanek and Apostolakis 1988; Möller and Blain 2001) in large part because few lenses require multiple lens galaxies at different redshifts with the exception of B2114+022 (Chae, Mao and Augusto 2001). In fact, we are not being as cavalier in making this approximation as it may seem.

The vast majority of strong lenses consist of a single lens galaxy perturbed by other objects. We can divide these objects into nearby objects, where a single screen is clearly appropriate, and objects distributed along the line of site for which a single screen may be inappropriate. Because the correlation function is so strong on small scales, the perturbations are dominated by objects within a correlation length of the lens galaxy (e.g. Keeton, Kochanek and Seljak 1997; Holder and Schechter 2003). The key to the relative safety of the single screen model is that weak perturbations from objects along the line of site, in the sense that in a multi-screen lens model they could be treated as a convergence and a shear, can be reduced to a single "effective" lens plane in which the true amplitudes of the convergence and shear are rescaled by distance ratios to convert them from their true redshifts to the redshift of the single screen (Kovner 1987b; Barkana 1996). The lens equation on the effective single screen takes the form

$$\boldsymbol{\beta} = (I + F_{OS})\,\boldsymbol{\theta} - (I + F_{LS})\,\boldsymbol{\alpha}\left[(I + F_{OL})\,\boldsymbol{\theta}\right], \tag{63}$$

where F_{OS}, F_{LS} and F_{OL} describe the shear and convergence due to perturbations between the observer and the source, the lens and the source and the observer and the lens respectively. For statistical calculations this can be simplified still further by making the coordinate transformation $\boldsymbol{\theta}' = (I + F_{OL})\boldsymbol{\theta}$ and $\boldsymbol{\beta}' = (I + F_{LS})\boldsymbol{\beta}$ to leave a lens equation identical to a single screen lens

$$\boldsymbol{\beta}' = (I + F_e)\,\boldsymbol{\theta}' - \boldsymbol{\alpha}\left[\boldsymbol{\theta}'\right] \tag{64}$$

in an effective convergence and shear of $F_e = F_{OL} + F_{LS} - F_{OS}$ (to linear order). In practice it will usually be safe to neglect the differences between (63) and (64) because the shearing terms affecting the deflections in (63) are easily mimicked by modest changes in the ellipticity and orientation of the primary lens. The rms amplitudes of these perturbations depend on the cosmological model and the amplitude of the non-linear power spectrum, but the general scaling is that the perturbations grow as $D_s^{3/2}$ with source redshift, and increase for larger σ_8 and Ω_M as shown in Fig. 19 from Keeton et al. (1997). The importance of these effects is very similar to concerns about the effects of lenses along the line-of-sight on the brightness of high redshift supernova being used to estimate the cosmological model (e.g. Dalal et al. 2003).

4.3 Constraining the Monopole

The most frustrating aspect of lens modeling is that it is very difficult to constrain the monopole. If we take a simple lens and fit it with any of the parametric models from the previous sub-section it will be possible to obtain a good fit provided the central surface density of the model is high enough to avoid the formation of a central image. As usual, it is simplest to begin understanding the problem with a circular, two-image lens whose images lie at radii θ_A and θ_B from the lens center (Fig. 20). The lens equation (4) constrains the deflections so that the two images correspond to the same source position,

$$\beta = \theta_A - \alpha(\theta_A) = -\theta_B + \alpha(\theta_B), \tag{65}$$

where the sign changes appear because the images are on opposite sides of the lens. Recall that for the power-law lens model, $\alpha(\theta) = b^{n-1}\theta^{2-n}$ (9), so we can easily solve the constraint equation to determine the Einstein radius of the lens,

$$b = \left[\frac{\theta_A + \theta_B}{\theta_A^{2-n} + \theta_B^{2-n}}\right]^{1/(n-1)} \tag{66}$$

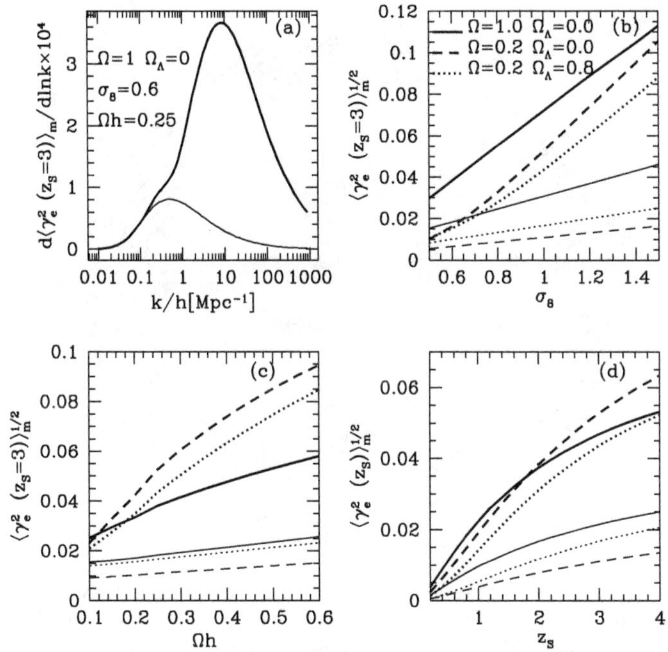

Fig. 19. Dependence of the shear generated by other objects along the line-of-sight for both linear (*light lines*) and non-linear (*heavy lines*) power spectra. (**a**) Shows the logarithmic contribution to the rms effective shear for a source at redshift $z_s = 3$ as a function of wave vector k. (**b**) Shows the dependence on σ_8 for a fixed power spectrum shape $\Omega_M h = 0.25$. (**c**) Shows the dependence on the shape $\Omega_M h$ with $\sigma_8 = 0.6$ for $\Omega_M = 1$ and $\sigma_8 = 1.0$ for $\Omega_M < 1$. (**d**) Shows the variation in the shear with source redshift for the models in (**c**) with $\Omega_M h = 0.25$

in terms of the image positions. In the limit of an SIS ($n = 2$) the Einstein radius is the arithmetic mean, $b = (\theta_A + \theta_B)/2$, and in the limit of a point source ($n \to 3$), it is the geometric mean, $b = (\theta_A \theta_B)^{1/2}$, of the image radii. More generally, for any deflection profile $bf(\theta)$, the two images simply determine the mass scale $b = (\theta_A + \theta_B)/(f(\theta_A) + f(\theta_B))$.

There are two important lessons here. First, the location of the tangential critical line is determined fairly accurately independent of the mass profile. We may only be able to determine the mass scale, but it is the most accurate measurement of galaxy masses available to astronomy. Second, it is going to be very difficult to determine radial mass distributions. In this example there is a perfect degeneracy between the exact location of the tangential critical line b and the exponent n. In theory, this is broken by the flux ratio of the images. However, a simple two-image lens has too few constraints even with perfectly measured flux ratios because a realistic lens model must also include some freedom in the angular structure of the lens. For a simple four-image lens, there begin to be enough constraints but the images all have similar

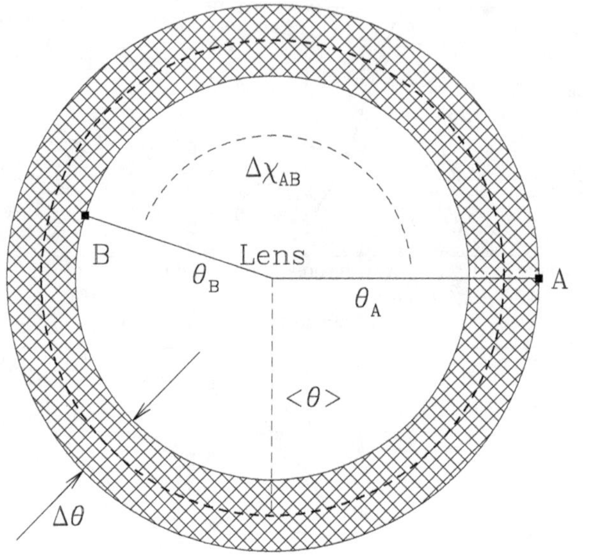

Fig. 20. A schematic diagram of a two-image lens. The lens galaxy lies at the origin with two images A and B at radii θ_A and θ_B from the lens center. The images define an annulus of average radius $\langle\theta\rangle = (\theta_A + \theta_B)/2$ and width $\delta\theta = \theta_A - \theta_B$, and they subtend an angle $\Delta\chi_{AB}$ relative to the lens center. For a circular lens $\Delta\chi_{AB} = 180°$ by symmetry

radii, making the flux ratios relatively insensitive to changes in the monopole. Combined with the systematic uncertainties in flux ratios, they are not useful for this purpose.

This example also leads to the major misapprehension about lens models and radial mass distributions, in that the constraints appear to lead to a degeneracy related to the global structure of the potential (i.e. the exponent n). This is not correct. The degeneracy is a purely local one that depends only on the structure of the lens in the annulus defined by the images, $\theta_B < \theta < \theta_A$, as shown in Fig. 20. To see this we will rewrite the expression for the bend angle (3) as

$$\alpha(\theta) = \frac{2}{\theta}\left[\int_0^{\theta_B} u du \kappa(u) + \int_{\theta_B}^{\theta} u du \kappa(u)\right] = \frac{1}{\theta}\left[b_B^2 + (\theta^2 - \theta_B^2)\langle\kappa\rangle(\theta, \theta_B)\right], \tag{67}$$

where $b_B^2 = 2\int_0^{\theta_B} u du \kappa(u)$ is the Einstein radius of the total mass interior to image B, and

$$\langle\kappa\rangle(\theta, \theta_B) = \frac{2}{\theta^2 - \theta_B^2}\int_{\theta_B}^{\theta} u du \kappa(u) \tag{68}$$

is the mean surface density in the annulus $\theta_B < u < \theta$. If we now solve the constraint (65) again, we find that

$$b_B^2 = \theta_A\theta_B - \langle\kappa\rangle_{AB}\theta_B(\theta_A - \theta_B), \tag{69}$$

where $\langle\kappa\rangle_{AB} = \langle\kappa\rangle(\theta_A, \theta_B)$ is the mean density in the annulus $\theta_B < \theta < \theta_A$ between the images. Thus, there is a degeneracy between the total mass interior to image B and the mean surface density (mass) between the two images. There is no dependence on the distribution of the mass interior to θ_B, the distribution of mass between the two images, or on either the amount or distribution of mass exterior to θ_A. This is Gauss' law for gravitational lens models.

If we normalize the mass scale at any point in the interior of the annulus then the result will depend on the distribution of the mass simply because the mass must be artificially divided. For example, suppose we model the surface density *locally* as a power law $\kappa \propto \theta^{1-n}$ with a mean surface density $\langle\kappa\rangle$ in the annulus $\theta_B < \theta < \theta_A$ between the images. The mass inside the mean image radius $\langle\theta\rangle$ is

$$\begin{aligned} b_{\langle\theta\rangle}^2 &= \theta_A\theta_B\,(1-\kappa_0) \\ &\quad + \delta\theta^2\langle\kappa\rangle\left[\frac{n}{4} + \left(\frac{\delta\theta}{\langle\theta\rangle}\right)^2 \frac{(4-n)(2-n)(1-n)}{192} + O\left(\left(\frac{\delta\theta}{\langle\theta\rangle}\right)^4\right)\right], \end{aligned} \tag{70}$$

where we have expanded the result in the ratio $\delta\theta/\langle\theta\rangle$ (in fact, the result as shown is exact for $n = 2/3$, 1, 2, 4 and 5). We included in this result an additional, global convergence κ_0 so that we can contrast the local degeneracies due to the distribution of matter between the images with the global degeneracies produced by a infinite mass sheet. The leading term $\theta_A\theta_B$ is the Einstein radius expected for a point mass lens (65). While the total enclosed mass $(\theta_A\theta_B)$ is fixed, the mass associated with the lens galaxy $b_{\langle\theta\rangle}^2$ must be modified in the presence of a global convergence by the usual $1-\kappa_0$ factor created by the mass sheet degeneracy (Falco, Gorenstein and Shapiro 1985). The structure of the lens in the annulus leads to fractional corrections to the mass of order $(\delta\theta/\langle\theta\rangle)^2$ that are proportional to $n\langle\kappa\rangle$ to lowest order.

Only if you have additional images inside the annulus can you begin to constrain the structure of the density in the annulus. The constraint is not, unfortunately, a simple constraint on the density. Suppose that we see an additional (pair) of images on the Einstein ring at θ_0, with $\theta_B < \theta_0 < \theta_A$. This case is simpler than the general case because it divides our annulus into two sub-annuli (from θ_B to θ_0 and from θ_0 to θ_A) rather than three. Since we put the extra image on the Einstein ring, we know that the mean surface density interior to θ_0 is unity (11). The A and B images then constrain a ratio

$$\frac{1-\langle\kappa\rangle_{B0}}{1-\langle\kappa\rangle_{A0}} = \frac{\theta_B}{\theta_A}\frac{\theta_A^2-\theta_0^2}{\theta_0^2-\theta_B^2} \simeq \frac{\theta_A-\theta_0}{\theta_0-\theta_B}\left[1 - \frac{\theta_A-\theta_B}{2\theta_0}\cdots\right] \tag{71}$$

of the average surface densities between the Einstein ring and image B ($\langle\kappa\rangle_{B0}$) and the Einstein ring and image A ($\langle\kappa\rangle_{A0}$). Since a physical distribution must have $0 < \langle\kappa\rangle_{A0} < \langle\kappa\rangle_{B0}$, the surface density in the inner sub-annulus must satisfy

$$\frac{\theta_A + \theta_B}{\theta_A} \frac{\theta_0^2 - \theta_A \theta_B}{\theta_0^2 - \theta_B^2} < \langle\kappa\rangle_{B0} < 1, \tag{72}$$

where the lower (upper) bound is found when the density in the outer sub-annulus is zero (when $\langle\kappa\rangle_{B0} = \langle\kappa\rangle_{A0}$). The term $\theta_0^2 - \theta_A\theta_B$ is the difference between the measured critical radius θ_0 and the critical radius implied by the other two images for a lens with no density in the annulus (e.g. a point mass), $(\theta_A\theta_B)^{1/2}$. Suppose we actually have images formed by an SIS, so $\theta_A = \theta_0(1+x)$ and $\theta_B = \theta_0(1-x)$ with $0 < x = \beta/\theta_0 < 1$, then the lower bound on the density in the inner sub-annulus is

$$\langle\kappa\rangle_{B0} > \frac{2x}{(2-x)(1+x)} \tag{73}$$

and the fractional uncertainly in the surface density is unity for images near the Einstein ring ($x \to 0$) and then steadily diminishes as the A and B images are more asymmetric. If you want to constrain the monopole, the more asymmetric the configuration the better. This rule becomes still more important with the introduction of angular structure.

Figure 21 illustrates these issues. We arbitrarily picked a model consisting of an SIS lens with two sources. One source is close to the origin and produces images at $\theta_A = 1''.1$ and $\theta_B = 0''.9$. The other source is farther from the origin with images at $\theta_A = 1''.5$ and $\theta_B = 0''.5$. We then modeled the lens with either a softened power law (57) or a three-dimensional cusp (58). We did not worry about the formation of additional images when the core radius becomes too large or the central cusp is too shallow – this would rule out models with very large core radii or shallow central cusps. If there were only a single source, either of these models can fit the data for any values of the parameters. Once, however, there are two sources, most of parameter space is ruled out except for degenerate tracks that look very different for the two mass models. Along these tracks, the models satisfy the additional constraint on the surface density given by (71). The first point to make about Fig. 21 is the importance of carefully defining parameters. The input SIS model has very different parameters for the two mass models – while the exponent $n = 2$ is the same in both cases, the SIS model is the limit $s \to 0$ for the core radius in the softened power law, but it is the limit $a \to \infty$ for the break radius in the cusp model. Similarly, models with an inner cusp $n = 0$ will closely resemble power law models whose exponent n matches the outer exponent m of the cuspy models. Our failure to explain these similarities is one reason why lens modeling seems so confusing. The second point to make about Fig. 21 is that the deflection profiles implied by these models are fairly similar over the annulus bounded by the images. Outside the annulus, particularly at smaller radii, they start to show very large fractional differences. Only if we were to add a third set of multiple images or measure a time delay with a known value of H_0 would the parameter degeneracy begin to be broken.

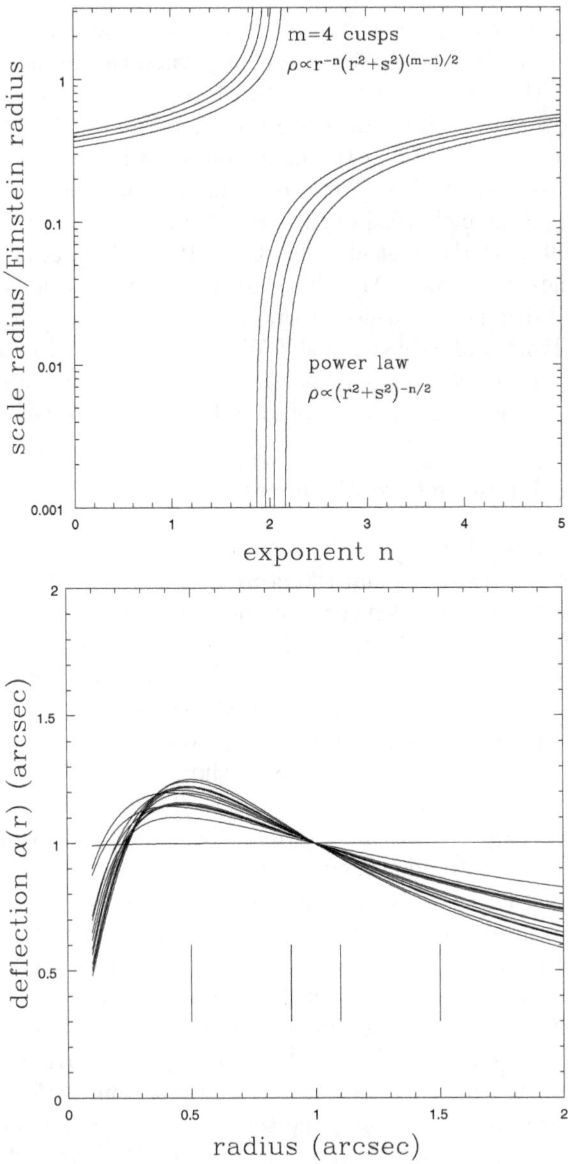

Fig. 21. Softened power law and cusped model fits to the images produced by an SIS lens with Einstein radius $b = 1\overset{''}{.}0$ and two source components located $0\overset{''}{.}1$ and $0\overset{''}{.}5$ from the lens center. In the top panel, the contours show the regions with astrometric fit residuals per image of $0\overset{''}{.}003$ and $0\overset{''}{.}010$. Models with $m = 3$ cusps so closely overly the $m = 4$ models that their error contours were not plotted. The bottom panel shows the deflection profiles of the best models at half-integer increments in the exponent n. The SIS model has a constant deflection, and the power-law and cusp models approach it in a sequence of slowly falling deflection profiles. All models agree with the SIS Einstein radius at $r = 1\overset{''}{.}0$. The positions of the images are indicated by the *vertical bars*

These general results suggest that studies of how lenses constrain the monopole need the ability to simultaneously vary the mass scale, the surface density of the annulus and possibly the slope of the density profile in the annulus to have the full range of freedom permitted by the data. Most parametric studies constraining the monopole have had two parameters, adjusting the mass scale and a correlated combination of the surface density and slope (e.g. Kochanek 1995a,b; Impey et al. 1998; Chae, Turnshek and Khersonsky 1998a, Barkana et al. 1999; Chae 1999; Cohn et al. 2001; Muñoz et al. 2001; Wucknitz et al. 2004), although there are exceptions using models with additional degrees of freedom (e.g. Bernstein and Fischer 1999; Keeton et al. 2000b; Trott and Webster 2002; Winn, Rusin and Kochanek 2003a). This limitation is probably not a major handicap, because realistic density profiles show a rather limited range of local logarithmic slopes.

4.4 The Angular Structure of Lenses

Assuming you have identified all the halos needed to model a particular lens, there are three sources of angular structure in the potential. The first source is the shape of the luminous lens galaxy, the second source is the dark matter in the halo of the lens, and the third source is perturbations from nearby objects or objects along the line-of-sight. Of these, the only one which is easily normalized is the contribution from the stars in the lens galaxy, since it must be tightly connected to the monopole deflection of the stars. The observed axis ratios of early-type galaxies show a deficit of round galaxies, a plateau for axis ratios from $q \sim 0.9$ to $q \sim 0.5$ and then a sharp decline beyond $q \sim 0.5$ (e.g. Khairul and Ryden 2002). Not surprisingly, the true elliptical galaxies are rounder than the lenticular (S0) galaxies even if both are grouped together as early-type galaxies. In three dimensions, the stellar distributions are probably close to oblate with very modest triaxialities (e.g. Franx et al. 1991). Theoretical models of galaxy formation predict ellipticities and triaxialities larger than observed for luminous galaxies (Dubinski 1992, 1994; Warren et al. 1992; Kazantzidis et al. 2004). Local estimates of the shape of dark matter halos are very limited (e.g. Olling and Merrifield 2001; Buote et al. 2002). Stellar isophotes also show deviations from perfect ellipses (e.g. Bender et al. 1989; Rest et al. 2001) and the deviations of simulated halos from ellipses have a similar amplitude (Heyl et al. 1994; Burkert and Naab 2003).

It is worth considering two examples to understand the relative importance of the higher order multipoles of a lens. The first is the singular isothermal ellipsoid (SIE) introduced in Sect. 3 (38-40). Let the major axis of the model lie on the θ_1 axis, in which case only the $\cos(m\chi)$ multipoles with $m = 2, 4, ...$ are non-zero. All non-zero poles also have the same radial dependence, with $\kappa_{cm} = A_m/\theta$ and $\Psi_{cm} = -2A_m\theta/(m^2-1)$. The ratio of the internal to the external multipole depends only on the index of the multipole, $\Psi_{cm,int}/\Psi_{cm,ext} = (m-1)/(m+1)$. Note, in particular, that the quadrupole moment of an SIE is dominated by the matter *outside* any given radius, with an internal quadrupole fraction of

$$f_{int} = \frac{\Psi_{c2,int}}{\Psi_{c2}} = \frac{1}{4}. \tag{74}$$

For lenses dominated by dark matter halos that have roughly flat global rotation curves, most of the quadrupole moment is generated outside the Einstein ring of the lens (i.e. by the halo !). This will hold provided any halo truncation radius is large compared to the Einstein ring radius. The tangential deflection is larger than the radial deflection, with $|\alpha_{cm,rad}/\alpha_{cm,tan}| = 1/m$. The final question is the relative amplitudes between the poles. The ratio of the angular deflection from the $m = 2$ quadrupole to the radial deflection of the monopole is

$$\frac{\alpha_{c2,tan}}{\alpha_{0,rad}} \simeq \frac{\epsilon}{3}\left[1 + \frac{1}{2}\epsilon + \frac{9}{32}\epsilon^2 \cdots\right], \tag{75}$$

while the ratio for the $m = 4$ quadrupole is

$$\frac{\alpha_{c4,tan}}{\alpha_{0,rad}} \simeq \frac{\epsilon^2}{20}\left[1 + \epsilon + \frac{19}{24}\epsilon^2 \cdots\right], \tag{76}$$

where the axis ratio of the ellipsoid is $q = 1 - \epsilon$. Each higher order multipole has an amplitude $\Psi_m \propto \epsilon^{m/2}$ to leading order.

The relative importance of the higher order poles can be assessed by computing the deflections for a typical lens with the monopole deflection (essentially the Einstein radius) fixed to be one arc second. Using the leading order scaling of the power-series, but setting the numerical value to be exact for an axis ratio $q = 1/2$, the angular deflection from the quadrupole is $0.''46\epsilon$ and that from the $m = 4$ pole is $0.''09\epsilon^2$, while the radial deflections will be smaller by a factors of 2 and 4 respectively. Since typical astrometric errors are of order $0.''005$, the quadrupole is quantitatively important for essentially any ellipticity while the $m = 4$ pole becomes quantitatively important only for $q \lesssim 0.75$ (and the $m = 6$ pole becomes quantitatively important for $q \lesssim 0.50$).

In Fig. 22 we compare the SIE to ellipsoidal de Vaucouleurs and NFW models. Unlike the SIE, these models are not scale free, so the multipoles depend on the distance from the lens center in units of the major axis scale length of the lens, R_{major}. The behavior of the de Vaucouleurs model will be typical of any ellipsoidal mass distribution that is more centrally concentrated than an SIE. Although the de Vaucouleurs model produces angular deflections similar to those of an SIE on small scales (for the same axis ratio), these are beginning to decay rapidly at the radii where we see lensed images (1–$2R_{major}$) because most of the mass is interior to the image positions and the amplitudes of the higher order multipoles decay faster with radius than the monopole (see (48)). Similarly, as more of the mass lies at smaller radii, the quadrupole becomes dominated by the internal quadrupole. The NFW model has a somewhat different behavior because on small scales it is less centrally concentrated than an SIE (a $\rho \propto 1/r$ central density cusp rather than $\propto 1/r^2$). It produces a somewhat bigger quadrupole for a given axis ratio,

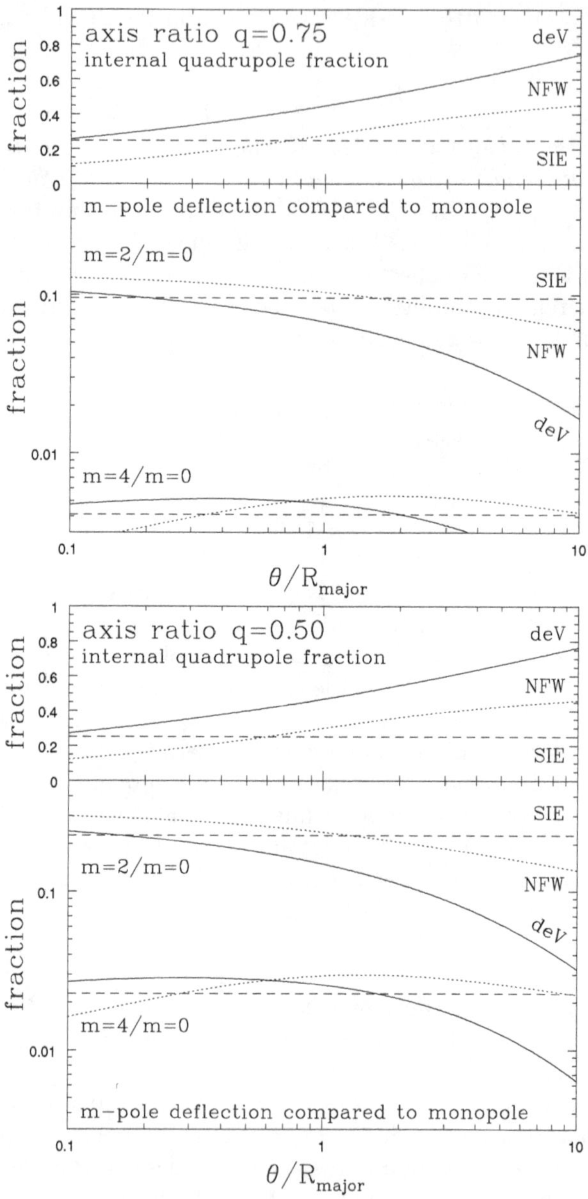

Fig. 22. Behavior of the angular multipoles for the de Vaucouleurs (*solid*), SIE (*dashed*) and NFW (*dotted*) models with axis ratios of either $q = 0.75$ (*top*) or $q = 0.5$ (*bottom*) as a function of radius from the lens center in units of the lens major axis scale R_{major}. For each axis ratio, the lower panel shows the ratio of the maximum angular deflections produced by the quadrupole ($m = 2$) and the $m = 4$ pole relative to the deflection produced by the monopole ($m = 0$). The upper panel shows the fraction of the quadrupole generated by the mass interior to each radius

and an even larger fraction of that quadrupole is generated on large scales. In a "standard" dark matter halo model, the region with $\theta < R_{major}$ is also where we see the lensed images. On larger scales, the NFW profile is more centrally concentrated than the SIE, so the quadrupole begins to decay and becomes dominated by the internal component.

It is unlikely that mass distributions are true ellipsoids producing only even poles ($m = 2, 4, \ldots$) with no twisting of the axes with radius. For model fits we need to consider the likely amplitude of these deviations and the ability of standard terms to absorb and mask their presence. It is clear from Fig. 22 that the amplitude of any additional terms must be of order the $m = 4$ deflections expected for an ellipsoid for them to be important. Here we illustrate the issues with the first few possible terms.

A dipole moment ($m = 1$) corresponds to making the galaxy lopsided with more mass on one side of the lens center than the other. Lopsidedness is not rare in disk galaxies (~30% at large radii, Zaritsky and Rix 1997), but is little discussed (and hence presumably small) for early-type galaxies. Certainly in the CASTLES photometry of lens galaxies we never see significant dipole residuals. It is difficult (impossible) to have an equilibrium system supported by random stellar motions with a dipole moment because the resulting forces will tend to eliminate the dipole. Similar considerations make it difficult to have a dark matter halo offset from the luminous galaxy. Only disks, which are supported by ordered rather than random motion, permit relatively long-lived lopsided structures. Where a small dipole exists, it will have little effect on the lens models unless the position of the lens galaxy is imposed as a stringent constraint. The reason is that a dipole adds terms to the effective potential of the form $\theta_1 G(\theta)$ whose leading terms are degenerate with a change in the unknown source position.

Perturbations to the quadrupole (relative to an ellipsoid) arise from variations in the ellipticity or axis ratio with radius. Since realistic lens models require an independent external shear simply to model the local environment, it will generally be very difficult to detect these types of perturbations or for these types of perturbations to significantly modify any conclusions. In essence, the amplitude and orientation of the external shear can capture most of their effects. Their actual amplitude is easily derived from perturbations. For example, if there is an isophote twist of $\Delta\chi$ between the region inside the Einstein ring and outside the Einstein ring, the fractional perturbations to the quadrupole will be of order $\Delta\chi$, or approximately $\epsilon\Delta\chi/3$ of the monopole – independent of the ability of the external shear to mimic the twist, the actual amplitude of the perturbation is approaching the typical measurement precision unless the twist is very large. Only in Q0957+561 have models found reasonably clear evidence for an effect arising from isophotal twists and ellipticity gradients, but both distortions are unusually large in this system (Keeton et al. 2000a). In general, in the CASTLES photometry of lens galaxies, deviations from simple ellipsoidal models are rare.

Locally we observe that the isophotes of elliptical galaxies are not perfect ellipses (e.g. Bender et al. 1989; Rest et al. 2001) and simulated halos show deviations of similar amplitude (Heyl et al. 1994; Burkert and Naab 2003). For lensing calculations it is useful to characterize these perturbations by a contribution to the lens potential and surface density of

$$\Psi = \frac{\epsilon_m}{m}\theta \cos m(\chi - \chi_m) \quad \text{and} \quad \kappa_m = \frac{\epsilon_m}{\theta}\frac{1-m^2}{m}\cos m(\chi - \chi_m), \tag{77}$$

respectively where the amplitude of the term is related to the usual isophote parameter $a_m = \epsilon_m |1-m^2|/mb$ for a lens with Einstein radius b. A typical early-type galaxy might have $|a_4| \sim 0.01$, so their fractional effect on the deflections, $|\epsilon_4|/b \sim |a_4|/4 \sim 0.003$, will be comparable to the astrometric measurement accuracy.

4.5 Constraining Angular Structure

The angular structure of lenses is usually simply viewed as an obstacle to understanding the monopole. This is a serious mistake. The reason angular structure is generally ignored is that the ability to accurately constrain the angular structure of the gravitational field is nearly unique to gravitational lensing. Since we have not emphasized the ability of lenses to measure angular structure and other methods cannot do so very accurately, there has been little theoretical work on the angular structure of galaxies with dark matter. Both theoretical studies of halos and modelers of gravitational lenses need to pay more attention to the angular structure of the gravitational potential.

We start by analyzing a simple two-image lens using our non-parametric model of the monopole (67) in an external shear (51). The two images are located at $\boldsymbol{\theta}_A = \theta_A(\cos\chi_A, \sin\chi_A)$, and $\boldsymbol{\theta}_B = \theta_B(\cos\chi_B, \sin\chi_B)$ as illustrated in Fig. 20. To illustrate the similarities and differences between shear and convergence, we will also include a global convergence κ_0 in the model. This corresponds to adding a term to the lens potential of the form $(1/2)\kappa_0\theta^2$. The model now has five parameters – two shear components, the mass and surface density of the monopole model and the additional global convergence. We have only two astrometric constraints, and so can solve for only two of the five parameters. Since the enclosed mass is always an interesting parameter, we can only solve for one of the two shear components. In general, we will find that the amplitude of γ_c depends on the amplitude of γ_s. There is, however, a special choice of the shear axis, $\chi_\gamma = (\chi_A + \chi_B)/2 + \pi/4$, such that the shear parameters become independent of each other. This allows us to determine the "invariant" shear associated with the images,

$$\gamma_1 = \frac{(1-\kappa_0 - \langle\kappa\rangle_{AB})\,(\theta_A^2 - \theta_B^2)\sin(\chi_A - \chi_B)}{\Delta\theta^2}, \tag{78}$$

where $\Delta\theta = |\boldsymbol{\theta}_A - \boldsymbol{\theta}_B|$ is the image separation. The monopole mass and the other shear component are degenerate,

$$b_B^2 + \gamma_2\theta_A\theta_B$$

$$= \frac{(1-\kappa_0)\left[\Delta\theta^2(\theta_A^2+\theta_B^2)-(\theta_A^2-\theta_B^2)^2\right]-\langle\kappa\rangle_{AB}\left(\theta_A^2-\theta_B^2\right)\left(\Delta\theta^2-\theta_A^2+\theta_B^2\right)}{2\Delta\theta^2}. \tag{79}$$

Several points are worth noting. First, the amplitude of the invariant shear γ_1 has the same degeneracy with the (local) surface density between the images $\langle\kappa\rangle_{AB}$ as it does with a global convergence κ_0. More centrally concentrated mass distributions with lower $\langle\kappa\rangle_{AB}$ require higher external shears to fit the same data. Second, the other component γ_2 introduces an uncertainty into the enclosed mass, with a series of somewhat messy trade offs between b_B^2, γ_1, $\langle\kappa\rangle_{AB}$ and κ_0. As a practical matter, the shear does not lead to an astronomically significant uncertainty in the mass, since $\gamma_2 \lesssim 0.1$ in all but the most extreme situations.

The external shear is only one component of the quadrupole. There is also an internal shear due to the mass interior to the images (52). The internal and external shears differ in their "handedness". For the same angular deflection ($d\Psi/d\chi$) they have opposite signs for the radial deflection ($d\Psi/d\theta$). The solution for two images is much the same as for an external shear. There is an invariant shear component, whose amplitude scales with $1-\kappa_0-\langle\kappa\rangle_{AB}$ but whose orientation differs from that of the external shear solution. The monopole mass b_B^2 is degenerate with the γ_2 shear component and the κ_0 and $\langle\kappa\rangle_{AB}$ surface densities. The actual expressions are far too complex to be illuminating. Figure 23 illustrates how the invariant shears combine to determine the overall structure of the quadrupole for the lens PG1115+080. For each image pair there is a line of permitted shears because of the degeneracy between the enclosed mass and the second shear component. The invariant shear component is the shear at the point where the line passes closest to the origin. If the quadrupole model is correct, the lines for all the image pairs will cross at a point, while if it is incorrect they will not. PG1115+080 is clearly going to be well modeled if the quadrupole is dominated by an external shear and poorly modeled if it is dominated by an internal shear. This provides a simple geometric argument for why full models of PG1115+080 are always dominated by an external shear (e.g. Impey et al. 1998). A failure of the curves to cross in both cases is primarily evidence for a mixture of external and internal quadrupoles or the presence of other multipoles rather than for a problem in the monopole mass distribution. In Fig. 23 we used an SIS for the monopole. For a point mass monopole, the figure looks almost the same provided we expand the scale – the invariant shear scales as $1-\langle\kappa\rangle_{AB}$ so in going from a SIS with $1-\langle\kappa\rangle_{AB} \simeq 1/2$ to a point mass with $1-\langle\kappa\rangle_{AB}=1$ the shear will double.

This scaling of the quadrupole with the surface density of the monopole provides an as yet unused approach to studying the monopole. Since the mass enclosed by the Einstein radius is nearly constant, the more centrally

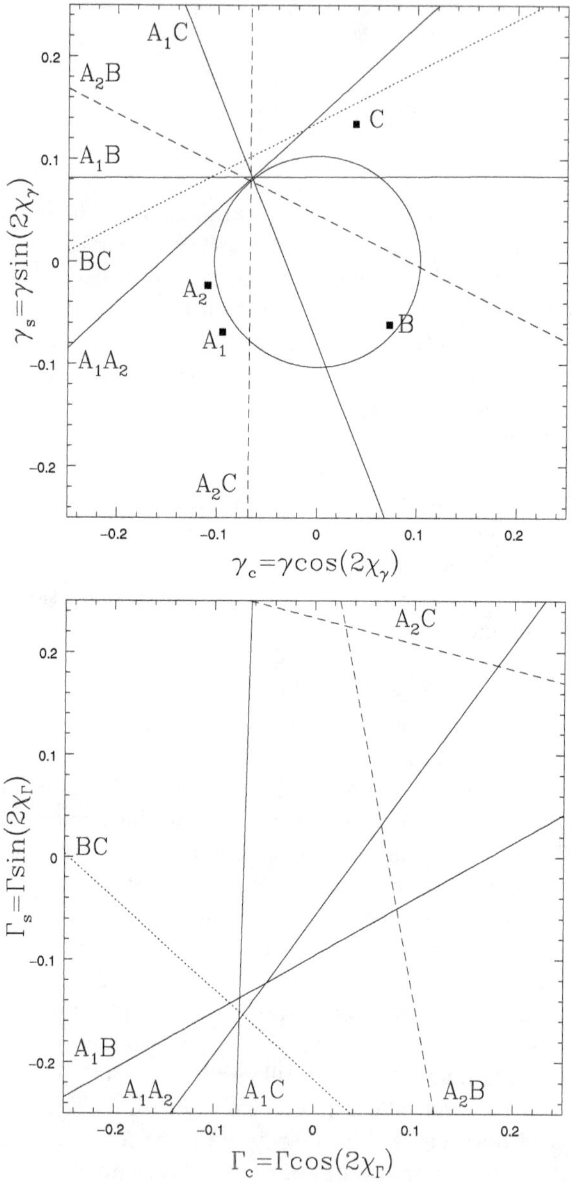

Fig. 23. The invariant shears for the lens PG1115+080 modeled using either an external (*top*) or an internal (*bottom*) quadrupole and an SIS monopole. Each possible image pair among the A_1, A_2, B and C images, constrains the quadrupole to lie on the labeled line. The amplitude and orientation of each invariant shear is given by the point where the corresponding line passes closest to the origin. Models of PG1115+080 show that the quadrupole is dominated by external (*tidal*) shear. Here we see that for the external quadrupole (*left*), the lines nearly cross at a point, so the data are consistent with an almost pure external shear. For an internal quadrupole (*right*), the A_2B and A_2C image pairs require shear parameters completely inconsistent with the other images

concentrated constant mass-to-light (M/L) ratio models must have lower surface mass densities near the images than the SIE model. As a result, they will require quadrupole amplitudes that are nearly twice those of models like the SIS with nearly flat rotation curves. Since the typical SIE model of a lens has an ellipticity that is comparable to the typical ellipticities of the visible galaxies, the more centrally concentrated monopole of a constant M/L model requires an ellipticity much larger than the observed ellipticity of the lens galaxy. The need to include an external tidal shear to represent the environment allows these models to produce acceptable fits, but the amplitudes of the required external shears are inconsistent with expectations from weak lensing (Part 3).

4.6 Model Fitting and the Mass Distribution of Lenses

Having outlined (in perhaps excruciating detail) how lenses constrain the mass distribution, we turn to the problem of actually fitting data. These days the simplest approach for a casual user is simply to download a modeling package, in particular the lensmodel package (Keeton 2001a,b) at http://cfa-www.harvard.edu/castles/, read the manual, try some experiments, and then apply it intelligently (i.e. read the previous sections about what you can extract and what you cannot !). Please publish results with a complete description of the models and the constraints using standard astronomical nomenclature.

In most cases we are interested in the problem of fitting the positions $\boldsymbol{\theta}_i$ of $i = 1, ..., n$ images where the image positions have been measured with accuracy σ_i. We may also know the positions and properties of one or more lens galaxies. Time delay ratios also constrain lens models but sufficiently accurate ratios are presently available for only one lens (B1608+656, Fassnacht et al. 2002), fitting them is already included in most packages, and they add no new conceptual difficulties. Flux ratios constrain the lens model, but we are so uncertain of their systematic uncertainties due to extinction in the ISM of the lens galaxy, microlensing (Part 4) and the effects of substructure (see Sect. 8) that we can never impose them with the accuracy needed to add a significant constraint on the model.

The basic issue with lens modeling is whether or not to invert the lens equations ("source plane" or "image plane" modeling). The lens equation supplies the source position

$$\boldsymbol{\beta}_i = \boldsymbol{\theta}_i - \boldsymbol{\alpha}(\boldsymbol{\theta}_i, \boldsymbol{p}) \tag{80}$$

predicted by the observed image positions $\boldsymbol{\theta}_i$ and the current model parameters $\boldsymbol{p}$. Particularly for parametric models it is easy to project the images on to the source plane and then minimize the difference between the projected source positions. This can be done with a χ^2 fit statistic of the form

$$\chi^2_{src} = \sum_i \left(\frac{\boldsymbol{\beta} - \boldsymbol{\beta}_i}{\sigma_i} \right)^2 , \tag{81}$$

where we treat the source position $\boldsymbol{\beta}$ as a model parameter. The astrometric uncertainties σ_i are typically a few milli-arcseconds. Moreover, where VLBI observations give significantly smaller uncertainties, they should be increased to approximately 0.″001–0.″005 because low mass substructures in the lens galaxy can produce systematic errors on this order (see Sect. 8). *You can impose astrometric constraints to no greater accuracy than the largest deflection scales produced by lens components you are not including in your models.* The advantage of χ^2_{src} is that it is fast and has excellent convergence properties. The disadvantages are that it is wrong, cannot be used to compute parameter uncertainties, and may lead to a model producing additional images that are not actually observed.

The reason it is wrong and cannot be used to compute parameter errors is that the uncertainty σ_i in the image positions does not have any meaning on the source plane. This is easily understood if we Taylor expand the lens equation near the projected source point $\boldsymbol{\beta}_i$ corresponding to an image

$$\boldsymbol{\beta} - \boldsymbol{\beta}_i = M_i^{-1}(\boldsymbol{\theta} - \boldsymbol{\theta}_i), \tag{82}$$

where M^{-1} is the inverse magnification tensor at the observed location of the image. In the frame where the tensor is diagonal, we have that $\Delta\beta_\pm = \lambda_\pm \Delta\theta_\pm$ so a positional error $\Delta\beta_\pm$ on the source plane corresponds to a positional error $\lambda_\pm^{-1}\Delta\beta_\pm$ on the image plane. Since the observed lensed images are almost always magnified (usually $\lambda_+ = 1+\kappa+\gamma \sim 1$ and $0.5 > |\lambda_- = 1+\kappa-\gamma| < 0.05$) there is always one direction in which small errors on the source plane are significantly magnified when projected back onto the image plane. Hence, if you find solutions with $\chi^2_{src} \sim N_{dof}$ where N_{dof} is the number of degrees of freedom, you will have source plane uncertainties $\Delta\beta \lesssim \sigma_i$. However, the actual errors on the image plane are $\mu = |M|$ larger and the χ^2 on the image plane is $\sim \mu^2 N_{dof}$ and you in fact have a terrible fit.

If you assume that in any interesting model you are close to having a good solution, then this Taylor expansion provides a means of using the easily computed source plane positions to still get a quantitatively accurate fitting statistic,

$$\chi^2_{int} = \sum_i \frac{(\boldsymbol{\beta} - \boldsymbol{\beta}_i) \cdot M^2 \cdot (\boldsymbol{\beta} - \boldsymbol{\beta}_i)}{\sigma_i^2} \tag{83}$$

in which the magnification tensor M is used to correct the error in the source position to an error in the image position. This procedure will be approximately correct provided the observed and model image positions are close enough for the Taylor expansion to be valid. Finally, there is the exact statistic where for the model source position $\boldsymbol{\beta}$ you numerically solve the lens equation to find the exact image positions $\boldsymbol{\theta}_i(\boldsymbol{\beta})$ and then compute the goodness of fit on the image plane

$$\chi^2_{img} = \sum_i \left(\frac{\boldsymbol{\theta}_i(\boldsymbol{\beta}) - \boldsymbol{\theta}_i}{\sigma_i} \right)^2 . \tag{84}$$

This will be exact even if the Taylor expansion of χ^2_{int} is breaking down, and if you find all solutions to the lens equations you can verify that the model predicts no additional visible images. Unfortunately, using the exact χ^2_{img} is also a much slower numerical procedure.

As we discussed earlier, even though lens models provide the most accurate mass normalizations in astronomy, they can constrain the mass distribution only if the source is more complex than a single compact component. Here we only show examples where there are multiple point-like components, deferring discussions of models with extended source structure to Sect. 10. The most spectacular example of a multi-component source is B1933+503 (Sykes et al. 1998, see Fig. 6) where a source consisting of a radio core and two radio lobes has 10 lensed images because the core and one lobe are quadruply imaged and the other lobe is doubly imaged. Since we have many images spread over roughly a factor of two in radius, this lens should constrain the radial mass distribution just as in our discussion for Sect. 4.3. Muñoz et al. (2001, also see Cohn et al. 2001 for softened power law models) fitted this system with cuspy models (55 with $\alpha = 2$ and $m = 4$), varying the inner density slope $n = \gamma$ ($\rho \propto r^{-n}$) and the break radius a. Figure 24 shows the resulting χ^2 as a function of the parameters and Fig. 24 illustrates the range of the acceptable monopole mass distributions – both are very similar to Fig. 21. The best fit is for $\gamma = 1.85$ with an allowed range of $1.6 < \gamma < 2.0$ that completely excludes the shallow $\gamma = 1$ cusps of the Hernquist and NFW profiles and is marginally consistent with the $\gamma = 2$ cusp of the SIS model. A second example, which illustrates how the distribution of mass well outside the region with images has little effect on the models are the Winn et al. (2003a,b,c) models of the three-image lens PMNJ1632–0033 shown in Fig. 25. In these models the outer slope η, with $\rho \propto r^{-\eta}$ asymptotically, of the density was also explored but has little effect on the results. Unless the break radius of the profile is interior to the B image, the mass profile is required to be close to isothermal $1.89 < \beta < 1.93$.

Unfortunately, systems like B1933+503 and PMNJ1632–0033 are a small minority of lens systems. For most lenses, obtaining information on the radial density profile requires some other information such as a dynamical measurement (Sect. 4.9), a time delay measurement (Sect. 5) or a lensed extended component of the source (Sect. 10). Even for these systems, it is important to remember that the actual constraints on the density structure really only apply over the range of radii spanned by the lensed images – the mass interior to the images is constrained but its distribution is not, while the mass exterior to the images is completely unconstrained. This is not strictly true when we include the angular structure of the gravitational field and the mass distribution is quasi-ellipsoidal.

It is also important to keep some problems with parametric models in mind. First, models that lack the degrees of freedom needed to describe the actual mass distribution can be seriously in error. Second, models with too many degrees of freedom can be nonsense. We can illustrate these two limiting problems with the sad history of Q0957+561 for the first problem and

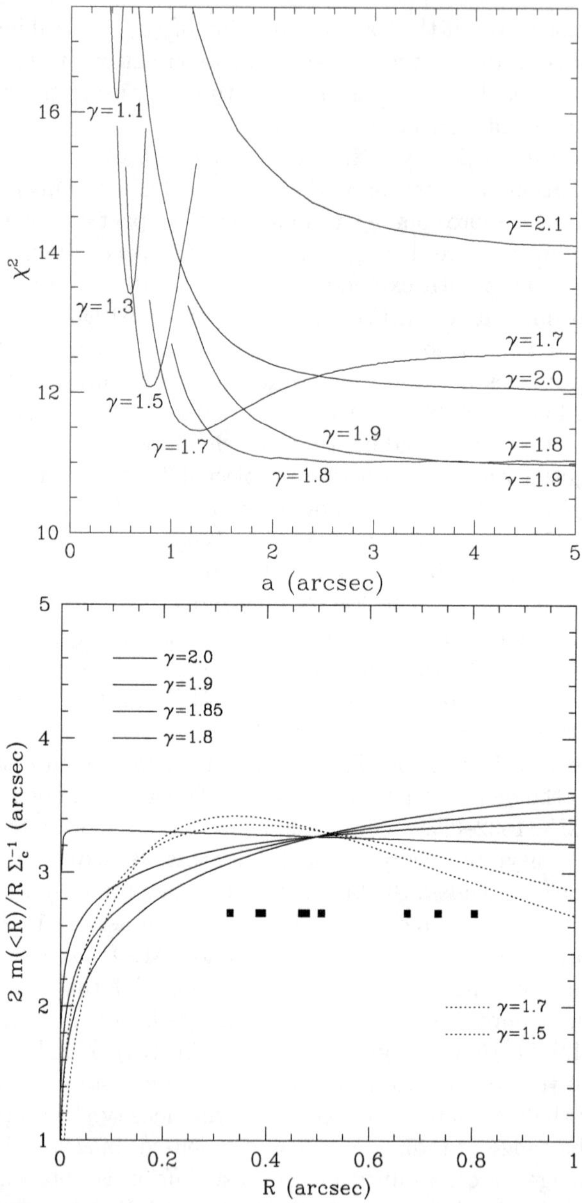

Fig. 24. (*Top*) Goodness of fit χ^2 for cuspy models of B1933+503 as a function of the inner density exponent γ ($\rho \propto r^{-\gamma}$) and the profile break radius a. Models with cusps significantly shallower or steeper than isothermal are ruled out, and acceptable models near isothermal must have break radii outside the region with the lensed images. (*Bottom*) The monopole deflections of the B1933+503 models for the range of permitted cusp exponents γ. The points show the radii of the lensed images, and the models only constrain the shape of the monopole in this region. The monopole deflection is closely related to the square of the rotation curve. Note the similarity to Fig. 21

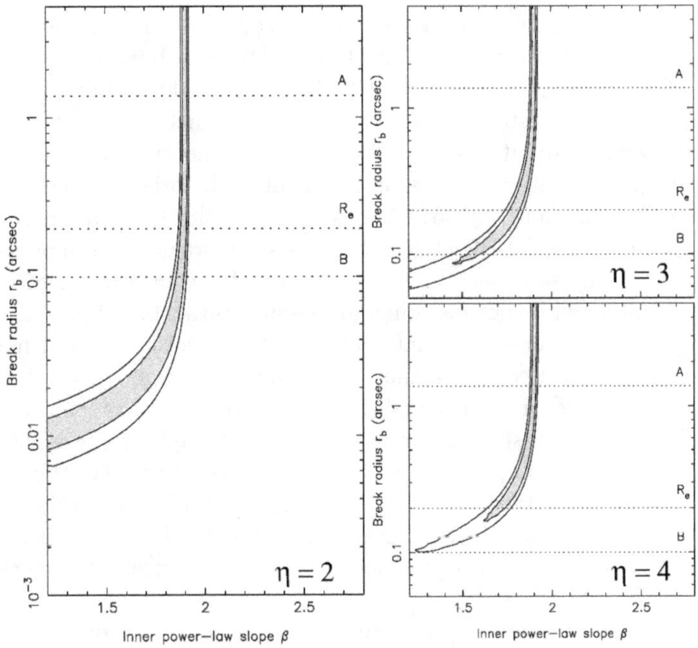

Fig. 25. (*Left*) Allowed parameters for cuspy models of PMNJ1632–0033 assuming that image C is a true third image. Each panel shows the constraints on the inner density cusp β ($\rho \propto r^{-\beta}$) and the break radius r_b for three different asymptotic density slopes $\rho \propto r^{-\eta}$. A Hernquist model has $\beta = 1$ and $\eta = 4$, an NFW model has $\beta = 1$ and $\eta = 3$, and a pseudo-Jaffe model has $\beta = 2$ and $\eta = 4$. Unless the break radius is placed interior to the B image, it is restricted to be close to isothermal ($\beta = 2$)

attempts to explain anomalous flux ratios (see Sect. 8) with complex angular structures in the density distribution for the dark matter.

Q0957+561, the first lens discovered (Walsh et al. 1979) and the first lens with a well measured time delay (see Sect. 5, Schild and Thomson 1995; Kundić et al. 1997 and references therein), is an ideal lens for demonstrating the trouble you can get into using parametric models without careful thought. The lens consists of a cluster and its brightest cluster galaxy with two lensed images of a radio source bracketing the galaxy. VLBI observations (e.g. Garrett et al. 1994) resolve the two images into thin, multi-component jets with very accurately measured positions (uncertainties as small as 0.1 mas, corresponding to deflections produced by a mass scale $\sim 10^{-8}$ of the primary lens !). Models developed along two lines. One line focused on models in which the cluster was represented as an external shear (e.g. Grogin and Narayan 1996; Chartas et al. 1998; Barkana et al. 1999; Chae 1999) while the other explored more complex models for the cluster (see Kochanek 1991c; Bernstein, Tyson and Kochanek 1993; Bernstein and Fischer 1999) and argued that external shear models had too few parameters to represent the mass distribution given the accuracy of the constraints. The latter view was born out by the morphology of the lensed host galaxy (Keeton et al. 2000a) and direct X-ray observations of the cluster (Chartas et al. 2002) which showed that the lens galaxy was within about one Einstein radius of the cluster center where a tidal shear approximation fails catastrophically. The origin of the problem is that as a two-image lens, Q0957+561 is critically short of constraints unless the fine details of the VLBI jet structures are included in the models. Many studies imposed these constraints to the limit of the measurements (tens of micro-arcseconds) while not including all possible terms in the potential which could produce a deflection on that scale (i.e. the precision should have been restricted to milli-arcseconds rather than micro-arcseconds). Models would adjust the positions and masses of the cluster and the lens galaxy in order to reproduce the small scale astrometric details of the VLBI jets without including less massive components of the mass distribution (e.g. the ellipticity gradient and isophote twist of the lens galaxy, Keeton et al. 2000a) that also affected the VLBI jet structure on these angular scales. Lens models must contain all reasonable structures producing deflections comparable to the scale of the measurement errors.

We are in the middle of an experiment exploring the second problem – if you include small scale structures but lack the constraints needed to measure them, their masses easily become unreasonable unless constrained by common sense, physical priors or additional data. Lately this has become an issue in studies (Evans and Witt 2003; Kochanek and Dalal 2004; Quadri, Möller and Natarajan 2003; Kawano et al. 2004) of whether the flux ratio anomalies in gravitational lenses could be due to complex angular structure in the lens galaxy rather than CDM substructure or satellites in the lens galaxy (see Sect. 8). The problem, as we discuss in the next section on non-parametric models (Sect. 4.7), is that lens modeling with large numbers of parameters is closely related to solving linear equations with more variables than constraints – as the matrix inversion necessary to finding a solution becomes singular, the parameters of the mass distribution show wild, large amplitude fluctuations even as the fit to the constraints becomes perfect. Thus, a model including enough unconstrained parameters is guaranteed to "solve" the anomalous flux ratio problem even if it should not. For example, Evans and Witt (2003) could match the flux ratios of Q2237+0305 even though for this lens we know from the time variability of the flux ratios that the flux ratio anomalies are created by microlensing rather than complex angular structures in the lens model (see Part 4).

If only the four compact images are modeled, then the flux ratio anomalies can be greatly reduced or eliminated in almost all lenses at the price of introducing deviations from an ellipsoidal density distribution far larger than expected (see Sect. 4.4). In some cases, however, you can test these solutions because the lens has extra constraints beyond the four compact images. We illustrate this in Fig. 26 where, by adding large amplitude $\cos 3\theta$ and $\cos 4\theta$

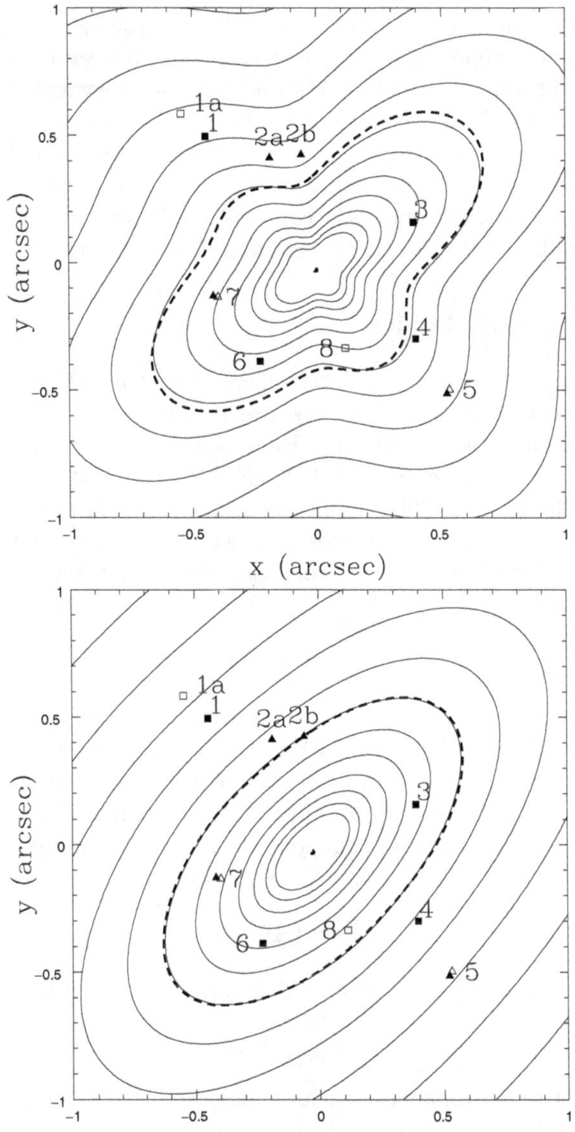

Fig. 26. Surface density contours for models of B1933+503 including misaligned a_3 and a_4 multipoles (*thin lines*). The model in the top panel is constrained only by the 4 compact images (images 1, 3, 4 and 6, *filled squares*). The model in the bottom panel is also constrained by the other images in the lens (the two-image system 1a/8, open squares; the four-image system 2a/2b/5/7 *filled triangles*; and the two-image system comprising parts of 5/7, *open triangles*). The tangential critical line of the model (*heavy dashed curve*) must pass between the merging images 2a/2b, but fails to do so in the first model (*top panel*)

perturbations to the surface density model for B1933+503, Kochanek and Dalal (2004) could reproduce the observed image flux ratios if they fit only the four compact sources. However, after adding the constraints from the other lensed components, the solution is driven back to being nearly ellipsoidal and the flux ratios cannot be fit. In every case, Kochanek and Dalal (2004) found that the extra constraints drove the solution back toward an ellipsoidal density distribution. In short, a sufficiently complex model can fit underconstrained data, but that does not mean it makes any sense to do so.

4.7 Non-Parametric Models

The basic idea behind non-parametric mass models is that the effective lens potential and the deflection equations are linear "functions" of the surface density. The surface density can be decomposed into multipoles (Kochanek 1991a; Trotter, Winn and Hewitt 2000; Evans and Witt 2003), pixels (see Saha and Williams 1997, 2004; Williams and Saha 2000), or any other form in which the surface density is represented as a linear combination of density functionals multiplied by unknown coefficients $\boldsymbol{\kappa}$. In any such model, the lens equation for image i takes the form

$$\boldsymbol{\beta} = \boldsymbol{\theta}_i - A_i \boldsymbol{\kappa}, \tag{85}$$

where A_i is the matrix that gives the deflection at the position of image i in terms of the coefficients of the surface density decomposition $\boldsymbol{\kappa}$. For a lens with $i = 1, ..., n$ images of the same source, such a system can be solved exactly if there are enough degrees of freedom in the description of the surface density. For simplicity, consider a two-image lens so that we can eliminate the source position by hand, leaving the system of equations

$$\boldsymbol{\theta}_1 - \boldsymbol{\theta}_2 = (A_1 - A_2)\boldsymbol{\kappa}, \tag{86}$$

which is easily solved by simply taking the inverse of the matrix $A_1 - A_2$ to find that

$$\boldsymbol{\kappa} = (A_1 - A_2)^{-1} (\boldsymbol{\theta}_1 - \boldsymbol{\theta}_2) . \tag{87}$$

Sadly, life is not that simple because as soon as the density decomposition has more degrees of freedom than there are constraints, the inverse $(A_1 - A_2)^{-1}$ of the deflection operators is singular.

The solution to this problem is to instead consider the problem as a more general minimization problem with a χ^2 statistic for the constraints and some form of regularization to restrict the results to plausible surface densities. One possibility is linear regularization, in which you minimize the function

$$F = \chi^2 + \lambda \boldsymbol{\kappa} \cdot H \cdot \boldsymbol{\kappa}, \tag{88}$$

where the χ^2 measures the goodness of fit to the lens constraints, H is a weight matrix and λ is a Lagrange multiplier. The Lagrange multiplier controls the relative weight given to fitting the lens constraints (minimizing the χ^2) versus producing a smooth density distribution (minimizing $\boldsymbol{\kappa} \cdot H \cdot \boldsymbol{\kappa}$). The simplest smoothing function is to minimize the variance of the surface density ($H = I$, the identity matrix), or, equivalently, ignore H and use the singular value decomposition for inverting a singular matrix. By using more complicated matrices you can minimize derivatives of the density (gradients, curvature etc.). Solutions are found by adjusting the multiplier λ until the goodness of fit satisfies $\chi^2 \simeq N_{dof}$ where N_{dof} is the number of degrees of freedom. Another solution is to use linear programming methods to impose constraints such as positive surface densities, negative density gradients from the lens center or density symmetries (Saha and Williams 1997, 2004; Williams and Saha 2000). Time delays, which are also linear functions of the surface density, are easily included. Flux ratios are more challenging because magnifications are quadratic rather than linear functions of the surface density except for the special case of the generalized singular isothermal models where $\Psi = b\theta F(\chi)$ (42), Witt, Mao and Keeton 2000; Kochanek et al. 2001a; Evans and Witt 2001). The best developed, publicly available non-parametric models are those by Saha and Williams (2004). These are available at http://ankh-morpork.maths.qmc.ac.uk/~saha/astron/lens/.

Personally, I am not a fan of the non-parametric models, essentially because almost all the additional degrees of freedom they include are irrelevant to the problem. As I have tried to outline in the preceding sections, there is no real ambiguity about the aspects of gravitational potentials either constrained or unconstrained by lens models. Provided the parametric models capture these degrees of freedom and you do not get carried away with the precision of the fits, you can ignore deviations of the $\cos(16\chi)$ term of the surface density from that expected for an ellipsoidal model. Similarly for the monopole profile, the distribution of mass interior and exterior to the images is irrelevant and for the most part only the mean surface density between the images has any physical effect. Nothing is gained by allowing arbitrary, fine-grained distributions.

There are also specific physical and mathematical problems with non-parametric models just as there are for parametric models. First, the trick of linearization only works if the lens equations are solved on the source plane. As we discussed when we introduced model fitting (Sect. 4.6), this makes it impossible to properly compute error bars on any parameters. The equations become non-linear if they include either the magnification tensor (83) or use the true image plane fit statistic (84), and this greatly reduces the attractiveness of these methods. Second, in many cases the non-parametric models are not constrained to avoid creating extra images not seen in the observations – the models reproduce the observed images exactly, but come with no guarantee that they are not producing 3 other images somewhere else. Third, it is very difficult to guarantee that the resulting models are physical. For example, consider a simple spherical lens constrained to have positive surface density. For the implied three-dimensional density to also be positive definite, the surface density must decline monotonically from the center of the lens. This constraint is usually applied by the Saha and Williams (2004) method. For the distribution function of the stars making up the galaxy to be positive definite, the three dimensional density must also decline monotonically – this implies a constraint on the second derivative of the surface density which is not imposed by any of these methods. For the distribution to be dynamically stable it must satisfy a criterion on the derivative of the distribution function with respect to the orbital energy, and this implies a criterion on the third derivative of the surface density which is also not imposed (see Binney and Tremaine 1987). Worse yet, for a non-spherical system we cannot even write down the constraints on the surface density required for the model to correspond to a stable galaxy with a positive definite distribution function. In short, most non-parametric models will be unphysical – they overestimate the degrees of freedom in the mass distribution. The critique being made, parametric models have a role because they define the outer limits of what is possible by avoiding the strong physical priors implicit in parametric models of galaxies.

4.8 Statistical Constraints on Mass Distributions

Where individual lenses may fail to constrain the mass distribution, ensembles of lenses may succeed. There are two basic ideas behind statistical constraints on mass distributions. The first idea is that models of individual lenses should be weighted by the likelihood of the observed configuration given the model parameters. The second idea is that the statistical properties of lens samples should be homogeneous.

An example of weighting models by the likelihood is the limit on the slopes of central density cusps from the observed absence of central images. Rusin and Ma (2001) considered 6 CLASS (see Sect. 6) survey radio doubles and computed the probability $p_i(n)$ that lens i would have a detectable third image in the core of the lens assuming power law mass densities $\Sigma \propto R^{1-n}$ and including a model for the observational sensitivities and the magnification bias (see Sect. 6.6) of the survey. They were only interested in the range $n < 2$, because as discussed in Sect. 3, density cusps with $n \geq 2$ never have central images. For most of the lenses they considered, it was possible to find models of the 6 lenses that lacked detectable central images over a broad range of density exponents. However, the shallower the cusp, the smaller the probability $p_i(n)$ of producing a lens without a visible central image. For any single lens, $p_i(n)$ varies too little to set a useful bound on the exponent, but the joint probability of the entire sample having no central images, $P = \Pi_i(1 - p_i(n))$, leads to a strong (one-sided) limit that $n > 1.78$ at 95% confidence (see Fig. 27). In practice, Keeton (2003b) demonstrated that the central stellar densities are sufficiently high to avoid the formation of visible central images in almost all lenses given the dynamic ranges of existing radio observations (i.e. stellar density distributions are sufficiently cuspy), and central black holes can also assist in suppressing the central image (Mao, Witt and Koopmans 2001).

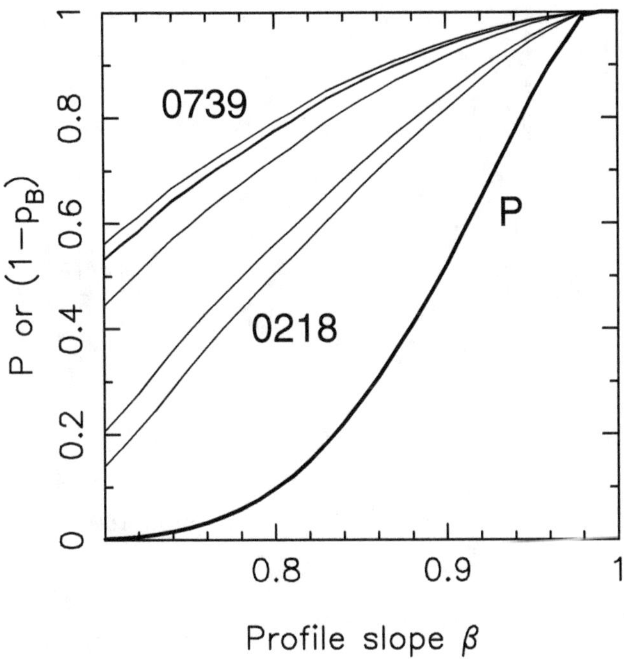

Fig. 27. Limits on the central density exponent for power-law density profiles $\rho \propto r^{-n} = r^{-1-\beta}$ from the absence of detectable central images in a sample of 6 CLASS survey radio doubles (Rusin and Ma 2001). The lighter curves show the limits for the individual lenses with the weakest constraint from B0739+366 and the strongest from B0218+357, and the heavy solid curve shows the joint probability P

However, the basic idea behind the Rusin and Ma (2001) analysis is important and underutilized.

An example of requiring the lenses to be homogeneous is the estimate of the misalignment between the major axis of the luminous lens galaxy and the overall mass distribution by Kochanek (2002a,b). Figure 28 shows the misalignment angle $\Delta\chi_{LM} = |\chi_L - \chi_M|$ between the major axis χ_L of the lens galaxy and the major axis χ_M of an ellipsoidal mass model for the lens. The particular mass model is unimportant because any single component model of a four-image lens will give a nearly identical value for χ_M (e.g. Kochanek 1991b; Wambsganss and Paczyński 1994). The distribution of the misalignment angle $\Delta\chi_{LM}$ is not consistent with the mass and the light being either perfectly correlated or uncorrelated. This is not surprising, because a simple ellipsoidal model determines the position angle of the mean quadrupole moment near the Einstein ring, which is a combination of the quadrupole moment of the lens galaxy, the halo of the lens galaxy, and the local tidal shear (see Sect. 4.4). Even if the lens galaxy and the halo were perfectly aligned, we would still find that the orientation of the mean quadrupole would differ from that of the light

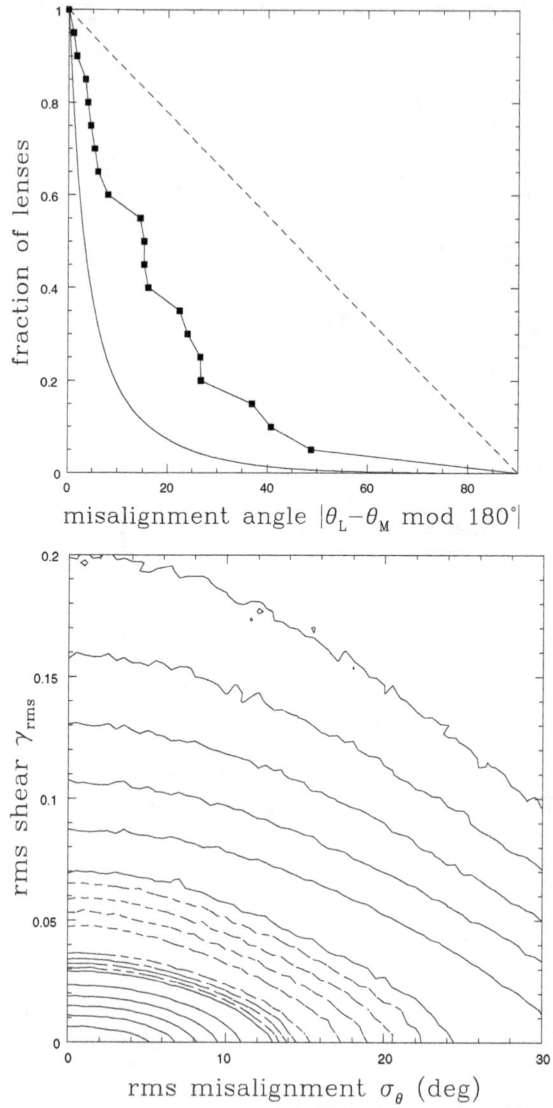

Fig. 28. (*Top*) The integral distribution of misalignment angles $\Delta\chi_{LM}$ between the major axes of the lens galaxy and an ellipsoidal lens model (*solid curve* with points for each lens). If the two angles were completely uncorrelated, the distribution would follow the dashed line. If the two angles were perfectly correlated they would follow the solid curve because of the measurement uncertainties in the two angles. (*Bottom*) Logarithmic contours of the probability for matching the distribution of misalignment angles as a function of the rms misalignment σ_θ between the mass and the light and the typical tidal shear γ_{rms}. Theoretically we expect tidal shears $\gamma_{rms} \simeq 0.06$. The solid contours are spaced by 0.5 dex and the dashed contours are spaced by 0.1 dex relative to the maximum likelihood contour. The differences between *dashed contours* are not statistically significant, while those between *solid contours* are statistically significant

because of the effects of the tidal shears. We can model this by estimating the probability of reproducing the observed misalignment distribution in terms of the strength of the local tidal shear γ_{rms} and the dispersion σ_χ in the angle between the major axis of the mass distribution and the light, as shown in Fig. 28. The observed mismatch can either be produced by having a typical tidal shear of $\gamma_{rms} \simeq 0.05$ or by having a typical misalignment between mass and light of $\sigma_\chi \simeq 20°$. We know, however, that the typical tidal shear cannot be zero because it can be estimated from the statistics of galaxies (e.g. Keeton, Kochanek and Seljak 1997; Holder and Schechter 2003). Keeton et al. (1997) obtained $\gamma_{rms} \simeq 0.05$, in which case mass must align with light and we obtain an upper limit of $\sigma_\chi \lesssim 10°$. Holder and Schechter (2003) argue for a much higher rms shear of $\gamma_{rms} = 0.15$ based on N-body simulations, which is too high to be consistent with the observed alignment of mass models and the luminous galaxy. One possible explanation (based on the results of White, Hernquist and Springel 2001) is that Holder and Schechter (2003) included parts of the lens galaxy's own halo in their estimate of the external shear. Alternatively, if lens galaxies are more compact than the SIE model used by Kochanek (2002a,b), then the lower surface density $\langle\kappa\rangle$ raises the required shear (since $\gamma \propto (1 - \langle\kappa\rangle)$, (78)). However, mass distributions similar to constant mass-to-light ratio models of the lenses would be required, which would be inconsistent with shear estimates from simulations in which galaxy masses are dominated by extended dark matter halos.

The trade-off between central concentration and shear leads to the interesting question of where the quadrupole structure of lenses originates. As we discussed in Sect. 4.4, we can break up the quadrupole of the mass distribution into the internal quadrupole due to the matter interior to the Einstein ring (52) and the exterior quadrupole due to the matter outside the Einstein ring (51). While the internal quadrupole is due only to the lens galaxy, the external quadrupole is a mixture of the quadrupole from the parts of the galaxy outside the Einstein ring (i.e. the dark matter halo) and the tidal shear from the environment. An important fact to remember is that for an isothermal ellipsoid, only $f_{int} = 25\%$ of the quadrupole is due to mass inside the Einstein ring (see Fig. 22, Sect. 4.4) ! Turner, Keeton and Kochanek (2004) explored this by fitting all the available four-image lenses with an SIS monopole combined with an internal and an external quadrupole. They then computed the fraction of the quadrupole f_{int} associated with the mass interior to the Einstein ring to find the distribution shown in Fig. 29. Most four-image lenses seem to be dominated by the external quadrupole, with internal quadrupole fractions below the $f_{int} = 0.25$ fraction expected for an isothermal ellipsoid. Lenses clearly in environments with very large tidal shears (e.g. RXJ0911+0551 which is near a massive cluster, Bade et al. 1997; Kneib et al. 2000; Morgan et al. 2001 or HE0435–1223 which is near a large galaxy, Wisotzki et al. 2002, see Fig. 4) show much smaller internal shear fractions. B1608+656 (Myers et al. 1995; Fassnacht et al. 1999), which has two lens galaxies inside the Einstein ring, shows a significantly higher internal quadrupole fraction. Combined with the

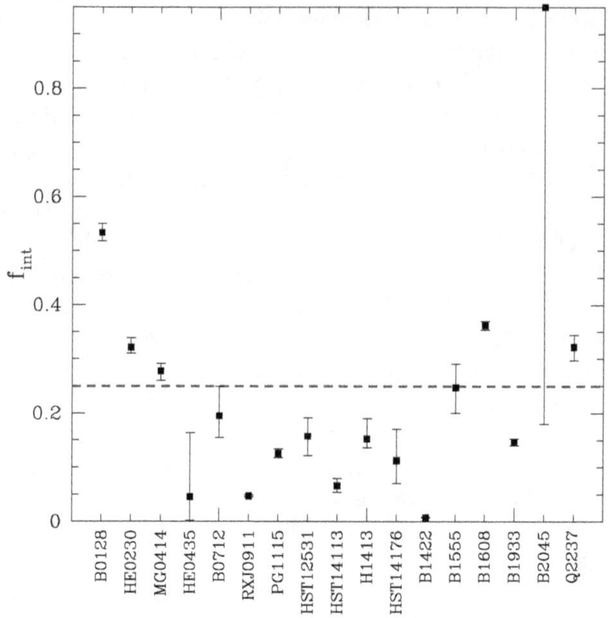

Fig. 29. The internal shear fraction f_{int} for the four-image lenses. Each system was fitted by an SIS combined with an internal shear and an external shear and $f_{int} = |\Gamma|/(|\Gamma| + |\gamma|)$ is the fraction of the quadrupole amplitude due to the internal shear. An SIE has $f_{int} = 1/4$ (see Fig. 22). Most of the quads have $f_{int} \lesssim 1/4$ as expected for an SIE in an additional external (*tidal*) shear field. Objects with very low f_{int} (e.g. HE0435–1223, RXJ0911+0551, B1422+231) have nearby galaxies or clusters generating anomalously large external shears, while objects with anomalously high f_{int} (B1608+656, HE0230–2130, MG0414+0534) tend to have additional lens components like the second lens galaxy of B1608+656. For some systems either the imaging data (e.g. B0128+437) or the models (e.g. B2045+265) do not allow a clear qualitative explanation

close correlation of mass model alignments with the luminous galaxies, this seems to argue for significant dark matter halos aligned with the luminous galaxy, but the final step of quantitatively assembling all the pieces has yet to be done.

The existence of the fundamental plane (see Sect. 9) strongly suggests that the structure of early-type galaxies is fairly homogeneous – in particular it is consistent with galaxies having self-similar mass distributions in the sense that the halo structure can be scaled from the structure of the visible galaxy. As a particular example based on our theoretical expectations, Rusin, Kochanek and Keeton (2003b) and Rusin and Kochanek (2005) modeled the visible galaxy with a Hernquist.

Equation 56 model scaled to match the observed effective radius of the lens galaxy, R_e, and then added a cuspy dark matter halo (59 with a variable inner cusp η, $\alpha = 2$ and $m = 3$) where the inner density cusp ($\rho \propto r^{-\eta}$), the halo break radius r_b and the dark matter fraction f_{CDM} inside $2R_e$ were kept as variables. The assumption of self-similarity enters by keeping the ratio r_b/R_e constant, the dark matter fraction f_{CDM} constant, and then scaling the mass-to-light ratio of the stars $\Upsilon \propto L^x$ with the luminosity. [3] We recover the fundamental plane in this model when $x \simeq 0.25$. Putting all the pieces together, the projected mass inside radius R is

$$M(< R) = \Upsilon_* L_* \left(\frac{L(0)}{L_*}\right)^{1+x} \left[g(R/R_e) + g(2)\frac{f_{CDM}}{1 - f_{CDM}} m_{CDM}(R/R_e)\right], \tag{89}$$

where Υ_* is the mass-to-light ratio of the stars in an L_* galaxy, $\log L(0) = \log L(z) - e(z)$ is the luminosity of the lens galaxy evolved to redshift zero (where we discuss estimates of the evolution rate $e(z)$ in Sect. 9), $g(x)$ is the fraction of the light inside dimensionless radius $x = R/R_e$ ($g(1) = 1/2$) and $m_{CDM}(x)$ is the dimensionless dark matter mass inside radius x with $m_{CDM}(2) = 1$ so that the CDM mass fraction inside $x = 2$ is f_{CDM}.

As we discussed earlier in Sect. 4.6, few lenses have sufficient constraints to estimate all the parameters in such a complex model. However, the assumption of self-similarity allows the average profile to be constrained statistically (Rusin et al. 2003a,b; Rusin and Kochanek 2005). Suppose we saw lensed images generated by the same galaxy at a range of different source and lens redshifts. Each observed lens only reliably measures an aperture mass $M_{ap}(R < R_{Ein})$ where R_{Ein} is the Einstein radius. But the physical scale R_{Ein} varies with redshift, so the ensemble of the lenses traces out the overall mass profile. Clearly we do not have ensembles of lenses generated by identical galaxies, but the assumption of self-similarity allows us to use the same idea for lenses with a range of luminosities and scale lengths. For 22 lenses with redshifts and accurate photometry we compared the measured aperture masses to the predicted aperture masses (the procedure for two-image lenses is a little more complicated, see Rusin et al. 2003a,b) to estimate all the model parameters. Figure 30 shows the results for the parameters associated with the dark matter halo. In the limit that $f_{CDM} \to 1$ we find that the mass distribution is consistent with a simple SIS model (the limit $f_{CDM} \to 1$ and $n \to 2$) almost independent of the break radius location. There is a slight trend with break radius because as the break to the steep $\rho \propto r^{-3}$ outer profile gets closer to the region with the lensed images the inner cusp can be shallower while keeping the overall profile over the region with images close to isothermal. As we reduce f_{CDM} and add mass to the stars, the inner cusp becomes shallower, such that for a NFW ($n = 1$) cusp the dark matter fraction inside $2R_e$ is $\sim 40\%$. It is interesting to note, however, that the total mass distribution (light + dark) changes little over the full range of allowed parameters (bottom panels of Fig. 30) – lensing constrains the global mass distribution not how

[3] They could also have allowed the CDM fraction to vary as $f_{CDM} \propto L^y$, but these led to degenerate models where only the combination $x + y$ was constrained.

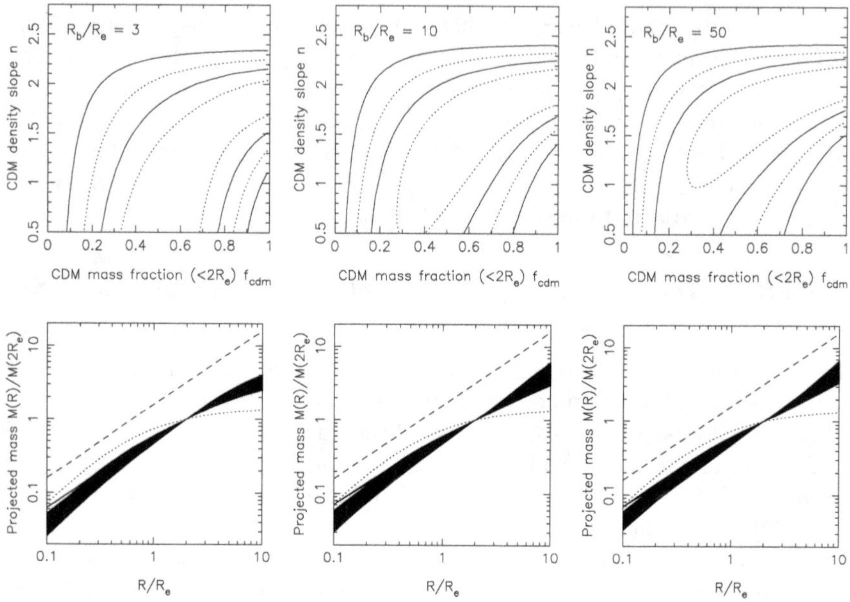

Fig. 30. The structure of lens galaxies in self-similar models. The top row shows the permitted region for the slope of the inner dark matter cusp ($\rho \propto r^{-n}$) and the fraction of the mass f_{CDM} inside $2R_e$ composed of dark matter. The results are shown for three ratios R_b/R_e between the break radius R_b of the dark matter profile and the effective radius R_e of the luminous galaxy. The solid (*dashed*) contours show the 68% and 95% confidence levels for two (one) parameter. Note that the estimates of n and f_{CDM} depend little on the location of the break radius relative to the effective radius. The Bottom row shows all the mass profiles lying with the (two parameter) 68% confidence region normalized to a fixed projected mass inside $2R_e$. For comparison we show the mass enclosed by a de Vaucouleurs model (*dotted line*) and an SIS (*offset dashed line*). While the allowed models exhibit a wide range of dark matter abundances, slopes and break radii, they all have roughly isothermal total mass profiles over the radial range spanned by the lensed images

it is divided into luminous and dark subcomponents. Note the resemblance of the statistical results to the results for detailed models of B1933+503 in Fig. 24.

4.9 Stellar Dynamics and Lensing

Stellar dynamical analyses of gravitational lenses have reached the level of studies of local galaxies approximately 15–20 years ago. The analyses are based on the spherical Jeans equations (see Binney and Tremaine 1987) with simple models of the orbital anisotropy and generally ignore both deviations from sphericity and higher order moments of the velocity distributions. The spherical Jeans equation

$$\frac{1}{\nu}\frac{d\nu\sigma_r^2}{dr} + \frac{2\beta(r)}{r}\sigma_r^2 = -\frac{GM(r)}{r^2} \tag{90}$$

relates the radial velocity dispersion $\sigma_r = \langle v_r^2 \rangle^{1/2}$, the isotropy parameter $\beta(r) = 1 - \sigma_\theta^2/\sigma_r^2$ characterizing the ratio of the tangential dispersion to the radial dispersion, the luminosity density of the stars $\nu(r)$ and the mass distribution $M(r)$. A well known result from dynamics is that you cannot infer the mass distribution $M(r)$ without constraining the isotropy $\beta(r)$ (e.g. Binney and Mamon 1982). Models with $\beta = 0$ are called isotropic models (i.e. $\sigma_r = \sigma_\theta$), while models with $\beta \to 1$ are dominated by radial orbits and models with $\beta \to -\infty$ are dominated by tangential orbits. These 3D components of the velocity dispersion must then be projected to measure the line-of-sight velocity dispersion $\langle v_{los}^2 \rangle^{1/2}$,

$$\Sigma(R)\langle v_{los}^2\rangle(R) = 2\int_0^\infty dz\nu\left[\frac{z^2}{r^2}\sigma_r^2 + \frac{R^2}{r^2}\sigma_\theta^2\right] = 2\int_0^\infty dz\nu\sigma_r^2\left[1 + \frac{R^2}{r^2}\beta(r)\right], \tag{91}$$

where $\Sigma(R) = 2\int_0^\infty dz\nu(r)$ is the projected surface brightness and z is the coordinate along the line-of-sight. Modern observations of local galaxies break the degeneracy between mass and isotropy by measuring higher order moments ($\langle v_{los}^n \rangle$) of the line-of-sight velocity distribution (LOSVD) because the shape of the LOSVD is affected by the isotropy of the orbits. Because the velocity dispersions are measured starting from a Gaussian fit to the LOSVD, the higher order moments are described by the amplitudes h_n of a decomposition of the LOSVD into Gauss–Hermite polynomials (e.g. van der Marel and Franx 1993). In general, the rms velocity (i.e. combining dispersion and rotation) and higher order moment profiles of early-type galaxies are fairly self-similar, with nearly flat rms velocity profiles, modest values of $h_4 \simeq 0.01 \pm 0.03$ and slightly radial orbits $\langle\beta\rangle \simeq 0.1$–$0.2$ (e.g. Romanowsky and Kochanek 1999; Gerhard et al. 2001).

Stellar dynamics is used for two purposes in lensing studies. The first is to provide a mass normalization for lens models used in studies of lens statistics. We will discuss this in Sect. 6. The second is to use comparisons between a mass estimated from the geometry of a lens and the velocity dispersion of the lens galaxy to constrain the mass distribution (e.g. Romanowsky and Kochanek 1999; Trott and Webster 2002; Koopmans and Treu 2002, 2003; Treu and Koopmans 2002a,b; Koopmans et al. 2003). It is important to understand that the systematic uncertainties in combining lensing and stellar dynamics to determine mass distributions are different from using either in isolation. For local galaxies we measure a velocity dispersion profile. The normalization of the profile sets the mass scale and the changes in the profile (and any higher order moments) with radius constrains the mass distribution. To lowest order, a simple scaling error in the velocity measurements will lead to errors in the mass scale rather than in the mass distribution. For lens galaxies, it is the comparison between the velocity dispersion and the mass determined by the geometry of the images that constrains the mass distribution. Thus, estimates of the mass distribution are directly affected by any calibration errors in the velocity dispersions.

We can understand the differences with a simple thought experiment. Suppose we have a mass distribution $M = M_0(R/R_0)^x$ in projection and we have mass estimates M_1 at R_1 and M_2 at R_2. Combining them we can solve for the exponent describing the mass distribution, $x = \ln(M_1/M_2)/\ln(R_1/R_2)$. In a dynamical observation the mass estimate is some sort of virial estimator $M \propto \sigma_v^2 R/G$ while in a lensing measurement it is a direct measurement of M. Standard velocity dispersion measurements start from the best fit Gaussian line width $\hat{\sigma}$ (uncertainties $\pm e_\sigma$) and then subtract an intrinsic line width σ_c (due to the instrument and the intrinsic line width of the star, uncertainties $\pm e_c$) in quadrature to estimate the portion of the line width due to the motions of the stars. Thus $\sigma_v^2 = f^2(\hat{\sigma}^2 - \sigma_c^2)$ where $f \simeq 1$ is a scale factor to account for deviations from spherical symmetry and non-Gaussian line-of-sight velocity distributions (LOSVDs). In a purely dynamical study, uncertainties in f and σ_c produce bigger fractional errors in the absolute mass scale M_0 than in the exponent x. For example, given measurements σ_1 and σ_2 at radii R_1 and R_2, the exponent, $x = 1 + \ln(\sigma_1^2/\sigma_2^2)/\ln(R_1/R_2)$, depends only on velocity dispersion ratios in which calibration errors tend to cancel. This is obvious for the scale factor f, which cancels exactly if it does not vary with radius. Since studies of lens dynamics use a comparison between a dynamical mass and a lensing mass to estimate the mass distribution, the results are more sensitive to calibration problems because these cancellations no longer occur. If we combine a velocity dispersion measurement σ_1 with a lensing mass measurement M_2 our estimate of the exponent becomes $x = \ln(\sigma_1^2 R_1/GM_2)/\ln(R_1/R_2)$ and the uncertainties are linear in the scale factor f rather than canceling. An error analysis for the effects of σ_c is messier, but you again find that the sensitivity in the mixed lensing and dynamics constraint to errors in σ_c is greater than in a purely dynamical study.

Velocity dispersions have now been measured for 10 lenses (0047–2808 Koopmans and Treu 2003; CFRS03.1077 Treu and Koopmans 2004; Q0957+561 Falco et al. 1997; Tonry and Franx 1999; PG1115+080 Tonry 1998; HST14176 +5226 Ohyama et al. 2002; Gebhardt et al. 2003; Treu and Koopmans 2004; HST15433+5352 Treu and Koopmans 2004; MG1549+3047 Lehár et al. 1996; B1608+656 Koopmans et al. 2003; MG2016+112 Koopmans and Treu 2002; Q2237+0305 Foltz et al. 1992). With the exception of Romanowsky and Kochanek (1999), who fitted for the distribution function of the lens, the analyses of the data have used the spherical Jeans equations with parameterized models for the isotropy $\beta(r)$. They include the uncertainties in σ_c about as well as any other dynamical study, although it is worth bearing in mind that this is tricky because we lack nearby stars with the appropriate metallicity and the problem of matching the spectral resolution for the galaxy and the template stars lacks direct checks of the success of the procedure. A useful rule of thumb to remember is that repeat measurements of velocity dispersions by different groups almost always show larger scatter than is consistent with the reported uncertainties. For example, the three velocity dispersion measurements for the lens HST14176+5226 (224 ± 15 km/s by Ohyama et al. 2002, 202 ± 9 km/s by Gebhardt et al. 2003, and 230 ± 14 km/s by Treu and Koopmans 2004) are mutually consistent only if the uncertainties are broadened by 30%.

In Fig. 31 we summarize the dynamical constraints for 9 of these systems using the self-similar mass distribution from Rusin and Kochanek (2005, (89)). This model is very similar to that used by Treu and Koopmans (2004). For most of the lenses, the region producing a good fit to the combined lensing and dynamical data overlaps the same region preferred by the Rusin and Kochanek (2005) self-similar models, shows the same general parameter degeneracy and is consistent with a simple SIS mass distribution ($f_{cdm} \rightarrow 1$ and $n = 2$). This is particularly true of 0047–2808, HST15433+5352, B1608+656, MG2016+112 and CFRS03.1077. Only Q2237+0305, where the lens is the bulge of a nearby spiral and we might not expect this mass model to be applicable, shows a very different trend (e.g. see the models of Trott and Webster 2002). PG1115+080 and to a lesser extent MG1549+3047 might have steeper than isothermal mass distributions (falling rotation curves) and the possibility of being consistent with a constant mass-to-light ratio model (Treu and Koopmans 2002a,b). HST14176+5226 and to a lesser extent HST15433+5352 could have shallower than isothermal mass distribution (rising rotation curves). Along the degeneracy direction for each lens we will find similar mass distributions with very different decompositions into luminous and dark matter, just as in Fig. 30. The problem raised by this panorama is whether it shows that the halo structure is largely homogeneous with some measurement outliers, or that the structure of early-type is heterogeneous with important implications for understanding time delays (Sect. 5) and galaxy evolution (Sect. 9).

My own view tends toward the first interpretation – that the dynamical data supports the homogeneity of early-type galaxy structure. The permitted bands in Fig. 31 show the 68% confidence regions given the formal measurement errors and the simple, spherical, isotropic Jeans equation models – this means that the true 68% confidence regions are significantly larger. We have already argued that the formal errors on dynamical measurements tend to be underestimates. For example, the need for HST14176+5226 to have a rising rotation curve would be considerably reduced if we used the higher velocity dispersion measurements from Ohyama et al. (2002) or Treu and Koopmans (2004) or if we broadened the uncertainties by the 30% needed to make the three estimates statistically consistent. Moreover, the existing analyses have also neglected the systematic uncertainties arising from the scaling factor f. There are two important issues that make $f \neq 1$. The first issue is that standard velocity dispersion measurements are the width of the best fit Gaussian model for the LOSVD, and this is not the same as the mean square velocity ($\langle v_{los}^2 \rangle^{1/2}$) appearing in the Jeans equations used to analyze the data unless the LOSVD is also a Gaussian. Stellar dynamics has adopted the dimensionless coefficients h_n of a Gauss–Hermite polynomial series to model the

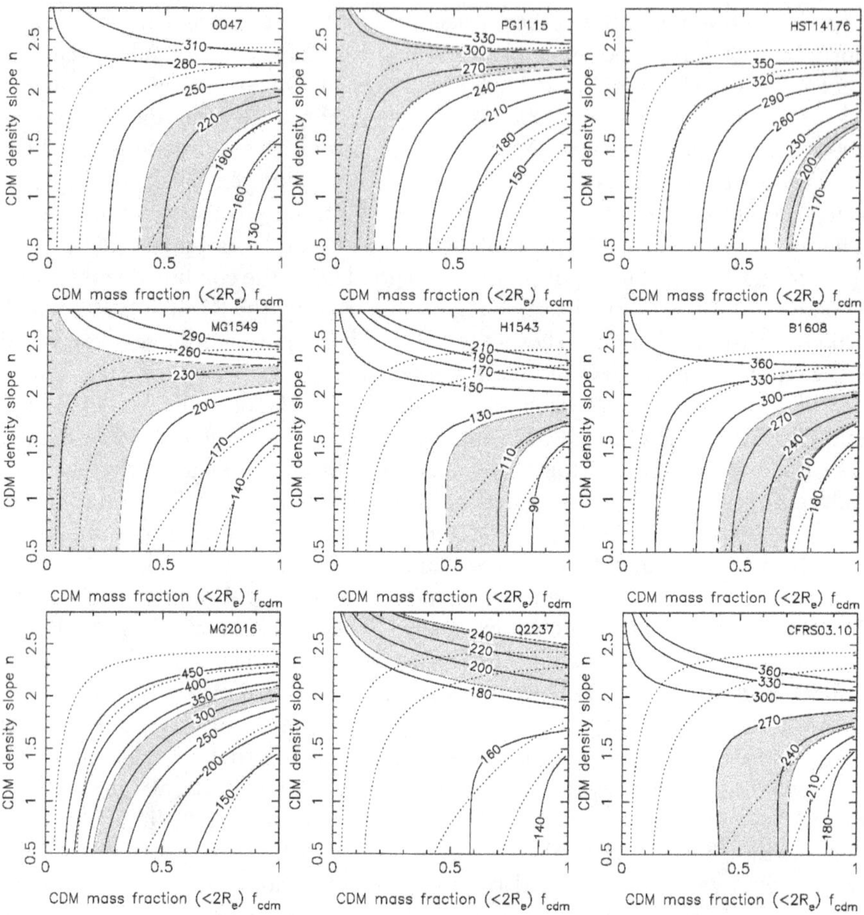

Fig. 31. Constraints from lens velocity dispersion measurements on the self-similar mass distributions of (89) and Fig. 30. The dotted contours show the 68% and 95% confidence limits from the self-similar models for $R_b/R_e = 50$. The shaded regions show the models allowed (68% confidence) by the formal velocity dispersion measurement errors, and the heavy solid lines show contours of the velocity dispersion in km/s. We used the low (Gebhardt et al. 2003) velocity dispersion for HST14176+5226 because it has the smallest formal error. These models assumed isotropic orbits, thereby underestimating the full uncertainties in the stellar dynamical models

deviations of the LOSVD from Gaussian, and a typical early-type galaxy has $|h_4| \lesssim 0.03$ (e.g. Romanowsky and Kochanek 1999). This leads to a systematic difference between the measured dispersions and the mean square velocity of $\langle v_{los}^2 \rangle^{1/2} \simeq \sigma(1 + \sqrt{6} h_4)$ (e.g. van der Marel and Franx 1993), so $|f - 1| \sim 7\%$ for $|h_4| \simeq 0.03$. Only the Romanowsky and Kochanek (1999) models of Q0957+561 and PG1115+080 have properly included this uncertainty.

In fact, Romanowsky and Kochanek (1999) demonstrated that there were stellar distribution functions in which the mass distribution of PG1115+080 is both isothermal and agrees with the measured velocity dispersion. While it is debatable whether these models allowed too much freedom, it is certainly true that models using the Jeans equations and ignoring the LOSVD have too little freedom and will overestimate the constraints.

The second issue is that lens galaxies are not spheres. Unfortunately there are few simple analytic results for oblate or triaxial systems like early-type galaxies in which the ellipticity is largely due to anisotropies in the velocity dispersion tensor rather than rotation. For the system as a whole, the tensor virial theorem provides a simple global relationship between the major and minor axis velocity dispersions

$$\frac{\sigma_{major}}{\sigma_{minor}} \simeq 1 + \frac{1}{5}e^2 + \frac{9}{70}e^4 + \cdots \tag{92}$$

for an oblate ellipsoid of axis ratio q and eccentricity $e = (1-q^2)^{1/2}$ (e.g. Binney and Tremaine 1987). The velocity dispersion viewed along the major axis is larger than that on the minor, and the correction can be quite large since a typical galaxy with $q = 0.7$ will have a ratio $\sigma_{major}/\sigma_{minor} \simeq 1.16$ that is much larger than typical measurement uncertainties. If galaxies are oblate, this provides no help for the case of PG1115+080 because making the line-of-sight dispersion too high requires a prolate galaxy. However, it is a very simple means of shifting HST14176+5226. Crudely, if we start with the low 209 km/s velocity dispersion and assume that the lens is a $q = 0.7$ galaxy viewed pole on, then $\sigma_{major}/\sigma_{minor} \simeq 1.14$ and the corrections for the shape are large enough to make HST14176+5226 consistent with the other systems.

A final caveat is that neglecting necessary degrees of freedom in your lens model can bias inferences from the stellar dynamics of lenses just as it can in pure lens modeling. For example, Sand et al. (2002, 2004) used a comparison of lensed arcs in clusters to velocity dispersion measurements of the central cluster galaxy to argue that the cluster dark matter distribution could not have the $\rho \propto 1/r$ cusp of the NFW model for CDM halos. However, Bartelmann and Meneghetti (2004) and Dalal and Keeton (2003) show that the data are consistent with an NFW cusp if the lens models include a proper treatment of the non-spherical nature of the clusters. This has not been an issue in the stellar dynamics of strong lenses where the lens models used to determine the mass scale have always included the effects of ellipticity and shear, but it is well worth remembering.

5 Time Delays

Nothing compares to the measurement of the Hubble constant in bringing out the worst in astronomers. As we discussed in the previous section on lens modeling, many discussions of lens models seem obfuscatory rather than illuminating, and the tendency in this direction increases when the models are used to estimate H_0. In this section we discuss the relationship between time delay measurements, lens models and H_0. All results in the literature are consistent with this discussion, although it may take you several days and a series of e-mails to confirm it for any particular paper. The basic idea is simple. We see images at extrema of the virtual time delay surface (e.g. Blandford and Narayan 1986, Part 1) so the propagation time from the source to the observer differs for each image. The differences in propagation times, known as time delays, are proportional to H_0^{-1} because the distances between the observer, the lens and the source depend on the Hubble constant (Refsdal 1964a,b). When the source varies, the variations appear in the images separated by the time delays and the delays are measured by cross-correlating the light curves. There are recent reviews of time delays and the Hubble constant by Courbin, Saha and Schechter (2002b) and Kochanek and Schechter (2004). Portions of this section are adapted from Kochanek and Schechter (2004) since we were completing that review at about the same time as we presented these lectures.

To begin the discussion we start with our standard simple model, the circular power law lens from Sect. 3. As a circular lens, we see two images at radii θ_A and θ_B from the lens center and we will assume that $\theta_A > \theta_B$ (Fig. 20). Image A is a minimum, so source variability will appear in image A first and then with a time delay Δt in the saddle point image B. We can easily fit this data with an SIS lens model (see (21) and (22)) to find that $\theta_A = \beta + b$ and $\theta_B = b - \beta$ where $b = (\theta_A + \theta_B)/2$ is the critical (Einstein) radius of the lens and $\beta = (\theta_A - \theta_B)/2$ is the source position. The light travel time for each image relative to a fiducial unperturbed ray is (see Part 1)

$$\tau(\boldsymbol{\theta}) = \frac{D_d D_s}{c D_{ds}} \left[\frac{1}{2} (\boldsymbol{\theta} - \boldsymbol{\beta})^2 - \Psi(\boldsymbol{\theta}) \right], \tag{93}$$

where the effective potential $\Psi = b\theta$ for the SIS lens. Remember that the distances are comoving angular diameter distances rather than the more familiar angular diameter distances and this leads to the vanishing of the extra $1 + z_l$ factor that appears in the numerator if you insist on using angular diameter distances. The propagation time scales as $H_0^{-1} \simeq 10h^{-1}$ Gyr because of the H_0^{-1} scalings of the distances. After substituting our lens model, and differencing the delays for the two images, we find that

$$\Delta t_{SIS} = \tau_B - \tau_A = \frac{1}{2} \frac{D_d D_s}{c D_{ds}} \left(\theta_A^2 - \theta_B^2 \right). \tag{94}$$

The typical deflection angle $b \sim 3 \times 10^{-6}$ radians (so $R_A^2 \sim 10^{-11}$) converts the $10h^{-1}$ Gyr propagation time into a time delay of months or years that can be measured by monitoring the light curves of the images. Naively, this result suggests that the problem of interpreting time delays to measure H_0 is a trivial problem in astrometry.

We can check this assumption by using our general power-law models from Sect. 3 instead of an SIS. The power-law models correspond to density distributions $\rho \propto r^{-n}$, surface densities $\kappa \propto R^{1-n}$ and circular velocities $v_c \propto r^{(2-n)/2}$ of which the SIS model is the special case with $n = 2$. These models have effective potentials

$$\Psi(\boldsymbol{\theta}) = \frac{b^2}{3-n}\left(\frac{\theta}{b}\right)^{3-n} . \tag{95}$$

As we discussed in Sect. 4.1 we can fit our simple, circular two-image lens with any of these models to determine $b(n)$ and $\beta(n)$ (66), which we can then substitute into the expression for the propagation time to find that the time delay between the images is

$$\Delta t(n) = (n-1)\Delta t_{SIS}\left[1 - \frac{(2-n)^2}{12}\left(\frac{\delta\theta}{\langle\theta\rangle}\right)^2 + \cdots\right], \tag{96}$$

where we have expanded the result as a series in the ratio between the mean radius of the images $\langle\theta\rangle = (\theta_A + \theta_B)/2$ and the thickness of the radial annulus separating them $\delta\theta = \theta_A - \theta_B$. While the expansion assumes that $\delta\theta/\langle\theta\rangle \sim \beta/b$ is small, we can usually ignore higher order terms even when $\delta\theta/\langle\theta\rangle$ is of order unity. We now see that the time delay depends critically on the density profile, with more centrally concentrated mass distributions (larger values of n) producing longer time delays or implying larger Hubble constants for a fixed time delay.

The other idealization of the SIS model, the assumption of a circular lens, turns out to be less critical. A very nice analytic example is to consider a singular isothermal model with arbitrary angular structure in which $\kappa = bF(\chi)/2\theta$ where $F(\chi)$ is an arbitrary function of the azimuthal angle. The singular isothermal ellipsoid (37) is an example of this class of potential. For this model family, $\Delta t = \Delta t_{SIS}$ independent of the actual angular structure $F(\chi)$ (Witt, Mao and Keeton 2000).

5.1 A General Theory of Time Delays

Just as for estimating mass distributions (Sect. 4), the aspect of modeling time delays that creates the greatest suspicion is the need to model the gravitational potential of the lens. Just as for mass distributions, this problem is largely of our own making, arising from poor communication, understanding and competition between groups. Here we will use simple mathematical expansions to show exactly what properties of the potential determine time delays. Any models which have these generic properties have all the degrees of freedom needed to properly interpret time delays. This does not, unfortunately, avoid the problem of degeneracies between the mass models and the Hubble constant.

The key to understanding time delays comes from Gorenstein, Falco and Shapiro (1988, Kochanek 2002a,b, see also Saha 2000) who showed that the time delay in a circular lens depends only on the image positions and *the surface density* $\kappa(\theta)$ *in the annulus between the images.* The two lensed images at radii $\theta_A > \theta_B$ define an annulus bounded by their radii, with an interior region for $\theta < \theta_B$ and an exterior region for $\theta > \theta_A$ (Fig. 20). As we discussed in Sect. 4.1, the mass in the interior region is implicit in the image positions and constrained by the astrometry. From Gauss' law we know that the distribution of the mass in the interior and the amount or distribution of mass in the exterior region is irrelevant. A useful approximation is to assume that the surface density in the annulus can be *locally* approximated by a power law, $\kappa(\theta) \propto \theta^{1-n}$ for $\theta_B < \theta < \theta_A$, with a mean surface density in the annulus of $\langle\kappa\rangle = \langle\Sigma\rangle/\Sigma_c$. The time delay between the images is then (Kochanek 2002a)

$$\Delta t = 2\Delta t_{SIS}\left[1 - \langle\kappa\rangle - \frac{1-n\langle\kappa\rangle}{12}\left(\frac{\delta\theta}{\langle\theta\rangle}\right)^2 + O\left(\left(\frac{\delta\theta}{\langle\theta\rangle}\right)^4\right)\right], \tag{97}$$

where $\langle\theta\rangle = (\theta_A + \theta_B)/2$ and $\delta\theta = \theta_A - \theta_B$ as before. Thus, the time delay is largely determined by the average surface density $\langle\kappa\rangle$ with only modest corrections from the local shape of the surface density distribution even when $\delta\theta/\langle\theta\rangle \sim 1$. This second order expansion is exact for an SIS lens ($\langle\kappa\rangle = 1/2$, $n = 2$), and it reproduces the time delay of a point mass lens ($\langle\kappa\rangle = 0$) to better than 1% even when $\delta\theta/\langle\theta\rangle = 1$. The local model also explains the scalings of the global power-law models. A $\kappa \propto \theta^{1-n}$ global power law has surface density $\langle\kappa\rangle = (3-n)/2$ near the Einstein ring, so the leading term of the time delay is $\Delta t = 2\Delta_{SIS}(1 - \langle\kappa\rangle) = (n-1)\Delta t_{SIS}$ just as in (96).

The role of the angular structure of the lens is easily incorporated into the expansion through the multipole expansion of Sect. 4. A quadrupole term in the potential with dimensionless amplitude ϵ_Ψ produces ray deflections of order $O(\epsilon_\Psi b)$ at the Einstein radius b of the lens. In a four-image lens, the quadrupole deflections are comparable to the fractional thickness of the annulus, $\epsilon_\Psi \simeq \delta\theta/\langle\theta\rangle$, while in a two-image lens they are smaller. For an ellipsoidal density distribution, the $\cos(2m\chi)$ multipole amplitude is smaller than the quadrupole amplitude by $\epsilon_{2m} \sim \epsilon_\Psi^m \lesssim (\delta\theta/\langle\theta\rangle)^m$. Hence, to lowest order in the expansion we only need to include the internal and external quadrupoles of the potential but not the changes of the quadrupoles in the annulus or any higher order multipoles. Remember that what counts is the angular structure of the potential rather than of the density, and that potentials are always much rounder than densities with a typical scaling of m^{-1}:m:1 between the potential, deflections and surface density for the $\cos m\chi$ multipoles (see Sect. 4.4)

While the full expansion for independent internal and external quadrupoles is too complex to be informative, the leading term for the case when the internal and external quadrupoles are aligned is informative. We have an internal shear of amplitude Γ and an external shear of amplitude γ with $\chi_\gamma = \chi_\Gamma$ as defined in (51) and (52). The leading term of the time delay is

$$\Delta t \simeq 2\Delta t_{SIS}\left(1 - \langle\kappa\rangle\right)\frac{\sin^2\left(\Delta\chi_{AB}/2\right)}{1 - 4f_{int}\cos^2\left(\Delta\chi_{AB}/2\right)}, \tag{98}$$

where $\Delta\chi_{AB}$ is the angle between the images (Fig. 20) and $f_{int} = \Gamma/(\Gamma + \gamma)$ is the internal quadrupole fraction we explored in Fig. 29. We need not worry about a singular denominator – successful models of the image positions do not allow such configurations.

A two-image lens has too few astrometric constraints to fully constrain a model with independent, misaligned internal and external quadrupoles. Fortunately, when the lensed images lie on opposite sides of the lens galaxy ($\Delta\chi_{AB} \simeq \pi + \delta$ with $|\delta| \ll 1$), the time delay becomes insensitive to the quadrupole structure. Provided the angular deflections are smaller than the radial deflections ($|\delta|\langle\theta\rangle \lesssim \delta\theta$), the leading term of the time delay reduces to the result for a circular lens, $\Delta t = 2\Delta t_{SIS}(1 - \langle\kappa\rangle)$ if we minimize the total shear of the lens. In the minimum shear solution the shear converges to the invariant shear (γ_1) and the other shear component $\gamma_2 = 0$ (see Sect. 4.5). If, however, you allow the other shear component to be non-zero, then you find that $\Delta t = 2\Delta t_{SIS}(1 - \langle\kappa\rangle - \gamma_2)$ to lowest order – the second shear component acts like a contribution to the convergence. In the absence of any other constraints, this adds a modest additional uncertainty (5–10%) to interpretations of time delays in two-image lenses. To first order its effects should average out in an ensemble of lenses because the extra shear has no preferred sign.

A four image lens has more astrometric constraints and can constrain a model with independent, misaligned internal and external quadrupoles – this was the basis of the Turner et al. (2004) summary of the internal to total quadrupole ratios shown in Fig. 29. If the external shear dominates, then $f_{int} \simeq 0$ and the leading term of the delay becomes $\Delta t = 2\Delta t_{SIS}(1 - \langle\kappa\rangle) \sin^2 \Delta\chi_{AB}/2$. If the model is isothermal, like the $\Psi = \theta F(\chi)$ model we introduced in (42), then $f_{int} = 1/4$ and we obtain the Witt et al. (2000) result that the time delay is independent of the angle between the images $\Delta t \simeq 2\Delta t_{SIS}(1 - \langle\kappa\rangle)$. Thus, delay ratios in a four-image lens are largely determined by the angular structure and provide a check on the potential model. Unfortunately, the only lens with precisely measured delay ratios, B1608+656 (Fassnacht et al. 2002), also has two galaxies inside the Einstein ring and is a poor candidate for a simple multipole treatment (although it is dominated by an internal quadrupole as expected, see Fig. 29). The delay ratios for PG1115+080 are less well measured (Schechter et al. 1997; Barkana 1997; Chartas et al. 2004), but should be dominated by external shear since the estimate from the image astrometry is that $f_{int} = 0.083$ ($0.055 < f_{int} < 0.111$ at 95% confidence). Figure 32 shows the dependence of the PG1115+080 delays on the leading angular dependence of the time delay (98) after scaling out the standard astrometry factor for the different radii of the images (94). Formally, the estimate from the time delays that $f_{int} = -0.02$ ($-0.09 < f_{int} < 0.03$ at 68% confidence) is a little discrepant, but the two estimates agree at the 95% confidence level and there are still some systematic uncertainties in the shorter optical delays of PG1115+080. Changes in f_{int} between lenses is one reason why (Saha 2004) found significant scatter between time delays scaled only by Δt_{SIS}, since the time delay lenses range

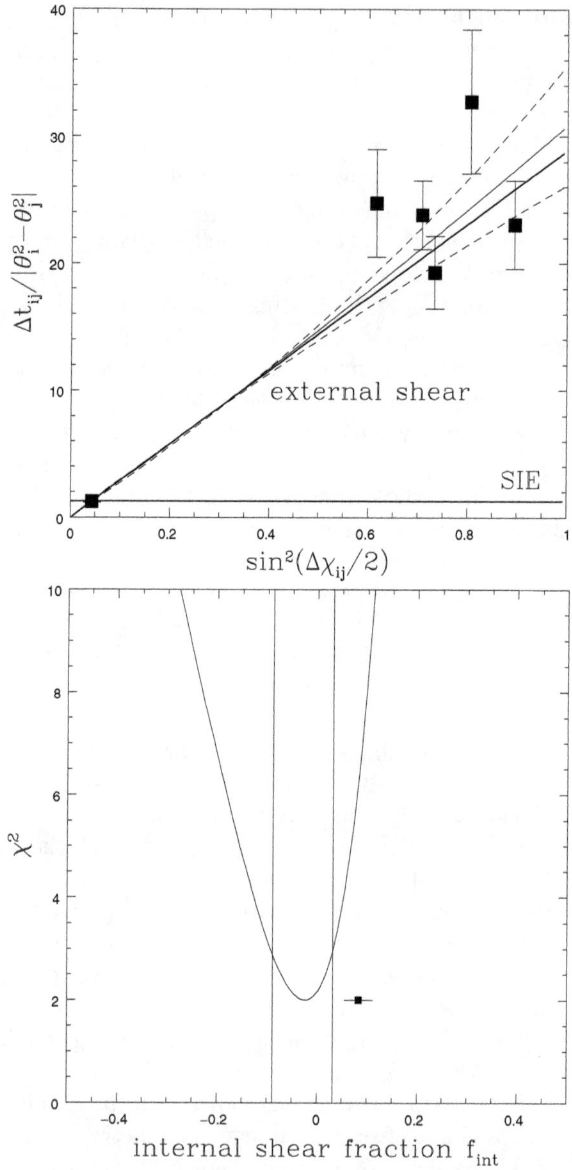

Fig. 32. (*Top*) The PG1115+080 time delays scaled by the astrometric factor $\theta_i^2-\theta_j^2$ appearing in Δt_{SIS} (94) as a function of the leading angular dependence of the time delay ($\sin^2 \Delta\chi_{ij}/2$) (98). The light solid curve and the dashed curves show the dependence for the best fit internal shear fraction f_{int} and its 68% confidence limits. A true external shear $f_{int} = 0$ is shown by the heavy solid curve inside the confidence limits, and the scaling for an SIE ($f_{int} = 1/4$) is shown by the horizontal line. (*Bottom*) The χ^2 goodness of fit for the internal shear fraction f_{int} from the time delay ratios is shown by the curve with the 68% confidence region bracketed by the vertical lines. The estimate of f_{int} from the image astrometry is shown by the point with an error bar

from external shear dominated systems like PG1115+080 to internal shear dominated systems like B1608+656.

5.2 Time Delay Lenses in Groups or Clusters

Most galaxies are not isolated, and many early-type lens galaxies are members of groups or clusters, so we need to consider the effects of the local environment on the time delays. Weak perturbations are easily understood since they will simply be additional contributions to the surface density (κ_c) and the external shear/quadrupole (γ_c) we discussed in Sect. 4. In general the effects of the external shear γ_c are minimal because they either have little effect on the delays (two-image lenses) or are tightly constrained by either the astrometry or delay ratios (four-image lenses or systems with lensed host galaxies see Sect. 10). The problems arise from either the degeneracies associated with the surface density κ_c or the need for a complete, complicated cluster model.

The problem with κ_c is the infamous *mass-sheet degeneracy* (Part 1, Falco, Gorenstein and Shapiro 1985). If we have a model predicting a time delay Δt_0 and add a sheet of constant surface density κ_c, then the time delay is changed to $(1-\kappa_c)\Delta t_0$ without changing the image positions, flux ratios, or time delay ratios. Its effects can be understood from Sect. 5.1 as a contribution to the annular surface density with $\langle\kappa\rangle = \kappa_c$ and $\eta = 1$. Its only observable effect other than that on the time delays is a reduction in the mass of the lens galaxy that could be detected given an independent estimate of the lens galaxy's mass such as a velocity dispersion (e.g. Sect. 4.9 see Romanowsky and Kochanek 1998 for an attempt to do this for Q0957+561). It can also be done given an independent estimate of the properties of the group or cluster using weak lensing (e.g. Fischer et al. 1997 in Q0957+561), cluster galaxy velocity dispersions (e.g., Angonin-Willaime, Soucail and Vanderriest 1994 for Q0957+561, Hjorth et al. 2002 for RXJ0911+0551) or X-ray temperatures/luminosities (e.g., Morgan et al. 2001 for RXJ0911+0551 or Chartas et al. 2002 for Q0957+561). The accuracy of these methods is uncertain at present because each suffers from its own systematic uncertainties, and they probably cannot supply the 10% or higher precision measurements of κ_c needed to strongly constrain models. When the convergence is due to an object like a cluster, there is a strong correlation between the convergence κ_c and the shear γ_c that is controlled by the density distribution of the cluster (for an isothermal model $\kappa_c = \gamma_c$). When the lens is in the outskirts of a cluster, as in RXJ0911+0551, it is probably reasonable to assume that $\kappa_c \leq \gamma_c$, as most mass distributions are more centrally concentrated than isothermal. Neglecting the extra surface density coming from nearby objects (galaxies, groups, clusters) leads to an overestimate of the Hubble constant, because these objects all have $\kappa_c > 0$. For most time delay systems this correction is probably $\lesssim 10\%$.

If the cluster or any member galaxies are sufficiently close, then we cannot ignore the higher-order perturbations in the expansion of (26). This is certainly true for Q0957+561 (as discussed in Sect. 4.6) where the lens galaxy is the brightest cluster galaxy and located very close to the center of the cluster. It is easy to gauge when they become important by simply comparing the deflections produced by any higher order moments of the cluster beyond the quadrupole with the uncertainties being used for the image positions. For a cluster of critical radius b_c at distance θ_c from a lens of Einstein radius b, these perturbations are of order $b_c(b/\theta_c)^2 \sim b\gamma_c(b/\theta_c)$. Because the astrometric precision of the measurements is so high, these higher order terms can be relatively easy to detect. For example, models of PG1115+080 (e.g. Impey et al. (1998)) find that using a group potential near the optical centroid of the nearby galaxies produces a better fit than simply using an external shear. In this case the higher order terms are fairly small and affect the results little, but results become very misleading if they are important but ignored.

5.3 Observing Time Delays and Time Delay Lenses

The first time delay measurement, for the gravitational lens Q0957+561, was reported in 1984 (Florentin-Nielsen 1984). Unfortunately, a controversy then developed between a short delay ($\simeq$ 1.1 years, Schild and Cholfin 1986; Vanderriest et al. 1989) and a long delay ($\simeq$ 1.5 years, Press, Rybicki, and Hewitt 1992a,b), which was finally settled in favor of the short delay only after 5 more years of effort (Kundić et al. 1997; also Schild and Thomson 1997 and Haarsma et al. 1999). Factors contributing to the intervening difficulties included the small amplitude of the variations, systematic effects, which, with hindsight, appear to be due to microlensing and scheduling difficulties (both technical and sociological).

While the long-running controversy over Q0957+561 led to poor publicity for the measurement of time delays, it allowed the community to come to an understanding of the systematic problems in measuring time delays, and to develop a broad range of methods for reliably determining time delays from typical data. Only the sociological problem of conducting large monitoring projects remains as an impediment to the measurement of time delays in large numbers. Even these are slowly being overcome, with the result that the last five years have seen the publication of time delays in 11 systems (see Table 1).

The basic procedures for measuring a time delay are simple. A monitoring campaign must produce light curves for the individual lensed images that are well sampled compared to the time delays. During this period, the source quasar in the lens must have measurable brightness fluctuations on time scales shorter than the monitoring period. The resulting light curves are cross correlated by one or more methods to measure the delays and their uncertainties (e.g., Press et al. 1992a,b; Beskin and Oknyanskij 1995; Pelt et al. 1996; references in Table 1). Care must be taken because there can be sources of uncorrelated variability between the images due to systematic errors in the photometry and real effects such as microlensing of the individual

Table 1. Time delay measurements

system	N_{im}	Δt (days)	astrometry	model	ref.
HE1104–1805	2	161 ± 7	+	"simple"	1
PG1115+080	4	25 ± 2	+	"simple"	2
SBS1520+530	2	130 ± 3	+	"simple"	3
B1600+434	2	51 ± 2	+/−	"simple"	4
HE2149–2745	2	103 ± 12	+	"simple"	5
RXJ0911+0551	4	146 ± 4	+	cluster/satellite	6
Q0957+561	2	417 ± 3	+	cluster	7
B1608+656	4	77 ± 2	+/−	satellite	8
B0218+357	2	10.5 ± 0.2	−	"simple"	9
PKS1830–211	2	26 ± 4	−	"simple"	10
B1422+231	4	(8 ± 3)	+	"simple"	11

N_{im} is the number of images. Δt is the longest of the measured delays and its 1σ error; delays in parenthesis require further confirmation. The "Astrometry" column indicates the quality of the astrometric data for the system: + (good), +/− (some problems), − (serious problems). The "Model" column indicates the type of model needed to interpret the delays. "Simple" lenses can be modeled as a single primary lens galaxy in a perturbing tidal field. More complex models are needed if there is a satellite galaxy inside the Einstein ring ("satellite") of the primary lens galaxy, or if the primary lens belongs to a cluster. References: (1) Ofek and Maoz 2003; Wyrzykowski et al. 2003; (2) Barkana 1997, based on Schechter et al. 1997; (3) Burud et al. 2002a,b; (4) Burud et al. 2000, also Koopmans et al. 2000a,b; (5) Burud et al. 2002a,b; (6) Hjorth et al. 2002; (7) Kundić et al. 1997, also Schild and Thomson 1997 and Haarsma et al. 1999; (8) Fassnacht et al. 2002; (9) Biggs et al. 1999, also Cohen et al. 2000; (10) Lovell et al. 1998; (11) Patnaik and Narasimha 2001.

images (e.g., Koopmans et al. 2000a,b; Burud et al. 2002a,b; Schechter et al. 2003). Figure 33 shows an example, the beautiful light curves from the radio lens B1608+656 by Fassnacht et al. (2002), where the variations of all four lensed images have been traced for over three years. One of the 11 systems, B1422+231, is limited by systematic uncertainties in the delay measurements.

We want to have uncertainties in the time delay measurements that are unimportant for the estimates of H_0. For the present, uncertainties of order 3%–5% are adequate (so improved delays are still needed for PG1115+080, HE2149–2745, and PKS1830–211). In a four-image lens we can measure three independent time delays, and the dimensionless ratios of these delays provide additional constraints on the lens models (see Sect. 5.1). These ratios are well measured in B1608+656 (Fassnacht et al. 2002), poorly measured in PG1115+080 (Barkana 1997; Schechter et al. 1997; Chartas et al. 2004) and unmeasured in either RXJ0911+0551 or B1422+231. Using the time delay lenses as very precise probes of H_0, dark matter and cosmology will eventually require still smaller delay uncertainties ($\sim 1\%$). Once a delay is known to 5%, it is relatively easy to reduce the uncertainties further because we can

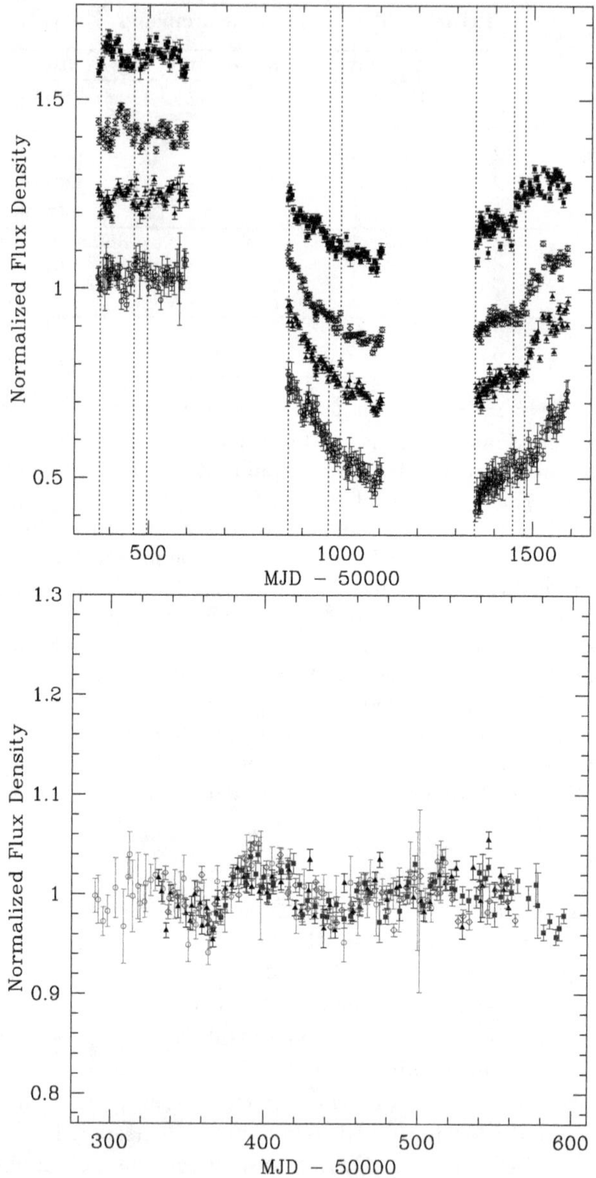

Fig. 33. VLA monitoring data for the four-image lens B1608+656. The top panel shows (from top to bottom) the normalized light curves for the B (*filled squares*), A (*open diamonds*), C (*filled triangles*) and D (*open circles*) images as a function of the Modified Julian Day (MJD). The bottom panel shows the composite light curve for the first monitoring season after cross correlating the light curves to determine the time delays ($\Delta t_{AB} = 31.5 \pm 1.5$, $\Delta t_{CB} = 36.0 \pm 1.5$ and $\Delta t_{DB} = 77.0 \pm 1.5$ days) and the flux ratios (from Fassnacht et al. 2002)

accurately predict when flux variations will appear in the other images and the lens will need to be monitored more intensively.

The expression for the time delay in an SIS lens (94) reveals the other data that are necessary to interpret time delays. First, the source and lens redshifts are needed to compute the distance factors that set the scale of the time delays. Fortunately, we know both redshifts for all 11 systems in Table 1 even though missing redshifts are a problem for the lens sample as a whole (see Sect. 2). The dependence of the distances D_d, D_s and D_{ds} on the cosmological model is unimportant until our total uncertainties approach 5%. Second, we require accurate relative positions for the images and the lens galaxy. These uncertainties are always dominated by the position of the lens galaxy relative to the images. For most of the lenses in Table 1, observations with radio interferometers (VLA, Merlin, VLBA) and HST have measured the relative positions of the images and lenses to accuracies $\lesssim 0\rlap{.}''005$. Sufficiently deep HST images can obtain the necessary data for almost any lens, but dust in the lens galaxy (as seen in B1600+434 and B1608+656) can limit the accuracy of the measurement even in a very deep image. For B0218+357 and PKS1830–211, however, the position of the lens galaxy relative to the images is not known to sufficient precision or determined only from models (see Biggs et al. 1999; Lehár et al. 2000; Courbin et al. 2002a,b; Winn et al. 2002a,b,c; Wucknitz, Biggs and Browne 2004; York et al. 2005).

We can also divide the systems by the complexity of the required lens model. We define eight of the lenses as "simple," in the sense that the available data suggests that a model consisting of a single primary lens in a perturbing shear (tidal gravity) field should be an adequate representation of the gravitational potential. In some of these cases, an external potential representing a nearby galaxy or parent group will improve the fits, but the differences between the tidal model and the more complicated perturbing potential are small (see Sect. 5.2). We include the quotation marks because the classification is based on an impression of the systems from the available data and models. While we cannot guarantee that a system is simple, we can easily recognize two complications that will require more complex models.

The first complication is that some primary lenses have less massive satellite galaxies inside or near their Einstein rings. This includes two of the time delay lenses, RXJ0911+0551 and B1608+656. RXJ0911+0551 could simply be a projection effect, since neither lens galaxy shows irregular isophotes. Here the implication for models may simply be the need to include all the parameters (mass, position, ellipticity, ...) required to describe the second lens galaxy, and with more parameters we would expect greater uncertainties in H_0. In B1608+656, however, the lens galaxies show the heavily disturbed isophotes typical of galaxies undergoing a disruptive interaction. How one should model such a system is unclear. If there was once dark matter associated with each of the galaxies, how is it distributed now? Is it still associated with the individual galaxies? Has it settled into an equilibrium configuration?

While B1608+656 can be well fit with standard lens models (Fassnacht et al. 2002; Koopmans et al. 2003), these complications have yet to be explored in detail.

The second complication occurs when the primary lens is a member of a more massive (X-ray) cluster, as in the time delay lenses RXJ0911+0551 (Morgan et al. 2001) and Q0957+561 (Chartas et al. 2002). The cluster model is critical to interpreting these systems (see Sect. 5.2). The cluster surface density at the position of the lens ($\kappa_c \gtrsim 0.2$) leads to large corrections to the time delay estimates and the higher-order perturbations are crucial to obtaining a good model. For example, models in which the Q0957+561 cluster was treated simply as an external shear were grossly incorrect (see Sect. 4.6, Keeton et al. 2000a). In addition to the uncertainties in the cluster model itself, we must also decide how to include and model the other cluster galaxies near the primary lens. Thus, lenses in clusters have many reasonable degrees of freedom beyond those of the "simple" lenses.

5.4 Results: The Hubble Constant and Dark Matter

With our understanding of the theory and observations of the lenses we will now explore their implications for H_0. We focus on the "simple" lenses PG1115+080, SBS1520+530, B1600+434, and HE2149–2745. We only comment on the interpretation of the HE1104–1805 delay because the measurement is too recent to have been interpreted carefully. We will briefly discuss the more complicated systems B0218+357, RXJ0911+0551, Q0957+561, and B1608+656, and we will not discuss the systems with problematic time delays or astrometry.

The most common, simple, realistic model of a lens consists of a singular isothermal ellipsoid (SIE) in an external (tidal) shear field (see Sect. 4). The model has 7 parameters (the lens position, mass, ellipticity, major axis orientation for the SIE, and the shear amplitude and orientation). It has many degrees of freedom associated with the angular structure of the potential, but the radial structure is fixed with $\langle\kappa\rangle \simeq 1/2$. For comparison, a two-image (four-image) lens supplies 5 (13) constraints on any model of the potential: 2 (6) from the relative positions of the images, 1 (3) from the flux ratios of the images, 0 (2) from the inter-image time delay ratios, and 2 from the lens position. With the addition of extra components (satellites/clusters) for the more complex lenses, this basic model provides a good fit to all the time delay lenses except Q0957+561. Although a naive counting of the degrees of freedom ($N_{dof} = -2$ and 6, respectively) suggests that estimates of H_0 would be under constrained for two-image lenses and over constrained for four-image lenses, the uncertainties are actually dominated by those of the time delay measurements and the astrometry in both cases. This is what we expect from Sect. 5.1 — the model has no degrees of freedom that change $\langle\kappa\rangle$ or η, so there will be little contribution to the uncertainties in H_0 from the model for the potential.

If we use a model that includes parameters to control the radial density profile (i.e., $\langle\kappa\rangle$), for example by adding a halo truncation radius a to the SIS profile (the pseudo-Jaffe model, $\rho \propto r^{-2}(r^2 + a^2)^{-1}$; e.g., Impey et al. 1998; Burud et al. 2002a)[4], then we find the expected correlation between a and H_0 — as we make the halo more concentrated (smaller a), the estimate of H_0 rises from the value for the SIS profile ($\langle\kappa\rangle = 1/2$ as $a \to \infty$) to the value for a point mass ($\langle\kappa\rangle = 0$ as $a \to 0$), with the fastest changes occurring when a is similar to the Einstein radius of the lens. We show an example of such a model for PG1115+080 in Fig. 34. This case is somewhat more complicated than a pure pseudo-Jaffe model because there is an additional contribution to the surface density from the group to which the lens galaxy belongs. As long as the structure of the radial density profile is fixed (constant a), the uncertainties are again dominated by the uncertainties in the time delay. Unfortunately, the goodness of fit, $\chi^2(a)$, shows too little dependence on a to determine H_0 uniquely. In general, two-image lenses have too few constraints, and the extra constraints supplied by a four-image lens constrain the angular structure rather than the radial structure of the potential. This basic problem holds for all existing models of the current sample of time delay lenses.

The inability of the present time delay lenses to directly constrain the radial density profile is the major problem for using them to determine H_0. Fortunately, it is a consequence of the available data on the current sample rather than a fundamental limitation. It is, however, a simple trade-off – models with less dark matter (lower $\langle\kappa\rangle$, more centrally concentrated densities) produce higher Hubble constants than those with more dark matter. We do have some theoretical limits on the value of $\langle\kappa\rangle$. In particular, we can be confident that the surface density is bounded by two limiting models. The mass distribution should not be more compact than the luminosity distribution, so a constant mass-to-light ratio (M/L) model should set a lower limit on $\langle\kappa\rangle \gtrsim \langle\kappa\rangle_{M/L} \simeq 0.2$, and an upper limit on estimates of H_0. We are also confident that the typical lens should not have a rising rotation curve at 1–2 optical effective radii from the center of the lens galaxy. Thus, the SIS model is probably the least concentrated reasonable model, setting an upper bound on $\langle\kappa\rangle \lesssim \langle\kappa\rangle_{SIS} = 1/2$, and a lower limit on estimates of H_0. Figure 35 shows joint estimates of H_0 from the four simple lenses for these two limiting mass distributions (Kochanek 2003b). The results for the individual lenses are mutually consistent and are unchanged by the new 0.149 ± 0.004 day delay for the A_1–A_2 images in PG1115+080 (Chartas et al. 2004). For galaxies with isothermal profiles we find $H_0 = 48 \pm 3$ km s^{-1} Mpc^{-1}, and for galaxies with constant M/L we find $H_0 = 71 \pm 3$ km s^{-1} Mpc^{-1}. While our best prior estimate for the mass distribution is the isothermal profile (see Sect. 4.6), the lens galaxies would have to have constant M/L to match Key Project estimate

[4] This is simply an example. The same behavior would be seen for any other parametric model in which the radial density profile can be adjusted.

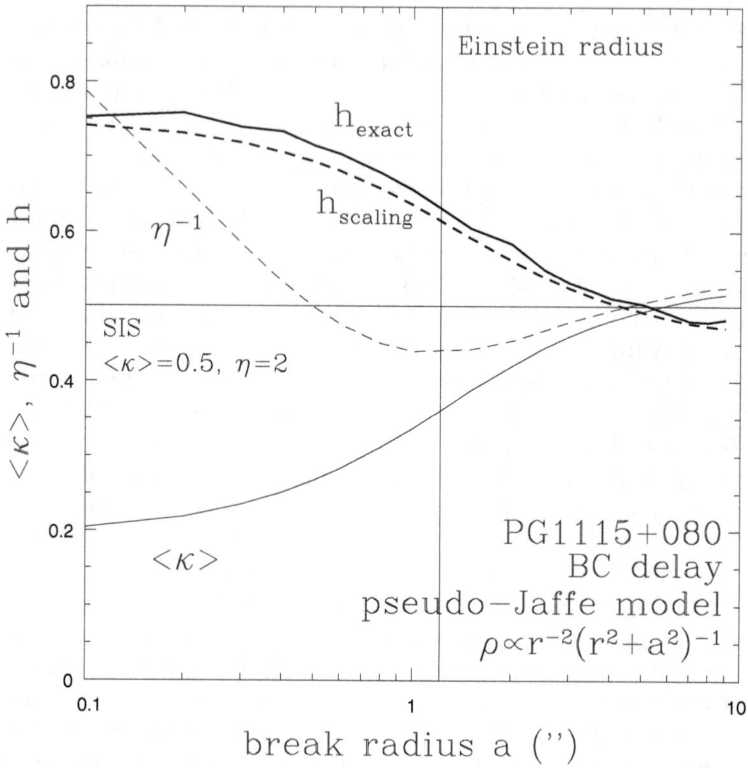

Fig. 34. H_0 estimates for PG1115+080. The lens galaxy is modeled as an ellipsoidal pseudo-Jaffe model, $\rho \propto r^{-2}(r^2+a^2)^{-1}$, and the nearby group is modeled as an SIS. As the break radius $a \rightarrow \infty$ the pseudo-Jaffe model becomes an SIS model, and as the break radius $a \rightarrow 0$ it becomes a point mass. The *heavy solid curve* (h_{exact}) shows the dependence of H_0 on the break radius for the exact, nonlinear fits of the model to the PG1115+080 data. The *heavy dashed curve* ($h_{scaling}$) is the value found using our simple theory (Sect. 5.1) of time delays. The agreement of the exact and scaling solutions is typical. The *light solid line* shows the average surface density $\langle\kappa\rangle$ in the annulus between the images, and the *light dashed line* shows the *inverse* of the logarithmic slope η in the annulus ($\kappa \propto \theta^{1-\eta}$). For an SIS model we would have $\langle\kappa\rangle = 1/2$ and $\eta^{-1} = 1/2$, as shown by the *horizontal line.* When the break radius is large compared to the Einstein radius (indicated by the *vertical line*), the surface density is slightly higher and the slope is slightly shallower than for the SIS model because of the added surface density from the group. As we make the lens galaxy more compact by reducing the break radius, the surface density decreases and the slope becomes steeper, leading to a rise in H_0. As the galaxy becomes very compact, the surface density near the Einstein ring is dominated by the group rather than the galaxy, so the surface density approaches a constant and the logarithmic slope approaches the value corresponding to a constant density sheet ($\eta = 1$)

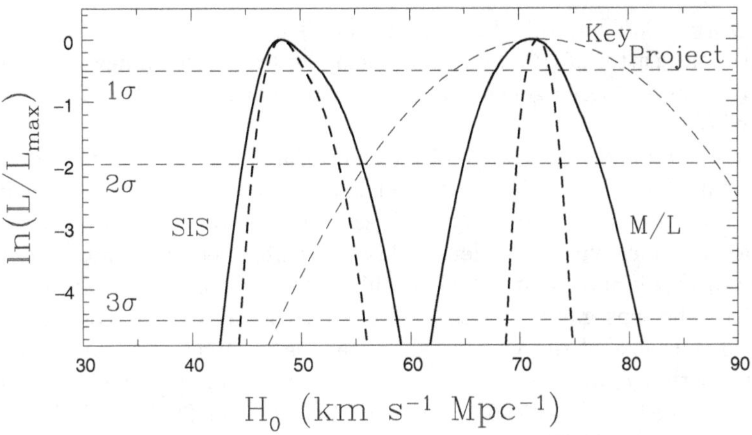

Fig. 35. H_0 likelihood distributions. The curves show the joint likelihood functions for H_0 using the four simple lenses PG1115+080, SBS1520+530, B1600+434, and HE2149–2745 and assuming either an SIS model (high $\langle\kappa\rangle$, *flat rotation curve*) or a constant M/L model (low $\langle\kappa\rangle$, *declining rotation curve*). The *heavy dashed curves* show the consequence of including the X-ray time delay for PG1115+080 from Chartas et al. (2004) in the models. The *light dashed curve* shows a Gaussian model for the Key Project result that $H_0 = 72 \pm 8$ km s^{-1} Mpc^{-1}

of $H_0 = 72 \pm 8$ km s^{-1} Mpc^{-1} (Freedman et al. 2001) or the WMAP estimate of $H_0 = 72 \pm 5$ km s^{-1} Mpc^{-1} for a flat universe with a cosmological constant (Spergel et al. 2003).

The difference between these two limits is entirely explained by the differences in $\langle\kappa\rangle$ and η between the SIS and constant M/L models. In fact, it is possible to reduce the H_0 estimates for each simple lens to an approximation formula, $H_0 = A(1 - \langle\kappa\rangle) + B\langle\kappa\rangle(\eta - 1)$. The coefficients, A and $|B| \approx A/10$, are derived from the image positions and the time delay using the simple theory from Sect. 5.1. These approximations reproduce numerical results using ellipsoidal lens models to accuracies of 3 km s^{-1} Mpc^{-1} (Kochanek 2002a,b). For example, in Fig. 34 we also show the estimate of H_0 computed based on the simple theory of Sect. 5.1 and the annular surface density ($\langle\kappa\rangle$) and slope (η) of the numerical models. The agreement with the full numerical solutions is excellent, even though the numerical models include both the ellipsoidal lens galaxy and a group. No matter what the mass distribution is, the five lenses PG1115+080, SBS1520+530, B1600+434, PKS1830–211,[5] and HE2149–2745 have very similar dark matter halos. For a fixed slope η, the five systems are consistent with a common value for the surface density of

$$\langle\kappa\rangle = 1 - 1.07h + 0.14(\eta - 1)(1 - h) \pm 0.04, \qquad (99)$$

[5] PKS1830–211 is included based on the Winn et al. (2002a,b,c) model of the *HST* imaging data as a single lens galaxy. Courbin et al. (2002a,b) prefer an interpretation with multiple lens galaxies which would invalidate the analysis.

where $H_0 = 100h$ km s^{-1} Mpc^{-1} and there is an upper limit of $\sigma_\kappa \lesssim 0.07$ on the intrinsic scatter of $\langle\kappa\rangle$. Thus, time delay lenses provide a new window into the structure and homogeneity of dark matter halos, regardless of the actual value of H_0.

There is an enormous range of parametric models that can illustrate how the extent of the halo affects $\langle\kappa\rangle$ and hence H_0 — the pseudo-Jaffe model we used above is only one example. It is useful, however, to use a physically motivated model where the lens galaxy is embedded in a standard NFW (Navarro, Frenk and White 1996) profile halo as we discussed at the end of Sect. 4.1. The lens galaxy consists of the baryons that have cooled to form stars, so the mass of the visible galaxy can be parameterized using the cold baryon fraction $f_{b,cold}$ of the halo, and for these CDM halo models the value of $\langle\kappa\rangle$ is controlled by the cold baryon fraction (Kochanek 2003a,b,c). A constant M/L model is the limit $f_{b,cold} \to 1$ (with $\langle\kappa\rangle \simeq 0.2$, $\eta \simeq 3$). Since the baryonic mass fraction of a CDM halo should not exceed the global fraction of $f_b \simeq 0.17 \pm 0.03$ (e.g., Spergel et al. 2003), we cannot use constant M/L models without also abandoning CDM. As we reduce $f_{b,cold}$, we are adding mass to an extended halo around the lens, leading to an increase in $\langle\kappa\rangle$ and a decrease in η. For $f_{b,cold} \simeq 0.02$ the model closely resembles the SIS model ($\langle\kappa\rangle \simeq 1/2$, $\eta \simeq 2$). If we reduce $f_{b,cold}$ further, the mass distribution begins to approach that of the NFW halo without any cold baryons. Figure 36 shows how $\langle\kappa\rangle$ and H_0 depend on $f_{b,cold}$ for PG1115+080, SBS1520+530, B1600+434 and HE2149–2745. When $f_{b,cold} \simeq 0.02$, the CDM models have parameters very similar to the SIS model, and we obtain a very similar estimate of $H_0 = 52 \pm 6$ km s^{-1} Mpc^{-1} (95% confidence). If all baryons cool, and $f_{b,cold} = f_b$, then we obtain $H_0 = 65 \pm 6$ km s^{-1} Mpc^{-1} (95% confidence), which is still lower than the Key Project estimates.

We excluded the lenses requiring significantly more complicated models with multiple lens galaxies or very strong perturbations from clusters. If we have yet to reach a consensus on the mass distribution of relatively isolated lenses, it seems premature to extend the discussion to still more complicated systems. We can, however, show that the clusters lenses require significant contributions to $\langle\kappa\rangle$ from the cluster in order to produce the same H_0 as the more isolated systems. As we discussed in Sect. 2 the three more complex systems are RXJ0911+0551, Q0957+561 and B1608+656.

RXJ0911+0551 is very strongly perturbed by the nearby X-ray cluster (Morgan et al. 2001; Hjorth et al. 2002). Kochanek (2003a,b,c) found $H_0 = 49 \pm 5$ km s^{-1} Mpc^{-1} if the primary lens and its satellite were isothermal and $H_0 = 67 \pm 5$ km s^{-1} Mpc^{-1} if they had constant mass-to-light ratios. The higher value of $H_0 = 71 \pm 4$ km s^{-1} Mpc^{-1} obtained by Hjorth et al. (2002) can be understood by combining Sect. 5.1 and Sect. 5.2 with the differences in the models. In particular, Hjorth et al. (2002) truncated the halo of the primary lens near the Einstein radius and used a lower mass cluster, both of which lower $\langle\kappa\rangle$ and raise H_0. The Hjorth et al. (2002) models also included many more cluster galaxies assuming fixed masses and halo sizes.

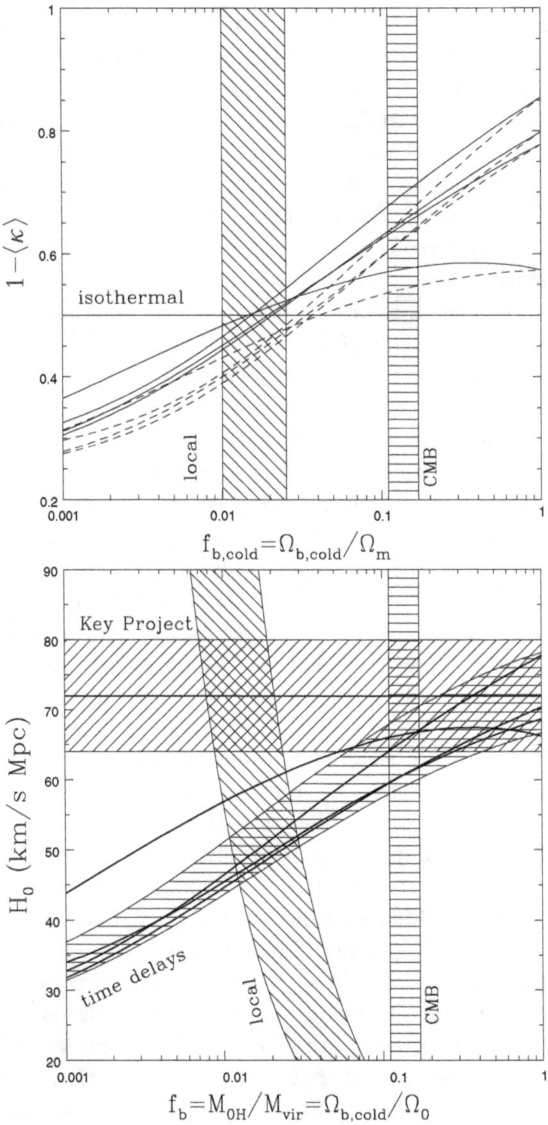

Fig. 36. H_0 in CDM halo models. The top panel shows $1 - \langle\kappa\rangle$ for the "simple" lenses (PG1115+080, SBS1520+530, B1600+434, and HE2149–2745) as a function of the cold baryon fraction $f_{b,cold}$. The solid (*dashed*) curves include (*exclude*) the adiabatic compression of the dark matter by the baryons. The *horizontal line* shows the value for an SIS potential. The *bottom panel* shows the resulting estimates of H_0, where the shaded envelope bracketing the curves is the 95% confidence region for the combined lens sample. The horizontal band shows the Key Project estimate. For larger $f_{b,cold}$, the density $\langle\kappa\rangle$ decreases and the local slope η steepens, leading to larger values of H_0. The *vertical bands* in the two panels show the lower bound on f_b from local inventories and the upper bound from the CMB

Q0957+561 is a special case because the primary lens galaxy is the brightest cluster galaxy and it lies nearly at the cluster center (Keeton et al. 2000a; Chartas et al. 2002). As a result, the lens modeling problems are particularly severe, and Keeton et al. (2000a,b) found that all previous models (most recently, Barkana et al. 1999; Bernstein and Fischer 1999; and Chae 1999, see Sect. 4.6) were incompatible with the observed geometry of the lensed host galaxy. While Keeton et al. (2000a) found models consistent with the structure of the lensed host, they covered a range of almost ±25% in their estimates of H_0. A satisfactory treatment of this lens remains elusive.

HE1104–1805 has the most recently measured time delay (Ofek and Maoz 2003; Wyrzykowski et al. 2003). Given the $\Delta t = 161 \pm 7$ day delay, a standard SIE model of this system predicts a very high $H_0 \simeq 90$ km s^{-1} Mpc^{-1}. The geometry of this system and the fact that the inner image is brighter than the outer image both suggest that HE1104–1805 lies in an anomalously high tidal shear field, while the standard model includes a prior to keep the external shear small. A prior is needed because a two-image lens supplies too few constraints to determine both the ellipticity of the main lens and the external shear simultaneously. Since the images and the lens in HE1104–1805 are nearly collinear, the anomalous H_0 estimate for the standard model may be an example of the shear degeneracy we briefly mentioned in Sect. 5.1. At present the model surveys needed to understand the new delay have not been made. Observations of the geometry of the host galaxy Einstein ring will resolve any ambiguities due to the shear in the near future (see Sect. 10).

The lens B1608+656 consists of two interacting galaxies, and, as we discussed in Sect. 2, this leads to a greatly increased parameter space. Fassnacht et al. 2002 used SIE models for the two galaxies to find $H_0 = 61 - 65$ km s^{-1} Mpc^{-1}, depending on whether the lens galaxy positions are taken from the H-band or I-band lens *HST* images (the statistical errors are negligible). The position differences are probably created by extinction effects from the dust in the lens galaxies. Like isothermal models of the "simple" lenses, the H_0 estimate is below local values, but the disagreement is smaller. These models correctly match the observed time delay ratios. Koopmans et al. (2003) obtain a still higher estimate of $H_0 = 75 \pm 7$ km s^{-1} Mpc^{-1} because the lens galaxy positions shift after they include extinction corrections. They use a foreground screen model to make the extinction corrections, which is a better approximation than no extinction corrections, but is unlikely to allow precise correction in a system like B1606+656 where the dust and stars are mixed and there is no simple relation between color excess and optical depth (e.g. Witt, Thronson and Capuano 1992).

Despite recent progress both in modeling the VLBI structure (Wucknitz et al. 2004) and obtaining deep images (York et al. 2005) it is unclear whether B0218+357 has escaped its problems with astrometry and models. While York et al. (2005) have clearly measured the position of the lens galaxy, the dependence of the position on the choice of the PSF model remains a significant source of uncertainty for estimates of H_0. Models of the system using power law models find a slope very close to isothermal $\eta = 2.04 \pm 0.02$ ($\rho \propto r^{-\eta}$). Unfortunately, these models have too few degrees of freedom given the small astrometric uncertainties in the VLBI structures providing the constraints (because the only angular structure in the model is the ellipsoidal potential used for the main lens galaxy), and this makes the limits on the power slope suspect (see Sect. 4.6). For example, while it is true that Lehár et al. (2000) estimated that the environmental shear near B0218+357 was small, even a $\gamma = 0.01$ external tidal shear produces deflections (3 milli-arcseconds) that are large compared to the accuracy of the constraints used for the models and so must be included for the models to be reliable. Within these caveats, B0218+357 (like the models of B1608+656 with significant extinction corrections) supports a nearly isothermal mass distribution with $H_0 = 73 \pm 8$ km s^{-1} Mpc^{-1}.

5.5 The Future of Time Delay Measurements

We understand the theory of time delays very well – the only important variable in the lens structure is the average surface density $\langle\kappa\rangle$ of the lens near the images for which the delay is measured. The angular structure of the potential has an effect on the delays, but it is either small or well-constrained by the observed image positions. Provided a lens does not lie in a cluster where the cluster potential cannot be described by a simple expansion, any lens model that includes the parameters needed to vary the average surface density of the lens near the images and to change the ratio between the quadrupole moment of the lens and the environment includes all the variables needed to model time delays, to estimate the Hubble constant, and to understand the systematic uncertainties in the results. Unfortunately, there is a tendency in the literature to confuse rather than to illuminate this understanding, even though all differences between estimates of the Hubble constant for the simple time delay lenses can be understood on this basis.

The problem with time delays lies with the confusing state of the data. The four simplest time delay lenses, PG1115+080, SBS1520+530, B1600+434 and HE2149–2745, can only match the currently preferred estimate of $H_0 \simeq 72 \pm 8$ km s^{-1} Mpc^{-1} (Freedman et al. 2001; Spergel et al. 2003) if they have nearly constant M/L mass distributions. If they have the favored quasi-isothermal mass distributions, then $H_0 \simeq 48 \pm 3$ km s^{-1} Mpc^{-1}. This leads to a conundrum: why do simple lenses with time delay measurements have falling rotation curves, while simple lenses with direct estimates of the mass profile do not? This is further confused by B1608+656 and B0218+357, which due to their observational complexity would be the last systems I would attempt to understand, but in current analyses can be both isothermal and have high H_0. In resolving this problem it is not enough to search for a "Golden Lens." *There is no such thing as a "Golden Lens"*. Chanting "my lens is better than your lens" may be satisfying but contributes little to understanding the basic problem.

The difficulty at the moment is that systems I would view as problematic (B0218+357 due to problems in astrometry or B1608+656 due to the interacting lens galaxies) allow both mass distributions with flat rotation curves and $H_0 = 72$ km s^{-1} Mpc^{-1}, while systems that should be simpler to interpret (the simple lenses in Table 1) do not. Yet the preponderance of evidence on the mass distributions of lens galaxies suggests that they are fairly homogeneous in structure and have roughly flat rotation curves (Sect. 4). The simplest way to clarify this problem is to measure accurate time delays for many more systems. At a fixed value of the Hubble constant we will either find significant scatter in the surface densities near the images or we will not.

6 Gravitational Lens Statistics

It is the opinion of the author that the statistics of lenses as a method for determining the cosmological model has largely ceased to be interesting. However, it is important to understand the underlying physics because it determines the types of lenses we detect. While most recent analyses have found cosmological results consistent with the concordance model (Chae et al. 2002; Chae 2003; Davis, Huterer and Krauss 2003, Mitchell et al. 2004) there are still large statistical uncertainties and some dangerous systematic assumptions. More importantly, there is little prospect at present of lens statistics becoming competitive with other methods. Gravitational lenses statistics arguably begins with Press and Gunn (1973), although the "modern" era begins with the introduction of magnification bias (Turner 1980), the basic statistics of normal galaxy lenses (Turner, Ostriker and Gott 1984), cross sections and optical depths for more general lenses (Blandford and Kochanek 1987a,b; Kochanek and Blandford 1987), explorations of the effects of general cosmologies (Fukugita et al. 1990; Fukugita and Turner 1991) and lens structure (Maoz and Rix 1993; Kochanek 1996a,b) and the development of the general methodology of interpreting observations (Kochanek 1993a,b,c, 1996a,b).

6.1 The Mechanics of Surveys

There are two basic approaches to searching for gravitational lenses depending on whether you start with a list of potentially lensed sources or a list of potential lens galaxies. Of the two, only a search of sources for lensed sources has a significant cosmological sensitivity – for a non-evolving population of lenses in a flat cosmological model we will find in Sect. 6.3 that the number of lensed sources scales with the volume between the observer and the source D_s^3. If you search potential lens galaxies for those which have actually lensed a source, then the cosmological dependence enters only through distance ratios, D_{ds}/D_s, and you require a precise knowledge of the source redshift distribution. Thus, while lenses found in this manner are very useful for many projects (mass distributions, galaxy evolution etc.), they are not very useful for determining the cosmological model. This changes for the case of cluster lenses where you may find multiple lensed sources at different redshifts behind the same lens (e.g. Soucail, Kneib and Golse 2004).

Most lenses have been found by searching for lensed sources because the number of targets which must be surveyed is considerably smaller. This is basically a statement about the relative surface densities of candidate sources and lenses. The typical lens is a galaxy with an Einstein radius of approximately $b \simeq 1\overset{''}{.}0$ so it has a cross section of order πb^2. If you search N lenses with such a cross section for signs of a lensed source, you would expect to find $N\pi b^2 \Sigma_{source}$ lenses where Σ_{source} is the surface density of detectable sources. If you search N sources for a lens galaxy in front of them, you would expect to find $N\pi b^2 \Sigma_{lens}$ lenses, where Σ_{lens} is the surface density of lens galaxies. Since the surface density of massive galaxies is significantly higher than the surface density of easily detectable higher redshift sources ($\Sigma_{lens} \gg \Sigma_{source}$), you need examine fewer sources than lens galaxies to find the same number of lensed systems. This is somewhat mitigated by the fact that the surface density of potential lens galaxies is high enough to allow you to examine many potential lenses in a single observation, while the surface density of sources is usually so low that they can be examined only one at a time.

For these reasons, we present a short synopsis of searches for sources behind lenses and devote most of this section to the search for lenses in front of sources. The first method for finding sources behind lenses is a simple byproduct of redshift surveys. Redshift surveys take spectra of the central regions of low redshift galaxies allowing the detection of spectral features from any lensed images inside the aperture used for the spectrum. Thus, the lens Q2237+0305 was found in the CfA redshift survey (Huchra et al. 1985) and SDSS0903+5028 (Johnston et al. 2003) was found in the SDSS survey. Theoretical estimates (Kochanek 1992a,b; Mortlock and Webster 2000a,b,c) suggest that the discovery rate should be one per 10^4–10^5 redshift measurements, but this does not seem to be borne out by the number of systems discovered in this age of massive redshift surveys (the origin of the lower rate in the 2dF survey is discussed by Mortlock & Webster 2001). Miralda-Escude and Lehár (1992) proposed searching for lensed optical (emission line) rings, a strategy successfully used by Warren et al. (1996) to find 0047–2808 and by Ratnatunga, Griffiths and Ostrander (1999) to find lenses in the HST Medium Deep Survey (MDS). There is also a hybrid approach whose main objective is simply to find lenses with minimal follow up observations by looking for high redshift radio lobes that have non-stellar optical counterparts (Lehár et al. 2001). Since radio lobes have no intrinsic optical emission, a lobe superposed on a galaxy is an excellent lens candidate. The present limitation on this method is the low angular resolution of the available all sky radio surveys (FIRST, NVSS) and the magnitude limits and star/galaxy separation problems of the current all-sky optical catalogs. Nonetheless, several systems have been discovered by this technique.

The vast majority of lens surveys, however, have focused on either optical quasars or radio sources because they are source populations known to lie at relatively high redshift ($z_s \gtrsim 1$) and that are easily detected even when there is an intervening lens galaxy. Surveys of lensed optical quasars (Crampton, McClure and Fletcher 1992; Yee, Fillipenko and Tang 1993; Maoz et al. 1993; Surdej et al. 1993; Kochanek, Falco and Schild 1995) have the advantage that the sources are bright, and the disadvantages that the bright sources can mask the lens galaxy and that the selection process is modified by dust in the lens galaxy and emission from the lens galaxy. We will discuss these effects in Sect. 9. While many more lensed quasars have been discovered since these efforts, none of the recent results have been presented as a survey. Surveys of all radio sources (the MIT/Greenbank survey, Burke, Lehár and Conner 1992) have the advantage that most lensed radio sources are produced by extended steep spectrum sources (see Kochanek and Lawrence 1990) but the disadvantage that the complex intrinsic structures of extended radio sources make the follow up observations difficult. Surveys of flat spectrum radio sources (the CLASS survey, Browne et al. 2003, the PANELS survey, Winn, Hewitt & Schechter 2001) have the advantage that the follow up observations are relatively simple because most unlensed flat spectrum sources are (nearly) point sources. There are disadvantages as well – because the source structure is so simple, flat spectrum lenses tend to provide fewer constraints on mass models than steep spectrum lenses. The radio sources tend to be optically faint, making it difficult to determine their redshifts in many cases.

The second issue for any survey is to understand the method by which the sources were originally identified. For example, it is important to know whether the source flux of a lens in the input catalog will be the total flux of all the images or only a part of the flux (e.g. the flux of the brightest image). This will have a significant effect on the statistical corrections for using a flux-limited catalog, a correction known in gravitational lensing as the "magnification bias" (see Sect. 6.6). All large, published surveys were essentially drawn from samples which would include the total flux of a lensed system. It is also important to know whether the survey imposed any criterion for the sources being point-like, since lensed sources are not, or any color criterion that might be violated by lensed sources with bright lens galaxies or significant extinction.

The third issue for any survey is to consider the desired selection function of the observations. This is some combination of resolution, dynamic range and field of view. These determine the range of lens separations that are detectable, the nature of any background sources, and the cost of any follow up observations. Any survey is a trade-off between completeness (what fraction of all lenses in sample that can be discovered), false positives (how many objects selected as lenses candidates that are not), and the cost of follow-up observations. The exact strategy is not critical provided it is well-understood. The primary advantages of the surveys of flat spectrum radio sources are the relatively low false positive rates and follow up costs produced by using a source population consisting almost entirely of point sources with no contaminating background population. This does not mean that the flat spectrum surveys are free of false positives – core-jet sources can initially look like asymmetric two-image lenses. On small angular scales ($\Delta\theta \lesssim 3''.0$) the quasar surveys share this advantage, but for wider separations there is contamination from binary quasars (see Sect. 7.2, Kochanek, Falco and Muñoz 1999; Mortlock, Webster and Francis 1999) and Galactic stars (see Kochanek 1993a,b; Kochanek 1993c).

6.2 The Lens Population

The probability that a source has an intervening lens requires a model for the distribution of the lens galaxies. In almost all cases these are based on the luminosity function of local galaxies combined with the assumption that the comoving density of galaxies does not evolve with redshift. Of course luminosity is not mass, so a model for converting the luminosity of a local galaxy into its deflection scale as a lens is a critical part of the process. For our purposes, the distributions of galaxies in luminosity are well-described by a Schechter (1976) function,

$$\frac{dn}{dL} = \frac{n_*}{L_*}\left(\frac{L}{L_*}\right)^{\alpha} \exp\left(-L/L_*\right). \tag{100}$$

The Schechter function has three parameters: a characteristic luminosity L_* (or absolute magnitude M_*), an exponent α describing the rise at low luminosity, and a comoving density scale n_*. All these parameters depend on the type of galaxy being described and the wavelength of the observations. In general, lens calculations have divided the galaxy population into two broad classes: late-type (spiral) galaxies and early-type (E/S0) galaxies. Over the period lens statistics developed, most luminosity functions were measured in the blue, where early and late-type galaxies showed similar characteristic luminosities. The definition of a galaxy type is a slippery problem – it may be defined by the morphology of the surface brightness (the traditional method), spectral classifications (the modern method since it is easy to do in redshift surveys), colors (closely related to spectra but not identical), and stellar kinematics (ordered rotational motions versus random motions). Each approach has advantages and disadvantages, but it is important to realize that the kinematic definition is the one most closely related to gravitational lensing and the one never supplied by local surveys. Figure 37 shows an example of a luminosity function, in this case K-band infrared luminosity function by Kochanek et al. (2001a,b, also Cole et al. 2001) where $M_{K*e} = -23.53 \pm 0.06$ mag, $n_{*e} = (0.45 \pm 0.06) \times 10^{-2} h^3$ Mpc^{-3}, and $\alpha_e = -0.87 \pm 0.09$ for galaxies which were morphologically early-type galaxies and $M_{K*l} = -22.98 \pm 0.06$ mag, $n_{*l} = (1.01 \pm 0.13) \times 10^{-2} h^3$ Mpc^{-3}, and $\alpha_l = -0.92 \pm 0.10$ for galaxies which were morphologically late-type galaxies. Early-type galaxies are less

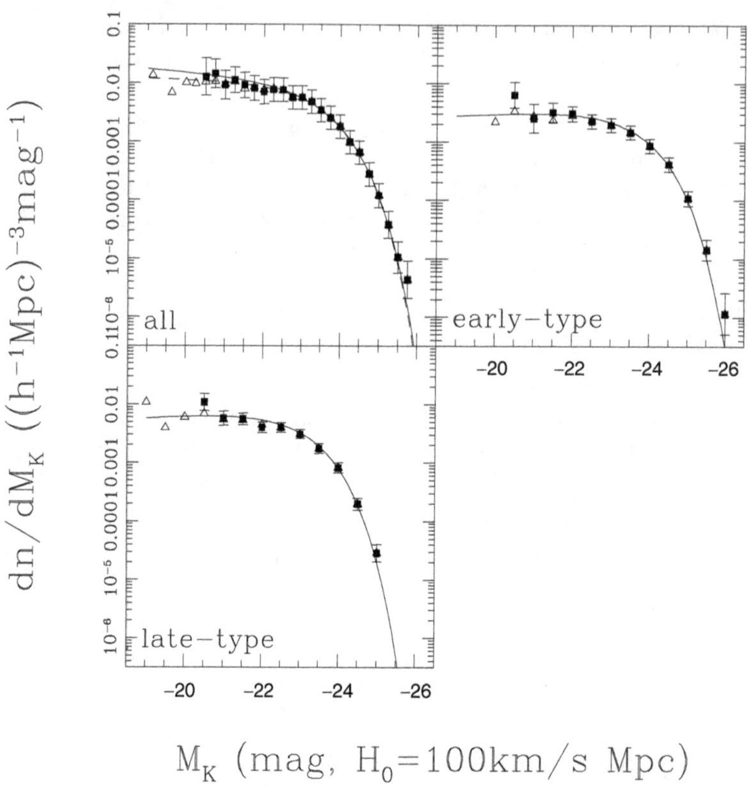

Fig. 37. Example of a local galaxy luminosity function. These are the K-band luminosity functions for either all galaxies or by morphological type from Kochanek et al. (2001a,b). The*curves* show the best fit Schechter models for the luminosity functions while the points with error bars show a non-parametric reconstruction

common but brighter than late-type galaxies at K-band. It is important to realize that the parameter estimates of the Schechter function are correlated, as shown in Fig. 38, and that it is dangerous to simply extrapolate them to fainter luminosities than were actually included in the survey.

However, light is not mass, and it is mass which determines lensing properties. One approach would simply be to assign a mass-to-light ratio to the galaxies and to the expected properties of the lenses. This was attempted only in Maoz and Rix (1993) who found that for normal stellar mass-to-light ratios it was impossible to reproduce the data (although it is possible if you adjust the mass-to-light ratio to fit the data, see Kochanek 1996a,b). Instead, most studies convert the luminosity functions dn/dL into a velocity functions dn/dv using the local kinematic properties of galaxies and then relate the stellar kinematics to the properties of the lens model. As Fig. 39 shows (for the same K-band magnitudes of our luminosity function example), both

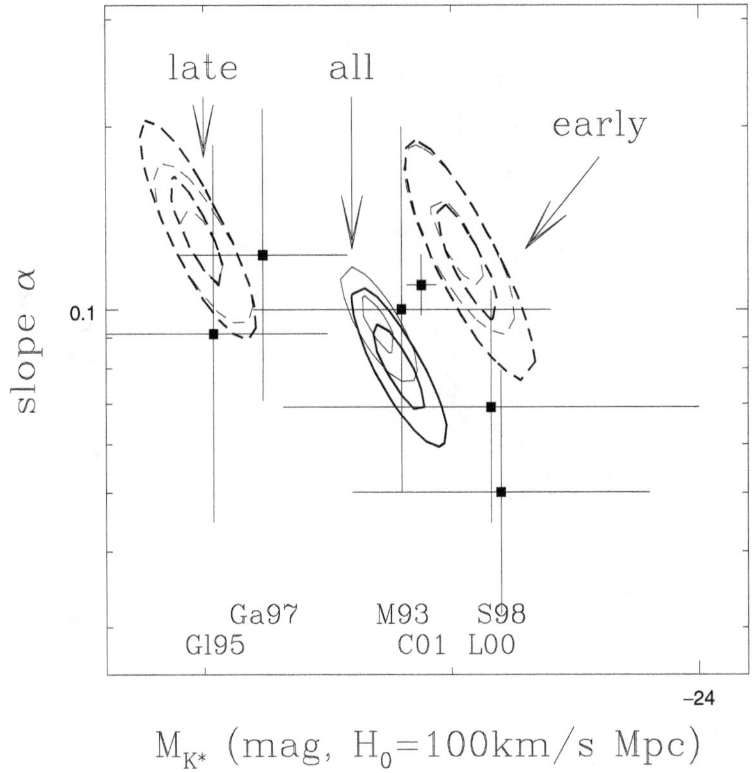

Fig. 38. Schechter parameters α and M_* for the 2MASS luminosity functions shown in Fig. 37. Note there is a significant correlations not only between α and M_* but also with the comoving density scale n_* that should be included in lens statistical calculations but generally are not

early-type and late-type galaxies show correlations between luminosity and velocity. For late-type galaxies there is a tight correlation known as the Tully–Fisher (1977) relation between luminosity L and circular velocity v_c and for early-type galaxies there is a loose correlation known as the Faber–Jackson (1976) relation between luminosity and central velocity dispersion σ_v. Early-type galaxies do show a much tighter correlation known as the fundamental plane (Dressler et al. 1987; Djorgovski and Davis 1987) but it is a three-variable correlation between the velocity dispersion, effective radius and surface brightness (or luminosity) that we will discuss in Sect. 9. While there is probably some effect of the FP correlation on lens statistics, it has yet to be found. For lens calculations, the circular velocity of late-type galaxies is usually converted into an equivalent (isotropic) velocity dispersion using $v_c = \sqrt{2}\sigma_v$. We can derive the kinematic relations for the same K-band-selected galaxies used in the Kochanek et al. (2001a,b) luminosity function,

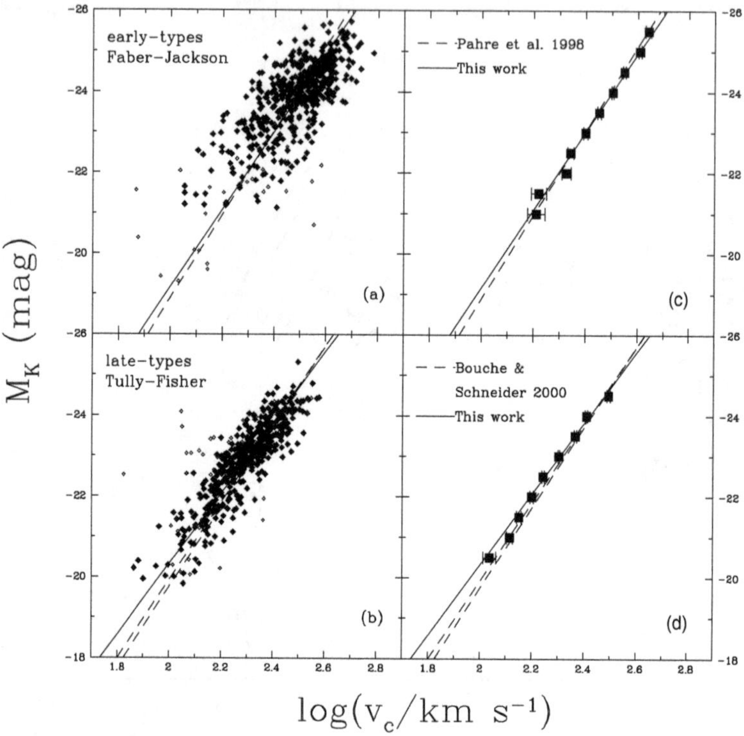

Fig. 39. K-band kinematic relations for 2MASS galaxies. The *top* panels show the Faber–Jackson relation and the *bottom* panels show the Tully–Fisher relations for 2MASS galaxies with velocity dispersions and circular velocities drawn from the literature. The left hand panels show the individual galaxies, while the right hand galaxies show the mean relations. Note the far larger scatter of the Faber–Jackson relation compared to the Tully–Fisher relation

finding the Faber–Jackson relation

$$M_k - 5\log h = (-23.83 \pm 0.03) - 2.5 \times (4.04 \pm 0.18)(\log v_c - 2.5) \qquad (101)$$

and the Tully–Fisher relation

$$M_k - 5\log h = (-22.92 \pm 0.02) - 2.5 \times (3.96 \pm 0.08)(\log v_c - 2.3). \qquad (102)$$

These correlations, when combined with the K-band luminosity function have the advantage that the magnitude systems for the luminosity function and the kinematic relations are identical, since magnitude conversions have caused problems for a number of lens statistical studies using older photographic luminosity functions and kinematic relations. For these relations, the characteristic velocity dispersion of an L_* early-type galaxy is $\sigma_{*e} \simeq 209$ km/s while that of an L_* late-type galaxy is $\sigma_{*l} \simeq 143$ km/s. These are fairly typical values even if derived from a completely independent set of photometric data.

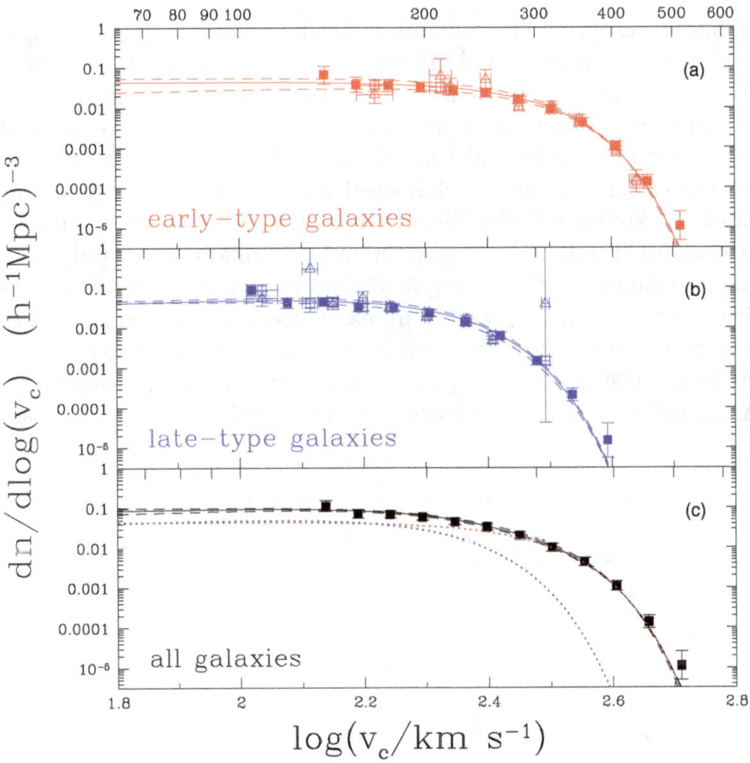

Fig. 40. The resulting velocity functions from combining the K-band luminosity functions (Fig. 37) and kinematic relations (Fig. 39) for early-type (*top*), late-type (*middle*) and all (*bottom*) galaxies. The points show partially non-parametric estimates of the velocity function based on the binned estimates in the right hand panels of Fig. 39 rather than power-law fits. Note that early-type galaxies dominate for high circular velocity

Both the Faber–Jackson and Tully–Fisher relations are power-law relations between luminosity and velocity, $L/L_* \propto (\sigma_v/\sigma_*)^{\gamma_{FJ}}$. This allows a simple variable transformation to convert the luminosity function into a velocity function,

$$\frac{dn}{dv} = \frac{dn}{dL}\left|\frac{dL}{dv}\right| = \gamma_{FJ}\frac{n_*}{\sigma_*}\left(\frac{\sigma_v}{\sigma_*}\right)^{(1+\alpha)\gamma_{FJ}-1} \exp\left(-(\sigma_v/\sigma_*)^{\gamma}_{FJ}\right) \tag{103}$$

as shown in Fig. 40.

There are three caveats to keep in mind about this variable change. First, we have converted to the distribution in stellar velocities, not some underlying velocity characterizing the dark matter distribution. Many early studies assumed a fixed transformation between the characteristic velocity of the stars and the lens model. In particular, Turner, Ostriker and Gott (1984) introduced the assumption $\sigma_{dark} = (3/2)^{1/2}\sigma_{stars}$ for an isothermal mass model based on the stellar dynamics (Jeans equation, (90) and Sect. 4.9) of a r^{-3} stellar density distribution in a r^{-2} isothermal mass distribution. Kochanek (1993a,b,c, 1994) showed that this oversimplified the dynamics and that if you embed a real stellar luminosity distribution in an isothermal mass distribution you actually find that the central stellar velocity dispersion is close to the velocity dispersion characterizing the dark matter halo. Figure 41 compares the stellar velocity dispersion to the dark matter halo dispersion for a Hernquist distribution of stars in an isothermal mass distribution. Such a normalization calculation is required in any calculation matching observed velocity functions with a particular mass model for the lenses. Second, in an ideal world, the luminosity function and the kinematic relations should be derived from a consistent set of photometric data, while in practice they rarely are. As we will see shortly, the cross section for lensing scales roughly as σ_*^4, so

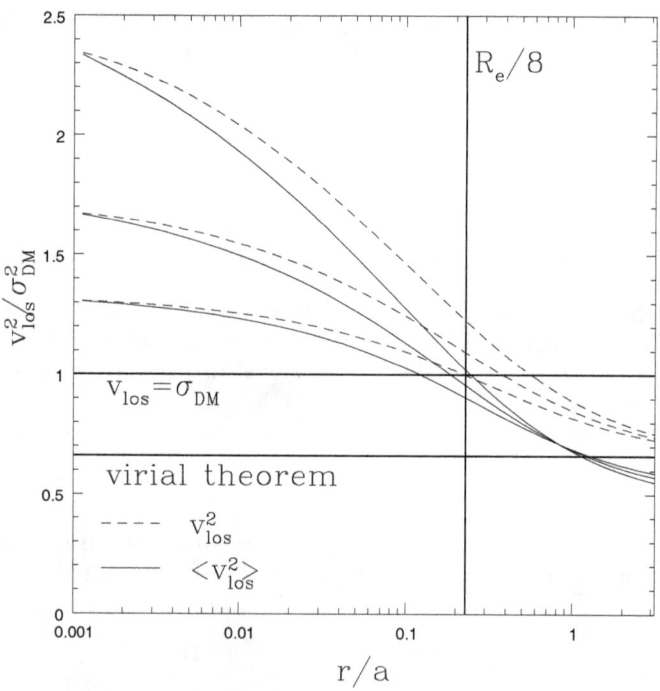

Fig. 41. Stellar velocity dispersions v_{los} for a Hernquist distribution of stars in an isothermal halo of dispersion σ_{DM}. The *solid curves* show the local value v_{los} and the *dashed curves* show the mean interior to the radius $\langle v^2_{los}\rangle$. Local velocity dispersions are typically measured on scales similar to $R_e/8$ where the stellar and dark matter dispersions are nearly equal rather than matching the viral theorem limit which would be reached in an infinite aperture. The upper, lower and middle curves are for stars with isotropies of $\beta = 0.2$ (somewhat radial), $\beta = 0$ (isotropic) and $\beta = -0.2$ (somewhat tangential)

small errors in estimates of the characteristic velocity have enormous impacts on the resulting cosmological results – a 5% velocity calibration error leads to a 20% error in the lens cross section. Since luminosity functions and kinematic relations are rarely derived consistently (the exception is Sheth et al. 2003), the resulting systematic errors creep into cosmological estimates. Finally, for the early-type galaxies where the Faber–Jackson kinematic relation has significant scatter, transforming the luminosity function using the mean relation as we did in (103) while ignoring the scatter underestimates the number of high velocity dispersion galaxies (Kochanek 1994; Sheth et al. 2003). This leads to underestimates of both the image separations and the cross sections. The fundamental lesson of all these issues is that the mass scale of the lenses should be "self-calibrated" from the observed separation distribution of the lenses rather than imposed using local observations (as we discuss below in Sect. 6.7).

Most lens calculations have assumed that the comoving density of the lenses does not evolve with redshift. For moderate redshift sources this only requires little evolution for $z_l < 1$ (mostly $z_l < 0.5$), but for higher redshift sources it is important to think about evolution as well. The exact degree of evolution is the subject of some debate, but a standard theoretical prediction for the change between now and redshift unity is shown in Fig. 42 (see Mitchell

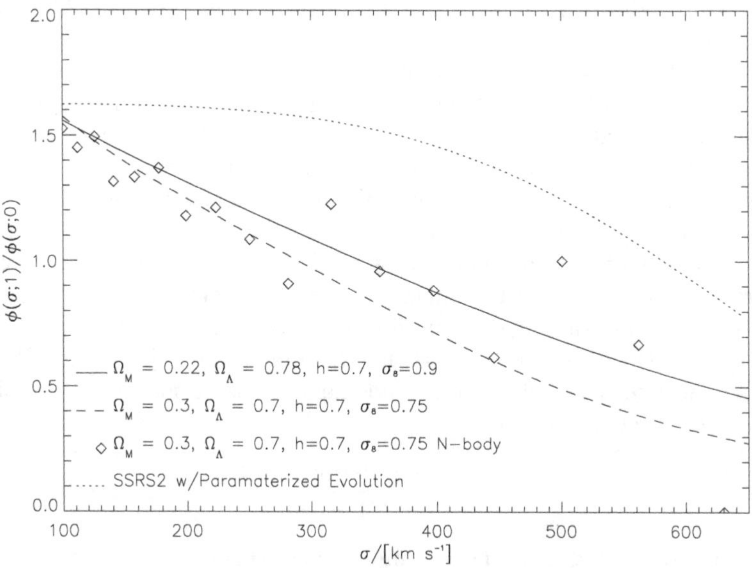

Fig. 42. The ratio of the velocity function of halos at $z = 1$ to that at $z = 0$ from Mitchell et al. (2004). The *solid curve* shows the expectation for an $\Omega_\Lambda \simeq 0.78$ flat cosmological model. The *points* show results from an N-body simulation with $\Omega_\Lambda \simeq 0.7$ and the *dashed curve* shows the theoretical expectation. For comparison, the *dotted curve* shows the evolution model used by Chae and Mao (2003)

et al. 2004 and references therein). Because lower mass systems merge to form higher mass systems as the universe evolves, low mass systems are expected to be more abundant at higher redshifts while higher mass systems become less abundant. For the $\sigma_v \sim \sigma_* \sim 200$ km/s galaxies which dominate lens statistics, the evolution in the number of galaxies is actually quite modest out to redshift unity, so we would expect galaxy evolution to have little effect on lens statistics. Higher mass systems evolve rapidly and are far less abundant at redshift unity, but these systems will tend to be group and cluster halos rather than galaxies and the failure of the baryons to cool in these systems is of greater importance to their lensing effects than their number evolution (see Sect. 7). There have been a number of studies examining lens statistics with number evolution (e.g. Mao 1991; Mao and Kochanek 1994; Rix et al. 1994) and several attempts to use the lens data to constrain the evolution (Ofek, Rix and Maoz 2003; Chae and Mao 2003; Davis, Huterer and Krauss 2003).

6.3 Cross Sections

The basic quantity we need for any statistical analysis is the cross section of the lens for producing the desired lensing effect (e.g. multiple images, two images, bright images...). The simplest cross section is the multiple imaging cross section of the SIS lens – the angular area on the source plane in which a source will produce two lensed images. We know from (21) and (22) that the source must lie within Einstein radius b of the lens center to produce multiple images, so the cross section is simply $\sigma_{SIS} = \pi b^2$. Since the Einstein radius $b = 4\pi(\sigma_v/c)^2 D_{ls}/D_s$ depends on the velocity dispersion and redshift of the lens galaxy, we will need a model for the distribution of lenses in redshift and velocity dispersion to estimate the optical depth for lensing. If we are normalizing directly to stellar dynamical measurements of lenses, then we will also need a dynamical model (e.g. the Jeans equations of Sect. 4.9) to relate the observed stellar velocity dispersions to the characteristic dark matter velocity dispersion σ_v appearing as a parameter of the SIS model. We can also compute cross sections for obtaining different image morphologies. For example, in (32) we calculated the caustic boundaries for the four-image region of an SIS in an external shear γ. If we integrate to find the area inside the caustic we obtain the four-image cross section

$$\sigma_4 = \frac{3\pi}{2} \frac{\gamma^2 b^2}{1 - \gamma^2}, \tag{104}$$

while (provided $|\gamma| < 1/3$) the two-image cross section is $\sigma_2 = \sigma_{SIS} - \sigma_4 \simeq \sigma_{SIS}$. If the shear is larger, then the tips of astroid caustic extend beyond the radial (pseudo-)caustic and the lens has regions producing two images, three images in the disk geometry (Fig. 18), and four images with no simple expression for the cross sections. There are no analytic results for the singular isothermal ellipsoid ((37) with $s = 0$), but we can power expand the cross section as a series in the ellipticity to find at lowest order that

$$\sigma_4 = \frac{\pi}{6} b^2 \epsilon^2 \tag{105}$$

for a lens with axis ratio $q = 1 - \epsilon$, while the total cross section is $\sigma_{SIS} = \pi b^2$ (e.g. Kochanek 1996a,b; Finch et al. 2002). As a general rule, a lens of ellipticity ϵ is roughly equivalent to a spherical lens in an external shear of $\gamma \simeq \epsilon/3$. According to the cross sections, the fraction of four-image lenses should be of order $\sigma_4/\sigma_{SIS} \sim \gamma^2 \sim (\epsilon/3)^2 \sim 0.01$ rather than the observed 30%. Most of this difference is a consequence of the different magnification biases of the two image multiplicities.

There is an important subtlety when studying lens statistics with models covering a range of axis ratios, namely that the definition of the critical radius b in (say) the SIE model (37) depends on the axis ratio and exactly what quantity you are holding fixed in your calculation (see Keeton, Kochanek and Seljak 1997; Keeton and Kochanek 1998; Rusin and Tegmark 2001; Chae 2003). For example, if we compare a singular isothermal sphere to a face on Mestel disk with the same equatorial circular velocity, the Einstein radius of the disk is $2/\pi$ smaller than the isothermal sphere because for the same circular velocity a disk requires less mass than a sphere. Since we usually count galaxies locally and translate these counts into a dynamical variable, this means that lens models covering a range of ellipticities must be normalized in terms of the same dynamical variables as were used to count the galaxies.

Much early effort focused on the effects of adding a finite core radius to these standard models (e.g. Blandford and Kochanek 1987a,b; Kochanek and Blandford 1987; Kovner 1987a; Hinshaw and Krauss 1987; Krauss and White 1992; Wallington and Narayan 1993; Kochanek 1996a,b). The core radius s leads to an evolution of the caustic structures (see Part 1, Blandford and Narayan 1986) with the ratio between the core radius and the critical radius s/b. Strong lenses with $s/b \ll 1$ act like singular models. Weak, or marginal, lenses with $s/b \sim 1$ have significantly reduced cross sections but higher average magnifications such that the rising magnification bias roughly balances the diminishing cross section to create a weaker than expected effect of core radii on the probability of finding a lens (see Kochanek 1996a,b). As the evidence that lenses are effectively singular has mounted, interest in these models has waned, and we will not discuss them further here. There is some interest in these models as a probe of large separation lenses due to groups and clusters where a finite core radius is replaced by effects of the shallow $\rho \propto r^{-1}$ NFW density cusp, and we will consider this problem in Sect. 7 where we discuss large separation lenses.

6.4 Optical Depth

The optical depth associated with a cross section is the fraction of the sky in which you can place a source and see the effect. This simply requires adding up the contributions from all the lens galaxies between the observer and the redshift of the source. For the SIS lens we simply need to know the comoving density of lenses per unit dark matter velocity dispersion $dn/d\sigma$ (which may be a function of redshift)

$$\tau_{SIS} = \int_0^{z_s} \frac{dV}{dz_l} dz_l \int_0^\infty \frac{dn}{d\sigma_v} \frac{\sigma_{SIS}}{4\pi} d\sigma_v, \tag{106}$$

where dV/dz_l is the comoving volume element per unit redshift (e.g. Turner, Ostriker and Gott 1984). For a flat cosmology, which we adopt from here on, the comoving volume element is simply $dV = 4\pi D_d^2 dD_d$ where D_d is the comoving distance to the lens redshift (2). The generalization to open or closed models can be found in Carroll et al. (1992). As with most lens calculations, this means that the expression simplifies if expressed in terms of the comoving angular diameter distances,

$$\tau_{SIS} = \int_0^{D_s} dD_d D_d^2 \left(\frac{D_{ds}}{D_s}\right)^2 \int_0^\infty \frac{dn}{d\sigma_v} 16\pi^2 \left(\frac{\sigma_v}{c}\right)^4 \tag{107}$$

(Gott, Park and Lee 1989; Fukugita, Futamase and Kasai 1990). If the comoving density of the lenses does not depend on redshift, the integrals separate to give

$$\tau_{SIS} = \frac{8\pi^2}{15} D_s^3 \int_0^\infty d\sigma_v \frac{dn}{d\sigma_v} \left(\frac{\sigma_v}{c}\right)^4 \tag{108}$$

(Fukugita and Turner 1991). If we now assume that the galaxies can be described by the combination of Schechter luminosity functions and kinematic relations described in Sect. 6.2, then we can do the remaining integral to find that

$$\tau_{SIS} = \frac{8\pi^2}{15} n_* \left(\frac{\sigma_*}{c}\right)^4 D_s^3 \Gamma[1+\alpha+6/\gamma] = \frac{1}{30} \tau_* r_H^{-3} D_s^3 \Gamma[1+\alpha+6/\gamma], \tag{109}$$

where $\Gamma[x]$ is a Gamma function, $r_H = c/H_0$ is the Hubble radius and the optical depth scale is

$$\tau_* = 16\pi^3 n_* r_H^3 \left(\frac{\sigma_*}{c}\right)^4 = 0.026 \left(\frac{n_*}{10^{-2} h^3 \mathrm{Mpc}^{-3}}\right) \left(\frac{\sigma_*}{200 \mathrm{km\ s}^{-1}}\right)^4 . \tag{110}$$

Thus, lens statistics are essentially a volume test of the cosmology (the D_s^3), predicated on knowing the comoving density of the lenses (n_*) and their average mass (σ_*). The result does not depend on the Hubble constant – all determinations of n_* scale with the Hubble constant such that $n_* D_s^3$ is independent of H_0.

Two other distributions, those in image separation and in lens redshift at fixed image separation, are easily calculated for the SIS model and useful if numerical for any other lens.

The SIS image separation is $\Delta\theta = 8\pi(\sigma_v/c)^2 D_{ds}/D_s$, so

$$\frac{d\tau_{SIS}}{d\Delta\theta} = \tfrac{1}{2} D_s^3 \hat{\Delta\theta}^2 \left(\Gamma\left[1+\alpha-2/\gamma_{FJ}, \xi\right] - 2\hat{\Delta\theta}\Gamma\left[1+\alpha-4/\gamma_{FJ}, \xi\right] \right.$$
$$\left. + \hat{\Delta\theta}^2 \Gamma\left[1+\alpha-6/\gamma_{FJ}, \xi\right] \right), \quad (111)$$

where $\xi = (\Delta\theta/\Delta\theta_*)^{\gamma_{FJ}/2}$ and

$$\Delta\theta_* = 8\pi \left(\frac{\sigma_*}{c}\right)^2 = 2''\!.3 \left(\frac{\sigma_*}{200\text{km s}^{-1}}\right)^2 \quad (112)$$

is the maximum separation produced by an L_* galaxy. The mean image separation,

$$\langle\Delta\theta\rangle = \frac{\Delta\theta_*}{2} \frac{\Gamma[1+\alpha+8/\gamma]}{\Gamma[1+\alpha+6/\gamma]^{1/2}}, \quad (113)$$

depends only on the properties of the lens galaxy and not on cosmology. If the cosmological model is not flat, a very weak dependence on cosmology is introduced (Kochanek 1993a,b,c). For a known separation $\Delta\theta$, the probability distribution for the lens redshift becomes

$$\frac{dP}{dz_l} \propto \frac{D_d^2}{D_s} \frac{dD_d}{dz_l} \exp\left[- \left(\frac{\Delta\theta}{\Delta\theta_*} \frac{D_s}{D_{ds}} \right)^{1/2} \right] \quad (114)$$

(we present the result only for Schechter function $\alpha = -1$ and Faber–Jackson $\gamma_{FJ} = 4$). The location of the exponential cut off introduced by the luminosity function has a strong cosmological dependence, so the presence or absence of lens galaxies at higher redshifts dominates the cosmological limits. The structure of this function is quite different from the total optical depth, which in a flat cosmology is a slowly varying function with a mean lens distance equal to one-half the distance to the source. The mean redshift changes with cosmology because of the changes in the distance-redshift relations, but the effect is not as dramatic as the redshift distributions for lenses of known separation.

We end this section by discussing the Keeton (2002) "heresy". Keeton (2002) pointed out that if you used a luminosity function derived at intermediate redshift rather than locally, then the cosmological sensitivity of the optical depth effectively vanishes when the median redshift of the lenses matches the median redshift of the galaxies used to derive the luminosity function. The following simple thought experiment shows that this is true at one level. Suppose there was only one kind of galaxy and we make a redshift survey and count all the galaxies in a thin shell at redshift z, finding N galaxies between z and $z+\Delta z$. The implied comoving density of the galaxies, $n = N/(\Delta z dV/dz)$, depends on the cosmological model with the same volume factor appearing in the optical depth calculation (106). To the extent that the redshift ranges and weightings of the galaxy survey and a lens survey are similar, the cosmological sensitivity of the optical depth vanishes because the volume factor cancels and the optical depth depends only on the number of observed galaxies N. This does not occur when we use a local luminosity function because changes in cosmology have no effect on the local volume element. The problem with the Keeton (2002) argument is that it basically says that if we could use galaxy number counts to determine the cosmological model then we would not need lensing to do so because the two are redundant. To continue our thought experiment, we also have local estimates n_{local} for the density of galaxies, and as we vary the cosmology we would find that n and n_{local} agree only for a limited range of cosmological models and this would restore the cosmological sensitivity. The problem is that the comparison of near and distant measurements of the numbers of galaxies is tricky because it depends on correctly matching the galaxies in the presence of galaxy evolution and selection effects – in essence, you cannot use this argument to eliminate the cosmological sensitivity of lens surveys unless you think you understand galaxy evolution so well that you can use galaxy number counts to determine the cosmological model, a program of research that has basically been abandoned.

6.5 Spiral Galaxy Lenses

Discussions of lens statistics, or even lenses in general, focus on early-type galaxies (E/S0). The reason is that spiral lenses are relatively rare. The only morphologically obvious spirals are B0218+357 (Sc, York et al. 2005), B1600+434 (S0/Sa, Jaunsen and Hjorth 1997), PKS1830–211 (Sb/Sc, Winn et al. 2002a,b,c), PMNJ2004–1349 (Sb/Sc, Winn, Hall and Schechter 2003c), and Q2237–0305 (Huchra et al. 1985). Other small separation systems may well be spiral galaxies, but we do not have direct evidence from imaging. There are studies of individual spiral lenses or the statistics of spiral lenses by Maller, Flores and Primack (1997); Keeton and Kochanek (1998); Koopmans et al. (1998); Maller et al. (2000); Trott and Webster (2002); Winn, Hall and Schechter (2003c).

The reason lens samples are dominated by early-type galaxies is that the early-type galaxies are more massive even if slightly less numerous (e.g. Fukugita and Turner 1991, see Sect. 6.2). The relative numbers of early-type and late-type lenses should be the ratio of their optical depths, $(n_l/n_e)(\sigma_l/\sigma_e)^4$, based on the comoving densities and characteristic velocity dispersions of the early and late-type galaxies. For example, in the Kochanek et al. (2001a,b) K-band luminosity function $n_l/n_e \simeq 2.2$ while the ratio of the characteristic velocity dispersions is $\sigma_{*l}/\sigma_{*e} = 0.68$ giving an expected ratio of spiral to early-type lenses of 0.47. Because the typical separation of the spiral lenses will also be smaller by a factor of $(\sigma_{*l}/\sigma_{*e})^2 = 0.46$, they will be much harder to resolve given the finite resolution of lens surveys. Thus, survey selections functions discriminate more strongly against late-type lenses than against early-type lenses. The higher prevalence of dust in late-type lenses adds a further bias against them in optical surveys.

6.6 Magnification Bias

The optical depth calculation suggests that the likelihood of finding that a $z_s \simeq 2$ quasar is lensed is very small ($\tau \sim 10^{-4}$) , while observational surveys of bright quasars typically find that of order 1% of bright quasars are lensed. The origin of the discrepancy is the effect known as "magnification bias" (Turner 1980), which is really the correction needed to account for the selection of survey targets from flux limited samples. Multiple imaging always magnifies the source, so lensed sources are brighter than the population from which they are drawn. For example, the mean magnification of all multiply imaged systems is simply the area over which we observe the lensed images divided by the area inside the caustic producing multiple images because the magnification is the Jacobian relating area on the image and source planes, $d^2\boldsymbol{\beta} = |\mu|^{-1} d^2\boldsymbol{\theta}$. For example, an SIS lens with Einstein radius b produces multiple images over a region of radius b on the source plane (i.e. the cross section is πb^2), and these images are observed over a region of radius $2b$ on the image plane, so the mean multiple-image magnification is $\langle \mu \rangle = (4\pi b^2)/(\pi b^2) = 4$.

Since fainter sources are almost always more numerous than brighter sources, magnification bias almost always increases your chances of finding a lens. The simplest example is to imagine a lens which always produces the same magnification μ applied to a population with number counts $N(F)$ with flux F. The number counts of the lensed population are then $N_{lens}(F) = \tau \mu^{-1} N(F/\mu)$, so the fraction of lensed objects (at flux F) is larger than the number expected from the cross section if fainter objects are more numerous than the magnification times the density of brighter objects. Where did the extra factor of magnification come from? It has to be there to conserve the total number of sources or equivalently the area on the source and lens planes – you can always check your expression for the magnification bias by computing the number counts of lenses and checking to make sure that the total number of lenses equals the total number of sources if the optical depth is unity.

Real lenses do not produce unique magnifications, so it is necessary to work out the magnification probability distribution $P(>\mu)$ (the probability of a magnification larger than μ) or its differential $dP/d\mu$ and then convolve it with the source counts. Equivalently we can define a magnification dependent cross section, $d\sigma/d\mu = \sigma dP/d\mu$ where σ is the total cross section. We can do this easily only for the SIS lens, where a source at β produces two images with a total magnification of $\mu = 2/\beta$ (21, 22) and $\mu > 2$ in the multiple image region, to find that $P(>\mu) = (2/\mu)^2$ and $dP/d\mu = 8/\mu^3$. The structure at low magnification depends on the lens model, but all sensible lens models have $P(>\mu) \propto \mu^{-2}$ at high magnification because this is generic to the statistics of fold caustics (Part 1, Blandford and Narayan 1986).

Usually people have defined a magnification bias factor $B(F)$ for sources of flux F so that the probability $p(F)$ of finding a lens with flux F is related to the optical depth by $p(F) = \tau B(F)$. The magnification bias factor is

$$B(F) = N(F)^{-1} \int \frac{d\mu}{\mu} \frac{dP}{d\mu} N\left(\frac{F}{\mu}\right) \tag{115}$$

for a source with flux F, or

$$B(m) = N(m)^{-1} \int d\mu \frac{dP}{d\mu} N\left(m + 2.5 \log \mu\right) \tag{116}$$

for a source of magnitude m. Note the vanishing of the extra $1/\mu$ factor when using logarithmic number counts $N(m)$ for the sources rather than the flux counts $N(F)$. Most standard models have magnification probability distributions similar to the SIS model, with $P(> \mu) \simeq (\mu_0/\mu)^2$ for $\mu > \mu_0$, in which case the magnification bias factor for sources with power law number counts $N(F) = dN/dF \propto F^{-\alpha}$ is

$$B(F) = \frac{2\mu_0^{\alpha-1}}{3-\alpha} \tag{117}$$

provided the number counts are sufficiently shallow ($\alpha < 3$). For number counts as a function of magnitude $N(m) = dN/dm \propto 10^{am}$ (where $a = 0.4(\alpha - 1)$) the bias factor is

$$B(F) = \frac{2\mu_0^{2.5a}}{2.5a - 2}. \tag{118}$$

The steeper the number counts and the brighter the source is relative to any break between a steep slope and a shallow slope, the greater the magnification bias. For radio sources a simple power law model suffices, with $\alpha \simeq 2.07 \pm 0.11$ for the CLASS survey (Rusin and Tegmark 2001), leading to a magnification bias factor of $B \simeq 5$. For quasars, however, the bright quasars have number counts steeper than this critical slope, so the location of the break from the steep slope of the bright quasars to the shallower slope for fainter quasars near $B \sim 19$ mag is critical to determining the magnification bias. Figure 46 shows an example of a typical quasar number counts distribution as compared to several (old) models for the distribution of lensed quasars. The changes in the magnification bias with magnitude are visible as the varying ratio between the lensed and unlensed counts, with a much smaller ratio for bright quasars (high magnification bias) than for faint quasars (low magnification bias) and a smooth shift between the two limits as you approach the break in the slope of the counts at $B \sim 19$ mag.

For optically-selected lenses, magnification bias is "undone" by extinction in the lens galaxy because extinction provides an effect that makes lensed quasars dimmer than their unlensed counterparts. Since the quasar samples were typically selected at blue wavelengths, the rest wavelength corresponding to the quasar selection band at the redshift of the lens galaxy where it encounters the dust is similar to the U-band. If we use a standard color excess

$E(B-V)$ for the amount of dust, then the images become fainter by of order $A_U E(B-V)$ magnitudes where $A_U \simeq 4.9$. Thus, if lenses had an average extinction of only $E(B-V) \simeq 0.05$ mag, the net magnification of the lensed images would be reduced by about 25%. If all lenses had the same demagnification factor $f < 1$ then the modifications to the magnification bias would be straight forward. For power-law number counts $N(F) \propto F^{-\alpha}$, the magnification bias is reduced by the factor f^α and a $E(B-V) = 0.05$ extinction leads to a 50% reduction in the magnification bias for objects with a slope $\alpha \simeq 2$ (faint quasars) and to still larger reductions for bright quasars. Some examples of the changes with the addition of a simple mean extinction are shown in the right panel of Fig. 46, although the levels of extinction shown there are larger than observed in typical lenses as we discuss in Sect. 9.1. Comparisons between the statistics of optically-selected and radio-selected samples can be used to estimate the magnitude of the correction. The only such comparison found estimated extinctions consistent with the direct measurements of Sect. 9.1 (Falco, Kochanek and Muñoz 1998). However, the ISM of real lenses is presumably far more complicated, with a distribution of extinctions and different extinctions for different images which may be a function of orientation and impact parameter relative to the lens galaxy, for which we have no good theoretical model.

The flux of the lens galaxy also can modify the magnification bias for faint quasars, although the actual sense of the effect is complex. The left panel Fig. 46 shows the effect of dropping lenses in which the lens galaxy represents some fraction of the total flux of the lensed images. The correction is unimportant for bright quasars because lens galaxies with $B < 19$ mag are rare. In this picture, the flux from the lens galaxy leads to the loss of lenses because the added flux from the lens galaxy makes the colors of faint lens galaxies differ from those of quasars so they are never selected as quasars to begin with. Alternatively, if one need not worry about color contamination, then the lens galaxy increases the magnification bias by supplying extra flux that makes lensed quasars brighter.

Any other selection effect, such as the dynamic range allowed for flux ratios between images as a function of their separation will also have an effect on the magnification bias. Exactly how the effect enters depends on the particular class of images being considered. For example, in the SIS lens (or more generally for two-image lenses), a limitation on the detectable flux ratio $0 < f_{min} < 1$ sets a minimum detectable magnification $\mu_{min} = 2(1+f_{min})/(1-f_{min}) > \mu_0 = 2$. Since most lens samples have significant magnification bias, which means that most lenses are significantly magnified, such flux limits have only modest effects. The other limit, which cannot be captured in the SIS model, is that almost all bright images are merging pairs on folds (or triplets on cusps) so the image separation decreases as the magnification increases. The contrast between the merging images and any other images also increases with increasing magnification – combined with limits on the detectability of images, these lead to selection effects against highly magnified images. This is also usually a modest effect – while magnification bias is important, the statistics are dominated by modestly magnified systems rather than very highly magnified images. In fact, there have been few attempts at complete studies of the complicated interactions between finding quasars, finding lenses, selection effects and magnification bias. There is an early general study by Kochanek (1991a,b,c) and a detailed practical application of many of these issues to the SDSS survey by Pindor et al. (2003). Unfortunately, Pindor et al. (2003) seem to arrive at a completeness estimate from their selection model that is too high given the number of lenses they found in practice. Some of this may be due to underestimating the luminosity of lens galaxies, the effects of the lens galaxy or extinction on the selection of quasars or the treatment of extended, multicomponent lenses compared to normal quasars in the photometric pipeline. These difficulties, as well as the larger size of the present radio-selected lens samples, are the reason that almost all recent statistical studies have focused exclusively on radio lenses.

The standard magnification bias expressions ((115) and (116)) are not always correct. They are correct for the statistics of lenses selected from source populations for which the total flux of the source (including all images of a lensed source) is defining F (or m). This is true of most existing surveys – for example the CLASS radio survey sources were originally selected from single dish observations with very poor resolution compared to typical image separations (see Browne et al. 2003). If, however, the separation of the images is large compared to the resolution of the observations and the fluxes of the images are considered separately, then the bias must be computed in terms of the bright image used to select sources to search for additional images. This typically reduces the bias. More subtle effects can also appear. For example, the SDSS survey selects quasar candidates based on the best fit point-source magnitudes, which will tend to be an underestimate of the flux of a resolved lens. Hence the magnification bias for lenses found in the SDSS survey will be less than in the standard theory. Samples selected based on more than one frequency can have more complicated magnification biases depending on the structure of the multidimensional number counts (Borgeest, von Linde and Refsdal 1991; Wyithe, Winn and Rusin 2003). The exact behavior is complex, but the magnification bias can be tremendously increased if the fluxes in the bands are completely uncorrelated or tightly but nonlinearly correlated. For example, if the luminosities in bands A and B are related by tight, nonlinear correlation of the form $L_A \propto L_B^{1/2}$, then the lensed examples of these objects will lie off the correlation. At present, there are too few deep, wide-area multiwavelength catalogs to make good use of this idea, but this is changing rapidly.

While most studies assume lenses are spherical when making statistical studies, there are significant and trivially observable consequences of ellipticity in lens statistics namely, the four image lenses, whose existence in observational samples is largely due to the differences in the magnification bias between quads and doubles. We noted earlier that the expectation from the cross section is that four-image lenses should represent order $\epsilon_\Psi^2 \sim \gamma^2 \sim 0.01$ of lenses where ϵ_Ψ is the ellipticity of the lens potential. Yet in Sect. 2 we saw that four-image lenses represent roughly one third of the observed population. Most of this difference is a consequence of the different magnification biases of the two image multiplicities. In general, the ellipticity of the lenses has little effect on the expected number of lenses, allowing the use of circular lens models for statistical studies that are uninterested in the morphologies of the images (e.g. Keeton, Kochanek and Seljak 1997; Rusin and Tegmark 2001; Chae 2003).

While simple models generally capture the total magnification bias of a sample, the magnification bias depends heavily on the number of images. Figure 43 shows the image magnification contours for an SIS lens in an external shear on both the image and source planes. The highly magnified regions are confined to lie near the critical line. If we Taylor expand the inverse magnification radially, then $\mu^{-1} = \Delta x |d\mu^{-1}/dx|$ where Δx is the distance from the critical line, so the magnification drops inversely with the distance from the critical line. If we Taylor expand the lens equations, then we find that the change in source plane coordinates is related to the change in image plane coordinates by $\Delta\beta = \mu^{-1}\Delta x \propto \mu^{-2}$. Thus, if L is the length of

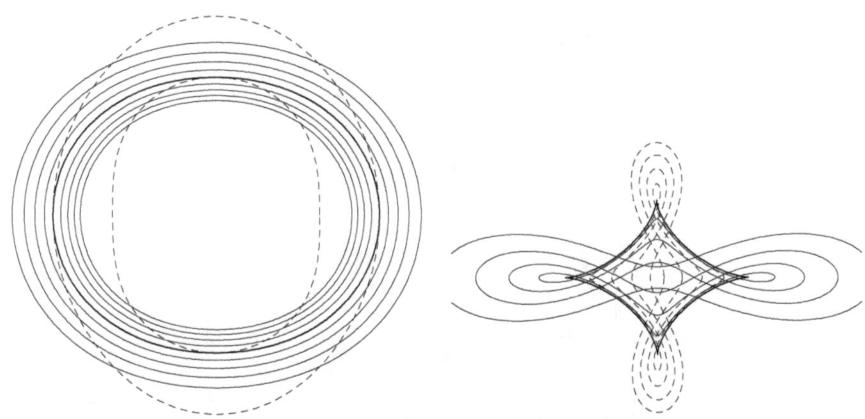

Fig. 43. Magnification contours on the image (*left*) and source (*right*) planes for an SIS in an external shear. The heavy solid contours show the tangential critical line (*left*) and its corresponding caustic (*right*). On the image plane (*left*), the light curves are magnification contours. These are positive outside the critical curve and negative inside the critical curve. The images found in a four-image lens are all found in the region between the two dashed contours – when two images are merging on the critical line, the other two images lie on these curves. On the source plane the solid (*dashed*) curves show the projections of the positive (*negative*) magnification contours onto the source plane. Note that the high magnification regions are dominated by the four-image systems with the exception of the small high magnification regions found just outside the tip of each cusp

the astroid curve, the probability of a magnification larger than μ scales as $P(>\mu) \propto \mu^{-2}L/|d\mu^{-1}/dx|$. This applies only to the four image region, because the only way to get a high magnification in the two-image region is for the source to lie just outside the tip of a cusp. The algebra is overly complex to present, but the generic result is that the region producing magnification μ extends μ^{-2} from the cusp tip but has a width that scales as $\mu^{-1/2}$, leading to an overall scaling that the asymptotic cross section declines as $P(>\mu) \propto \mu^{-7/2}$ rather than $P(>\mu) \propto \mu^{-2}$. This can all be done formally (see Blandford and Narayan 1986) so that asymptotic cross sections can be derived for any model (e.g. Kochanek and Blandford 1987; Finch et al. 2002), but a reasonable approximation for the four-image region is to compute the magnification, μ_0, for the cruciform lens formed when the source is directly behind the lens and then use the estimate that $P(>\mu) = (\mu_0/\mu)^2$. Unfortunately, such simple estimates are not feasible for the two-image region. These distributions are relatively easy to compute numerically, as in the example shown in Fig. 44.

Because the minimum magnification increases $\propto \gamma^{-1}$ even as the cross section decreases as $\propto \gamma^2$, the expected number of four-image lenses in a sample varies much more slowly with ellipticity than expected from the cross section. The product $\sigma_4 B(F) \propto \gamma^2 \mu_0^{\alpha-1}$, of the four-image cross section, σ_4, and the

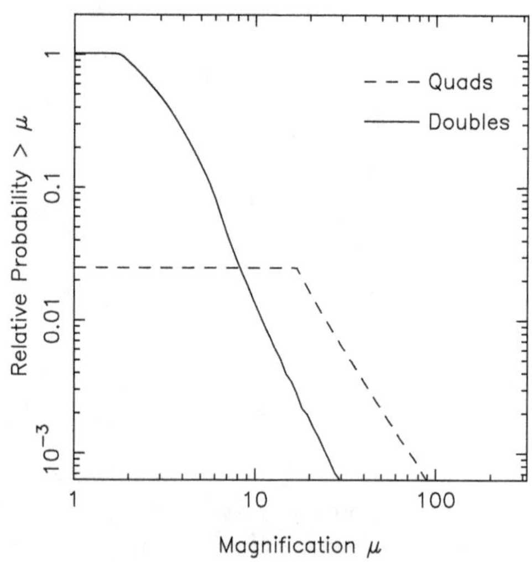

Fig. 44. The integral magnification probability distributions for a singular isothermal ellipsoid with an axis ratio of $q = 0.7$ normalized by the total cross section for finding two images. Note that the total four-image cross section is only of order $\epsilon_\Psi^2 \sim (\epsilon/3)^2 \sim 0.01$ of the total, but that the minimum magnification for the four-image systems ($\mu_{min} \sim 1/\epsilon \sim 10$) is much larger than that for the two-image systems ($\mu_{min} \sim 2$ just as for an SIS). The entire four-image probability distribution is well approximated by the $P(>\mu) \propto \mu^{-2}$ power law expected for fold caustics, while the two-image probability distribution is steeper since highly magnified images can only be created by the cusps. Figure courtesy of D. Rusin

magnification bias, $B(F)$, scales as $\gamma^{3-\alpha} \propto \gamma$ for the CLASS survey ($\alpha \simeq 2$), which is a much more gentle dependence on ellipticity than the quadratic variation expected from the cross section. There is a limit, however, to the fraction of four-image lenses, as shown in Fig. 45. If the potential becomes too flat, the astroid caustic extends outside the radial caustic (Fig. 18), to produce three-image systems in the "disk" geometry rather than additional four-image lenses. In the limit that the axis ratio goes to zero (the lens becomes a line), only the disk geometry is produced. The existence of a maximum four-image lens fraction, and its location at an axis ratio inconsistent with the observed axis ratios of the dominant early-type lenses has made it difficult to explain the observed fraction of four image lenses (King and Browne 1996; Kochanek 1996a,b; Keeton, Kochanek and Seljak 1997; Keeton and Kochanek 1998; Rusin and Tegmark 2001). Recently, Cohn and Kochanek (2004) argued that satellite galaxies of the lenses provide the explanation by somewhat boosting the fraction of four-image lenses while at the same time explaining the existence of the more complex lenses like B1359+154 (Myers et al. 1999; Rusin et al. 2001) and PMNJ0134–0931 (Winn et al. 2002a,b,c; Keeton and Winn 2003) formed by having multiple lens galaxies with more complex caustic structures. It is not, however, clear in the existing data that four-image systems are more likely to have satellites to the lens galaxy than two-image systems as one would expect for this explanation.

Gravitational lenses can produce highly magnified images without multiple images only if they are highly elliptical or have a low central density. The SIS lens has a single-image magnification probability distribution

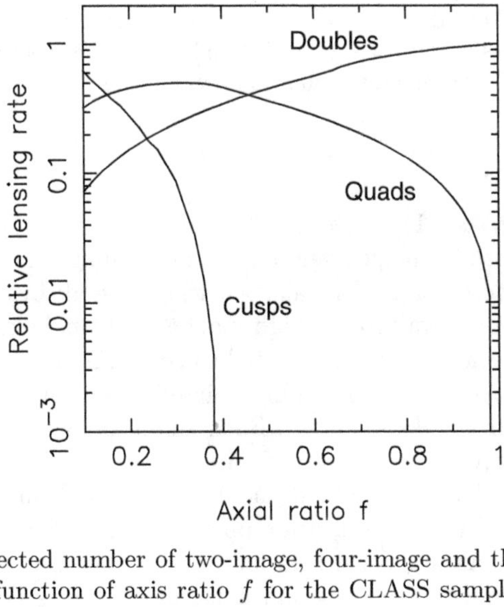

Fig. 45. The expected number of two-image, four-image and three-image (disk or cusp) lenses as a function of axis ratio f for the CLASS sample. From Rusin and Tegmark (2001)

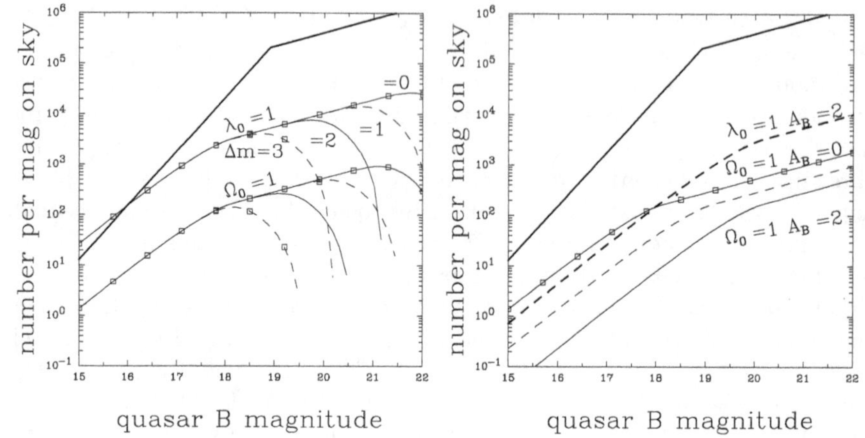

Fig. 46. Examples of selection effects on optically selected lens samples. The *heavy solid curves* in the two panels shows a model for the magnitude distribution of optically-selected quasars. The *light curves* labeled $\Omega_M = 1$ and $\lambda_0 = 1$ show the distribution of lensed quasars for flat cosmologies that are either pure matter or pure cosmological constant. The change in the ratio between the lensed curves and the unlensed curves illustrates the higher magnification bias for bright quasars where the number count distribution is steeper than for faint quasars. In the left panel the *truncated curves* show the effect of losing the lensed systems where the lens galaxy is $\Delta m = 1$, 2 or 3 magnitudes fainter than the quasars. Once surveys are searching for lensed quasars with $B \gtrsim 20$ mag, the light from the lens galaxy becomes an increasing problem, particularly since the systems with the brightest lens galaxies will also have the largest image separations that would otherwise make them easily detected. In the right panel we illustrate the effect of adding a net extinction of $A_B = 1$ or 2 mag from dust in the lens galaxies. These correspond to larger than expected color excesses of $E(B-V) \simeq 0.2$ and 0.4 mag respectively. Note how the extinction "undoes" the magnification bias by shifting the lensed distributions to fainter magnitudes

of $\tau dP/d\mu = 2\pi b^2/(\mu-1)^3$ with $\mu < 2$ compared to $\tau dP/d\mu = 2\pi b^2/\mu^3$ with $\mu \geq 2$ for the multiply imaged region, so single images are never magnified by more than a factor of 2. For galaxies, where we always expect high central densities, the only way to get highly magnified single images is when the astroid caustic extends outside the radial caustic (Fig. 18). A source just outside an exposed cusp tip can be highly magnified with a magnification probability distribution $dP/d\mu \propto \mu^{-7/2}$. Such single image magnifications have recently been a concern for the luminosity function of high redshift quasars (e.g. Wyithe 2004; Keeton, Kuhlen and Haiman 2005) and will be the high magnification tail of any magnification perturbations to supernova fluxes (e.g. Dalal et al. 2003). As a general rule for galaxies, the probability of a single image being magnified by more than a factor of two is comparable to the probability of being multiply imaged.

6.7 Cosmology With Lens Statistics

The statistics of lenses, in the sense of the number of lenses expected in a sample of sources as a function of cosmology, is a volume test of the cosmological model because the optical depth (at least for flat cosmologies) is proportional to D_s^3. However, the number of lenses also depends on the comoving density and mass of the lenses (n_*, σ_* and α in the simple SIS model). While n_* could plausibly be estimated locally, the σ_*^4 dependence on the mass scale makes it very difficult to use local estimates of galaxy kinematics or masses to normalize the optical depth. The key step to eliminating this problem is to note that there is an intimate relation between the cross section, the observed image separations and the mass scale. While this will hold for any mass model, the SIS model is the only simple analytic example. The mean image separation for the lenses should be independent of the cosmological model for flat cosmologies (and only weakly dependent on it otherwise). Thus, in any lens sample you can eliminate the dependence on the mass scale by replacing it with the observed mean image separation, $\tau_{SIS} \propto n_* \langle \Delta\theta \rangle^2 D_s^3$. Full calculations must include corrections for angular selection effects. Most odd results in lens cosmology arise in calculations that ignore the close coupling between the image separations and the cross section.

In practice, real calculations are based on variations of the maximum likelihood method introduced by Kochanek (1993a,b,c, 1996a,b). For each lens i you compute the probability p_i that it is lensed including magnification bias and selection effects. The likelihood of the observations is then

$$\ln L_0 = \sum_{lenses} \ln p_i + \sum_{unlensed} \ln(1 - p_i) \simeq \sum_{lenses} \ln p_i - \sum_{unlensed} p_i, \qquad (119)$$

where $\ln(1 - p_i) \simeq -p_i$ provided $p_i \ll 1$. This simply encodes the likelihood of finding the observed number of lenses given the individual probabilities that the objects are lensed. Without further information, this likelihood could determine the limits on the cosmological model only to the extent we had accurate prior estimates for n_* and σ_*.

If we add, however, a term for the probability that each detected lens has its observed separation (including any selection effects)

$$\ln L = \ln L_0 + \sum_{lensed} \ln \left(\frac{p_i(\Delta\theta_i)}{p_i} \right), \qquad (120)$$

then the lens sample itself can normalize the typical mass scale of the lenses (Kochanek 1993a,b,c). This has two advantages. First, it eliminates any systematic problems arising from the dynamical normalization of the lens model and its relation to the luminosity function. Second, it forces the cosmological estimates from the lenses to be consistent with the observed image separations – it makes no sense to produce cosmological limits that imply image separations inconsistent with the observations. In theory the precision exceeds that of any local calibration very rapidly. The fractional spread of the separations about the mean is ~ 0.7, so the fractional uncertainty in the mean separation scales as $0.7/N^{1/2}$ for a sample of N lenses. Since the cross section goes as the square of the mean separation, the uncertainty in the mean cross section $1.4/N^{1/2}$ exceeds any plausible accuracy of a local normalization for σ_* (10% in σ_*, or 20% in $\langle\theta\rangle \propto \sigma_*^2$, or 40% in $\tau \propto \sigma_*^4$) with only $N \simeq 10$ lenses.

Any other measurable property of the lenses can be added to the likelihood, but the only other term that has been seriously investigated is the probability of the observed lens redshift given the image separations and the source redshift (Kochanek 1992a,b, 1996a,b; Helbig and Kayser 1996; Ofek, Rix and Maoz 2003). In general, cosmologies with a large cosmological constant predict significantly higher lens redshifts than those without, and in theory this is a very powerful test because of the exponential cutoff in (114). The biggest problem in actually using the redshift test, in fact so big that it probably cannot be used at present, is the high incompleteness of the lens redshift measurements (Sect. 2). There will be a general tendency, even at fixed separation, for the redshifts of the higher redshift lens galaxies to be the ones that are unmeasured. Complete samples could be defined for a separation range, usually by excluding small separation systems, but a complete analysis needs to include the effects of groups and cluster boosting image separations beyond the splitting produced by an isolated galaxy. For example, how do we include Q0957+561 with its separation of 6.″2 that is largely due to the lens galaxy but has significant contributions from the surrounding cluster?

6.8 The Current State

Recent analyses of lens statistics have focused exclusively on the CLASS flat spectrum radio survey (Browne et al. 2003). Chae et al. (2002); Chae (2003) and Mitchell et al. (2004) focus on estimating the cosmological model and find results in general agreement with estimates from Type Ia supernovae (e.g. Riess et al. 2004). The general approach of both groups is to use variants of the maximum likelihood methods described above in Sect. 6.7. Chae (2003) uses an obsolete estimate of the galaxy luminosity function combined with a Faber–Jackson relation and the variable transformation of (103) but normalized the velocity scale using the observed distribution of lens separations. Mitchell et al. (2004) use the true velocity dispersion function from the SDSS survey (Sheth et al. 2003) and incorporate a Press-Schechter (1974) model for the evolution of the velocity function. Chae (2003) used ellipsoidal galaxies, although this has little cosmological effect, while Mitchell et al. (2004) considered only SIS models. Figure 47 shows the cosmological limits from Mitchell et al. (2004), which are typical of the recent results. There are also attempts to use lens statistics to constrain dark energy (e.g. Chae et al. 2004; Kuhlen, Keeton and Madau 2004), but far larger, well-defined samples are needed before the resulting constraints will become interesting.

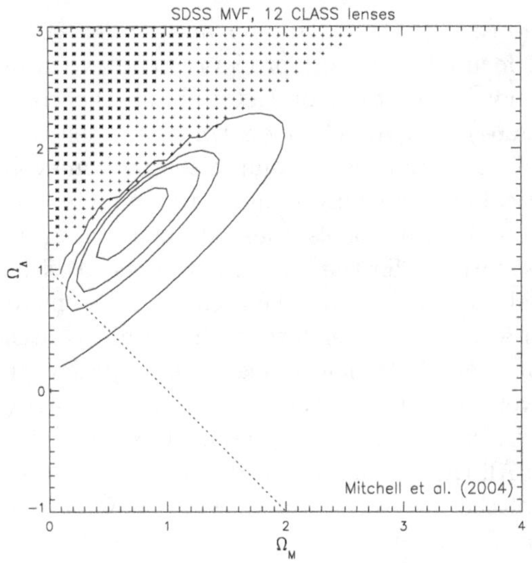

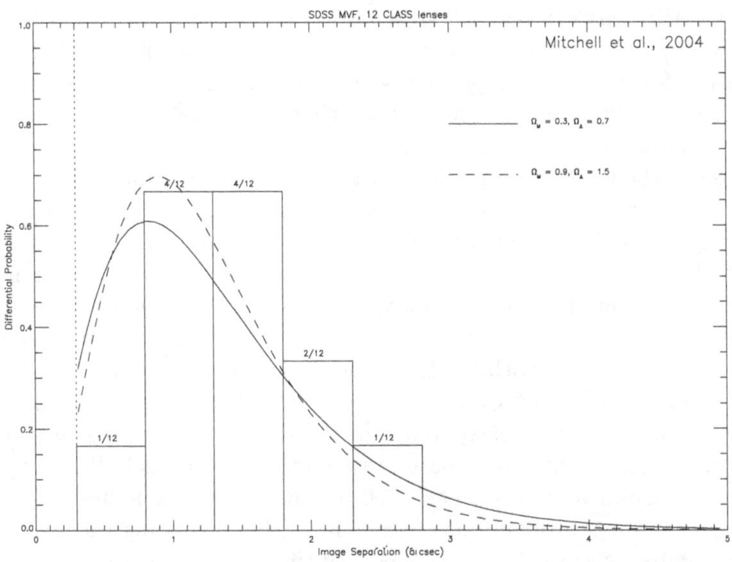

Fig. 47. (*Top*) Likelihood functions for the cosmological model from Mitchell et al. (2004) using the velocity function of galaxies measured from the SDSS survey and a sample of 12 CLASS lenses. The contours show the 68, 90, 95 and 99% confidence intervals on the cosmological model. In the shaded regions the cosmological distances either become imaginary or there is no big bang. (*Bottom*) The histogram shows the separation distribution of the 12 CLASS lenses used in the analysis and the curve shows the distribution predicted by the maximum likelihood model including selection effects

Chae and Mao (2003); Davis, Huterer and Krauss (2003) and Ofek, Rix and Maoz (2003) focused on galaxy properties and evolution in a fixed, concordance cosmology rather than on determining the cosmological models. Mitchell et al. (2004) compared models where the lenses evolved following the predictions of CDM models in comparison to non-evolving models. Because lens statistical estimates are unlikely to compete with other means of estimating the cosmological models, these are more promising applications of gravitational lens statistics for the future. Attempts to estimate the evolution of the lens population usually allow the n_* and σ_* parameters of the velocity function (103) to evolve as power laws with redshift. Mitchell et al. (2004), Fig. 42) point out that CDM halo models make specific predictions for the evolution of the velocity function that have a different structure from simple power laws in redshift, but with the present data the differences are probably unimportant. All these evolution studies came to the conclusion that the number density of the $\sigma_v \sim \sigma_*$ galaxies which dominate lens statistics has changed little ($\sim \pm 50\%$) between the present day and redshift unity.

I have three concerns about these analyses and their focus on the "complete" CLASS lens samples. First, a basic problem with the CLASS survey is that we lack direct measurements of the redshift distribution of the source population forming the lenses (e.g Marlow et al. 2000; Muñoz et al. 2003). In particular, Muñoz et al. (2003) note that the radio source population is changing radically from nearly all quasars to mostly galaxies as you approach the fluxes of the CLASS source population. This makes it dangerous to extrapolate the source population redshifts from the brighter radio fluxes where the redshift samples are nearly complete to the fainter samples where they are not. The second problem is that no study has a satisfactory treatment of the lenses with satellites or associated with clusters. All the analyses use isolated lens models and then either include lenses with satellites but ignore the satellites or drop lenses with satellites and ignore the fact that they have been dropped. The analysis by Cohn and Kochanek (2004) of lens statistics with satellites shows that neither approach is satisfactory – dropping the satellites biases the results to underestimate cross sections while including them does the reverse. Cohn and Kochanek (2004) concluded that including the systems with satellites probably has fewer biases than dropping them. A similar problem probably arises from the effects of the group halos to which many of the lenses belong (e.g. Keeton et al. 2000b; Fassnacht and Lubin 2002). My third concern is that the separations of the radio lenses seem to be systematically smaller than the optically selected lenses even though the Optical HST Snapshot Lens Survey (Maoz et al. 1993) had the greatest sensitivity to small separation systems. It is possible that this is simply due to selection effects in the optical samples, but I have seen no convincing scenario for producing such a selection effect. We see no clear correlation of extinction with image separation (see Sect. 9.1), emission from the lens galaxy is less important for small separation systems than for large separation systems, and the selection function due to the resolution of the observations is fairly simple to model.

On the other hand, the various lens samples may all be consistent. One way to compare the different data sets is to non-parametrically construct the velocity function from the observed image separations of the samples. To do this we assume an SIS lens model for the conversion from image separations to circular velocities, and then adopt the standard non-parametric methods used to construct luminosity functions from redshift surveys to construct the velocity function from the image separations (Kochanek 2003a,b,c). The results for the flat-spectrum lens surveys (CLASS, JVAS, PANELS), all radio surveys and all radio surveys plus the quasar lenses are shown in Fig. 48. We normalized the estimates to the density at $v_c = 300$ km/s to eliminate any dependence on the cosmological model. The lens data can estimate the velocity function from roughly $v_c \sim 100$ km/s to 500 km/s. At lower velocities the finite resolution of the observations makes the uncertainties in the density explode, and at higher velocities the surveys have not searched large enough

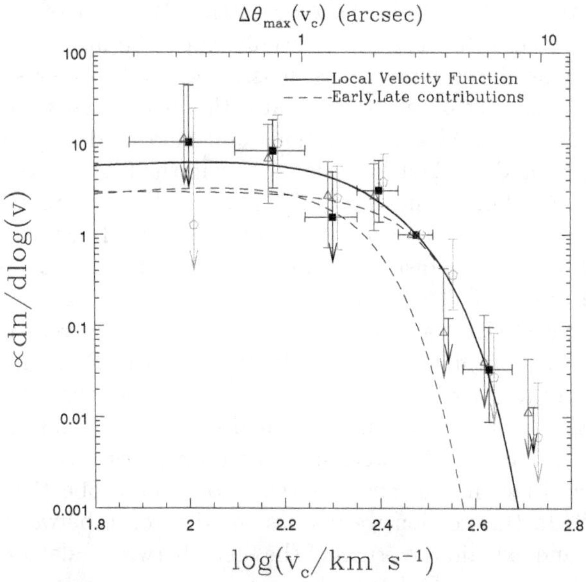

Fig. 48. Non-parametric reconstructions of the velocity function from the observed separations of gravitational lenses assuming an SIS lens model. The velocity functions are all normalized to the bin centered at 300 km/s. The *filled squares* use only the lenses in the flat spectrum radio surveys, the *triangles* use all radio-selected lenses and the *pentagons* include all radio lenses and all quasar lenses. The *horizontal error bars* on the filled squares show the bin widths. The triangles and pentagons are horizontally offset from the squares to make them more visible. The *curves* show the velocity function estimated from the 2MASS sample from Fig. 40. The horizontal scale at the top of the figure shows the maximum separation produced by a lens of the corresponding circular velocity. The mean separation produced by such a lens will be one-half the maximum

angular regions around the lens galaxies. The shape of the velocity function is consistent with local estimates (Fig. 40) except in the highest circular velocity bin where we begin to see the contribution from clusters we will consider in Sect. 7. Figure 48 also makes it clear why constraints on the evolution of the lenses are so weak – evolution estimates basically try to compare the low-redshift separation distribution to the high redshift separation distribution, and we simply do not have large enough lens samples to begin subdividing them in redshift (to say nothing of dealing with unmeasured redshifts) and still have small statistical uncertainties.

7 What Happened to the Cluster Lenses?

One would think from the number of conference proceeding covers featuring HST images of cluster arcs that these are by far the most common type of lens. In fact, this is an optical delusion created by the ease of finding the rich clusters even though they are exponentially rare. The most common kind of lens is the one produced by a typical massive galaxy – as we saw in Fig. 48. For a comparison, Fig. 49 shows several estimates of the velocity function based on standard CDM mass functions and halo models (from Kochanek and White 2001 and Kochanek 2003a,b,c, using the Sheth and Tormen 1999 mass function combined with the NFW halo model from Sect. 4.1). We see for high masses or circular velocities that the predicted distribution of halos agrees with the observed distribution of clusters. At the velocities typical of galaxies, the observed density of galaxies is nearly an order of magnitude higher than expected for a CDM halo mass function. At very low velocities we expect many more halos than we observe galaxies. The velocity function estimated from the observed image separations matches that of galaxies with the beginnings of a tail extending onto the distribution of clusters at the high velocity end (Fig. 49). At low velocities the limited resolution of the present surveys means that the current lens data does not probe the low velocity end very well. In this section we discuss the difference between cluster and galaxy lenses and explain the origin of the break between galaxies and clusters observed. In Sect. 8 on CDM substructure we will discuss the divergence at low circular velocities.

The standard halo mass function is roughly a power law with $dn/dM \sim M^{-1.8}$ combined with an exponential cutoff at the mass scale corresponding to the largest clusters that could have formed at any epoch (e.g. the Sheth and Tormen 1999 halo mass function). Typically these rich clusters have internal velocity dispersions above 1000 km/s and can produce image splittings of ~ 30 arcsec. If halo structure was independent of mass, then we would expect the separation distribution of gravitational lenses to show a similar structure – a power law out to the mass scale of rich clusters followed by an exponential cutoff. In Fig. 50 we compare the observed distribution of radio lenses to that expected from the halo mass function assuming either NFW halos or NFW

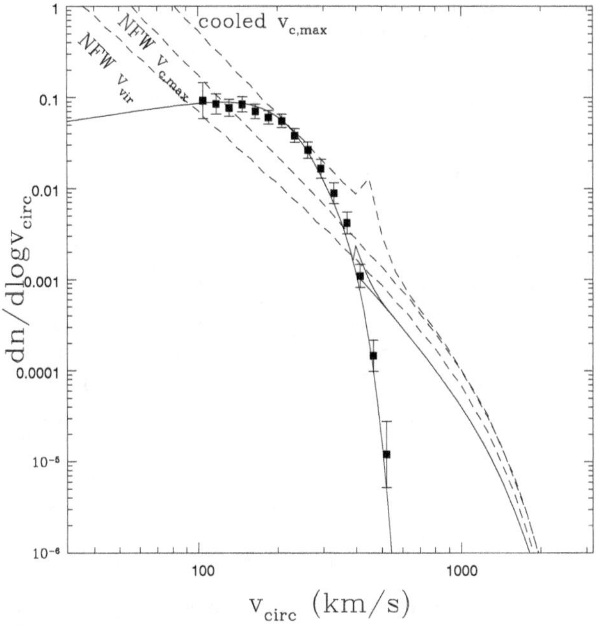

Fig. 49. The expected circular velocity function $dn/d\log v_c$ of CDM halos. The *lowest dashed curve* labeled NFW v_{vir} shows the velocity function using the NFW halo virial velocity v_{vir} for the circular velocity (see Sect. 4.1). The *middle dashed curve* labeled NFW $v_{c,max}$ shows the velocity function if the peak circular velocity of the halo is used rather than the virial velocity. The *upper dashed curve* is a model in which the baryons of halos with $M \lesssim 10^{13} M_\odot$ cool, raising the central density and circular velocity. The *solid curve* with the points shows the estimate of the local velocity function of galaxies (Fig. 40) and the solid curve extending to higher velocities is an estimate of the local velocity function of groups and clusters

halos in which the baryons, representing 5% of the halo mass has cooled and condensed into the centers of the halos (Kochanek and White 2001). We would find similar curves if we used simple SIS models rather than these more complex CDM-based models (Keeton et al. 1998; Porciani and Madau 2000). In practice, the most complete survey for multiply imaged sources, the CLASS survey, found a largest separation of 4.″5 (B2108+213) despite carefully checking candidates out to separations of 15.″0 (Phillips et al. 2001). The largest lens found in a search for multiply imaged sources has an image separation of roughly 15 arcsec (SDSS1004+4112, Inada et al. 2003). The overall separation distribution (see Fig. 50) has a sharp cutoff on scales of 3 arcsec corresponding to galaxies with velocity dispersions of $\sim$ 250 km/s. The principal searches for wide separation lenses are Maoz et al. (1997); Ofek et al. (2001) and Phillips et al. (2001), although most surveys searched for

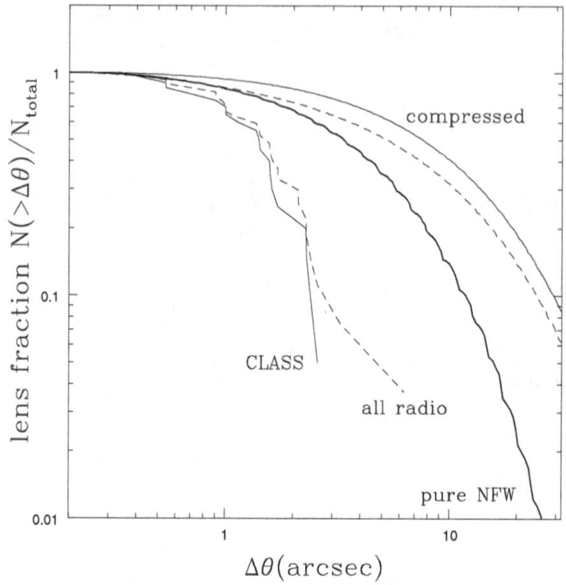

Fig. 50. Predicted image separation distributions assuming the structure of halos does not change with halo mass. The *heavy solid line* shows the prediction for pure NFW models while the *light solid (dashed) curves* shows the predictions after 5% of the baryons have cooled into a disk (a disk plus a bulge with 10% of the baryonic mass in the bulge). The curves labeled CLASS (for the CLASS survey lenses) and all radio (for all radio selected lenses) show the observed distributions

image separations of at least 6.″0. A large number of studies focused only on the properties of lenses produced by CDM mass functions (e.g. Narayan and White 1988; Wambsganss et al. 1995, 1998; Kochanek 1995a,b; Maoz et al. 1997; Flores and Primack 1996; Mortlock and Webster 2000a,b,c; Li and Ostriker 2002; Keeton and Madau 2001b; Wyithe, Turner and Spergel 2001). We will not discuss these in detail because such models cannot reproduce the observed separation distributions of lenses. Most recent analyses allow for changes in the density distributions between galaxies and clusters.

Physically the important difference between galaxies and clusters is that the baryons in the galaxies have cooled and condensed into the center of the halo to form the visible galaxy. As the baryons cool, they also drag some of the dark matter inward through a process known as adiabatic compression (Blumenthal et al. 1986), although this is less important than the cooling. As we show in Fig. 51, standard dark matter halos are terrible lenses because their central cusps ($\rho \propto r^{-\gamma}$ and $1.5 \geq \gamma \geq 1$) are too shallow. In this case, a standard NFW halo with a total mass of $10^{12} M_\odot$ and a concentration of $c = 8$ (see (60–62)) at a redshift of $z_l = 0.5$ is unable to produce multiple images of a source at redshift $z_s = 2$ despite having an asymptotic circular

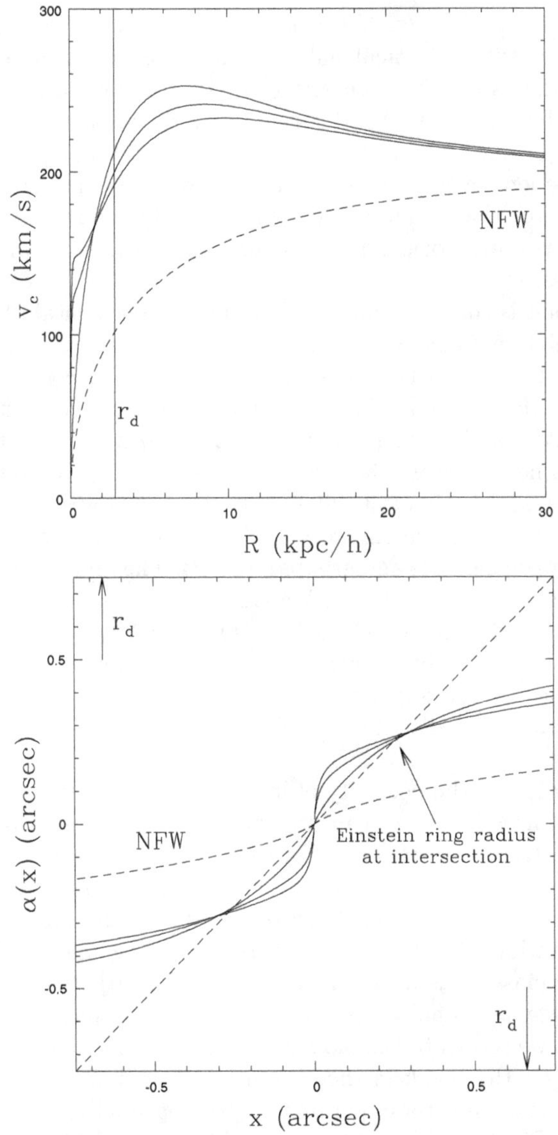

Fig. 51. (*Top*) The rotation curve and (*bottom*) the bending angle $\alpha(x)$ for a $10^{12} M_\odot$ halo at $z_l = 0.5$ with a concentration of $c = 8$ lensing a source at $z_s = 2.0$. The *dashed curves* show the results for the initial NFW halo, while the *solid curves* show the results after allowing 5% of the mass to cool conserving angular momentum (spin parameter $\lambda = 0.04$) and adiabatically compressing the dark matter. The three solid curves show the effect of putting 0%, 10% or 20% of the baryonic mass into a central bulge. Higher bulge masses raise the central circular velocity and steepen the central deflection profile. The final disk scale length is r_d. Compare these to the bending angles of our simple models in Figs. 10–14

velocity of nearly 200 km/s. If we now assume that 5% of the mass is in baryons starting with a typical halo angular momentum and then cooling into a disk of radius r_d while conserving angular momentum we see that the rotation curve becomes flatter and the galaxy is now able to produce multiple images. Putting some fraction of the mass into a still more compact, central bulge make the lens even more supercritical and the bending angle diagram begins to resemble that of an SIS lens (see Fig. 11). Thus, the cooling of the baryons converts a sub-critical dark matter halo into one capable of producing multiple images.

The key point is that only intermediate mass halos contain baryons which have cooled. High mass halos (groups and clusters) have cooling times longer than the Hubble time so they have not had time too cool (e.g. Rees and Ostriker 1977). Most low mass halos also probably resemble dark matter halos more than galaxies with large quantities of cold baryons because they lost their baryons due to heating from the UV background during the initial period of star formation (e.g. Klypin et al. 1999; Bullock, Kravtsov and Weinberg 2000; see Sect. 8). Here we ignore the very low mass halos and consider only the distinction between galaxies and groups/clusters. The fundamental realization in recent studies (e.g. Porciani and Madau 2000; Kochanek and White 2001; Kuhlen, Keeton and Madau 2004; Li and Ostriker 2003) is that introducing a cooling mass scale M_c below which the baryons cool to form galaxies and above which they do not supplies the explanation for the difference between the observed separation distribution of lenses and naive estimates from halo mass functions.

Once we recognize the necessity of introducing a distinction between cluster and galaxy mass halos, we can use the observed distribution of lens separations to constrain the mass scale of the break and the physics of cooling. Figure 52 shows the most common version of these studies, where separation distributions are computed as a function of the cooling mass scale M_c. We show the separation distributions for various cooling mass scales assuming that 5% of the mass cools into a disk plus a bulge with 10% of the baryonic mass in the bulge for all halos with $M < M_c$. If the cooling mass is either too low or too high we return to the models of Fig. 50, while at some intermediate mass scale we get the break in the separation distribution to match the observed angular scale. For these parameters, the optimal cooling mass scale is $M_c \simeq 10^{13} M_\odot$ (Fig. 52). This agrees reasonably well with Porciani and Madau (2000) and Kuhlen, Keeton and Madau (2004) who found a somewhat higher mass scale $M_c \simeq 3 \times 10^{13} M_\odot$ using SIS models for galaxies. Cosmological hydrodynamic simulations by Pearce et al. (1999) also found that approximately 50% of the baryons had cooled on mass scales near $10^{13} M_\odot$. Note, however, that the mass scale needed to fit the data depends on the assumed fraction of the mass in cold baryons. With fewer cold baryons a halo becomes a less efficient lens producing smaller image separations so M_c must increase to keep the break at the observed scale. If the cold baryon fraction is too low ($\lesssim 1\%$), it becomes impossible to explain the data at all. Crudely, the

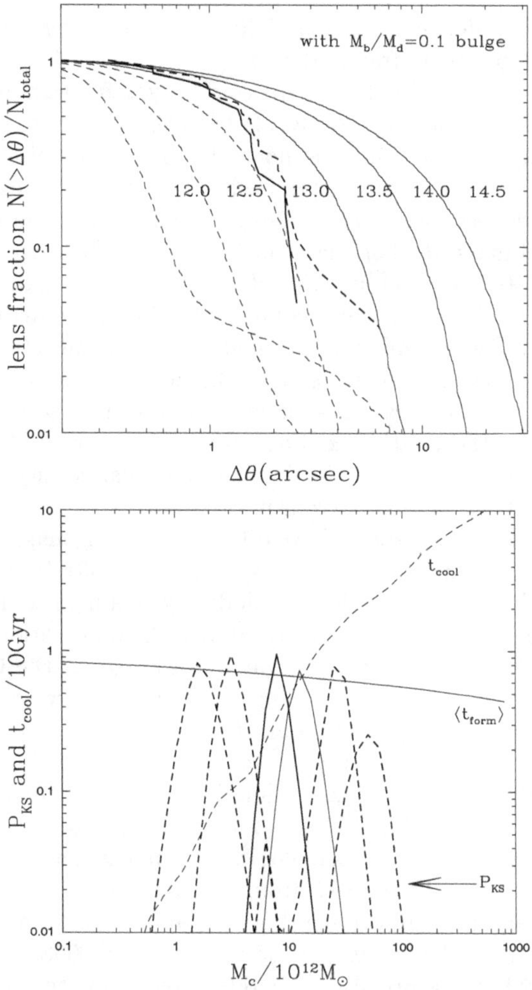

Fig. 52. (*Top*) Predicted separation distributions as a function of the cooling mass scale M_c in which 5% of the mass cools with 90% of the cooled material in a disk and 10% in a bulge. The *dashed curves* show the distributions for $M_c = 10^{12} M_\odot$, $3 \times 10^{12} M_\odot$ and $10^{13} M_\odot$, while the *solid curves* show the distributions for $M_c = 3 \times 10^{13} M_\odot$, $10^{14} M_\odot$ and $3 \times 10^{14} M_\odot$. The heavy solid (*dashed*) curves shows the observed distribution of the CLASS (all radio-selected) lenses. (*Bottom*) The Kolmogorov–Smirnov probability, P_{KS}, of fitting the observed distribution of CLASS lenses as a function of the cooling mass scale M_c. The *heavy solid curves* show the results when 5% of the mass cools without (with) 10% of that mass in a bulge. The *heavy dashed curves* show the results for models where lower (1% and 2%) or higher (10% and 20%) halo mass fractions cool, where the optimal cooling mass scale M_c decreases as the cold baryon fraction increases. For comparison, the light dashed line shows the cooling time t_{cool} in units of 10 Gyr for the radius enclosing 50% of the baryonic mass in the standard model. The *light solid line* shows the average formation epoch, $\langle t_{form} \rangle$, also in units of 10 Gyr

cooling mass scale depends exponentially on the cold baryon fraction with $\log M_c/M_\odot \simeq 13.6 - (\text{cold fraction})/0.15$.

The mass scale of the break and the cold baryon fraction are not independent parameters and should be derivable from the physics of the cooling gas. In its full details this must include not only the cooling of the gas but also reheating of the gas in galaxies due to feedback from star formation. Figure 52 also shows the dependence of the cooling time scale and the formation time scale for halos of mass M_c. For this model (based on the semi-analytic models of Cole et al. 2000), the cooling time becomes shorter than the age of the halo very close to the mass scale required to explain the distribution of image separations. These semi-analytic models suggest an alternate approach in where the cooling mass scale need not be added as an *ad hoc* parameter. We could instead follow the semi-analytic models and use the cooling function to determine the relative cooling rates of halos with different masses. We leave as the free parameter, the final cosmological density in cold baryons $\Omega_{b,cool} \leq \Omega_b \simeq 0.04$ (i.e. some baryons may never cool or cool and are reheated by feedback). Low $\Omega_{b,cool}$ models have difficulty cooling, making them equivalent to models with a high cooling mass scale (see Fig. 53). High $\Omega_{b,cool}$ models cool easily, making them equivalent to models with a high cooling mass scale. Models with $0.015 \lesssim \Omega_{b,cool} \lesssim 0.025$ agree with the observations (see Fig. 53). The result depends little on whether we add a bulge, fit the CLASS sample or all radio lenses or adjust the cooling curve by a factor of two. Thus, the characteristic scale of the gravitational lens separation distribution is a probe of the cosmological baryon density Ω_b and the fraction of those baryons that cool in the typical massive galaxy. While it would be premature to use this as a method for determining Ω_b, it is interesting to note that our estimate is significantly below current cosmological estimates that $\Omega_b \simeq 0.04$ which would be consistent with feedback from star formation and other processes preventing all baryons from cooling, but well above the estimates of the cold baryon fraction in local galaxies ($0.0045 \lesssim \Omega_{b,cool} \lesssim 0.0068$, Fukugita, Hogan and Peebles 1998). These are also the models generating the velocity function estimate with baryonic cooling in Fig. 49. The cooling of the baryons shifts the more numerous low velocity halos to higher circular velocities so that the models match the observed density of σ_* galaxies. The models do not correctly treat the break region because they allow "over-cooled" massive groups, but then merge back onto the peak circular velocity distribution of the CDM halos at higher velocities. Since the models allow all low mass halos to cool, there is still a divergence at low circular velocities which is closely related to the problem of CDM substructure we discuss in Sect. 8.

7.1 The Effects of Halo Structure and the Power Spectrum

Estimating the structure of clusters using gravitational lensing is primarily a topic for Part 3, so we include only an abbreviated discussion of lensing by clusters here. For a fixed cosmological model, two parameters largely control

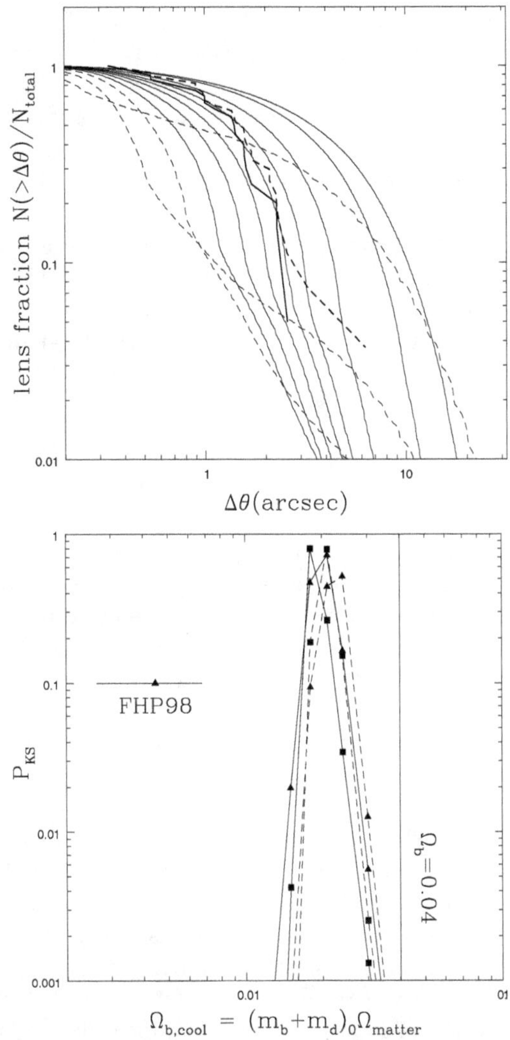

Fig. 53. (*Top*) Predicted separation distributions as a function of the cosmological cold baryon density $\Omega_{b,cool}$. The *dashed curves* show the results for $\Omega_{b,cool} = 0.003$, 0.006 and 0.009 (right to left at large separation) and the *solid curves* show the results for $\Omega_{b,cool} = 0.0012$, 0.015, 0.018, 0.021, 0.024, 0.030, 0.045 and 0.060 (from left to right at large separation). The models have 10% of the cold baryons in a bulge. The heavy solid (*dashed*) curves show the observed distribution of CLASS (all radio) lenses. (*Bottom*) The Kolmogorov–Smirnov probability, P_{KS}, of fitting the observed distribution of lenses as a function of the cold baryon density $\Omega_{b,cool}$. The squares (*triangles*) indicate models with no bulge (10% of the cooled material in a bulge), and the solid (*dashed*) lines correspond to fitting the CLASS (all radio) lenses. For comparison, the horizontal error bar is the estimate by Fukugita, Hogan and Peebles 1998 for the cold baryon (*stars, remnants, cold gas*) content of local galaxies. The vertical line marks the total baryon content of the concordance model

the abundance of cluster lenses. First, the abundance of clusters varies nearly exponentially with the standard normalization $\sigma_8 \simeq 1$ of the power spectrum on $8h^{-1}$ Mpc scales. Second, the cross sections of the individual clusters depend strongly on the exponent of the central density cusp of the cluster. There are recent studies of these issues by Li and Ostriker (2002, 2003); Huterer and Ma (2004); Kuhlen, Keeton and Madau (2004); Oguri et al. (2004), and Oguri and Keeton (2004).

We can understand the general effects of halo structure very easily from our simple power law model in (9). In Sect. 3 we normalized the models to have the same Einstein radius, but we now want to normalize them so that all have the same total mass interior to some much larger radius R_0. This is roughly what happens when we keep the virial mass and break radius of the halo constant but vary the central density exponent $\rho \propto r^{-n}$. The deflection profile becomes

$$\alpha(\theta) = \frac{b_0^2}{R_0}\left(\frac{\theta}{R_0}\right)^{2-n}, \tag{121}$$

where $b_0 \ll R_0$ sets the mass interior to R_0 and we recover our old example if we let $b = b_0 = R_0$. The typical image separation is determined by the tangential critical line at $\theta_t = R_0(b_0/R_0)^{2/(n-1)}$, so more centrally concentrated lenses (larger n) produce larger image separations when $b_0/R_0 \ll 1$. The radial caustic lies at $\beta_r = f(n)\theta_t$ where $f(n)$ is a not very interesting function of the index n, so the cross section for multiple imaging $\sigma \propto \beta_r^2 \propto R_0^2(b_0/R_0)^{4/(n-1)}$ – for an SIS profile $\sigma \propto b^4/R_0^2$, while the cross section for a Moore profile ($n = 3/2$) $\sigma \propto b^8/16R_0^6$ is significantly smaller.

We cannot go to the limit of an NFW profile ($n = 1$) because our power law model has a constant surface density rather than a logarithmically divergent surface density in the limit as $n \to 1$, but we can see that as the density profile becomes shallower the multiple image cross section drops rapidly when the models have constant mass inside a radius which is much larger than their Einstein radius. As a result, the numbers of group or cluster lenses depends strongly on the central exponent of the density distribution even when the mass function of halos is fixed. Magnification bias will weaken the dependence on the density slope because the models with shallower slopes and smaller cross sections will generally have higher average magnifications. The one caveat to these calculations is that many groups or clusters will have central galaxies, and the higher surface density of the galaxy can make the central density profile effectively steeper than the CDM halo in isolation.

7.2 Binary Quasars

Weedman et al. (1982) reported the discovery of the third "gravitational" lens, Q2345+007, a pair of $z = 2.15$ quasars separated by 7″.3. The optical spectra of the two images are impressively similar (e.g. Small et al. 1997), but repeated attempts to find a lens have failed in both the optical (e.g. Pello et al. 1996) and with X-rays (Green et al. 2002). Q2345+007 is the founding member of a class of objects seen in the optical as a pair of quasars with very similar spectra, small velocity differences and separations $3.''0 \lesssim \Delta\theta \lesssim 15.''0$. The most recent compilation contained 15 examples (Mortlock, Webster and Francis 1999). The incidence of these quasar pairs in surveys is roughly 2 per 1000 LBQS quasars (see Hewett et al. 1998) and 1 per 14000 CLASS radio sources (Koopmans et al. 2000a,b). The separations of these objects correspond to either very massive galaxies or groups/clusters. Obvious lenses on these scales, in the sense that we see the lens, are rare but have an incidence consistent with theoretical expectations (see Fig. 50). If, however, even a small fraction of the objects like Q2345–007 are actually gravitational lenses, then dark lenses outnumber normal groups and clusters and dominate the halo population on mass scales above $M \gtrsim 10^{13} M_\odot$.

If the criterion of possessing a visible lens is dropped, so as to allow for dark lenses, proving objects are lenses becomes difficult. There are two unambiguous tests – measuring a time delay between the images, which is very difficult given the long time delays expected for lenses with such large separations, or using deep imaging to show that the host galaxies of the quasars show the characteristic arcs or Einstein rings of lensed hosts (Figs. 3 and 4). The latter test is feasible with HST[6] and will be trivial with JWST. Spectral comparisons have been the main area of debate. In the optical, many of the pairs have alarmingly similar spectra if they are actually binary quasars (e.g. Q2345+007 or Q1634+267, see Small et al. 1997) – indeed, some of these dark lens candidates have more similar spectra than genuinely lensed quasars (see Mortlock, Webster and Francis 1999). The clearest examples of dark lens candidates that have to be binary quasars are the cases in which only one quasar is radio loud. These objects, such as PKS1145–071 (Djorgovski et al. 1987) or MGC2214+3550 (Muñoz et al. 1998), represent 4 of the 15 candidates. Similarly, the dramatic difference in the flux ratio between optical and X-ray wavelengths of Q2345+007 is the strongest direct argument for this object being a binary quasar (Green et al. 2002).

Two statistical arguments provide the strongest evidence that these objects must be binary quasars independent of any weighting of spectral similarities. The first argument, due to Kochanek, Falco and Muñoz 1999), is that the existence of binary quasars like MGC2214+3550 in which only one of the quasars is radio loud predicts the incidence of pairs in which both are radio quiet. We can label the quasar pairs as either O^2R^2, where both quasars are seen in the optical (O) and the radio (R), O^2R, where only one quasar is seen in the radio, or O^2 where neither quasar is seen in the radio. Lenses must be either O^2R^2 or O^2 pairs. Surveys of quasars find that only $P_R \simeq 10\%$ of quasars are radio sources with 3.6 cm fluxes above 1 mJy (e.g. Bischof and

[6] We detected the host galaxies of the Q2345–007 quasars in the CASTLES H-band image. Their morphology is probably inconsistent with the lens hypothesis, but we viewed the data as too marginal to publish the result.

Becker 1997). If all the quasar pairs were binary quasars and the probability of being radio loud is independent of whether a quasar is in a binary, then the relative number of O^2, O^2R and O^2R^2 binaries should be 1 to $2P_R = 0.2$ to $P_R^2 = 0.01$. Given that we observed 4 O^2R binaries we should observe 20 O^2 binaries and 0.2 O^2R^2 binaries. This statistical pattern matches the data, and Kochanek, Falco and Muñoz (1999) found that the most probable solution was that all quasar pairs were binary quasars with an upper limit of only 8% (68% confidence) on the fraction that could be dark lenses. With the subsequent expansion of the quasar pair sample and the discovery of the first O^2R^2 binary (B0827+525, Koopmans et al. 2000a,b), these limits could be improved.

The second statistical argument is that the dark lens candidates do not have the statistical properties expected for lenses. Three aspects of the quasar pairs make them unlikely to be lenses simply given the properties of gravitational lensing. First, there are no four-image dark lens candidates even though a third of the normal lenses are quads. Second, many of the dark lens candidates have very high flux ratios between the images – 4 of the 9 ambiguous quasar pairs considered by Rusin (2002) have flux ratios of greater than 10:1. Magnification bias makes such large flux ratios very improbable for true gravitational lenses (Sect. 6.6, Kochanek 1995a,b). Third, the suppression of central/third/odd images in the lens population is a consequence of baryonic cooling and the resulting increase of the central surface density. Standard dark matter halos with their shallow central cusps, $\rho \propto r^{-1}$, generally produce detectable third images. Since it is probably a requirement for a lens to remain dark that the baryons in the halo cannot cool (or they would form stars), you would expect the typical dark lens to resemble APM08279+5255 and have an easily detectable third image (Rusin 2002). Thus, in the context of CDM we would expect dark lenses to be standard cuspy density distributions like the NFW model (60). Rusin (2002) evaluated the likelihood of the quasar pairs assuming that dark lenses have the structure of CDM halos and found that the observed flux ratios and the lack of three-image dark lenses were extremely unlikely. Only the real lens APM08279+5255 had a significant probability of being produced by a dark CDM halo (also see Muñoz, Kochanek and Keeton 2001), although for this case I think the exposed cusp/disk lens explanation for the morphology is more likely.

The evidence overwhelmingly favors interpreting the quasar pairs as binary quasars. However, as originally pointed out by Djorgovski (1991), the one problem with the binary hypothesis is that the incidence of the quasar pairs is two orders of magnitude above that expected from an extrapolation of the quasar-quasar correlation function on scales of Mpc. As discussed in Kochanek, Falco and Muñoz 1999 and Mortlock, Webster and Francis (1999) the incidence can be increased if the incipient merger of the two host galaxies is triggering the quasar activity. The separation distribution of the binary quasars is crudely compatible with tidally triggered activity when the merger starts followed by a coalescence of the host galaxies driven by tidal friction.

Small separation binary quasars ($\Delta\theta < 3''\!.0$) are rare because the decay of the host galaxy orbits accelerates as their separation diminishes. Well-measured angular distributions of binary quasars, potentially obtainable from SDSS, might allow detailed explorations of the triggering and merging physics.

8 The Role of Substructure

Simulations of CDM halos predicted many more small satellites than were actually observed in the Milky Way (e.g. Kauffmann et al. 1993; Moore et al. 1999; Klypin et al. 1999). Crudely 5–10% of the mass was left in satellites with perhaps 1–2% at the projected separations of 1–$2R_e$ where we see most lensed images (e.g. Zentner and Bullock 2003; Mao et al. 2004). This is far larger than the observed fraction of 0.01–0.1% in observed satellites (e.g. Chiba 2002). Solutions were proposed in three broad classes: hide the satellites by preventing star formation so they are present but dark (e.g. Klypin et al. 1999; Bullock et al. 2000), destroy them using self-interacting dark matter (e.g. Spergel and Steinhardt 2000), or avoid forming them by changing the power spectrum to something similar to warm dark matter with significantly less power on the relevant mass scales (e.g. Bode et al. 2001). These hypotheses left the major observational challenge of distinguishing dark satellites from non-existent ones. This became known as the CDM substructure problem.

It was well known in the lensing community that the fluxes of lensed images were usually poorly fit by lens models. There was a long litany of reasons for ignoring them arising from possible systematic errors which can corrupt image fluxes. Differential effects between the images from the interstellar medium of the lens can corrupt the fluxes (dust in the optical/IR, scatter broadening in the radio, see Sect. 9.1). Time delays combined with source variability can corrupt any single-epoch measurement. Microlensing by the stars in the lens galaxy can modify the fluxes of any sufficiently compact component of the source (at a minimum the quasar accretion disk, see Part 4). The most peculiar problem was the anomalous flux ratios in radio lenses. Radio sources are essentially unaffected by the ISM of the lens galaxy in low resolution observations that minimize the effects of scatter broadening (VLA rather than VLBI), true absorption appears to be rare, radio sources generally show little variability even when monitored, and most of the flux should come from regions too large to be affected by microlensing. Yet in B1422+231, for example, the three cusp images violated the cusp relation for their fluxes (that the sum of the signed magnifications of the three images should be zero, see Metcalf and Zhao 2002; Keeton, Gaudi and Petters 2003; or Schneider, Ehlers and Falco 1992).[7]

[7] In specific models there can also be global invariants relating image positions and magnifications (e.g. Witt and Mao 2000; Hunter and Evans 2001; Evans and Hunter 2002). These results are usually for simple softened power law models

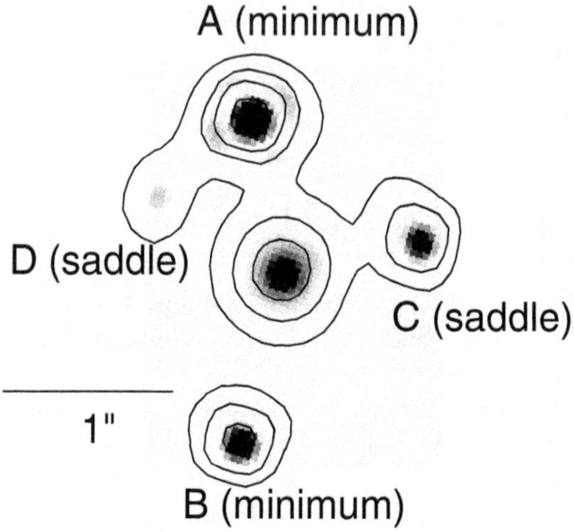

Fig. 54. The most spectacular example of an anomalous flux ratio, SDSS0924+0219 (Inada et al. 2003). In this CASTLES infrared HST image, the D image should be comparable in brightness to the A image, but is actually an order of magnitude dimmer. The A and B images are minima, while C and D are saddle points. The contours are spaced by factors of two from the peak of the A image. The lens galaxy is seen at the center. At present we do not know whether the suppression of the saddle point in this lens is due to microlensing or substructure. If it is microlensing, ongoing monitoring programs should see it return to its expected flux within approximately 10 years

It is easier to outline the problem of anomalous flux ratios near a fold caustic (such as images A and D in SDSS0924+0219, see Fig. 54), than a cusp caustic. Near a fold, the lens equations can be reduced to a one-dimensional model with

$$\beta = \theta \left(1 - \Psi''\right) - \frac{1}{2}\Psi'''\theta^2 \rightarrow -\frac{1}{2}\Psi'''\theta^2 \tag{122}$$

and inverse magnification

$$\mu^{-1} = (1 - \Psi'') - \Psi'''\theta \rightarrow -\Psi'''\theta, \tag{123}$$

where we choose our coordinates such that there is a critical line at $\theta = 0$ (i.e. $1 - \Psi'' = 0$) and the primes denote derivatives of the potential. These equations are easily solved to find that you have images at $\theta_\pm = \pm(-2\beta/\Psi''')^{1/2}$ if the argument of the square root is positive and no solutions otherwise – as you cross the fold caustic ($\beta = 0$) two images are created or destroyed on the critical line at $\theta = 0$. Their inverse magnifications of $\mu_\pm^{-1} = \mp(-2\beta\Psi''')^{1/2}$

using either ellipsoidal potentials or an external shear rather than ellipsoidal cuspy density distributions with an external shear, so their applicability to the observed lenses is unclear.

are equal in magnitude but reversed in sign. Hence, if the assumptions of the Taylor expansion hold, the images merging at a fold should have identical fluxes. Either by guessing or by tedious algebra you can determine that the fractional correction to the magnification from the next order term is of order $\theta_\pm \Psi^{(4)}/\Psi'''$. For any reasonable central potential where the images are at radius θ_0 from the lens center, the fractional correction will be of order $\theta_\pm/\theta_0 \sim 0.1$ for the typical pair of anomalous images. Hence, using gravity to produce the anomalous flux ratios requires terms in the potential with a length scale comparable to the separation of the images to significantly violate the rule that they should have similar fluxes. Mao and Schneider (1998) pointed out that a very simple way of achieving this was to put a satellite near the images, and they found that this could explain the anomaly in B1422+231. Metcalf and Madau (2001, also see Bradac et al. 2002 for images of the magnification patterns expected from a CDM halo) put these two pieces together, pointing out that if normal satellite galaxies were too rare to make anomalous flux ratios common, the missing CDM substructure was not. They predicted that in CDM, anomalous flux ratios should be common.

If we add a population of satellites with surface density $\kappa_{sat} = \Sigma_{sat}/\Sigma$ near the images we can estimate the nature of the perturbations. If we model them as pseudo-Jaffe potentials with critical radius b and break radius[8] $a = (bb_0)^{1/2}$, then the satellites produce a deflection perturbation of order

$$\langle \delta\theta^2 \rangle^{1/2} \sim 10^{-3} b_0 \left(\frac{10\Sigma_{sat}}{\Sigma_c} \right)^{1/2} \left(\frac{10^3 b}{b_0} \right)^{3/4} . \tag{124}$$

Only massive satellites will be able to produce deflection perturbations large enough to be detected given typical astrometric errors. Because the astrometric constraints for lenses are so accurate, generally better than 0.″005, satellites with deflection scales larger than $b \gtrsim 10^{-2} b_0$ will usually have observable effects on model fits and must be included in the basic lens model. The shear perturbation

$$\langle \delta\gamma^2 \rangle^{1/2} \sim 0.1 \left(\frac{10\Sigma_{sat}}{\Sigma_c} \right)^{1/2} \left(\frac{10^3 b}{b_0} \right)^{1/4} \left(\frac{\ln \Lambda}{10} \right)^{1/2} , \tag{125}$$

where $\ln \Lambda = \ln(a/s)$ is a Coulomb logarithm required to make the integral converge at small separations, is significantly larger. The effects of substructure gain on those from the primary lens as we move to quantities requiring more derivatives of the potential because the substructure has less mass but shorter length scales. For example most astronomical objects have masses and sizes that scale with internal velocity σ_v as $M \propto \sigma_v^4$ and $R \propto \sigma_v^2$. So time delays, which depend on the potential $\phi \propto M \propto \sigma_v^4$, will be completely unaffected by substructure. Deflections, which require one spatial derivative

[8] This is the tidal truncation radius for an SIS of critical radius b orbiting in an SIS of critical radius $b_0 > b$. The total satellite mass is $\simeq \pi a b \Sigma_c$.

of the potential, $\alpha \propto \phi/R \propto \sigma_v^2$, are affected only be the more massive substructures. Magnifications, which require two spatial derivatives of the potential, $\kappa \sim \gamma \sim \phi/R^2 \propto \sigma_v^0$, are affected equally by all mass scales provided the Einstein radius of the object is larger than the characteristic size of the source. Substructure will also affect brighter images more than fainter images because the magnifications of the brighter images are more unstable to small perturbations. Recall that the magnification $\mu = (\lambda_+\lambda_-)^{-1}$ where one of the eigenvalues $\lambda_\pm = 1 - \kappa \pm \gamma$, usually λ_-, is small for a highly magnified image. If we now add a shear perturbation $\delta\gamma$, the perturbation to the magnification is of order $\delta\gamma/\lambda_-$ so you have a bigger fractional perturbation to the magnification for the same shear perturbation if the image is more highly magnified. The last important effect from substructure, for which I know of no simple, qualitative explanation, is that substructure discriminates between saddle points and minima when it is a small fraction of the total surface density (Schechter and Wambsganss 2002; Keeton 2003b). In this regime, the magnification distributions for the saddle points develop an extended tail toward demagnification that is not present for the minima.

It turns out that anomalous flux ratios are very common – a fact which had been staring us in the face but was ignored because most people (including the author !) were mainly just annoyed that the flux ratios could not be used to constrain the potential of the primary lens so as to determine the radial mass profile. When Dalal and Kochanek (2002) collected the available four-image radio lenses to estimate the abundance of substructure, they found that 5 of 6 systems showed anomalies. In order to estimate the abundance of substructure Dalal and Kochanek 2002 developed a Bayesian Monte Carlo method which estimated the likelihood that adding substructure would significantly improve models of seven four-image lenses including the fact that the model for the primary lens would have to be adjusted each time any substructure was added. Figure 55 illustrates some tests of the method. Under the assumption that the uncertainties in flux measurements (systematic as well as statistical) were 10%, they found a substructure mass fraction of $0.006 < f_{sat} < 0.07$ (90% confidence) with a median estimate of $f_{sat} = 0.02$. This is consistent with expectations from CDM simulations, including estimates of the destruction of the satellites in the inner regions of galaxies (Zentner and Bullock 2003; Mao et al. 2004), and too high to be explained by normal satellite populations. Because the result is driven by the flux anomalies, which do not depend on the mass of the substructures, rather than astrometric anomalies, which do depend on the mass, the results had almost no ability to estimate the mass scale associated with the substructure.

While substructure with approximately the surface density expected from CDM is consistent with the data, it is worth examining other possibilities. We would expect any effect from the ISM to be strongly frequency dependent (whether in the radio or in the optical). At least for radio lenses, Kochanek and Dalal (2004) found that the optical depth function needed to explain the radio flux anomalies would have to be gray, ruling out all the standard radio suspects. We would also expect propagation effects at radio frequencies to

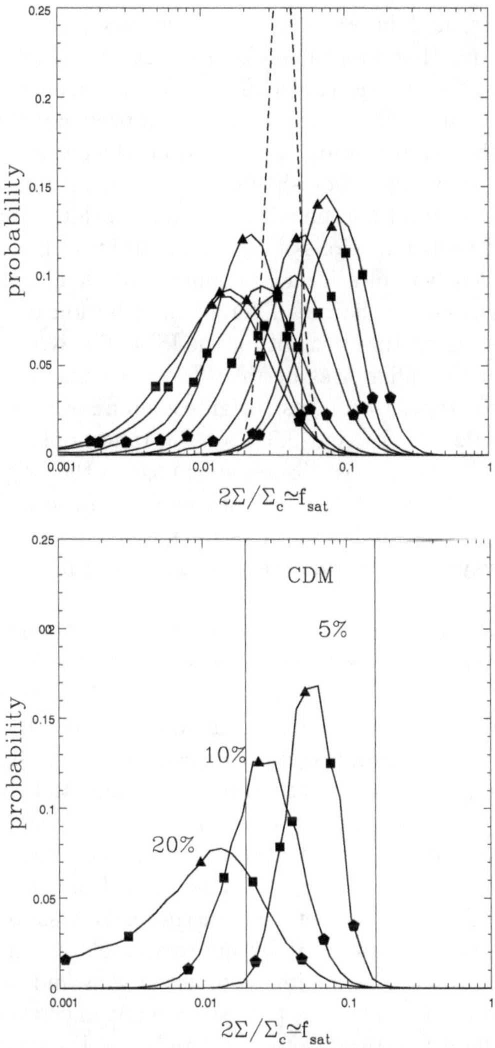

Fig. 55. (*Top*) A Monte Carlo test for estimating substructure surface densities. The *heavy curves* show the estimated probability distribution for the substructure surface density fraction in a sample of 7 four-image lenses in which the input fraction was 5% (marked by the *vertical line*). The points on the curve show the median, 1σ and 2σ confidence limits. The output distributions are consistent with the true input fraction. The dashed line shows how the accuracy would improve given a sample of 56 lenses (i.e. multiplying the 8 trials of 7 images each). (*Bottom*) The same method applied to the real data. The three distributions show the effects of changing assumptions on the actual flux measurement errors – the greater the measurement uncertainties the less substructure surface density is required to explain the flux ratio anomalies. The middle case (10%) is probably slightly too conservative (20% is ridiculously conservative and 5% is probably too optimistic)

preferentially affect the faintest images because they have the smallest angular sizes – remember that more magnified images are always bigger even if you cannot resolve the change in size. The ISM also cannot discriminate between images based on parity – the ISM is a local property of the lens and the parity is not, so they cannot show a correlation. Hence, if radio propagation effects created the anomalies they should be the same for minima and saddle points and more important for the fainter than the brighter images. Figure 56 shows the cumulative distributions of flux residuals for radio, optical and combined four-image lens samples from Kochanek and Dalal (2004). The bright saddle point images clearly have a different distribution in each case, as we would expect for substructure but not for the ISM. The Kolmogorov–Smirnov test significance of the differences between the most magnified saddle points and the other three types of images (brightest minimum, faintest minimum, faintest saddle) is 0.04%, 5% and 0.3% for the radio, optical and joint samples respectively. The next most discrepant image is the brightest minimum, also as expected for substructure, but with less significance. Various statistical games (bootstrap resampling methods of estimating significance or testing for anomalies) always give the same results. Thus, the ISM is ruled out as an explanation.

Even though simple Taylor series arguments make it unlikely that changes to the central potential are a solution (see Sect. 4.4), it still has its advocates (Evans and Witt 2003; Quadri et al. 2003; Möller, Hewett and Blain 2003; Kawano et al. 2004). The basic answer is that it is possible to create flux anomalies by making the deviations of the central potential from ellipsoidal sufficiently large for the angular structure of the potential to change rapidly enough between nearby images to produce the necessary magnification changes. There are three basic problems with this solution (see Sect. 4.6 as well).

The first problem is that the required deviations from an ellipsoidal profile are far too large. This is true even though the biggest survey of such models allowed image positions to shift by approximately 10 times their actual uncertainties in order to alter the image fluxes (Evans and Witt 2003) – had they forced the models to match the true astrometric uncertainties they would have needed even larger perturbations. Kochanek and Dalal (2004) found that models fitting the flux anomalies required $|a_4| \gg 0.01$ compared to the typical values observed for galaxies and simulated halos ($|a_4| \sim 0.01$, see Sect. 4.4). It is fair to say, however, that the quantitative results on the multipole structure of simulated halos are limited.

The second problem is that when we test these solutions in lenses for which we have additional model constraints, the models are forced back toward the standard ellipsoidal models. The basic problem, as Evans and Witt (2003) show, is that the problem of fitting image positions and fluxes with potentials of the form $rF(\theta)$ can be reduced to a problem in linear algebra if $F(\theta)$ is expanded as a multipole series – by adding enough terms it is possible to fit any four-image lens exactly. The reasons go back to the lack of constraints we discussed in Sect. 4.6. Figure 26 illustrates this point using the lens B1933+503.

Kochanek and Dalal (2004) first fit the four compact images with a model including deviations from an ellipsoidal surface density. With sufficiently strong deviations there were models that could eliminate the flux anomalies in this system. However, this lens, B1933+503, actually has three components to its source – a compact core forming the four-image system with the anomaly but also to radio lobes lensed into another four-image system and a two-image system for 10 images in all (Fig. 6). When we add the constraints from these other images the model is forced back to being a standard ellipsoidal model with a flux ratio anomaly. In the future, the degree to which lens galaxy potentials are ellipsoidal could be thoroughly tested in the lenses with Einstein ring images of their host galaxies.

The third problem with using the central potential to produce flux ratio anomalies is that it does not lead to the discrimination between saddle points and minima shown in Fig. 56. Kochanek and Dalal (2004) demonstrate this with Monte Carlo simulations, but the basic reason is simple. Consider a lens like PG1115+080 with two images merging at a saddle point. The sense with which the saddle point and minima are perturbed depends on the phase of the higher order multipoles relative to the images and the critical line, but for any fixed lens potential, that phase varies depending on the source position, so the average effect cannot make the bright saddle points show a significantly different set of properties from the bright minima. Every observed flux anomaly could be explained by adding complex angular structures to the main lens, but the inability of these models to differentiate between saddle points and minima would still rule them out.

For the moment there are two barriers to improving estimates of the substructure mass fraction. First, radio lens surveys have run out of sources bright enough to conduct efficient surveys. This will only change as upgrades to existing radio arrays are completed. The proposed Merlin and VLA upgrades will provide both sensitivity and resolution improvements that will make the next generation of radio lens surveys easier than the last. Second, searches for substructure using optical quasars need to separate the effects of microlensing and substructure. With simple imaging this can be done by finding parts of the quasar which are sufficiently extended to avoid significant contamination from microlensing. Emission line (e.g. Moustakas and Metcalf 2003) and dust emission regions should both be large enough to filter out the effects of the stars. Studying emission line ratios is now relatively easy because of the new generation of small-pixel integral field spectrographs on 8m-class telescopes. Mid-infrared flux ratios for the dusty regions remain difficult, but they have been obtained for one lens (Q2237+0305, Agol et al. 2000) and could be extended to several more.

The gold standard, however, would be astrometric detection of dark substructure so that we would obtain a direct, mass estimate. In all the present analyses, the most massive substructures were included as part of the model. They were not, however, dark substructures because they matched to satellites visible in HST images of the lenses. For example, Object X in MG0414

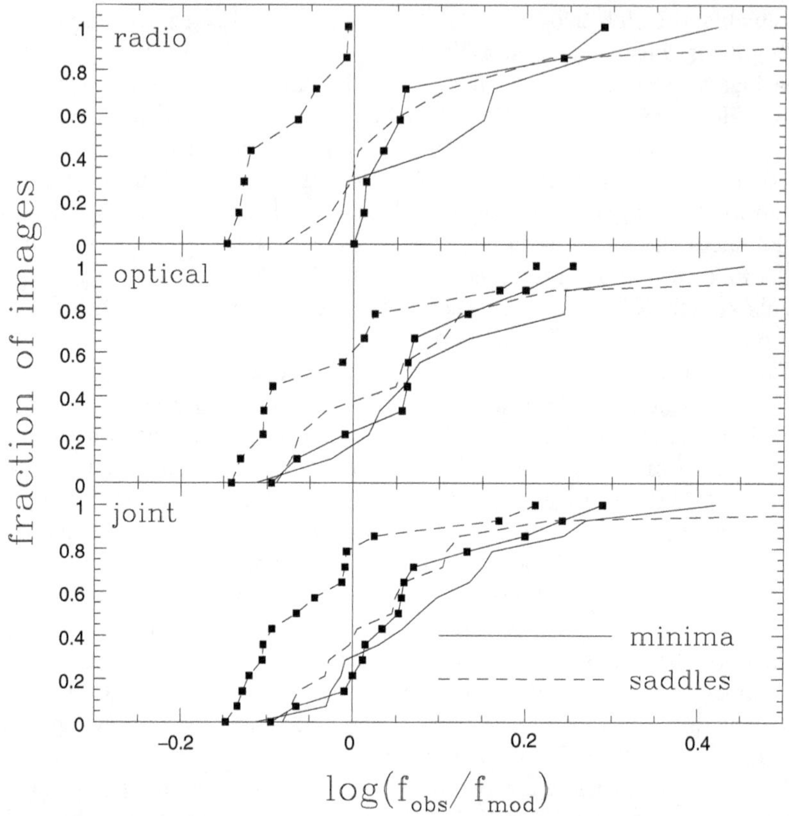

Fig. 56. Saddle point suppression in lenses. The three panels show the cumulative distributions of model flux residuals, $\log(f_{obs}/f_{mod})$, in the real data, assuming constant fractional flux errors for each image. The solid (*dashed*) lines are for minima (*saddle points*), with squares (*no squares*) for the distribution corresponding to the most (least) magnified image. From top to bottom the distributions are shown for samples of 8 radio, 10 optical or 15 total four-image lenses. If the flux residuals are created by propagation effects we would not expect the distributions to depend on the image parity or magnification, while if they are due to low optical depth substructure we would expect the distribution for the brightest saddle points to be shifted to lower observed fluxes

+0534 (Fig. 7) has effects on the image positions that are virtually impossible to reproduce with changes in the potential of the central lens galaxy (Trotter, Winn and Hewitt 2000), while models with it easily fit the data (Ros et al. 2000). Figure 57 shows the dependence of the goodness of fit to MG0414+0534 on the location of an additional lens component, with a deep minimum located at the observed position of Object X. The deflections produced by an object of mass M generally scale as $M^{1/2}$, so it is relatively easy to detect the deflection perturbations from objects only 1% the mass of the

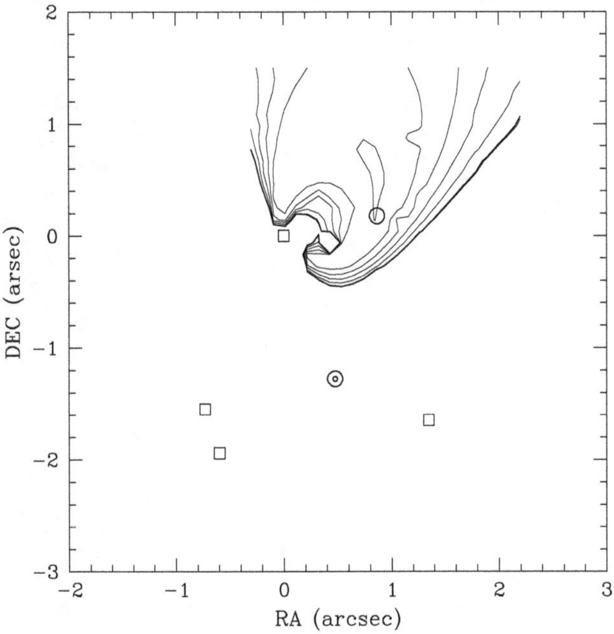

Fig. 57. The improvement in the fit to the Ros et al. (2000) VLBI data on MG0414+0534 from adding an additional lens with an Einstein radius 15% that of the primary lens galaxy as a function of its position. The *squares* show the location of the quasar images, the *central circles* mark the position of the main lens galaxy and the *single circle* marks the position of object X (see Fig. 7). The heavy contour has the same $\chi^2 = 123$ as single component models, and they then drop a factor of 0.2 per lighter contour to a minimum of $\chi^2 = 0.6$ almost exactly at the position of Object X

primary lens. One approach is to search lenses with VLBI structures for signs of perturbations. This has been attempted for B1152+199 by Metcalf (2002), but the case for substructure is not very solid given the limited nature of the data. The cleanest example of astrometric detection of something small, but sadly not dark, is in the VLBI structure of image C in MG2016+112 (Koopmans et al. 2002). The asymmetry in the VLBI component separations of image C on either side of the critical line (see Fig. 58) is due to a very faint galaxy 0.″8 South of the image with a deflection scale ∼ 10% of the primary lens (see Fig. 7). This is in reasonable agreement with the prediction from the H-band magnitude difference of 4.6 mag and the (lens) Faber–Jackson relation between magnitudes and deflections. In this case, we even know that the satellite is at the same redshift as the lens because Koopmans and Treu (2002) accidentally measured its redshift in the course of their observations to measure the velocity dispersion of the lens galaxy.

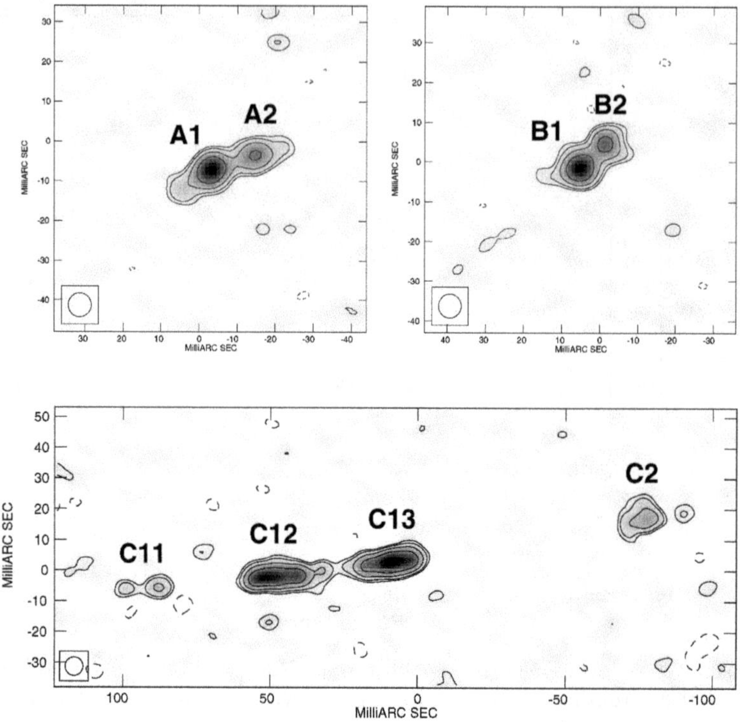

Fig. 58. VLBI maps of MG2016+112 (Koopmans et al. 2002). The large difference in the C_{11}/C_{12} separation as compared to the C_{13}/C_2 separation is the clearest example of an "astrometric" anomaly in a lens. The critical line passes between C_{12} and C_{13} and by symmetry we would expect the separations of the subcomponents on either side of the critical line to be similar. In this case the cause of the asymmetry seems to be a galaxy D about 0.″8 South of the C image (see Fig. 7). Galaxy D has the same redshift as the primary lens (Koopmans and Treu 2002)

8.1 Low Mass Dark Halos

When we are examining a particular lens, almost all the substructure will consist of satellites associated with the lens with only $\sim 10\%$ contamination from other small halos along the line-of-sight to the source (Chen, Kravtsov and Keeton 2003). However, the excess of low mass halos in CDM mass functions relative to visible galaxies is a much more general problem because the low mass CDM satellites should exist everywhere, not just as satellites of massive galaxies (Fig. 49, Gonzalez et al. 2000; Kochanek 2003a,b,c). Crudely, luminosity functions diverge as $dn/dL \sim 1/L \sim 1/M$ while CDM mass functions diverge as $dn/dM \sim M^{-1.8}$ so the fraction of low mass halos that must be dark increases $\sim M^{-0.8}$ at low masses. Figure 49 illustrates this assuming that all low mass halos have baryons which have cooled (e.g. Gonzalez et al. 2000; Kochanek 2003a,b,c). In the context of CDM, the solution to this general problem is presumably the same as for the satellites responsible for anomalous flux ratio – they exist but lost their baryons before they could form stars. Such processes are implicit in semianalytic models which can reproduce galaxy luminosity function (e.g. Benson et al. 2003) but can be modeled empirically in much the same way was employed for the break between galaxies in clusters in Sect. 7 (e.g. Kochanek 2003a,b,c). In any model, the probability of the baryons cooling to form a galaxy has to drop rapidly for halo masses below $\sim 10^{11} M_\odot$ just as it has to drop rapidly for halo masses above $\sim 10^{13} M_\odot$. Unlike groups and clusters, where we still expect to be able to detect the halos from either their member galaxies or X-ray emission from the hot baryons trapped in the halo, these low mass halos almost certainly cannot be detected in emission.

Unlike substructures in the halo of a massive galaxy that can be detected from their influence on the fluxes of lensed images, we can only detect isolated, low-mass dark halos if they multiply image background sources. For SIS lenses the distribution of image separations for small separations ($\Delta\theta/\Delta\theta_* \ll 1$, (111)) scales as

$$\frac{d\tau_{SIS}}{d\Delta\theta} \propto \Delta\theta^{1+\gamma_{FJ}(1+\alpha)/2}, \tag{126}$$

where α describes the divergence of the mass/luminosity function at low mass and γ_{FJ} is the conversion from mass to velocity dispersion (see Sect. 6.2). For the standard parameters of galaxies, $\alpha \simeq -1$ and $\gamma_{FJ} \simeq 4$, the separation distribution is $d\tau_{SIS}/d\Delta\theta \propto \Delta\theta$. In practice we do not observe this distribution because the surveys have angular selection effects that prevent the detection of small image separations (below 0.″25 for the radio surveys), so the observed distributions show a much sharper cutoff (Fig. 1). Even without a cutoff, there would be few lenses to find – the CLASS survey found 9 lenses between $0.''3 \leq \Delta\theta \leq 1.''0$ in which case we expect only one lens with $\Delta\theta < 0.''3$ even in the absence of any angular selection effects. A VLBI survey of 3% of the CLASS sources with milli-arcsecond resolution found no lenses (Wilkinson et al. 2001), nor would it be expected to for normal galaxy populations. Our non-parametric reconstruction of the velocity function including selection effects confirms that the existing lens samples are consistent with this standard model (Fig. 48).

The result is very different if we extrapolate to low mass with the $\alpha \simeq -1.8$ slope of the CDM halo mass function. The separation distribution becomes integrably divergent, $d\tau_{SIS}/d\Delta\theta \propto \Delta\theta^{-0.6}$, and we would expect 15 lenses with $\Delta\theta < 0.''3$ given 9 between $0.''3 \leq \Delta\theta \leq 1.''0$. Unfortunately, the Wilkinson et al. (2001) VLBI survey is too small to rule out such a model. A larger VLBI survey could easily do so, allowing the lenses to confirm the galaxy counting argument for the existence of second break in the density structure of halos at low mass (Kochanek 2003a,b,c; Ma 2003) similar to the one between galaxies and high mass halos (Sect. 7). If the baryons in the low mass halos either fail to cool, or cool and are then ejected by feedback, then their density distributions should revert to those of their CDM halos. If they are standard NFW halos, Ma (2003) shows that such low mass dark lenses will be very difficult to detect even in far larger surveys than are presently possible. Nonetheless, improving the scale of searches for very small separations from the initial attempt by Wilkinson et al. (2001) would provide valuable limits on their existence.

The resulting small, dark lenses would be the same as the dark lenses we discussed in Sect. 7.2 for binary quasars and explored by Rusin (2002). They will also create the same problems about proving or disproving the lens hypothesis as was raised by the binary quasars with the added difficulty that they will be far more difficult to resolve. Time delays, while short enough to be easily measured, will also be on time scales where quasars show little variability. Confirmation of any small dark lens will probably require systems with three or four images, rather than two images, and the presence of resolvable (VLBI) structures.

9 The Optical Properties of Lens Galaxies

The optical properties of lens galaxies and the properties of their interstellar medium (ISM) are important for two reasons. First, statistical calculations such as those in Sect. 6 rely on lens galaxies obeying the same scaling relations as nearby galaxies and the selection effects depend on the properties of the ISM. Thus, measuring the scaling relations of the observed lenses and the properties of their ISM are an important part of validating these calculations. Second, lenses have a unique advantage for studying the evolution of galaxies because they are the only sample of galaxies selected based on mass rather than luminosity, surface brightness or color. Evolution studies using optically-selected samples will always be subject to strong biases arising from the difficulty of matching nearby galaxies to distant galaxies. Selection by mass rather than light makes the lens samples almost immune to these biases.

Most lens galaxies are early-type galaxies with relatively red colors and few signs of significant on-going star formation (like the 3727Å or 5007Å Oxygen lines). The resulting need to measure absorption line redshifts is one of the reasons that the completeness of the lens redshift measurements is so poor. Locally, early-type galaxies follow a series of correlations which also exist for the lens galaxies and have been explored by Im, Griffiths and Ratnatunga (1997); Keeton, Kochanek and Falco (1998); Kochanek et al. (2000a,b); Rusin et al. (2003a); Rusin, Kochanek and Keeton (2003b); van de Ven, van Dokkum and Franx (2003); Rusin and Kochanek (2005).

The first, crude correlation is the Faber–Jackson relation between velocity dispersion and luminosity used in most lens statistical calculations. A typical local relation is that from Sect. 6.2 and shown in Fig. 39. Most lenses lack directly measured velocity dispersions, but all lenses have a well-determined image separation $\Delta\theta$. For specific mass models the image separation can be converted into an estimate of a velocity dispersion, such as the

$\Delta\theta = 8\pi(\sigma_v/c)^2 D_{ds}/D_s$ relation of the SIS, but the precise relationship depends on the mass distribution, the orbital isotropy, the ellipticity and so forth (see Sect. 4.9). For the lenses, there is a close relationship between the Faber–Jackson relation and aperture mass-to-light ratios. The image separation, $\Delta\theta$, defines the aperture mass interior to the Einstein ring,

$$M_{ap} = \frac{\pi}{4}\Sigma_c \Delta\theta^2, \tag{127}$$

where $\Sigma_c = c^2 D_s/4\pi G D_{ds} D_d$ is the critical surface density. By image separation we usually mean either twice the mean distance of the images from the lens galaxy or twice the critical radius of a simple lens model rather than a directly measured image separation because these quantities will be less sensitive to the effects of shear and ellipticity. If we measure the luminosity in the aperture L_{ap} using (usually) HST, then we know the aperture mass-to-light (M/L) ratio $\Upsilon_{ap} = M_{ap}/L_{ap}$.

If the mass-to-light ratio varies with radius or with mass, then to compare values of Υ_{ap} from different lenses we must correct them to a common radius and common mass. If these scalings can be treated as power laws, then we can define a corrected aperture mass-to-light ratio $\Upsilon_* = \Upsilon_{ap}(D_d^{ang}\Delta\theta/2R_0)^x$ where R_0 is a fiducial radius and x is an unknown exponent, and we would expect to find a correlation of the form

$$\log \Upsilon_* = 2(1+a)\log\Delta\theta + 0.4M_{abs} + \text{constant}, \tag{128}$$

where M_{abs} is the absolute magnitude of the lens (in some band) and a value $a \neq 0$ indicates that the mass-to-light ratio varies either with mass or with radius. We can then rewrite this in a more familiar form as

$$M_{abs} = M_{abs,0} + \gamma_{EV} z_l - 1.25\gamma_{FJ}\log\left(\frac{\Delta\theta}{\Delta\theta_0}\right), \tag{129}$$

where $\Delta\theta_0$ sets an arbitrary separation scale, γ_{EV} (or a more complicated function) determines the evolution of the luminosity with redshift, and $\gamma_{FJ} = 4(1+a)$ sets the scaling of luminosity with normalized separation defined so that for an SIS lens (where $\Delta\theta \propto \sigma_v^2$) the exponent γ_{FJ} will match the index of the Faber–Jackson relation (102). Figure 59 shows the resulting relation converted to the rest frame B band at redshift zero. The relation is slightly tighter than local estimates of the Faber–Jackson relation, but the scatter is still twice that expected from the measurement errors. The best fit exponent $\gamma_{FJ} = 3.29 \pm 0.58$ (Fig. 59) is consistent with local estimates and implies a scaling exponent $a = -0.18 \pm 0.14$ that is marginally non-zero. If the mass-to-light ratio of early-type galaxies increases with mass as $\Upsilon \propto M^x$, then $x = -a = 0.18 \pm 0.14$ is consistent with estimates from the fundamental plane that more massive early-type galaxies have higher mass-to-light ratios. The solutions also require evolution with $\gamma_{EV} = -0.41 \pm 0.21$, so that early-type galaxies were brighter in the past. These scalings can also be done in terms of observed magnitudes rather than rest frame magnitudes to provide

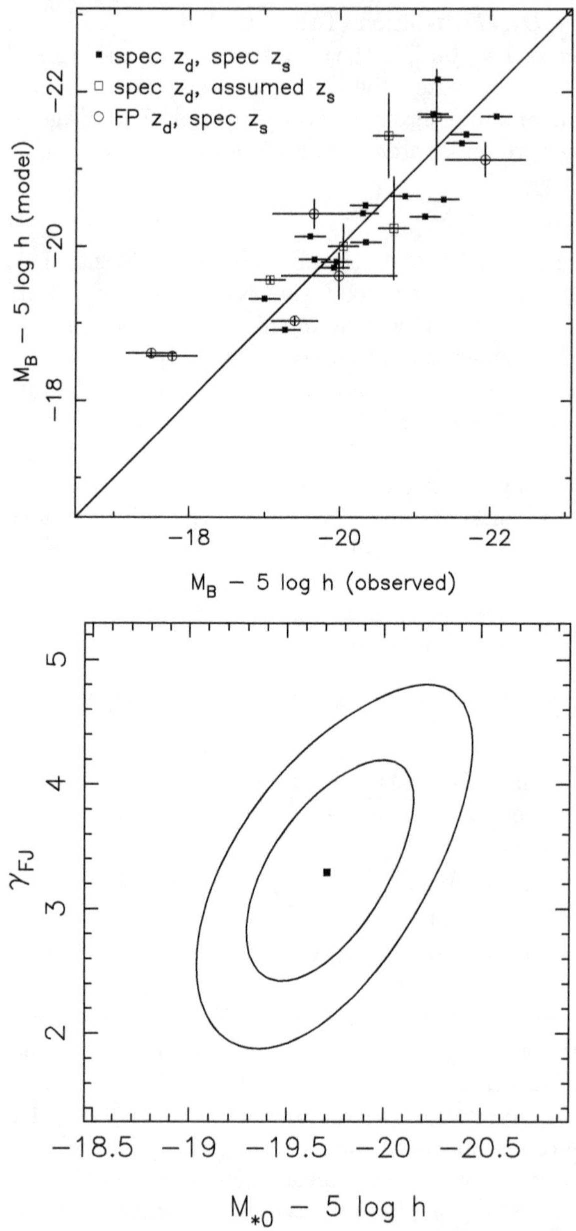

Fig. 59. (*Top*) The "Faber–Jackson" relation for gravitational lenses. The figure compares the observed absolute B magnitude corrected for evolution to that predicted from the equivalent of the Faber–Jackson relation for gravitational lenses (129). The different point styles indicate whether the lens and source redshifts were directly measured or estimated. From Rusin et al. (2003a,b). (*Bottom*) The redshift zero absolute B-band magnitude and effective exponent of the "Faber–Jackson" relation $L \propto \Delta\theta^{\gamma_{FJ}/2}$ for gravitational lenses

simple estimation formulas for the apparent magnitudes of lens galaxies in various bands as a function of redshift and separation to an rms accuracy of approximately 0.5 mag (see Rusin et al. 2003a,b).

The significant scatter of the Faber–Jackson relation makes it a crude tool. Early-type galaxies also follow a far tighter correlation known as the fundamental plane (FP, Dressler et al. 1987; Djorgovski and Davis 1987) between the central, stellar velocity dispersion σ_c, the effective radius R_e and the mean surface brightness inside the effective radius $\langle \mathrm{SB}_e \rangle$ of the form

$$\log\left(\frac{R_e}{h^{-1}\mathrm{kpc}}\right) = \alpha \log\left(\frac{\sigma_c}{\mathrm{km\ s}^{-1}}\right) + \beta\left(\frac{\langle \mathrm{SB}_e \rangle}{\mathrm{mag\ arcsec}^{-2}}\right) + \gamma, \tag{130}$$

where the slope α and the zero-point γ depend on wavelength but the slope $\beta \simeq 0.32$ does not (e.g. Scodeggio et al. 1998; Pahre, de Carvalho and Djorgovski 1998). Local estimates for the rest frame B-band give $\alpha = 1.25$ and $\gamma_0 = -8.895 - \log(h/0.5)$ (e.g. Bender et al. 1998). In principle both the zero points and the slopes may evolve with redshift, but all existing studies have assumed fixed slopes and studied only the evolution of the zero point with redshift. For galaxies with velocity dispersion measurements, the basis of the method is that measurement of R_e and σ_v provides an estimate of the surface brightness the galaxy will have at redshift zero. The difference between the measured surface brightness at the observed redshift and the surface brightness predicted for $z = 0$ measures the evolution of the stellar populations between the two epochs as a shift in the zero-point $\Delta\gamma$. The change in the zero-point is related to the change in the luminosity by $\Delta L = -0.4\Delta\mathrm{SB}_e = \Delta\gamma/(2.5\beta)$. While these estimates are always referred to as a change in the mass-to-light ratio, no real mass measurement enters operationally. If, however, we assume a non-evolving virial mass estimate $M = c_M \sigma_v^2 R_e/G$ for some constant c_M, then the FP can be rewritten in terms of a mass-to-light ratio,

$$\log \Upsilon = \log\left(\frac{M}{L}\right) \propto \left(\frac{10\beta - 2\alpha}{5\beta}\right)\log\sigma_c + \left(\frac{2-5\beta}{5\beta}\right)\log R_e - \frac{\gamma}{2.5\beta}, \tag{131}$$

so that if both α and β do not evolve, the evolution of the mass-to-light ratio is $d\log\Upsilon/dz = -(d\gamma/dz)/(2.5\beta)$. Either way of thinking about the FP, either as an empirical estimator of the redshift zero surface brightness or an implicit estimate of the virial mass, leads to the same evolution estimates but alternate ways of thinking about potential systematic errors.

Confusion about applications of lenses to the FP and galaxy evolution usually arise because most gravitational lenses lack direct measurements of the central velocity dispersion. Before addressing this problem, it is worth considering what is done for distant galaxies with direct measurements. The central dispersion appearing in the FP has a specific definition – usually either the velocity dispersion inside the equivalent of a 3.″0 aperture in the Coma cluster or the dispersion inside $R_e/8$. Measurements for particular galaxies almost never exactly match these definitions, so empirical corrections are applied to adjust the velocity measurements in the observed aperture to the standard aperture. As we explore more distant galaxies, resolution problems mean that the measurement apertures become steadily larger than the standard apertures. The corrections are made with a single, average local relation for all galaxies – implicit in this assumption is that the dynamical structure of the galaxies is homogeneous and non-evolving. This seems reasonable since the minimal scatter around the FP seems to require homogeneity, but says nothing about evolution. These are also the same assumptions used in the lensing analyses.

If early-type galaxies are homogeneous and have mass distributions that are homologous with the luminosity distributions, then there is no difference between the lens FP and the normal kinematic FP, independent of the actual mass distribution of the galaxies (Rusin and Kochanek 2005). If the mass distributions are homologous, then the mass and velocity dispersion are related by $M = c_M \sigma_c^2 R_e/G$ where c_M is a constant, σ_c is the central velocity dispersion (measured in a self-similar aperture like the $R_e/8$ aperture used in many local FP studies), and R_e is the effective radius. If we allow the mass-to-light ratio to scale with luminosity as $\Upsilon \propto L^x$, then the normal FP can be written as

$$\log R_e = \frac{2}{2x+1} \log \sigma_c + \frac{0.4(x+1)}{2x+1} \langle SB_e \rangle + \frac{\log c_M}{2x+1}, \qquad (132)$$

which looks like the local FP (130) if $\alpha = 2/(2x+1)$ and $\beta = 0.4(x+1)/(2x+1)$ (see Faber et al. 1987). Thus, the lens galaxy FP will be indistinguishable from the FP provided early-type galaxies are homologous and the slopes can be reproduced by a scaling of the mass-to-light ratio (as they can for $x \simeq 0.3$ given $\alpha \simeq 1.2$ and $\beta \simeq 0.3$, e.g., Jorgensen, Franx and Kjaergaard 1996 or Bender et al. 1998). All the details about the mass distribution, orbital isotropies and the radius interior to which the velocity dispersion is measured enter only through the constant c_M or equivalently from differences between the FP zero point γ measured locally and with gravitational lenses. In practice, Rusin and Kochanek (2005) show that the zero point must be measured to an accuracy significantly better than $\Delta\gamma = 0.1$ before there is any sensitivity to the actual mass distribution of the lenses from the FP. Thus, there is no difference between the aperture mass estimates for the FP and its evolution and the normal stellar dynamical approach unless the major assumption underlying both approaches is violated. It also means, perhaps surprisingly, that measuring central velocity dispersions adds almost no new information once these conditions are satisfied.

Rusin and Kochanek (2005) used the self-similar models we described in Sect. 4.8 to estimate the evolution rate and the star formation epoch of the lens galaxies while simultaneously estimating the mass distribution. Thus, the models for the mass include the uncertainties in the evolution and the reverse. Figure 60 shows (top) the estimated evolution rate, and shows (bottom) how this is related to a limit on the average star formation epoch $\langle z_f \rangle$ based on

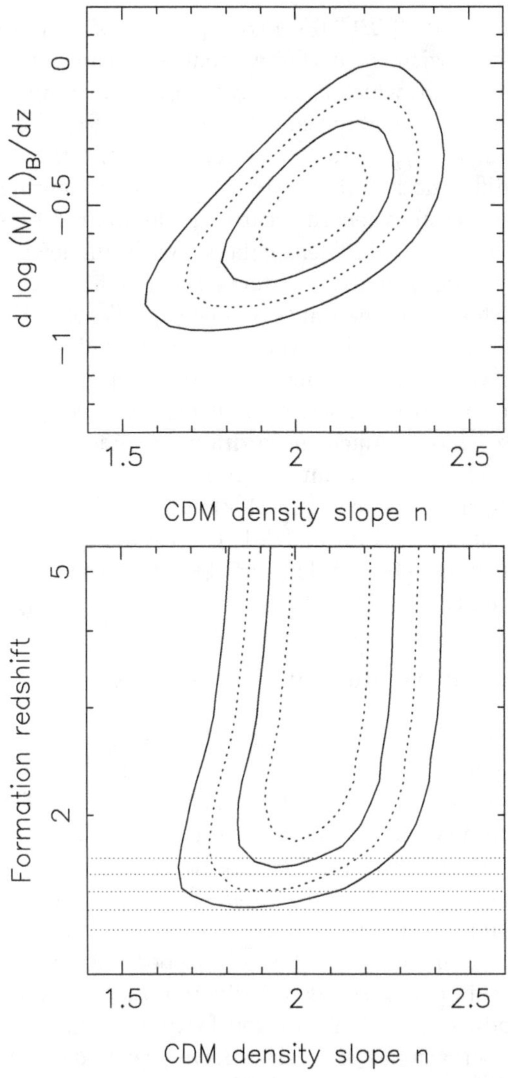

Fig. 60. (*Top*) Constraints on the B-band luminosity evolution rate $d\log(M/L)_B/dz$ as a function of the logarithmic density slope n ($\rho \propto r^{-n}$) of the galaxy mass distribution. Solid (*dashed*) contours are the 68% and 95% confidence limits on two parameter (one parameter). These use the self-similar mass models of (89) and are closely related to the fundamental plane. From Rusin and Kochanek (2005). (*Bottom*) Constraints on the mean star formation epoch $\langle z_f \rangle$ as a function of the logarithmic density slope n ($\rho \propto r^{-n}$) of the galaxy mass distribution. Solid (*dashed*) contours are the 68% and 95% confidence limits on two parameter (one parameter). The *horizontal dotted lines* mark $\langle z_f \rangle = 1.3$, 1.4, 1.5, 1.6 and 1.7. The lens sample favors older stellar populations with $\langle z_f \rangle > 1.5$ at 95% confidence. These use the self-similar mass models of (89) and are closely related to the fundamental plane. From Rusin and Kochanek (2005)

Bruzual and Charlot (1993, BC96 version) population synthesis models. This estimate is consistent with the earlier estimates by Kochanek et al. (2000a,b) and Rusin et al. (2003a,b) which used only isothermal lens models, as we would expect. van de Ven, van Dokkum and Franx (2003) found a somewhat lower star formation epoch ($\langle z_f \rangle = 1.8^{+1.4}_{-0.5}$) when analyzing the same data, which can be traced to differences in the analysis. First, by weighting the galaxies by their measurement errors when the scatter is dominated by systematics and by dropping two higher redshift lens galaxies with unknown source redshifts, van de Ven et al. (2003) analysis reduces the weight of the higher redshift lens galaxies, which softens the limits on low $\langle z_f \rangle$. Second, they used a power law approximation to the stellar evolution tracks which underestimates the evolution rate as you approach the star formation epoch, thereby allowing lower star formation epochs. These two effects leverage a small difference in the evolution rate[9] into a much more dramatic difference in the estimated star formation epoch. These evolution rates are consistent with estimates for cluster or field ellipticals by (e.g. van Dokkum and Franx 1996; van Dokkum et al. 2001; van Dokkum and Franx 2001; van Dokkum and Ellis 2003; Kelson et al. 1997; Kelson et al. 2000), and inconsistent with the much faster evolution rates found by Treu et al. (2001, 2002) or Gebhardt et al. (2003).

9.1 The Interstellar Medium of Lens Galaxies

As well as studying the emission by the lens galaxy we can study its absorption of emission from the quasar as a probe of the interstellar medium (ISM) of the lens galaxies. The most extensively studied effect of the ISM is dust extinction because of its effects on estimating the cosmological model from optically-selected lenses and because it allows unique measurements of extinction curves outside the local Group. There are also broad band effects on the radio continuum due to free-free absorption, scatter broadening and Faraday rotation. While all three effects have been observed, they have been of little practical importance so far. Finally, in both the radio and the optical, the lens can introduce narrow absorption features. While these are observed in some lenses, observational limitations have prevented them from being as useful as the are in other areas of astrophysics.

As we mentioned in Sect. 6, extinction is an important systematic problem for estimating the cosmological model using the statistics of optically selected lenses. It modifies the results by changing the effective magnification bias of the sample because it provides an effect to make lensed quasars dimmer than their unlensed counterparts. Because we see multiple images of the same quasar, it is relatively easy to estimate the differential extinction between

[9] Rusin and Kochanek (2005) obtained $d\log(M/L)_B/dz = -0.50 \pm 0.19$ including the uncertainties in the mass distribution, Rusin et al. (2003a,b) obtained -0.54 ± 0.09 for a fixed SIS model, and van de Ven et al. (2003) obtained -0.62 ± 0.13 for a fixed SIS model.

lensed images under the assumption that the quasar spectral shapes are not varying on the time scale corresponding to the time delay between the images and that microlensing effects are not significantly changing the slope of the quasar continuum. The former is almost certainly valid, while for the latter we simply lack the necessary data to check the assumption (although we have a warning sign from the systems where the continuum and emission line flux ratios differ, see Part 4). Under these assumptions, the magnitude difference at wavelength λ between two images A and B

$$m_A(\lambda) - m_B(\lambda) = -2.5 \log \left| \frac{\mu_A}{\mu_B} \right| + R\left(\frac{\lambda}{1+z_l}\right) \Delta E(B-V) \qquad (133)$$

depends on the ratio of the image magnifications μ_A/μ_B, the differential extinction $\Delta E(B-V) = E_A - E_B$ between the two images and the extinction law $R(\lambda/(1+z_l))$ of the dust in the rest frame of the dust. We have the additional assumption that either the extinction law is the same for both images or that one image dominates the total extinction (Nadeau et al. 1991). Because it is a purely differential measurement that does not depend on knowing the intrinsic spectrum of the quasar, it provides a means of determining extinctions and extinction laws that is otherwise only achievable locally where we can obtain spectra of individual stars (the pair method, e.g. Cardelli, Clayton and Mathis 1989). The total extinction cannot be determined to any comparable accuracy because estimates of the total extinction require an estimate of the intrinsic spectrum of the quasar. Figure 61 shows the distribution of differential extinctions found in the Falco et al. (1999) survey of extinction in 23 gravitational lenses. Only 7 of the 23 systems had colors consistent with no extinction, and after correcting for measurement errors and excluding the two outlying, heavily extincted systems the data are consistent with a one-sided Gaussian distribution of extinctions starting at 0 and with a dispersion of $\sigma_{\Delta E} \simeq 0.1$ mag. The two outlying systems, B0218+357 and PKS1830–211, were both radio-selected and both have one image that lies behind a molecular cloud of the late type lens galaxy (see below).

For lenses that have the right amount of dust, so that the image flux ratio can be measured accurately over a broad range of wavelengths, it is possible to estimate the extinction curve $R(\lambda/(1+z_l))$ of the dust (Nadeau et al. 1991) or to estimate the dust redshift under the assumption that the extinction curve is similar to those measured locally (Jean and Surdej 1998). Starting with Nadeau et al. (1991), there have been many estimates of extinction curves in lens galaxies (Falco et al. 1999; Toft, Hjorth and Burud 2000; Motta et al. 2002; Muñoz et al. 2004). The most interesting of these are for systems where the region near the 2175Å extinction feature is visible. This requires source and lens redshifts that put the feature at long enough wavelengths to be easily observed (i.e. higher lens redshifts) with a quasar UV continuum extending to shorter wavelengths (i.e. lower source redshifts). Motta et al. (2002) achieved the first cosmological detection of the feature in the $z_l = 0.83$ lens SBS0909+532, as shown in Fig. 62. The

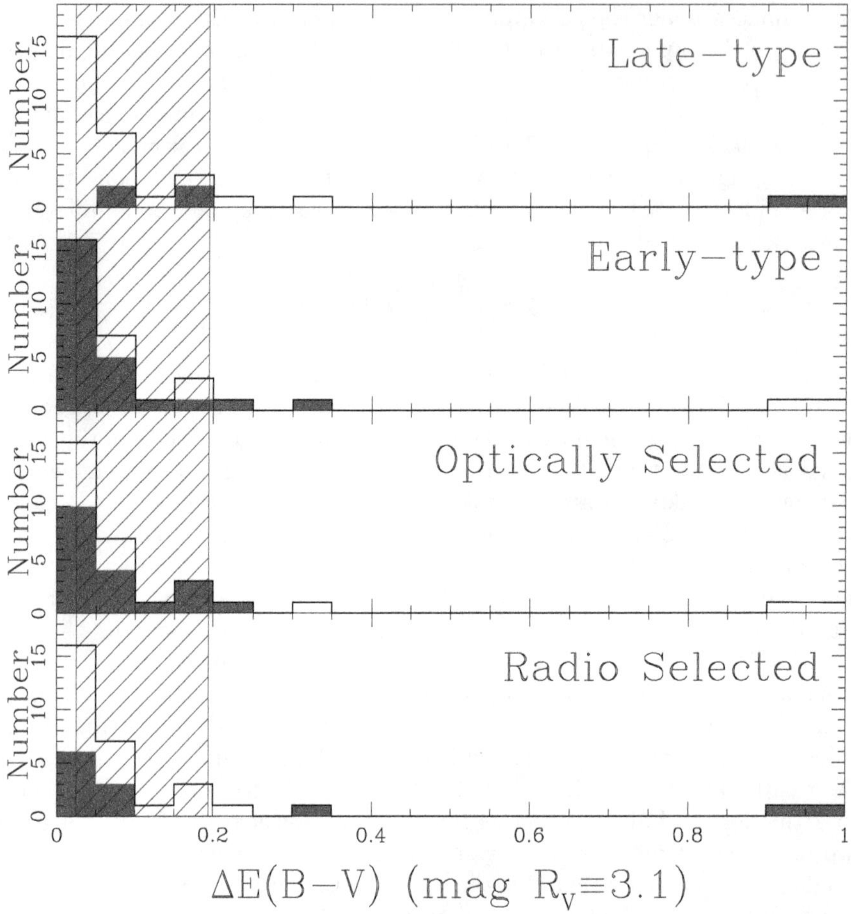

Fig. 61. Histograms of the differential extinction in various lens subsamples from Falco et al. (1999). In each panel the solid histogram shows the full sample of 37 differential extinctions measured in 23 lenses while the shaded histogram shows the distributions for different selection methods (radio/optical) or galaxy types (early/late). The hatched region shows the extinction range consistent with the Falco, Kochanek and Muñoz 1998 analysis of the difference between the statistics of radio-selected and optically-selected lens samples (see Sect. 6.6). Note that the most highly extincted systems, PKS1830–211 and B0218+357, are both radio-selected and late-type galaxies. The lowest differential extinction bins are contaminated by the effects of finite measurement errors

overall extinction curve is marginally consistent with a standard Galactic $R_V = 3.1$ extinction curve. Other cosmologically distant extinction curves are very different from normal Galactic models ranging for an anomalously low R_V curve in MG0414+0534 at $z_l = 0.96$ (Falco et al. 1999), probably an SMC

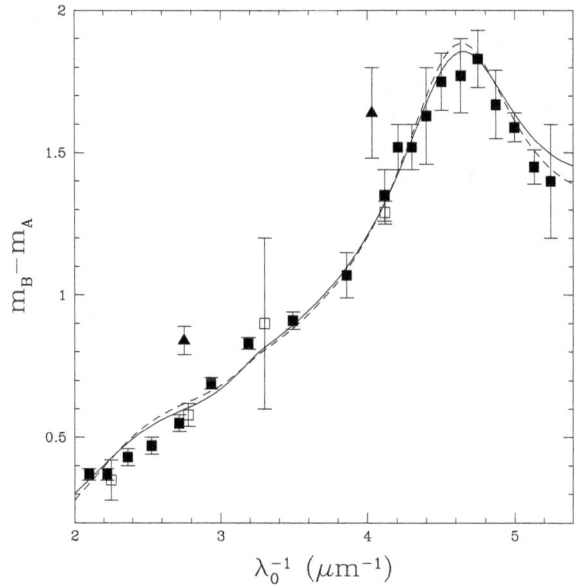

Fig. 62. The extinction curve of the dust in SBS0909+532 at $z_l = 0.83$ by Motta et al. (2002). The *solid squares* show the magnitude difference as a function of inverse rest wavelength derived from integral field spectra of the continuum of the quasars. The *open squares* are broad band measurements from earlier HST imaging and the *filled triangles* are the flux ratios in the quasar emission lines. The *solid curve* shows the best fit $R_V = 2.1 \pm 0.9$ Cardelli, Clayton and Mathis (1989) extinction curve while the *dashed curve* shows a standard $R_V = 3.1$ curve. The offset between the continuum and emission line flux ratios seems not to depend on wavelength and is probably due to microlensing

extinction curve in LBQS1009–252 at an estimated redshift of $z_l \simeq 0.88$ (Muñoz et al. 2004), and a anomalously high R_V extinction curve for the dust in the molecular cloud of the $z_l = 0.68$ lens galaxy in B0218+357. The Jean and Surdej (1998) idea of using the shape of the extinction curve to estimate the redshift of the dust also seems to work given a reasonable amount of dust and wavelength coverage (see Falco et al. 1999; Muñoz et al. 2004), but too few lenses with unknown redshifts satisfy the requirements for widespread use of the method.

For broad band radio emission from the source, the three observed propagation effects are free-free absorption, scatter broadening and Faraday rotation. For example, in PMNJ1632–0033, the candidate third image of the lens (C) has the same radio spectrum as the other two images except at the lowest frequency observed (1.4 GHz) where it is fainter than expected. This can be interpreted as free-free absorption by electrons at the center of the lens galaxy but the interpretation needs to be confirmed by measurements at additional frequencies to demonstrate that the dependence of the optical depth on wavelength is consistent with the free-free process (Winn, Rusin and Kochanek 2004). Scatter broadening is observed in many radio lenses (e.g. PMN0134–0931, Winn et al. 2003a,b,c; B0128+437, Biggs et al. 2004; PKS1830–211, Jones et al. 1996; B1933+503, Marlow et al. 1999) primarily as changes in the fluxes of images between high resolution VLBI observations and lower resolution VLA observations or apparently finite sizes for compact source components in VLBI observations. Aside from its effects in altering radio fluxes determined in VLBI observations, it seems to have practical consequences. In the presence of a magnetic field, the scattering medium will also rotate polarization vectors (e.g. MG1131+0456, Chen and Hewitt 1993). This is only of practical importance if maps which depend on the polarization vector are used to constrain the lens potential. In short, these effects are observed but have so far been of little practical consequence.

More surprisingly, absorption by atoms and molecule has also been of little practical import for lens physics as yet. Wiklind and Alloin (2002) provide an extensive review of molecular absorption and emission in gravitational lenses. The two systems with the strongest absorption systems are B0218+357 and PKS1830–211 (see Gerin et al. 1997 and references therein) where one of the two images lies behind a molecular cloud of the spiral galaxy lens. These two systems also show the highest extinction of any lensed images (Falco et al. 1999). Molecular absorption systems can be used to determine time delays (Wiklind and Alloin 2002), measure the redshift of lens galaxies (the lens redshift in PKS1830-211 is measured using molecular absorption lines, Wiklind and Combes 1996), and potentially to determine the rotation velocity of the lens galaxy (e.g. Koopmans and de Bruyn 2003). These studies at centimeter and millimeter wavelengths are heavily limited by the resolution and sensitivity of existing instruments, and the importance of these radio absorption features will probably rise dramatically with the completion of the next generation of telescopes (e.g. ALMA, LOFAR, SKA).

Similar problems face studies of metal absorption lines in the optical. Since most lenses are at modest redshifts, the strongest absorption lines expected from the lens galaxies tend to be observable only from space because they lie at shorter wavelengths than the atmospheric cutoff. For most lenses only the MgII (2800Å) lines are potentially observable from the ground since you only require a lens redshift $z_l \gtrsim 0.26$ to get the redshifted absorption lines longwards of 3500Å. The other standard metal line, CIV (1549Å), is only visible for $z_l \gtrsim 1.25$, and we have no confirmed lens redshifts in this range. Spectroscopy with HST can search for metal lines in the UV, but the integration times tend to be prohibitively long unless the quasar images are very bright. Thus, while absorption lines either associated with the lens galaxy or likely to be associated with the lens galaxy are occasionally found (e.g. SDSS1650+4251, Morgan, Snyder and Reens 2003; or HE1104–1805, Lidman et al. 2000), there have been no systematic studies of metal absorption in gravitational lenses. Nonetheless, some very bright quasar lenses are favored targets for very high dispersion studies of their Lyα forest, particularly the four-image lens B1422+231 and the three image lens APM08279+5255, because the lens magnification makes these systems anomalously bright for quasars at $z_s > 3$.

10 Extended Sources and Quasar Host Galaxies

As we saw in Figs. 3, 4, and 8, we frequently see lensed emission from extended components of the source. These arcs and rings are important because they can supply the extra constraints needed to determine the radial mass distribution that we lack in a simple two-image or four-image lens (Sect. 4.1). The magnification produced by gravitational lensing also allows us to study far fainter quasar host galaxies than is otherwise possible. Comparisons of the luminosities and colors of high and low redshift host galaxies and the relative luminosities of the host and the quasar are important for understanding the growth of supermassive black holes and their relationships with their parent halos.

Modeling extended emission is more difficult than modeling point sources largely because of the complications introduced by the finite resolution of the observations. In this section we first discuss a simple theory of Einstein ring images, then some methods for modeling extended emission, and finally some results about the mass distributions of lenses and the properties of quasar host galaxies. All models of extended lenses sources start from the fact that lensing preserves the surface brightness of the source – what we perceive as magnification is only an artifact of the finite resolution of our observations. This can be modified by absorption in the ISM of the lens galaxy (e.g. see, Koopmans et al. 2003), but we will neglect this complication in what follows. We start with a simple analytic model for the formation of Einstein rings, then discuss numerical reconstructions of lenses sources and their ability to constraint mass distributions, and end with a survey of the properties of quasar host galaxies.

10.1 An Analytic Model for Einstein Rings

Most of the lensed extended sources we see are dominated by an Einstein ring – this occurs when the size of the source is comparable to the size of the astroid caustic associated with producing four-image lenses. When the Einstein ring is fairly thin, there is a simple analytic model for the formation of Einstein rings (Kochanek, Keeton and McLeod 2001a). The important point to understand is that the ring is a pattern rather than a simple combination of multiple images. Mathematically, what we identify as the ring is the peak of surface brightness as a function of angle around the lens galaxy. We can identify the peak by finding the maximum intensity $\lambda(\chi)$ along radial spokes in the image plane, $\boldsymbol{\theta}(\lambda) = \boldsymbol{\theta}_0 + \lambda(\cos\chi, \sin\chi)$. At a given azimuth χ we find the extremum of the surface brightness of the image $f_D(\boldsymbol{\theta})$ along each spoke, and these lie at the solutions of

$$0 = \partial_\lambda f_D(\boldsymbol{\theta}) = \nabla_{\boldsymbol{\theta}} f_D(\boldsymbol{\theta}) \cdot \frac{d\boldsymbol{\theta}}{d\lambda}. \tag{134}$$

The next step is to translate the criterion for the ring location onto the source plane. In real images, the observed image $f_D(\boldsymbol{\theta})$ is related to the actual surface density $f_I(\boldsymbol{\theta})$ by a convolution with the beam (PSF), $f_D(\boldsymbol{\theta}) = B * f_I(\boldsymbol{\theta})$, but for the moment we will assume we are dealing with a true surface brightness map. Under this assumption $f_D(\boldsymbol{\theta}) = f_I(\boldsymbol{\theta}) = f_S(\boldsymbol{\beta})$ because of surface brightness conservation. When we change variables the criterion for the peak brightness becomes

$$0 = \nabla_{\boldsymbol{\beta}} f_S(\boldsymbol{\beta}) \cdot M^{-1} \cdot \frac{d\boldsymbol{\theta}}{d\lambda}, \tag{135}$$

where the inverse magnification tensor $M^{-1} = d\boldsymbol{\theta}/d\boldsymbol{\beta}$ is introduced by the variable transformation. Geometrically we must find the point where the tangent vector of the curve, $M^{-1} \cdot d\boldsymbol{\theta}/d\lambda$ is perpendicular to the local gradient of the surface brightness $\nabla_{\boldsymbol{\beta}} f_S(\boldsymbol{\beta})$. These steps are illustrated in Fig. 63.

This result is true in general but not very useful. We next assume that the source has ellipsoidal surface brightness contours, $f_S(m^2)$, with $m^2 = \Delta\boldsymbol{\beta} \cdot S \cdot \Delta\boldsymbol{\beta}$ where $\Delta\boldsymbol{\beta} = \boldsymbol{\beta} - \boldsymbol{\beta}_0$ is the distance from the center of the source, $\boldsymbol{\beta}_0$, and the matrix S is defined by the axis ratio $q_s = 1 - \epsilon_s \leq 1$ and position angle χ_s of the source. We must assume that the surface brightness declines monotonically, $df_s(m^2)/dm^2 < 0$, but require no additional assumptions about the actual profile. With these assumptions the Einstein ring curve is simply the solution of

$$0 = \Delta\boldsymbol{\beta} \cdot S \cdot \mu^{-1} \cdot \frac{d\boldsymbol{\theta}}{d\lambda}. \tag{136}$$

The ring curve traces out a four (two) lobed cloverleaf pattern when projected on the source plane if there are four (two) images of the center of the source (see Fig. 63). These lobes touch the tangential caustic at their maximum ellipsoidal distance from the source center, and these cyclic variations in the ellipsoidal radius produce the brightness variations seen around the ring. The surface brightness along the ring is defined by $f_I(\lambda(\chi), \chi)$ for a spoke at azimuth χ and distance $\lambda(\chi)$ found by solving (135). The extrema in the surface brightness around the ring are located at the points where $\partial_\chi f_I(\lambda(\chi), \chi) = 0$, which occurs only at extrema of the surface brightness of the source (the center of the source, $\Delta\boldsymbol{\beta} = 0$ in the ellipsoidal model), or when the ring crosses a critical line of the lens and the magnification tensor is singular ($|M|^{-1} = \mu^{-1} = 0$) for the minima. These are general results that do not depend on the assumption of ellipsoidal symmetry.

For an SIE lens in an external shear field we can derive some simple properties of Einstein rings to lowest order in the various axis ratios. Let the SIE have critical radius b, axis ratio $q_l = 1 - \epsilon_l$ and put its major axis along θ_1. Let the external shear have amplitude γ and orientation θ_γ. We let the source

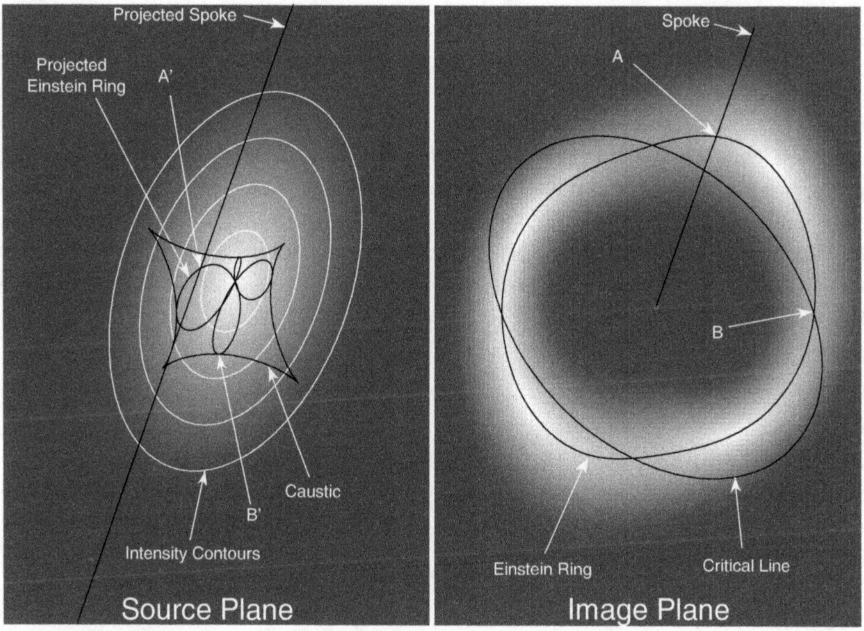

Fig. 63. An illustration of ring formation by an SIE lens. An ellipsoidal source (*left gray-scale*) is lensed into an Einstein ring (*right gray-scale*). The source plane is magnified by a factor of 2.5 relative to the image plane. The tangential caustic (*astroid on left*) and critical line (*right*) are superposed. The Einstein ring curve is found by looking for the peak brightness along radial spokes in the image plane. For example, the spoke in the illustration defines point A on the ring curve. The long line segment on the left is the projection of the spoke onto the source plane. Point A corresponds to point A$'$ on the source plane where the projected spoke is tangential to the intensity contours of the source. The ring in the image plane projects into the four-lobed pattern on the source plane. Intensity maxima along the ring correspond to the center of the source. Intensity minima along the ring occur where the ring crosses the critical curve (e.g. point B). The corresponding points on the source plane (e.g. B$'$) are where the astroid caustic is tangential to the intensity contours

be an ellipsoid with axis ratio $q_s = 1 - \epsilon_s$ and a major axis angle χ_s located at position $(\beta \cos\chi_0, \beta \sin\chi_0)$ from the lens center. The tangential critical line of the lens lies at radius

$$r_{crit}/b = 1 + \frac{\epsilon_l}{2} \cos 2\chi - \gamma \cos 2(\chi - \chi_\gamma), \tag{137}$$

while the Einstein ring lies at

$$\frac{r_E}{b} = 1 + \frac{\beta}{b} \cos(\chi - \chi_0) - \frac{\epsilon_l}{6} \cos 2\chi + \gamma \cos 2(\chi - \chi_\gamma). \tag{138}$$

At this order, the Einstein ring is centered on the source position rather than the lens position. The orientation of the ring is generally perpendicular to that of the critical curve, although it need not be exactly so when the SIE and the shear are misaligned due to the differing coefficients of the shear and ellipticity terms in the two expressions. These results lead to a false impression that the results do not depend on the shape of the source. In making the expansion we assumed that all the terms were of the same order ($\beta/b \sim \gamma \sim \epsilon_l \sim \epsilon_s$), but we are really doing an expansion in the ellipticity of the potential of the lens $e_\Psi \sim e_l/3$ rather than the ellipticity of the density distribution of the lens, so second order terms in the shape of the source are as important as first order terms in the ellipticity of the potential. For example in a circular lens with no shear ($\epsilon_l = 0$, $\gamma = 0$) the ring is located at

$$\frac{r_E}{b} = 1 + \frac{\beta}{b} \frac{(2-\epsilon_s)\cos(\chi - \chi_0) + \epsilon_s \cos(2\chi_s - \chi - \chi_0)}{2 - \epsilon_s + \epsilon_s \cos 2(\chi_s - \chi)}, \tag{139}$$

which has only odd terms in its multipole expansion and converges slowly for flattened sources. In general, the ring shape is a weak function of the source shape only if the potential is nearly round and the source is almost centered on the lens. The structure of the lens potential dominates the even multipoles of the ring shape, while the structure of the source dominates the odd multipoles.

In fact, the shape of the ring can be used to simply "read off" the amplitudes of the higher order multipoles of the lens potential. This is nicely illustrated by an isothermal potential with arbitrary angular structure, $\Psi = rbF(\chi)$ with $\langle F(\chi)\rangle = 1$ (see Zhao and Pronk 2001; Witt et al. 2000; Kochanek et al. 2001a,b; Evans and Witt 2001) in the absence of any shear. The tangential critical line of the lens is

$$\frac{r_{crit}}{b} = F(\chi) + F''(\chi). \tag{140}$$

If $\hat{e}_\chi$ and $\hat{e}_\theta$ are tangential and radial unit vectors relative to the lens center and $\boldsymbol{\beta}_0$ is the distance of the source from the lens center, then the Einstein ring curve is

$$\frac{r_E}{b} = F(\chi) + F'(\chi)\frac{\hat{e}_\chi \cdot S \cdot \hat{e}_\theta}{\hat{e}_\theta \cdot S \cdot \hat{e}_\theta} + \frac{\boldsymbol{\beta}_0 \cdot S \cdot \hat{e}_\theta}{\hat{e}_\theta \cdot S \cdot \hat{e}_\theta} \rightarrow F(\chi) + \boldsymbol{\beta}_0 \cdot \hat{e}_\theta \tag{141}$$

with the limit showing the result for a circular source.

Thus, by analyzing the multipole structure of the ring curve one can deduce the multipole structure of the potential. While this has not been done non-parametrically, the ability of standard ellipsoidal models to reproduce ring curves strongly suggests that higher order multipoles cannot be significantly different from the ellipsoidal scalings. Figure 64 shows two examples of fits to the ring curves in PG1115+080 and B1938+666 using SIE plus external shear lens models. The major systematic problem with fitting the real data are that bright quasar images must frequently be subtracted from the image before the ring curve can be extracted, and this can lead to artifacts like the wiggle in the

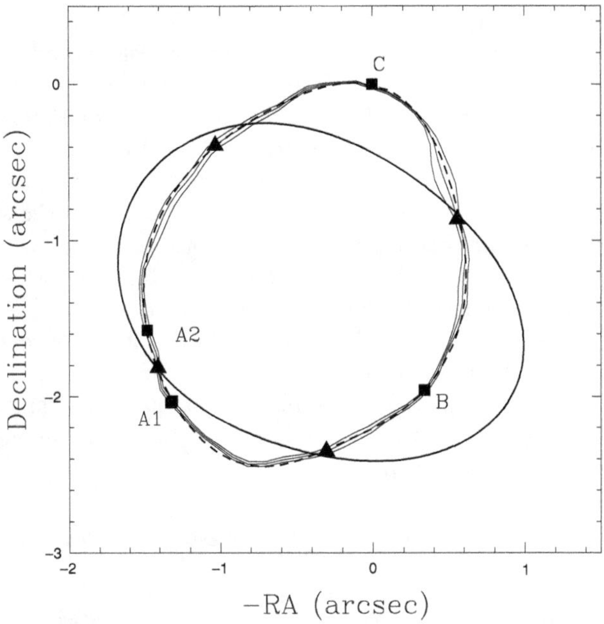

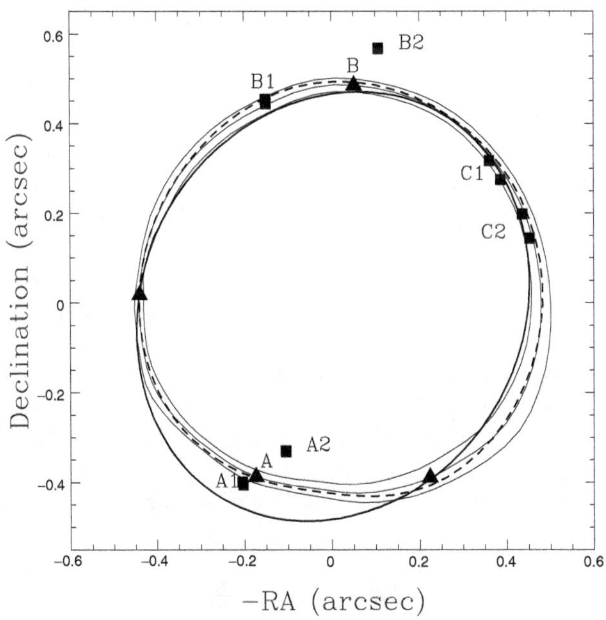

Fig. 64. The Einstein ring curves in PG1115+080 (*top*) and B1938+666 (*bottom*). The black squares mark the lensed quasar or compact radio sources. The *light black lines* show the ring curve and its uncertainties. The *black triangles* show the intensity minima along the ring curve (but not their uncertainties). The best fit model ring curve is shown by the *dashed curve*, and the *heavy solid curve* shows the critical line of the best fit model. The model was not constrained to fit the critical line crossings

curve between the bright A_1/A_2 images of PG1115+080. Other than that, the accuracy with which the ellipsoidal (plus shear) models reproduce the curves is consistent with the uncertainties. In both cases the host galaxy is relatively flat ($q_s = 0.58 \pm 0.02$ for PG1115+080 and 0.62 ± 0.14 for B1938+666). The flatness of the host explains the "boxiness" of the PG1115+080 ring, while the B1938+666 host galaxy shape is poorly constrained because the center of the host is very close to the center of the lens galaxy so the shape of the ring is insensitive to the shape of the source. Unless the source is significantly offset from the center of the lens as we might see for the host galaxy of an asymmetric two-image lens, it does not constrain the radial density profile of the lens very well – after considerable algebraic effort you can show that the dependence on the radial structure scales as $|\Delta\boldsymbol{\beta}|^4$. It can, however, help considerably in this circumstance because it eliminates the angular degrees of freedom in the potential that make it impossible for two-image lenses to constrain the radial density profile at all.

10.2 Numerical Models of Extended Lensed Sources

Obviously the ring curve and its extrema are an abstraction of the real structure of the lensed source. Complete modeling of extended sources requires a real model for the surface brightness of the source. In many cases it is sufficient to simply use a parameterized model for the source, but in other cases it is not. The basic idea in any non-parametric method is that there is an optimal estimate of the source structure for any given lens model. This is most easily seen if we ignore the smearing of the image by the beam (PSF) and assume that our image is a surface brightness map. Since surface brightness is conserved by lensing, $f_I(\boldsymbol{\theta}) = f_S(\boldsymbol{\beta})$. For any lens model with parameters $\boldsymbol{p}$, the lens equations define the source position $\boldsymbol{\beta}(\boldsymbol{\theta}, \boldsymbol{p})$ associated with each image position. If we had only single images of each source point, this would be useless for modeling lenses. However, in a multiply imaged region, more than one point on the image plane is mapped to the same point on the source plane. In a correct lens model, all image plane points mapped to the same source plane position should have the same surface brightness, while in an incorrect model, points with differing surface brightnesses will be mapped to the same source point. This provided the basis for the first non-parametric method, sometimes known as the "Ring Cycle" method (Kochanek et al. 1989; Wallington, Kochanek and Koo 1995). Suppose source plane pixel j is associated with image plane pixels $i = 1, ..., n_j$ with surface brightness f_i and uncertainties σ_i. The goodness of fit for this source pixel is

$$\chi_j^2 = \sum_{i=1}^{n_j} \left(\frac{f_i - f_s}{\sigma_i} \right)^2 , \tag{142}$$

where f_s will be our estimate of the surface brightness on the source plane. For each lens model we compute $\chi^2(\boldsymbol{p}) = \sum \chi_j^2$ and then optimize the lens parameters to minimize the surface brightness mismatches.

The problem with this algorithm is that we never have images that are true surface brightness maps – they are always the surface brightness map convolved with some beam (PSF). We can generalize the simple algorithm into a set of linear equations. Although the source and lens plane are two-dimensional, the description is simplified if we simply treat them as a vector $\boldsymbol{f}_S$ of source plane surface brightness and a vector $\boldsymbol{f}_I$ of image plane flux densities (i.e. including any convolution with the beam). The two images are related by a linear operator $A(\boldsymbol{p})$ that depends on the parameters of the current lens model and the PSF. In the absence of a lens, A is simply the real-space (PSF) convolution operator. In either case, the fit statistic

$$\chi^2 = \frac{|\boldsymbol{f}_I - A(\boldsymbol{p})\boldsymbol{f}_S|^2}{\sigma^2} \tag{143}$$

(with uniform uncertainties here, but this is easily generalized) must first be solved to determine the optimal source structure for a given lens model and then minimized as a function of the lens model. The optimal source structure $d\chi^2/d\boldsymbol{f}_S = 0$ leads to the equation that $\boldsymbol{f}_S = A^{-1}(\boldsymbol{p})\boldsymbol{f}_I$. The problem, which is the same as we discussed for non-parametric mass models in Sect. 4.7, is that a sufficiently general source model when combined with a PSF will lead to a singular matrix for which $A(\boldsymbol{p})^{-1}$ is ill-defined – physically, there will be wildly oscillating source models for which it is possible to obtain $\chi^2(\boldsymbol{p}) = 0$.

Three approaches have been used to solve the problem. The first is LensClean (Kochanek and Narayan 1992; Ellithorpe, Kochanek and Hewitt 1996; Wucknitz 2004), which is based on the Clean algorithm of radio astronomy. Like the normal Clean algorithm, LensClean is a non-linear method using a prior that radio sources can be decomposed into point sources for determining the structure of the source. The second is LensMEM (Wallington, Kochanek and Narayan 1996), which is based on the Maximum Entropy Method (MEM) for image processing. The determination of the source structure is stabilized by minimizing $\chi^2 + \lambda \int d^2\beta f_S \ln(f_S/f_0)$ while adjusting the Lagrange multiplier λ such that at the minimum $\chi^2 \sim N_{dof}$ where N_{dof} is the number of degrees of freedom in the model. Like Clean/LensClean, MEM/LensMEM is a non-linear algorithm in which solutions must be solved iteratively. Both LensClean and LensMEM can be designed to produce only positive-definite sources. The third approach is linear regularization where the source structure is stabilized by minimizing $\chi^2 + \lambda \boldsymbol{f}_S \cdot H \cdot \boldsymbol{f}_S$ (Warren and Dye 2003; Koopmans et al. 2003). The simplest choice for the matrix H is the identity matrix, in which case the added criterion is to minimize the sum of the squares of the source flux. More complicated choices for H will minimize the gradients or curvature of the source flux. The advantage of this scheme is that the solution is simply a linear algebra problem with $(A^T(\boldsymbol{p})A(\boldsymbol{p}) + \lambda H)\boldsymbol{f}_S = A^T(\boldsymbol{p})\boldsymbol{f}_I$.

In all three of these methods there are two basic systematic issues which need to be addressed. First, all the methods have some sort of adjustable parameter – the Lagrange multiplier λ in LensMEM or the linear regularization methods and the stopping criterion in the LensClean method. As the lens model changes, the estimates of the parameter errors will be biased if the treatment of the multiplier or the stopping criterion varies with changes in the lens model in some poorly understood manner. Second, it is difficult to work out the accounting for the number of degrees of freedom associated with the model for the source when determining the significance of differences between different lens models. Both of these problems are particularly severe when comparing models where the size of the multiply imaged region depends on the lens model. Since only multiply imaged regions supply any constraints on the model, one way to improve the goodness of fit is simply to shrink the multiply imaged region so that there are fewer constraints. Since changes in the radial mass distribution have the biggest effect on the multiply imaged region, this makes estimates of the radial mass distribution particularly sensitive to controlling these biases. It is fair to say that all these algorithms lack a completely satisfactory understanding of this problem. For radio data there are added complications arising from the nature of interferometric observations, which mean that good statistical models must work with the raw visibility data rather than the final images (see Ellithorpe et al. 1996).

These methods, including the effects of the PSF, have been applied to determining the mass distributions in 0047-2808 (Dye and Warren 2005), B0218 +357 (Wucknitz, Biggs and Browne 2004), MG1131+0456 (Chen, Kochanek and Hewitt 1995, and MG1654+134 (Kochanek 1995a,b). We illustrate them with the Dye and Warren (2005) results for 0047-2808 in Fig. 65. The mass distribution consists of the lens galaxy and a cuspy dark matter halo, where Fig. 65 shows the final constraints on the mass-to-light ratio of the stars in the lens galaxy and the exponent of the central dark matter density cusp ($\rho \propto r^{-\gamma}$). The allowed parameter region closely resembles earlier results

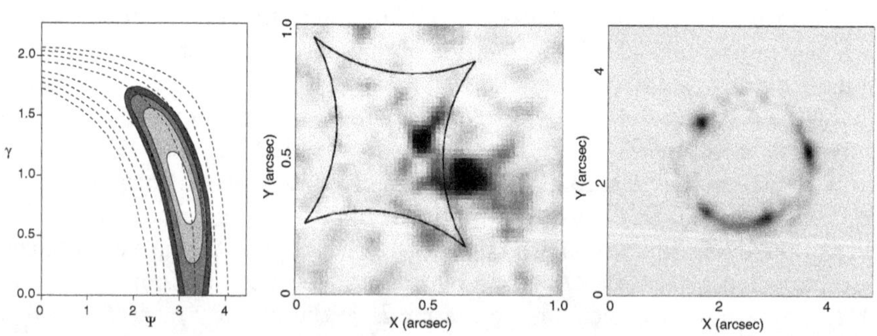

Fig. 65. Models of 0047–2808 from Dye and Warren (2005). The *right panel* shows the lensed image of the quasar host galaxy after the foreground lens has been subtracted. The *middle panel* shows the reconstructed source and its position relative to the tangential (astroid) caustic. The *left panel* shows the resulting constraints on the central exponent of the dark matter halo ($\rho \propto r^{-\gamma}$) and the stellar mass-to-light ratio of the lens galaxy. The *dashed contours* show the constraints for the same model using the central velocity dispersion measurement from Koopmans and Treu (2003)

using either statistical constraints (Fig. 30 or stellar dynamics (Fig. 31). In fact, the results using the stellar dynamical constraint from Koopmans and Treu (2003) are superposed on the constraints from the host in Fig. 65, with the host providing a tighter constraint on the mass distribution than the central velocity dispersion. The one problem with all these models is that they have too few degrees of freedom in their mass distributions by the standards we discussed in Sect. 4.6. In particular, we know that four-image lenses require both an elliptical lens and an external tidal shear in order to obtain a good fit to the data (e.g. Keeton, Kochanek and Seljak 1997), while none of these models for the extended sources allows for multiple sources for the angular structure in the potential. In fact, the lack of an external shear probably drives the need for dark matter in the 0047–2808 models. Without dark matter, the decay of the stellar quadrupole and the low surface density at the Einstein ring means that the models generate too small a quadrupole moment to fit the data in the absence of a halo. The dark matter solves the problem both through its own ellipticity and the reduction in the necessary shear with a higher surface density near the ring (recall that $\gamma \propto 1 - \langle\kappa\rangle$). Again see the need for a greater focus on the angular structure of the potential.

10.3 Lensed Quasar Host Galaxies

The advantage of studying lensed quasars is that the lens magnification enormously enhances the visibility of the quasar host. A typical HST PSF makes the image of a point source have a mean surface brightness that declines as R^{-3} with distance R from the quasar. Compared to an unlensed quasar, the host galaxy of a lensed quasar is stretched along the Einstein ring leading to an improvement in the contrast between the host in the quasar of μ^2 for an image magnified by μ – you gain μ^3 by stretching the host away from the quasar and lose μ because the quasar is magnified. Perpendicular to the Einstein ring, the contrast becomes a factor of μ worse than for an unlensed quasar. Since the alignment of the magnification tensor relative to the host changes with each image, the segment of the host where contrast is lost will correspond to a segment where it is gained for another image leading to a net gain for almost all parts of the source when you consider all the images. The distortions produced by lensing also mean the host structure is more easily distinguished from the PSF. In a few cases, like SDSS0924+0219 in Fig. 54, microlensing or substructure may provide a natural coronograph that suppresses the flux from the quasar but not that from the host. Despite naive expectations (and TAC comments), the distortions have little consequence for understanding the structure of the host even though a lens model is required to produce a photometric model of the host.

The only extensive survey of lensed quasar hosts is that of Peng (2004). Figure 66 shows the example of PG1115+080, a $z_s = 1.72$ radio-quiet quasar (RQQ). The Einstein ring image is easily visible even in a short, one-orbit exposure. For comparison, we also took the final model for the quasar and

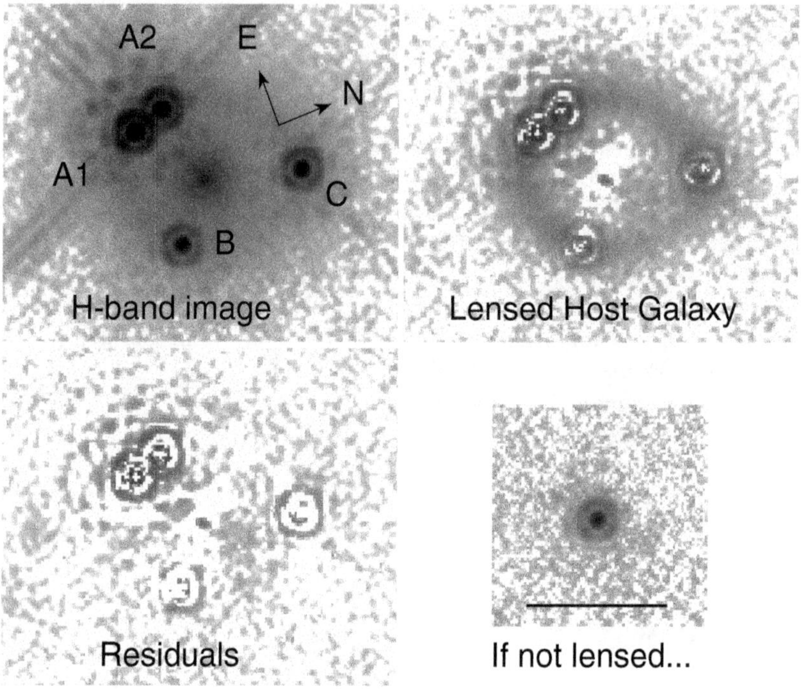

Fig. 66. The host galaxy in PG1115+080. The *top left* panel shows the 1-orbit NICMOS image from Impey et al. (1998). The *top right* panel shows the lensed host galaxy after subtracting the quasar images and the lens galaxy, The *lower left* panel shows the residuals after subtracting the host as well. For comparison, the lower right panel shows what an image of an unlensed PG1115+080 quasar and host would look like in the same integration time. The host galaxy is an H= 20.8 mag late-type galaxy (Sersic index $n = 1.4$) with a scale length of $R_e = 1.5h^{-1}$ kpc. The demagnified magnitude of the quasar is H= 19.0 mag. The axis ratio of the source, $q_s = 0.65 \pm 0.04$ is consistent with the estimate of $q_s = 0.58 \pm 0.02$ from the simpler ring curve analysis (Sect. 10.1, Fig. 64, Kochanek, Keeton and McLeod 2001a)

the source and produced the image that would be obtained in the same time if we observed the quasar in the absence of lensing. It is quite difficult to see the host, and this problem will carry through in any numerical analysis.

At low redshifts ($z < 1$), quasar host galaxies tend to be massive early-type galaxies (e.g. Mclure et al. 1999; Dunlop et al. 2003). Over 80% of quasars brighter than $M_V < -23.5$ mag are in early-type galaxies with $L \gtrsim 2L_*$ and effective radii of $R_e \sim 10$ kpc for $z \lesssim 0.5$. Radio quiet quasars (RQQ) tend to be in slightly lower luminosity hosts than radio loud quasars (RLQ) but only by factors of ~ 2 at redshift unity. Far fewer unlensed host galaxies have been detected above redshift unity (e.g. Kukula et al. 2001; Ridgway et al. 2001) with the surprising result that the host galaxies are 2–3 mag brighter than the typical host galaxy at low redshift and corresponded to $\sim 4L_*$ galaxies. Given that the low redshift hosts were already very massive galaxies it was expected that higher redshift hosts would have lower masses because they were still in the process of being assembled and forming stars (e.g. Kauffmann and Haehnelt 2000). One simple explanation was that by selecting from bright radio sources, these samples picked quasars with more massive black holes as the redshift increased, creating a bias in favor of more massive hosts. The key to checking for such a bias is to be able to detect far less luminous hosts, and the improved surface brightness contrast provided by lensing the host galaxies provides the means.

Figure 67 shows the observed H-band magnitudes of the lensed hosts as compared to low redshift host galaxies and other studies of high redshift host

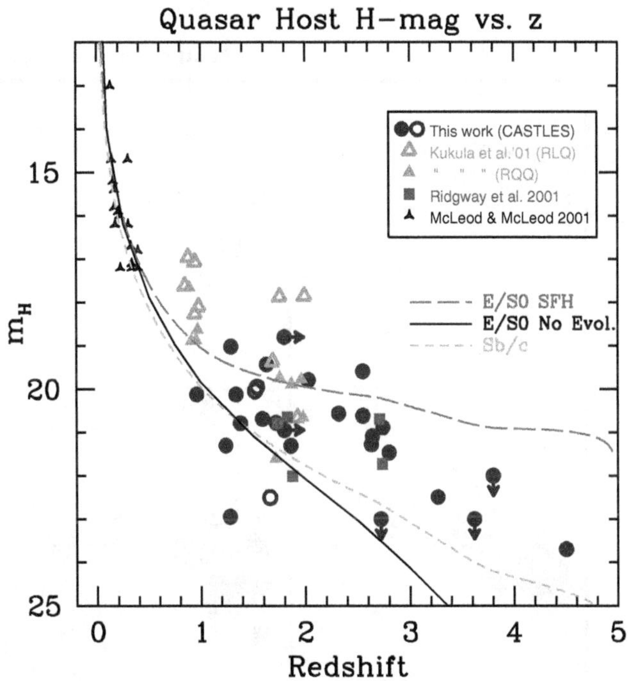

Fig. 67. Observed H-band magnitudes of quasar host galaxies. The solid (*open*) points are secure (more questionable) hosts detected in the CASTLES survey of lensed hosts. The low redshift points are from McLeod and McLeod (2001). All the Ridgway et al. (2001) systems are radio quiet. For comparison we superpose the evolutionary tracks for a non-evolving E/S0 galaxy (*solid curve*), an evolving E/S0 galaxy which starts forming stars at $z_f = 5$ with a 1 Gyr exponentially decaying star formation rate (*long dashed line*) and a star forming Sb/c model (*short dashed line*). The evolution models are matched to the luminosity of an L_* early-type galaxy at redshift zero. The CASTLES observations can reliably detect hosts about 4 magnitudes fainter than the quasar

galaxies. Although 30% of the lensed quasars are radio-loud, they have luminosities similar to the lensed (or unlensed) radio-quiet hosts. There are no hosts as bright as the Kukula et al. (2001) radio-loud quasar hosts. Once the luminosities of the quasar and the host galaxy are measured we can compare them to the theoretical expectations (Fig. 68). While the models agree with the data at low redshift, they are nearly disjoint by $z \sim 3$ in the sense that the observed quasars and hosts are significantly more luminous than predicted. The same holds for the Kukula et al. (2001) and Ridgway et al. (2001) samples, suggesting that black holes masses grow more rapidly than predicted by the theoretical models or that accretion efficiencies were higher in the past. Vestergaard (2004) makes a similar argument based on estimates of black hole masses from emission line widths.

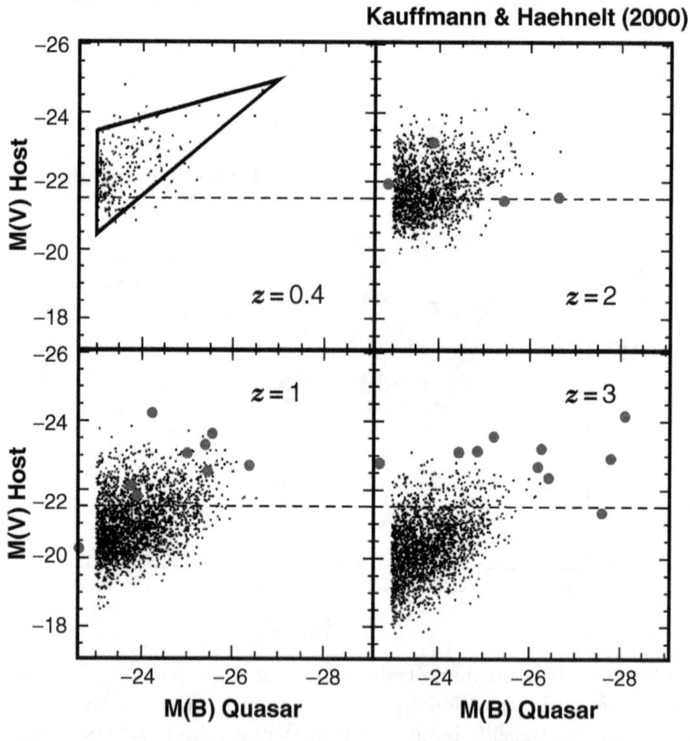

Fig. 68. A comparison of the estimated rest frame absolute magnitudes of the quasars and hosts as compared to the theoretical models for the evolution of galaxies and the growth of *black holes* as a function of redshift by Kauffmann and Haehnelt (2000). The low redshift quasars from McLeod and McLeod (2001) occupy the triangle in the upper left panel. At intermediate redshift the lensed host galaxies occupy a region similar to the models, but the two distributions are nearly disjoint by $z \simeq 3$. Both the hosts and the quasars are significantly more luminous than predicted. The horizontal line marks the luminosity of an L_* galaxy at $z = 0$

11 Does Strong Lensing Have a Future?

Well, you can hardly expect an answer of "No !" at this point, can you? Since we have just spent nearly 200 pages on the astrophysical uses of lenses, there is no point in reviewing all the results again here. Instead I suggest some goals for the future.

Our first goal is to expand the sample of lenses from ~ 100 to ~ 200. While 80 lenses seems like a great many compared to even a few years ago, it is still too few to pursue many interesting questions. The problem worsens if the analysis must be limited to lenses meeting other criteria (radio lenses, lenses found in a well-defined survey, lenses outside the cores of clusters,...) or if the sample must be subdivided into bins (redshift, separation, luminosity,...). For example, one of the most interesting applications of lenses will be to map out the halo mass function. This is difficult to do with any other approach because no other selection method works homogeneously on dark low-mass halos, galaxies of different types, groups and clusters. Unlike any other sample in astronomy, gravitational lenses are selected based on mass rather than luminosity, so the same search method works for all halos – the separation distribution of lenses is a direct mapping of the halo mass function. It is not a trivial mapping because the structure of halos changes with mass, but the systematics are far better than those of any other approach. The fact that lenses are mass-selected also gives them an enormous advantage in studying the evolution of galaxies with redshift over optically-selected samples where it will be virtually impossible to select galaxies in the same manner at both low and high redshift. The upgraded VLA and Merlin radio arrays are the most promising tools for this objective.

Our second goal is to systematically monitor the variability of as many lenses as possible. Time delays, if measured in large numbers and measured accurately, can resolve most of the remaining issues about the mass distributions of lenses. This is true even if you regard the H_0 as unmeasured or uncertain – the Hubble constant is the same number for all lenses, so as the number of time delay systems increases, the contribution of the actual value of the Hubble constant to constraining the mass distribution diminishes. At the present time, we are certain that the typical early-type galaxy has a substantial dark matter halo, but we are uncertain how it merges with the luminous galaxy. Steady monitoring of microlensing of the source quasars by the stars in the lens galaxy will also help to resolve this problem because the patterns of the microlensing variability constrain both the stellar surface density near the lensed images and the total density (Part 4, Schechter and Wambsganss 2002). The constraints from time delays and microlensing will be complemented by the continued measurement of central velocity dispersions.

Our third goal should be to obtain ultra-deep, high resolution radio maps of the lenses to search for central images in order to measure the central surface densities of galaxies and to search for supermassive black holes. Keeton (2003a) showed that the dynamic ranges of the existing radio maps of lenses are 1–2 orders of magnitude too small to routinely detect central images given the expected central surface densities of galaxies. Only very asymmetric doubles like PMN1632–0033, where Winn et al. (2004) have detected a central image, are likely to show central images with the present data. Once we reach the sensitivity needed to detect central images, we will also either find central black holes or set strict limits on their existence (Mao, Witt and Koopmans 2001). This is the only approach that can directly detect even quiescent black holes and determine their masses at cosmological distances. The existing limits could be considerably improved simply by co-adding the existing radio maps either for individual lenses or even for multiple lenses in order to obtain statistical limits.

Our fourth goal should be to unambiguously identify a "dark" satellite of a lens galaxy. For starters we need to conduct complete statistical analyses of lens galaxy satellites in general, by determining their mass functions and radial distributions. As part of such an analysis we can obtain upper bounds on the number of dark satellites. Then, with luck, we will find an example of a lens that requires a satellite at a specific location for which there is no optical counterpart. This may be too conservative a condition. For example, Peng (2004) argues that much of the flux of Object X in MG0414+0534 (Fig. 7) is actually coming from lensed images of the quasar host galaxy rather than the satellite.

Finally, lens magnification already means that it is far easier to do photometry of a lensed quasar host galaxy than an unlensed galaxy. The next frontier is to measure the kinematics of cosmologically distant host galaxies. This might already be doable for the host galaxy of Q0957+561 at $z_s = 1.41$, but will generally require either JWST or the next generation of ground based telescopes. With larger lens samples we may also find more cases like SDSS0924+0219 where gravitational lensing provides a natural coronograph for the quasar.

Acknowledgments

I would like to thank E.E. Falco, C.R. Keeton, D. Rusin, D.H. Weinberg and J.N. Winn for commenting on this manuscript, and S. Dye, C.D. Fassnacht, D.R. Marlow, J.L. Mitchell, C.Y. Peng, D. Rusin and J.N Winn for supplying figures. G. Meylan showed considerable patience with the author. This research has been supported by the NASA ATP grant NAG5-9265, and by grants HST-GO-7495, 7887, 8175, 8804, and 9133 from the Space Telescope Science Institute. The continuing HST observations of gravitational lenses are an absolutely essential part of converting gravitational lenses from curiosities into astrophysical tools.

References

Agol, E., Jones, B., & Blaes, O., 2000, ApJ, 545, 657
Angonin-Willaime, M.-C., Soucail, G. & Vanderriest, C. 1994, A&A 291, 411

Bade, N., Siebert, J., Lopez, S., Voges, W. & Reimers, D. 1997, A&A 317, 13
Barkana, R. 1996, ApJ 468, 17
Barkana, R. 1997, ApJ 489, 21
Barkana, R. 1998, ApJ 502, 531
Barkana, R., Lehár, J., Falco, E.E., Grogin, N.A., Keeton, C.R. & Shapiro, I.I. 1999, ApJ 523, 54
Bartelmann, M. & Meneghetti, M. 2004, A&A 418, 413, also astro-ph/0312011
Bender, R. et al. 1989, A&A 217, 35
Bender, R., Saglia, R.P., Ziegler, B., Belloni, P., Greggio, L., Hopp, U. & Bruzual, G. 1998, ApJ 493, 529
Benson, A.J., Bower, R.G., Frenk, C.S., Lacey, C.G., Baugh, C.M. & Cole, S. 2003, ApJ 599, 38
Bernstein, G.M., Tyson, J.A. & Kochanek, C.S. 1993, AJ 105, 816
Bernstein, G. & Fischer, P. 1999, AJ 118, 14
Beskin, G.M. & Oknyanskij, V.L. 1995, A&A 304, 341
Biggs, A.D., Browne, I.W.A, Helbig, P., Koopmans, L.V.E., Wilkinson, P.N. & Perley, R.A. 1999, MNRAS 304, 349
Biggs, A.D., Browne, I.W.A., Jackson, N.J., York, T., Norbury, M.A., McKean, J.P. & Phillips, P.M. 2004, MNRAS submitted astro-ph/0402128
Binney, J.J. & Mamon, G.A. 1982, MNRAS 200, 361
Binney, J.J. & Tremaine, S. 1987, Galactic Dynamics (Princeton University Press: Princeton)
Bischof, O.B. & Becker, R.H. 1997, AJ 113, 2000
Blandford, R.D. & Narayan, R. 1986, ApJ 310, 568
Blandford, R.D. & Kochanek, C.S. 1987a, in *Dark Matter in the Universe*, J. Bahcall, T. Piran & S. Weinberg (eds.), (World Scientific: Singapore) p. 133
Blandford, R.D. & Kochanek, C.S. 1987b, ApJ 321, 658
Blandford, R.D. & Narayan, R. 1992, ARA&A 30, 311
Blumenthal G.R., Faber S.M., Flores R., Primack J.R. 1986, ApJ 301, 27
Bode, P., Ostriker, J.P. & Turok, N. 2001, ApJ 556, 93
Borgeest, U., von Linde, J., & Refsdal, S. 1991, A&A 251, L35
Bradac, M., Schneider, P., Steinmetz, M., Lombardi, M. & King, L.J. 2002, A&A 388, 373
Browne, I.W.A., Wilkinson, P.N., Jackson, N.J.F. et al. 2003, MNRAS 341, 13
Bruzual, G., & Charlot, S. 1993, ApJ 405, 538
Bullock, J.S., Kravtsov, A.V. & Weinberg, D.H. 2000, ApJ 539, 517
Bullock J.S., Dekel, A., Kolatt, T.S., Primack, J.R. & Somerville, R.S. 2001a, ApJ 550, 21
Bullock J.S., Kolatt, T.S., Sigad, Y., Somerville, R.S., Kravtsov, A.V., Klypin, A.A., Primack, J.R. & Dekel, A. 2001b, MNRAS 321, 559
Buote, D.A., Jeltema, T.E., Canizares, C.R., & Garmire, G.P. 2002, ApJ 577, 183

Burke, B.F., Lehár, J. & Conner, S.R. 1992, in Gravitational Lenses, R.Kayser, T.Schramm & L.Nieser (eds.), (Springer-Verlag: Berlin:), p. 237
Burkert, A. & Naab, T. 2003, in Galaxies & Chaos, G. Contopulos & N. Voglis (eds.), (Springer-Verlag: Berlin) p. 327 also astro-ph/0301385
Burud, I. et al. 2000, ApJ 544, 117
Burud, I. et al. 2002a, A&A 383, 71
Burud, I. et al. 2002b, A&A 391, 481
Byun, Y.-I., Grillmair, C.J., Faber, S.M., Ajhar, E.A., Dressler, A., Kormendy, J., Lauer, T.R., Richstone, D. & Tremaine, S. 1996, AJ 111, 1889
Cardelli, J.A., Clayton, G.C. & Mathis, J.S. 1989, ApJ 345, 245
Carroll, S.M., Press, W.H. & Turner, E.L. 1992, ARA&A 30, 499
Chae, K.-H., Turnshek, D.A. & Khersonsky, V.K. 1998a, ApJ 495, 609
Chae, K.-H., Khersonsky, V.K. & Turnshek, D.A. 1998b, ApJ 506, 80
Chae, K.-H. 1999, ApJ 524, 582
Chae, K.-H., Mao, S. & Augusto, P. 2001, MNRAS 326, 1015
Chae, K.-H. 2002, AJ 568, 500
Chae, K.-H., Biggs, A.D., Blandford, R.D. et al. 2002, PhRvL 89, 1301
Chae, K.-H. 2003, MNRAS 346, 746
Chae, K.-H., & Mao, S. 2003, ApJ 599, L61
Chae, K.-H., Chen, G., Ratra, B. & Lee, D.-W. 2004, ApJ 607, 71
Chartas, G., Chuss, D., Forman, W., Jones, C. & Shapiro, I. 1998, ApJ 504, 661
Chartas, G., Bautz, M., Garmire, G., Jones, C. & Schneider, D.P. 2001, ApJ 550, L163
Chartas, G., Gupta, V., Garmire, G., Jones, C., Falco, E.E., Shapiro, I.I. & Tavecchio, F. 2002, ApJ 565, 96
Chartas, G., Dai, X. & Garmire, G.P. 2004, in Measuring and Modeling the Universe, W.L. Freedman (ed.), Carnegie Observatories Astrophysics Series, Vol. 2 (Carnegie Observatories: Pasadena), http://www.ociw.edu/ociw/symposia/series/symposium2/proceedings.html
Chen, G.H. & Hewitt, J.N. 1993, AJ 106, 1719
Chen, G.H., Kochanek,C.S. & Hewitt, J.N. 1995, ApJ 447, 62
Chen, J., Kravtsov, A.V. & Keeton, C.R. 2003, ApJ 592, 24
Chiba, M. 2002, ApJ 565, 17
Claeskens, J. & Surdej, J. 2002, A&ARv 10, 263
Cohen, A.S., Hewitt, J.N., Moore, C.B. & Haarsma, D.B. 2000, ApJ 545, 578
Cohn, J.D., Kochanek, C.S., McLeod, B.A. & Keeton, C.R. 2001, ApJ 554, 1216
Cohn, J.D., & Kochanek, C.S. 2004, ApJ 608, 25, also astro-ph/0306171
Cole, S., Lacey, C.G., Baugh, C.M. & Frenk, C.S. 2000, MNRAS 319, 168
Cole, S. et al. 2001, MNRAS 326, 255
Courbin, F., Meylan, G., Kneib, J.-P. & Lidman, C. 2002a, ApJ 575, 95
Courbin, F., Saha, P., & Schechter, P., 2002b, in Gravitational Lensing: An Astrophysical Tool, F. Courbin & D. Minniti, (eds.), (Springer-Verlag: Berlin), LRP, 608, 1

Crampton, D., McClure, R.D. & Fletcher, J.M. 1992, ApJ 392, 23

Dalal, N. & Kochanek, C.S. 2002, ApJ 572, 25

Dalal, N. & Keeton, C.R. 2003, astro-ph/0312072

Dalal, N., Holz, D.E., Chen, X. & Frieman, J.A. 2003, ApJ 585, L11

Davis, A.N., Huterer, D. & Krauss, L.M. 2003, MNRAS 344, 1029

de Vaucouleurs, G. 1948, Ann. Astrophys. 11, 247

Djorgovski, S., Perley, R., Meylan, G., & McCarthy, P. 1987, ApJ 321, L17

Djorgovski, S.G. & Davis, M. 1987, ApJ 313, 59

Djorgovski, S. 1991, in The Space Distribution of Quasars, D. Crampton (ed.), (ASP Conf. Series: San Francisco), ASP Conf. Ser. # 21, p. 349

Dressler, A., Lynden-Bell, D., Burstein, D., Davies, R.L., Faber, S.M., Terlevich, R.J. & Wegner, G. 1987, ApJ 313, 42

Dubinski, J. 1992, ApJ 401, 441

Dubinski, J. 1994, ApJ 431, 617

Dunlop, J.S., McLure, R.J., Kukula, M.J., Baum, S.A., O'Dea, C.P. & Hughes, D.H. 2003, MNRAS 340, 1095

Dye, S. & Warren, S. 2005, ApJ 623, 31, also astro-ph/0311185

Ellithorpe, J.D., Kochanek, C.S. & Hewitt, J.N. 1996, ApJ 464, 556

Evans, N.W. & Wilkinson, M.I. 1998, MNRAS 296, 800

Evans, N.W. & Witt, H.J. 2001, MNRAS 327, 1260

Evans, N.W. & Hunter, C. 2002, ApJ 575, 68

Evans, N.W. & Witt, H.J. 2003, MNRAS 345, 1351

Faber, S.M. & Jackson, R.E. 1976, ApJ 204, 668

Faber, S.M., Dressler, A., Davies, R.L, Burstein, D. & Lynden-Bell, D. 1987, in Nearly Normal Galaxies, S.M. Faber (ed.) (Springer Verlag: New York) p. 175

Falco, E.E., Gorenstein, M.V. & Shapiro, I.I. 1985, ApJ 289, L1

Falco, E.E., Shapiro, I.I., Moustakas, L.A. & Davis, M. 1997, ApJ 484, 70

Falco, E.E., Kochanek, C.S. & Muñoz, J.A. 1998, ApJ 494, 47

Falco, E.E., Impey, C.D., Kochanek, C.S., Lehár, J., McLeod, B.A., Rix, H.-W., Keeton, C.R., Muñoz, J.A. & Peng, C.Y. 1999, ApJ 523, 617

Fassnacht, C.D., Pearson, T.J., Readhead, A.C.S., Browne, I.W.A., Koopmans, L.V.E., Myers, S.T. & Wilkinson, P.N. 1999, ApJ 527, 498

Fassnacht, C.D., Xanthopoulos, E., Koopmans, L.V.E. & Rusin, D. 2002, ApJ 581, 823

Fassnacht, C.D. & Lubin, L.M. 2002, AJ 123, 627

Finch, T., Carlivati, L., Winn,J.N. & Schechter, P.L. 2002, ApJ 577, 51

Fischer, P., Bernstein, G., Rhee, G. & Tyson, J.A. 1997, AJ 113, 521

Florentin-Nielsen, R. 1984, A&A 138, L19

Flores, R. & Primack, J.R. 1996, ApJ 457, L5

Foltz, C.B, Hewitt, P.C., Webster, R.L., Lewis, G.F. 1992, ApJ 386, L43

Franx, M., Illingworth, G. & de Zeeuw, T. 1991, ApJ 383, 112

Freedman, W. et al. 2001, ApJ 553, 47

Fukugita, M., Futamase, T. & Kasai, M. 1990, MNRAS 246, 24

Fukugita, M. & Turner, E.L. 1991, MNRAS 253, 99

Fukugita M., Hogan C.J., Peebles P.J.E. 1998, ApJ 503, 518
Garrett, M.A., Calder, R.J., Porcas, R.W., King, L.J., Walsh, D. & Wilkinson, P.N. 1994, MNRAS 270, 457
Gebhardt, K. et al. 2003, ApJ 597, 239, also astro-ph/0307242
Gerhard, O., Kronawitter, A., Saglia, R.P. & Bender, R. 2001, AJ 121, 1936
Gerin, M., Phillips, T.G., Benford, D.J., Young, Y.H., Menten, K.M. & Frye, B. 1997, ApJ 488, L31
Gonzalez, A.H., Williams, K.A., Bullock, J.S., Kolatt, T.S. & Primack, J.R. 2000, ApJ 528, 145
Gorenstein, M.V., Falco, E.E. & Shapiro, I.I. 1988, ApJ 327, 693
Gott, J.R., Park, M.-G. & Lee, H.M 1989, ApJ 338, 1
Green, P.J., Kochanek, C.S., Siemiginowska, A., Kim, D.-W., Markevitch, M., Silverman, J., Dosaj, A., Jannuzi, B.T. & Smith, C. 2002, ApJ 571, 721
Grogin, N.A. & Narayan, R. 1996, ApJ 464, 92; erratum 1996, ApJ 473, 570
Haarsma, D.B., Hewitt, J.N., Lehár, J., Burke, B.F. 1999, ApJ 510, 64
Helbig, P. & Kayser, R. 1996, A&A 308, 359
Hernquist, L. 1990, ApJ 356, 359
Hewett, P.C., Foltz, C.B., Harding, M.E. & Lewis, G.F. 1998, AJ 115, 383
Hewitt, J.N., Turner, E.L., Lawrence, C.R., Schneider, D.P. & Brody, J.P. 1992, AJ 104, 968
Heyl, J.S., Hernquist, L. & Spergel, D.N. 1994, ApJ 427, 165
Hinshaw, G. & Krauss, L.M. 1987, ApJ 320, 468
Hjorth, J. et al. 2002, ApJ 572, L11
Holder, G.P. & Schechter, P.L. 2003, ApJ 589, 688
Huchra, J., Gorenstein, M., Kent, S., Shapiro, I., Smith, G., Horine, E. & Perley, R. 1985, AJ 90, 691
Hunter, C. & Evans, N.W. 2001, ApJ 554, 1227
Huterer, D. & Ma, C.-P. 2004, ApJ 600, 7
Ibata, R.A., Lewis, G.F., Irwin, M.J., Lehár, J. & Totten, E.J. 1999, AJ 118, 1922
Im, M., Griffiths, R.E. & Ratnatunga, K.U. 1997, ApJ 475, 457
Impey, C.D., Falco, E.E., Kochanek, C.S., Lehár, J., McLeod, B.A., Rix, H.-W., Peng, C.Y. & Keeton, C.R. 1998, ApJ 509, 551
Inada, N., Oguri, M., Pindor, B. et al. 2003, Nature 426, 810
Irwin, M.J., Ibata, R.A., Lewis, G.F. & Totten, E.J. 1998, ApJ 505, 529
Jaffe, W. 1983, MNRAS 202, 995
Jaunsen, A.O. & Hjorth, J. 1997, A&A 317, L39
Jean, C. & Surdej, J. 1998, A&A 339, 729
Johnston, D.E., Richards, G.T., Frieman , J.A. et al. 2003, AJ 126, 2281
Jones, D.L. et al. 1996, ApJ 470, 23
Jorgensen, I., Franx, M. & Kjaergaard, P. 1996, MNRAS 280, 167
Kassiola, A. & Kovner, I. 1993, ApJ 417, 450
Kauffmann, G., White, S.D.M. & Guiderdoni, B. 1993, MNRAS 264, 201
Kauffmann, G. & Haehnelt, M. 2000, MNRAS 311, 576

Kawano, Y., Oguri, M., Matsubara, T. & Ikeuchi, S. 2004, PASJ 56, 253, also astro-ph/0404013
Kazantzidis, S., Kravtsov, A.V., Zentner, A.R., Algood, B., Nagai, D. & Moore, B. 2004, ApJL 611, L73, also astro-ph/0405189
Keeton, C.R., Kochanek, C.S. & Seljak, U. 1997, ApJ 482, 604
Keeton C.R. & Kochanek C.S. 1998, ApJ 495, 157
Keeton C.R., Kochanek C.S., & Falco E.E. 1998, ApJ 509, 561
Keeton, C.R., Falco, E.E., Impey, C.D., Kochanek, C.S., Lehár, J., McLeod, B.A.,Rix, H.-W.; Muñoz, J.A. & Peng, C.Y. 2000a, ApJ 542, 74
Keeton, C.R., Christlein, D. & Zabludoff, A.I. 2000b ApJ, 545, 129
Keeton, C.R. 2001a, astro-ph/0102340
Keeton, C.R. & Madau, P. 2001b, ApJ 549, 25
Keeton, C.R. 2001b, ApJ 46, 60
Keeton, C.R. 2002, ApJ 575, L1
Keeton, C.R., Gaudi, B.S. & Petters, A.O. 2003, ApJ 598, 138
Keeton, C.R. 2003a, ApJ 582, 17
Keeton, C.R. 2003b, ApJ 584, 664
Keeton, C.R. & Winn, J.N. 2003, ApJ 590, 39
Keeton, C.R., Kuhlen, M. & Haiman, Z. 2005, ApJ 621, 559, also astro-ph/0405143
Kelson, D.D., van Dokkum, P, Franx, M., Illingworth, G.& Fabricant, D. 1997, ApJ 478, L13
Kelson, D.D., Illingworth, G.D., van Dokkum, P. & Franx, M. 2000, ApJ 531, 184
Khairul, A.S.M. & Ryden, B.S. 2002, ApJ 570, 610
King, L. & Browne, I.W.A. 1996, MNRAS 282, 67
Klypin, A., Kravtsov, A.V., Valenzuela, O. & Prada, F. 1999, ApJ 522, 82
Kneib, J.-P., Ellis, R.S., Smail, I., Couch, W.J. & Sharples, R.M. 1996, ApJ 471, 643
Kneib, J.-P., Cohen, J.G & Hjorth, J. 2000, ApJ 544, L35
Kochanek, C.S. & Blandford, R.D. 1987, ApJ 321, 676
Kochanek, C.S. & Apostolakis, J. 1988, MNRAS 235, 1073
Kochanek, C.S., Blandford, R.D., Lawrence, C.R. & Narayan, R. 1989, MNRAS 238, 43
Kochanek, C.S. & Lawrence, C.R. 1990, AJ 99, 1700
Kochanek, C.S. 1991a, ApJ 373, 354
Kochanek, C.S. 1991b, ApJ 379, 517
Kochanek, C.S. 1991c, ApJ 382, 58
Kochanek, C.S. 1992a, ApJ 384, 1
Kochanek, C.S. 1992b, ApJ 397, 381
Kochanek, C.S. & Narayan, R. 1992, ApJ 401, 461
Kochanek, C.S. 1993a, ApJ 419, 12
Kochanek, C.S. 1993b, MNRAS 261, 453
Kochanek, C.S. 1993c, ApJ 417, 438

Kochanek, C.S. 1994, ApJ 436, 56
Kochanek, C.S., Falco, E.E. & Schild, R. 1995, ApJ 452, 109
Kochanek, C.S. 1995a, ApJ 445, 559
Kochanek, C.S. 1995b, ApJ 453, 545
Kochanek, C.S. 1996b, ApJ 466, 638
Kochanek, C.S. 1996a, ApJ 473, 595
Kochanek, C.S., Falco, E.E. & Muñoz, J.A. 1999, ApJ 510, 590
Kochanek, C.S., et al. 2000a, ApJ 543, 131
Kochanek, C.S., Falco, E.E., Impey, C.D., Lehár, J., McLeod, B.A., Rix, H.-W., Keeton, C.R., Peng, C.Y. & Muñoz, J.A. 2000b, ApJ 535, 692
Kochanek, C.S., Keeton, C.R. & McLeod, B.A. 2001a, ApJ 547, 50
Kochanek, C.S., & White, M. 2001, ApJ 559, 531
Kochanek, C.S., Pahre, M.A., Falco, E.E., Huchra, J.P., Mader, J., Jarrett, T.H., Chester, T., Cutri, R. & Schneider, S.E. 2001b, ApJ 560, 566
Kochanek, C.S. 2002a, ApJ 578, 25
Kochanek, C.S. 2002b, in Shapes of Galaxies and Their Dark Halos, P. Natarajan (ed.), (World Scientific: Singapore), 62
Kochanek, C.S. 2003a, ApJ 583, 49
Kochanek, C.S. 2003b, astro-ph/0204043
Kochanek, C.S. 2003c, in The Dark Universe: Matter Energy and Gravity, M. Livio (ed.), (Cambridge Univ. Press: Cambridge), p. 139
Kochanek, C.S. & Schechter, P. 2004, in Measuring and Modeling the Universe, W.L Freedman (ed.), Carnegie Observatories Centennial Symposia, Cambridge Univ. Press: Cambridge), p. 117
Kochanek, C.S. 2004, ApJ 605, 58
Kochanek, C.S. & Dalal, N. 2004, ApJ 610, 69, also astro-ph/0302036
Koopmans, L.V.E., de Bruyn, A.G. & Jackson, N. 1998, MNRAS 295, 534
Koopmans, L.V.E., de Bruyn, A.G., Xanthopoulos, E. & Fassnacht, C.D. 2000a, A&A 356, 391
Koopmans, L.V.E., de Bruyn, A.G., Fassnacht, C.D., Marlow, D.R. et al. 2000b, A&A 361, 815
Koopmans, L.V.E., Garrett, M.A., Blandford, R.D., et al. 2002, MNRAS 334, 39
Koopmans, L.V.E. & Treu, T. 2002, ApJ 568, 5
Koopmans, L.V.E. & Treu, T. 2003, ApJ 583, 606
Koopmans, L.V.E., Treu, T., Fassnacht, C.D., Blandford, R.D., Surpi, G. 2003, ApJ 599, 70
Koopmans, L.V.E. & de Bruyn, A.G. 2003, MNRAS 360, L6, also astro-ph/0311567
Kormann, R., Schneider, P. & Bartelmann, M. 1994, A&A 284, 285
Kovner, I. 1987a, ApJ 312, 22
Kovner, I. 1987b, ApJ 316, 52
Krauss, L.M. & White, M. 1992, ApJ 394, 385
Kuhlen, M., Keeton, C.R. & Madau, P. 2004, ApJ 601, 104

Kukula, M.J., Dunlop, J.S., McLure, R.J, Miller, L., Percival, W.J., Baum, S.A. & O'Dea, C.P. 2001, MNRAS 326, 1533
Kundić, T., Turner, E.L., Colley, W.N., Gott, J.R., Rhoads, J.E., Wang, Y., Bergeron, L.E., Gloria, K.A., Long, D.C., Malhotra, S., Wambsganss, J. 1997, ApJ 482, 75
Kuzmin, G. 1956, Astr.Zh. 33, 27
Lawrence, C.R., Schneider, D.P., Schmidt, M., Bennett, C.L., Hewitt, J.N., Burke, B.F., Turner, E.L. & Gunn, J.E. 1984, Science 223, 46
Lehár, J., Cooke, AJ, Lawrence, C.R., Silber, A.D. & Langston, G.I. 1996, AJ 111, 1812
Lehár, J., Falco, E.E., Kochanek, C.S., McLeod, B.A, Muñoz, J.A., Impey, C.D., Rix, H.-W., Keeton, C.R. & Peng, C.Y. 2000, ApJ 536, 584
Lehár, J., Buchalter, A., McMahon, R.G., Kochanek, C.S. & Muxlow, T.W.B. 2001, ApJ 547, 60
Li, L.-X. & Ostriker, J.P. 2002, ApJ 566, 652
Li, L.-X. & Ostriker, J.P. 2003, ApJ 595, 603
Lidman, C., Courbin, F., Kneib, J.-P., Golse, G., Castandar, F. & Soucail, G. 2000, A&A 364, L62
Loewenstein, M. & Mushotzky, R.F. 2003, ApJ submitted
Lovell, J.E.J., Jauncey, D.L., Reynolds, J.E., Wieringa, M.H., King, E.A., Tzioumis, A., McCulloch, P.M., Edwards, P.G. 1998, ApJ 508, L51
Ma, C.-P. 2003, ApJ 584, L1
Maller, A.H., Flores, R.A. & Primack, J.R. 1997, ApJ 486, 681
Maller, A.H., Simard, L., Guhathakurta, P., Hjorth, J., Jaunsen, A.O., Flores, R.A. & Primack, J.R. 2000, ApJ 533, 194
Mao, S.D. 1991, ApJ 380, 9
Mao S.D. & Kochanek, C.S. 1994, MNRAS 268, 569
Mao, S. & Schneider, P. 1998, MNRAS 295, 587
Mao, S., Witt, H.J. & Koopmans, L.V.E. 2001 MNRAS 323, 301
Mao, S., Jing, Y., Ostriker, J.P. & Weller, J. 2004, ApJ, 604, 5L
Maoz, D., Bahcall, J.N., Schneider, D.P. et al., 1993 ApJ 409, 28
Maoz, D. & Rix, H.-W. 1993, ApJ 416, 425
Maoz D., Rix H.-W., Gal-Yam A., Gould A. 1997, ApJ 486, 75
Marlow, D.R., Browne, I.W.A., Jackson, N., & Wilkinson, P.N. 1999, MNRAS 305, 15
Marlow, D.R., Rusin,¡ D., Jackson, N., Wilkinson, P.N., Browne, I.W.A. & Koopmans, L. 2000, AJ 119, 2629
McKay, T.A. et al. 2002, ApJ 571, L85
McLeod, K.K. & McLeod, B.A. 2001, ApJ 546, 782
McLure, R.J., Kukula, M.J., Dunlop, J.S., Baum, S.A., O'Dea, C.P. & Hughes, D.H. 1999, MNRAS 308, 377
Mestel, L. 1963, MNRAS 126, 553
Metcalf, R.B. & Madau, P. 2001, ApJ 563, 9
Metcalf, R.B. 2002, ApJ 580, 696
Metcalf, R.B. & Zhao, H. 2002, ApJ 567, L5

Miralda-Escude, J. & Lehár, J. 1992, MNRAS 259, 31
Mitchell, J.L., Keeton, C.R., Frieman, J.A. & Sheth, R.K. 2004, ApJ 622, 81, also astro-ph/0401138
Mo H.J., Mao S. & White S.D.M. 1998, MNRAS 295, 319
Möller, O., Hewett, P., & Blain, A.W. 2003, MNRAS 345, 1
Möller, O. & Blain, A.W. 2001, MNRAS 327, 339
Moore B. et al. 1998, ApJ 499, L5
Moore, B., Quinn, T., Governato, F., Statel, J. & Lake, G. 1999, MNRAS 310, 1147
Moore, B., Ghigna, S., Governato, F., Lake, G., Quinn, T., Stadel, J. & Tozzi, P. 1999, ApJ 524, L19
Morgan, N.D., Chartas, G., Malm, M., Bautz, M.W., Burud, I., Hjorth, J., Jones, S.E. & Schechter, P.L. 2001, ApJ 555, 1
Morgan, N.D., Snyder, J.A. & Reens, L.H. 2003, AJ 126, 2145
Mortlock, D.J., Webster, R.L. & Francis, P.J. 1999, MNRAS 309, 836
Mortlock, D.J. & Webster, R.L. 2000a, MNRAS 319, 879
Mortlock, D.J. & Webster, R.L. 2000b, MNRAS 319, 860
Mortlock, D.J. & Webster, R.L. 2000c, MNRAS 319, 872
Mortlock, D.J. & Webster, R.L. 2001, MNRAS 321, 629
Motta, V., Mediavilla, E., Muñoz, J.A., Falco, E.E. et al. 2002, ApJ 574, 719
Moustakas, L.A. & Metcalf, R.B. 2003, MNRAS 339, 607
Muñoz, J.A., Falco, E.E., Kochanek, C.S., Lehár, J., Herold, L.K., Fletcher, A.B. & Burke, B.F. 1998, ApJ 492, L9
Muñoz, J.A., Kochanek, C.S. & Falco, E.E. 1999, ApJ 521, L17
Muñoz, J.A., Kochanek, C.S. & Keeton, C.R. 2001, ApJ 558, 657
Muñoz, J.A., Falco, E.E., Kochanek, C.S., Lehár, J. & Mediavilla, E. 2003, ApJ 594, 684
Muñoz, J.A., Falco, E.E., Kochanek, C.S., McLeod, B.A. & Mediavilla, E. 2004, ApJ 605, 614
Myers, S.T., Fassnacht, C.D., Djorgovski, S.G. et al. 1995, ApJ 447, L5
Myers, S.T., Rusin, D., Fassnacht, C.D., Blandford, R.D., Pearson, T.J., Readhead, A.C.S., Jackson, N., Browne, I.W.A., Marlow, D.R., Wilkinson, P.N., Koopmans, L.V.E. & de Bruyn, A.G. 1999, AJ 117, 2565
Nadeau, D., Yee, H.K.C., Forrest, W.J., Garnett, J.D., Ninkov, Z. & Pipher, J.L. 1991, ApJ 376, 430
Narayan R. & White S.D.M. 1988, MNRAS 231, P97
Narayan, R. & Bartelmann, M. 1999, in Formation of Structure in the Universe, A. Dekel & J. Ostriker (eds.), (Cambridge University Press: Cambridge) p. 360
Navarro J., Frenk C.S., & White S.D.M. 1996, ApJ 462, 563
Ofek, E.O., Maoz, D., Prada, F., Kolatt, T. & Rix, H.-W. 2001, MNRAS 324, 463
Ofek, E.O. & Maoz, D. 2003, ApJ 594, 101
Ofek, E.O., Rix, H.-W. & Maoz, D. 2003, MNRAS 343, 639
Oguri, M. & Keeton, C.R. 2004, ApJ submitted, astro-ph/0403633

Oguri, M., Inada, N., Keeton, C.R., Pindor, B., Hennawi, J.F., Gregg, M.D., Becker, R.H. et al. 2004, ApJ 604, 78
Ohyama, Y. et al. 2002, AJ 123, 2903
Olling, R.P. & Merrifield, M. 2001, MNRAS 311, 361
Pahre, M.A., de Carvalho, R.R. & Djorgovski, S.G. 1998, AJ 116, 1606
Patnaik, A.R. & Narasimha, D. 2001, MNRAS 326, 1403
Pearce F.R., Jenkins A., Frenk C.S., Colberg J.M., White S.D.M., Thomas P.A., Couchman H.M.P., Peacock J.A., & Efstathiou G. 1999, ApJ 521, L99
Pello, R., Miralles, J.M., Le Borgne, J.-F., Picat, J.-P., Soucail, G. & Bruzual, G. 1996, A&A 314, 73
Pelt, J., Kayser, R., Refsdal, S. & Schramm, T. 1996, A&A 305, 97
Peng, C.Y., 2004, PhD Thesis, Steward Observatory, University of Arizona
Phillips, P.M., Browne, I.W.A., Jackson, N.J., Wilkinson, P.N., Mao, S., Rusin, D., Marlow, D.R., Snellen, I. & Nesser, M., 2001, MNRAS 328, 1001
Pindor, B., Turner, E.L., Lupton, R.H. & Brinkmann, J. 2003, AJ 125, 2325
Porciani C., & Madau P. 2000, ApJ 532, 679
Press, W.H. & Gunn, J.E. 1973, ApJ 185, 397
Press, W.H. & Schechter P. 1974, ApJ 187, 425
Press, W.H., Rybicki, G.B. & Hewitt, J.N. 1992a, ApJ 385, 404
Press, W.H., Rybicki, G.B. & Hewitt, J.N. 1992b, ApJ 385, 416
Quadri, R., Möller, O. & Natarajan, P. 2003, ApJ 597, 659
Ratnatunga, K.U., Griffiths, R.E. & Ostrander, E.J. 1999, AJ 117, 2010
Rees M.J. & Ostriker J.P. 1977, MNRAS 179, 541
Refsdal, S. 1964a, MNRAS 128, 295
Refsdal, S. 1964b, MNRAS 128, 307
Refsdall, S. & Surdej, J. 1994, RPPh 57, 117
Rest, A., van den Bosch, F.C., Jaffe, W., Tran, H., Tsvetanov, Z., Ford, H.C., Davies, J. & Schafer, J.L. 2001, AJ 121, 2431
Ridgway, S.E., Heckman, T.M., Calzetti, D. & Lehnert, M. 2001, ApJ 550, 122
Riess, A.G., Strolger, L.-G., Tonry, J., Casertano, S. et al. 2004, ApJ 607, 665
Rix H.-W., Maoz D., Turner E.L. & Fukugita M. 1994, ApJ 435, 49
Romanowsky, A.J. & Kochanek, C.S. 1998, ApJ 493, 64
Romanowsky, A.J. & Kochanek, C.S. 1999, ApJ 516, 18
Ros, E., Guirado, J.C., Marcaide, J.M., Perez-Torres, M.A., Falco, E.E., Muñoz, J.A., Alberdi, A. & Lara L. 2000, A&A 362, 845
Rusin, D. & Ma, C.-P. 2001, ApJ 549, L33
Rusin, D., Kochanek, C.S., Norbury, M. et al. 2001, ApJ 557, 594
Rusin, D. & Tegmark, M. 2001, ApJ 553, 709
Rusin, D. et al. 2002, MNRAS 330, 205
Rusin, D. 2002, ApJ 572, 705
Rusin, D., Kochanek, C.S., Falco, E.E., Keeton, C.R., McLeod, B.A., Impey, C.D., Lehár, J., Muñoz, J.A., Peng, C.Y. & Rix, H.-W. 2003a, ApJ 587, 143

Rusin, D., Kochanek, C.S. & Keeton, C.R. 2003b, ApJ 595, 29
Rusin, D. & Kochanek, C.S. 2005, ApJ 623, 666
Saha, P. & Williams, L.R.L. 1997, MNRAS 292, 148
Saha, P. 2000, AJ 120, 1654
Saha, P. & Williams, L.L.R. 2001, AJ 122, 585
Saha, P. 2004, A&A 414, 425
Saha, P. & Williams, L.L.R. 2004, AJ 127,2604, also astro-ph/0402135
Sand, D.J., Treu, T. & Ellis, R.S. 2002, ApJ 574, L129
Sand, D.J., Treu, T., Smith, G.P. & Ellis, R.S. 2004, ApJ 604, 88
Schechter, P. 1976, ApJ 203, 297
Schechter, P.L. & Moore, C.B. 1993, AJ 105, 1
Schechter, P.L., Bailyn, C.D., Barr, R. et al. 1997, ApJ 475, L85
Schechter, P.L. & Wambsganss, J. 2002, ApJ 580, 685
Schechter, P.L., Udalski, A., Szymanski, M., Kubiak, M., Pietrzynski, G., Soszynski, I., Wozniak, P., Zebrun, K., Szewczyk, O. & Wyrzykowki, L. 2003, ApJ 584, 657
Schild, R. & Cholfin, B. 1986, ApJ 300, 209
Schild, R. & Thomson, D.J. 1995, AJ 109, 1970
Schild, R.E. & Thomson, D.J. 1997, AJ 113, 130
Schneider, P., Ehlers, J. & Falco, E.E. 1992, Gravitational Lenses, (Springer-Verlag: Berlin)
Scodeggio, M., Giovanelli, R. & Haynes, M.P. 1998, ApJ 116, 2738
Sersic, J.-L., 1968, Atlas de Galaxias Australes (Obs. Astron.: Cordoba)
Sheth, R.K. & Tormen, G. 1999, MNRAS 308, 119
Sheth, R.K., Bernardi, M., Schechter, P.L. et al. 2003, ApJ 594, 225
Sluse, D., Surdej, J., Claeskens, J.-F., Hutsemekers, D., Jean, C., Courbin, F., Nakos, T., Billeres, M. & Khmil, S.V. 2003, A&A 406, L43
Small, T.A., Sargent, W.L.W. & Steidel, C.C. 1997, AJ 114, 2254
Soucail, G., Kneib, J.-P. & Golse, G. 2004, A&A 417, L33
Spergel, D. & Steinhardt, P. 2000, PRL 84, 3760
Spergel, D.N., Verde, L., Peiris, H.V. et al. 2003, ApJS 148, 175
Surdej, J., Claeskens, J.F., Crampton, D. et al. 1993, AJ 105, 2064
Sykes, C.M., Browne, I.W.A., Jackson, N.J., Marlow, D.R., Nair, S., Wilkinson, P.N., Blandford, R.D., Cohen, J., Fassnacht, C.D., Hogg, D., Pearson, T.J., Readhead, A.C.S., Womble, D.S., Myers, S.T., de Bruyn, A.G., Bremer, M., Miley, G.K. & Schilizzi, R.T. 1998, MNRAS 301, 310
Toft, S., Hjorth, J., & Burud, I. 2000, A&A 357, 115
Tonry, J.L. 1998, AJ 115, 1
Tonry, J.L. & Franx, M. 1999, ApJ 515, 512
Treu, T., Stiavelli, M., Bertin, G., Casertano, S. & Møller, P. 2001, MNRAS 326, 327
Treu, T., Stiavelli, M., Casertano, S., Møller, P. & Bertin, G. 2002, ApJ 564, L13
Treu, T. & Koopmans, L.V.E. 2002a, MNRAS 337, P6

Treu, T. & Koopmans, L.V.E. 2002b, ApJ 575, 87
Treu, T. & Koopmans, L.V.E. 2004, ApJ 611, 739, also astro-ph/0401373
Trott, C.M. & Webster, R.L. 2002, MNRAS 334, 621
Trotter, C.S., Winn, J.N. & Hewitt, J.N. 2000, ApJ 535, 671
Tully, R.B. & Fisher, B. 1977, A&A 54, 661
Turner, C.M., Keeton, C.R. & Kochanek, C.S. 2004, in preparation
Turner, E.L. 1980, ApJ 242, L135
Turner E.L., Ostriker J.P., Gott J.R. 1984, ApJ 284, 1
Vanderriest, C., Schneider, J., Herpe, G., Chevreton, M., Moles, M. & Wlérick, G. 1989, A&A 215, 1
van de Ven, G., van Dokkum, P.G. & Franx, M. 2003, MNRAS 344, 924
van Dokkum, P.G. & Franx M. 1996, MNRAS 281, 985
van Dokkum, P.G. & Franx, M. 2001, ApJ 553, 90
van Dokkum, P.G., Franx, M., Kelson, D.D. & Illingworth, G.D., 2001, ApJ 553, L39
van Dokkum, P.G. & Ellis, R.S. 2003, ApJ 592, L53
van der Marel, R.P. & Franx, M. 1993, ApJ 407, 525
Vestergaard, M. 2004, ApJ 601, 676
Wallington, S. & Narayan, R. 1993, ApJ 403, 517
Wallington, S., Kochanek, C.S. & Koo, D.C. 1995, ApJ 441, 58
Wallington, S., Kochanek, C.S. & Narayan, R. 1996, ApJ 465, 64
Walsh, D., Carswell, R.F. & Weymann, R.J. 1979, Nature 279, 381
Wambsganss, J., & Paczyński, B., 1994, AJ 108, 1156
Wambsganss, J., Cen R., Ostriker, J.P. & Turner, E.L. 1995, Science 268, 274
Wambsganss, J., Cen R. & Ostriker, J.P. 1998, ApJ 494, 29
Wambsganss, J. 1998, LRR 1, 12
Warren, M.S., Quinn, P.J., Salmon, J.K. & Zurek, W.H. 1992, ApJ 399, 405
Warren, S.J., Hewett, P.C., Lewis, G.F., Møller, P., Iovino, A. & Shaver, P.A. 1996, MNRAS 278, 139
Warren, S.J. & Dye, S. 2003, ApJ 590, 673
Weedman, D.W., Weymann, R.J., Green, R.F. & Heckman, T.M. 1982, ApJ 255, L5
Weymann, R.J., Latham, D., Roger, J., Angel, P., Green, R.F., Liebert, J., Turnshek, D.A., Turnshek, D.E. & Tyson, J.A. 1980, Nature 285, 641
White M., Hernquist L. & Springel V. 2001, astro-ph/0107023
Wiklind, T. & Combes, F. 1996, Nature 379, 139
Wiklind, T. & Alloin, D. 2002, in Gravitational Lensing: An Astrophysical Tool, F. Courbin & D. Minniti (eds.), (Springer-Verlag: Berlin), Lecture Notes in Physics 608, 124
Wilkinson, P.N., Henstock, D.R., Browne, I.W.A., Readhead, A.C.S., Pearson, T.J., Taylor, G.B. & Vermeulen, R.C. 2001, in Gravitational Lensing: Recent Progress Future Goals T.G. Brainerd & C.S. Kochanek (eds.), ASP Conf. Ser. # 237, (ASP: San Francisco) p. 37
Williams, L.L.R. & Saha, P. 2000, AJ 119, 439

Winn, J.N., Hewitt, J.N. & Schechter, P.L. 2001, Gravitational Lensing: Recent Progress and Future Goals, T.G. Brainerd & C.S. Kochanek (eds.), ASP Conf. Ser. # 237, (San Francisco: ASP) p. 61
Winn, J.N. et al. 2002a, AJ 123, 10
Winn, J.N., Lovell, J.E.J., Chen, H., Fletcher, A., Hewitt, J.N., Patnaik, A.R. & Schechter, P.L. 2002b, ApJ 564, 143
Winn, J.N., Kochanek, C.S., McLeod, B.A., Falco, E.E., Impey, C.D. & Rix, H.-W. 2002c, ApJ 575, 103
Winn, J.N., Rusin, D. & Kochanek, C.S. 2003a, ApJ 587, 80
Winn, J.N., Kochanek, C.S., Keeton, C.R. & Lovell, J.E.J. 2003b, ApJ 590, 26
Winn, J.N., Hall, P.B. & Schechter, P.L. 2003c, ApJ 597, 672
Winn, J.N., Rusin, D.S. & Kochanek, C.S. 2004, Nature 427, 613
Wisotzki, L., Kohler, T., Kayser, R. & Reimers, D. 1993, A&A 278, L15
Wisotzki, L., Schechter, P.L., Bradt, H.V., Heinmuller, J. & Reimers, D. 2002, A&A 395, 17
Witt, A.N., Thronson, H.A. & Capuano, J.M. 1992, ApJ 393, 611
Witt, H.J. & Mao, S. 2000, MNRAS 311, 689
Witt, H.J., Mao, S. & Keeton, C.R. 2000, ApJ 544, 98
Wucknitz, O. 2002, MNRAS 332, 951
Wucknitz, O. 2004, MNRAS 349, 1
Wucknitz, O., Biggs, A.D. & Browne, I.W.A. 2004, MNRAS 349, 14
Wyithe, J.S.B., Turner, E.L. & Spergel, D.N. 2001, ApJ 555, 504
Wyithe, J.S.B., Winn, J.N. & Rusin, D. 2003, ApJ 583, 58
Wyithe, J.S.B. 2004, MNRAS 351, 1266
Wyrzykowski, L., Udalski, A., Schechter, P.L., Szewczyk, O., Szymanski, M., Kubiak, M., Piertrzynski, G., Soszynski, I. & Zebrun, K. 2003, AcA 53, 229
Yee, H.K.C., Fillipenko, A.V. & Tang, D. 1993, AJ 105, 7
York, T., Jackson, N., Browne, I.W.A., Wucknitz, O. & Skelton, J.E. 2005, MNRAS 357, 124, also astro-ph/0405115
Zaritsky, D. & Rix, H.-W. 1997, AJ 477, 118
Zentner, A.R. & Bullock, J.S. 2003, ApJ 598, 49
Zhao, H. 1997, MNRAS 287, 525
Zhao, H. & Pronk, D. 2001, MNRAS 320, 401

Part 3: Weak Gravitational Lensing

P. Schneider

1 Introduction

Multiple images, microlensing (with appreciable magnifications) and arcs in clusters are phenomena of *strong lensing*. In *weak gravitational lensing*, the Jacobi matrix $\mathcal{A}$ is very close to the unit matrix, which implies weak distortions and small magnifications. Those cannot be identified in individual sources, but only in a statistical sense. Because of that, the accuracy of any weak lensing study will depend on the number of sources which can be used for the weak lensing analysis. This number can be made large either by having a large number density of sources, or to observe a large solid angle on the sky, or both. Which of these two aspects is more relevant depends on the specific application. Nearly without exception, the sources employed in weak lensing studies up to now are distant galaxies observed in the optical or near-IR passband, since they form the densest population of distant objects in the sky (which is a statement both about the source population in the Universe and the sensitivity of detectors employed in astronomical observations). To observe large number densities of sources, one needs deep observations to probe the faint (and thus more numerous) population of galaxies. Faint galaxies, however, are small, and therefore their observed shape is strongly affected by the Point Spread Function, caused by atmospheric seeing (for ground-based observations) and telescope effects. These effects need to be well understood and corrected for, which is the largest challenge of observational weak lensing studies. On the other hand, observing large regions of the sky quickly leads to large data sets, and the problems associated with handling them. We shall discuss some of the most important aspects of weak lensing observations in Sect. 3.

The effects just mentioned have prevented the detection of weak lensing effects in early studies with photographic plates (e.g., Tyson et al. 1984); they are not linear detectors (so correcting for PSF effects is not reliable), nor are they sensitive enough for obtaining sufficiently deep images. Weak lensing research came through a number of observational and technical advances.

Schneider P (2006), Weak gravitational lensing. In: Meylan G, Jetzer Ph and North P (eds) Gravitational lensing: Strong, weak, and micro. Saas-Fee Adv Courses vol 33, pp 269–451
DOI 10.1007/3-540-30309-X_3

Soon after the first giant arcs in clusters were discovered (see Sect. 1.2 of Schneider, this volume; hereafter referred to as IN) by Soucail et al. (1987) and Lynds and Petrosian (1989), Fort et al. (1988) observed objects in the lensing cluster Abell 370 which were less extremely stretched than the giant arc, but still showed a large axis ratio and was aligned in the direction tangent to its separation vector to the cluster center; they termed these images ‘arclets’. Indeed, with the spectroscopic verification (Mellier et al. 1991) of the arclet A5 in Abell 370 being located at much larger distance from us than the lensing cluster, the gravitational lens origin of these arclets was proven. When the images of a few background galaxies are deformed so strongly that they can be identified as distorted by lensing, there should be many more galaxy images where the distortion is much smaller, and where it can only be detected by averaging over many such images. Tyson et al. (1990) reported this statistical distortion effect in two clusters, thereby initiating the weak lensing studies of the mass distribution of clusters of galaxies. This very fruitful field of research was put on a rigorous theoretical basis by Kaiser and Squires (1993) who showed that from the measurement of the (distorted) shapes of galaxies one can obtain a parameter-free map of the projected mass distribution in clusters.

The flourishing of weak lensing in the past ten years was mainly due to three different developments. First, the potential of weak lensing was realized, and theoretical methods were worked out for using weak lensing measurements in a large number of applications, many of which will be described in later sections. This realization, reaching out of the lensing community, also slowly changed the attitude of time allocation committees, and telescope time for such studies was granted. Second, returning to the initial remark, one requires large fields-of-views for many weak lensing application, and the development of increasingly large wide-field cameras installed at the best astronomical sites has allowed large observational progress to be made. Third, quantitative methods for the correction of observations effects, like the blurring of images by the atmosphere and telescope optics, have been developed, of which the most frequently used one came from Kaiser et al. (1995). We shall describe this technique, its extensions, tests and alternative methods in Sect. 3.5.

We shall start by describing the basics of weak lensing in Sect. 2, namely how the shear, or the projected tidal gravitational field of the intervening matter distribution can be determined from measuring the shapes of images of distant galaxies. Practical aspects of observations and the measurements of image shapes are discussed in Sect. 3. The next two sections are devoted to clusters of galaxies; in Sect. 4, some general properties of clusters are described, and their strong lensing properties are considered, whereas in Sect. 5 weak lensing by clusters is treated. As already mentioned, this allows us to obtain a parameter-free map of the projected (2-D) mass distribution of clusters.

We then turn to lensing by the inhomogeneously distributed matter distribution in the Universe, the large-scale structure. Starting with Gunn (1967), the observation of the distortion of light bundles by the inhomogeneously distributed matter in the Universe was realized as a unique probe to study the properties of the cosmological (dark) matter distribution. The theory of this cosmic shear effect, and its applications, was worked out in the early 1990's (e.g., Blandford et al. 1991). In contrast to the lensing situations studied in the rest of this book, here the deflecting mass is manifestly three-dimensional; we therefore need to generalize the theory of geometrically-thin mass distributions and consider the propagation of light in an inhomogeneous Universe. As will be shown, to leading order this situation can again be described in terms of an 'equivalent' surface mass density. The theoretical aspects of this large-scale structure lensing, or cosmic shear, are contained in Sect. 6. Although the theory of cosmic shear was well in place for quite some time, it took until the year 2000 before it was observationally discovered, independently and simultaneously by four groups. These early results, as well as the much more extensive studies carried out in the past few years, are presented and discussed in Sect. 7. In Sect. 8, we consider the weak lensing effects of galaxies, which can be used to investigate the mass profile of galaxies. As we shall see, this galaxy-galaxy lensing, first detected by Brainerd et al. (1996), is directly related to the connection between the galaxy distribution in the Universe and the underlying (dark) matter distribution; this lensing effect is therefore ideally suited to study the biasing of galaxies; we shall also describe alternative lensing effects for investigating the relation between luminous and dark matter. In the final Sect. 9 we discuss higher-order cosmic shear statistics and how lensing by the large-scale structure affects the lens properties of localized mass concentrations. Some final remarks are given in Sect. 10.

Until very recently, weak lensing has been considered by a considerable fraction of the community as 'black magic' (or to quote one member of a PhD examination committee: "You have a mass distribution about which you don't know anything, and then you observe sources which you don't know either, and then you claim to learn something about the mass distribution ?"). Most likely the reason for this is that weak lensing is indeed weak. One cannot 'see' the effect, nor can it be graphically displayed easily. Only by investigating many faint galaxy images can a signal be extracted from the data, and the human eye is not sufficient to perform this analysis. This is different even from the analysis of CMB anisotropies which, similarly, need to be analyzed by statistical means, but at least one can display a temperature map of the sky. However, in recent years weak lensing has gained a lot of credibility, not only because it has contributed substantially to our knowledge about the mass distribution in the Universe, but also because different teams, with different data set and different data analysis tools, agree on their results.

Weak lensing has been reviewed before; we shall mention only five extensive reviews. Mellier (1999) provides a detailed compilation of the weak lensing results before 1999, whereas Bartelmann and Schneider (2001; hereafter BS01) present a detailed account of the theory and technical aspects of weak lensing.[1]

[1] We follow here the notation of BS01, except that we denote the angular diameter distance explicitly by $D^{\rm ang}$, whereas D is the comoving angular diameter distance,

More recent summaries of results can also be found in Wittman (2002) and Refregier (2003a), as well as the cosmic shear review by van Waerbeke and Mellier (2003).

The coverage of topics in this review has been a subject of choice; no claim is made about completeness of subjects or references. In particular, due to the lack of time during the lectures, the topic of weak lensing of the CMB temperature fluctuations has not been covered at all, and is also not included in this written version. Apart from this increasingly important subject, I hope that most of the currently actively debated aspects of weak lensing are mentioned, and the interested reader can find her/his way to more details through the references provided.

2 The Principles of Weak Gravitational Lensing

2.1 Distortion of Faint Galaxy Images

Images of distant sources are distorted in shape and size, owing to the tidal gravitational field through which light bundles from these sources travel to us. Provided the angular size of a lensed image of a source is much smaller than the characteristic angular scale on which the tidal field varies, the distortion can be described by the linearized lens mapping, i.e., the Jacobi matrix $\mathcal{A}$. The invariance of the surface brightness by gravitational light deflection, $I(\boldsymbol{\theta}) = I^{(\mathrm{s})}[\boldsymbol{\beta}(\boldsymbol{\theta})]$, together with the locally linearized lens equation,

$$\boldsymbol{\beta} - \boldsymbol{\beta}_0 = \mathcal{A}(\boldsymbol{\theta}_0) \cdot (\boldsymbol{\theta} - \boldsymbol{\theta}_0) , \tag{1}$$

where $\boldsymbol{\beta}_0 = \boldsymbol{\beta}(\boldsymbol{\theta}_0)$, then describes the distortion of small lensed images as

$$I(\boldsymbol{\theta}) = I^{(\mathrm{s})}[\boldsymbol{\beta}_0 + \mathcal{A}(\boldsymbol{\theta}_0) \cdot (\boldsymbol{\theta} - \boldsymbol{\theta}_0)] . \tag{2}$$

We recall (see IN) that the Jacobi matrix can be written as

$$\mathcal{A}(\boldsymbol{\theta}) = (1-\kappa) \begin{pmatrix} 1-g_1 & -g_2 \\ -g_2 & 1+g_1 \end{pmatrix} , \text{ where } g(\boldsymbol{\theta}) = \frac{\gamma(\boldsymbol{\theta})}{[1-\kappa(\boldsymbol{\theta})]} \tag{3}$$

is the reduced shear, and the g_α, $\alpha = 1, 2$, are its Cartesian components. The reduced shear describes the shape distortion of images through gravitational light deflection. The (reduced) shear is a 2-component quantity, most conveniently written as a complex number,

$$\gamma = \gamma_1 + \mathrm{i}\gamma_2 = |\gamma| \, \mathrm{e}^{2\mathrm{i}\varphi} ; \quad g = g_1 + \mathrm{i}g_2 = |g| \, \mathrm{e}^{2\mathrm{i}\varphi} ; \tag{4}$$

---

which we also write as f_K, depending on the context; see Sect. 4.3 of IN for more details. In most cases, the distance ratio $D_{\mathrm{ds}}/D_{\mathrm{s}}$ is used, which is the same for both distance definitions.

its amplitude describes the degree of distortion, whereas its phase φ yields the direction of distortion. The reason for the factor '2' in the phase is the fact that an ellipse transforms into itself after a rotation by 180°. Consider a circular source with radius R (see Fig. 1); mapped by the local Jacobi matrix, its image is an ellipse, with semi-axes

$$\frac{R}{1-\kappa-|\gamma|} = \frac{R}{(1-\kappa)(1-|g|)} \quad ; \quad \frac{R}{1-\kappa+|\gamma|} = \frac{R}{(1-\kappa)(1+|g|)}$$

and the major axis encloses an angle φ with the positive θ_1-axis. Hence, if sources with circular isophotes could be identified, the measured image ellipticities would immediately yield the value of the reduced shear, through the axis ratio

$$|g| = \frac{1-b/a}{1+b/a} \quad \Leftrightarrow \quad \frac{b}{a} = \frac{1-|g|}{1+|g|}$$

and the orientation of the major axis φ. In these relations it was assumed that $b \le a$, and $|g| < 1$. We shall discuss the case $|g| > 1$ later.

However, faint galaxies are not intrinsically round, so that the observed image ellipticity is a combination of intrinsic ellipticity and shear. The strategy to nevertheless obtain an estimate of the (reduced) shear consists in locally averaging over many galaxy images, assuming that the intrinsic ellipticities are *randomly oriented*. In order to follow this strategy, one needs to clarify first how to define 'ellipticity' for a source with arbitrary isophotes (faint galaxies are not simply elliptical); in addition, seeing by the atmospheric turbulence will blur – and thus circularize – observed images, together with other effects related to the observation procedure. We will consider these issues in turn.

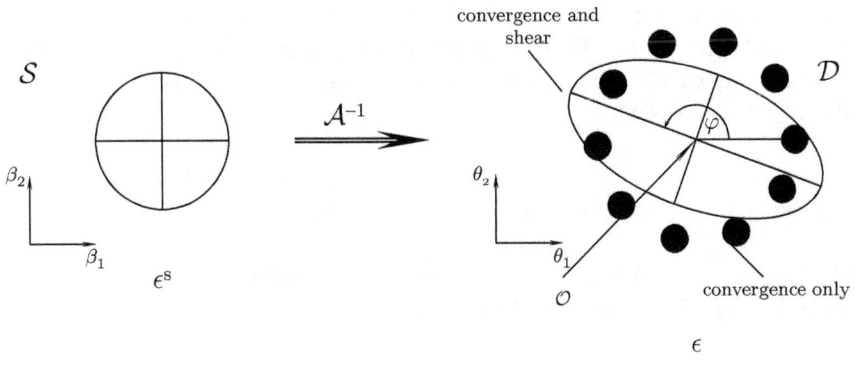

Fig. 1. A circular source, shown at the left, is mapped by the inverse Jacobian $\mathcal{A}^{-1}$ onto an ellipse. In the absence of shear, the resulting image is a circle with modified radius, depending on κ. Shear causes an axis ratio different from unity, and the orientation of the resulting ellipse depends on the phase of the shear (source: M. Bradac)

2.2 Measurements of Shapes and Shear

Definition of Image Ellipticities

Let $I(\boldsymbol{\theta})$ be the brightness distribution of an image, assumed to be isolated on the sky; the center of the image can be defined as

$$\bar{\boldsymbol{\theta}} \equiv \frac{\int \mathrm{d}^2\theta \, I(\boldsymbol{\theta}) \, q_I[I(\boldsymbol{\theta})] \, \boldsymbol{\theta}}{\int \mathrm{d}^2\theta \, I(\boldsymbol{\theta}) \, q_I[I(\boldsymbol{\theta})]} , \tag{5}$$

where $q_I(I)$ is a suitably chosen weight function; e.g., if $q_I(I) = \mathrm{H}(I - I_{\mathrm{th}})$, where $\mathrm{H}(x)$ is the Heaviside step function, $\bar{\boldsymbol{\theta}}$ would be the center of light within a limiting isophote of the image. We next define the tensor of second brightness moments,

$$Q_{ij} = \frac{\int \mathrm{d}^2\theta \, I(\boldsymbol{\theta}) \, q_I[I(\boldsymbol{\theta})] \, (\theta_i - \bar{\theta}_i) \, (\theta_j - \bar{\theta}_j)}{\int \mathrm{d}^2\theta \, I(\boldsymbol{\theta}) \, q_I[I(\boldsymbol{\theta})]} , \quad i,j \in \{1,2\} . \tag{6}$$

Note that for an image with circular isophotes, $Q_{11} = Q_{22}$, and $Q_{12} = 0$. The trace of Q describes the size of the image, whereas the traceless part of Q_{ij} contains the ellipticity information. From Q_{ij}, one defines two complex ellipticities,

$$\chi \equiv \frac{Q_{11} - Q_{22} + 2\mathrm{i}Q_{12}}{Q_{11} + Q_{22}} \quad \text{and} \quad \epsilon \equiv \frac{Q_{11} - Q_{22} + 2\mathrm{i}Q_{12}}{Q_{11} + Q_{22} + 2(Q_{11}Q_{22} - Q_{12}^2)^{1/2}} . \tag{7}$$

Both of them have the same phase (because of the same numerator), but a different absolute value. Figure 2 illustrates the shape of images as a function of their complex ellipticity χ. For an image with elliptical isophotes of axis ratio $r \le 1$, one obtains

$$|\chi| = \frac{1 - r^2}{1 + r^2} \quad ; \quad |\epsilon| = \frac{1 - r}{1 + r} . \tag{8}$$

Which of these two definitions is more convenient depends on the context; one can easily transform one into the other,

$$\epsilon = \frac{\chi}{1 + (1 - |\chi|^2)^{1/2}} , \quad \chi = \frac{2\epsilon}{1 + |\epsilon|^2} . \tag{9}$$

In fact, other (but equivalent) ellipticity definitions have been used in the literature (e.g., Kochanek 1990; Miralda-Escudé 1991; Bonnet and Mellier 1995), but the two given above appear to be most convenient.

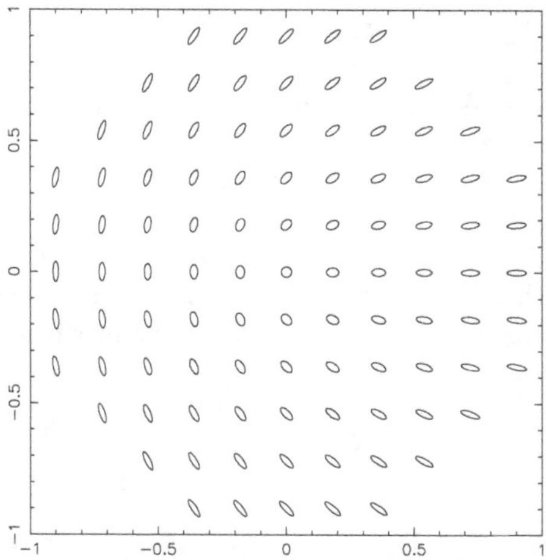

Fig. 2. The shape of image ellipses for a circular source, in dependence on their two ellipticity components χ_1 and χ_2; a corresponding plot in term of the ellipticity components ϵ_i would look quite similar. Note that the ellipticities are rotated by 90° when $\chi \to -\chi$ (source: D. Clowe)

From Source to Image Ellipticities

In total analogy, one defines the second-moment brightness tensor $Q_{ij}^{(\mathrm{s})}$, and the complex ellipticities $\chi^{(\mathrm{s})}$ and $\epsilon^{(\mathrm{s})}$ for the unlensed source. From

$$Q_{ij}^{(\mathrm{s})} = \frac{\int \mathrm{d}^2\beta \, I^{(\mathrm{s})}(\boldsymbol{\theta}) \, q_I[I^{(\mathrm{s})}(\boldsymbol{\beta})] \, (\beta_i - \bar{\beta}_i) \, (\beta_j - \bar{\beta}_j)}{\int \mathrm{d}^2\beta \, I^{(\mathrm{s})}(\boldsymbol{\theta}) \, q_I[I^{(\mathrm{s})}(\boldsymbol{\beta})]} \, , \quad i,j \in \{1,2\} \, , \tag{10}$$

one finds with $\mathrm{d}^2\beta = \det \mathcal{A} \, \mathrm{d}^2\theta$, $\boldsymbol{\beta} - \bar{\boldsymbol{\beta}} = \mathcal{A}\left(\boldsymbol{\theta} - \bar{\boldsymbol{\theta}}\right)$, that

$$Q^{(\mathrm{s})} = \mathcal{A} \, Q \, \mathcal{A}^T = \mathcal{A} \, Q \, \mathcal{A} \, , \tag{11}$$

where $\mathcal{A} \equiv \mathcal{A}(\bar{\boldsymbol{\theta}})$. Using the definitions of the complex ellipticities, one finds the transformations (e.g., Schneider and Seitz 1995; Seitz and Schneider 1997)

$$\chi^{(\mathrm{s})} = \frac{\chi - 2g + g^2\chi^*}{1 + |g|^2 - 2\mathcal{R}\mathrm{e}(g\chi^*)} \, ; \quad \epsilon^{(\mathrm{s})} = \begin{cases} \dfrac{\epsilon - g}{1 - g^*\epsilon} & \text{if } |g| \le 1 \, ; \\[2ex] \dfrac{1 - g\epsilon^*}{\epsilon^* - g^*} & \text{if } |g| > 1 \, . \end{cases} \tag{12}$$

The inverse transformations are obtained by interchanging source and image ellipticities, and $g \to -g$ in the foregoing equations.

Estimating the (Reduced) Shear

In the following we make the assumption that the intrinsic orientation of galaxies is random,

$$\mathrm{E}\left(\chi^{(\mathrm{s})}\right) = 0 = \mathrm{E}\left(\epsilon^{(\mathrm{s})}\right) , \tag{13}$$

which is expected to be valid since there should be no direction singled out in the Universe. This then implies that the expectation value of ϵ is [as obtained by averaging the transformation law (12) over the intrinsic source orientation]

$$\mathrm{E}(\epsilon) = \begin{cases} g & \text{if } |g| \le 1 \\ 1/g^* & \text{if } |g| > 1 . \end{cases} \tag{14}$$

This is a remarkable result (Schramm and Kayser 1995; Seitz and Schneider 1997), since it shows that each image ellipticity provides an unbiased estimate of the local shear, though a very noisy one. The noise is determined by the intrinsic ellipticity dispersion

$$\sigma_\epsilon = \sqrt{\left\langle \epsilon^{(\mathrm{s})} \epsilon^{(\mathrm{s})*} \right\rangle} ,$$

in the sense that, when averaging over N galaxy images all subject to the same reduced shear, the 1-σ deviation of their mean ellipticity from the true shear is $\sigma_\epsilon/\sqrt{N}$. A more accurate estimate of this error is

$$\sigma = \sigma_\epsilon \left[1 - \min\left(|g|^2, |g|^{-2}\right)\right] /\sqrt{N} \tag{15}$$

(Schneider et al. 2000). Hence, the noise can be beaten down by averaging over many galaxy images; however, the region over which the shear can be considered roughly constant is limited, so that averaging over galaxy images is always related to a smoothing of the shear. Fortunately, we live in a Universe where the sky is 'full of faint galaxies', as was impressively demonstrated by the Hubble Deep Field images (Williams et al. 1996) and previously from ultra-deep ground-based observations (Tyson 1988). Therefore, the accuracy of a shear estimate depends on the local number density of galaxies for which a shape can be measured. In order to obtain a high density, one requires deep imaging observations. As a rough guide, on a 3 hour exposure with a 4-meter class telescope, about 30 galaxies per arcmin^2 can be used for a shape measurement.

In fact, considering (14) we conclude that the expectation value of the observed ellipticity is the same for a reduced shear g and for $g' = 1/g^*$. Schneider and Seitz (1995) have shown that one cannot distinguish between these two values of the reduced shear from a purely local measurement, and term this fact the 'local degeneracy'; this also explains the symmetry between $|g|$ and $|g|^{-1}$ in (15). Hence, from a local weak lensing observation one cannot tell the case $|g| < 1$ (equivalent to $\det \mathcal{A} > 0$) from the one of $|g| > 1$ or $\det \mathcal{A} < 0$.

This local degeneracy is, however, broken in large-field observations, as the region of negative parity of any lens is small (the Einstein radius inside of which $|g| > 1$ of massive lensing clusters is typically $\lesssim 30''$, compared to data fields of several arcminutes used for weak lensing studies of clusters), and the reduced shear must be a smooth function of position on the sky.

Whereas the transformation between source and image ellipticity appears simpler in the case of χ than ϵ – see (12), the expectation value of χ cannot be easily calculated and depends explicitly on the intrinsic ellipticity distribution of the sources. In particular, the expectation value of χ is not simply related to the reduced shear (Schneider and Seitz 1995). However, in the weak lensing regime, $\kappa \ll 1$, $|\gamma| \ll 1$, one finds

$$\gamma \approx g \approx \langle\epsilon\rangle \approx \frac{\langle\chi\rangle}{2} \,. \tag{16}$$

2.3 Tangential and Cross Component of Shear

Components of the Shear

The shear components γ_1 and γ_2 are defined relative to a reference Cartesian coordinate frame. Note that the shear is *not* a vector (though it is often wrongly called that way in the literature), owing to its transformation properties under rotations: Whereas the components of a vector are multiplied by $\cos\varphi$ and $\sin\varphi$ when the coordinate frame is rotated by an angle φ, the shear components are multiplied by $\cos(2\varphi)$ and $\sin(2\varphi)$, or simply, the complex shear gets multiplied by $\mathrm{e}^{-2\mathrm{i}\varphi}$. The reason for this transformation behavior of the shear traces back to its original definition as the traceless part of the Jacobi matrix $\mathcal{A}$. This transformation behavior is the same as that of the linear polarization; the shear is therefore a *polar*. In analogy with vectors, it is often useful to consider the shear components in a rotated reference frame, that is, to measure them w.r.t. a different direction; for example, the arcs in clusters are tangentially aligned, and so their ellipticity is oriented tangent to the radius vector in the cluster.

If ϕ specifies a direction, one defines the *tangential* and *cross components* of the shear *relative to this direction* as

$$\gamma_\mathrm{t} = -\mathcal{R}\mathrm{e}\left[\gamma\,\mathrm{e}^{-2\mathrm{i}\phi}\right] \quad , \quad \gamma_\times = -\mathcal{I}\mathrm{m}\left[\gamma\,\mathrm{e}^{-2\mathrm{i}\phi}\right] \;; \tag{17}$$

For example, in case of a circularly-symmetric matter distribution, the shear at any point will be oriented tangent to the direction toward the center of symmetry. Thus in this case choose ϕ to be the polar angle of a point; then, $\gamma_\times = 0$. In full analogy to the shear, one defines the tangential and cross components of an image ellipticity, ϵ_t and $\epsilon_\times$. An illustration of these definitions is provided in Fig. 3.

The sign in (17) is easily explained (and memorized) as follows: consider a circular mass distribution and a point on the θ_1-axis outside the Einstein

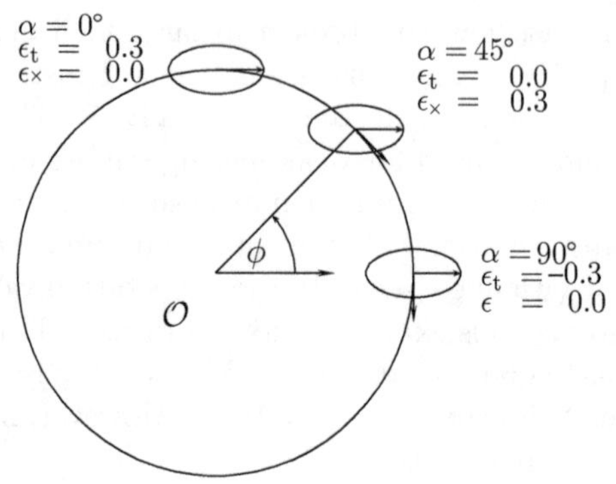

Fig. 3. Illustration of the tangential and cross-components of the shear, for an image with $\epsilon_1 = 0.3$, $\epsilon_2 = 0$, and three different directions ϕ with respect to a reference point (source: M. Bradac)

radius. The image of a circular source there will be stretched in the direction of the θ_2-axis. In this case, $\phi = 0$ in (17), the shear is real and negative, and in order to have the tangential shear positive, and thus to define tangential shear in accordance with the intuitive understanding of the word, a minus sign is introduced. Negative tangential ellipticity implies that the image is oriented in the radial direction. We warn the reader that sign conventions and notations have undergone several changes in the literature, and the current author had his share in this.

Minimum Lens Strength for its Weak Lensing Detection

As a first application of this decomposition, we consider how massive a lens needs to be in order that it produces a detectable weak lensing signal. For this purpose, consider a lens modeled as an SIS with one-dimensional velocity dispersion σ_v. In the annulus $\theta_{\rm in} \le \theta \le \theta_{\rm out}$, centered on the lens, let there be N galaxy images with positions $\boldsymbol{\theta}_i = \theta_i(\cos\phi_i, \sin\phi_i)$ and (complex) ellipticities ϵ_i. For each one of them, consider the tangential ellipticity

$$\epsilon_{{\rm t}i} = -\mathcal{R}{\rm e}\left(\epsilon_i\,{\rm e}^{-2{\rm i}\phi_i}\right) \ . \tag{18}$$

The weak lensing signal-to-noise for the detection of the lens obtained by considering a weighted average over the tangential ellipticity is (see BS01, Sect. 4.5)

$$\begin{aligned} \frac{\rm S}{\rm N} &= \frac{\theta_{\rm E}}{\sigma_\epsilon}\sqrt{\pi n}\,\sqrt{\ln(\theta_{\rm out}/\theta_{\rm in})} \\ &= 8.4\left(\frac{n}{30\,{\rm arcmin}^{-2}}\right)^{1/2}\left(\frac{\sigma_\epsilon}{0.3}\right)^{-1}\left(\frac{\sigma_v}{600\,{\rm km\,s}^{-1}}\right)^2 \\ &\times\left(\frac{\ln(\theta_{\rm out}/\theta_{\rm in})}{\ln 10}\right)^{1/2}\left\langle\frac{D_{\rm ds}}{D_{\rm s}}\right\rangle , \end{aligned} \tag{19}$$

where $\theta_{\rm E} = 4\pi(\sigma_v/c)^2(D_{\rm ds}/D_{\rm s})$ is the Einstein radius of an SIS, n the mean number density of galaxies, and the average of the distance ratio is taken over the source population from which the shear measurements are obtained. Hence, the S/N is proportional to the lens strength (as measured by $\theta_{\rm E}$), the square root of the number density, and inversely proportional to σ_ϵ, as expected. From this consideration we conclude that clusters of galaxies with $\sigma_v \gtrsim 600\,{\rm km/s}$ can be detected with sufficiently large S/N by weak lensing, but individual galaxies ($\sigma_v \lesssim 200\,{\rm km/s}$) are too weak as lenses to be detected individually. Furthermore, the final factor in (19) implies that, for a given source population, the cluster detection will be more difficult for increasing lens redshift.

Mean Tangential Shear on Circles

In the case of axi-symmetric mass distributions, the tangential shear is related to the surface mass density $\kappa(\theta)$ and the mean surface mass density $\bar\kappa(\theta)$ inside the radius θ by $\gamma_{\rm t} = \bar\kappa - \kappa$, as can be easily shown by the relation in Sect. 3.1 of IN. It is remarkable that a very similar expression holds for general matter distributions. To see this, we start from Gauss' theorem, which states that

$$\int_0^\theta {\rm d}^2\vartheta\; \nabla\cdot\nabla\psi = \theta \oint {\rm d}\varphi\; \nabla\psi\cdot\mathbf{n}\,,$$

where the integral on the left-hand side extends over the area of a circle of radius θ (with its center chosen as the origin of the coordinate system), ψ is an arbitrary scalar function, the integral on the right extends over the circle with radius θ, and $\mathbf{n}$ is the outward directed normal on this circle. Taking ψ to be the deflection potential and noting that $\nabla^2\psi = 2\kappa$, one obtains

$$m(\theta) \equiv \frac{1}{\pi}\int_0^\theta {\rm d}^2\vartheta\; \kappa(\boldsymbol{\vartheta}) = \frac{\theta}{2\pi}\oint {\rm d}\varphi\; \frac{\partial\psi}{\partial\theta}\,, \tag{20}$$

where we used that $\nabla\psi\cdot\mathbf{n} = \psi_{,\theta}$. Differentiating this equation with respect to θ yields

$$\frac{{\rm d}m}{{\rm d}\theta} = \frac{m}{\theta} + \frac{\theta}{2\pi}\oint {\rm d}\varphi\; \frac{\partial^2\psi}{\partial\theta^2}\,. \tag{21}$$

Consider a point on the θ_1-axis; there, $\psi_{,\theta\theta} = \psi_{11} = \kappa + \gamma_1 = \kappa - \gamma_{\rm t}$. This last expression is independent on the choice of coordinates and must therefore hold for all φ. Denoting by $\langle\kappa(\theta)\rangle$ and $\langle\gamma_{\rm t}(\theta)\rangle$ the mean surface mass density and mean tangential shear on the circle of radius θ, (21) becomes

$$\frac{{\rm d}m}{{\rm d}\theta} = \frac{m}{\theta} + \theta\left[\langle\kappa(\theta)\rangle - \langle\gamma_{\rm t}(\theta)\rangle\right]\,. \tag{22}$$

The dimensionless mass $m(\theta)$ in the circle is related to the mean surface mass density inside the circle $\bar{\kappa}(\theta)$ by

$$m(\theta) = \theta^2 \,\bar{\kappa}(\theta) = 2 \int_0^\theta \mathrm{d}\vartheta \; \vartheta \; \langle \kappa(\vartheta) \rangle \; . \tag{23}$$

Together with $\mathrm{d}m/\mathrm{d}\theta = 2\theta \,\langle \kappa(\theta) \rangle$, (22) becomes, after dividing through θ,

$$\langle \gamma_\mathrm{t} \rangle = \bar{\kappa} - \langle \kappa \rangle \; , \tag{24}$$

a relation which very closely matches the result mentioned above for axi-symmetric mass distributions (Bartelmann 1995). One important immediate implication of this result is that from a measurement of the tangential shear, averaged over concentric circles, one can determine the azimuthally-averaged mass profile of lenses, even if the density is not axi-symmetric.

2.4 Magnification Effects

Recall from IN that a magnification μ changes source counts according to

$$n(>S, \boldsymbol{\theta}, z) = \frac{1}{\mu(\boldsymbol{\theta}, z)} \, n_0 \left(> \frac{S}{\mu(\boldsymbol{\theta}, z)}, z \right) \; , \tag{25}$$

where $n(>S, z)$ and $n_0(>S, z)$ are the lensed and unlensed cumulative number densities of sources, respectively. The first argument of n_0 accounts for the change of the flux (which implies that a magnification $\mu > 1$ allows the detection of intrinsically fainter sources), whereas the prefactor in (25) stems from the change of apparent solid angle. In the case that $n_0(S) \propto S^{-\alpha}$, this yields

$$\frac{n(>S)}{n_0(>S)} = \mu^{\alpha - 1} \; , \tag{26}$$

and therefore, if $\alpha > 1$ (<1), source counts are enhanced (depleted); the steeper the counts, the stronger the effect. In the case of weak lensing, where $|\mu - 1| \ll 1$, one probes the source counts only over a small range in flux, so that they can always be approximated (locally) by a power law. Provided that $\kappa \ll 1$, $|\gamma| \ll 1$, a further approximation applies,

$$\mu \approx 1 + 2\kappa \; ; \quad \text{and} \quad \frac{n(>S)}{n_0(>S)} \approx 1 + 2(\alpha - 1)\kappa \; . \tag{27}$$

Thus, from a measurement of the local number density $n(>S)$ of galaxies, κ can in principle be inferred directly. It should be noted that $\alpha \sim 1$ for galaxies in the B-band, but in redder bands, $\alpha < 1$ (e.g., Ellis 1997); therefore, one expects a depletion of their counts in regions of magnification $\mu > 1$. Broadhurst et al. 1995 have discussed in detail the effects of magnification in weak lensing. Not only are the number counts affected, but since this is a redshift-dependent effect (since both κ and γ depend, for a given physical surface mass density, on the source redshift), the redshift distribution of galaxies is locally changed by magnification.

Since magnification is merely a stretching of solid angle, Bartelmann and Narayan (1995) pointed out that magnified images at fixed surface brightness have a larger solid angle than unlensed ones; in addition, the surface brightness of a galaxy is expected to be a strong function of redshift [$I \propto (1+z)^{-4}$], owing to the Tolman effect. Hence, if this effect could be harnessed, a (redshift-dependent) magnification could be measured statistically. Unfortunately, this method is hampered by observational difficulties; it seems that estimating a reliable estimate for the surface brightness from seeing-convolved images (see Sect. 3.5) is even more difficult than determining image shapes.

3 Observational Issues and Challenges

Weak lensing, employing the shear method, relies on the shape measurements of faint galaxy images. Since the noise due to intrinsic ellipticity dispersion is $\propto \sigma_\epsilon/\sqrt{n}$, one needs a high number density n to beat this noise component down. However, the only way to increase the number density of galaxies is to observe to fainter magnitudes. As it turns out, galaxies at faint magnitudes are small, in fact typically smaller than the size of the point-spread function (PSF), or the seeing disk (see Fig. 4). Hence, for them one needs usually large correction factors between the true ellipticity and that of the

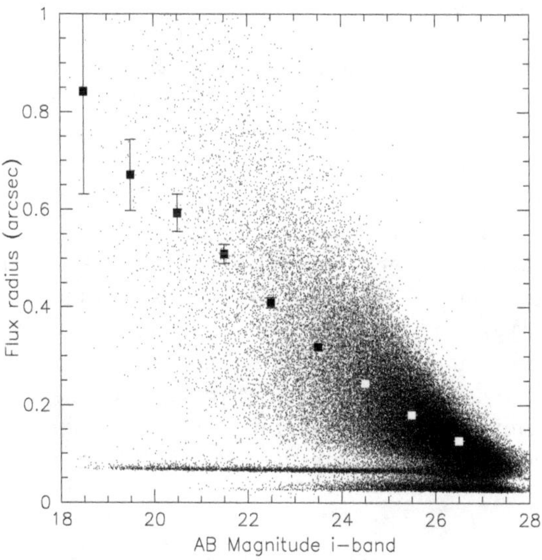

Fig. 4. The size of galaxies observed with the ACS camera on-board HST. Small dots denote the half-light radius of individual galaxies, bigger points with error bars show the mean size in a magnitude bin. The horizontal line of point at $r_\mathrm{h} \approx 0\farcs08$ correspond to stellar images in the ACS fields, as they have all the same size but vary in magnitude, and points at even smaller size are noise artefacts which are not used for any lensing analysis (source: T. Schrabback)

seeing-convolved image. On the other hand, fainter galaxies tend to probe higher-redshift galaxies, which increases the lensing signal due to $D_{\rm ds}/D_{\rm s}$-dependence of the 'lensing efficiency'.

3.1 Strategy

In the present observational situation, only the optical sky is densely populated with sources; therefore, weak lensing observations are performed with optical (or near-IR) CCD-cameras (photographic plates are not linear enough to measure these subtle effects). In order to substantiate this comment, note that the Hubble Deep Field North contains about 3000 galaxies, but only seven radio sources are detected in a very deep integration with the VLA (Richards et al. 1998).[2] In order to obtain a high number density of sources, long exposures are needed: as an illustrative example, to get a number density of useful galaxies (i.e., those for which a shape can be measured reliably) of $n \sim 20\,\mathrm{arcmin}^{-2}$, one needs ~2 hours integration on a 4-m class telescope in good seeing $\sigma \lesssim 1''$.

Furthermore, large solid angles are desired, either to get large areas around clusters for their mass reconstruction, or to get good statistics of lenses on blank field surveys, such as they are needed for galaxy–galaxy lensing and cosmic shear studies. It is now possible to cover large area in reasonable amounts of observing time, since large format CCD cameras have recently become available; for example, the Wide-Field Imager (WFI) at the ESO/MPG 2.2-m telescope at La Silla has $(8\mathrm{K})^2$ pixels and covers an area of $\sim(0.5\,\mathrm{deg})^2$. Until recently, the CFH12K camera with 8K × 12K pixels and field $\sim 30' \times 45'$ was mounted at the Canada-French-Hawaii Telescope (CFHT) on Mauna Kea and was arguably the most efficient wide-field imaging instrument hitherto. In 2003, MegaCam has been put into operation on the CFHT which has $(18\mathrm{K})^2$ pixels and covers $\sim 1\,\mathrm{deg}^2$. Several additional cameras of comparable size will become operational in the near future, including the 1 deg^2 instrument OmegaCAM on the newly built VLT Survey Telescope on Paranal. The largest field camera on a 10-m class telescope is Suprime-Cam, a $34' \times 27'$ multi-chip camera on the Subaru 8.2-meter telescope. Unfortunately, many optical astronomers (and decision making panels of large facilities) consider the prime use of large telescopes to be spectroscopy; for example, although the four ESO VLT unit telescopes are equipped with a total of ten instruments, the largest imagers on the VLT are the two FORS instruments, with a $\sim 6\overset{\prime}{.}7$ field-of-view.[3]

[2] The source density on the radio sky will become at least comparable to that currently on the optical sky with the future Square Kilometer Array (SKA).

[3] Nominally, the VIMOS instrument has a four times larger f.o.v., but our analysis of early VIMOS imaging data indicates that it is totally useless for weak lensing observations, owing to its highly anisotropic PSF, which even seems to show discontinuities on chips, and its large variation of the seeing size across chips. It

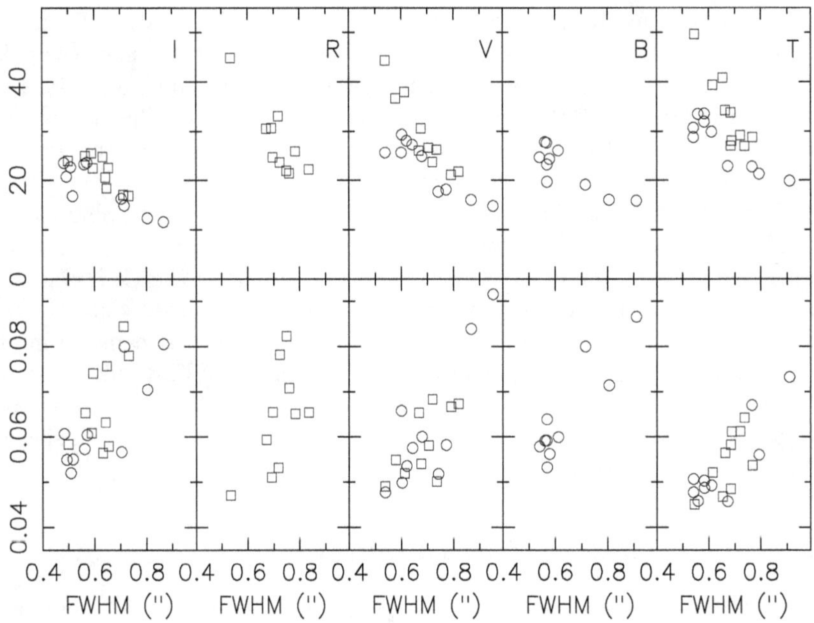

Fig. 5. Mean number density of galaxy images for which a shape can be measured (*upper row*) and the r.m.s. noise of a shear measurement in an area of 1 arcmin^2 as a function of the full width at half maximum (FWHM) of the point-spread function (PSF) – i.e., the seeing. The data were taken on 20 different fields with the FORS2 instrument at the VLT, with different filters (I, R, V and R). Squares show data taken with about 2 hours integration time, circles those with ∼45 min exposure. The right-most panels show the coadded data of I,R,V for the long exposures, and I,V,B for the 45 min fields. The useful number of galaxy images is seen to be a strong function of the seeing, except for the I-band (which is related to the higher sky brightness and the way objects are detected). But even more dramatically, the noise due to intrinsic source ellipticity decreases strongly for better seeing conditions, which is due to (1) higher number density of galaxies for which a shape can be measured, and (2) smaller corrections for PSF blurring, reducing the associated noise of this correction. In fact, this figure shows that seeing is a more important quantity than the total exposure time (from Clowe et al. 2004b)

The typical pixel size of these cameras is ∼0.″2, which is needed to sample the seeing disk in times of good seeing. From Fig. 4 one concludes immediately that the seeing conditions are absolutely critical for weak lensing: an image with 0.″6 is substantially more useful than one taken under the more typical condition of 0.″8 (see Fig. 5). There are two separate reasons why the seeing is such an important factor. First, seeing blurs the images and make

may be hoped that some of these image defects are improved after a complete overhaul of the instrument which occurred recently.

them rounder; accordingly, to correct for the seeing effect, a larger correction factor is needed in the worse seeing conditions. In addition, since the galaxy images from which the shear is to be determined are faint, a larger seeing smears the light from these galaxies over a larger area on the sky, reducing its contrast relative to the sky noise, and therefore leads to noisier estimates of the ellipticities even before the correction.

Deep observations of a field require multiple exposures. As a characteristic number, the exposure time for an R-band image on a 4-m class telescope is not longer than $\sim$10 min to avoid the non-linear part of the CCD sensitivity curve (exposures in shorter wavelength bands can be longer, since the night sky is fainter in these filters). Therefore, these large-format cameras imply a high data rate; e.g., one night of observing with the WFI yields $\sim$30 GB of science and calibration data. This number will increase by a factor $\sim$6 for MegaCam. Correspondingly, handling this data requires large disk space for efficient data reduction.

3.2 Data Reduction: Individual Frames

We shall now consider a number of issues concerning the reduction of imaging data, starting here with the steps needed to treat individual chips on individual frames, and later consider aspects of combining them into a coadded image.

Flatfielding

The pixels of a CCD have different sensitivity, i.e., they yield different counts for a given amount of light falling onto them (Fig. 6). In order to calibrate the pixel sensitivity, one needs flatfielding. Three standard methods for this are in use:

1. Dome-flats: a uniformly illuminated screen in the telescope dome is exposed; the counts in the pixels are then proportional to their sensitivity. The problem here is that the screen is not really of uniform brightness.
2. Twilight-flats: in the period of twilight after sunset, or before sunrise, the cloudless sky is nearly uniformly bright. Short exposures of regions of the sky without bright stars are then used to calibrate the pixel sensitivity.
3. Superflats: if many exposures with different pointings are taken with a camera during a night, then any given pixel is not covered by a source for most of the exposures (because the fraction of the sky at high galactic latitudes which is covered by objects is fairly small, as demonstrated by the deep fields taken by the HST). Hence, the (exposure-time normalized) counts of any pixel will show, in addition to a little tail due to those exposures when a source has covered it, a distribution around its sensitivity to the uniform night-sky brightness; from that distribution, the flat-field can be constructed, by taking its mode or its median.

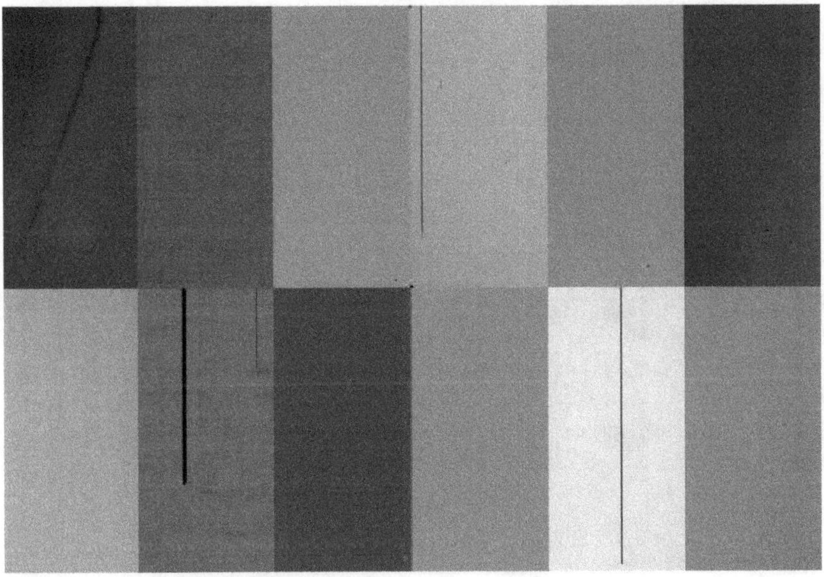

Fig. 6. A flat field for the CFH12K camera, showing the sensitivity variations between pixels and in particular between chips. Also, bad columns are clearly seen

Bad Pixels

Each CCD has defects, in that some pixels are dead or show a signal unrelated to their illumination. This can occur as individual pixels, or whole pixel columns (Fig. 6). No information of the sky image is available at these pixel positions. One therefore employs dithering: several exposures of the same field, but with slightly different pointings (dither positions) are taken. Then, any position of the field falls on bad pixels only in a small fraction of exposures, so that the full two-dimensional brightness distribution can be recovered.

Cosmic Rays

Those mimic groups of bad pixels; they can be removed owing to the fact that a given point of the image will most likely be hit by a cosmic only once, so that by comparison between the different exposures, cosmic rays can be removed (or more precisely, masked). Another signature of a cosmic ray is that the width of its track is typically much smaller than the seeing disk, the minimum size of any real source.

Bright Stars

Those cause large diffraction spikes, and depending on the optics and the design of the camera, reflection rings, ghost images and other unwanted features.

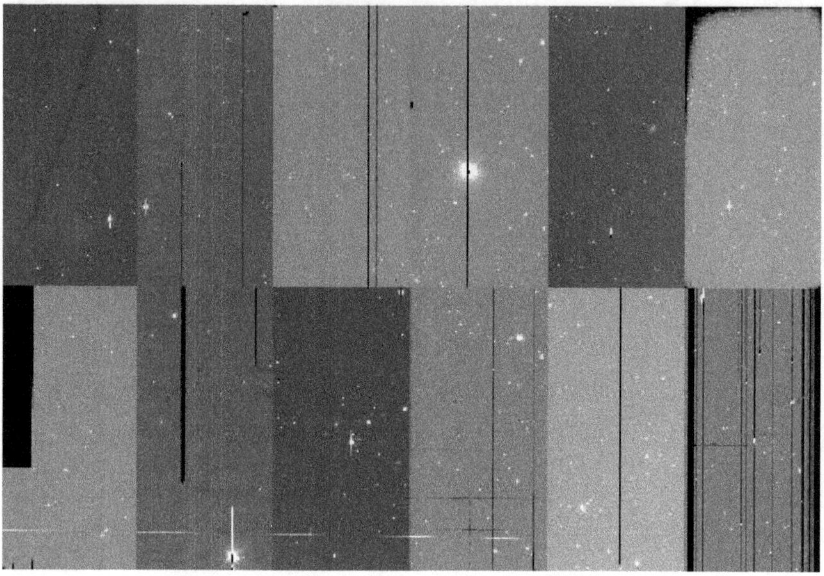

Fig. 7. A raw frame from the CFH12K camera, showing quite a number of effects mentioned in the text: bad column, saturation of bright stars, bleeding, and sensitivity variations across the field and in particular between chips

It is therefore best to choose fields where no or very few bright stars are present. The diffraction spikes of stars need to be masked, as well as the other features just mentioned (Fig. 7).

Fringes

Owing to light reflection within the CCD, patterns of illumination across the field can be generated (see Fig. 8); this is particularly true for thin chips when rather long wavelength filters are used. In clear nights, the fringe pattern is stable, i.e., essentially the same for all images taken during the night; in that case, it can be deduced from the images and subtracted off the individual exposures. However, if the nights are not clear, this procedure no longer works well; it is then safer to observe at shorter wavelength. For example, for the WFI, fringing is a problem for I-band images, but for the R-band filter, the amplitude of fringing is small. For the FORS instruments at the VLT, essentially no fringing occurs even in the I band (Maoli et al. 2001).

Gaps

The individual CCDs in multi-chip cameras cannot be brought together arbitrarily close; hence, there are gaps between the CCDs (see Fig. 9 for an

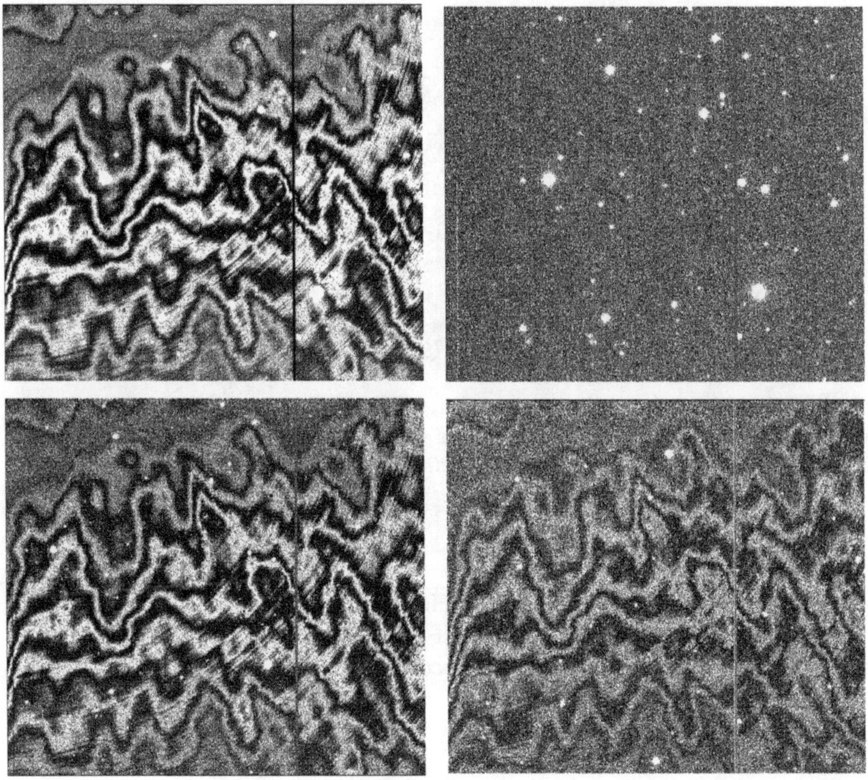

Fig. 8. The two left panels show the fringe patterns of images taken with the WFI in the I-band; the upper one was taken during photometric conditions, the lower one under non-photometric conditions. Since the fringe pattern is spatially stable, it can be corrected for (*right panels*), but the result is satisfactory only in the former case (source: M. Schirmer and T. Erben)

example). In order to cover the gaps, the dither pattern can be chosen such as to cover the gaps, so that they fall on different parts of the sky in different exposures. As we shall see, such relatively large dither patterns also provide additional advantages.

Satellite Trails, Asteroid Trails

Those have to be identified, either by visual inspection (currently the default) or by image recognition software which can detect these linear features which occur either only once, or at different positions on different exposures. These are then masked, in the same way as some of the other features mentioned above.

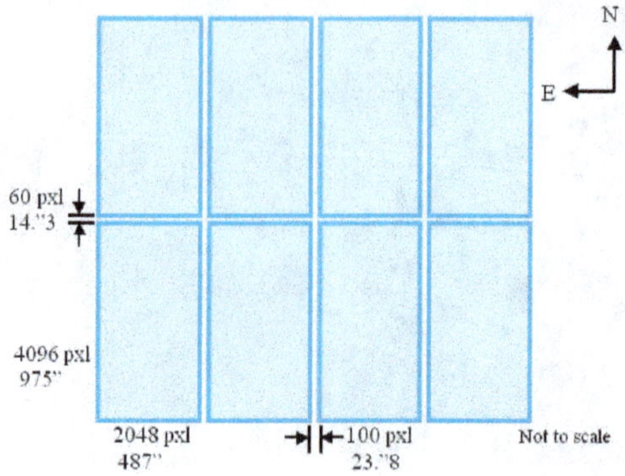

Fig. 9. Layout of the Wide Field Imager (WFI) at the ESO/MPG 2.2m telescope at La Silla. The eight chips each have $\sim 2048 \times 4096$ pixels and cover $\sim 7\overset{\prime}{.}5 \times 15'$

3.3 Data Reduction: Coaddition

After taking several exposures with slightly different pointing positions (for the reasons given above), frames shall be coadded to a sum-frame; some of the major steps in this coaddition procedure are:

Astrometric Solution

One needs to coadd data from the same true (or sky) position, not the same pixel position. Therefore, one needs a very precise mapping from sky coordinates to pixel coordinates. Field distortions, which occur in every camera (and especially so in wide-field cameras), make this mapping non-linear (see Fig. 10). Whereas the distortion map of the telescope/camera system is to a large degree constant and therefore one of the known features, it is not stable to the sub-pixel accuracy needed for weak lensing work, owing to its dependence on the zenith angle (geometrical distortions of the telescope due to gravity), temperature etc. Therefore, the pixel-to-sky mapping has to be obtained from the data itself. Two methods are used to achieve this: one of them makes use of an external reference catalog, such as the US Naval Observatory catalogue for point sources; it contains about 2 point sources per arcmin^2 (at high Galactic latitudes) with $\sim$0.3 arcsec positional accuracy. Matching point sources on the exposures with those in the USNO catalog therefore yields the mapping with sub-arcsecond accuracy. Far higher accuracy of the relative astrometry is achieved (and needed) from internal astrometry, which is obtained by matching objects which appear at different pixel coordinates, and in particular, on different CCDs for the various dithering positions. Whereas the sky coordinates are constant, the pixel coordinates change between dithering positions. Since the distortion map can be described by a low-order polynomial, the comparison of many objects appearing at (substantially) different

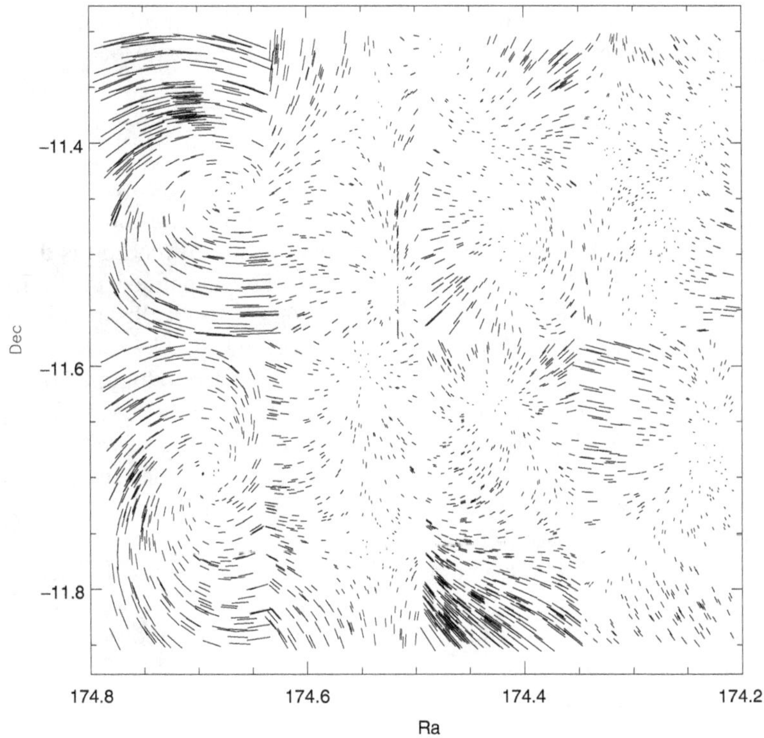

Fig. 10. This figure shows the geometric distortion of the WFI. Plotted is the difference of the positions of stars as obtained from a simple translation, and a third-order astrometric correction obtained in the process of image reduction. The patterns in the two left chips is due to their rotation relative to the other six chips. Whereas this effect looks dramatic at first sight, the maximum length of the sticks corresponds to about 6 pixels, or $1''\!.2$. Given that the WFI covers a field of $\sim 33'$, the geometrical distortions are remarkably small – however, they are sufficiently large that they have to be taken into account in the coaddition process (source: T. Erben and M. Schirmer)

pixel positions yield many more constraints than the free parameters in the distortion map and thus yields the distortion map with much higher relative accuracy than external data. The corresponding astrometric solution can routinely achieve an accuracy of 0.1 pixel, or typically $0''\!.02$ – compared with a typical field size of $\sim 30'$.

Photometric Solution

Flatfielding corrects for the different sensitivities of the pixels and therefore yields accurate relative photometry across individual exposures. The different exposures are tied together by matching the brightness of joint objects, in particular across chip boundaries. To achieve an absolute photometric calibration, one needs external data (e.g., standard star observations).

The Coaddition Process

Coaddition has to happen with sub-pixel accuracy; hence, one cannot just shift pixels from different exposures on top of each other, although this procedure is still used by some groups. The by-now standard method is drizzling (Fruchter and Hook 2002), in which a new pixel frame is defined which usually has smaller pixel size than the original image pixels (typically by a factor of two) and which is linearly related to the sky coordinates. The astrometrically and photometrically calibrated individual frames are now remapped onto this new pixel grid, and the pixel values are summed up into the sub-pixel grid, according to the overlap area between exposure pixel and drizzle pixel (see Fig. 11). By that, drizzling automatically is flux conserving. In the coaddition process, weights are assigned, accounting for the noise properties of the individual exposures (including the masks, of course).

The result of the coaddition procedure is then a science frame, plus a weight map which contains information about the pixel noise, which is of course spatially varying, owing to the masks, CCD gaps, removed cosmic rays and bad pixels. Figure 12 shows a typical example of a coadded image and its corresponding weight map.

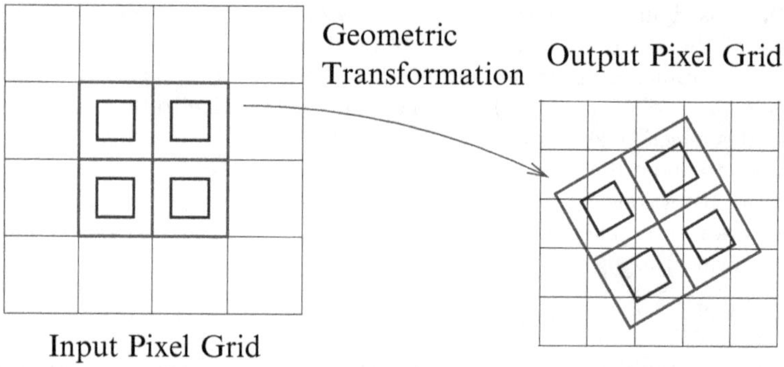

Fig. 11. The principle of drizzling in the process of coaddition is shown. The pixel grid of each individual exposure is mapped onto an output grid, where the shifts and geometric distortions obtained during the astrometric solutions are applied. The counts of the input pixel, multiplied by the relative weight of this pixel, are then dropped onto the output pixels, according to the relative overlap area, where the output pixels can be chosen smaller than the input pixels. The same procedure is applied to the weight maps of the individual exposures. If many exposures are coadded, the input pixel can also be shrunk before dropping onto the output pixel. After processing all individual exposures in this way, a coadded image and a coadded weight map is obtained (source: T. Schrabback)

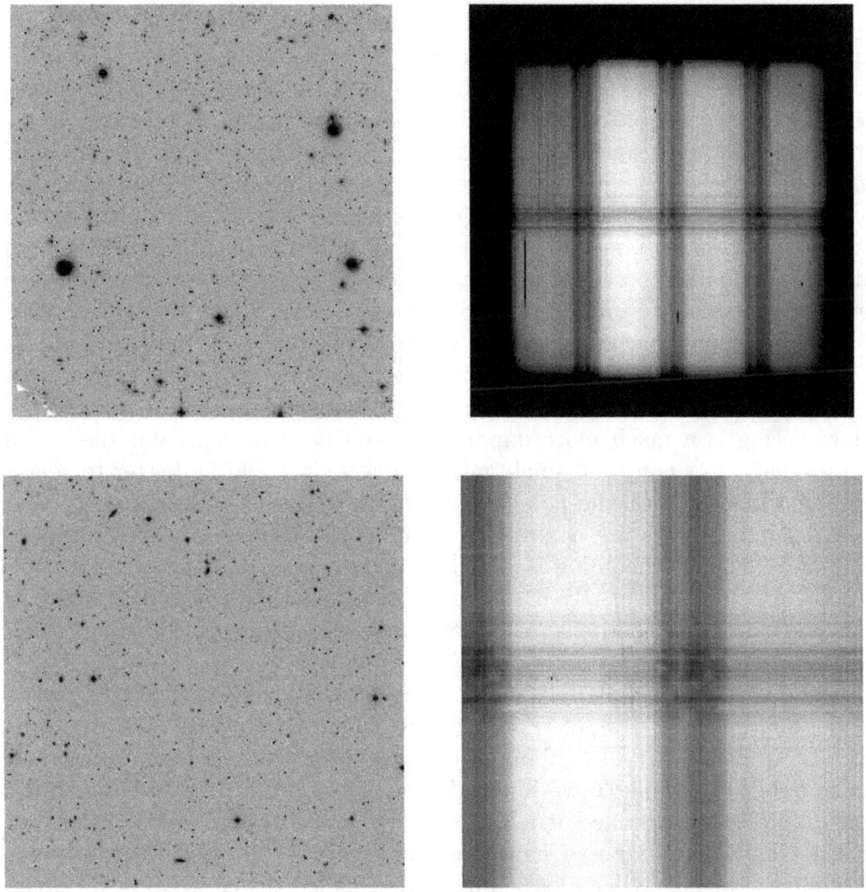

Fig. 12. A final coadded frame from a large number of individual exposures with the WFI is shown in the upper left panel, with the corresponding weight map at the upper right. The latter clearly shows the large-scale inhomogeneity of the chip sensitivity and the illumination, together with the different number of exposures contributing to various regions in the output image due to dithering and the gaps between CCDs. The two lower panels show a blow-up of the central part. Despite the highly inhomogeneous weight, the coadded image apparently shows no tracer of the gaps, which indicates that a highly accurate relative photometric solution was obtained (source: T. Erben and M. Schirmer)

The quality of the coadded image can be checked in a number of ways. Coaddition should not erase information contained in the original exposures (except, of course, the variability of sources). This means that the PSF of the coadded image should not be larger than the weighted mean of the PSFs of the individual frames. Insufficient relative astrometry would lead to a blurring of images in the coaddition. Furthermore, the anisotropy of the PSF should be similar to the weighted mean of the PSF anisotropies of the individual frames; again, insufficient astrometry could induce an artificial anisotropy of the PSF in the coaddition (which can be easily visualized, by adding two round images with a slight center offset, where a finite ellipticity would be induced).

Probably, there does not exist the 'best' coadded image from a given set of individual exposures. This can be seen by considering a set of exposures with fairly different individual seeing. If one is mainly interested in photometric properties of rather large galaxies, one would prefer a coaddition which puts all the individual exposures together, in order to maximize the total exposure time and therefore to minimize the photometric noise of the coadded sources. For weak lensing purposes, such a coaddition is certainly not optimal, as adding exposures with bad seeing together with those of good seeing creates a coadded image with a seeing intermediate between the good and the bad. Since seeing is a much more important quantity than depth for the shape determination of faint and small galaxy images, it would be better to coadd only the images with the good seeing. In this respect, the fact that large imaging instruments are operated predominantly in service observing more employing queue scheduling is a very valuable asset: data for weak lensing studies are then taken only if the seeing is better than a specified limit; in this way one has a good chance to get images of homogeneously good seeing conditions.

As a specific example, we show in Fig. 13 the 'deepest wide-field image in the Southern sky', targeted toward the Chandra Deep Field South, one of regions in the sky in which all major observatories have agreed to obtain, and make publically available, very deep images for a detailed multi-band study. For example, the Hubble Ultra Deep Field (Beckwith et al. 2003) is located in the CDFS, the deepest Chandra X-ray exposures are taken in this field, as well as two ACS@HST mosaic images, one called the GOODS field (Great Observatories Origins Deep Survey; cf. Giavalisco and Mobasher 2004), the other the GEMS survey (Rix et al. 2004).

3.4 Image Analysis

The final outcome of the data reduction steps described above is an image of the sky, together with a weight map providing the noise properties of the image. The next step is the scientific exploitation of this image, which in the case of weak lensing includes the identification of sources, and to measure their magnitude, size and shape.

As a first step, individual sources on the image need to be identified, to obtain a catalog of sources for which the ellipticities, sizes and magnitudes are to be determined later. This can done with by-now standard software, like SExtractor (Bertin and Arnouts 1996), or may be part of specialized software packages developed specifically for weak lensing, such as IMCAT, developed by Nick Kaiser (see below). Although this first step seems straightforward at first glance, it is not: images of sources can be overlapping, the brightness

Chandra Deep Field South (CDF-S)
(MPG/ESO 2.2-m + WFI)
ESO PR Photo 02a/03 (10 January 2003) ©European Southern Observatory

Fig. 13. A multi-color WFI image of the CDFS; the field is slightly larger than one-half degree on the side. To obtain this image, about 450 different WFI exposures were combined, resulting in a total exposure time of 15.8 hours in B, 15.6 hours in V, and 17.8 hours in R. The data were obtained in the frame of three different projects – the GOODS project, the public ESO Imaging Survey, and the COMBO-17 survey. These data were reduced and coadded by Mischa Schirmer and Thomas Erben; more than 2 TB of disk space were needed for the reduction

distribution of many galaxies (in particular those with active star formation) tends to be highly structured, with a collection of bright spots, and therefore the software must be taught whether or not these are to be split into different sources, or be taken as one (composite) source. This is not only a software problem; in many cases, even visual inspection cannot decide whether a given light distribution corresponds to one or several sources. The shape and size of the images are affected by the point-spread function (PSF), which results from the telescope optics, but for ground-based images, is dominated by the blurring caused by the atmospheric turbulence; furthermore, the PSF may be affected by telescope guiding and the coaddition process described earlier.

The Point-Spread Function

Atmospheric turbulence and the other effects mentioned above smear the image of the sky, according to

$$I^{\rm obs}(\boldsymbol{\theta}) = \int {\rm d}^2\vartheta \; I(\boldsymbol{\vartheta}) \, P(\boldsymbol{\theta} - \boldsymbol{\vartheta}) \,, \tag{28}$$

where $I(\boldsymbol{\vartheta})$ is the brightness profile outside the atmosphere, $I^{\rm obs}(\boldsymbol{\vartheta})$ the observed brightness profile, and P is the PSF; it describes how point sources would appear on the image. To first approximation, the PSF is a bell-shaped function; its full width at half maximum (FWHM) is called the 'seeing' of the image. At excellent sites, and excellent telescopes, the seeing has a median of $\sim 0\farcs7$–$\sim 0\farcs8$; exceptionally, images with a seeing of $\sim 0\farcs5$ can be obtained. Recall that typical faint galaxies are considerably smaller than this seeing size, hence their appearance is dominated by the PSF.

The main effect of seeing on image shapes is that it makes an elliptical source rounder: a small source with a large ellipticity will nevertheless appear as a fairly round image if its size is considerably smaller than the PSF. If not properly corrected for, this smearing effect would lead to a serious underestimate of ellipticities, and thus of the shear estimates. Furthermore, the PSF is not fully isotropic; small anisotropies can be introduced by guiding errors, the coaddition, the telescope optics, bad focusing etc. An anisotropic PSF makes round sources elliptical, and therefore mimics a shear. Also here, the effect of the PSF anisotropy depends on the image size and is strongest for the smallest sources. PSF anisotropies of several percent are typical; hence, if not corrected for, its effect can be larger than the shear to be measured.

The PSF can be measured at the position of stars (point sources) on the field; if it is a smooth function of position, it can be fitted by a low-order polynomial, which then yields a model for the PSF at all points, in particular at every image position, and one can correct for the effects of the PSF. A potential problem occurs if the PSF jumps between chips boundaries in multichip cameras, since then the coaddition produces PSF jumps on the coadded frame; this happens in cameras where the chips are not sufficiently planar, and thus not in focus simultaneously. For the WFI@ESO/MPG 2.2-m, this however is not a problem, but for some other cameras this problem exists and is severe. There is an obvious way to deal with that problem, namely to coadd data only from the same CCD chip. In this case, the gaps between chips cannot be closed in the coadded image, but for most weak lensing purposes this is not a very serious issue. In order not to lose too much area in this coaddition, the dither pattern, i.e., the pointing differences in the individual exposures, should be kept small; however, it should not be smaller than, say, 20″, since otherwise some pixels may always fall onto a few larger galaxies in the field, which then causes problems in constructing a superflat. Furthermore, small shifts between exposures means that the number of objects falling onto different chips in different exposures is small, thus reducing the accuracy of the astrometric solution. In any case, the dither strategy shall be constructed for each camera individually, taken into account its detailed properties.

3.5 Shape Measurements

Specific software has been developed to deal with the issues mentioned above; the one that is most in use currently has been developed by Kaiser et al. (1995; hereafter KSB), with substantial additions by Luppino and Kaiser (1997), and later modifications by Hoekstra et al. (1998). The numerical implementation of this method is called IMCAT and is publically available. The basic features of this method shall be outlined next.

First one notes that the definition (6) of the second-order moments of the image brightness is not very practical for applying it to real data. As the effective range of integration depends on the surface brightness of the image (through the weight function q_I) the presence of noise enters the definition of the Q_{ij} in a non-linear fashion. Furthermore, neighboring images can lead to very irregularly shaped integration ranges. In addition, this definition is hampered by the discreteness of pixels. For these reasons, the definition is modified by introducing a weight function $q_\theta(\boldsymbol{\theta})$ which depends explicitly on the image coordinates,

$$Q_{ij} = \frac{\int d^2\theta \, q_\theta(\boldsymbol{\theta}) \, I(\boldsymbol{\theta}) \, (\theta_i - \bar{\theta}_i) \, (\theta_j - \bar{\theta}_j)}{\int d^2\theta \, q_\theta(\boldsymbol{\theta}) \, I(\boldsymbol{\theta})} \; , \quad i,j \in \{1,2\} \; , \tag{29}$$

where the size of the weight function q_θ is adapted to the size of the galaxy image (for optimal S/N measurement). One typically chooses q_θ to be circular Gaussian. The image center $\bar{\boldsymbol{\theta}}$ is defined as before, but also with the new weight function $q_\theta(\boldsymbol{\theta})$, instead of $q_I(I)$. However, with this definition, the transformation between image and source brightness moments is no longer simple; in particular, the relation (11) between the second-order brightness moments of source and image no longer holds. The explicit spatial dependence of the weight, introduced for very good practical reasons, destroys the convenient relations that we derived earlier – welcome to reality.

In KSB, the anisotropy of the PSF is characterized by its (complex) ellipticity q, measured at the positions of the stars, and fitted by a low-order polynomial. Assume that the (reduced) shear g and the PSF anisotropy q are small; then they both will have a small effect on the measured ellipticity. Linearizing these two effects, one can write (employing the Einstein summation convention)

$$\hat{\chi}^{\rm obs}_\alpha = \chi^0_\alpha + P^{\rm sm}_{\alpha\beta} q_\beta + P^g_{\alpha\beta} g_\beta \; . \tag{30}$$

The interpretation of the various terms is found as follows: First consider an image in the absence of shear and the case of an isotropic PSF; then $\hat{\chi}^{\rm obs} = \chi^0$; thus, χ^0 is the image ellipticity one would obtain for $q = 0$ and $g = 0$; it is the source smeared by an isotropic PSF. It is important to note that $\mathrm{E}(\chi^0) = 0$, due to the random orientation of sources. The tensor P^{sm} describes how the image ellipticity responds to the presence of a PSF anisotropy; similarly, the tensor P^g describes the response of the image ellipticity to shear in the presence of smearing by the seeing disk. Both, $P^{\rm sm}$ and P^g have to be calculated for each image individually; they depend on higher-order moments of the brightness distribution and the size of the PSF. A full derivation of the explicit equations can be found in Sect. 4.6.2 of BS01.

Given that $\langle \chi^0 \rangle = 0$, an estimate of the (reduced) shear is provided by

$$\epsilon = (P^g)^{-1} \left(\hat{\chi}^{\rm obs} - P^{\rm sm} q \right) . \tag{31}$$

If the source size is much smaller than the PSF, the magnitude of P^g can be very small, i.e., the correction factor in (31) can be very large. Given that the measured ellipticity $\hat{\chi}^{\rm obs}$ is affected by noise, this noise then also gets multiplied by a large factor. Therefore, depending on the magnitude of P^g, the error of the shear estimates differ between images; this can be accounted for by specifically weighting these estimates when using them for statistical purposes (e.g., in the estimate of the mean shear in a given region). Different authors use different weighting schemes when applying KSB. Also, the tensors P^{sm} and P^g are expected to depend mainly on the size of the image and their signal-to-noise; therefore, it is advantageous to average these tensors over images having the same size and S/N, instead of using the individual tensor values which are of course also affected by noise. Erben et al. (2001) and Bacon et al. (2001) have tested the KSB scheme on simulated data and in particular investigated various schemes for weighting shear estimates and for determining the tensors in (30); they concluded that simulated shear values can be recovered with a systematic uncertainty of about 10%.

Maybe by now you are confused – what is 'real ellipticity' of an image, independent of weights etc. ? Well, this question has no answer, since only images with conformal elliptical isophotes have a 'real ellipticity'. By the way, not necessarily the one that is the outcome of the KSB procedure. The KSB process does not aim toward measuring 'the' ellipticity of any individual galaxy image; it tries to measure 'a' ellipticity which, when averaged over a random intrinsic orientation of the source, yields an unbiased estimate of the reduced shear.

Given that the shape measurements of faint galaxies and their correction for PSF effects is central for weak lensing, several different schemes for measuring shear have been developed (e.g., Valdes et al. 1983; Bonnet and Mellier 1995; Kuijken 1999; Kaiser 2000; Refregier 2003b; Bernstein and Jarvis 2002). In the shapelet method of Refregier (2003b; see also Refregier and Bacon 2003), the brightness distribution of galaxy images is expanded in a set of basis functions ('shapelets') whose mathematical properties are particularly convenient. With a corresponding decomposition of the PSF (the shape of stars) into these shapelets and their low-order polynomial fit across the image, a partial deconvolution of the measured images becomes possible, using linear algebraic relations between the shapelet coefficients. The effect of a shear on the shapelet coefficients can be calculated, yielding then an estimate of the reduced shear. In contrast to the KSB scheme, higher-order brightness moments, and not just the quadrupoles, of the images are used for the shear estimate.

These alternative methods for measuring image ellipticities (in the sense mentioned above, namely to provide an unbiased estimate of the local reduced shear) have not been tested yet to the same extent as is true for the KSB method. Before they become a standard in the field of weak lensing, several groups need to independently apply these techniques to real and synthetic data sets to evaluate their strengths and weaknesses. In this regard, one needs to note that weak lensing has, until recently, been regarded by many researchers as a field where the observational results are difficult to 'believe' (and sure, not all colleagues have given up this view, yet). The difficulty to display the directly measured quantities graphically so that they can be directly 'seen' makes it difficult to convince others about the reliability of the measurements. The fact that the way from the coadded imaging data to the final result is, except for the researchers who actually do the analysis, close to a black box with hardly any opportunity to display intermediate results (which would provide others with a quality check) implies that the methods employed should be standardized and well checked.

Surprisingly enough, there are very few (published) attempts where the same data set is analyzed by several groups independently, and intermediate and final results being compared. Kleinheinrich (2003) in her dissertation has taken several subsets of the data that led to the deep image shown in Fig. 13 and compared the individual image ellipticities between the various subsets. If the subsets had comparable seeing, the measured ellipticities could be fairly well reproduced, with an rms difference of about 0.15, which is small compared to the dispersion of the image ellipticities $\sigma_\epsilon \sim 0.35$. Hence, these differences, which presumably are due to the different noise realizations on the different images, are small compared to the 'shape noise' coming from the finite intrinsic ellipticities of galaxies. If the subsets had fairly different seeing, the smearing correction turns out to lead to a systematic bias in the measured ellipticities. From the size of this bias, the conclusions obtained from the simulations are confirmed – measuring a shear with better than $\sim$10% accuracy will be difficult with the KSB method, where the main problem lies in the smearing correction.

Shear Observations from Space

We conclude this section with a few comments on weak lensing observations from space. Since the PSF is the largest problem in shear measurements, one might be tempted to use observations from space which are not affected by the atmosphere. At present, the Hubble Space Telescope (HST) is the only spacecraft that can be considered for this purpose. Weak lensing observations have been carried out using two of its instruments, WFPC2 and STIS. The former has a field-of-view of about 5 arcmin^2, whereas STIS has a field of $51''$. These small fields imply that the number of stars that can be found on any given exposure at high galactic latitude is very small, in fact typically zero for STIS. Therefore, the PSF cannot be measured from these exposures themselves. Given that an instrument in space is expected to be much more stable than one on the ground, one might expect that the PSF is stable in time; then, it can be investigated by analyzing exposures which contain many stars (e.g., from a star cluster). In fact, Hoekstra et al. (1998) and Hämmerle et al. (2002) have shown that the PSFs of WFPC2 and STIS are approximately constant in time. The situation is improved with the new camera ACS onboard HST, where the field size of $\sim 3\overset{\prime}{.}4$ is large enough to contain about a dozen stars even for high galactic latitude, and where some control over the PSF behavior on individual images is obtained. We shall discuss the PSF stability of the ACS in Sect. 7.3 below.

The PSF of a diffraction-limited telescope is much more complex than that of the seeing-dominated one for ground-based observations. The assumption underlying the KSB method, namely that the PSF can be described by a axi-symmetric function convolved with a small anisotropic kernel, is strongly violated for the HST PSF; it is therefore less obvious how well the shear measurements with the KSB method work in space. In addition, the HST PSF in not well sampled with the current imaging instruments, even though STIS and ACS have a pixel scale of $0\overset{\prime\prime}{.}05$. The number density of cosmic rays is much larger in space, so their removal can be more cumbersome than for ground-based observations. The intense particle bombardment also leads to aging of the CCD, which lose their sensitivity and attain charge-transfer efficiency problems. Despite these potential problems, a number of highly interesting weak lensing results obtained with the HST have been reported, in particular on clusters, and we shall discuss some of them in later sections. The new Advanced Camera for Surveys (ACS) on-board HST has a considerably larger field-of-view than previous instruments and will most likely become a highly valuable tool for weak lensing studies.

4 Clusters of Galaxies: Introduction, and Strong Lensing

4.1 Introduction

Galaxies are not distributed randomly, but they cluster together, forming groups and clusters of galaxies. Those can be identified as overdensities of galaxies projected onto the sky, and this has of course been the original method for the detection of clusters, e.g., leading to the famous and still heavily used Abell (1958) catalog and its later Southern extension (Abell et al. 1989; ACO).

Only later – with the exception of Zwicky's early insight in 1933 that the Coma cluster must contain a lot of missing mass – it was realized that the visible galaxies are but a minor contribution to the clusters since they are dominated by dark matter. From X-ray observations we know that clusters contain a very hot intracluster gas which emits via free-free and atomic line radiation. Many galaxies are members of a cluster or a group; indeed, the Milky Way is one of them, being one of two luminous galaxies of the Local Group (the other one is M31, the Andromeda galaxy), of which ∼35 member galaxies are known, most of them dwarfs.

In the first part of this section we shall describe general properties of galaxy clusters, in particular methods to determine their masses, before turning to their strong lensing properties, such as show up in the spectacular giant luminous arcs. Very useful reviews on clusters of galaxies are from Sarazin (1986) and in a recent proceedings volume (Mulchaey et al. 2004).

4.2 General Properties of Clusters

Clusters of galaxies contain tens to hundreds of bright galaxies; their galaxy population is dominated by early-type galaxies (E's and S0's), i.e. galaxies without active star formation. Often a very massive cD galaxy is located at their center; these galaxies differ from normal ellipticals in that they have a much more extended brightness profile – they are the largest galaxies. The morphology of clusters as seen in their distribution of galaxies can vary a lot, from regular, compact clusters (often dominated by a central cD galaxy) to a bimodal distribution, or highly irregular morphologies with strong substructure. Since clusters are at the top of the mass scale of virialized objects, the hierarchical merging scenario of structure growth predicts that many of them have formed only recently through the merging of two or more lower-mass sub-clusters, and so the irregular morphology just indicates that this happened.

X-ray observations reveal the presence of a hot (several keV) intracluster medium (ICM) which is highly enriched in heavy elements; hence, this gas has been processed through star-formation cycles in galaxies. The mass of the ICM surpasses that of the baryons in the cluster galaxies; the mass balance in clusters is approximately as follows: stars in cluster galaxies contribute ∼3% of the total mass, the ICM another ∼15%, and the rest ($\gtrsim$80%) is dark matter. Hence, clusters are dominated by dark matter; as discussed below (Sect. 4.3), the mass of clusters can be determined with three vastly different methods which overall yield consistent results, leading to the aforementioned mass ratio.

We shall now quote a few characteristic values which apply to rich, massive clusters. Their virial radius, i.e., the radius inside of which the mass distribution is in approximate virial equilibrium (or the radius inside of which the mean mass density of clusters is ∼200 times the critical density of the Universe – cf. Sect. 4.5 of IN) is $r_{\rm vir} \sim 1.5\,h^{-1}\,$Mpc. A typical value for the one-dimensional velocity dispersion of the member galaxies is $\sigma_v \sim 1000\,\mathrm{km/s}$. In equilibrium, this equals the thermal velocity of the ICM, corresponding to a temperature of $T \sim 10^{7.5}\,\mathrm{K} \sim 3\,\mathrm{keV}$. The mass of massive clusters within the virial radius (i.e., the *virial mass*) is $\sim 10^{15} M_\odot$. The mass-to-light ratio of clusters (as measured from the B-band luminosity) is typically of order $(M/L) \sim 300h^{-1}\,(M_\odot/L_\odot)$. Of course, the much more numerous typical clusters have smaller masses (and temperatures).

Cosmological Interest for Clusters

Clusters are the most massive bound and virialized structures in the Universe; this, together with the (related) fact that their dynamical time scale (e.g., the crossing time $\sim r_{\mathrm{vir}}/\sigma_v$) is not much smaller than the Hubble time H_0^{-1} – so that they retain a 'memory' of their formation – render them of particular interest for cosmologists. The evolution of their abundance, i.e., their comoving number density as a function of mass and redshift, is an important probe for cosmological models and traces the growth of structure; massive clusters are expected to be much rarer at high redshift than today. Their present-day abundance provides one of the measures for the normalization of the power spectrum of cosmological density fluctuations. Furthermore, they form (highly biased) signposts of the dark matter distribution in the Universe, so their spatial distribution traces the large-scale mass distribution in the Universe. Clusters act as laboratories for studying the evolution of galaxies and baryons in the Universe. Since the galaxy number density is highest in clusters, mergers of their member galaxies and, more importantly, other interactions between them occur frequently. Therefore, the evolution of galaxies with redshift is most easily studied in clusters. For example, the Butcher–Oemler effect (the fact that the fraction of blue galaxies in clusters is larger at higher redshifts than today) is a clear sign of galaxy evolution which indicates that star formation in galaxies is suppressed once they have become cluster members. More generally, there exists a density-morphology relation for galaxies, with an increasing fraction of early-types with increasing spatial number density, with clusters being on the extreme for the latter. Finally, clusters were (arguably) the first objects for which the presence of dark matter has been concluded (by Zwicky 1933). Since they are so large, and present the gravitational collapse of a region in space with initial comoving radius of $\sim 8h^{-1}\,\mathrm{Mpc}$, one expects that their mixture of baryonic and dark matter is characteristic for the mean mass fraction in the Universe (White et al. 1993). With the baryon fraction of $\sim 15\%$ mentioned above, and the density parameter in baryons determined from big-bang nucleosynthesis in connection to the determination of the deuterium abundance in Lyα QSO absorption systems, $\Omega_{\mathrm{b}} \approx 0.02h^{-2}$, one obtains a density parameter for matter of $\Omega_{\mathrm{m}} \sim 0.3$, in agreement with results from other methods, most noticeably from the recent WMAP CMB measurements (e.g., Spergel et al. 2003).

4.3 The Mass of Galaxy Clusters

Cosmologists can predict the abundance of clusters as a function of their mass (e.g., using numerical simulations); however, the mass of a cluster is not directly observable, but only its luminosity, or the temperature of the X-ray emitting intra-cluster medium. Therefore, in order to compare observed clusters with the cosmological predictions, one needs a way to determine their masses. Three principal methods for determining the mass of galaxy clusters are in use:

- Assuming virial equilibrium, the observed velocity distribution of galaxies in clusters can be converted into a mass estimate, employing the virial theorem; this method typically requires assumptions about the statistical distribution of the anisotropy of the galaxy orbits.
- The hot intra-cluster gas, as visible through its Bremsstrahlung in X-rays, traces the gravitational potential of the cluster. Under certain assumptions (see below), the mass profile can be constructed from the X-ray emission.
- Weak and strong gravitational lensing probes the projected mass profile of clusters, with strong lensing confined to the central regions of clusters, whereas weak lensing can yield mass measurements for larger radii.

All three methods are complementary; lensing yields the line-of-sight projected density of clusters, in contrast to the other two methods which probe the mass inside spheres. On the other hand, those rely on equilibrium (and symmetry) conditions; e.g., the virial method assumes virial equilibrium (that the cluster is dynamically relaxed) and the degree of anisotropy of the galaxy orbit distribution.

Dynamical Mass Estimates

Estimating the mass of clusters based on the virial theorem,

$$2E_{\rm kin} + E_{\rm pot} = 0 \;, \tag{32}$$

has been the traditional method, employed by Zwicky (1933) to find strong hints for the presence of dark matter in the Coma cluster. The specific kinetic energy of a galaxy is $v^2/2$, whereas the potential energy is determined by the cluster mass profile, which can thus be determined using (32). One should note that only the line-of-sight component of the galaxy velocities can be measured; hence, in order to derive the specific kinetic energy of galaxies, one needs to make an assumption on the distribution of orbit anisotropies in the cluster potential. Assuming an isotropic distribution of orbits, the l.o.s. velocity distribution can then be related to the 3-D velocity dispersion, which in turn can be transformed into a mass estimate if spherical symmetry is assumed. This method requires many redshifts for an accurate mass estimate, which are available only for a few clusters. However, a revival of this method is expected and already seen by now, owing to the new high-multiplex optical spectrographs.

X-ray Mass Determination of Clusters

The intracluster gas emits via Bremsstrahlung; the emissivity depends on the gas density and temperature, and, at lower T, also on its chemical composition, since at $T \lesssim 1\,\mathrm{keV}$ the line radiation from highly ionized atomic species starts to dominate the total emissivity of a hot gas. Investigating the properties of the ICM with X-ray observations have revealed a wealth of information on the properties of clusters (see Sarazin 1986). Assuming that the gas is in hydrostatic equilibrium in the potential well of the cluster, the gas pressure P must balance gravity, or

$$\nabla P = -\rho_{\rm g}\, \nabla \Phi \,,$$

where $\rho_{\rm g}$ is the gas density. In the case of spherical symmetry, this becomes

$$\frac{1}{\rho_{\rm g}}\frac{{\rm d}P}{{\rm d}r} = -\frac{{\rm d}\Phi}{{\rm d}r} = -\frac{G\,M(r)}{r^2} \,.$$

From the X-ray brightness profile and temperature measurement, $M(r)$, the mass inside r, both dark and luminous, can then be determined,

$$M(r) = -\frac{k_{\rm B}Tr^2}{G\mu m_{\rm p}}\left(\frac{{\rm d}\ln\rho_{\rm g}}{{\rm d}r} + \frac{{\rm d}\ln T}{{\rm d}r}\right) \,, \tag{33}$$

where $\mu m_{\rm p}$ is the mean particle mass in the gas. Only for relatively few clusters are detailed X-ray brightness and temperature profile measurements available. In the absence of a temperature profile measurement, one often assumes that T does not vary with distance form the cluster center. In this case, assuming that the dark matter particles also have an isothermal distribution (with velocity traced by the galaxy velocities), one can show that

$$\rho_{\rm g}(r) \propto \left[\rho_{\rm tot}(r)\right]^{\beta} \;; \quad \text{with} \quad \beta = \frac{\mu m_{\rm p}\sigma_v^2}{k_{\rm B}T_{\rm g}} \,. \tag{34}$$

Hence, β is the ratio between kinetic and thermal energy. The mass profile corresponding to the isothermality assumption follows from the Lame–Emden equation which, however, has no closed-form solution. In the King approximation, the density and X-ray brightness profile (which is obtained by a line-of-sight integral at projected distance R from the cluster center over the emissivity, which in turn is proportional to the square of the electron density, or $\propto \rho_{\rm g}^2$, for an isothermal gas) become

$$\rho_{\rm g}(r) = \rho_{\rm g0}\left[1 + \left(\frac{r}{r_{\rm c}}\right)^2\right]^{-3\beta/2} \;; \quad I(R) \propto \left[1 + \left(\frac{R}{r_{\rm c}}\right)^2\right]^{-3\beta/2+1/2} \,,$$

where $r_{\rm c}$ is the core radius. The observed brightness profile can now be fitted with these β-models, yielding estimates of β and $r_{\rm c}$ from which the cluster mass follows. Typical values for r_c range from 0.1 to $0.3h^{-1}$ Mpc; and $\beta = \beta_{\rm fit} \sim 0.65$. On the other hand, one can determine β from the temperature T and the galaxy velocity dispersion using (34), which yields $\beta_{\rm spec} \approx 1$. The discrepancy between these two estimates of β is not well understood and probably indicates that one of assumptions underlying this 'β-models' fails in many clusters, which is not too surprising (see below).

The hot ICM loses energy through its thermal radiation; the cooling time $t_{\rm cool}$ of the gas, i.e., the ratio between the thermal energy density and the X-ray emissivity, is larger than the Hubble time $\sim H_0^{-1}$ for all but the innermost regions. In the center of clusters, the gas density can be high enough to have $t_{\rm cool} < H_0^{-1}$, so that there the gas can no longer be in hydrostatic equilibrium. One expects that the gas flows toward the cluster center, thereby being compressed and therefore maintain approximate pressure balance. Such 'cooling flows' (see, e.g., Fabian 1994) are observed indirectly, through highly peaked X-ray emission in cluster centers which indicates a strong increase of the gas density; furthermore, these cooling-flow clusters show a decrease of T toward the center. The mass-flow rate in these clusters can be as high as $100 M_\odot\,{\rm yr}^{-1}$ or even more, so that the total cooled mass can be larger than the baryonic mass of a massive galaxy. However, the fate of the cooled gas is unknown.

New Results from Chandra and XMM

The two X-ray satellites Chandra and XMM, launched in 1999, have greatly increased our view of the X-ray Universe, and have led to a number of surprising results about clusters. X-ray spectroscopy verified the presence of cool gas near the center of cooling-flow clusters, but no indication for gas with temperature below $\sim$1 keV has been seen, whereas the cooling is expected to rapidly proceed to very low temperatures, as the cooling function increases for lower T where atomic transitions become increasingly important. Furthermore, the new observations have revealed that at least the inner regions of clusters often show a considerably more complicated structure than implied by hydrostatic equilibrium. In some cases, the intracluster medium is obviously affected by a central AGN, which produces additional energy and entropy input, which might explain why no sub-keV gas has been detected. As the AGN activity of a galaxy may be switched on and off, depending on the fueling of the central black hole, even in clusters without a currently active AGN such heating might have occurred in the recent past, as indicated in some cases by radio relics. Cold fronts with very sharp edges (discontinuities in density and temperature, but such that $P \propto \rho T$ is approximately constant across the front), and shocks have been discovered, most likely showing ongoing or recent merger events. In many clusters, the temperature and metallicity appears to be strongly varying functions of position which invalidates the assumption of isothermality underlying the β-model. Therefore, mass estimates of central parts of clusters from X-ray observations require special care, and one needs to revise the simplified models used in the pre-Chandra era. In fact, has there ever been the believe that the β-model provides an adequate description of the gas in a cluster, the results from Chandra and XMM show that this is unjustified. The physics of the intracluster gas appears to be considerably more complicated than that.

4.4 Luminous Arcs and Multiple Images

Strong lensing effects in cluster show up in the form of giant luminous arcs, strongly distorted arclets, and multiple images of background galaxies. Since strong lensing only occurs in the central part of clusters, it can be used only to probe their inner mass structure. However, strong lensing yields by far the most accurate central mass determinations in those cases where several strong lensing features can be identified. For a detailed account of strong lensing in clusters, the reader is referred to the review by Fort and Mellier (1994).

Furthermore, clusters thus act as a 'natural telescope'; many of the most distant galaxies have been found by searching behind clusters, employing the lensing magnification. For example, the recently discovered very high redshift galaxies at $z \approx 7$ (Kneib et al. 2004) and $z = 10$ (Pelló et al. 2004) were found through a search in the direction of the high-magnification region in the clusters A 2218 and A1835, respectively. In the first of these two cases, the multiple imaging of the background galaxy provides not only the magnification, but also an estimate of the redshift of the source (which is not determined by any spectral line), whereas in the latter case, only the implied high magnification makes the source visible on deep HST images and allows its spectroscopy, yielding a spectral line which most likely is due to Lyα. The magnification is indeed a very important asset, as can be seen from a simple example: a value of $\mu = 5$ reduces the observing time for obtaining a spectrum by a factor 25 (in the case where the noise is sky background dominated) – which is the difference of being doable or not. Recognizing the power of natural telescopes, the deepest SCUBA surveys for faint sub-millimeter sources have been conducted (e.g., Blain et al. 1999) around clusters with well-constrained (from lensing) mass distribution to reach further down the (unlensed) flux scale.

First go: $M(\leq \theta_{\rm E})$

Giant arcs occur where the distortion (and magnification) is very large, that is near critical curves. To a first approximation, assuming a spherical mass distribution, the location of the arc from the cluster center (which usually is assumed to coincide with the brightest cluster galaxy) yields the Einstein radius of the cluster, so that the mass estimate (see IN, Eq. 43) can be applied:

$$M(\theta_{\rm arc}) \approx \pi \, (D_{\rm d}^{\rm ang} \, \theta_{\rm arc})^2 \, \Sigma_{\rm cr} \; . \tag{35}$$

Fig. 14. The galaxy cluster Abell 1689 is the most impressive lensing cluster yet found. This image has been taken with the new Advanced Camera for Surveys (ACS) onboard HST. Numerous arcs are seen. A simple estimate for the mass of the center of the cluster, obtained by identifying the arcs radius with the Einstein radius, yields an extremely large equivalent velocity dispersion. The distribution of the arcs shown here indicates that such a simple assumption is misleading, and more detailed modeling required

Therefore, this simple estimate yields the mass inside the arc radius. However, this estimate is not very accurate, perhaps good to within ~30% (Bartelmann and Steinmetz 1996). Its reliability depends on the level of asymmetry and substructure in the cluster mass distribution. Furthermore, it is likely to overestimate the mass in the mean, since arcs preferentially occur along the major axis of clusters. Of course, the method is very difficult to apply if the center of the cluster is not readily identified or if the cluster is obviously bimodal. For these reasons, this simple method for mass estimates is not regarded as particularly accurate.

Detailed Modeling

The mass determination in cluster centers becomes much more accurate if several arcs and/or multiple images are present, since in this case, detailed modeling can be done. This typically proceeds in an interactive way: First, multiple images have to be identified (based on their colors and/or detailed morphology, as available with HST imaging). Simple (plausible) mass models are then assumed, with parameters fixed by matching the multiple images, and requiring the distortion at the arc location(s) to be strong and to have the correct orientation. This model then predicts the presence of possible further multiple images; they can be checked for through morphology, surface brightness (in particular if HST images of the cluster are available) and color. If confirmed, a new, refined model is constructed including these new additional strong lensing constraints, which yields further strong lensing predictions etc. As is the case for galaxy lensing (see SL), the components of the mass models are not arbitrary, but chosen to be physically motivated. Typically, as major component an ellipsoidal isothermal or NWF distribution is used to describe the overall mass distribution of the cluster. Refinements of the mass distribution are introduced as mass components centered on bright cluster member galaxies or on subgroups of such galaxies, describing massive subhalos which survived a previous merger. Such models have predictive power and can be trusted in quite some detail; the accuracy of mass estimates in some favorable cases can be as high as a few percent.

In fact, these models can be used to predict the redshift of arcs and arclets. As an example, we mention the strong lensing analysis of the cluster Abell 2390 based on HST imaging (Pelló et al. 1999). Two pairs of multiple images were identified (see Fig. 15) which then implies that the critical curve has to pass between the individual components. The location of the critical curves depends, however, on the source redshift. As shown in the figure, the sources have to be at a high redshift in order for the corresponding critical curves to have the correct location. In fact, spectroscopy placed the two sources at $z_s = 4.04$ and $z_s = 4.05$, as predicted by the lens model.

Since the distortion of a lens also depends on the source redshift, once a detailed mass model is available from arcs with known redshifts for at least some of them, one can estimate the value of the lens strength $\propto D_{ds}/D_s$ and thus infer the redshift of arclets. This method has been successfully applied to HST observations of clusters (Ebbels et al. 1998). Of course, having spectroscopic redshifts of the arcs available increases the calibration of the mass models; they are therefore very useful.

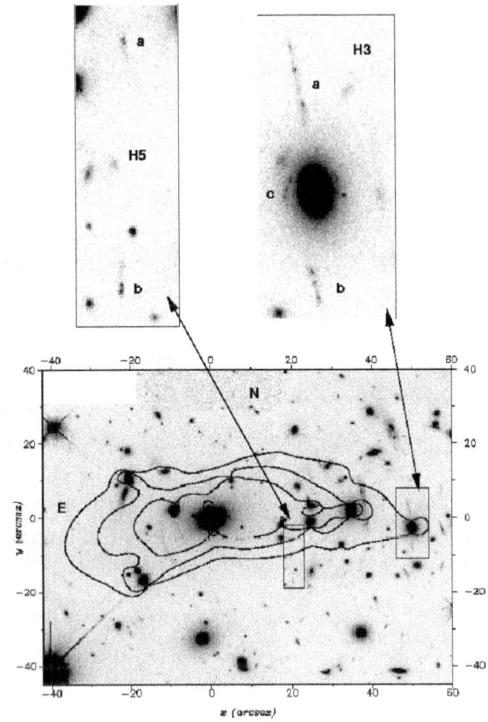

Fig. 15. The lower panel shows the critical curves of the cluster A 2390 (cluster redshift $z_{\rm d} = 0.231$), for three different source redshifts of $z_{\rm s} = 1$, 2.5 and 4 (from inner to outer). The lens model is based on the detailed HST image shown here. Identified are two sets of multiple images, shown in the upper two panels, which obviously need to be at very high redshift. Indeed, spectroscopy shows that they have $z_{\rm s} = 4.04$ and $z_{\rm s} = 4.05$ (from Pelló et al. 1999)

Lens Properties from Fourier Transforms

Before discussing results from these detailed models, a brief technical section shall be placed here, related to calculating lens properties of general mass distributions. A general method to obtain the lensing quantities of a mass distribution is through Fourier transformation. We assume that we have a mass distribution of finite mass; this is not a serious restriction even for models with formally infinite total mass, because we can truncate them on large scales, thus making the total mass finite, without affecting any lensing properties at smaller scales. We define the Fourier transform $\hat\kappa(\boldsymbol{\ell})$ of the surface mass density as[4]

$$\hat\kappa(\boldsymbol{\ell}) = \int_{\mathbb{R}^2} \mathrm{d}^2\theta \; \kappa(\boldsymbol{\theta}) \, \exp\left(\mathrm{i}\boldsymbol{\ell}\cdot\boldsymbol{\theta}\right) \; , \tag{36}$$

[4] We denote the Fourier variable of three-dimensional space as **k**, that of angular position by $\boldsymbol{\ell}$.

and its inverse by

$$\kappa(\boldsymbol{\theta}) = \frac{1}{(2\pi)^2} \int_{\mathbb{R}^2} \mathrm{d}^2\ell \, \hat{\kappa}(\boldsymbol{\ell}) \exp\left(-\mathrm{i}\boldsymbol{\ell}\cdot\boldsymbol{\theta}\right) \; . \tag{37}$$

Similarly, we define the Fourier transforms of the deflection potential, $\hat{\psi}(\boldsymbol{\ell})$, of the deflection angle, $\hat{\boldsymbol{\alpha}}(\boldsymbol{\ell})$, and of the complex shear, $\hat{\gamma}(\boldsymbol{\ell})$. Differentiation by θ_i in real space is replaced by multiplication by $-\mathrm{i}\ell_i$ in Fourier space. Therefore, the Fourier transform of $\partial\psi/\partial\theta_j$ is $-\mathrm{i}\ell_j\hat{\psi}(\boldsymbol{\ell})$. Hence, the Poisson equation as given in Sect. 2.2 of IN becomes in Fourier space

$$-|\boldsymbol{\ell}|^2\hat{\psi}(\ell) = 2\hat{\kappa}(\boldsymbol{\ell}) \; . \tag{38}$$

Thus, for $\boldsymbol{\ell} \neq \mathbf{0}$, the Fourier transform of the potential which satisfies the Poisson equation can be readily determined. The $\boldsymbol{\ell} = \mathbf{0}$ mode remains undetermined; however, since this mode corresponds to a constant in ψ, it is unimportant and can be set to zero. Once $\hat{\psi}$ is determined, the Fourier transform of the deflection angle and the shear follows from their definitions in terms of the deflection potential, given in Sect. 2.2 of IN,

$$\hat{\boldsymbol{\alpha}}(\boldsymbol{\ell}) = -\mathrm{i}\boldsymbol{\ell}\hat{\psi}(\boldsymbol{\ell}) \; , \tag{39}$$

$$\hat{\gamma}(\boldsymbol{\ell}) = -\left(\frac{\ell_1^2 - \ell_2^2}{2} + \mathrm{i}\ell_1\ell_2\right)\hat{\psi}(\boldsymbol{\ell}) \; . \tag{40}$$

Thus, in principle, one determines the relevant quantities by Fourier transforming κ, then calculating the Fourier transforms of the potential, deflection, and shear, whose real-space counterparts are then obtained from an inverse Fourier transform, like in (37).

Up to now we have not gained anything; the Fourier transforms as defined above are two-dimensional integrals, as are the real-space relations between deflection angle and shear, and the surface-mass density. However, provided κ becomes 'small enough' for large values of $|\boldsymbol{\theta}|$, the integral in (36) may be approximated by one over a finite region in $\boldsymbol{\theta}$-space. This finite integral is further approximated as a sum over gridpoints, with a regular grid covering the lens plane. Consider a square in the lens plane of side L, and let N be the number of gridpoints per dimension, so that $\Delta\theta = L/N$ is the size of a gridcell. The inverse grid, i.e., the $\boldsymbol{\ell}$-grid, has a gridcell of size $\Delta\ell = 2\pi/L$. The discrete Fourier transform then uses the values of κ on the $\boldsymbol{\theta}$-grid to calculate $\hat{\kappa}$ on the $\boldsymbol{\ell}$-grid. The latter, in fact, is then the Fourier transform of the periodic continuation of the mass distribution in $\boldsymbol{\theta}$-space. Because of this periodic continuation, the deflection angle as calculated from the discrete Fourier transform, which is performed by the Fast Fourier Transform (FFT) method, is the sum of the input mass distribution, plus all of its periodic continuation. Here, finally, is why we have considered the Fourier method: the FFT is a very efficient and quick procedure (see, e.g., Press et al. 1986), and arguably the best one in cases of mass distributions for which no analytical progress can be made. The lensing properties are calculated on a grid; if needed, they can be obtained for other points by interpolation.

Because of the periodic continuation, the mass distribution has to decrease sufficiently quickly for large $|\boldsymbol{\theta}|$, or be truncated at large radii. In any case, L should be taken sufficiently large to minimize these periodicity effects.

Another point to mention is that a periodic mass distribution, each element of which has positive total mass, has an infinite mass, so that the deflection potential has to diverge; on the other hand, the deflection potential is enforced to be periodic. This apparent contradiction can be resolved by noting that the $\boldsymbol{\ell} = \mathbf{0}$ mode of $\hat{\kappa}$ is not used in the calculation of $\hat{\boldsymbol{\alpha}}$ and $\hat{\gamma}$. Indeed, if $\hat{\psi}$ and ψ are calculated from the above equations, then the resulting ψ *does not satisfy the Poisson equation*; the ψ resulting from this procedure is the one corresponding to $\kappa - \bar{\kappa}$, where $\bar{\kappa}$ is the average of κ on the $\boldsymbol{\theta}$-grid. A similar remark is true for the deflection angle. Thus, at the end, one has to add a term $\bar{\kappa} |\boldsymbol{\theta}|^2 /2$ to ψ, and a term $\bar{\kappa}\boldsymbol{\theta}$ to $\boldsymbol{\alpha}$.

Since the FFT is very fast, one can choose N and L large, and then consider only the central part of the $\boldsymbol{\theta}$-grid needed for the actual lens modeling.

4.5 Results from Strong Lensing in Clusters

The main results of the strong lensing investigations of clusters can be summarized as follows:

- The mass in cluster centers is much more concentrated than predicted by (simple) models based on X-ray observations. The latter usually predict a relatively large core of the mass distribution. These large cores would render clusters sub-critical to lensing, i.e., they would be unable to produce giant arcs or multiple images. In fact, when arcs were first discovered they came as a big surprise because of these expectations. By now we know that the intracluster medium is much more complicated than assumed in these 'β-model' fits for the X-ray emission.
- The mass distribution in the inner region of clusters often shows strong substructure, or multiple mass peaks. These are also seen in the galaxy distribution of clusters, but with the arcs can be verified to also correspond to mass peaks (examples of this include the cluster Abell 2218 where arcs also curve around a secondary concentration of bright galaxies, clearly indicating the presence of a mass concentration, or the obviously bimodal cluster Abell 370). These are easily understood in the frame of hierarchical mergers in a CDM model; the merged clusters retain their multiple peaks for a dynamical time or even longer, and are therefore not in virial equilibrium.
- The orientation of the (dark) matter appears to follow closely the orientation of the light in the cD galaxy; this supports the idea that the growth of the cD galaxy is related to the cluster as a whole, through repeated accretion of lower-mass member galaxies. In that case, the cD galaxy 'knows' the orientation of the cluster.

- There is in general good agreement between lensing and X-ray mass estimates (e.g., Ettori and Lombardi 2003; Donahue et al. 2003) for those clusters where a 'cooling flow' indicates that they are in dynamical equilibrium, provided the X-ray analysis takes the presence of the cooling flow into account (Allen 1998).

Probably our 'favorite' clusters in which strong lensing effects are investigated in detail are biased in favor of having strong substructure, as this increases the lensing cross section for the occurrence of giant arcs (see below). Hence, it may be that the most detailed results obtained from strong lensing in clusters apply to a class of clusters which are especially selected because of their ability to produce spectacular arcs, and thus of their asymmetric mass distribution. Therefore, one must be careful in generalizing conclusions drawn from the 'arc clusters' to the cluster population as a whole.

Discrepancies

There are a few clusters where the lensing results and those obtained from analyzing the X-ray observations or cluster dynamics are in stong apparent conflict. Two of the most prominent ones shall be mentioned here. The cluster A 1689 (see Fig. 14) has arcs more than $\sim 40''$ away from the cluster center, which would imply a huge mass in this cluster center. This high mass is apparently confirmed by the high velocity dispersion of its member galaxies, although their distribution in redshift makes it likely that the cluster consists of several subcomponents (see Clowe and Schneider 2001 for a summary of these results). Several weak lensing results of this cluster have been published, and they are not all in agreement: whereas Tyson and Fischer (1995) from weak shear, and Taylor et al. (1998) and Dye et al. (2001) from the magnification method (that will be discussed in the next section) find also a very high mass for this cluster, the weak lensing analysis of Clowe and Schneider (2001; see also King et al. 2002b), based on deep wide-field imaging data of this cluster, finds a more moderate mass (or equivalent velocity dispersion) for this cluster. A new XMM-Newton X-ray observation of this cluster (Andersson and Madejski 2004) lends support for the smaller mass; in fact, their estimate of the virial mass of the cluster agrees with that obtained by Clowe and Schneider (2001). However, the disrepancy with the strong lensing mass in the cluster center remains at present; a quantitative analysis of the ACS data shown in Fig. 14 will hopefully shed light on this issue.

A second clear example for discrepant results in the cluster Cl 0024+17. It has a prominent arc system, indicating an Einstein radius of $\sim 30''$, and thus a high mass. The X-ray properties of this cluster, however, indicate a much smaller mass (Soucail et al. 2000), roughly by a factor of three. This discrepancy has been reaffirmed by recent Chandra observations, which confirmed this factor-of-three problem (Ota et al. 2004). The resolution of this discrepancy has probably been found by Czoske et al. (2001, 2002), who performed an extensive spectroscopic survey of cluster galaxies. Their result is best interpreted such that Cl 0024+17 presents a merger of two clusters along our line-of-sight, which implies that the measured velocity dispersion cannot be easily turned into a mass, as this system is not in virial equailibrium, and that the X-ray data cannot be converted to a mass either, due to the likely strong deviation from spherical symmetry and equilibrium. A wide field sparsely sampled HST observation of this cluster (Kneib et al. 2003) also indicates the presence of a second mass concentration about $3'$ away from the main peak. As will be mentioned below, clusters undergoing mergers have particularly high cross sections for producing arcs (Torri et al. 2004); hence, our 'favorites' are most likely selected for these non-equilibrium clusters.

Arc Statistics

The abundance of arcs is expected to be a strong function of the cosmological parameters: they not only determine the abundance of massive clusters (through the mass function discussed in Sect. 4.5 of IN), but also the degree of relaxation of clusters, which in turn affects their strong lensing cross section (Bartelmann et al. 1998). It is therefore interesting to consider the expected abundance of arcs as a function of cosmological parameters and compare this to the observed abundance. In a series of papers, M. Bartelmann and his colleagues have studied the expected giant arc abundance, using analytical as well as numerical techniques (e.g., Bartelmann and Weiss 1994; Bartelmann et al. 1995, 1998, 2002; Meneghetti et al. 2005; see also Dalal et al. 2003; Oguri et al. 2003; Wambsganss et al. 2004). Some of the findings of these studies can be summarized as follows:

- The formation of arcs depends very sensitively on the deviation from spherical symmetry and the detailed substructure of the mass distribution in the cluster. For this reason, analytical models which cannot describe this substructure with sufficient realism (see Bergmann and Petrosian 1993) do not provide realiable predictions for the arc statistics (in particular, axisymmetric mass models are essentially useless for estimating arc statistics), and one needs to refer to numerical simulations of structure formation. Since the substructure and triaxiality plays such an important role, these simulations have to be of high spatial and mass resolution.
- The frequency of arcs depends of course on the abundance of clusters, which in turn depends on the cosmological model and the fluctuation spectrum of the matter, in particular its normalization σ_8. Furthermore, clusters at a given redshift have different mean ages in different cosmological models, as the history of structure growth, and thus the merging history, depends on $\Omega_{\rm m}$ and Ω_Λ. Since the age of a cluster is one of the determining parameters for its level of substructure – younger clusters do not have had enough time to fully relax – this affects the lensing cross section of the clusters for arc formation. In fact, during epochs of mergers, the arc cross-section can have temporary excursions by large factors. Even the same cluster at the same epoch can have arc forming cross sections that vary by

more than an order-of-magnitude between different projection directions of the cluster. For fixed cluster abundance today, low-density models form clusters earlier than high-density models.

- Since the largest contribution of the total cross section for arc formation comes from clusters at intermediate redshift ($z \sim 0.4$), also the equation-of-state of the dark energy matters; as shown in Meneghetti et al. (2005), what matters is the dark energy density at the epoch of cluster formation. In addition, the earlier clusters form, the higher their characteristic density, which then makes them more efficient lenses for arc formation.

Taking these effects together, a low-density open model produces a larger number of arcs than a flat low-density model, which in turn has more arcs than a high-density model, for a given cluster abundance today. Whereas the differences between these models obtained by Meneghetti et al. (2005) are smaller than claimed in Bartelmann et al. (1998), they in principle allow constraining the cosmological parameters, provided they can be compared with the observed number of arcs.

Unfortunately, there are only a few systematic studies of clusters with regards to their strong lensing contents. Luppino et al. (1999) report on 8 giant arcs in their sample of the 38 most massive clusters found in the Einstein Medium Sensitivity Survey. Zaritsky and Gonzalez (2003) surveyed clusters in the redshift range $0.5 \lesssim z \lesssim 0.7$ over $69\,\mathrm{deg}^2$ and found two giant arcs with $R < 21.5$ and a length $\theta_1 > 10''$. Gladders at al. (2003) found 5 arc candidates in their Red Cluster Sequence survey of $90\,\mathrm{deg}^2$, all of them being associated with high-redshift clusters. In contrast to the claim by Bartelmann et al. (1998), these observered arc frequencies can be accounted for in a standard ΛCDM Universe, as shown by Dalal et al. (2003). There are several differences between these two studies, which are based on different assumptions about the number density of clusters and the source redshift distribution, which Dalal et al. (2003) took from the Hubble Deep Field, whereas Bartelmann et al. (1998) assumed all sources having $z_\mathrm{s} = 1$.

The strong dependence on the source redshift distribution has been pointed out by Wambsganss et al. (2004). In contrast to the other studies, they investigated the arc statistics using ray tracing through a three-dimensional mass distribution obtained from cosmological simulations, where- as the other studies mentioned considered the lensing effect of individual clusters found in these simulations. Although the former approach is more realistic, the assumption of Wambsganss et al. (2004) that the magnification of a light ray is a good measure for the length-to-width ratio of a corresponding arc is certainly not justified in detail, as shown in Dalal et al. (2003). The agreement of the lensing probability between Wambsganss et al. (2004) and Bartelmann et al. (1998) for all $z_\mathrm{s} = 1$ is therefore most likely a coincidence.

There are further difficulties in obtaining realistic predictions for the occurrence of giant arcs that can be compared with observations. First, the question of whether an image counts as an arc depends on a combination of source size, lens magnification, and seeing. Seeing makes arcs rounder and therefore reduces their length-to-width ratio. An impressive demonstration of this effect is provided by the magnificent system of arcs in the cluster A 1689 observed with the ACS onboard the HST, as shown in Fig. 14, compared to earlier ground-based images of this cluster. Second, several of the above-mentioned papers assume the source size to be $\theta = 1''$, whereas many arcs observed with HST are essentially unresolved in width, implying much smaller source sizes (and accordingly, a much higher sensitivity to seeing effects). Third, magnification bias is usually not taken into account in these theoretical studies. In fact, accounting properly for the magnification bias is quite difficult, as the surveys reporting on arc statistics are not really flux-limited. One might argue that they are surface brightness-limited, but even if this were true, the surface brightness of an arc coming from a small source depends very much on the seeing.

Therefore at present, the abundance of arcs seem to be not in conflict with a ΛCDM model, but more realistic simulations which take the aforementioned effects into account are certainly needed for a definite conclusion on this issue. On the observational side, increasing the number of clusters for which high-quality imaging is performed is of great importance, and the survey of luminous X-ray clusters imaged either with the ACS@HST or with ground-based telescopes during periods of excellent seeing would improve the observational situation dramatically. Blank-field surveys, such as they are conducted for cosmic shear research (see Sect. 7), could be used for blind searches of arcs (that is, not restricted to regions around known clusters). It may turn out, however, that the number of 'false positives' is unacceptably high, e.g., by misidentification of edge-on spirals, or blends of sources that yield apparent images with a high length-to-width ratio.

Constraints on Collisional Dark Matter

Spergel and Steinhardt (2000) suggested the possibility that dark matter particles are not only weakly interacting, but may have a larger elastic scattering cross-section. If this cross-section of such self-interacting dark matter is sufficiently large, it may help to explain two of the remaining apparent discrepancies between the predictions of the Cold Dark Matter model and observations: The slowly rising rotation curves of dwarf galaxies (e.g., de Blok et al. 2001) and the substructure of galaxy-scale dark matter halos (see Sect. 8 of SL). Self-interacting may soften the strength of the central density concentration as compared to the NFW profile, and could destroy most of the subclumps. However, there are other consequence of such an interaction, in that the shapes of the inner parts of dark matter halos tend to be more spherical. Meneghetti et al. (2001) have investigated the influence of self interaction of dark matter particles on clusters of galaxies, in particular their ability to form giant arcs. From their numerical simulations of clusters with varying cross-sections of particles, they showed that even a relatively small cross-section is sufficient to reduce the ability of clusters to produce giant arcs by an order of magnitude.

This is mainly due to two effects, the reduced asymmetry of the resulting mass distribution and the shallower central density profile. Furthermore, self-interactions destroy the ability of clusters to form radial arcs. Therefore, the 'desired' effect of self-interaction – to smooth the mass distribution of galaxies – has the same consequence for clusters, and can therefore probably be ruled out as a possible mechanism to cure the aforementioned apparent problems of the CDM model. From combining X-ray and lensing data of the cluster 0657−56, Markevitch et al. (2004) obtained upper limits on the self-interaction cross section of dark matter.

Do Clusters Follow the Universal NFW Profile ?

The CDM paradigm of structure formation predict a universal density profile of dark matter halos. One might therefore investigate whether the strong lensing properties of clusters are compatible with this mass profile. Of particular value for such an investigation are clusters which contain several strong lensing features, and in particular a radial arc, as it probes the inner critical curve of the cluster. Sand et al. (2004; see also Sand et al. 2002) claim from a sample of three clusters with radial arcs, that the slope of the inner mass profile must be considerably flatter than predicted by the NFW model. However, this conclusion is derived under the assumption of an axially-symmetric lens model. As is true for strong lensing by galaxies (see SL), axisymmetric mass model are not generic, and therefore conclusions derived from them are prone to the systematic of the symmetry assumption. That was demonstrated by Bartelmann and Meneghetti (2004) who showed that, as expected, the conclusion about the inner slope changes radically once a finite ellipticity of the mass distribution is allowed for, removing the apparent discrepancy with the predictions from CDM models.

Cosmological Parameters from Strong Lensing Systems

The lens strength, at given physical surface mass density Σ, depends on the redshifts of lens and source, as well as on the geometry of the Universe which enters the distance-redshift relation. Therefore, it has been suggested that a cluster which contains a large number of strong lensing features can be used to constrain cosmological parameters, provided the sources of the arcs and multiple image systems cover a large range of redshifts (Link and Pierce 1998). Simulations of this effect, using realistic cluster models, confirmed that such purely geometrical constraints can in principle be derived (Golse et al. 2002). One of the best studied strong-lensing cluster up to now is A 2218, for which four multiple-image systems with measured (spectroscopic) redshift have been identified which allows very tight constraints on the mass distribution in this cluster. Soucail et al. (2004) applied the aforementioned method to this cluster and obtained first constraints on the density parameter $\Omega_{\rm m}$, assuming a flat cosmological model. This work can be viewed as a proof of concept; the new

ACS camera onboard HST will allow the identification of even richer strong lensing systems in clusters, of which the one in A 1689 (see Fig. 14) is a particularly impressive example.

5 Mass Reconstructions from Weak Lensing

Whereas strong lensing probes the mass distribution in the inner part of clusters, weak lensing can be used to study the mass distribution at much larger angular separations from the cluster center. In fact, as we shall see, weak lensing can provide a parameter-free reconstruction of the projected two-dimensional mass distribution in clusters – and hence offers the prospect of mapping the dark matter distribution of clusters directly. This discovery (Kaiser and Squires 1993) can be viewed to mark the beginning of quantitative weak lensing research. But even before this discovery, weak lensing by clusters has been observed in a number of cases. Fort et al. (1988) found that in addition to the giant arc in Abell 370, there are a number of images stretched in the direction tangent to the center of the cluster, but with much less spectacular axis ratios than the giant arc in this cluster; they termed these new features 'arclets'. Tyson et al. (1990) found a statistically significant tangential alignment of faint galaxy images relative to the center of the clusters Abell 1689 and Cl 1409+52, and obtained a mass profile from these lens distortion maps. Comparison with numerical simulations yielded an estimate of the cluster velocity dispersion, assuming an isothermal sphere profile.

In this section we consider the parameter-free mass reconstruction technique, first the original Kaiser and Squires method, and then a number of improvements of this method. We then turn to the magnification effects; the change of the number density of background sources, as predicted from (26), can be turned into a local estimate of the surface mass density, and this method has been employed in a number of clusters. Next we shall consider inverse methods for the reconstruction of the mass distribution, which on the one hand are more difficult to apply than the 'direct' methods, but on the other hand are expected to yield more satisfactory results. Whereas the two-dimensional maps yield a good visual impression on the mass distribution in clusters, it is hard to extract quantitative information from them. In order to get quantities that describe the mass and that can be compared between clusters, often parameterized mass models are more useful, which are considered next. Finally, we consider aperture mass measures, which have been introduced originally to obtain a mass quantity that is unaffected by the mass-sheet degeneracy, but as will be shown, has a number of other useful features. In particular, employing the aperture mass, one can device a method to systematically search for mass concentrations on cluster-mass scales, using their shear properties only, i.e. without referring to their luminous properties.

5.1 The Kaiser–Squires Inversion

Weak lensing yields an estimate of the local (reduced) shear, as discussed in Sect. 2.2. Here we shall discuss how to derive the surface mass density from a measurement of the (reduced) shear. Recalling (IN-26), the relation between shear and surface mass density is

$$\begin{aligned} \gamma(\boldsymbol{\theta}) &= \frac{1}{\pi}\int_{\mathbb{R}^2} \mathrm{d}^2\theta'\,\mathcal{D}(\boldsymbol{\theta}-\boldsymbol{\theta}')\,\kappa(\boldsymbol{\theta}')\,, \quad \text{with} \\ \mathcal{D}(\boldsymbol{\theta}) &\equiv -\frac{\theta_1^2-\theta_2^2+2\mathrm{i}\theta_1\theta_2}{|\boldsymbol{\theta}|^4} = \frac{-1}{(\theta_1-\mathrm{i}\theta_2)^2}\,. \end{aligned} \tag{41}$$

Hence, the complex shear γ is a convolution of κ with the kernel $\mathcal{D}$, or, in other words, $\mathcal{D}$ describes the shear generated by a point mass. This relation can be inverted: in Fourier space this convolution becomes a multiplication,

$$\hat{\gamma}(\boldsymbol{\ell}) = \pi^{-1}\hat{\mathcal{D}}(\boldsymbol{\ell})\,\hat{\kappa}(\boldsymbol{\ell}) \quad \text{for} \quad \boldsymbol{\ell} \neq \mathbf{0}\,,$$

which can be inverted to yield

$$\hat{\kappa}(\boldsymbol{\ell}) = \pi^{-1}\hat{\gamma}(\boldsymbol{\ell})\,\hat{\mathcal{D}}^*(\boldsymbol{\ell}) \quad \text{for} \quad \boldsymbol{\ell} \neq \mathbf{0}\,, \tag{42}$$

where the Fourier transform of $\mathcal{D}$ is[5]

$$\hat{\mathcal{D}}(\boldsymbol{\ell}) = \pi\frac{\left(\ell_1^2-\ell_2^2+2\mathrm{i}\ell_1\ell_2\right)}{|\boldsymbol{\ell}|^2}\,; \tag{43}$$

note that this implies that $\hat{\mathcal{D}}(\boldsymbol{\ell})\hat{\mathcal{D}}^*(\boldsymbol{\ell}) = \pi^2$, which has been used in obtaining (42). It is obvious that $\hat{\mathcal{D}}$ is undefined for $\boldsymbol{\ell} = \mathbf{0}$, which has been indicated in the foregoing equations. Fourier back-transformation of (42) then yields

$$\begin{aligned} \kappa(\boldsymbol{\theta}) - \kappa_0 &= \frac{1}{\pi}\int_{\mathbb{R}^2} \mathrm{d}^2\theta'\,\mathcal{D}^*(\boldsymbol{\theta}-\boldsymbol{\theta}')\,\gamma(\boldsymbol{\theta}') \\ &= \frac{1}{\pi}\int_{\mathbb{R}^2} \mathrm{d}^2\theta'\,\mathcal{R}\mathrm{e}\left[\mathcal{D}^*(\boldsymbol{\theta}-\boldsymbol{\theta}')\,\gamma(\boldsymbol{\theta}')\right]\,. \end{aligned} \tag{44}$$

Note that the constant κ_0 occurs since the $\boldsymbol{\ell} = \mathbf{0}$-mode is undetermined. Physically, this is related to the fact that a uniform surface mass density yields no shear. Furthermore, it is obvious (physically, though not so easily seen mathematically) that κ must be real; for this reason, the imaginary part

[5] The form of $\hat{\mathcal{D}}$ can be obtained most easily by using the relations between the surface mass density and the shear components in terms of the deflection potential ψ, given in (IN-18). Fourier transforming those immediately yields $\hat{\kappa} = -|\boldsymbol{\ell}|^2\hat{\psi}/2$, $\hat{\gamma}_1 = -(\ell_1^2-\ell_2^2)\hat{\psi}/2$, $\hat{\gamma}_2 = -\ell_1\ell_2\hat{\psi}$. Eliminating $\hat{\psi}$ from the foregoing relations, the expression for $\hat{\mathcal{D}}$ is obtained.

of the integral should be zero, and taking the real-part only [as in the second line of (44)] makes no difference. However, in practice this is different, since noisy data, when inserted into the inversion formula, will produce a non-zero imaginary part. What (44) shows is that if γ can be measured, κ can be determined.

Before looking at this in more detail, we briefly mention some difficulties with the inversion formula as given above:

- Since γ can at best be estimated at discrete points (galaxy images), smoothing is required. One might be tempted to replace the integral in (44) by a discrete sum over galaxy positions, but as shown by Kaiser and Squires (1993), the resulting mass density estimator has infinite noise (due to the θ^{-2}-behavior of the kernel $\mathcal{D}$).
- It is not the shear γ, but the reduced shear g that can be determined from the galaxy ellipticities; hence, one needs to obtain a mass density estimator in terms of g. In the case of 'weak' weak lensing, i.e., where $\kappa \ll 1$ and $|\gamma| \ll 1$, then $\gamma \approx g$.
- The integral in (44) extends over $\mathbb{R}^2$, whereas data are available only on a finite field; therefore, it needs to be seen whether modifications allow the construction of an estimator for the surface mass density from finite-field shear data.
- To get absolute values for the surface mass density, the additive constant κ_0 is of course a nuisance. As will be explained soon, this indeed is the largest problem in mass reconstructions, and is the *mass-sheet degeneracy* discussed in Sect. 2.5 of IN.

5.2 Improvements and Generalizations

Smoothing

Smoothing of data is needed to get a shear field from discrete data points. Consider first the case that we transform (44) into a sum over galaxy images (ignoring the constant κ_0 for a moment, and also assuming the weak lensing case, $\kappa \ll 1$, so that the expectation value of ϵ is the shear γ),

$$\kappa_{\rm disc}(\boldsymbol{\theta}) = \frac{1}{n\,\pi} \sum_i \mathcal{R}\mathrm{e}\left[\mathcal{D}(\boldsymbol{\theta} - \boldsymbol{\theta}_i)\,\epsilon_i\right] \; , \tag{45}$$

where the sum extends over all galaxy images at positions $\boldsymbol{\theta}_i$ and complex ellipticity ϵ_i, and n is the number density of background galaxies. As shown by Kaiser and Squires (1993), the variance of this estimator for κ diverges. However, one can smooth this estimator, using a weight function $W(\Delta\theta)$ (assumed to be normalized to unity), to obtain

$$\kappa_{\rm smooth}(\boldsymbol{\theta}) = \int \mathrm{d}^2\theta' \; W(|\boldsymbol{\theta} - \boldsymbol{\theta}'|)\,\kappa_{\rm disc}(\boldsymbol{\theta}') \; , \tag{46}$$

which now has a finite variance. One might expect that, since (i) smoothing can be represented by a convolution, (ii) the relation between κ and γ is a convolution, and (iii) convolution operations are transitive, it does not matter whether the shear field is smoothed first and inserted into (44), or one uses (46) directly. This statement is true if the smoothing of the shear is performed as

$$\gamma_{\rm smooth;1}(\boldsymbol{\theta}) = \frac{1}{n} \sum_i W(|\boldsymbol{\theta} - \boldsymbol{\theta}_i|) \, \epsilon_i \; . \tag{47}$$

If this expression is inserted into (44), one indeed recovers the estimate (46). However, this is not a particularly good method for smoothing, as can be seen as follows: the background galaxy positions will at least have Poisson noise; in fact, since the angular correlation function even of faint galaxies is non-zero, local number density fluctuations will be larger than predicted from a Poisson distribution. However, in the estimator (45) and in the smoothing procedure (47), these local variations of the number density are not taken into account. A much better way (Seitz and Schneider 1995a,b) to smooth the shear is given by

$$\gamma_{\rm smooth;2}(\boldsymbol{\theta}) = \left[\sum_i W(|\boldsymbol{\theta} - \boldsymbol{\theta}_i|) \right]^{-1} \sum_i W(|\boldsymbol{\theta} - \boldsymbol{\theta}_i|) \, \epsilon_i \; , \tag{48}$$

which takes these local number density fluctuations into account. Lombardi and Schneider (2001) have shown that the expectation value of the smoothed shear estimate (48) is not exactly the shear smoothed by the kernel W, but the deviation (i.e., the bias) is very small provided the effective number of galaxy images inside the smoothing function W is substantially larger than unity, which will always be the case for realistic applications. Lombardi and Schneider (2002) then have demonstrated that the variance of (48) is indeed substantially reduced compared to that of (47), in agreement with the finding of Seitz and Schneider (1995a,b).

When smoothed with a Gaussian kernel of angular scale $\theta_{\rm s}$, the covariance of the resulting mass map is finite, and given by (Lombardi and Bertin 1998; van Waerbeke 2000)

$$\mathrm{Cov}\left(\kappa(\boldsymbol{\theta}), \kappa(\boldsymbol{\theta}')\right) = \frac{\sigma_\epsilon^2}{4\pi\theta_{\rm s}^2 n} \exp\left(-\frac{|\boldsymbol{\theta} - \boldsymbol{\theta}'|^2}{2\theta_{\rm s}^2} \right) \; . \tag{49}$$

Thus, the larger the smoothing scale, the less noisy is the corresponding mass map; on the other hand, the more are features washed out. Choosing the appropriate smoothing scale is not easy; we shall come back to this issue in Sect. 5.3 below.

The Non-linear Case, $g \neq \gamma$

Noting that the reduced shear $g = \gamma/(1-\kappa)$ can be estimated from the ellipticity of images (assuming that we avoid the potentially critical inner region of the cluster, where $|g| > 1$; indeed, this case can also be taken into account, at the price of somewhat increased complexity), one can write:

$$\kappa(\boldsymbol{\theta}) - \kappa_0 = \frac{1}{\pi} \int_{\mathbb{R}^2} \mathrm{d}^2\theta' \left[1 - \kappa(\boldsymbol{\theta}')\right] \mathcal{R}\mathrm{e}\left[\mathcal{D}^*(\boldsymbol{\theta} - \boldsymbol{\theta}')\, g(\boldsymbol{\theta}')\right] ; \tag{50}$$

this integral equation for κ can be solved by iteration, and it converges quickly (Seitz and Schneider 1995a,b). Note that in this case, the undetermined constant κ_0 no longer corresponds to adding a uniform mass sheet. What the arbitrary value of κ_0 corresponds to can be seen as follows: The transformation

$$\begin{aligned} \kappa(\boldsymbol{\theta}) \to \kappa'(\boldsymbol{\theta}) &= \lambda\kappa(\boldsymbol{\theta}) + (1 - \lambda) \quad \text{or} \\ [1 - \kappa'(\boldsymbol{\theta})] &= \lambda[1 - \kappa(\boldsymbol{\theta})] \end{aligned} \tag{51}$$

changes the shear $\gamma \to \gamma' = \lambda\gamma$, and thus leaves g invariant; this is the mass-sheet degeneracy ! It can be broken if magnification information can be obtained, since $\mathcal{A} \to \mathcal{A}' = \lambda\mathcal{A}$, so that

$$\mu \to \mu' = \lambda^{-2}\mu \, .$$

Magnification information can be obtained from the number counts of images (Broadhurst et al. 1995), owing to the magnification bias, provided the unlensed number density is sufficiently well known. In principle, the mass sheet degeneracy can also be broken if redshift information of the source galaxies is available and if the sources are widely distributed in redshift; this can be seen as follows: let

$$Z(z_\mathrm{s}) = \frac{D_\mathrm{ds}/D_\mathrm{s}}{\lim_{z_\mathrm{s}\to\infty} D_\mathrm{ds}/D_\mathrm{s}} \mathrm{H}(z_\mathrm{s} - z_\mathrm{d}) \tag{52}$$

(H being the Heaviside step function) be the ratio of the lens strength of a source at z_s to that of a fiducial source at infinite redshift (see Fig. 16); then, if κ_∞ and γ_∞ denote the surface mass density and shear for such a fiducial source, the reduced shear for a source at z_s is

$$g = \frac{Z\gamma_\infty}{1 - Z\kappa_\infty} \, , \tag{53}$$

and there is no global transformation of κ_∞ that leaves g invariant for sources at all redshifts, showing the validity of the above statement. However, even in this case the mass-sheet degeneracy is only mildly broken (see Bradac et al. 2004). In particular, only those regions in the cluster where the non-linearity (i.e., the difference between γ and g) is noticeable can contribute to the degeneracy breaking, that is, the region near the critical curves where $|g| \sim 1$.

In the non-linear case ($\gamma \neq g$) the reduced shear needs to be obtained from smoothing the galaxy ellipticities in the first place. Since the relation between

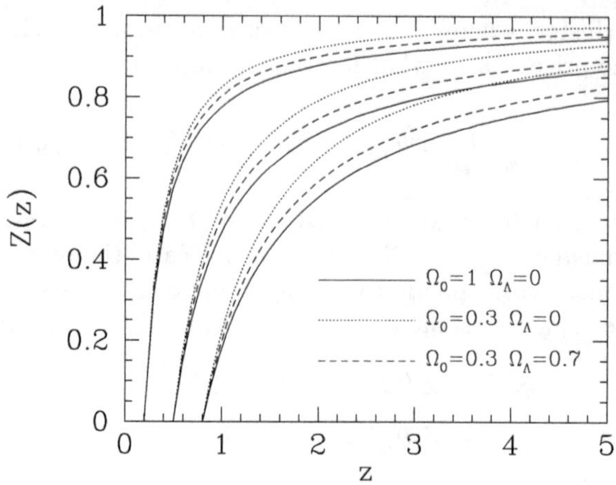

Fig. 16. The redshift weight function $Z(z_{\rm s})$, defined in (52), for three different values of the lens redshift $z_{\rm d} = 0.2$, 0.5, and 0.8, and three different geometries of the Universe, as indicated in the labels (here, $\Omega_{\rm m}$ is denoted as Ω_0). Asymptotically for $z_{\rm s} \to \infty$, all curves tend to $Z = 1$ (from Bartelmann and Schneider 2001)

g and κ is non-linear, the 'transitivity of convolutions' no longer applies; one thus cannot start from a discretization of an integral over image ellipticities and smooth the resulting mass map later. We also note that the accuracy with which the (reduced) shear is estimated can be improved provided redshift estimates of individual source galaxies are available (see Fig. 17). In particular for high-redshift clusters, redshift information on individual source galaxies becomes highly valuable. This can be understood by considering a high-redshift lens, where an appreciable fraction of faint 'source' galaxies are located in front of the lens, and thus do not contribute to the lensing signal. However, they do contribute to the noise of the measurement. Redshift information allows the elimination of these foreground galaxies in the shear estimate and thus the reduction of noise.

Finite-Field Mass Reconstruction

In order to obtain a mass map from a finite data field, one starts from the relation (Kaiser 1995)

$$\nabla\kappa = \begin{pmatrix} \gamma_{1,1} + \gamma_{2,2} \\ \gamma_{2,1} - \gamma_{1,2} \end{pmatrix} \equiv \mathbf{u}_\gamma(\boldsymbol{\theta}) \;, \tag{54}$$

which is a *local* relation between shear and surface mass density; it can easily be derived from the definitions of κ and γ in terms of $\psi_{,ij}$. A similar relation can be obtained in terms of reduced shear,

$$\nabla K(\boldsymbol{\theta}) = \frac{-1}{1 - g_1^2 - g_2^2} \begin{pmatrix} 1 - g_1 & -g_2 \\ -g_2 & 1 + g_1 \end{pmatrix} \begin{pmatrix} g_{1,1} + g_{2,2} \\ g_{2,1} - g_{1,2} \end{pmatrix} \equiv \mathbf{u}_g(\boldsymbol{\theta}) \;, \tag{55}$$

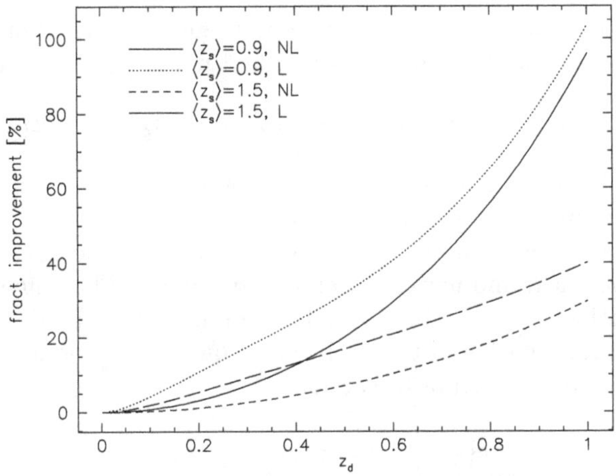

Fig. 17. The fractional gain in accuracy of the shear estimate when using redshift information of individual source galaxies, relative to the case where only the redshift distribution of the population is known, plotted as a function of the lens redshift. It is assumed that the sources have a broad redshift distribution, with a mean of $\langle z_{\rm s}\rangle = 0.9$ (*solid and dotted curves*) or $\langle z_{\rm s}\rangle = 1.5$ (*short-dashed and long-dashed curves*). The gain of accuracy also depends on the lens strength; the dotted and long-dashed curves assume local lens parameters of $\gamma_\infty = 0.3 = \kappa_\infty$, whereas the solid and short-dashed curves assume only very weak lensing, here approximated by $\gamma_\infty = 0 = \kappa_\infty$. One sees that the gain is dramatic once the lens redshift becomes comparable to the mean redshift of the source galaxies and is therefore of great importance for high-redshift clusters (from Bartelmann and Schneider 2001)

where

$$K(\boldsymbol{\theta}) \equiv \ln[1 - \kappa(\boldsymbol{\theta})] \tag{56}$$

is a non-linear function of κ. Based on these local relations, finite-field inversion relations can be derived, and several of them appeared in the literature right after the foregoing equations have been published. For example, it is possible to obtain finite-field mass maps from line integrations (Schneider 1995; for other methods, see Squires and Kaiser 1996). Of all these finite-field methods, one can be identified as optimal, by the following reasoning: in the case of noise-free data, the imaginary part of (44) should vanish. Since one is always dealing with noisy data (at least coming from the finite intrinsic ellipticity distribution of the sources), in real life the imaginary part of (44) will not be zero. But being solely a noise component, one can choose that finite-field inversion which yields a zero imaginary component when averaged over the data field (Seitz and Schneider 1996). One way of deriving this mass map is obtained by a further differentiation of (54); this then yields a von Neumann boundary-value problem on the data field $\mathcal{U}$ (Seitz and Schneider 2001),

$$\nabla^2\kappa = \nabla \cdot \mathbf{u}_\gamma \quad \text{with} \quad \mathbf{n}\cdot\nabla\kappa = \mathbf{n}\cdot\mathbf{u}_\gamma \quad \text{on} \quad \partial\mathcal{U} \; , \tag{57}$$

where $\mathbf{n}$ is the outward-directed normal on the boundary $\partial\mathcal{U}$ of $\mathcal{U}$. The analogous equation holds for K in terms of g and $\mathbf{u}_g$,

$$\nabla^2 K = \nabla \cdot \mathbf{u}_g \quad \text{with} \quad \mathbf{n} \cdot \nabla K = \mathbf{n} \cdot \mathbf{u}_g \quad \text{on} \quad \partial\mathcal{U} \, . \tag{58}$$

Note that (57) determines the solution κ only up to an additive constant, and (58) determines K only up to an additive constant, i.e., $(1-\kappa)$ up to a multiplicative factor. Hence, in both cases we recover the mass-sheet degeneracies for the linear and non-linear case, respectively. The numerical solution of these equations is fast, using overrelaxation (see Press et al. 1992). In fact, the foregoing formulation of the problem is equivalent (Lombardi and Bertin 1998) to the minimization of the action

$$A = \int_{\mathcal{U}} \mathrm{d}^2\theta \, |\nabla\kappa(\boldsymbol{\theta}) - \mathbf{u}_\gamma(\boldsymbol{\theta})|^2 \, , \tag{59}$$

from which the von Neumann problem can be derived as the Euler equation of the variational principle $\delta A = 0$. Furthermore, Lombardi and Bertin (1998) have shown that the solution of (57) is 'optimal', in that for this estimator the variance of κ is minimized.

Since (57) provides a linear relation between the shear and the surface mass density, one expects that it can also be written in the form

$$\kappa(\boldsymbol{\theta}) = \int_{\mathcal{U}} \mathrm{d}^2\theta' \, \mathbf{H}(\boldsymbol{\theta};\boldsymbol{\theta}') \cdot \mathbf{u}_\gamma(\boldsymbol{\theta}') \, , \tag{60}$$

where the vector field $\mathbf{H}(\boldsymbol{\theta};\boldsymbol{\theta}')$ is the Green's function of the von Neumann problem (57). Accordingly,

$$K(\boldsymbol{\theta}) = \int_{\mathcal{U}} \mathrm{d}^2\theta' \, \mathbf{H}(\boldsymbol{\theta};\boldsymbol{\theta}') \cdot \mathbf{u}_g(\boldsymbol{\theta}') \, . \tag{61}$$

Seitz and Schneider (1996) gave explicit expression for $\mathbf{H}$ in the case of a circular and rectangular data field.

One might ask how important the changes in the resulting mass maps are compared to the Kaiser–Squires formula applied to a finite data field. For that we note that applying (44) or (50) to a finite data field is equivalent to setting the shear outside the data field to zero. Hence, the resulting mass distribution will be such as to yield a zero shear outside the data field, despite the fact that we have no indication from data that the shear indeed is zero there. This induces features in the mass map, in form of a pillow-like overall mass distribution. The amplitude of this feature depends on the strength of the lens, its location inside the data field, and in particular the size of the data field. Whereas for large data fields this amplitude is small compared to the noise amplitude of the mass map, it is nevertheless a systematic that can easily be avoided, and should be avoided, by using the finite-field inversions, which cause hardly any additional technical problems.

Various tests have been conducted in the literature as to the accuracy of the various inversions. For those, one generates artificial shear data from a known mass distribution, and compares the mass maps reconstructed with the various methods with the original (e.g., Seitz and Schneider 1996, 2001; Squires and Kaiser 1996). One of the surprising results of such comparisons is that in some cases, the Kaiser and Squires original reconstruction faired better than the explicit finite-field inversions, although it is known to yield systematics. The explanation for this apparent paradox is, however, easy: the mass models used in these test consisted of one or more localized mass peaks well inside the data field, so the shear outside the data field is very small. Noting that the KS formula applied to a finite data field is equivalent to setting $\gamma = 0$ outside the data field, this methods provides 'information' to the reconstruction process which is not really there, but for the mass models used in the numerical tests is in fact close to the truth. Of course, by adding this nearly correct 'information' to the mass reconstruction, the noise can be lowered relative to the finite-field reconstructions where no assumptions about the shear field outside the data field is made.

Constraints on the Geometry of the Universe from Weak Lensing Mass Reconstructions

The strength of the lensing signal depends, for a given lens redshift, on the redshift of the sources, through the function $Z(z_{\rm s})$ (52). Suppose that the surface mass density of a cluster was well known, and that the redshifts of background sources can be determined. Then, by comparing the measured shear signal from sources at a given redshift $z_{\rm s}$ with the one expected from the mass distribution, the value of $Z(z_{\rm s})$ can be determined. Since $Z(z)$ depends on the geometry of the Universe, parameterized through $\Omega_{\rm m}$ and Ω_Λ, these cosmological parameters can in principle be determined. A similar strategy for strong lensing clusters was described at the end of Sect. 4.

Of course, the surface mass density of the cluster cannot be assumed to be known, but needs to be reconstructed from the weak lensing data itself. Consider for a moment only the amplitude of the surface mass density, assuming that its shape is obtained from the reconstruction. Changing the function $Z(z)$ by a multiplicative factor would be equivalent to changing the surface mass density Σ of the cluster by the inverse of this factor, and hence such a constant factor in Z is unobservable due to the mass-sheet degeneracy. Hence, not the amplitude of the function $Z(z)$ shown in Fig. 16 is important here, but its shape.

Lombardi and Bertin (1999) have suggested a method to perform cluster mass reconstructions and at the same time determine the cosmological parameters by minimizing the difference between the shear predicted from the reconstructed mass profile and the observed image ellipticities, where the former depends on the functional form of $Z(z)$. A nice and simple way to illustrate such a method was given in Gautret et al. (2000), called the 'triplet method'. Consider three background galaxies which have a small separation on the sky, and assume to know the three source redshifts. Because of their closeness, one might assume that they all experience the same tidal field and surface mass density from the cluster. In that case, the shear of the three galaxies is described by five parameters, the two components of γ_∞, κ, and $\Omega_{\rm m}$ and Ω_Λ. From the six observables (two components of three galaxy ellipticities), one can minimize the difference between the predicted shear and the observed ellipticities with respect to these five parameters, and in particular obtain an estimate for the cosmological parameters. Repeating this process for a large number of triplets of background galaxies, the accuracy on the Ω's can be improved, and results from a large number of clusters can be combined.

This procedure is probably too simple to be applied in practice; in particular, it treats κ_∞ and γ_∞ for each triplet as independent numbers, whereas the mass profile of the cluster is described by a single scalar function. However, it nicely illustrates the principle. Lombardi and Bertin (1999) have used a single density profile $\kappa_\infty(\boldsymbol{\theta})$ of the cluster, but assumed that the mass-sheet degeneracy is broken by some other means. Jain and Taylor (2003) suggested a similar technique for employing the lensing strength as a function of redshifts and cosmological parameters to infer constraints on the latter. Clearly, more work is needed in order to turn these useful ideas into a practically applicable method.

5.3 Inverse Methods

In addition to these 'direct' methods for determining κ, inverse methods have been developed, such as a maximum-likelihood fit (Bartelmann et al. 1996; Squires and Kaiser 1996) to the data. There are a number of reasons why these are in principle preferable to the direct method discussed above. First, in the direct methods, the smoothing scale is set arbitrarily, and in general kept constant. It would be useful to obtain an objective way how this scale should be chosen, and perhaps, that the smoothing scale be a function of position: e.g., in regions with larger number densities of sources, the smoothing scale could be reduced. Second, the direct methods do not allow additional input coming from complementary observations; for example, if both shear and magnification information are available, the latter could not be incorporated into the mass reconstruction. The same is true for clusters where strong lensing constraints are known.

The Shear Likelihood Function

In the inverse methods, one tries to fit a (very general) lens model to the observational data, such that the data agree within the estimated errors with the model. In the maximum-likelihood methods, one parameterizes the lens by the deflection potential ψ on a grid and then minimizes the regularized log-likelihood

$$-\ln \mathcal{L} = \sum_{i=1}^{N_{\rm g}} \frac{|\epsilon_i - g(\boldsymbol{\theta}_i, \{\psi_n\})|^2}{\sigma_i^2(\boldsymbol{\theta}_i, \{\psi_n\})} + 2\ln\sigma_i(\boldsymbol{\theta}_i, \{\psi_n\}) + \lambda_{\rm e} S(\{\psi_n\}) , \qquad (62)$$

where $\sigma_i \approx \sigma_\epsilon \left(1 - |g(\boldsymbol{\theta}_i, \{\psi_n\})|^2\right)$ [see (15) for the case $|g| < 1$ that was assumed here], with respect to these gridded ψ-values; this specific form of the likelihood assumes that the intrinsic ellipticity distribution follows a Gaussian with width σ_ϵ.[6] In order to avoid overfitting, one needs a regularization term S; entropy regularization (Seitz et al. 1998) seems very well suited (see Bridle et al. 1998; Marshall et al. 2002 for alternative regularizations). The entropy term S gets large if the mass distribution has a lot of structure; hence, in minimizing (62) one tries to match the data as closely as permitted by the entropic term (Narayan and Nityananda 1986). As a result, one obtains a model as smooth as compatible with the data, but where structure shows up where the data require it. The parameter $\lambda_{\rm e}$ is a Langrangean multiplier which sets the relative weight of the likelihood function and the regularization; it should be chosen such that the χ^2 per galaxy image is about unity, i.e.,

$$\sum_{i=1}^{N_{\rm g}} \frac{|\epsilon_i - g(\boldsymbol{\theta}_i, \{\psi_n\})|^2}{\sigma_i^2(\boldsymbol{\theta}_i, \{\psi_n\})} \approx N_{\rm g} ,$$

since then the deviation of the observed galaxy ellipticities from their expectation value g is as large as expected from the ellipticity dispersion. This choice of the regularization parameter $\lambda_{\rm e}$ then fixes the effective smoothing used for the reconstruction.

Strong lensing constraints can be incorporated into the inverse method by adding a term to the log-likelihood function which forces the minimum to satisfy these strong constraints nearly precisely. E.g., if a pair of multiple images at $\boldsymbol{\theta}_1$ and $\boldsymbol{\theta}_2$ is identified, one could add the term

$$\lambda_{\rm s} \left|\boldsymbol{\beta}(\boldsymbol{\theta}_1) - \boldsymbol{\beta}(\boldsymbol{\theta}_2)\right|^2 = \lambda_{\rm s} \left|[\boldsymbol{\theta}_1 - \boldsymbol{\alpha}(\boldsymbol{\theta}_1)] - [\boldsymbol{\theta}_2 - \boldsymbol{\alpha}(\boldsymbol{\theta}_2)]\right|^2$$

to the log-likelihood; by turning up the parameter $\lambda_{\rm s}$, its minimum is guaranteed to correspond to a solution where the multiple image constraint is satisfied. Note that the form of this 'source-plane minimization' is simplified – see Sect. 4.6 of SL – but in the current context this approach suffices.

[6] This specific form (62) of the likelihood function assumes that the sheared ellipticity probability distribution follows a two-dimensional Gaussian with mean g and dispersion σ; note that this assumption is not valid in general, not even when the intrinsic ellipticity distribution is Gaussian (see Geiger and Schneider 1999 for an illustration of this fact). The exact form of the lensed ellipticity distribution follows from the intrinsic distribution $p_{\rm s}(\epsilon^{(\rm s)})$ and the transformation law (12) between intrinsic and lensed ellipticity, $p(\epsilon) = p_{\rm s}\left(\epsilon^{(\rm s)}(\epsilon; g)\right) \det\left(\partial\epsilon^{(\rm s)}/\partial\epsilon\right)$. However, in many cases the Gaussian approximation underlying (62) is sufficient and convenient for analytical considerations.

Magnification Likelihood

Similarly, when accurate number counts of faint background galaxies are available, the magnification information can be incorporated into the log-likelihood function. If the number counts behave (locally) as a power law, $n_0(>S) \propto S^{-\alpha}$, the expected number of galaxies on the data field $\mathcal{U}$ then is

$$\langle N \rangle = n_0 \int_{\mathcal{U}} \mathrm{d}^2\theta \, |\mu(\boldsymbol{\theta})|^{\alpha-1} \; ; \tag{63}$$

see (26). The likelihood of observing N galaxies at the positions $\boldsymbol{\theta}_i$ can then be factorized into a term that yields the probability of observing N galaxies when the expected number is $\langle N \rangle$, and one that the N galaxies are at their observed locations. Since the probability for a galaxy to be at $\boldsymbol{\theta}_i$ is proportional to the expected number density there, $n = n_0\, \mu^{\alpha-1}$, the likelihood function becomes (Seitz et al. 1998)

$$\mathcal{L}_\mu = P_N\left(\langle N \rangle\right) \prod_{i=1}^{N} |\mu(\boldsymbol{\theta}_i)|^{\alpha-1} \; , \tag{64}$$

with the first factor yielding the Poisson probability. Note that this expression assumes that the background galaxies are unclustered on the sky; in reality, where (even faint) galaxies cluster, this factorization does not strictly apply.

It should be pointed out that the deflection potential ψ, and not the surface mass density κ, should be used as variable on the grid, for two reasons: first, shear and κ depend locally on ψ, and are thus readily calculated by finite differencing from ψ, whereas the relation between γ and κ is non-local and requires summation over all gridpoints, which is of course more time consuming. Second, and more important, the surface mass density on a finite field *does not* determine γ on this field, since mass outside the field contributes to γ as well. In fact, one can show (Schneider and Bartelmann 1997) that the shear inside a circle is fully determined by the mass distribution inside the circle and the multipole moments of the mass distribution outside the circle; in principle, the latter can thus be determined from the shear measurement.

Despite these reasons, some authors prefer to construct inverse methods in which the surface mass density on a grid serves as variables (e.g., Bridle et al. 1998; Marshall et al. 2002). The fact that the mass density on a finite field does not describe the shear in this field is accounted for in these methods by choosing a reconstruction grid that is larger than the data field and by allowing the surface mass density in this outer region to vary as well. Whereas the larger numerical grid requires a larger numerical effort, in addition to the non-local relation between κ and γ, this is of lesser importance, provided the numerical resources are available. Worse, however, is the view that the mass distribution outside the data field obtained by this method has any physical significance ! It has not. This mass distribution is solely one of infinitely many that can approximately generate the shear in the data field from mass outside the data field. The fact that numcrical tests show that one can indeed recover some of the mass distribution outside the data field is again a fluke, since these models are usually chosen such that all mass distribution outside the field in contained in a boundary region around the data field which is part of the numerical grid – and hence, the necessary 'external' shear must be generated by a mass distribution in this boundary zone which by construction is where it is. In real life, however, there is no constraint on where the 'external' shear contribution comes from.

5.4 Parameterized Mass Models

Whereas the parameter-free mass maps obtained through one of the methods discussed above provide a direct view of the mass distribution of a cluster, their quantitative interpretation is not straightforward. Peaks in the surface mass density can indicate the presence of a mass concentration, or else be a peak caused by the ellipticity noise of the galaxies. Since the estimated values for κ at different locations $\boldsymbol{\theta}$ are correlated [see (49)], it is hard to imagine 'error bars' attached to each point. Therefore, it is often preferable to use parameterized mass models to fit the observed data; for example, fitting shear (and/or magnification) data to an NFW mass profile (see IN, Sect. 6.2) yields the virial mass M_{200} of the cluster and its concentration index c. There are basically two methods which have been used to obtain such parameterized models. The first one, assuming a spherical mass model, orders the tangential component of the observed image ellipticities into radial bins and fits a parameterized shear profile through these bins, by minimizing a corresponding χ^2-function. One of the disadvantages of this method is that the result of the fitting process can depend on the selected binning, but this can be largely avoided by choosing the bins fine enough. This then essentially corresponds to minimizing the first term in (62).

Alternatively, a likelihood method can be used, in which the log-likelihood function (62) – without the regularization term – is minimized, with the values of the potential on the grid $\{\psi_n\}$ replaced by a set of parameters which describe the mass profile. Schneider et al. (2000) have used this likelihood method to investigate with which accuracy the model parameters of a mass profile can be obtained, using both the shear information as well as magnification information from number counts depletion. One of the surprising findings of this study was that the slope of the fitted mass profile is highly degenerate if only shear information is used; indeed, the mass-sheet degeneracy strikes again and causes even fairly different mass profiles to have very similar reduced shear profiles, as is illustrated for a simple example in Fig. 18. In Fig. 19, the resulting degeneracy of the profile slope is seen. This degeneracy can be broken if number count information is used in addition. As seen in the middle panel of Fig. 18, the magnification profiles of the four models displayed are quite different and thus the number counts sensitive to the profile slope. Indeed, the confidence regions in the parameter fits, shown in Fig. 19, obtained from the magnification information are highly inclined relative to

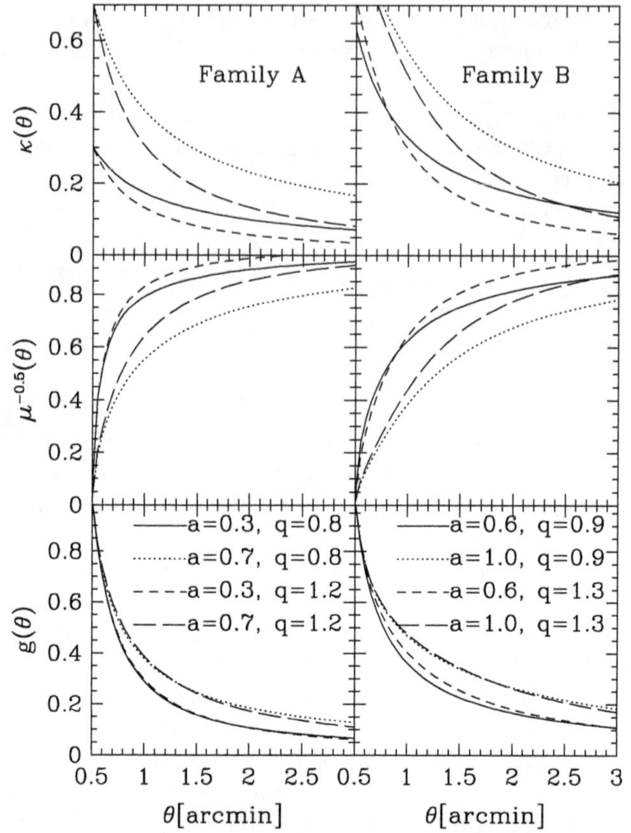

Fig. 18. The Einstein radius of a spherical mass distribution was assumed to be $\theta_{\rm E} = 0\overset{\prime}{.}5$, and the density profile outside the Einstein radius was assumed to follow a power law, $\kappa(\theta) = a(\theta/\theta_{\rm E})^{-q}$; an SIS would have $a = 1/2$ and $q = 1$. The figure displays for four combinations of model parameters the surface mass density $\kappa(\theta)$, the function $\mu^{-1/2}$, which would be the depletion factor for source counts of slope $\beta = 1/2$, and the reduced shear $g(\theta)$. As can be seen, whereas the density profiles of the four models are quite different, the reduced shear profiles are pairwise almost fully degenerate. This is due to the mass-sheet degeneracy; it implies that it will be difficult to determine the slope q of the profiles from shear measurements alone, unless much larger fields around the cluster are used (from Schneider et al. 2000)

those from the shear measurements, implying that the combination of both methods yields much better constraints on the model parameters. Of course, as mentioned before, the mass-sheet degeneracy can also be broken if redshift information of individual background galaxies is available.

However, in order for the magnification information to yield significant constraints on the mass parameters, one needs to know the unlensed number density n_0 of sources quite accurately. In fact, even an uncertainty of less than

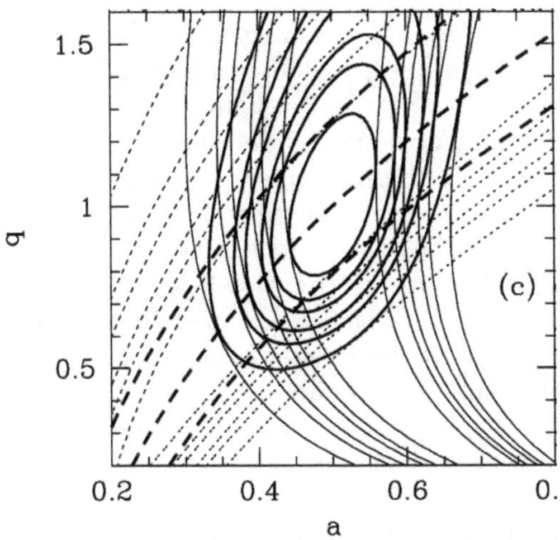

Fig. 19. For the power-law models of Fig. 18, confidence regions in the slope q and amplitude a are drawn, as derived from the shear (*thin solid contours*), the magnification (*dotted*) and their combination (*thick solid*). A number density of $30/\text{arcmin}^2$ for shear measurements and $120/\text{arcmin}^2$ for number counts was assumed. Thick dashed curves show models with constant total number of galaxies in the field, demonstrating that most of the constraint from magnification is due to the total counts, with little information about the detailed profile. It was assumed here that the unlensed number density of background galaxies is perfectly known; the fact that most of the magnification information comes from the total number of galaxies in the field implies that any uncertainty in the unlensed number density will quickly remove most of the magnification information (from Schneider et al. 2000)

$\sim$10% in the value of n_0 renders the magnification information in relation to the shear information essentially useless (in the frame of parameterized models). Note that an accurate determination of n_0 is difficult to achieve: since n_0 corresponds to the unlensed number density of faint galaxies at the same flux limit as used for the actual data field, one requires an accurate photometric calibration. A flux calibration uncertainty of 0.1 mag corresponds to an uncertainty in n_0 of about $\sim$5% for a slope of $\alpha = 0.5$, and such uncertainties are likely at the very faint flux limits needed to achieve a high number density of sources. In addition, the presence of bright cluster galaxies renders the detection and accurate brightness measurement of background galaxies difficult and requires masking of regions around them. Nevertheless, in cases where only magnification information is available, it can provide information on the mass profile by itself. Such a situation can occur for observing conditions with seeing above $\sim 1''$, when the shear method is challenged by the smallness of faint galaxies.

The result shown in Fig. 19 implies that the shape of the mass profile cannot be very well determined from the shear method, owing to the mass sheet degeneracy. This result extends to more general mass profiles than power-law models; e.g., King and Schneider 2001 considered NFW models with their two parameters c and r_{200}. A fairly strong degeneracy between these two parameters was found. Furthermore, the mass-sheet degeneracy renders it surprisingly difficult to distinguish an isothermal mass model from an NFW profile. The ability to distinguish these two families of models increases with a larger field-of-view of the observations. This expectation was indeed verified in King et al. 2002b where the wide-field imaging data of the cluster A 1689 were analyzed with the likelihood method. Although the field size is larger than $30'$, so that the shear profile up to $\sim 15'$ from the cluster center can be measured, an NFW profile is preferred with less than 90% confidence over a power-law mass model. The determination of the mass profiles is likely to improve when strong lensing constraints are taken into account as well.

The likelihood method for obtaining the parameters of a mass model is robust in the sense that the result is only slightly affected by substructure, as has been shown by King et al. (2001) using numerically generated cluster models. However, if a 'wrong' parameterization of the mass distribution is chosen, the interpretation of the resulting best-fit model must proceed carefully, and the resulting physical parameters, such as the total mass, may be biased. The principal problems with parameterized models are the same as for lens galaxies in strong lensing: unless the parameters have a well-defined physical meaning, one does not learn much, even if they are determined with good accuracy (see Sect. 4.7 of SL).

5.5 Problems of Weak Lensing Cluster Mass Reconstruction and Mass Determination

In this section, some of the major problems of determining the mass profile of clusters from weak lensing techniques are summarized. The finite ellipticity dispersion of galaxies generates a noise which provides a fundamental limit to the accuracy of all shear measurements. We will mention a number of additional issues here.

Number 1: The Mass-Sheet Degeneracy

As mentioned several times, the major problem is the mass-sheet degeneracy, which implies that there is always one arbitrary constant that is undetermined from the shear data. Number count depletion can in principle lift this degeneracy, but this magnification effect has been observed in only a few clusters yet, and as mentioned above, this method has its own problems. Employing redshift information of individual source galaxies can also break this degeneracy (Bradac et al. 2004). Note that the mass-sheet degeneracy causes quite different mass profiles to have very similar reduced shear profiles.

Source Redshift Distribution

Since the critical surface mass density $\Sigma_{\rm cr}$ depends on the source redshift, a quantitative interpretation of the weak lensing mass reconstruction requires the knowledge of the redshift distribution of the galaxy sample used for the shear measurements. Those are typically so faint (and numerous) that it is infeasible to obtain individual spectroscopic redshifts for them. There are several ways to deal with this issue: probably the best is to obtain multi-color photometry of the fields and employ photometric redshift techniques (e.g. Connolly et al. 1995; Benítez 2000; Bolzonella et al. 2000). In order for them to be accurate, the number of bands needs to be fairly large; in addition, since much of the background galaxy population is situated at redshifts above unity, one requires near-IR images, as optical photometry alone cannot be used for photometric redshifts above $z \gtrsim 1.3$ (where the 4000 Å-break is redshifted out of the optical window). The problem with near-IR photometry is, however, that currently near-IR cameras have a substantially smaller field-of-view than optical cameras; in addition, due to the much higher sky brightness for ground-based near-IR observations, they extend to brighter flux limits (or smaller galaxy number densities) than optical images, for the same observing time. Nevertheless, upcoming wide-field near-IR cameras, such as theVISTA project on Paranal or WIRCAM at the CFHT, will bring great progress in this direction.

The alternative to individual redshift estimates of background galaxies is to use the redshift distribution obtained through spectroscopic (or detailed photometric redshift) surveys in other fields, and identify this with the faint background galaxy population at the same magnitude. In this way, the redshift distribution of the galaxies can be estimated. The issues that need to be considered here is that neither the targets for a spectroscopic survey, nor the galaxy population from which the shear is estimated, are strictly magnitude selected. Very small galaxies, for example, cannot be used for a shear estimate (or are heavily downweighted) owing to their large smearing corrections from the PSF. Similarly, for low-surface brightness galaxies it is much harder to determine a spectroscopic redshift. Hence, in these redshift identifications, care needs to be excersized.

For cluster mass reconstructions, the physical mass scale is obtained from the average $\beta := \langle D_{\rm ds}/D_{\rm s} \rangle$ over all source galaxies. This average is fairly insensitive to the detailed redshift distribution, as long as the mean source redshift is substantially larger than the lens redshift. This is typically the case for low-redshift ($z \lesssim 0.3$) clusters. However, for higher-redshift lenses, determining β requires a good knowledge of the galaxy redshift distribution.

Contamination of the Source Sample

Next on the list is the contamination of the galaxy sample from which the shear is measured by cluster galaxies; a fraction of the faint galaxies will be foreground objects or faint cluster members. Whereas the foreground population is automatically taken into account in the normal lensing analysis (i.e., in determining β), the cluster members constitute an additional population of galaxies which is not included in the statistical redshift distribution. The galaxy sample used for the shear measurement is usually chosen as to be substantially fainter than the brighter cluster member galaxies; however, the abundance of dwarf galaxies in clusters (or equivalently, the shape of the cluster galaxy luminosity function) is not well known, and may vary substantially from cluster to cluster (e.g., Trentham and Tully 2002, and references therein). Including cluster members in the population from which the shear is measured weakens the lensing signal, since they are not sheared. As a consequence, a smaller shear is measured, and a lower cluster mass is derived. In addition, the dwarf contamination varies as a function of distance from the cluster center, so that the shape of the mass distribution will be affected. Color selection of faint galaxies can help in the selection of background galaxies, i.e., to obtain a cleaner set of true background galaxies. Of course, cluster dwarfs, if not properly accounted for, will also affect the magnification method. One method to deal with this problem is to use only galaxies redder than the Red Cluster Sequence of the cluster galaxies in the color-magnitude diagram, as this sequence indicates the reddest galaxies at the corresponding redshift.

Accuracy of Mass Determination via Weak Lensing

Comparing the 'true' mass of a cluster with that measured by weak lensing is not trivial, as one has to define what the true mass of a cluster is. Using clusters from numerical simulations, the mass is defined as the mass inside a sphere of radius r_{200} around the cluster center within which the overdensity is 200 times the critical density of the universe at the redshift considered. When comparing this mass with the projected mass inside a circle of radius $R = r_{200}$, one should not be surprised that the latter is larger (Metzler et al. 2001), since one compares apples (the mass inside a sphere) with oranges (the mass within a cylinder). Metzler et al. ascribed this to the mass in dark matter filaments at the intersection of which massive clusters are located, but it is most likely mainly an effect of the mass definitions.

The mass-sheet degeneracy tell us there is little hope to measure the 'total' mass of a cluster without further assumptions. Therefore, one natural strategy is to assume a parameterized mass profile and see how accurately one can determine these parameters. The effect of ellipticity noise has already been described in Sect. 5.4. Using simulated clusters, Clowe et al. (2004a) have studied the effect of asphericity and substructure of clusters on these mass parameters, by analyzing the shear field obtained from independent projection of the clusters. They find that the non-spherical mass distribution and substructure induce uncertainties in the two parameters (r_{200} and the concentration c) of an NWF profile which are larger than those from the ellipticity noise under very good observing conditions. Among different projections of the same cluster, the value of r_{200} has a spread of 10–15%, corresponding to a spread in virial mass of ~40 %. Averaging over the different projections, they find that there is little bias in the mass determination, except for clusters with very large ellipticity.

Lensing by the Large-Scale Structure

Lensing by foreground and background density inhomogeneities (i.e., the LSS), yields a fundamental limit to the accuracy of cluster mass estimates. Since lensing probes the projected density, these foreground and background inhomogeneities are present in the lensing signal. Hoekstra (2003) has investigated this effect in the determination of the parameters of an NFW mass profile; we shall return to this issue in Sect. 9.2 below when we consider lensing by the large-scale structure. In principle, the foreground and background contributions can be eliminated if the individual redshifts of the source galaxies are known, since in this case a three-dimensional mass reconstruction becomes possible (see Sect. 7.6); however, the resulting cluster mass map will be very noisy.

5.6 Results

After the first detection of a coherent alignment of galaxy images in two clusters by Tyson et al. (1990) and the development of the Kaiser and Squires (1993) mass reconstruction method, the cluster MS 1224+20 was the first for which a mass map was obtained (Fahlman et al. 1994). This investigation of the X-ray selected cluster yielded a mass map centered on the X-ray centroid of the cluster, but also a surprisingly high M/L-ratio of $\sim 800\,h$ (here and in following we quote mass-to-light ratios always in Solar units). This high M/L ratio has later been confirmed in an independent analysis by Fischer (1999). This mass estimate is in strong conflict with that obtained from a virial analysis (Carlberg et al. 1994); however, it is known that this cluster has a very complex structure, is not relaxed, and most likely a superposition of galaxy concentrations in redshift.

Since this pioneering work,mass reconstructions of many clusters have been performed; see Mellier (1999) and Sect. 5.4 of BS. Here, only a few recent results shall be mentioned, followed by a summary.

Wide-Field Mass Reconstructions

The advent of large mosaic CCD cameras provides an opportunity to map large regions around clusters to be used for a mass reconstruction, and thus to measure the shear profile out to the virial radius of clusters. These large-scale observations offer the best promise to investigate the outer slope of the mass profile, and in particular distinguish between isothermal distributions

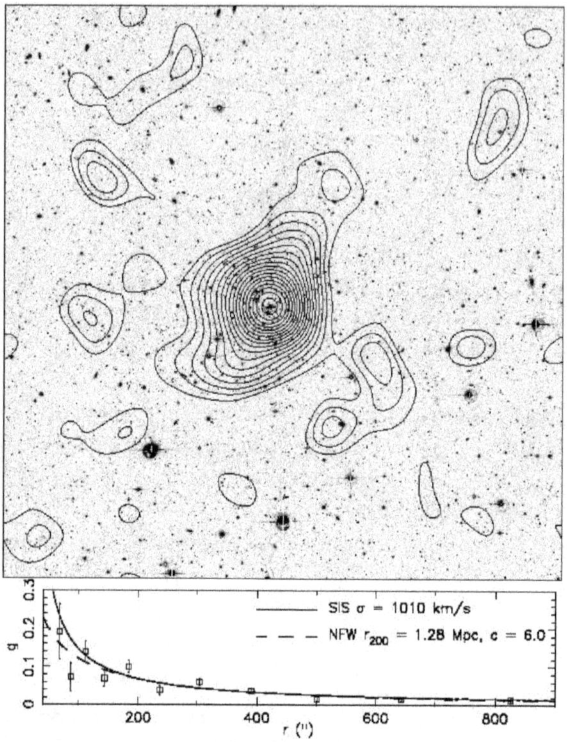

Fig. 20. Contours show the mass reconstruction of the cluster A 1689, obtained from data taken with the WFI at the ESO/MPG 2.2m telescope. The image is $\sim 33'$ on a side, corresponding to $\sim 4.3\,h^{-1}$ Mpc at the cluster redshift of $z_{\rm d} = 0.18$. In the lower panel, the reduced shear profile is shown, together with the best fitting SIS and NFW models. The mass reconstruction has been smoothed by a $1\rlap{.}'15$ Gaussian, and contour spacing is $\Delta\kappa = 0.01$. No corrections have been applied to account for contamination of the lensing signal by cluster dwarf galaxies – that would increase the mass of the best fit models by ~ 25 % (taken from Clowe and Schneider 2001)

and those following the NWF profile. Figure 20 shows an example of such a mass reconstruction, that of the cluster Abell 1689 with $z_{\rm d} = 0.182$. A significant shear is observed out to the virial radius. The mass peak is centered on the brightest cluster galaxy, and the overall lens signal is significant at the 13.4-σ level. The shear signal is fit with two models, as shown in the lower panel of Fig. 20; the NWF profile yields a better fit than an SIS profile. Two more clusters observed with the WFI by Clowe and Schneider (2002) yield similar results, i.e., a detection of the lensing signal out to the virial radius, and a preference for an NWF mass profile, although in one of the two cases this preference is marginal. The lensing signal of such rich clusters could be contaminated by faint cluster member galaxies; correcting for this effect would increase the estimate of the lensing strength, but requires multi-color imaging for source selection.

The cluster A 1689 is (one of) the strongest lensing clusters known (see Fig. 14); in fact, it is strong enough so that a weak lensing signal can be significantly detected from near-IR images (King et al. 2002a) despite the fact that the usable number density of (background) galaxies is only $\sim 3\,\mathrm{arcmin}^{-2}$. The estimate of its velocity dispersion from weak lensing yields an Einstein radius well below the distance of the giant arcs from the cluster center. Hence, in this cluster we see a discrepancy between the strong and weak lensing results, which cannot be easily explained by redshift differences between the arc sources and the mean redshift of the faint galaxies used for the weak lensing analysis. On the other hand, A 1689 is known to be not a relaxed cluster, due to the redshift distribution of its member galaxies. This may explain the fact that the weak lensing mass estimates is also lower than that obtained from X-ray studies.

Filaments Between Clusters

One of the predictions of CDM models for structure formation is that clusters of galaxies are located at the intersection points of filaments formed by the dark matter distribution. In particular, this implies that a physical pair of clusters should be connected by a bridge or filament of (dark) matter, and weak lensing mass reconstructions can in principle be used to search for them. In the investigation of the $z = 0.42$ supercluster MS 0302+17, Kaiser et al. (1998) found an indication of a possible filament connecting two of the three clusters, with the caveat (as pointed out by the authors) that the filament lies just along the boundary of two CCD chips; in fact, an indepedent analysis of this supercluster (Gavazzi et al. 2004) failed to confirm this filament. Gray et al. (2002) saw evidence for a filament connecting the two clusters A 901A/901B in their mass reconstruction of the A 901/902 supercluster field. Another potential filament has been found in the wide-field mass reconstruction of the field containing the pair of clusters A 222/223 (Fig. 21, Dietrich et al. 2004). Spectroscopy shows that there are also galaxies at the same redshift as the two clusters present in the 'filament' (Dietrich et al. 2002).

One of the problems related to the unambiguous detection of filaments is the difficulty to define what a 'filament' is, i.e. to device a statistics to quantify the presence of a mass bridge. The eye easily picks up a pattern and identifies it as a 'filament', but quantifying such a pattern turns out to be very difficult, as shown by Dietrich et al. (2004). Because of that, it is difficult to distinguish between noise in the mass maps, the 'elliptical' extension of two clusters pointing toward each other, and a true filament. However, this problem is not specific to the weak investigation: even if the true projected mass distribution of a pair of clusters were known (e.g., from a cluster pair in numerical simulations), it is not straightforward to define what a filament would be.

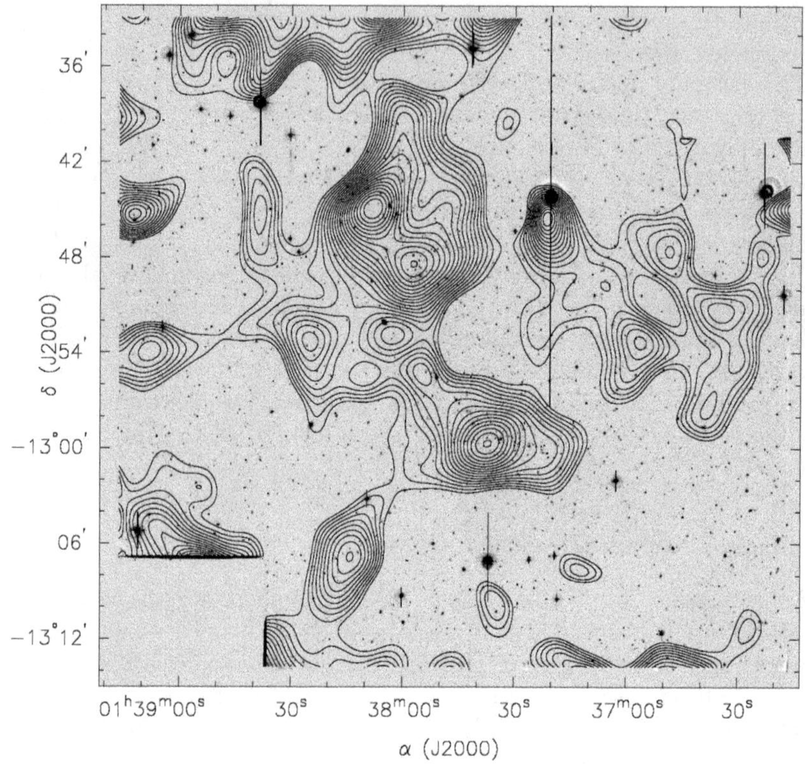

Fig. 21. A deep R-band image of the cluster pair Abell 222/223, obtained from two different pointings with the WFI@ESO/MPG 2.2m, with contours showing the reconstructed κ-map. The two clusters are in the region where the pointings overlap and thus deep imaging is available there. Both clusters are obviously detected in the mass map, with A223 (the Northern one) clearly split up into two subclusters. The mass reconstruction shows a connection between the two clusters which can be interpreted as a filament; galaxies at the clusters' redshift are present in this inter-cluster region. A further mass concentration is seen about 13′ to the South-East of A222, which is significant at the 3.5σ level and where a clear concentration of galaxies is visible. A possible red cluster sequence indicates a substantially higher redshift for this cluster, compared to $z \approx 0.21$ of the double cluster (from Dietrich et al. 2004)

Correlation Between Mass and Light

Mass reconstructions on wide-fields, particularly those covering supercluster regions, are ideally suited to investigate the relation between mass and galaxy light. For example, a smoothed light map of the color-selected early-type galaxies can be correlated with the reconstructed κ-map; alternatively, assuming that light traces mass, the expected shear map can be predicted from the early-type galaxies and compared to the observed shear, with the mass-to-light ratio being the essential fit parameter. Such studies have been carried out on the aforementioned supercluster fields, as well as on blank fields (Wilson et al. 2001). These studies yield very consistent results, in that the mass of clusters is very well traced by the distribution of early-type galaxies, but the mass-to-light ratio seems to vary between different fields, with $\sim 400h$ (in solar units) for the 0302 supercluster (Gavazzi et al. 2004), $\sim 200h$ for the A 901/902 supercluster (Gray et al. 2002), and $\sim 300h$ for empty fields (Wilson et al. 2001) in the rest-frame B-band. When one looks in more detail at these supercluster fields, interesting additional complications appear. The three clusters in the 0302 field, as well as the three clusters in the A 901/902 field (A 901 is indeed a pair of clusters) have quite different properties. In terms of number density of color-selected galaxies, A 901a and A 902 dominate the field, whereas only A 901b seems to be detected inX-rays. Considering early-type galaxies' luminosity, A 901a is the most prominent of the three clusters. In contrast to this, A 902 seems to be most massive as judged from the weak lensing reconstruction. Similar differences between the three clusters in the 0302 field are also seen. It therefore appears that the mass-to-light properties of clusters cover quite a range.

Cluster Mass Reconstructions from Space

The exquisite image quality that can be achieved with the HST – imaging without the blurring effects of atmospheric seeing – suggests that such data would be ideal for weak lensing studies. This is indeed partly true: from space, the shape of smaller galaxy images can be measured than from the ground where the size of the seeing disk limits the image size of galaxies that can be used for ellipticity measurements in practice. Figure 22 shows an HST image of the cluster A 851 ($z_\mathrm{d} = 0.41$), together with a mass reconstruction. The agreement between the mass distribution and the angular distribution of bright cluster galaxies is striking. A detailed X-ray observation of this cluster with XMM-Newton (De Filippis et al. 2003) finds two extended X-ray components coinciding with the two maxima of the bright galaxy distribution, and thus of the mass map shown in Fig. 22, in addition to several compact X-ray sources inside the HST field. Clearly, this cluster is a dynamically young system, as also seen by the inhomogeneities of the X-ray temperature and metallicity of the intracluster gas.

The drawback of cluster weak lensing studies with the HST is the small field-of-view of its WFPC2 camera, which precludes imaging of large regions around the cluster center. To compensate for this, one can use multiple pointings to tile a cluster. For example, Hoekstra and collaborators have observed three X-ray selected clusters with HST mosaics; the results from this survey are summarized in Hoekstra et al. (2002d). One example is shown in Fig. 23, the high-redshift cluster MS 1054$-$03 at $z_\mathrm{d} = 0.83$. Also in this cluster one detects clear substructure, here consisting of three mass peaks, which is matched by the distribution of bright cluster galaxies. The shape of the

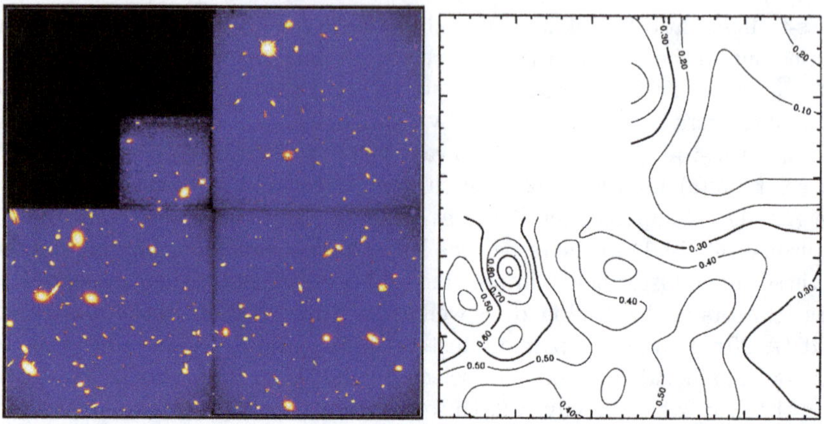

Fig. 22. The left panel shows an WFPC2@HST image of the cluster Cl0939+4713 (=Abell 851; taken from Seitz et al. 1996; the field is about 2.′5 on a side), whereas the right panel shows a mass reconstruction obtained by Geiger and Schneider (1999); this was obtained using the entropy-regularized maximum likelihood method of Seitz et al. (1998). One notices the increased spatial resolution of the resulting mass map near the center of the cluster, which this method yields 'automatically' in those regions where the shear signal is large. Indeed, this mass map predicts that the cluster is critical in the central part, in agreement with the finding of Trager et al. (1997) that strong lensing features (multiple images plus an arc) of sources with $z \sim 4$ are seen there. The strong correlation between the distribution of mass and that of the bright cluster galaxies is obvious: Not only does the peak of the mass distribution coincide with the light center of the cluster, but also a secondary maximum in the surface mass density corresponds to a galaxy concentration (seen in the lower middle), as well as a pronounced minimum on the left where hardly any bright galaxies are visible

mass maps indicates that this cluster is not relaxed, but perhaps in a later stage of merging, a view also supported by its hot X-ray temperature. In fact, new observations with Chandra and XMM-Newton of MS 1054 have shown that this cluster has a much lower temperature than measured earlier with ASCA (Gioia et al. 2004). Only two of the three components seen in the galaxy distribution and the mass reconstruction are seen in X-rays, with the central weak lensing component being the dominant X-ray source. The newly determined X-ray temperature is consistent with the velocity dispersion of cluster galaxies.

Magnification Effects

As mentioned in Sect. 2.4, the magnification of a lens can also be used to reconstruct its surface mass density (Broadhurst et al. 1995). Provided a population of background source galaxies is identified whose number count slope α – see (26) - differs significantly from unity, local counts of these sources

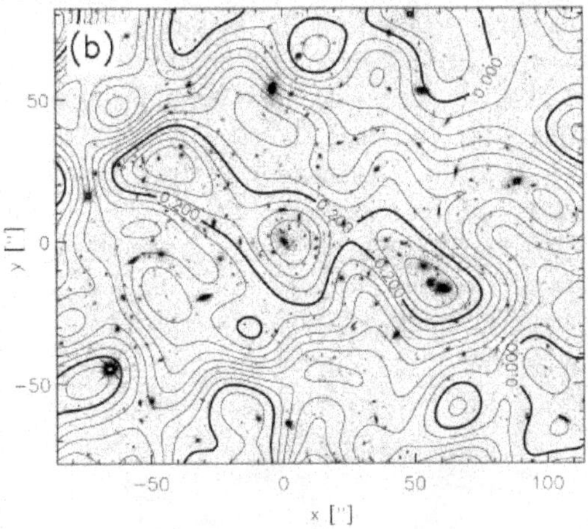

Fig. 23. Mass reconstruction (contours) of the inner part of the high-redshift ($z_{\rm d} = 0.83$) cluster MS 1054−03, based on a mosaic of six pointings obtained with the WFPC2@HST (from Hoekstra et al. 2000). The splitting of the cluster core into three subcomponents, also previously seen from ground-based images by Clowe et al. (2000), shows that this cluster is not yet relaxed

can be turned into an estimator of the local magnification. If the lens is weak, (27) provides a relation between the local number counts and the local surface mass density. If the lens is not weak, this relation no longer suffices, but one needs to use the full expression

$$|\mu|^{-1} = \left|(1-\kappa)^2 - |\gamma|^2\right| \;, \tag{65}$$

where we have written absolute values to account for the fact that the sign of the magnification cannot be observed. There are two obvious difficulties with (65): the first comes from the sign ambiguities, namely whether μ is positive or negative, and whether $\kappa < 1$ or >1. Assuming that we are in the region of the cluster where $\mu > 0$ and $\kappa < 1$ (that is, outside the outer critical curve), then (65) can be rewritten as

$$\kappa = 1 - \sqrt{\mu^{-1} + |\gamma|^2} \;, \tag{66}$$

which shows the second difficulty: in order to estimate κ from μ, one needs to know the shear magnitude $|\gamma|$.

There are various ways to deal with this second problem. Consider first the case that the (reduced) shear is also observed, in which case one better writes

$$\kappa = 1 - \left[\mu\left(1 - |g|^2\right)\right]^{-1/2} \;; \tag{67}$$

but of course, if shear measurements are available, they should be combined with magnification observations in a more optimized way. A second method, using magnification only, is based on the fact that γ depends linearly on κ (ignoring finite-field problems here), and so (66) can be turned into a quadratic equation for the κ field (Dye and Taylor 1998). From numerical models of clusters, van Kampen (1998) claimed that the shear in these clusters approximately follows on average a relation of the form $|\gamma| = (1-c)\sqrt{\kappa/c}$, with $c \sim 0.7$; however, there is (as expected) large scatter around this mean relation which by itself has little theoretical justification. Figure 24 shows the mass reconstruction of the cluster Cl 0024+17 using galaxy number counts and the two reconstruction methods just mentioned.

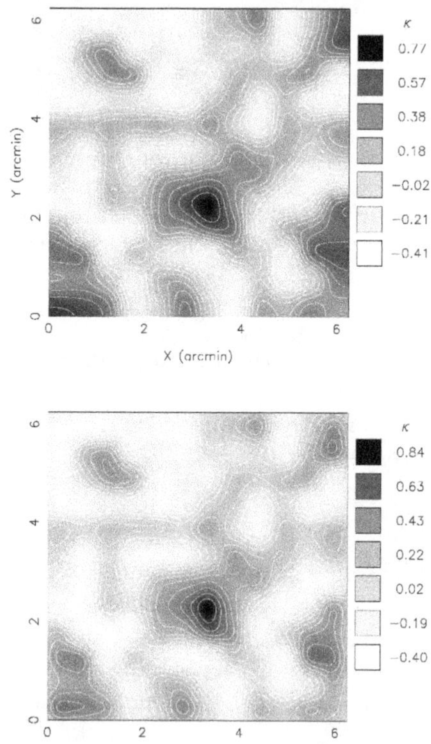

Fig. 24. Mass reconstruction of the cluster Cl 0024+17 from the magnification method. The two different reconstructions are based on two different ways to turn the magnification signal – number count depletions – into a surface mass density mass, as described in the text: in the upper panel, a local relation between surface mass density and shear magnitude has been used, whereas in the lower panel, the magnification was transformed into a κ map using the (non-local) quadratic dependence of the inverse magnification on the surface mass density field. Overall, these two reconstructions agree very well. To account for the presence of bright foreground galaxies, the data field had to be masked before local number densities of background galaxies were estimated – the mask is shown in Fig. 25 (from Dye et al. 2002)

Magnification effects have been observed for a few clusters, most noticeably Cl 0024+17 (Fort et al. 1997; Rögnvaldsson et al. 2001; Dye et al. 2002) and A 1689 (Taylor et al. 1998; Dye et al. 2001). We shall describe some of the results obtained for Cl 0024+17 as an example (Dye et al. 2002). Since the cluster galaxies generate a local overdensity of galaxy counts, they need to be removed first, which can be done based on a color and magnitude criterion. Comparison with extensive spectroscopy of this cluster (Czoske et al. 2001) shows that this selection is very effective for the brighter objects. For the fainter galaxies – those from which the lensing signal is actually measured – a statistical subtraction of foreground and cluster galaxies needs to be performed, which is done by subtracting galaxies according to the field luminosity function with $z < z_d$ and cluster galaxies according to the cluster luminosity function. The latter is based on the assumption that the luminosity distribution of cluster galaxies is independent from the distance to the cluster center. Next, the field of the cluster needs to be masked for bright objects, near which the photometry of fainter galaxies becomes inaccurate or impossible; Fig. 25 shows the masked data field. The number density of sources is then determined from the unmasked area. The resulting mass reconstruction is shown in Fig. 24. The results confirm the earlier finding from strong lensing (see Sect. 4.4) that the mass in the inner part of this cluster is larger by a factor ~3 than estimated from its X-ray emission (Soucail et al. 2000).

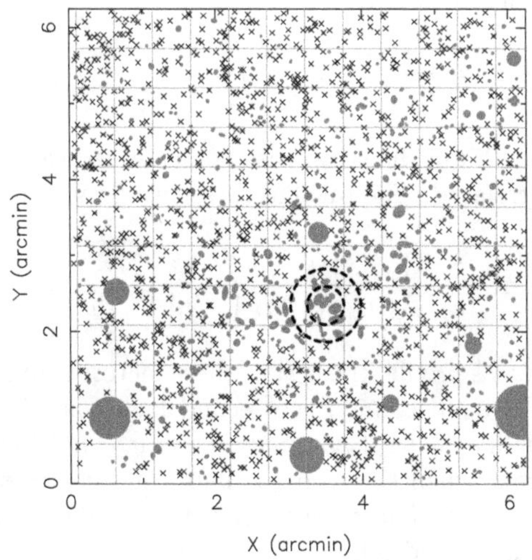

Fig. 25. The mask of the data field of the cluster Cl 0024+17 (*grey circles*) and the location of putative background objects (*crosses*). The inner dashed circle shows the critical curve of the cluster as derived from the multiply imaged arc system (from Dye et al. 2002)

Magnification and Shear Method Compared

It is interesting to consider the relative merits of shear and magnification methods for weak lensing studies. The number of clusters that have been investigated with either method are quite different, with less than a handful for which the magnification effect has been seen. The reason for this is multifold. First, the shear method does not need external calibration, as it is based on the assumption of random source ellipticity; in contrast to this, the magnification method requires the number counts of unlensed sources. Whereas this can be obtained from the same dataset, provided it covers a sufficiently large area, this self-calibration removes one of the strongest appeals of the magnification effect, namely its potential to break the mass-sheet degeneracy. Second, the magnification method is affected by the angular correlation of galaxies, as clearly demonstrated by Athreya et al. (2002) in their study of the clusterMS 1008−1224, where the background number counts revealed the presence of a background cluster which, if not cut out of the data, would contaminate the resulting mass profile substantially. Third, the removal of foreground galaxies, and more seriously, of faint cluster members introduces an uncertainty in the results which is difficult to control. Finally, the number count method yields a lower lensing signal-to-noise than the shear method: If we consider N_γ and N_μ galaxies in a given patch of the sky, such that for the former ones the ellipticities have been measured, and for the latter ones accurate photometry is available and the galaxies are above the photometric completeness brightness, the signal-to-noise ratio from the shear – see (15) – and number count methods are

$$\left(\frac{\mathrm{S}}{\mathrm{N}}\right)_\gamma = \frac{|\gamma|}{\sigma_\epsilon}\sqrt{N_\gamma}\;;\;\left(\frac{\mathrm{S}}{\mathrm{N}}\right)_\mu = 2\kappa|\alpha-1|\sqrt{N_\mu}\;, \tag{68}$$

where we employed (27) in the latter case and assumed that the source galaxy positions are uncorrelated. The ratio of these two S/N values is

$$\frac{(\mathrm{S/N})_\gamma}{(\mathrm{S/N})_\mu} = \frac{|\gamma|}{\kappa}\,\frac{1}{2\sigma_\epsilon|1-\alpha|}\sqrt{\frac{N_\gamma}{N_\mu}}\;. \tag{69}$$

For an isothermal mass profile, the first of these factors is unity. With $\sigma_\epsilon \approx 0.4$ and $\alpha \approx 0.75$ for R-band counts, the second factor is $\sim$5. The final factor depends on the quality of the data: in good seeing conditions, this ratio is of order unity. However, when the seeing is bad, the photometric completeness level can be considerably fainter than the magnitude for which the shape of galaxies can be measured reliably. Therefore, for data with relatively bad seeing, the magnification effect may provide a competitive means to extract weak lensing information. Having said all of this, the magnification method will keep its position as an alternative to shear measurements, in particular for future multi-color datasets where the separation of foreground and cluster galaxies from the background population can be made more cleanly.

Summary

The mass reconstruction of clusters using weak lensing has by now become routine; quite a few cameras at excellent sites yield data with sub-arcsecond image quality to enable this kind of work. Overall, the reconstructions have shown that the projected mass distribution is quite similar to that of the projected galaxy distribution and the shape of the X-ray emission, at least for clusters that appear relaxed. There is no strong evidence for a discrepancy between the mass obtained from weak lensing and that from X-rays, again with exceptions like for Cl0024+16 mentioned above (which most likely is not a single cluster). The weak lensing mass profiles are considered more reliable than the ones obtained from X-ray studies, since they do not rely on symmetry or equilibrium assumptions. On the other hand, they contain contributions from foreground and background mass inhomogeneities, and are affected by the mass-sheet degeneracy. What is still lacking is a combined analysis of clusters, making use of weak lensing, X-ray, Sunyaev–Zeldovich, and galaxy dynamics measurements, although promising first attempts have been published (e.g., Zaroubi et al. 1998, 2001; Reblinsky 2000; Doré et al. 2001; Marshall et al. 2003).

5.7 Aperture Mass and Other Aperture Measures

In the weak lensing regime, $\kappa \ll 1$, the mass-sheet degeneracy corresponds to adding a uniform surface mass density κ_0. However, one can define quantities in terms of the surface mass density which are invariant under this transformation. In addition, several of these quantities can be determined directly in terms of the locally measured shear. In this section we shall present the basic properties of the aperture measures, whereas in the following section we shall demonstrate how the aperture mass can be used to find mass concentrations based solely on their weak lensing properties.

Aperture Mass

Let $U\left(|\boldsymbol{\theta}|\right)$ be a compensated weight (or filter) function, meaning $\int \mathrm{d}\theta\, \theta\, U(\theta) = 0$, then the *aperture mass*

$$M_{\mathrm{ap}}(\boldsymbol{\theta}_0) = \int \mathrm{d}^2\theta\, \kappa(\boldsymbol{\theta})\, U(|\boldsymbol{\theta} - \boldsymbol{\theta}_0|) \tag{70}$$

is independent of κ_0, as can be easily seen. For example, if U has the shape of a Mexican hat, M_{ap} will have a maximum if the filter center is centered on a mass concentration. The important point to notice is that M_{ap} can be written directly in terms of the shear (Kaiser et al. 1994; Schneider 1996)

$$M_{\mathrm{ap}}(\boldsymbol{\theta}_0) = \int \mathrm{d}^2\theta\, Q(|\boldsymbol{\theta}|)\, \gamma_{\mathrm{t}}(\boldsymbol{\theta};\boldsymbol{\theta}_0)\,, \tag{71}$$

where we have defined the *tangential component* γ_t of the shear relative to the point $\boldsymbol{\theta}_0$ [cf. 17], and

$$Q(\theta) = \frac{2}{\theta^2}\int_0^\theta d\theta'\,\theta'\,U(\theta') - U(\theta)\,. \tag{72}$$

These relations can be derived from (54), by rewriting the partial derivatives in polar coordinates and subsequent integration by parts (see Schneider and Bartelmann 1997); it can also be derived directly from the Kaiser and Squires inversion formula (44), as shown in Schneider (1996). Perhaps easiest is the following derivation (Squires and Kaiser 1996): We first rewrite (70) as

$$\begin{aligned} M_{ap} &= 2\pi\int_0^{\theta_u} d\vartheta\,\vartheta\,U(\vartheta)\,\langle\kappa(\vartheta)\rangle \\ &= 2\pi\left[X(\vartheta)\,\langle\kappa(\vartheta)\rangle\right]_0^{\theta_u} - 2\pi\int_0^{\theta_u} d\vartheta\,X(\vartheta)\,\frac{d\,\langle\kappa\rangle}{d\vartheta}\,, \end{aligned} \tag{73}$$

where θ_u is the radius of the aperture, and we have defined

$$X(\theta) = \int_0^\theta d\vartheta\,\vartheta\,U(\vartheta)\,.$$

This definition and the compensated nature of U implies that the boundary terms in (73) vanish. Making use of (24), one finds that

$$\frac{d\,\langle\kappa\rangle}{d\vartheta} = \frac{d\bar\kappa}{d\vartheta} - \frac{d\,\langle\gamma_t\rangle}{d\vartheta} = -\frac{2}{\vartheta}\,\langle\gamma_t\rangle - \frac{d\,\langle\gamma_t\rangle}{d\vartheta}\,,$$

where we used (23) and (24) to obtain $d\bar\kappa/d\vartheta = -2\,\langle\gamma_t\rangle/\vartheta$. Inserting the foregoing equation into (73), one obtains

$$\begin{aligned} M_{ap} &= 2\pi\int_0^{\theta_u} d\vartheta\,\vartheta\,\frac{2X(\vartheta)}{\vartheta^2}\,\langle\gamma_t(\vartheta)\rangle \\ &\quad + 2\pi\left[X(\vartheta)\,\langle\gamma_t(\vartheta)\rangle\right]_0^{\theta_u} - 2\pi\int_0^{\theta_u} d\vartheta\,\frac{dX}{d\vartheta}\,\langle\gamma_t(\vartheta)\rangle\,. \end{aligned} \tag{74}$$

The boundary term again vanishes, and one sees that the last equation has the form of (71), with the weight function $Q = 2X/\vartheta^2 - U$, reproducing (72).

We shall now consider a few properties of the aperture mass, which follow directly from (72).

- If U has finite support, then Q has finite support, which is due to the compensated nature of U. This implies that the aperture mass can be calculated on a finite data field, i.e., from the shear in the same circle where $U \neq 0$.
- If $U(\theta) = \text{const.}$ for $0 \le \theta \le \theta_{in}$, then $Q(\theta) = 0$ for the same interval, as is seen directly from (72). Therefore, the strong lensing regime (where γ deviates appreciably from g) can be avoided by properly choosing U (and Q).

- If $U(\theta) = (\pi\theta_{\rm in}^2)^{-1}$ for $0 \le \theta \le \theta_{\rm in}$, $U(\theta) = -[\pi(\theta_{\rm out}^2 - \theta_{\rm in}^2)]^{-1}$ for $\theta_{\rm in} < \theta \le \theta_{\rm out}$, and $U = 0$ for $\theta > \theta_{\rm out}$, then $Q(\theta) = \theta_{\rm out}^2\,\theta^{-2}\left[\pi(\theta_{\rm out}^2 - \vartheta_{\rm in}^2)\right]^{-1}$ for $\theta_{\rm in} \le \theta \le \theta_{\rm out}$, and $Q(\theta) = 0$ otherwise. For this special choice of U,

$$M_{\rm ap} = \bar\kappa(\theta_{\rm in}) - \bar\kappa(\theta_{\rm in}, \theta_{\rm out}) \; , \tag{75}$$

  the mean mass density inside $\theta_{\rm in}$ minus the mean density in the annulus $\theta_{\rm in} \le \theta \le \theta_{\rm out}$ (Kaiser 1995). Since the latter is non-negative, this yields a lower limit to $\bar\kappa(\theta_{\rm in})$, and thus to $M(\theta_{\rm in})$.

The aperture mass can be generalized to the case where the weight function U is constant on curves other than circles, e.g., on ellipses, in the sense that the corresponding expressions can be rewritten directly in terms of the shear on a finite region (see Squires and Kaiser 1996 for the case where U is constant on a set of self-similar curves, and Schneider and Bartelmann 1997 for a general set of nested curves). In general, $M_{\rm ap}$ is not a particularly good measure for the total mass of a cluster – since it employs a compensated filter – but it has been specifically designed that way to be immune against the mass-sheet degeneracy. However, $M_{\rm ap}$ is a very convenient measure for mass concentrations (see Sect. 5.8) and, as shown above, yields a robust lower limit on cluster masses.

Aperture Multipoles

The aperture method can also be used to calculate multipoles of the mass distribution: define the multipoles

$$Q^{(n)} := \int {\rm d}^2\theta \; |\boldsymbol\theta|^n \; U(|\boldsymbol\theta|)\, {\rm e}^{n{\rm i}\varphi}\, \kappa(\boldsymbol\theta) \; , \tag{76}$$

then the $Q^{(n)}$ can again be expressed as an integral over the shear. Here, U is a radial weight function for which certain restrictions apply (see Schneider and Bartelmann 1997 for details), but is not required to be compensated for $n > 0$. A few cases of interest are: a weight function U which is non-zero only within an annulus $\theta_{\rm in} \le \theta \le \theta_{\rm out}$ and which continuously goes to zero as $\theta \to \theta_{\rm in,out}$; in this case, the shear is required only within the same annulus. Likewise, if U is constant for $0 \le \theta \le \theta_{\rm in}$ and then decreases smoothly to zero at $\theta_{\rm out}$, only the shear within the annulus is required to calculate the multipoles. Aperture multipoles can be used to calculate the multipole moments of mass concentrations like clusters directly from the shear, i.e., without obtaining first a mass map, which allows a more direct quantification of signal-to-noise properties.

The Cross Aperture

We have seen that the Kaiser and Squires inversion, given by the first expression in (44), must yield a real result; the imaginary part of the integral in (44) vanishes in the absence of noise. Suppose one would multiply the complex shear by $\mathrm{i} = \mathrm{e}^{2\mathrm{i}\pi/4}$; this would transform the real part of the integral into the imaginary part and the imaginary part into the negative of the real part. Geometrically, multiplication by this phase factor corresponds to rotating the shear at every point by 45°. Hence, if all shears are rotated by $\pi/4$, the real part of the Kaiser and Squires inversion formula (44) yields zero. This 45-degree test has been suggested by A. Stebbins; it can be used on real data to test whether typical features in the mass map are significant, as those should have larger amplitude than spurious features obtained from the mass reconstruction in which the shear has been rotated by $\pi/4$ (the corresponding 'mass map' then yields a good indication of the typical noise present in the real mass map).

One can define in analogy to (71) the cross aperture by replacing the tangential component of the shear by its cross component. According to the 45-degree test, the resulting cross aperture should be exactly zero. Hence, if we define for $\boldsymbol{\theta}_0 = \mathbf{0}$

$$M := M_{\rm ap} + \mathrm{i}M_{\perp} = \int \mathrm{d}^2\theta \; Q(|\boldsymbol{\theta}|) \left[\gamma_{\rm t}(\boldsymbol{\theta}) + \mathrm{i}\gamma_{\times}(\boldsymbol{\theta})\right]$$

$$= -\int \mathrm{d}^2\theta \, Q(|\boldsymbol{\theta}|) \, \gamma(\boldsymbol{\theta}) \, \mathrm{e}^{-2\mathrm{i}\phi} \; , \tag{77}$$

where ϕ is the polar angle of $\boldsymbol{\theta}$ as in (17), then M is expected to be purely real. We shall make use of this definition and the interpretation of M in later sections.

5.8 Mass Detection of Clusters

Motivation

If a weak lensing mass reconstruction of a cluster has been performed and a mass peak is seen, it can also be quantified by applying the aperture mass statistics to it: placing the center of the aperture on the mass peak, and choosing the radius of the aperture to match the extent of the mass peak will give a significant positive value of $M_{\rm ap}$. Now consider to observe a random field in the sky, and to determine the shear in this field. Then, one can place apertures on this field and determine $M_{\rm ap}$ at each point. If $M_{\rm ap}$ attains a significant positive value at some point, it then corresponds to a point around which the shear is tangentially oriented. Such shear patterns are generated by mass peaks according to (70) – hence, a significant peak in the $M_{\rm ap}$-map corresponds to a mass concentration (which can, in principle at least, be a mass concentration just in two-dimensional projection, not necessarily in 3D). Hence, the aperture mass statistics allows us to search for mass concentrations on blank fields, using weak lensing methods (Schneider 1996). From the estimate (19), we see that the detectable mass concentrations have to have typical cluster masses.

The reason why this method is interesting is obvious: As discussed in Sect. 6 of IN, the abundance of clusters as a function of mass and redshift is an important cosmological probe. Cosmological simulations are able to predict the abundance of massive halos for a given choice of cosmological parameters. To compare these predictions with observations, cluster samples are analyzed. However, clusters are usually detected either as an overdensity in the galaxy number counts (possibly in connection with color information, to employ the red cluster sequence – see Gladders and Yee 2000), or from extended X-ray sources. In both cases, one makes use of the luminous properties of the clusters, and cosmologists find it much more difficult to predict those, as the physics of the baryonic component of the matter is much harder to handle than the dark matter. Hence, a method for cluster detection that is independent of their luminosity would provide a clean probe of cosmology. From what was said above, the aperture mass provides such method (Schneider 1996).

To illustrate this point, we show in Fig. 26 the projected mass and the corresponding shear field as it results from studying the propagation of light rays through a numerically generated cosmological matter distribution (Jain et al. 2000; we shall return to such simulations in Sect. 6.6). From the comparison of these two panels, one sees that for each large mass concentration there is a tangential shear pattern centered on the mass peak. Thus, a systematic search for such shear patterns can reveal the presence and abundance of peaks in the mass map.

The Method

The search for mass concentrations can thus be carried out by calculating the aperture mass on a grid over the data field and to identify significant peaks. A practical estimator for $M_{\rm ap}$ is obtained by replacing the integral in (71) by a finite sum over image ellipticities:

$$\hat{M}_{\rm ap}(\boldsymbol{\theta}_0) = \frac{1}{n} \sum_i \epsilon_{{\rm t}i}(\boldsymbol{\theta}_0)\, Q(|\boldsymbol{\theta}_i - \boldsymbol{\theta}_0|) \;, \tag{78}$$

where n is the mean number density of galaxy images, and $\epsilon_{{\rm t}i}(\boldsymbol{\theta}_0)$ is the ellipticity component of a galaxy at $\boldsymbol{\theta}_i$ tangent to the center $\boldsymbol{\theta}_0$ of the aperture. This estimator has easy-to-quantify signal-to-noise properties. In the absence of a lensing signal, $\left\langle \hat{M}_{\rm ap} \right\rangle \equiv 0$, and the dispersion of $\hat{M}_{\rm ap}(\boldsymbol{\theta}_0)$ is

$$\sigma^2(\boldsymbol{\theta}_0) = \frac{\sigma_\epsilon^2}{2n^2} \sum_i Q^2(|\boldsymbol{\theta}_i - \boldsymbol{\theta}_0|) \;; \tag{79}$$

hence, the signal-to-noise of $\hat{M}_{\rm ap}(\boldsymbol{\theta}_0)$ is

$$\frac{\rm S}{N} = \frac{\sqrt{2}}{\sigma_\epsilon} \frac{\sum_i \epsilon_{{\rm t}i}(\boldsymbol{\theta}_0)\, Q(|\boldsymbol{\theta}_i - \boldsymbol{\theta}_0|)}{\sqrt{\sum_i Q^2(|\boldsymbol{\theta}_i - \boldsymbol{\theta}_0|)}} \;. \tag{80}$$

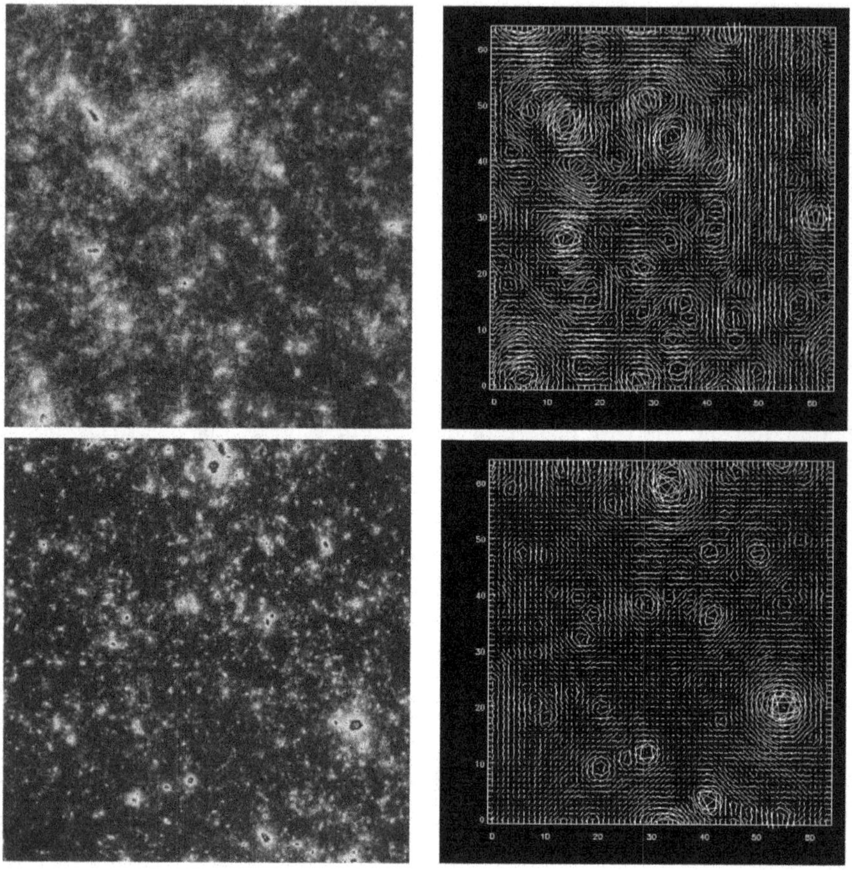

Fig. 26. Projected mass distribution of the large-scale structure (*left*), and the corresponding shear field (*right*), where the length and orientation of the sticks indicate the magnitude and direction of the local shear. The top panels correspond to an Einstein–de Sitter model of the Universe, whereas the bottom panels are for a low density open model. The size of the field is one degree on the side, and the background galaxies are assumed to all lie at the redshift $z_{\rm s} = 1$. Note that each mass concentration seen in the left-hand panels generates a circular shear pattern at this position; this forms the basic picture of the detection of mass concentrations from a weak lensing observation (from Jain et al. 2000)

The noise depends on $\boldsymbol{\theta}_0$, as the image number density can vary of data field. The size (or radius) of the aperture shall be adapted to the mass concentrations excepted: too small aperture radii miss most of the lensing signal of real mass concentrations, but is more susceptible to noise peaks, whereas too large aperture radii include regions of very low signal which may be swamped again by noise. In addition, the shape of the filter function Q can be adapted to the expected mass profiles of mass concentrations; e.g., one can design filters which are particularly sensitive to NFW-like density profiles. In order not to prejudice the findings of a survey, it may be advantageous to use a 'generic' filter function, e.g., of the form

$$U(\vartheta) = \frac{9}{\pi\theta^2}\left(1 - \frac{\vartheta^2}{\theta^2}\right)\left(\frac{1}{3} - \frac{\vartheta^2}{\theta^2}\right) \; ; \; Q(\vartheta) = \frac{6}{\pi\theta^2}\frac{\vartheta^2}{\theta^2}\left(1 - \frac{\vartheta^2}{\theta^2}\right) . \tag{81}$$

The relation between the two expressions for $M_{\rm ap}$ given by (70) and (71) is only valid if the aperture lies fully inside the data field. If it does not, i.e., if the aperture crosses the boundary of the data field, these two expressions are no longer equivalent; nevertheless, the estimator (78) still measures a tangential shear alignment around the aperture center and thus signifies the presence of a mass concentration.

There are superior estimates of the significance of a detected mass peak than using the signal-to-noise ratio (80). One consists in bootstrapping; there one calculates $M_{\rm ap}$ at a given point (where N galaxies are in the aperture) many times by randomly drawing – with replacement – N galaxies and tests how often is signal negative. The fraction of cases with negative values corresponds to the error level of having a positive detection of $M_{\rm ap}$. Alternatively, one can conduct another Monte-Carlo experiment, by randomizing all galaxy image orientations and calculating $M_{\rm ap}$ from these randomized samples, and ask in which fraction of realizations is the value of $M_{\rm ap}$ larger than the measured value ? As the randomized galaxies should show no lensing signal, this fraction is again the probability of getting a value as large as that measured from random galaxy orientations. In fact, from the central limit theorem one expects that the probability distribution of $M_{\rm ap}$ from randomizing the image orientations will be a Gaussian of zero mean, and its dispersion can be calculated directly from (78) to be

$$\sigma^2(\boldsymbol{\theta}_0) = \frac{1}{2n^2}\sum_i |\epsilon_i|^2 \, Q^2(|\boldsymbol{\theta}_i - \boldsymbol{\theta}_0|) \; , \tag{82}$$

which is similar to (79), but accounts for the moduli of the ellipticity of the individual galaxy images.

Both of the aforementioned methods take the true ellipticity distribution of galaxy images into account, and should yield very similar results for the significance. Highly significant peaks signify the presence of a mass concentration, detected solely on the basis of its mass, and therefore, it is a very promising search method for clusters.

There is nothing special about the weight function (81), except mathematical simplicity. It is therefore not clear whether these filter functions are most efficient to detect cluster-mass matter concentrations. In fact, as shown in Schneider (1996), the largest S/N is obtained if the filter function U follows the true mass profile of the lens or, equivalently, if Q follows its radial shear profile. Hennawi and Spergel (2005) and Schirmer (2004) tested a large range of filter functions, including (81), Gaussians, and those approximating an NFW profile. Based on numerical ray-tracing simulations, Hennawi and Spergel conclude that the 'truncated' NFW filter is most efficient for cluster detections; the same conclusion has been achieved by Schirmer (2004) based on wide-field imaging data.

Furthermore, Hennawi and Spergel have complemented their cluster search by a 'tomographic' component, assuming that the source galaxies have (photometric) redshift estimates available. Since the lens strength is a function of source redshift, the expected behavior of the aperture mass signal as a function of estimated source redshift can be used as an additional search criteria. They shown that this additional information increases the sensitivity of weak lensing to find mass concentrations, in particular for higher-redshift ones; in fact, the cluster search by Wittman et al. (described below) has made use of redshift information. As an additional bonus, this method also provides an estimate of the lens redshift.

Results

In the past few years, a number of clusters and/or cluster candidates have been detected by the weak lensing method, and a few of them shall be discussed here. The right-hand panel of Fig. 27 shows the mass reconstruction

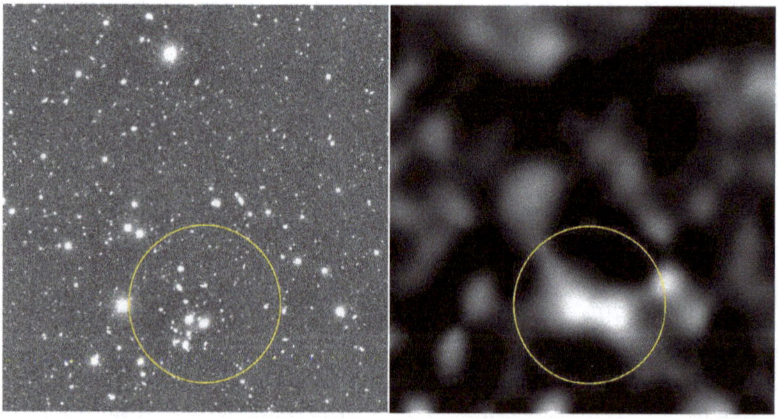

Fig. 27. A cosmic shear survey was carried out with the FORS1 instrument on the VLT (see Maoli et al. 2001 and Sect. 7.1 below). The left panel shows one of the 50 fields observed in the course of this survey, whereas the right panel shows a weak-lensing mass reconstruction of this field. Obviously, a strong mass peak is detected in this reconstruction, indicated by the circle. At the same position, one finds a strong overdensity of relatively bright galaxies on the VLT image; therefore, this mass peak corresponds to a cluster of galaxies. A reanalysis of all 50 VLT fields (Hetterscheidt 2003) yielded no further significant cluster candidate; however, with a field size of only $\sim 6\overset{\prime}{.}5$, detecting clusters in them is difficult unless these are positioned close to the field centers

of one of the 50 FORS1@VLT fields observed in the course of a cosmic shear survey (see Sect. 7.1). This reconstruction shows an obvious mass peak, indicated by a circle. The left panel shows the optical image, and it is obvious that the location of the mass peak coincides with a concentration of bright galaxies – this certainly is a cluster, detected by its weak lensing signal. However, no follow-up observations have been conducted yet to measure its redshift.

Wittman et al. (2001, 2003) reported on the discovery of two clusters from their wide-field weak lensing survey; one of them is shown in Fig. 28 and discussed here. First, a peak in their mass reconstruction was identified which has a significance of 4.5σ. The location of the mass peak is identified with a concentration of red elliptical galaxies, with the two centers separated by about $1'$ (which is about the accuracy with which the centers of mass concentrations are expected to be determined from mass reconstructions). Follow-up spectroscopy confirmed the galaxy concentration to be a cluster at redshift $z_{\rm d} = 0.28$, with a velocity dispersion of $\sigma_v \sim 600\,{\rm km/s}$. Since multi-color photometry data are available, photometric redshift estimates of the faint galaxy population have been obtained, and the tangential shear around the mass peak has been investigated as a function of this estimated redshift. The lens signal rises as the redshift increases, as expected due to the lensing efficiency factor $D_{\rm ds}/D_{\rm s}$. In fact, from the source redshift dependence of the lens signal, the lens redshift can be estimated, and yields a result within $\sim$0.03 of the spectroscopically measured $z_{\rm d}$. Hence, in this case not only can the presence of a cluster be inferred from weak lensing, but at the same time a cluster redshift has been obtained from lensing observations alone. This is one example of using source redshift information to investigate the redshift structure of the lensing matter distribution; we shall return to a more general discussion of this issue in Sect. 7.6.

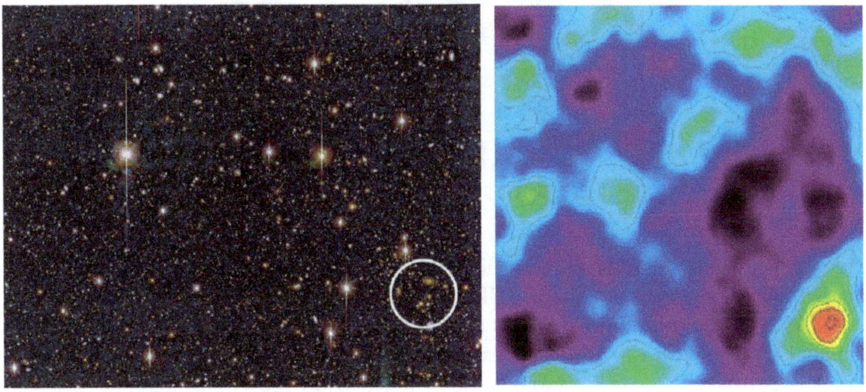

Fig. 28. Left: BTC image of a blank field, right: mass reconstruction, showing the presence of a (mass-selected) cluster near the lower right corner – spectroscopically verified to be at $z = 0.276$ (from Wittman et al. 2001)

In a wide-field imaging weak lensing survey of galaxy clusters, Dahle et al. (2003) detected three significant mass peaks away from the clusters that were targeted. One of these cases is illustrated in Fig. 29, showing the mass reconstruction in the field of the cluster Abell 1705. The mass peak South-West of the cluster coincides with a galaxy concentration at $z \sim 0.55$, as estimated from their color, and an arc is seen near the brightest galaxy of this cluster. A further cluster was detected in the wide-field image of the A222/223 double cluster field (Dietrich et al. 2004) which coincides with an overdensity of galaxies. Hence, by now of order ten cluster-mass matter concentrations have been discovered by weak lensing techniques and verified as genuine clusters from optical photometry and, for some of them, spectroscopy.

Miyazaki et al. (2002) used a 2.1 deg^2 deep image taken with the Suprime-Cam wide-field imager on Subaru to search for mass peaks. They compared their peak statistics with both, the expected peak statistics from a noise field created by intrinsic galaxy ellipticities (Jain and van Waerbeke 2000) as well as from N-body simulations, and found a broader distribution in the actual data. They interpret this as statistical evidence for the presence of mass peaks; however, their interpretation of the significant dips in the mass map as evidence for voids cannot hold, as the density contrast of voids is too small (since the fractional density contrast $\delta > -1$) to be detectable with weak

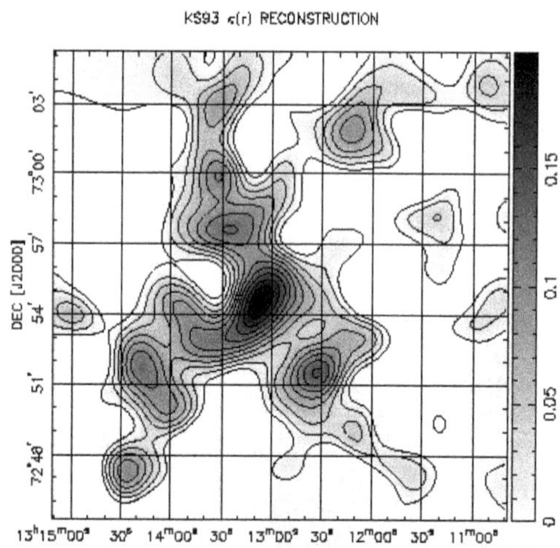

Fig. 29. Shown is the mass reconstruction of the field containing the cluster A 1705, located near the center of this field. The peak $\sim 4'$ to the North-East of A 1705 appears to be associated with galaxies at the same redshift as A 1705. However, the peak $\sim 4'$ South-West of A 1705 seems to be associated with galaxies at considerably larger redshift, at $z \sim 0.55 \pm 0.05$, as determined from the $V - I$ colors of the corresponding galaxy concentration. Indeed, an arc curving around the central galaxy of this newly detected cluster candidate is observed (from Dahle et al. 2003)

lensing. They find a number density of $> 5\sigma$ peaks of about $5\,\mathrm{deg}^{-2}$, well in agreement with predictions from Kruse and Schneider (1999) and Reblinsky et al. (1999). Schirmer (2004) investigated about $16\,\mathrm{deg}^2$ of images taken with the WFI@ESO/MPG 2.2m, and detected $100 > 4\sigma$-peaks, again in good agreement with theoretical expectations.

Dark Clusters?

In addition, however, this method has the potential to discover mass concentrations with very large mass-to-light ratio, i.e., clusters which are very faint optically and which would be missed in more conventional surveys for clusters. Two potential 'dark clusters' have been reported in the literature.[7] Umetsu and Futamase (2000), using the WFPC2 onboard HST detected a highly significant (4.5σ) mass concentration $1\farcm7$ away from the cluster Cl 1604+4304, also without an apparent overdensity of associated galaxies.

In the course of a wide-field weak lensing analysis of the cluster A 1942, Erben et al. (2000) detected a mass peak which, using the aperture mass statistics introduced previously, has been shown to be highly significant ($\sim 4.7\sigma$ on the V-band image), with the significance being obtained from the randomization and bootstrapping techniques described above. An additional I-band image confirmed the presence of a mass peak at the same location as on the V-band image, though with somewhat lower significance. No concentration of galaxies is seen near the location of the mass peak, which indicates that it either is a very dark mass concentration, or a cluster at a fairly high redshift (which, however, would imply an enormous mass for it), or, after all, a statistical fluke. It is important to note that the signal in $M_{\rm ap}$ comes from a range of radii (see Fig. 30); it is not dominated by a few highly flattened galaxies which happen to have a fortuitous orientation. Gray et al. (2000) have used near-IR images to search for a galaxy concentration in this direction, without finding an obvious candidate. Therefore, at present it is unclear whether the 'dark clump' is indeed a very unusual cluster. A low-significance X-ray source near its position, as obtained in a ROSAT observation of Abell 1942, certainly needs confirmation by the more sensitive X-ray observatory XMM.[8] Of course, if there are really dark clusters, their confirmation by methods other than weak lensing would be extremely difficult; but even if we are dealing with a statistical fluke, it would be very important to find the cause for it. An HST mosaic observation of this field has been conducted; a first analysis of these data was able to confirm the findings of Erben et al., in the sense that the shear signal from galaxies seen in both, the HST images and the

---

[7] A third case reported in Miralles et al. (2002) has in the meantime been considerably weakened (Erben et al. 2003).

[8] Judging from the results of several proposal submissions, people on X-ray TACs seem not to care too much about dark cluster candidates.

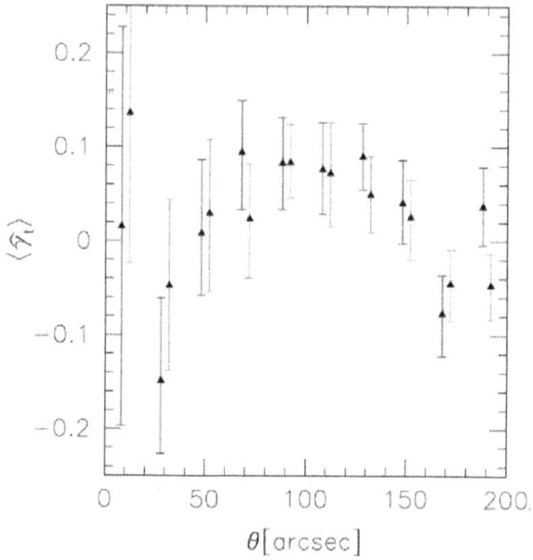

Fig. 30. Tangential shear profile from both (V- and I-band) images around the 'dark cluster' candidate near the cluster A1942. For each angular scales, two points (and corresponding error bars) are plotted, which are derived from two different images of the field in the V- and I-band. It can be seen that the tangential shear signal extends over quite a range in radius (from Erben et al. 2000)

ground-based data, have a significant tangential alignment (von der Linden 2004). However, contrary to expectations if this was truly a lensing mass signal, there is hardly any tangential alignment from fainter galaxies, although they are expected to be located at higher redshift and thus should show a stronger shear signal. However, as a word of caution, the PSF anisotropy of WFPC2 cannot be controlled from stars on the image, owing to the small field-of-view, and no stellar cluster has been observed with the filter with which the dark clump observations were conducted, so that the PSF anisotropy cannot be accurately inferred from such calibration images. The existence of dark clusters would be highly unexpected in view of our current understanding of structure formation and galaxy evolution, and would require revisions of these models.

The search for clusters by weak lensing will certainly continue, due to the novel properties of the cluster samples obtained that way. The observational data required are the same as those used for cosmic shear studies, and several very wide-field surveys are currently conducted, as will be described in Sect. 7. Hence, we can expect to have a sizable sample of shear-selected clusters in the near future. The search for mass concentrations by weak lensing techniques is affected by foreground and background inhomogeneities, which impose fundamental limits on the reliability and completeness of such searches; we shall return to this issue in Sect. 9.2.

Expectations

Kruse and Schneider (1999) have calculated the expected number density of lensing-detected clusters, using the aperture-mass method, for different cosmological parameters; these have been verified in numerical simulations of the large-scale structure by Reblinsky et al. (1999). Depending on the cosmological model, a few clusters per $\deg^2$ should be detected at about the 5σ level. The dependence of the expected number density of detectable mass peaks on the cosmological parameters can be used as a cosmological probe; in particular, Bartelmann et al. (2002) and Weinberg and Kamionkowski (2003) demonstrate that the observed abundance of weak lensing clusters can probe the equation-of-state of the dark energy. Bartelmann et al. (2001) argued that the abundance of weak lensing detected clusters strongly depends on their mass profile, with an order-of-magnitude difference between NFW profiles and isothermal spheres. Weinberg and Kamionkowski (2002) argued, based on the spherical collapse model of cluster formation, that a considerable fraction of such detections are expected to be due to non-virialized mass concentrations, which would then be considerably weaker X-ray emitters and may be candidates for the 'dark clusters'.

6 Cosmic Shear – Lensing by the LSS

Up to now we have considered the lensing effect of localized mass concentrations, like galaxies and clusters. In addition to that, light bundles propagating through the Universe are continuously deflected and distorted by the gravitational field of the inhomogeneous mass distribution, the large-scale structure (LSS) of the cosmic matter field. This distortion of light bundles causes shape and size distortions of images of distant galaxies, and therefore, the statistics of the distortions reflect the statistical properties of the LSS (Gunn 1967; Blandford et al. 1991; Miralda-Escudé 1991; Kaiser 1992).

Cosmic shear deals with the investigation of this connection, from the measurement of the correlated image distortions to the inference of cosmological information from this distortion statistics. As we shall see, cosmic shear has become a very important tool in observational cosmology. From a technical point-of-view, it is quite challenging, first because the distortions are indeed very weak and therefore difficult to measure, and second, in contrast to 'ordinary' lensing, here the light deflection does not occur in a 'lens plane' but by a 3-D matter distribution, implying the need for a different description of the lensing optics. We start by looking at the description of light propagating through the Universe, and then consider the second-order statistical properties of the cosmic shear which reflect the second-order statistical properties of the cosmic matter field, i.e., the power spectrum. Observational results from cosmic shear surveys are presented in Sect. 7, whereas higher-order statistical properties of the shear field will be treated in Sect. 9.

6.1 Light Propagation in an Inhomogeneous Universe

In this brief, but rather technical section, we outline the derivation of the lensing effects of the three-dimensional mass distribution between the faint background galaxy population and us; the reader is referred to Bartelmann and Schneider (2001) for a more detailed discussion. The final result of this consideration has a very simple interpretation: in the lowest-order approximation, the 3-D cosmological mass distribution can be considered, for sources at a single redshift z_s, as an effective surface mass density κ, just like in ordinary lensing. The resulting κ is obtained as a line-of-sight integral of the density contrast $\Delta\rho$, weighted by the usual geometrical factor entering the lens equations.

The laws of light propagation follow from Einstein's General Relativity; according to it, light propagates along the null-geodesics of the space-time metric. As shown in Schneider et al. (1992, hereafter SEF; see also Seitz et al. 1994), one can derive from General Relativity that the governing equation for the propagation of thin light bundles through an arbitrary space-time is the equation of geodesic deviation,

$$\frac{\mathrm{d}^2\boldsymbol{\xi}}{\mathrm{d}\lambda^2} = \mathcal{T}\,\boldsymbol{\xi}\,, \tag{83}$$

where $\boldsymbol{\xi}$ is the separation vector of two neighboring light rays, λ the affine parameter along the central ray of the bundle, and $\mathcal{T}$ is the *optical tidal matrix* which describes the influence of space-time curvature on the propagation of light. $\mathcal{T}$ can be expressed directly in terms of the Riemann curvature tensor (Fig. 31).

For the case of a weakly inhomogeneous Universe, the tidal matrix can be explicitly calculated in terms of the peculiar Newtonian potential. For that, we write the slightly perturbed metric of the Universe in the form

$$\mathrm{d}s^2 = a^2(\tau)\left[\left(1+\frac{2\Phi}{c^2}\right)c^2\mathrm{d}\tau^2 - \left(1-\frac{2\Phi}{c^2}\right)\left(\mathrm{d}w^2 + f_K^2(w)\mathrm{d}\omega^2\right)\right]\,, \tag{84}$$

where w is the comoving radial distance, $a = (1+z)^{-1}$ the scale factor, normalized to unity today, τ is the conformal time, related to the cosmic time

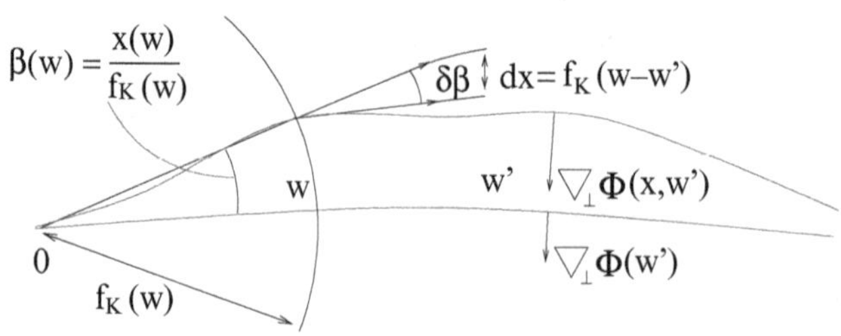

Fig. 31. Illustration of the evolution of the separation between two light rays in a curved space-time (source: T. Schrabback)

t through $\mathrm{d}t = a\,\mathrm{d}\tau$, $f_K(w)$ is the comoving angular diameter distance, which equals w in a spatially flat model, and $\Phi(\mathbf{x}, w)$ denotes the Newtonian peculiar gravitational potential which depends on the comoving position vector $\mathbf{x}$ and cosmic time, here expressed in terms of the comoving distance w (see Sect. 4 of IN for a more detailed description of the various cosmological terms). In this metric, the tidal matrix $\mathcal{T}$ can be calculated in terms of the Newtonian potential Φ, and correspondingly, the equation of geodesic deviation (83) yields the evolution equation for the comoving separation vector $\mathbf{x}(\boldsymbol{\theta}, w)$ between a ray separated by an angle $\boldsymbol{\theta}$ at the observer from a fiducial ray (Fig. 31)

$$\frac{\mathrm{d}^2\mathbf{x}}{\mathrm{d}w^2} + K\,\mathbf{x} = -\frac{2}{c^2}\left[\nabla_\perp \Phi\left(\mathbf{x}(\boldsymbol{\theta}, w), w\right) - \nabla_\perp \Phi^{(0)}\left(w\right)\right], \tag{85}$$

where $K = (H_0/c)^2\,(\Omega_\mathrm{m} + \Omega_\Lambda - 1)$ is the spatial curvature of the Universe, $\nabla_\perp = (\partial/\partial x_1, \partial/\partial x_2)$ is the transverse *comoving* gradient operator, and $\Phi^{(0)}(w)$ is the potential along the fiducial ray.[9] The formal solution of this transport equation is obtained by the method of Green's function, to yield

$$\mathbf{x}(\boldsymbol{\theta}, w) = f_K(w)\boldsymbol{\theta} - \frac{2}{c^2}\int_0^w \mathrm{d}w'\, f_K(w-w')\left[\nabla_\perp \Phi\left(\mathbf{x}(\boldsymbol{\theta}, w'), w'\right) - \nabla_\perp \Phi^{(0)}\left(w'\right)\right]. \tag{86}$$

A source at comoving distance w with comoving separation $\mathbf{x}$ from the fiducial light ray would be seen, in the absence of lensing, at the angular separation $\boldsymbol{\beta} = \mathbf{x}/f_K(w)$ from the fiducial ray (this statement is nothing but the definition of the comoving angular diameter distance). Hence, $\boldsymbol{\beta}$ is the unlensed angular position in the 'comoving source plane' at distance w, where the origin of this source plane is given by the intersection point with the fiducial ray. Therefore, in analogy with standard lens theory, we define the Jacobian matrix

$$\mathcal{A}(\boldsymbol{\theta}, w) = \frac{\partial \boldsymbol{\beta}}{\partial \boldsymbol{\theta}} = \frac{1}{f_K(w)}\frac{\partial \mathbf{x}}{\partial \boldsymbol{\theta}}, \tag{87}$$

and obtain from (86)

$$\mathcal{A}_{ij}(\boldsymbol{\theta}, w) = \delta_{ij} - \frac{2}{c^2}\int_0^w \mathrm{d}w'\,\frac{f_K(w-w')f_K(w')}{f_K(w)}\,\Phi_{,ik}\left(\mathbf{x}(\boldsymbol{\theta}, w'), w'\right)\,\mathcal{A}_{kj}(\boldsymbol{\theta}, w'), \tag{88}$$

which describes the locally linearized mapping introduced by LSS lensing. To derive (88), we noted that $\nabla_\perp\Phi^{(0)}$ does not depend on $\boldsymbol{\theta}$, and used the chain

[9] In some of the literature, this transport equation is written without the term accounting for the potential along the fiducial ray. The idea behind this is to compare a light ray in the inhomogeneous universe with one in the homogeneous, unperturbed universe. Apart from the conceptual difficulty, this 'first-order expansion' is not justified, as the light rays in an inhomogeneous universe can deviate quite significantly from straight rays in the homogeneous reference universe – much more than the length scale of typical density fluctuations. These difficulties are all avoided if one starts from the exact equation of geodesic deviation, as done here.

rule in the derivative of Φ. This equation still is exact in the limit of validity of the weak-field metric. Next, we expand $\mathcal{A}$ in powers of Φ, and truncate the series after the linear term:

$$\mathcal{A}_{ij}(\boldsymbol{\theta}, w) = \delta_{ij} - \frac{2}{c^2} \int_0^w \mathrm{d}w' \, \frac{f_K(w-w') f_K(w')}{f_K(w)} \, \Phi_{,ij}\left(f_K(w')\boldsymbol{\theta}, w'\right) . \quad (89)$$

Hence, to linear order, the distortion can be obtained by integrating along the unperturbed ray $\mathbf{x} = f_K(w)\,\boldsymbol{\theta}$; this is also called the Born approximation. Corrections to the Born approximation are necessarily of order Φ^2. Throughout this article, we will employ the Born approximation; later, we will comment on its accuracy. If we now define the deflection potential

$$\psi(\boldsymbol{\theta}, w) := \frac{2}{c2} \int_0^w \mathrm{d}w' \, \frac{f_K(w-w')}{f_K(w)\, f_K(w')} \, \Phi\left(f_K(w')\boldsymbol{\theta}, w'\right) , \quad (90)$$

then $\mathcal{A}_{ij} = \delta_{ij} - \psi_{,ij}$, just as in ordinary lens theory. *In this approximation, lensing by the 3-D matter distribution can be treated as an equivalent lens plane with deflection potential* ψ*, mass density* $\kappa = \nabla^2\psi/2$*, and shear* $\gamma = (\psi_{,11} - \psi_{,22})/2 + \mathrm{i}\psi_{,12}$.

6.2 Cosmic Shear: The Principle

The Effective Surface Mass Density

Next, we relate κ to fractional density contrast δ of matter fluctuations in the Universe; this is done in a number of steps:

1. To obtain $\kappa = \nabla^2\psi/2$, take the 2-D Laplacian of ψ, and add the term $\Phi_{,33}$ in the resulting integrand; this latter term vanishes in the line-of-sight integration, as can be seen by integration by parts.
2. We make use of the 3-D Poisson equation in comoving coordinates

$$\nabla^2\Phi = \frac{3H_0^2\Omega_\mathrm{m}}{2a}\delta \quad (91)$$

   to obtain

$$\kappa(\boldsymbol{\theta}, w) = \frac{3H_0^2\Omega_\mathrm{m}}{2c^2} \int_0^w \mathrm{d}w' \, \frac{f_K(w') f_K(w-w')}{f_K(w)} \, \frac{\delta\left(f_K(w')\boldsymbol{\theta}, w'\right)}{a(w')} . \quad (92)$$

   Note that κ is proportional to Ω_m, since lensing is sensitive to $\Delta\rho \propto \Omega_\mathrm{m}\,\delta$, not just to the density contrast $\delta = \Delta\rho/\bar{\rho}$ itself.
3. For a redshift distribution of sources with $p_z(z)\,\mathrm{d}z = p_w(w)\,\mathrm{d}w$, the effective surface mass density becomes

$$\begin{aligned}\kappa(\boldsymbol{\theta}) &= \int \mathrm{d}w \; p_w(w)\, \kappa(\boldsymbol{\theta}, w) \\ &= \frac{3H_0^2\Omega_\mathrm{m}}{2c^2} \int_0^{w_\mathrm{h}} \mathrm{d}w \; g(w)\, f_K(w) \, \frac{\delta\left(f_K(w)\boldsymbol{\theta}, w\right)}{a(w)} ,\end{aligned} \quad (93)$$

with

$$g(w) = \int_w^{w_\mathrm{h}} \mathrm{d}w'\, p_w(w') \frac{f_K(w'-w)}{f_K(w')} , \tag{94}$$

which is the source-redshift weighted lens efficiency factor $D_\mathrm{ds}/D_\mathrm{s}$ for a density fluctuation at distance w, and w_h is the comoving horizon distance, obtained from $w(a)$ by letting $a \to 0$.

The expression (92) for the effective surface mass density can be interpreted in a very simple way. Consider a redshift interval of width $\mathrm{d}z$ around z, corresponding to the proper radial distance interval $\mathrm{d}D_\mathrm{prop} = |c\,\mathrm{d}t| = H^{-1}(z)(1+z)^{-1}\,c\,\mathrm{d}z$. The surface mass density in this interval is $\Delta\rho\,\mathrm{d}D_\mathrm{prop}$, where only the density contrast $\Delta\rho = \rho - \bar\rho$ acts as a lens (the 'lensing effect' of the mean matter density of the Universe is accounted for by the relations between angular diameter distance and redshift; see Schneider and Weiss 1988a). Dividing this surface mass density by the corresponding critical surface mass density, and integrating along the line-of-sight to the sources, one finds

$$\kappa = \int_0^{z_\mathrm{s}} \mathrm{d}z\, \frac{4\pi G}{c^2} \frac{D_\mathrm{d}^\mathrm{ang} D_\mathrm{ds}^\mathrm{ang}}{D_\mathrm{s}^\mathrm{ang}} \frac{\mathrm{d}D_\mathrm{prop}}{\mathrm{d}z} \Delta\rho \, . \tag{95}$$

This expression is equivalent to (92), as can be easily shown (by the way, this is a good exercise for practicing the use of cosmological quantities like redshift, distances etc.).

Limber's Equation

The density field δ is assumed to be a realization of a random field. It is the properties of the random field that cosmologists can hope to predict, and not a specific realization of it. In particular, the second-order statistical properties of the density field are described in terms of the power spectrum (see IN, Sect. 6.1). We shall therefore look at the relation between the quantities relevant for lensing and the power spectrum $P_\delta(k)$ of the matter distribution in the Universe. The basis of this relation is formed by Limber's equation. If δ is a homogeneous and isotropic 3-D random field, then the projections

$$g_i(\boldsymbol{\theta}) = \int \mathrm{d}w\, q_i(w)\, \delta\left(f_K(w)\boldsymbol{\theta}, w\right) \tag{96}$$

also are (2-D) homogeneous and isotropic random fields, where the q_i are weight functions. In particular, the correlation function

$$C_{12} = \langle g_1(\boldsymbol{\varphi}_1)\, g_2(\boldsymbol{\varphi}_2)\rangle \equiv C_{12}(|\boldsymbol{\varphi}_1 - \boldsymbol{\varphi}_2|) \tag{97}$$

depends only on the modulus of the separation vector. The original form of the Limber (1953) equation relates C_{12} to the correlation function of δ which is a line-of-sight projection. Alternatively, one can consider the Fourier-space analogy of this relation: The power spectrum $P_{12}(\ell)$ – the Fourier transform of $C_{12}(\theta)$ – depends linearly on $P_\delta(k)$ (Kaiser 1992, 1998),

$$P_{12}(\ell) = \int \mathrm{d}w\, \frac{q_1(w)\, q_2(w)}{f_K^2(w)}\, P_\delta\left(\frac{\ell}{f_K(w)}, w\right) , \tag{98}$$

if the largest-scale structures in δ are much smaller than the effective range Δw of the projection. Hence, we obtain the (very reasonable) result that the 2-D power at angular scale $1/\ell$ is obtained from the 3-D power at length scale $f_K(w)\,(1/\ell)$, integrated over w.

Comparing (93) with (98), one sees that $\kappa(\boldsymbol{\theta})$ is such a projection of δ with the weights $q_1(w) = q_2(w) = (3/2)(H_0/c)^2\Omega_{\mathrm{m}} g(w) f_K(w)/a(w)$, so that

$$P_\kappa(\ell) = \frac{9H_0^4\Omega_{\mathrm{m}}^2}{4c^4} \int_0^{w_{\mathrm{h}}} \mathrm{d}w\, \frac{g^2(w)}{a^2(w)}\, P_\delta\left(\frac{\ell}{f_K(w)}, w\right) . \tag{99}$$

The power spectrum P_κ, if observable, can therefore be used to constrain the 3-D power spectrum P_δ. For a number of cosmological models, the power spectrum $P_\kappa(\ell)$ is plotted in Fig. 32. Predictions of P_κ are plotted both for assuming linear growth of the density structure (see Sect. 6.1 of IN), as well as the prescription of the fully nonlinear power spectrum as given by the fitting formulae of Peacock and Dodds (1996). From this figure one infers that the nonlinear evolution of the density fluctuations becomes dominant for values of $\ell \gtrsim 200$, corresponding to an angular scale of about 30′; the precise values depend on the cosmological model and the redshift distribution of the sources. Furthermore, the dimensionless power spectrum $\ell^2\, P_\kappa(\ell)$, that is, the power per logarithmic bin, peaks at around $\ell \sim 10^4$, corresponding to an angular scale of $\sim$1′, again somewhat depending on the source redshift distribution. Third, one notices that the shape and amplitude of P_κ depends on the values of the cosmological parameters; therefore, by measuring the power spectrum, or quantities directly related to it, one can constrain the values of the cosmological parameters. We consider next appropriate statistical measures of the cosmic shear which are directly and simply related to the power spectrum P_κ.

6.3 Second-Order Cosmic Shear Measures

We will now turn to statistical quantities of the cosmic shear field which are quadratic in the shear, i.e., to second-order shear statistics. Higher-order statistical properties, which already have been detected in cosmic shear surveys, will be considered in Sect. 9. As we shall see, all second-order statistics of the cosmic shear yield (filtered) information about, and are fully described in terms of P_κ. The most-often used second-order statistics are:

- The two-point correlation function(s) of the shear, $\xi_\pm(\theta)$,
- the shear dispersion in a (circular) aperture, $\left\langle|\bar\gamma|^2\right\rangle(\theta)$, and
- the aperture mass dispersion, $\left\langle M_{\mathrm{ap}}^2\right\rangle(\theta)$.

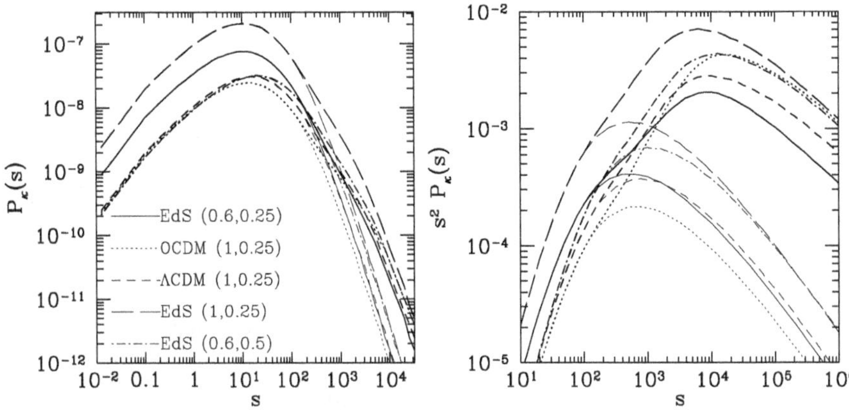

Fig. 32. The power spectrum $P_\kappa(\ell)$ (*left panel*) and its dimensionless form $\ell^2\,P_\kappa(\ell)$ (*right panel*) for several cosmological models (where here, ℓ is denoted by s). Specifically, EdS denotes an $\Omega_\mathrm{m}=1$, $\Omega_\Lambda=0$ Einstein-de Sitter model, OCDM an open $\Omega_\mathrm{m}=0.3$, $\Omega_\Lambda=0$ Universe, and ΛCDM a flat, low-density $\Omega_\mathrm{m}=0.3$, $\Omega_\Lambda=0.7$ model. Numbers in parenthesis indicate $(\Gamma_\mathrm{spect},\sigma_8)$, where Γ_spect is the shape parameter of the power spectrum (see IN, Sect. 6.1) and σ_8 is the power-spectrum normalization. For these power spectra, the mean redshift of the galaxy distribution was assumed to be $\langle z_\mathrm{s}\rangle=1.5$. Thin curves show the power spectra assuming linear evolution of the density fluctuations in the Universe, and thick curves use the fully non-linear evolution, according to the prescription of Peacock and Dodds (1996). For angular scales below $\sim 30'$, corresponding to $\ell \geq 200$, the non-linear evolution of the power spectrum becomes very important (from Schneider et al. 1998a)

Those will be discussed next, and their relation to $P_\kappa(\ell)$ shown. As a preparation, consider the Fourier transform of κ,

$$\hat\kappa(\boldsymbol{\ell}) = \int \mathrm{d}^2\theta\,\mathrm{e}^{\mathrm{i}\boldsymbol{\ell}\cdot\boldsymbol{\theta}}\,\kappa(\boldsymbol{\theta})\ ; \tag{100}$$

then,

$$\left\langle \hat\kappa(\boldsymbol{\ell})\hat\kappa^*(\boldsymbol{\ell}')\right\rangle = (2\pi)^2\,\delta_\mathrm{D}(\boldsymbol{\ell}-\boldsymbol{\ell}')\,P_\kappa(\ell)\ , \tag{101}$$

which provides another definition of the power spectrum P_κ [compare with (123) of IN]. The Fourier transform of the shear is

$$\hat\gamma(\boldsymbol{\ell}) = \left(\frac{\ell_1^2-\ell_2^2+2\mathrm{i}\ell_1\ell_2}{|\boldsymbol{\ell}|^2}\right)\hat\kappa(\boldsymbol{\ell}) = \mathrm{e}^{2\mathrm{i}\beta}\,\hat\kappa(\boldsymbol{\ell})\ , \tag{102}$$

where β is the polar angle of the vector $\boldsymbol{\ell}$; this follows directly from (42) and (43). Equation (102) implies that

$$\left\langle \hat\gamma(\boldsymbol{\ell})\hat\gamma^*(\boldsymbol{\ell}')\right\rangle = (2\pi)^2\,\delta_\mathrm{D}(\boldsymbol{\ell}-\boldsymbol{\ell}')\,P_\kappa(\ell). \tag{103}$$

Hence, the power spectrum of the shear is the same as that of the surface mass density.

Shear Correlation Functions

Consider a pair of points (i.e., galaxy images); their separation direction φ (i.e. the polar angle of the separation vector $\boldsymbol{\theta}$) is used to define the tangential and cross-component of the shear at these positions *for this pair*, $\gamma_{\rm t} = -\mathcal{R}{\rm e}\left(\gamma\,{\rm e}^{-2{\rm i}\varphi}\right)$, $\gamma_\times = -\mathcal{I}{\rm m}\left(\gamma\,{\rm e}^{-2{\rm i}\varphi}\right)$, as in (17). Using these two shear components, one can then define the correlation functions $\langle\gamma_{\rm t}\gamma_{\rm t}\rangle(\theta)$ and $\langle\gamma_\times\gamma_\times\rangle(\theta)$, as well as the mixed correlator. However, it turns out to be more convenient to define the following combinations,

$$\xi_\pm(\theta) = \langle\gamma_{\rm t}\gamma_{\rm t}\rangle(\theta) \pm \langle\gamma_\times\gamma_\times\rangle(\theta)\;, \quad \xi_\times(\theta) = \langle\gamma_{\rm t}\gamma_\times\rangle(\theta)\;. \tag{104}$$

Due to parity symmetry, $\xi_\times(\theta)$ is expected to vanish, since under such a transformation, $\gamma_{\rm t} \to \gamma_{\rm t}$, but $\gamma_\times \to -\gamma_\times$. Next we relate the shear correlation functions to the power spectrum P_κ: Using the definition of $\xi_\pm$, replacing γ in terms of $\hat\gamma$, and making use of relation between $\hat\gamma$ and $\hat\kappa$, one finds (e.g., Kaiser 1992)

$$\xi_+(\theta) = \int_0^\infty \frac{{\rm d}\ell\,\ell}{2\pi}\,{\rm J}_0(\ell\theta)\,P_\kappa(\ell)\;;\quad \xi_-(\theta) = \int_0^\infty \frac{{\rm d}\ell\,\ell}{2\pi}\,{\rm J}_4(\ell\theta)\,P_\kappa(\ell)\;, \tag{105}$$

where ${\rm J}_n(x)$ is the n-th order Bessel function of first kind. $\xi_\pm$ can be measured as follows: on a data field, select all pairs of faint galaxies with separation within $\Delta\theta$ of θ and then take the average $\langle\epsilon_{{\rm t}i}\,\epsilon_{{\rm t}j}\rangle$ over all these pairs; since $\epsilon_i = \epsilon_i^{\rm (s)} + \gamma(\boldsymbol{\theta}_i)$, the expectation value of $\langle\epsilon_{{\rm t}i}\,\epsilon_{{\rm t}j}\rangle$ is $\langle\gamma_{\rm t}\gamma_{\rm t}\rangle(\theta)$, provided source ellipticities are uncorrelated. Similarly, the correlation for the cross-components is obtained. It is obvious how to generalize this estimator in the presence of a weight factor for the individual galaxies, as it results from the image analysis described in Sect. 3.5.

The Shear Dispersion

Consider a circular aperture of radius θ; the mean shear in this aperture is $\bar\gamma$. Averaging over many such apertures, one defines the shear dispersion $\left\langle|\bar\gamma|^2\right\rangle(\theta)$. It is related to the power spectrum through

$$\left\langle|\bar\gamma|^2\right\rangle(\theta) = \frac{1}{2\pi}\int {\rm d}\ell\,\ell\,P_\kappa(\ell)\,W_{\rm TH}(\ell\theta)\;, \quad \text{where } W_{\rm TH}(\eta) = \frac{4{\rm J}_1^2(\eta)}{\eta^2} \tag{106}$$

is the top-hat filter function (see, e.g., Kaiser 1992). A practical unbiased estimator of the mean shear in the aperture is $\hat{\bar\gamma} = N^{-1}\sum_{i=1}^N \epsilon_i$, where N is the number of galaxies in the aperture. However, the square of this expression is *not* an unbiased estimator of $\left\langle|\bar\gamma|^2\right\rangle$, since the diagonal terms of the resulting double sum yield additional terms, since ${\rm E}\left(\epsilon_i\epsilon_i^*\right) = |\gamma(\boldsymbol{\theta}_i)|^2 + \sigma_\epsilon^2$.

An unbiased estimate for the shear dispersion is obtained by omitting the diagonal terms,

$$\widehat{\left\langle |\bar{\gamma}|^2 \right\rangle} = \frac{1}{N(N-1)} \sum_{i \neq j}^{N} \epsilon_i \, \epsilon_j^* \, . \tag{107}$$

This expression is then averaged over many aperture placed on the data field. Again, the generalization to allow for weighting of galaxy images is obvious. Note in particular that this estimator is not positive semi-definite.

The Aperture Mass

Consider a circular aperture of radius θ; for a point inside the aperture, define the tangential and cross-components of the shear relative to the center of the aperture (as before); then define

$$M_{\rm ap}(\theta) = \int {\rm d}^2\vartheta \; Q(|\boldsymbol{\vartheta}|) \, \gamma_{\rm t}(\boldsymbol{\vartheta}) \, , \tag{108}$$

where Q is a weight function with support $\vartheta \in [0, \theta]$. If we use the function Q given in (81), the dispersion of $M_{\rm ap}(\theta)$ is related to power spectrum by (Schneider et al. 1998a)

$$\left\langle M_{\rm ap}^2 \right\rangle (\theta) = \frac{1}{2\pi} \int_0^\infty {\rm d}\ell \; \ell \, P_\kappa(\ell) \, W_{\rm ap}(\theta\ell) \, , \quad \text{with} \;\; W_{\rm ap,1}(\eta) := \frac{576 {\rm J}_4^2(\eta)}{\eta^4} \, . \tag{109}$$

Crittenden et al. (2002) suggested a different pair U and Q of filter functions,

$$U(\vartheta) = \frac{1}{2\pi\,\theta^2} \left[1 - \left(\frac{\vartheta^2}{2\theta^2} \right) \right] \exp\left(-\frac{\vartheta^2}{2\theta^2} \right) \; ; \; Q(\vartheta) = \frac{\vartheta^2}{4\pi\theta^4} \exp\left(-\frac{\vartheta^2}{2\theta^2} \right) \, . \tag{110}$$

These function have the disadvantage of not having finite support; however, due to the very strong fall-off for $\vartheta \gg \theta$, for many practical purposes the support can be considered effectively as finite. This little drawback is compensated by the convenient analytic properties of these filter functions, as we shall see later. For example, the relation of the corresponding aperture mass dispersion is again given by the first of (109), but the filter function simplifies to

$$W_{\rm ap,2}(\eta) = \frac{\eta^4}{4} \, {\rm e}^{-\eta^2} \, . \tag{111}$$

Whereas the filter functions which relate the power spectrum to the shear correlation functions, i.e., the Bessel function appearing in (105), and to the shear dispersion, given by $W_{\rm TH}$, are quite broad filters, implying that these statistics at a given angular scale depend on the power spectrum over a wide range of ℓ, the two filter functions $W_{\rm ap,1,2}$ are very localized and thus the aperture mass dispersion yields highly localized information about the power spectrum (see Bartelmann and Schneider 1999, who showed that replacing the filter function W by a delta-'function' causes an error of only ~10%). Hence, the shape of $\langle M_{\rm ap}^2 \rangle(\theta)$ directly reflects the shape of the power spectrum as can also be seen in Fig. 35 below.

Interrelations

These various 2-point statistics all depend linearly on the power spectrum P_κ; therefore, one should not be too surprised that they are all related to each other (Crittenden et al. 2002). The surprise perhaps is that these interrelations are quite simple. First, the relations between $\xi_\pm$ and P_κ can be inverted, making use of the orthonormality relation of Bessel functions:

$$P_\kappa(\ell) = 2\pi \int_0^\infty {\rm d}\theta\, \theta\, \xi_+(\theta)\, {\rm J}_0(\ell\theta) = 2\pi \int_0^\infty {\rm d}\theta\, \theta\, \xi_-(\theta)\, {\rm J}_4(\ell\theta) \,. \tag{112}$$

Next, we take one of these and plug them into the relation (105) between the other correlation function and P_κ, to find:

$$\xi_+(\theta) = \xi_-(\theta) + \int_\theta^\infty \frac{{\rm d}\vartheta}{\vartheta} \xi_-(\vartheta) \left(4 - 12\frac{\theta^2}{\vartheta^2}\right) \,; \tag{113}$$

$$\xi_-(\theta) = \xi_+(\theta) + \int_0^\theta \frac{{\rm d}\vartheta\, \vartheta}{\theta^2} \xi_+(\vartheta) \left(4 - 12\frac{\vartheta^2}{\theta^2}\right) \,. \tag{114}$$

These equations show that the two shear correlation functions are not independent of each other, the reason for that being that the shear (which itself is a two-component quantity) is derived from a single scalar field, namely the deflection potential ψ. We shall return to this issue further below. Using (112) in the equation for the shear dispersion, one finds

$$\left\langle |\bar\gamma|^2 \right\rangle(\theta) = \int_0^{2\theta} \frac{{\rm d}\vartheta\, \vartheta}{\theta^2}\, \xi_+(\vartheta)\, S_+\left(\frac{\vartheta}{\theta}\right) = \int_0^\infty \frac{{\rm d}\vartheta\, \vartheta}{\theta^2}\, \xi_-(\vartheta)\, S_-\left(\frac{\vartheta}{\theta}\right) \,,$$

where the $S_\pm$ are simple functions, given explicitly in Schneider et al. (2002a) and plotted in Fig. 33. Finally, the same procedure for the aperture mass dispersion lets us write

$$\langle M_{\rm ap}^2 \rangle(\theta) = \int_0^{2\theta} \frac{{\rm d}\vartheta\, \vartheta}{\theta^2}\, \xi_+(\vartheta)\, T_+\left(\frac{\vartheta}{\theta}\right) = \int_0^{2\theta} \frac{{\rm d}\vartheta\, \vartheta}{\theta^2}\, \xi_-(\vartheta)\, T_-\left(\frac{\vartheta}{\theta}\right) \,, \tag{115}$$

again with analytically known functions $T_\pm$, given for the filter function (81) in Schneider et al. (2002a), and for the filter function (110) in Jarvis et al. (2003). Hence, all these 2-point statistics can be evaluated from the correlation

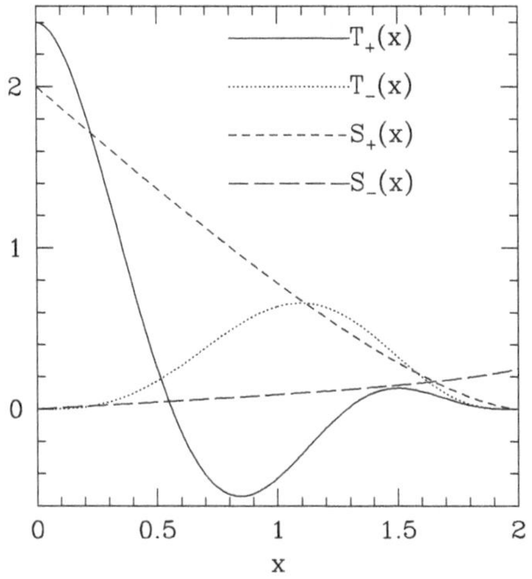

Fig. 33. The function $S_{\pm}(x)$ and $T_{\pm}(x)$ which relate the shear and aperture mass dispersion to the correlation functions. Note that S_- does not vanish for $x > 2$, as is the case for the other three functions (from Schneider et al. 2002a)

functions $\xi_{\pm}(\theta)$, which is of particular interest, since they can be measured best: Real data fields contain holes and gaps (like CCD defects; brights stars; nearby galaxies, etc.) which makes the placing of apertures difficult; however, the evaluation of the correlation functions is not affected by gaps, as one uses all pairs of galaxy images with a given angular separation. Furthermore, it should be noted that the aperture mass dispersion at angular scale θ can be calculated from $\xi_{\pm}(\vartheta)$ in the finite range $\vartheta \in [0, 2\theta]$, and the shear dispersion can be calculated from ξ_+ on $\vartheta \in [0, 2\theta]$, but not from ξ_- on a finite interval; this is due to the fact that ξ_- on small scales does not contain the information of the power spectrum on large scales, because of the filter function J_4 in (105).

We also note that from a cosmic shear survey, the power spectrum P_κ can be determined directly, as has been investigated by Kaiser (1998), Seljak (1998) and Hu and White (2001). This is *not* done by applying (112), as these relations would require the determination of the correlation function for all separation, but by more sophisticated methods. A simple example (though not optimal) is to consider the measured shear field on the square; Fourier transforming it and binning modes in $|\boldsymbol{\ell}|$ then yields an estimate of the power spectrum, once the power from the intrinsic ellipticity dispersion is subtracted. Better methods aim at minimizing the variance of the reconstructed power spectrum (Seljak 1998; Hu and White 2001). As mentioned before, the aperture mass dispersion is a filtered version of the power spectrum with such a narrow filter, that it contains essentially the same information as P_κ over the corresponding angular scale and at $\ell \sim 5/\theta$, provided P_κ has no sharp features.

6.4 Cosmic Shear and Cosmology

Why Cosmology From Cosmic Shear ?

Before continuing, it is worth to pause for a second and ask the question why one tries to investigate cosmological questions by using cosmic shear – since it is widely assumed that the CMB will measure 'all' cosmological quantities with high accuracy. Partial answers to this question are:

- Cosmic shear measures the mass distribution at much lower redshifts ($z \lesssim 1$) and at smaller physical scales [$R \sim 0.3\,h^{-1}\,(\theta/1')\,\mathrm{Mpc}$] than the CMB; indeed, it is the only way to map out the dark matter distribution directly without any assumptions about the relation between dark and baryonic matter.
- Cosmic shear measures the non-linearly evolved mass distribution and its associated power spectrum $P_\delta(k)$; hence, in combination with the CMB it allows us to study the evolution of the power spectrum and in particular, provide a very powerful test of the gravitational instability paradigm for structure growth.
- As was demonstrated by the recent results from the WMAP satellite (Bennett et al. 2003), the strongest constraints are derived when combining CMB measurements (constraining the power spectrum on large spatial scales) with measurements on substantially smaller scales, to break parameter degeneracies remaining from the CMB results alone (see Spergel et al. 2003). Hu and Tegmark (1999) have explicitly demonstrated how much the accuracy of estimates of cosmological parameters is improved when the CMB results from missions like WMAP and later Planck is complemented by cosmic shear measurements (see Fig. 34). In fact, as we shall see later, combinations of CMB anisotropy measurements have already been combined with cosmic shear measurements (see Fig. 47) and lead to substantially improved constraints on the cosmological parameters.
- It provides a fully independent way to probe the cosmological model; given the revolutionary claims coming from the CMB, SN Ia, and the LSS of the galaxy distribution, namely that more than 95% of the contents in the Universe is in a form that we have not the slightest idea about what it is (the names 'dark matter' and 'dark energy' reflect our ignorance about their physical nature), an additional independent verification of these claims is certainly welcome.
- For a foreseeable future, astronomical observations will provide the only possibility to probe the dark energy empirically. The equation of state of the dark energy can be probed best at relatively low redshifts, that is with SN Ia and cosmic shear observations, whereas CMB anisotropy measurements are relatively insensitive to the properties of the dark energy, as the latter was subdominant at the epoch of recombination.
- As we have seen in Sect. 5.8, cosmic shear studies provide a new and highly valuable search method for cluster-scale matter concentrations.

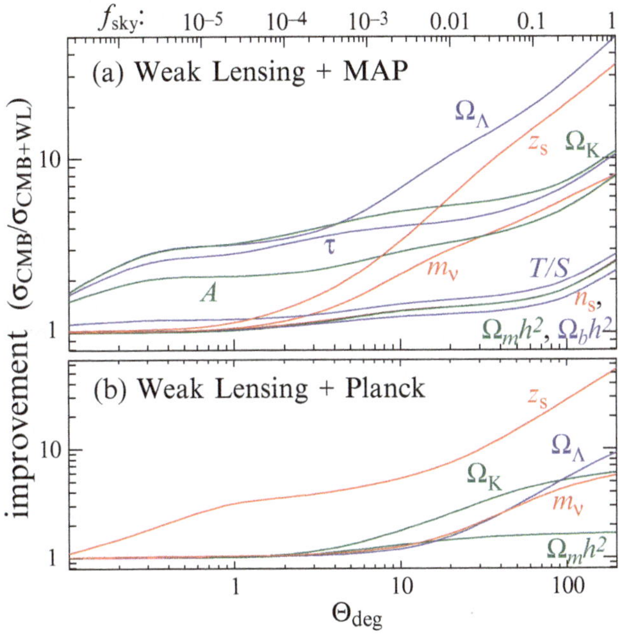

Fig. 34. The improvement of the accuracy of cosmological parameters when supplementing CMB data from WMAP (*upper panel*) and the Planck satellite (*lower panel*) by a cosmic shear survey of solid angle $\theta^2\pi$. The accuracies are significantly improved, certainly when combined with WMAP, but even in combination with Planck, the accuracies of the density parameters can be increased, when using next-generation cosmic shear surveys with hundreds of square degrees (from Hu and Tegmark 1999)

Expectations

The cosmic shear signal depends on the cosmological model, parameterized by Ω_m, Ω_Λ, and the shape parameter Γ_{spect} of the power spectrum, the normalization of the power spectrum, usually expressed in terms of σ_8, and the redshift distribution of the sources. By measuring $\xi_\pm$ over a significant range of angular scales one can derive constraints on these parameters. To first order, the amplitude of the cosmic shear signal depends on the combination $\sim\sigma_8\,\Omega_m^{0.5}$, very similar to the cluster abundance. Furthermore, the cosmic shear signal shows a strong dependence on the source redshift distribution. These dependencies are easily understood qualitatively: A higher normalization σ_8 increases P_δ on all scales, thus increasing P_κ. The increase with Ω_m is mainly due to the prefactor in (99), i.e. due to the fact that the light deflection depends on $\Delta\rho$, not just merely on $\delta = \Delta\rho/\bar{\rho}$, as most other cosmological probes. Finally, increasing the redshift of sources has two effects: first, the lens efficiency $D_{\rm ds}/D_{\rm s} = f_K(w_{\rm s} - w)/f_K(w_{\rm s})$ at given distance w increases as the sources are moved further away, and second, a larger source redshift implies a longer ray path through the inhomogeneous matter distribution.

In Fig. 35 the predictions of the shear dispersion and the aperture mass dispersion are shown as a function of angular scale, for several cosmological models. The dependencies of the power spectrum P_κ on cosmological parameters and ℓ is reflected in these cosmic shear measures. In particular, the narrow filter function which relates the aperture mass dispersion to the power spectrum implies that the peak in $\ell^2 P_\kappa(\ell)$ at around $\ell \sim 10^4$ (see Fig. 32) translates into a peak of $\langle M_{\rm ap}^2 \rangle$ at around $\theta \sim 1'$. The non-linear evolution of the power spectrum is dominating the cosmic shear result for scales below

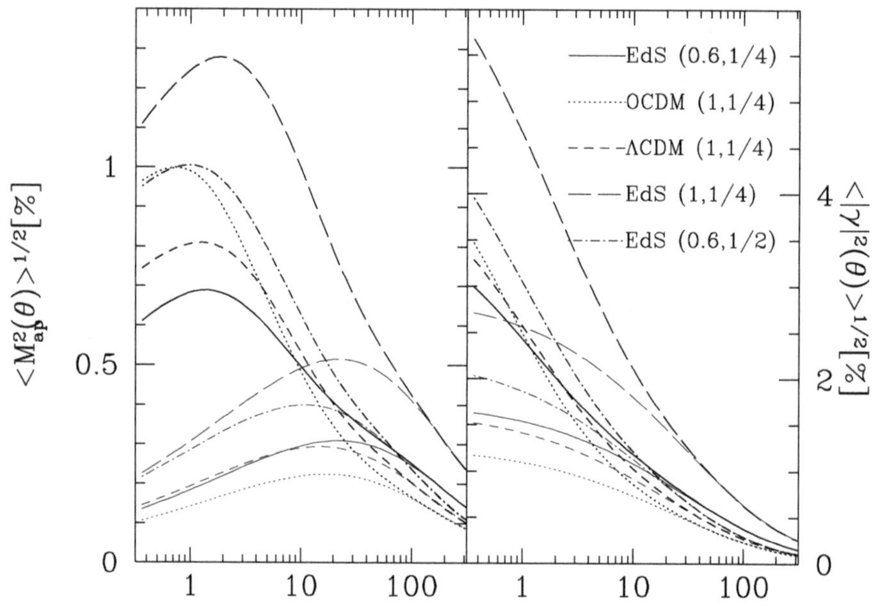

Fig. 35. The square root of the aperture mass dispersion (*left*) and of the shear dispersion (*right*), for the same cosmological models as were used for Fig. 32, again with results from assuming linear growth of structure in the Universe shown as thin curves, whereas the fully non-linear evolution was taken into account for the thick curves. One sees that the aperture mass signal is considerably smaller than that of the shear dispersion; this is due to the fact that the filter function $W_{\rm ap}$ is much narrower than $W_{\rm TH}$; hence, at a given angular scale, $\langle M_{\rm ap}^2 \rangle$ samples less power than $\langle |\bar\gamma|^2 \rangle$. However, this also implies that the aperture mass dispersion provides much more localized information about the power spectrum than the shear dispersion and is therefore a more useful statistics to consider. Other advantages of $\langle M_{\rm ap}^2 \rangle$ will be described further below. For scales below $\sim 30'$, the non-linear evolution of the power spectrum becomes very important (from Schneider et al. 1998a)

$\sim 30'$; the fact that the non-linear prediction approach the linear ones at somewhat smaller scales for the shear dispersion $\langle|\bar{\gamma}|^2\rangle$ is due to the fact that this statistics corresponds to a broad-band filter $W_{\rm TH}$ (106) of P_κ which includes the whole range of small ℓ values, which are less affected by non-linear evolution.

Deriving Constraints

From the measured correlation functions $\xi_\pm(\theta)$ (or any other measure of the cosmic shear, but we will concentrate on the statistics which is most easily obtained from real data), obtaining constraints on cosmological parameters can proceed through maximizing the likelihood $\mathcal{L}(p|\xi^{\rm obs})$, which yields the probability for the set of cosmological parameters being p, given the observed correlation function $\xi^{\rm obs}$. This likelihood is given by the probability $P(\xi^{\rm obs}|p)$ that the observed correlation function is $\xi^{\rm obs}$, given the parameters p. For a given set of parameters p, the correlation function $\xi(p)$ is predicted. If one assumes that the observed correlations $\xi^{\rm obs}$ are drawn from a (multi-variate) Gaussian probability distribution, then

$$P(\xi^{\rm obs}|p) = \frac{1}{(2\pi)^{n/2}\sqrt{\det {\rm Cov}}} \exp\left(\frac{-\chi^2(p,\xi^{\rm obs})}{2}\right) ,$$

with

$$\chi^2(p,\xi^{\rm obs}) = \sum_{ij} \left(\xi_i(p) - \xi_i^{\rm obs}\right) {\rm Cov}_{ij}^{-1} \left(\xi_j(p) - \xi_j^{\rm obs}\right) . \tag{116}$$

Here, the $\xi_i = \xi(\theta_i)$ are the values of the correlation function(s) (i.e., either $\xi_\pm$, or using both) in angular bins, n is the number of angular bins in case either one of the $\xi_\pm$ is used, or if both are combined, twice the number of angular bins, and ${\rm Cov}_{ij}^{-1}$ is the inverse of the covariance matrix, which is defined as

$${\rm Cov}_{ij} = \left\langle \left[\xi_i(p) - \xi_i^{\rm obs}\right] \left[\xi_j(p) - \xi_j^{\rm obs}\right] \right\rangle , \tag{117}$$

where the average is over multiple realizations of the cosmic shear survey under consideration. ${\rm Cov}_{ij}$ can be determined either from the $\xi_\pm$ itself, from simulations, or estimated from the data in terms of the $\xi_\pm^{\rm obs}$ (see Schneider et al. 2002b; Kilbinger and Schneider 2004, Simon et al. 2004). Nevertheless, the calculation of the covariance is fairly cumbersome, and most authors have used approximate methods to derive it, such as the field-to-field variations of the measured correlation. In fact, this latter approach may be more accurate than using the analytic expressions of the covariance in terms of the correlation function, which are obtained by assuming that the shear field is Gaussian, so that the four-point correlation function can be factorized as products of two-point correlators. As it turns out, $\xi_+(\theta)$ is strongly correlated across angular bins, much less so for $\xi_-(\theta)$; this is due to the fact that the filter function that describes ξ in terms of the power spectrum P_κ is much broader for ξ_+ (namely J_0) than J_4 which applies for ξ_-.

The accuracy with which $\xi_\pm$ can be measured, and thus the covariance matrix, depends on the number density of galaxies (that is, depth and quality of the images), the total solid angle covered by the survey, and its geometrical arrangement (compact survey vs. widely separated pointings). The accuracy is determined by a combination of the intrinsic ellipticity dispersion and the cosmic (or sampling) variance. The likelihood function then becomes

$$\mathcal{L}(p|\xi^{\rm obs}) = \frac{1}{(2\pi)^{n/2}\sqrt{\det {\rm Cov}}} \exp\left(\frac{-\chi^2(p,\xi^{\rm obs})}{2}\right) P_{\rm prior}(p) \;, \tag{118}$$

where $P_{\rm prior}(p)$ contains prior information (or prejudice) about the parameters to be determined. For example, the redshift distribution of the sources (at given apparent magnitude) is fairly well known from spectroscopic redshift surveys, and so the prior probability for $z_{\rm s}$ would be chosen to be a fairly narrow function which describes this prior knowledge on the redshifts. One often assumes that all but a few parameters are known precisely, and thus considers a restricted space of parameters; this is equivalent to replacing the prior for those parameters which are fixed by a delta-'function'. If m parameters are assumed to be undetermined, but one is mainly interested in the confidence contours of $m' < m$ parameters, then the likelihood function is integrated over the remaining $m - m'$ parameters; this is called marginalization and yields the likelihood function for these m' parameters.

There are two principal contributions to the 'noise' of cosmic shear measurements. One is the contribution coming from the finite intrinsic ellipticity dispersion of the source galaxies, the other due to the finite data fields of any survey. This latter effect implies that only a *typical* part of the sky is mapped, whose properties will in general deviate from the *average* properties of such a region in the sky for a given cosmology. This effect is called cosmic variance, or sample variance. Whereas the noise from intrinsic ellipticity dispersions dominates at small angular scales, at scales beyond a few arcminutes the cosmic variance is always the dominating effect (e.g., Kaiser 1998; White and Hu 2000).

Of course, all of what was said above can be carried over to the other second-order shear statistics, with their respective covariance matrices. The first cosmic shear measurements were made in terms of the shear dispersion and compared to theoretical prediction from a range of cosmological models. As is true for the correlation functions, the shear dispersion is strongly correlated between different angular scales. This is much less the case for the aperture mass dispersion, where the correlation quickly falls off once the angular scales differ by more than a factor $\sim$1.5 (see Schneider et al. 2002b). Even less correlated is the power spectrum itself. These properties are of large interest if the results from a cosmic shear survey are displayed as a curve with error bars; for the aperture mass dispersion and the power spectrum estimates, these errors are largely uncorrelated. However, for deriving cosmological constraints, the correlation function $\xi_\pm$ are most useful since they contain all second-order information in the data, in addition of being the primary observable.

6.5 E-Modes, B-Modes

In the derivation of the lensing properties of the LSS, we ended up with an equivalent surface mass density. In particular, this implied that $\mathcal{A}$ is a symmetric matrix, and that the shear can be obtained in terms of κ or ψ. Now, the shear is a 2-component quantity, whereas both κ and ψ are scalar fields. This then obviously implies that the two shear components are not independent of each other !

Recall that (54) yields a relation between the gradient of κ and the first derivatives of the shear components; in particular, (54) implies that $\nabla \times \mathbf{u}_\gamma \equiv 0$, yielding a local constraint relation between the second derivatives of the shear components. The validity of this constraint equation guarantees that the imaginary part of (44) vanishes. This constraint is also present at the level of 2-point statistics, since one expects from (112) that

$$\int_0^\infty \mathrm{d}\theta\,\theta\,\xi_+(\theta)\mathrm{J}_0(\theta\ell) = \int_0^\infty \mathrm{d}\theta\,\theta\,\xi_-(\theta)\mathrm{J}_4(\theta\ell) \ .$$

Hence, the two correlation functions $\xi_\pm$ are not independent. The observed shear field is not guaranteed to satisfy these relations, due to noise, remaining systematics, or other effects. Therefore, searching for deviations from this relation allows a check for these effects. However, there might also be a 'shear' component present that is not due to lensing (by a single equivalent thin matter sheet κ). Shear components which satisfy the foregoing relations are called E-modes; those which don't are B-modes – these names are exported from the polarization of the CMB, which has the same mathematical properties as the shear field, namely that of a polar.

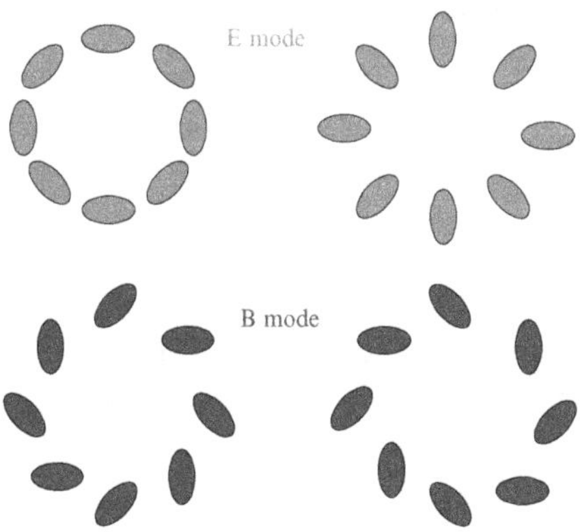

Fig. 36. Sketch of the distinction between E- and B-modes of the shear. The upper row shows a typical E-mode shear pattern coming from a mass overdensity (*left*) or underdensity (*right*), yielding tangential and radial alignment of the shear, respectively. The lower row shows a B-mode pattern, which is obtained from the E-mode pattern by rotating all shears by 45°. Those cannot be produced from gravitational lensing (from van Waerbeke and Mellier 2003)

The best way to separate these modes locally is provided by the aperture measures: $\langle M_{\rm ap}^2(\theta)\rangle$ is sensitive *only* to E-modes. If one defines in analogy – recall (77)

$$M_\perp(\theta) = \int {\rm d}^2\vartheta\; Q(|\boldsymbol{\vartheta}|)\,\gamma_\times(\boldsymbol{\vartheta}) \;, \tag{119}$$

then $\langle M_\perp^2(\theta)\rangle$ is sensitive *only* to B-modes. In fact, one can show that for a pure E-mode shear field, $M_\perp \equiv 0$, and for a pure B-mode field, $M_{\rm ap} \equiv 0$. Furthermore, in general (that is, even if a B-mode is present), $\langle M_{\rm ap}\rangle = 0$, since $\langle\kappa\rangle = 0$, and $\langle M_\perp\rangle = 0$, owing to parity invariance: a non-zero mean value of $M_\perp$ would introduce a net orientation into the shear field. Using the same argument, one finds that $\langle M_{\rm ap}^m M_\perp^n\rangle = 0$ for n odd (Schneider 2003).

E/B-mode Decomposition of a Shear Field

There are a number of (equivalent) ways to decompose a shear field into its two modes. One is provided by the Kaiser and Squires mass reconstruction (44), which yields, for a general shear field, a complex surface mass density $\kappa = \kappa^{\rm E} + {\rm i}\kappa^{\rm B}$. Another separation is obtained by considering the vector field $\mathbf{u}_\gamma(\boldsymbol{\theta})$ (54) obtained from the first derivatives of the shear components. This vector will in general not be a gradient field; its gradient component corresponds to the E-mode field, the remaining one to the B-mode. Hence one defines

$$\nabla^2\kappa^{\rm E} = \nabla\cdot\mathbf{u}_\gamma \quad ; \quad \nabla^2\kappa^{\rm B} = \nabla\times\mathbf{u}_\gamma \;. \tag{120}$$

In full analogy with the 'lensing-only' case (i.e., a pure E-mode), one defines the (complex) potential $\psi(\boldsymbol{\theta}) = \psi^{\rm E}(\boldsymbol{\theta}) + {\rm i}\psi^{\rm B}(\theta)$ by the Poisson equation $\nabla^2\psi = 2\kappa$, and the shear is obtained in terms of the complex ψ in the usual way,

$$\begin{aligned}\gamma &= \gamma_1 + {\rm i}\gamma_2 = \left(\psi_{,11} - \psi_{,22}\right)/2 + {\rm i}\psi_{,12} \\ &= \left[\frac{1}{2}\left(\psi^{\rm E}_{,11} - \psi^{\rm E}_{,22}\right) - \psi^{\rm B}_{,12}\right] + {\rm i}\left[\psi^{\rm E}_{,12} + \frac{1}{2}\left(\psi^{\rm B}_{,11} - \psi^{\rm B}_{,22}\right)\right] \;.\end{aligned} \tag{121}$$

On the level of second-order statistics, one considers the Fourier transforms of the E- and B-mode convergence, and defines the two power spectra $P_{\rm E}$, $P_{\rm B}$, and the cross-power spectrum $P_{\rm EB}$ by

$$\begin{aligned}\left\langle\hat\kappa^{\rm E}(\boldsymbol{\ell})\hat\kappa^{\rm E*}(\boldsymbol{\ell}')\right\rangle &= (2\pi)^2\,\delta_{\rm D}(\boldsymbol{\ell}-\boldsymbol{\ell}')\,P_{\rm E}(\ell) \;, \\ \left\langle\hat\kappa^{\rm B}(\boldsymbol{\ell})\hat\kappa^{\rm B*}(\boldsymbol{\ell}')\right\rangle &= (2\pi)^2\,\delta_{\rm D}(\boldsymbol{\ell}-\boldsymbol{\ell}')\,P_{\rm B}(\ell) \;, \\ \left\langle\hat\kappa^{\rm E}(\boldsymbol{\ell})\hat\kappa^{\rm B*}(\boldsymbol{\ell}')\right\rangle &= (2\pi)^2\,\delta_{\rm D}(\boldsymbol{\ell}-\boldsymbol{\ell}')\,P_{\rm EB}(\ell) \;.\end{aligned} \tag{122}$$

From what was said above, the cross power $P_{\rm EB}$ vanishes for parity-symmetric shear fields, and we shall henceforth ignore it. The shear correlation functions now depend on the power spectra of both modes, and are given as (Crittenden et al. 2002; Schneider et al. 2002a)

$$\xi_+(\theta) = \int_0^\infty \frac{\mathrm{d}\ell\,\ell}{2\pi}\, \mathrm{J}_0(\ell\theta)\,[P_\mathrm{E}(\ell) + P_\mathrm{B}(\ell)] \;,$$

$$\xi_-(\theta) = \int_0^\infty \frac{\mathrm{d}\ell\,\ell}{2\pi}\, \mathrm{J}_4(\ell\theta)\,[P_\mathrm{E}(\ell) - P_\mathrm{B}(\ell)] \;.$$

Hence, in the presence of B-modes, the ξ_- correlation function cannot be obtained from ξ_+, as was the case for a pure E-mode shear field. The inverse relation (112) now gets modified to

$$P_\mathrm{E}(\ell) = \pi \int_0^\infty \mathrm{d}\theta\,\theta\,\left[\xi_+(\theta)\mathrm{J}_0(\ell\theta) + \xi_-(\theta)\mathrm{J}_4(\ell\theta)\right] \;,$$

$$P_\mathrm{B}(\ell) = \pi \int_0^\infty \mathrm{d}\theta\,\theta\,\left[\xi_+(\theta)\mathrm{J}_0(\ell\theta) - \xi_-(\theta)\mathrm{J}_4(\ell\theta)\right] \;. \tag{123}$$

Hence, the two power spectra can be obtained from the shear correlation functions. However, due to the infinite range of integration, one would need to measure the correlation functions over all angular scales to apply the previous equations for calculating the power spectra. Much more convenient for the E/B-mode decomposition is the use of the aperture measures, since one can show that

$$\left\langle M_\mathrm{ap}^2\right\rangle(\theta) = \frac{1}{2\pi}\int_0^\infty \mathrm{d}\ell\,\ell\,P_\mathrm{E}(\ell)\,W_\mathrm{ap}(\theta\ell) \;,$$

$$\left\langle M_\perp^2\right\rangle(\theta) = \frac{1}{2\pi}\int_0^\infty \mathrm{d}\ell\,\ell\,P_\mathrm{B}(\ell)\,W_\mathrm{ap}(\theta\ell) \;, \tag{124}$$

so that these two-point statistics clearly separate E- and B-modes. In addition, as mentioned before, they provide a highly localized measure of the corresponding power spectra, since the filter function $W_\mathrm{ap}(\eta)$ involved is very narrow. As was true for the E-mode only case, the aperture measures can be expressed as finite integrals over the correlation functions,

$$\left\langle M_\mathrm{ap}^2\right\rangle(\theta) = \frac{1}{2}\int \frac{\mathrm{d}\vartheta\,\vartheta}{\theta^2}\left[\xi_+(\vartheta)\,T_+\!\left(\frac{\vartheta}{\theta}\right) + \xi_-(\vartheta)\,T_-\!\left(\frac{\vartheta}{\theta}\right)\right] ,$$

$$\left\langle M_\perp^2\right\rangle(\theta) = \frac{1}{2}\int \frac{\mathrm{d}\vartheta\,\vartheta}{\theta^2}\left[\xi_+(\vartheta)\,T_+\!\left(\frac{\vartheta}{\theta}\right) - \xi_-(\vartheta)\,T_-\!\left(\frac{\vartheta}{\theta}\right)\right] , \tag{125}$$

where the two functions $T_\pm$ are the same as in (115) and have been given explicitly in Schneider et al. (2002a) for the weight function Q given in (81), and in Jarvis et al. (2003) for the weight function (110). Hence, the relations (125) remove the necessity to calculate the aperture measures by placing apertures on the data field which, owing to gaps and holes, would make this an inaccurate and biased determination. Instead, obtaining the correlation functions from the data is all that is needed.

The relations given above have been applied to recent cosmic shear surveys, and significant B-modes have been discovered (see Sect. 7); the question now is what are they due to ? As mentioned before, the noise, which contributes to both E- and B-modes in similar strengths, could be underestimated, the cosmic variance which also determines the error bars on the aperture measures and which depends on fourth-order statistical properties of the shear field could also be underestimated, there could be remaining systematic effects, or B-modes could indeed be present. There are two possibilities known to generate a B-mode through lensing: The first-order in Φ (or 'Born') approximation may not be strictly valid, but as shown by ray-tracing simulations through cosmic matter fields (e.g., Jain et al. 2000), the resulting B-modes are expected to be very small. Clustering of sources also yields a finite B-mode (Schneider et al. 2002a), but again, this effect is much smaller than the observed amplitude of the B-modes (see Fig. 37).

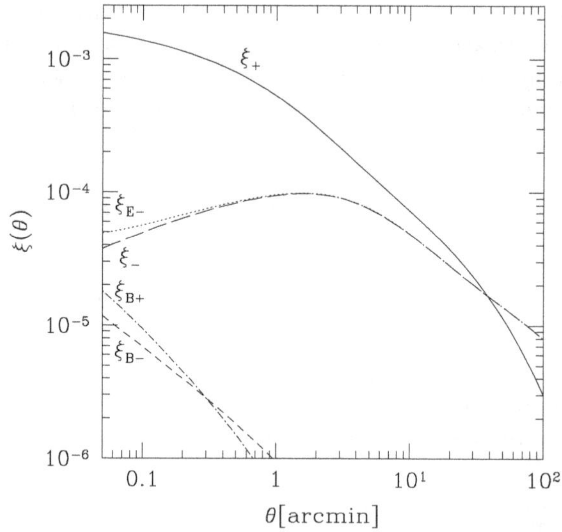

Fig. 37. The correlation functions $\xi_\pm(\theta)$ for a ΛCDM model with $\Gamma_{\rm spect} = 0.21$ and $\sigma_8 = 1$, and a source population with mean redshift of $\langle z_{\rm s}\rangle = 1.5$. Also plotted are the corresponding correlation functions that arise separately from the E- and B-modes, with the ξ_{E+} mode curve coinciding within the line thickness with ξ_+. In this calculation, the clustering of the faint galaxy population was taken into account, and they give rise to a very small B-mode contribution, as can be seen from the $\xi_{B\pm}$ curves. The smallness of the B-mode due to intrinsic source clustering renders this effect not viable to explain the B-modes observed in some of the cosmic shear surveys (from Schneider et al. 2002a)

Intrinsic Alignment of Source Galaxies

Currently the best guess for the generation of a finite B-mode are intrinsic correlations of galaxy ellipticities. Such intrinsic alignments of galaxy ellipticities can be caused by tidal gravitational fields during galaxy formation, owing to tidal interactions between galaxies, or between galaxies and clusters. Predictions of the alignment of the projected ellipticity of the galaxy mass can be made analytically (e.g. in the frame of tidal torque theory) or from numerical simulations; however, the predictions from various groups differ by large factors (e.g., Croft and Metzler 2001; Crittenden et al. 2001; Heavens et al. 2000; Jing 2002) which means that the process is not well understood at present. For example, the results of these studies depend on whether one assumes that the light of a galaxy is aligned with the dark matter distribution, or aligned with the angular momentum vector of the dark halo. This is related to the question of whether the orientation of the galaxy light (which is the issue of relevance here) is the same as that of the mass.

If intrinsic alignments play a role, then

$$\xi_+ = \left\langle \epsilon_i \, \epsilon_j^* \right\rangle = \left\langle \epsilon_i^{(\mathrm{s})} \, \epsilon_j^{(\mathrm{s})*} \right\rangle + \xi_+^{\mathrm{lens}} \,, \tag{126}$$

and measured correlations $\xi_\pm$ contain both components, the intrinsic correlation and the shear. Of course, there is no reason why intrinsic correlations should have only a B-mode. If a B-mode contribution is generated through this process, then the measured E-mode is most likely also contaminated by intrinsic alignments. Given that intrinsic alignments yield ellipticity correlations only for spatially close sources (i.e., close in 3-D, not merely in projection), it is clear that the deeper a cosmic shear survey is, and thus the broader the redshift distribution of source galaxies, the smaller is the relative amplitude of an intrinsic signal. Most of the theoretical investigations on the strength of intrinsic alignments predict that the deep cosmic shear surveys (say, with mean source redshifts of $\langle z_\mathrm{s} \rangle \sim 1$) are affected at a $\sim$10% level, but that shallower cosmic shear surveys are more strongly affected; for them, the intrinsic alignment can be of same order or even larger than the lensing signal.

However, the intrinsic signal can be separated from the lensing signal if redshift information of the sources is available, owing to the fact that $\left\langle \epsilon_i^{(\mathrm{s})} \, \epsilon_j^{(\mathrm{s})*} \right\rangle$ will be non-zero only if the two galaxies are at essentially the same redshift. Hence, if z-information is available (e.g., photometric redshifts), then galaxy pairs which are likely to have similar redshifts are to be avoided in estimating the cosmic shear signal (King and Schneider 2002, Heymans and Heavens 2003, Takada and White 2004). This will change the expectation value of the shear correlation function, but in a controllable way, as the redshifts are assumed to be known. Indeed, using (photometric) redshifts, one can simultaneously determine the intrinsic and the lensing signal, essentially providing a cosmic shear tomography (King and Schneider 2003). This again is accomplished by employing the fact that the intrinsic correlation can only come from galaxies very close in redshift. Hence, in the presence of intrinsic alignments, the redshift dependent correlation functions $\xi_\pm(z_1, z_2; \theta)$ between galaxies with estimated redshifts z_i are expected to show a strong peak over the range $|z_1 - z_2| \lesssim \Delta z$, where Δz is the typical uncertainty in photometric redshifts. It is this peak that allows one to identify and subtract the intrinsic signal from the correlation functions. An efficient method to calculate the covariance of the redshift-dependent correlation functions has been developed by Simon et al. (2004), where the improvement in the constraints on cosmological parameters from redshift information has been studied, confirming the earlier results by Hu (1999) which were based on considerations of the power spectrum.

Brown et al. (2003) obtained a measurement of the intrinsic ellipticity correlation from the Super-COSMOS photographic plate data, where the galaxies are at too low a redshift for cosmic shear playing any role. Heymans et al. (2004) used the COMBO-17 data set (that will be described in Sect. 7.3 below) for which accurate photometric redshifts are available to measure the intrinsic alignment. The results from both studies is that the models predicting a large intrinsic amplitude can safely be ruled out. Nevertheless, intrinsic alignment affects cosmic shear measurements, at about the 2% level for a survey with the depth of the VIRMOS-DESCART survey, and somewhat more for the slightly shallower COMBO-17 survey. Hence, to obtain precision measurements of cosmic shear, very important for constraining the equation of state of dark energy, these physically close pairs of galaxies need to be identified in the survey, making accurate photometric redshifts mandatory.

Correlation Between Intrinsic Ellipticity and Shear

The relation (126) above implicitly assumes that the shear is uncorrelated with the intrinsic shape of a neighboring galaxy. However, as pointed out by Hirata and Seljak (2004), this is not necessarily the case. Hence consider galaxies at two significantly different redshifts $z_i < z_j$. For them, the first term in (126) vanishes. However, making use of $\epsilon = \epsilon^{(s)} + \gamma$, one finds

$$\left\langle \epsilon_i \epsilon_j^* \right\rangle = \left\langle \epsilon_i^{(s)} \gamma_j^* \right\rangle + \xi_+^{\rm lens} , \tag{127}$$

where the first term on the right-hand side describes the correlation between the intrinsic ellipticity of the lower-redshift galaxy with the shear along the l.o.s. to the higher-redshift one. The correlation can in principle be non-zero: if the intrinsic alignment of the light of a galaxy is determined by the large-scale tidal gravitational field, then this tidal field at the redshift z_i causes both, an alignment of the nearer galaxy and a contribution to the shear of the more distant one (see Fig. 38). This alignment effect can therefore not be removed by considering only pairs at different redshifts.

The importance of this effect depends on the nature of the alignment of galaxies relative to an external tidal field. If the alignment is linear in the tidal

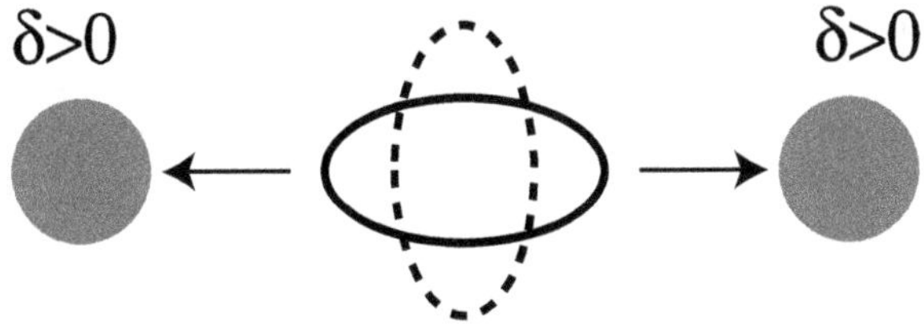

Fig. 38. A tidal gravitational field, for example caused by two matter concentrations, can produce an alignment of a galaxy situated at the same redshift (indicated by the *solid ellipse*), as well as contributing to the shear toward a more distant galaxy (as indicated by the *dashed ellipse*) (from Hirata and Seljak 2004)

field strength, then this effect can be a serious contaminant of the cosmic shear signal, in particular for relatively shallow surveys (where the mean source redshift is small); in particular, this effect can yield much larger contaminations than the intrinsic alignment given by the first term in (126). As can be seen from Fig. 38, the resulting contribution is negative, hence decreases the lensing signal. If, however, the intrinsic alignment depends quadratically on the tidal field, as is suggested by tidal torque theory, than this effect is negligible. Whether or not this effect is relevant needs to be checked from observations. Assuming that the matter density field is represented approximately by the galaxy distribution, the latter can be used to estimate the tidal gravitational field, in particular its direction. Alternatively, since the correlation between the intrinsic alignment and the shear toward more distant galaxies has a different redshift dependence than the lensing shear signal, these two contributions can be disentangled from the z-dependence of the signal.

It should be noted that the use of photometric redshifts also permits to study the cosmic shear measures as a function of source redshift; hence, one can probe various redshift projections $P_\kappa(\ell)$ of the underlying power spectrum $P_\delta(k;z)$ separately. This is due to the fact that the cosmic shear signal from different populations of galaxies (i.e., with different redshift distributions) lead to different weight functions $g(w)$ [see (94)], and thus to different weighting in the projection (99) of the power spectrum. Not surprisingly, uncertainties of cosmological parameters are thereby reduced (Hu 1999; Simon et al. 2004). Also, as shown by Taylor (2001), Hu and Keeton (2002) and Bacon and Taylor (2003), in principle the three-dimensional mass distribution $\delta(\boldsymbol{x})$ can be reconstructed if the redshifts of the source galaxies are known (see Sect. 7.6).

6.6 Predictions; Ray-Tracing Simulations

The power spectrum of the convergence P_κ can be calculated from the power spectrum of the cosmological matter distribution P_δ, using (99); the latter in turn is determined by the cosmological model. However, since the non-linear evolution of the power spectrum is essential for making accurate quantitative predictions for the shear properties, there is no analytic method known how to calculate the necessary non-linear P_δ. As was mentioned in Sect. 6.1 of IN, fairly accurate fitting formulae exist which yield a closed-form expression for P_δ and which can be used to obtain P_κ (see, e.g., Jain and Seljak 1997). Nevertheless, there are a number of reasons why this purely analytic approach should at least be supplemented by numerical simulations.

- First, the fitting formulae for P_δ (Peacock and Dodds 1996; Smith et al. 2003) have of course only a finite accuracy, and are likely to be insufficient for comparison with results from the ongoing cosmic shear surveys which are expected to yield very accurate measurements, owing to their large solid angle.
- A second reason why simulations are needed is to test whether the various approximations that enter the foregoing analytical treatment are in fact accurate enough. To recall them, we employed the Born approximation, i.e., neglected terms of higher order than linear in the Newtonian potential when deriving the convergence, and we assumed that the shear everywhere is small, so that the difference between shear and reduced shear can be neglected, at least on average. This, however, is not guaranteed: regions in the sky with large shear are most likely also those regions where the convergence is particularly large, and therefore, there one expects a correlation between γ and κ, which can affect the dispersion of $g = \gamma/(1-\kappa)$.
- Third, whereas fairly accurate fitting formulae exist for the power spectrum, this is not the case for higher-order statistical properties of the matter distribution; hence, when considering higher-order shear statistics (Sect. 9), numerical simulations will most likely be the only way to obtain accurate predictions.
- The covariance of the shear correlations (and all other second-order shear measures) depends on fourth-order statistics of the shear field, for which hardly any useful analytical approximations are available. The analytical covariance estimates are all based on the Gaussian assumption for the fourth-order correlators. Therefore, simulations are invaluable for the calculation of these covariances, which can be derived for arbitrary survey geometries.

Ray-Tracing Simulations: The Principle

The simulations proceed by following light rays through the inhomogeneous matter distribution in the Universe. The latter is generated by cosmological simulations of structure evolution. Those start at an early epoch by generating a realization of a Gaussian random field with a power spectrum according to the cosmological model considered, and follow the evolution of the density and velocity field of the matter using Newtonian gravity in an expanding Universe.

The mass distribution is represented by discrete particles whose evolution in time is followed. A finite volume of the Universe is simulated this way, typically a box of comoving side-length L, for which periodic boundary conditions are applied. This allows one to use Fast Fourier Transforms (FFT) to evaluate the gravitational potential and forces from the density distribution. The box size L should be chosen such that the box contains a representative part of the real Universe, and must therefore be larger than the largest scales on which structure is expected, according to the power spectrum; a reasonable choice is $L \gtrsim 100h^{-1}\,\mathrm{Mpc}$. The number of grid points and the number of particles that can be distributed in this volume is limited by computer memory; modern simulations work typically with 256^3 points and the same number of particles, though larger simulations have also been carried out; this immediately yields the size of grid cells, of order $0.5h^{-1}\,\mathrm{Mpc}$. This comoving length, if located at a redshift of $z \sim 0.3$ (which is about the most relevant for cosmic shear), subtends an angle of roughly $2'$ on the sky. The finite number of particles yields the mass resolution of the simulations, which is typically $\sim 10^{10}h^{-1}M_\odot$, depending on cosmological parameters.

In order to obtain higher spatial resolution, force calculations are split up into near-field and far-field forces. The gravitational force due to the distant matter distribution is obtained by grid-based FFT methods, whereas the force from nearby masses is calculated from summing up the forces of individual particles; such simulations yield considerably higher resolution of the resulting mass distribution. Since the matter in these simulations is represented by massive particles, these can undergo strong interactions, leading to (unphysical) large orbital deflections. In order to avoid these unphysical strong collisions, the force between pairs of particles is modified at short distances, typically comparable to the mean separation of two particles in the simulation. This softening length defines the minimum length scale on which the results from numerical simulations can be considered reliable. Cosmological simulations consider either the dark matter only or, more recently, the hydrodynamics effects of baryons have been incorporated as well.

The outcome of such simulations, as far as they are relevant here, are the 3-dimensional positions of the matter particles at different (output) times or redshifts. In order to study the light propagation through this simulated mass distribution, one employs multiple lens-plane theory. First, the volume between us and sources at some redshift z_s is filled with boxes from the cosmological simulations. That is, the comoving distance $w_\mathrm{s} = w(z_\mathrm{s})$ is split up into n intervals of length L, and the mass distribution at an output time close to $t_i = t(w = (i - 1/2)L)$ is considered to be placed at this distance. In this way, one has a light cone covered by cubes containing representative matter distributions. Since the mass distributions at the different times t_i are not independent of each other, but one is an evolved version of the earlier one, the resulting mass distribution is highly correlated over distances much larger than L. This can be avoided by making use of the statistical homogeneity and isotropy of the mass distribution: each box can be translated by an arbitrary two-dimensional vector, employing the periodicity of the mass distribution, and rotated by an arbitrary angle; furthermore, the three different projections of the box can be used for its orientation. In this way – a kind of recycling of numerical results – the worst correlations are removed.

Alternatively, one can combine the outputs from several simulations with different realizations of the initial conditions. In this case, one can use simulation boxes of different spatial extent, to match the comoving size of a big light cone as a function of redshift. That is, for a given light-cone size, only relatively small boxes are needed at low redshifts, and bigger ones at higher redshift (see White and Hu 2000).

Second, the mass in each of these boxes is projected along the line-of-sight, yielding a surface mass density at the appropriate comoving distance $w_i = (i - 1/2)L$. Each of these surface mass densities can now be considered a lens plane, and the propagation of light can be followed from one lens plane to the next; the corresponding theory was worked out in detail by Blandford and Narayan(1986; see also Chap. 9 of SEF), but applied as early as 1970 by Refsdal (1970) for a cosmological model consisting of point masses only (see also Schneider and Weiss 1988a,b). Important to note is that the surface mass density Σ in each lens plane is the projection of $\Delta\rho = \rho - \bar{\rho}$ of a box, so that for each lens plane, $\langle \Sigma \rangle = 0$. As has been shown in Seitz et al. (1994), this multiple lens-plane approach presents a well-defined discretization of the full 3-dimensional propagation equations. Light bundles are deflected and distorted in each lens plane and thus represented as piecewise straight rays. The resulting Jacobi matrix $\mathcal{A}$ is then obtained as a sum of products of the tidal matrices in the individual lens planes, yielding a discretized version of the form (88) for $\mathcal{A}$. The result of such simulations is then the matrix $\mathcal{A}(\boldsymbol{\theta})$ on a predefined angular grid, as well as the positions $\boldsymbol{\beta}(\boldsymbol{\theta})$ in the source plane. The latter will not be needed here, but have been used in studies of multiple images caused by the LSS (see Wambsganss et al. 1998).

One needs special care in applying the foregoing prescription; in particular, in the smoothing process to obtain a mass distribution from the discrete particles; Jain et al. (2000) contains a detailed discussion on these issues.[10] The finite spatial resolution in the simulations translates into a redshift-dependent angular resolution, which degrades for the low redshift lens planes; on the other hand, those have a small impact on the light propagation due to the large value of $\Sigma_{\rm cr}$ for them [see (10) of IN]. The discreteness of particles gives rise to a shot-noise term in the mass distribution, yielding increased power on small angular scales.

[10] For other recent ray-tracing simulations related to cosmic shear, see e.g. Barber et al. (2000); Hamana and Mellier (2001); Premadi et al. (2001); Taruya et al. (2002); Fluke et al. (2002); Barber (2002); Vale and White (2003).

Results From Ray-Tracing Simulations

We shall summarize here some of the results from ray-tracing simulation:

- Whereas the Jacobi matrix in this multi-deflection situation is no longer symmetric, the contribution from the asymmetry is very small. The power spectrum of the asymmetric part of $\mathcal{A}$ is at least three orders of magnitude smaller than the power spectrum P_κ, for sources at $z_s = 1$ (Jain et al. 2000). This result is in accord with analytical expectations (e.g., Bernardeau et al. 1997; Schneider et al. 1998a), i.e., that terms quadratic in the Newtonian potential are considerably smaller than first-order terms, and supports the validity of the Born approximation. Furthermore, this result suggests that a simpler method for predicting cosmic shear distributions from numerical simulations may be legitimate, namely to project the mass distribution of all lens planes along the grid of angular positions, with the respective weighting factors, according to (92), i.e., employing the Born approximation. Of course this simplified method is computationally much faster than the full ray-tracing.
- The power spectra obtained reproduce the ones derived using (99), over the range of wavevectors which are only mildly affected by resolution and discreteness effects. This provides an additional check on the accuracy of the fitting formulae for the non-linear power spectrum.
- The simulation results give the full two-dimensional shear map, and thus can be used to study properties other than the second-order ones, e.g., higher-order statistics, or the occurrence of circular shear patters indicating the presence of strong mass concentrations. An example of such maps is shown in Fig. 26. These shear maps can be used to simulate real surveys, e.g, including the holes in the data field resulting from masking or complicated survey geometries, and thus to determine the accuracy with which the power spectra can be determined from such surveys. Note that in order to quantify the error (or covariance matrix) of any second-order statistics, one needs to know the fourth-order statistics, which in general cannot be obtained analytically when outside the linear (Gaussian) regime. Simulations are also used to obtain good survey strategies.

Higher-Order Correction Terms

Up to now we have considered the lowest-order approximation of the Jacobi matrix (88) and have argued that this provides a sufficiently accurate description. Higher-order terms in Φ were neglected since we argued that, because the Newtonian potential is very small, these should play no important role. However, this argument is not fully correct since, whereas the potential certainly is small, its derivatives are not necessarily so. Of course, proper ray-tracing simulation take these higher-order terms automatically into account.

We can consider the terms quadratic in Φ when expanding (88) to higher order. There are two such terms, one containing the product of second-order derivatives of Φ, the other a product of first derivatives of Φ and its third derivatives. The former is due to lens-lens coupling. The shear and surface mass densities from different redshifts (or lens planes, in the discretized approximation) do not simply add, but multi lens plane theory shows that the tidal matrices from different lens planes get multiplied. The latter term comes from dropping the Born approximation and couples the deflection of a light ray (first derivative of Φ) with the change of the tidal matrix with regards to the position (third derivatives of Φ). These terms are explicitly given in the appendix of Schneider et al. (1998a), in Bernardeau et al. (1997) and in Cooray and Hu (2002) and found to be indeed small, providing corrections of at most a few percent. Furthermore, Hamana (2001) has shown that the magnification bias caused by the foreground matter inhomogeneities on the selection of background galaxies has no practical effect on second-order cosmic shear statistics.

Another effect that affects the power spectrum P_κ is the difference between shear and reduced shear, the latter being the observable. Since the correlation function of the reduced shear is the correlation function of the shear plus a term containing a product of two shears and one surface mass density, this correction depends linearly on the third-order statistical properties of the projected mass κ. Also this correction turns out to be very small; moreover, it does not give rise to any B-mode contribution (Schneider et al. 2002a).

7 Large-Scale Structure Lensing: Results

After the theory of cosmic shear was considered in some detail in the previous section, we shall summarize here the observational results that have been obtained so far. In fact, as we will see, progress has been incredibly fast over the past ~four years, with the first detections reported in 2000, and much larger surveys being available by now, with even larger ones ongoing or planned. Already by now, cosmic shear is one of the pillars on which our cosmological model rests.

The predictions discussed in the previous section have shown that the rms value of cosmic shear is of the order of ~2 % on angular scales of ~1′, and considerably smaller on larger scales. These small values make the measurements of cosmic shear particularly challenging, as the observational and instrumental effects described in Sect. 3 are expected to be larger than the cosmic shear signal, and thus have to be understood and removed with great precision. For example, the PSF anisotropy of nearly all wide-field cameras is considerably larger than a few percent and thus needs to be corrected for. But, as also discussed in Sect. 3, methods have been developed and thoroughly tested which are able to do so.

7.1 Early Detections of Cosmic Shear

Whereas the theory of cosmic shear was worked out in the early 1990's (Blandford et al. 1991; Miralda-Escudé 1991; Kaiser 1992), it took until the year 2000 before this effect was first discovered.[11] The reason for this evolution must be seen by a combination of instrumental developments, i.e. the wide-field CCD mosaic cameras, and the image analysis software, like IMCAT (the software package encoding the KSB method discussed in Sect. 3.5), with which shapes of galaxies can be accurately corrected for PSF effects. Then in March 2000, four groups independently announced their first discoveries of cosmic shear (Bacon et al. 2000; Kaiser et al. 2000; van Waerbeke et al. 2000, Wittman et al. 2000). In these surveys, of the order of 10^5 galaxy images have been analyzed, covering about 1 deg^2. Later that year, Maoli et al. (2001) reported a significant cosmic shear measurement from 50 widely separated FORS1@VLT images, each of size $\sim 6\overset{\prime}{.}5 \times 6\overset{\prime}{.}5$, which also agreed with the earlier results. The fact that the results from four independent teams agreed within the respective error bars immediately lend credit to this new window of observational cosmology. This is also due to the fact that 4 different telescopes, 5 different cameras (the UH8K and CFH12K at CFHT, the $8' \times 16'$-imager on WHT, the BTC at the 4m-CTIO telescope and FORS1 at the VLT), independent data reduction tools and at least two different image analysis methods have been used. These early results are displayed in Fig. 39, where the (equivalent) shear dispersion is plotted as a function of effective circular aperture radius, together with the predictions for several cosmological models. It is immediately clear that a high-normalization Einstein-de Sitter model can already be excluded from these early results, but the other three models displayed are equally valid approximations to the data.

Maoli et al. (2001) considered the constraints one obtains by combining the results from these five surveys, in terms of the normalization parameter σ_8 of the power spectrum. The confidence contours in the $\Omega_{\rm m} - \sigma_8$-plane are shown in Fig. 40. There is clearly a degeneracy between these two parameters from the data sets considered, roughly tracing $\sigma_8 \sim 0.59\Omega_{\rm m}^{-0.47}$; although the best fitting model is defined by $\Omega_{\rm m} = 0.26$, $\sigma_8 = 1.1$, it cannot be significantly distinguished from, e.g., a $\Omega_{\rm m} = 1$, $\sigma_8 = 0.62$ model since the error bars displayed in Fig. 39 are too large and the range of angular scales over which the shear was measured is too small. In Fig. 40, the solid curve displays the normalization as obtained from the abundance of massive clusters, which is seen to follow pretty much the valley of degeneracy from the cosmic shear analysis. This fact should not come as a surprise, since the cluster abundance probes the power spectrum on a comoving scale of about $8h^{-1}$ Mpc, which is

---

[11] An early heroic attempt by Mould et al. (1994) to detect cosmic shear on a single $\sim 9' \times 9'$ field only yielded an upper limit, and the putative detection of a shear signal by Schneider et al. (1998b; see also Fort et al. 1996) in three $2' \times 2'$ fields is, due to the very small sky area, of no cosmological relevance.

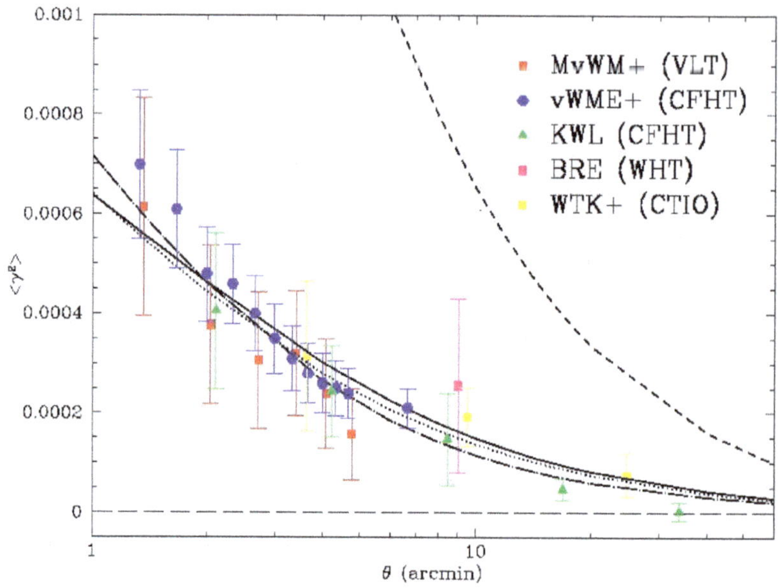

Fig. 39. Shear dispersion as a function of equivalent circular aperture radius as obtained from the first five measurements of cosmic shear (MvWM+: Maoli et al. 2001; vWME+: van Waerbeke et al. 2000; KWL: Kaiser, Wilson and Luppino 2000; BRE: Bacon, Refregier and Ellis 2000; WTK: Wittman et al. 2000). The data points within each team are not statistically independent, due to the fairly strong covariance of the shear dispersion on different angular scales, but points from different teams are independent (see text). The error bars contain the noise from the intrinsic ellipticity dispersion and, for some of the groups, also an estimate of cosmic variance. The four curves are predictions from four cosmological models; the upper-most one corresponds to an Einstein-de Sitter Universe with normalization $\sigma_8 = 1$, and can clearly be excluded by the data. The other three models are cluster normalized – see Sect. 4.4 of IN – and all provide equally good fits to these early data (courtesy: Y. Mellier)

comparable to the median scale probed by the cosmic shear measurements. However, the predictions of the cluster abundance rely on the assumption that the initial density field was Gaussian, whereas the cosmic shear prediction is independent of this assumption, which therefore can be tested by comparing the results from both methods.

7.2 Integrity of the Results

As mentioned before, the cosmic shear effects are smaller than many observational effects (like an anisotropic PSF) that could mimic a shear; it is therefore

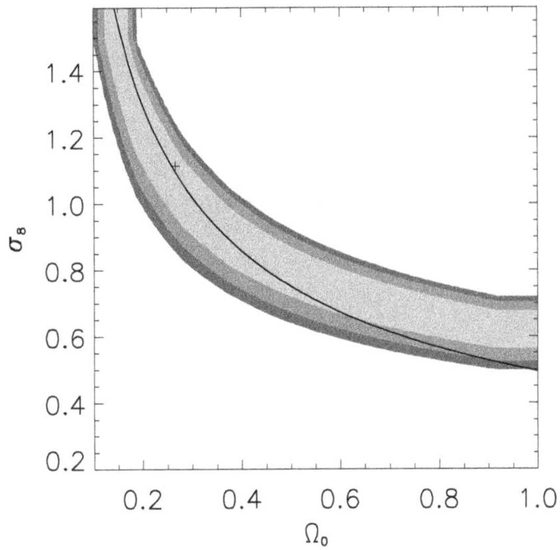

Fig. 40. Constraints on $\Omega_{\rm m}$ and σ_8 from the five surveys shown in Fig. 39; shown are 1, 2 and 3-σ confidence regions. The cross denotes the best-fitting model, but as can be seen, these two parameters are highly degenerate with the data used. The solid curve displays the constraint from cluster normalization (from Maoli et al. 2001)

necessary to exclude as much as possible such systematics from the data. The early results described above were therefore accompanied by quite a large number of tests; they should be applied to all cosmic shear surveys as a sanity check. A few of those shall be mentioned here.

Stellar Ellipticity Fits

The ellipticity of stellar objects should be well fitted by a low-order function, so one is able to predict the PSF anisotropy at galaxy locations. After subtracting this low-order fit from the measured stellar ellipticities, there should be no coherent spatial structure remaining, and the ellipticity dispersion of the corrected ellipticities should be considerably smaller that the original ones, essentially compatible with measurement noise.

Correlation of PSF Anisotropy with Corrected Galaxy Ellipticities

After correcting for the anisotropy of the PSF, there should remain no correlation between the corrected galaxy ellipticities and the ellipticity of the PSF. This correlation can be measured by considering $\langle \epsilon\, \epsilon^* \rangle$, where ϵ is the corrected galaxy ellipticity (31), and ϵ^* the uncorrected stellar ellipticity (i.e., the PSF anisotropy). Bacon et al. (2000) found that for fairly low signal-to-noise galaxy images, this correlation was significantly different from zero, but for galaxies with high S/N (only those entered their cosmic shear analysis), no significant correlation remained. The same was found in van Waerbeke et al. (2000), except that the average $\langle\epsilon_1\rangle$ was slightly negative, but independent of ϵ_1^*. The level of $\langle\epsilon_1\rangle$ was much smaller than the estimated cosmic shear, and does not affect the latter by more than 10 %.

Spatial Dependence of Mean Galaxy Ellipticity

When a cosmic shear survey consist of many uncorrelated fields, the mean galaxy ellipticity at a given position on the CCD chips should be zero, due to the assumed statistical isotropy of the shear field. If, on the other hand, the shear averaged over many fields shows a dependence on the chip position, most likely optical distortions and/or PSF effects have not been properly accounted for.

Parity Invariance

The two-point correlation function $\xi_\times(\theta) = \langle\gamma_t\gamma_\times\rangle(\theta)$ is expected to vanish for a density distribution that is parity symmetric. More generally, every astrophysical cause for a 'shear' signal (such as intrinsic galaxy alignments, or higher-order lensing effects) is expected to be invariant under parity transformation. A significant cross-correlation $\xi_\times$ would therefore indicate systematic effects in the observations and/or data analysis.

7.3 Recent Cosmic Shear Surveys

Relatively soon after the announcement of the first cosmic shear detections, additional results were published. These newer surveys can roughly be classified as follows: deep surveys, shallower, but much wider surveys, and special surveys, such as obtained with the Hubble Space Telescope. We shall mention examples of each of these classes here, without providing a complete list.

Deep Surveys

Currently the largest of the deep surveys from which cosmic shear results have been published is the VIRMOS-DESCART survey, carried out with the CFH12K camera at the CFHT; this camera covers about $45' \times 30'$ in one exposure. The exposure time of the images, taken in the I-band, is one hour. The survey covers four fields of $2° \times 2°$ each, of which roughly $8.5\,\mathrm{deg}^2$ have been used for a weak lensing analysis up to now (van Waerbeke et al. 2001, 2002). About 20% of the area is masked out, to account for diffraction spikes, image defects, bright and large foreground objects etc. The number density of galaxy images used for the cosmic shear analysis is about $17\,\mathrm{arcmin}^{-2}$. A small part of this survey was used for the early cosmic shear detection (van Waerbeke et al. 2000). Compared to the earlier results, the error bars on the shear

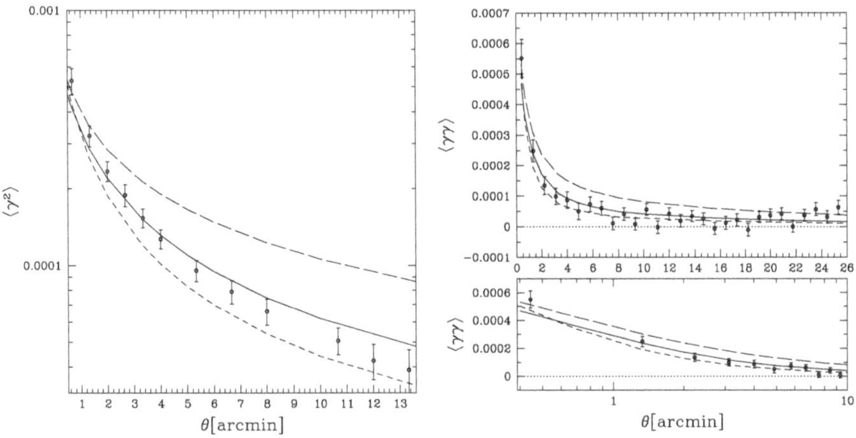

Fig. 41. The shear dispersion as a function of aperture radius (*left*) and the shear correlation function $\xi_+(\theta)$ (*right*) as measured from the VIRMOS-DESCART survey (van Waerbeke et al. 2001). The lower panel on the right shows an enlargement with logarithmic axis of the larger figure. The error bars were calculated from simulations in which the galaxy images have been randomized in orientation. The curves show predictions from three different cosmological models, corresponding to $(\Omega_\mathrm{m}, \Omega_\Lambda, \sigma_8) = (0.3, 0, 0.9)$ (open model, *short-dashed curves*), $(0.3, 0.7, 0.9)$ (low-density flat model, *solid curves*), and $(1, 0, 0.6)$ (Einstein-de Sitter Universe, long-dashed curves). In all cases, the shape parameter of the power spectrum was set to $\Gamma_\mathrm{spect} = 0.21$. The redshift distribution of the sources was assumed to follow the law (128), with $\alpha = 2$, $\beta = 1.5$ and $z_0 = 0.8$, corresponding to a mean redshift of $\bar{z} \approx 1.2$

measurements are greatly reduced, owing to the much better statistics. We show in Fig. 41 the shear dispersion and the correlation function as measured from this survey. Furthermore, this survey yielded the first detection of a significant $\langle M_\mathrm{ap}^2 \rangle$-signal; we shall come back to this later. In order to compare the measured shear signal with cosmological predictions, one needs to assume a redshift distribution for the galaxies; a frequently used parameterization for this is

$$p(z) = N \left(\frac{z}{z_0} \right)^\alpha \exp\left[-\left(\frac{z}{z_0} \right)^\beta \right] , \tag{128}$$

where α and β determine the shape of the redshift distribution, z_0 the characteristic redshift, and N is a normalization factor, chosen such as $\int \mathrm{d}z\, p(z) = 1$.

Another example of a deep survey is the Suprime-Cam survey (Hamana et al. 2003), a $2.1\,\mathrm{deg}^2$ survey taken with the wide-field camera Suprime-Cam (with a $34' \times 27'$ field-of-view) at the 8.2-m Subaru telescope. With an exposure time of 30 min, the data is considerably deeper than the VIRMOS-DESCART survey, due to the much larger aperture of the telescope. After cuts in the object catalog, the resulting number density of objects used for the weak lensing analysis is $\approx 30\,\mathrm{arcmin}^{-2}$. Figure 42 shows how small the

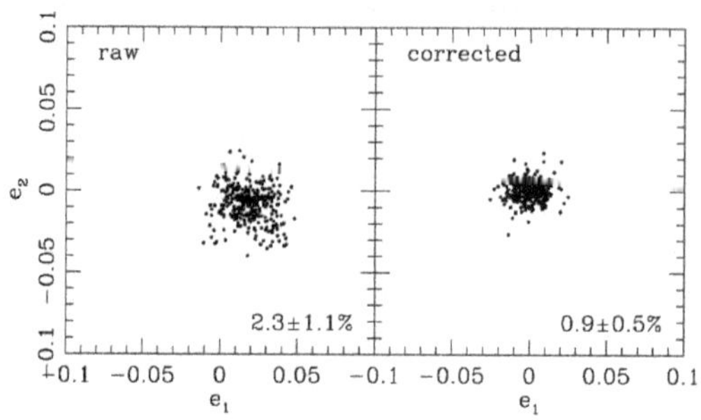

Fig. 42. Stellar ellipticities before and after correction for PSF anisotropies in the Suprime-Cam survey. Numbers give mean and dispersion of stellar ellipticities $|\chi|$ (from Hamana et al. 2003)

PSF anisotropy is, and that the correction with a fifth-order polynomial over the whole field-of-view in fact reduces the remaining stellar ellipticities considerably. This survey has detected a significant cosmic shear signal, as measured by the shear correlation functions and the aperture mass dispersion, over angular scales $2' \lesssim \theta \lesssim 40'$. The shear signal increases as fainter galaxies are used in the analysis, as expected, since fainter galaxies are expected to be at larger mean redshift and thus show a stronger shear signal.

Bacon et al. (2003) combine images taken at the Keck II telescope and the WHT. For the former, 173 fields were used, each having a f.o.v. of $2' \times 8'$; and the data from WHT were obtained from 20 different fields, covering about $1\,\mathrm{deg}^2$ in total. The large number of fields minimizes the sample variance of this particular survey, and the two instruments used allowed a cross-check of instrumental systematics.

Very Wide Surveys

Within a given observing time, instead of mapping a sky region to fairly deep magnitudes, one can also map larger regions with smaller exposure time; since most of the surveys have been carried out with goals in addition to cosmic shear, the survey strategy will depend on these other considerations. We shall mention two very wide surveys here.

Hoekstra et al. (2002a; also Hoekstra et al. 2002b) used the Red Cluster Sequence (RCS) survey, a survey designed to obtain a large sample of galaxy clusters using color selection techniques (Gladders and Yee 2000). The cosmic shear analysis is based on $53\,\mathrm{deg}^2$ of R_C-band data, spread over 13 patches on the sky and observed with two different instruments, the CFH12K@CFHT for Northern fields, and the Mosaic II camera at the CTIO 4m telescope in the South. The integration times are 900 s and 1200 s, respectively. The shear dispersion as measured with the two instruments are in satisfactory agreement and thus can be safely combined. Owing to the shallower magnitude, the detected shear is smaller than in the deeper surveys mentioned above: on a scale of 2.5 arcmin, the shear dispersion is $\langle|\bar{\gamma}|^2\rangle \sim 4 \times 10^{-5}$ in the RCS survey, compared to $\sim 2 \times 10^{-4}$ in the deeper VIRMOS-DESCART survey (see Fig. 41), in accordance with expectations.

Jarvis et al. (2003) presented a cosmic shear survey of 75 deg^2, taken with the BTC camera and the Mosaic II camera on the CTIO 4m telescope, with about half the data taken with each instrument. The survey covers 12 fields, each with sidelength of $\sim 2.5°$. For each pointing, three exposures of 5 min were taken, making the depth of this survey comparable to the RCS. A total of $\sim 2 \times 10^6$ galaxies with $R \leq 23$ were used for the shear analysis. Since this survey has some peculiar properties which are very educational, it will be discussed in somewhat more detail. The first point to notice is the large pixel size of the BTC, of 0.″43 per pixel – for comparison, the CFH12K has ∼0.″20 per pixel. With a median seeing of 1.″05, the PSF is slightly undersampled with the BTC. Second, the PSF anisotropy on the BTC is very large, as shown in Fig. 43 – a large fraction of the exposures has stellar images with ellipticities higher than 10%. Obviously, this renders the image analysis and the correction for PSF effects challenging. As shown on the right-hand part of Fig. 43, this challenge is indeed met. This fact is very nicely illustrated in Fig. 44, where the corrected stellar ellipticities are shown as a function of the PSF anisotropy; in essence, the correction reduces the PSF anisotropy by nearly a factor of 300 !

The third point to notice is that the image analysis for this survey has not been carried out with IMCAT (as for most of the other surveys), but by a different image analysis method described in Bernstein and Jarvis (2002).

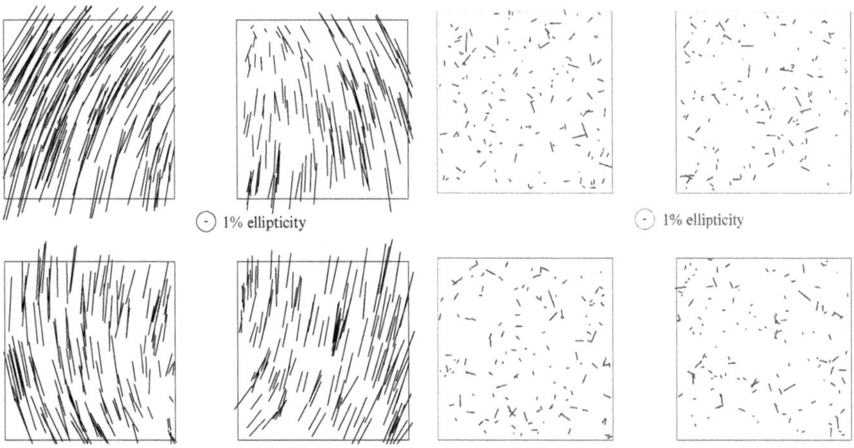

Fig. 43. On the left-hand side, the raw ellipticities of stars are shown for the four CCDs of the BTC instrument; for reference, a 1% ellipticity is indicated. After correcting for the PSF anisotropy, the remaining stellar ellipticities (shown on the right) are of order 1–2%, and essentially uncorrelated with position on the chip, i.e., they are compatible with measurement noise (from Jarvis et al. 2003)

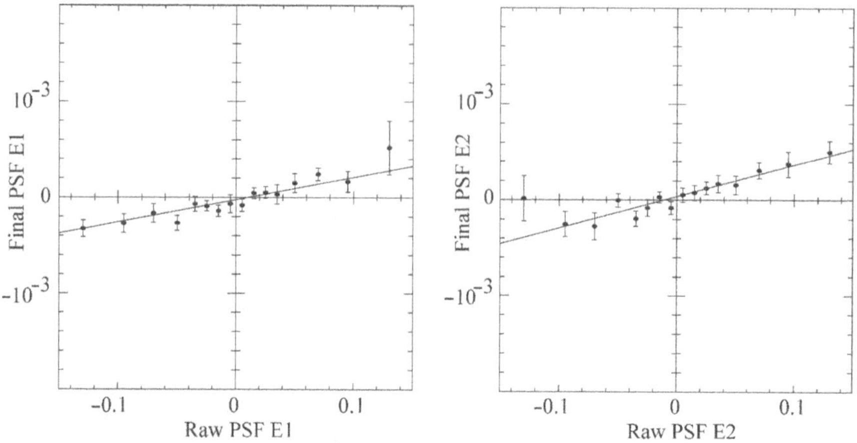

Fig. 44. These two plots show the two components of the stellar ellipticities as measured on the data (x-axis) and after correction, from the Jarvis et al. (2003) survey. The slope of the straight line is about $1/300$, meaning that the strong PSF anisotropy can be corrected for up to this very small residual. The final PSF anisotropy is well below 5×10^{-4}. This figure, together with Fig. 43, demonstrates how well the procedures for PSF corrections work (from Jarvis et al. 2003)

In this respect, this survey is independent of all the others described in this section; it is important to have more than one image analysis tool to check potential systematics of either one.

One of the amazing results from the CTIO cosmic shear survey is that the shear dispersion can be measured with about a 3σ significance on each of the 12 fields. Hence, this provides a shear dispersion measurement on scales larger than 1 degree (the radius of a circle with area of the mean area of the 12 fields of $\sim$6.2 deg^2); the shear dispersion on these scales is $\left\langle |\bar{\gamma}|^2 \right\rangle = 0.0012 \pm 0.0003$.

Special Surveys

There are a number of cosmic shear surveys which cover a much smaller total area than the ones mentioned above, and are thus not competitive in terms of statistical accuracy, but which have some special properties which give them an important complementary role. One example are surveys carried out with the Hubble Space Telescope. Since for them the PSF is much smaller than for ground-based observations, PSF corrections in measuring galaxy ellipticities are expected to be correspondingly smaller. The drawback of HST observations is that its cameras, at least before the installment of the ACS, have a small field-of-view, less than 1 arcmin^2 for the STIS CCD, and about 5 arcmin^2 for WFPC2. This implies that the total area covered by HST surveys are smaller than those achievable from the ground, and that the number of stars per field are very small, so that PSF measurements are typically not possible on those frames which are used for a cosmic shear analysis. Hence, the PSF needs to be measured on different frames, e.g., taken on star clusters, and one needs to assume (this assumption can be tested, of course) that the PSF is fairly stable in time. In fact, this is not really true, as the telescopes moves in and out the Earth's shadow every orbit, thereby changing its temperature and thus changing its length (an effect called breathing). A further potential problem of HST observations is that the WFPC2 has a pixel scale of $0\farcs1$ and thus substantially undersamples the PSF; this is likely to be a serious problem for very faint objects whose size is not much larger than the PSF size.

Cosmic shear surveys from two instruments onboard HST have been reported in the literature so far. One of the surveys uses archival data from the Medium Deep Survey, a mostly parallel survey carried out with the WFPC2. Refregier et al. (2002) used 271 WFPC2 pointings observed in the I-band, selected such that each of them is separated from the others by at least $10'$ to have statistically independent fields. They detected a shear dispersion on the scale of the WFC-chips (which is equivalent to a scale $\theta \sim 0\farcm72$) of $\left\langle|\bar{\gamma}|^2\right\rangle \sim 3.5 \times 10^{-4}$, which is a 3.8σ detection. The measurement accuracy is lower than that, owing to cosmic variance and uncertainties in the redshift distribution of the sources. Hämmerle et al. (2002) used archival parallel data taken with STIS; from the 121 fields which are deep enough, have multiple exposures, and are at sufficiently high galactic latitude, they obtained a shear dispersion of $\left\langle|\bar{\gamma}|^2\right\rangle \sim 15 \times 10^{-4}$ on an effective scale of $\sim 30''$, a mere 1.5σ detection. This low significance is due to the small total area covered by this survey. On the other hand, since the pixel scale of STIS is half of that of WFPC2, the undersampling problem is much less in this case. A larger set of STIS parallel observations were analyzed with respect to cosmic shear by Rhodes et al. (2004) and Miralles et al. (2005). Whereas Rhodes et al. obtained a significant ($\sim 5\sigma$) detection on an angular scale of $\sim 30''$, Miralles et al. concluded that the degradation of the STIS CCD in orbit regarding the charge transfer efficiency prevents a solid measurement of weak lensing. The discrepancies between these two works, which are based to a large degree on the same data set, is unclear at present. Personally I consider this discrepancy as a warning sign that weak lensing measurement based on small fields-of-view, and correspondingly too few stars to control the PSF on the science exposures, need to be regarded with extreme caution.

The new ACS onboard HST offers better prospects for cosmic shear measurements, since it has a substantially larger field-of-view. A first result was derived by Schrabback (2004), again based on parallel data. He found that the PSF is not stable in time, but that the anisotropy pattern changes among only a few characteristic patterns. He used those as templates, and the (typically a dozen) stars in the science frames to select a linear combination of these templates for the PSF correction of individual frames, thereby obtaining a solid detection of cosmic shear from the early ACS data.

A further survey that should be mentioned here is the one conducted on COMBO-17 fields (Brown et al. 2003). COMBO-17 is a one square degree survey, split over four fields, taken with the WFI at the ESO/MPG 2.2m telescope on La Silla, in 5 broad-band and 12 medium-band filters. In essence, therefore, this multi-band survey produces low-resolution spectra of the objects and thus permits to determine very accurate photometric redshifts of the galaxies taken for the shear analysis. Therefore, for the analysis of Brown et al., the redshift distribution of the galaxies is assumed to be very well known and not a source of uncertainty in translating the cosmic shear measurement into a constraint on cosmological parameters. We shall return to this aspect in Sect. 7.6. The data set was reanalyzed by Heymans et al. (2004) where special care has been taken to identify and remove the signal coming from intrinsic alignment of galaxy shapes.

7.4 Detection of B-Modes

The recent cosmic shear surveys have measured the aperture mass dispersion $\langle M_{\rm ap}^2(\theta)\rangle$, as well as its counterpart $\langle M_\perp^2(\theta)\rangle$ for the B-modes (see Sect. 6.5). These aperture measures are obtained in terms of the directly measured shear correlation functions, using the relations (125). As an example, we show in Fig. 45 the aperture measures as obtained from the Red Cluster Sequence survey (Hoekstra et al. 2002a). A significant measurement of $\langle M_{\rm ap}^2(\theta)\rangle$ is obtained over quite a range of angular scales, with a peak around a few arcminutes, as predicted from CDM power spectra (see Fig. 35). In addition to that, however, a significant detection of $\langle M_\perp^2(\theta)\rangle$ signifies the presence of B-modes. As discussed in Sect. 6.5, those cannot be due to cosmic shear. The only plausible explanation for them, apart from systematics in the observations and data analysis, is an intrinsic alignment of galaxies. If this is the cause of the B-modes, then one would expect that the relative contribution of the B-mode signal decreases as higher-redshift galaxies are used for the shear measurement. In fact, this expectation is satisfied, as shown in Fig. 45, where the galaxy sample is split into a bright and faint part, and the relative amplitude of the B-mode signal is smaller for the fainter (and thus presumably more distant) sample.

Similar detections of a B-mode signal have been obtained by the other surveys. For example, van Waerbeke et al. (2001) reported a significant B-mode signal on angular scales of a few arcminutes. In the reanalysis of the VIRMOS-DESCART data, van Waerbeke et al. (2002) reported that the B-mode on these scales was caused by the polynomial PSF anisotropy fit: the third-order function (fitted for each chip individually) has its largest amplitude near the boundary of the chips and is least well constrained there, unless one finds stars close to these edges. If a second-order polynomial fit is used, the B-modes on a few arcminute scales disappear. van Waerbeke et al. (2002) calculate the aperture statistics from the uncorrected stellar ellipticities in their survey and found that the 'E- and B-modes' of the PSF anisotropy have very similar amplitude and shape (as a function of θ). This similarity is unlikely to change in the course of the PSF correction procedure. Thus,

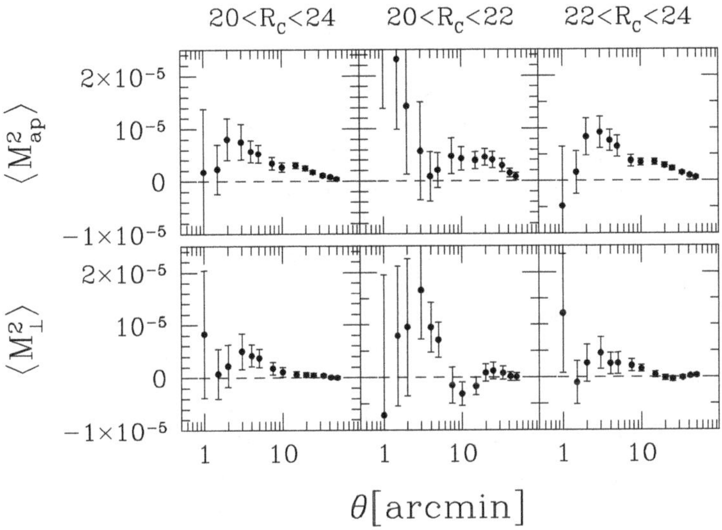

Fig. 45. The aperture mass dispersion $\langle M_{\rm ap}^2(\theta)\rangle$ (*top panels*) and the cross aperture dispersion $\langle M_\perp^2(\theta)\rangle$ (*bottom panels*) from the RCS survey (Hoekstra et al. 2002a). In the left panels, all galaxies with apparent magnitude $20 \le R_C \le 24$ are used, the middle and right panels show the same statistics for the brighter and fainter subsamples of background galaxies, respectively. Error bars in the former are larger, owing to the smaller number of bright galaxies

they argue, that if the B-mode is due to systematics in the data analysis, a systematic error of very similar amplitude will also affect the E-mode. Jarvis et al. (2003) found a significant B-mode signal on angular scales below $\sim 30'$; hence, despite their detection of an E-mode signal over a large range of angular scales $1' \lesssim \theta \lesssim 100'$, one suspects that part of this signal might be due to non-lensing effects.

Given our lack of understanding about the origin of the B-mode signal, and the associated likelihood that any effect causing a B-mode signal also contributes a non-lensing part to the E-mode signal, one needs a prescription on how to use the detected E-mode signal for a cosmological analysis. Depending on what one believes the B-modes are due to, this prescription varies. For example, if the B-mode is due to a residual systematic, one would add its signal in quadrature to the error bars of the E-mode signal, as done in van Waerbeke et al. (2002). On the other hand, if the B-mode signal is due to intrinsic alignments of galaxies, as is at least suggested for the RCS survey from Fig. 45 owing to its dependence on galaxy magnitudes, then it could be more reasonable to subtract the B-mode signal from the E-mode signal, if one assumes that intrinsic alignments produce similar amplitudes of both modes [which is far from clear, however; Mackey et al. (2002) find that the E-mode signal from intrinsic alignments is expected to be ~ 3.5 times higher than the corresponding B-mode signal].

Owing to the small size of the fields observed with the early HST instruments, no E/B-mode decomposition can be carried out from these surveys – the largest size of these fields is smaller than the angular scale at which the aperture mass dispersion is expected to peak (see Fig. 35). However, future cosmic shear studies carried out with ACS images will most likely be able to detect, or set upper bounds on the presence of B-modes.

In fact, it is most likely that (most of) the B-mode signal seen in the cosmic shear surveys is due to remaining systematics. Hoekstra (2004) investigated the PSF anisotropy of the CFH12k camera using fields with a high number density of stars. Randomly selecting about 100 stars per CCD, which is the typical number observed in high galactic latitute fields, he fitted a second-order polynomial to these stars representing the PSF anisotropy. Correcting with this model all the stars in the field, the remaining stellar ellipticities carry substantial E- and B-mode signals, essentially on all angular scales, but peaking at about the size of a CCD. A substantially smaller residual is obtained if the ellipticities of stars in one of the fields is corrected by a more detailed model of the PSF anisotropy as measured from a different field; this improvement indicates that the PSF anisotropy pattern in the data set used by Hoekstra is fairly stable between different exposures. This, however, is not necessarily the case in other datasets. Nevertheless, if one assumes that the PSF anisotropy is a superposition of two effects, one from the properties of the telescope and instrument itself, the other from the specific observation procedure (e.g., tracking, wind shake, etc.), and further assuming that the latter one affects mainly the large-scale properties of the anisotropy pattern, then a superposition of a PSF model (obtained from a dense stellar field and describing the small-scale properties of the anisotropy pattern) plus a low-order polynomial can be a better representation of the PSF anisotropy. This indeed was verified in the tests made by Hoekstra (2004). In their reanalysis of the VIRMOS-DESCART survey, van Waerbeke et al. (2005) have fitted the PSF anisotropy with a rational function, instead of a polynomial. This functional form was suggested by the study of Hoekstra (2004). When correcting the galaxy ellipticities with this new PSF model, essentially no more B-modes in the VIRMOS-DESCART survey are detected. Further studies on PSF anisotropy corrections need to be conducted; possibly the optimal way of dealing with them will be instrument-specific.

7.5 Cosmological Constraints

The measured cosmic shear signal can be translated into constraints on cosmological parameters, by comparing the measurements with theoretical predictions. In Sect. 6.4 we have outlined how such a comparison can be made; there, we have concentrated on the shear correlation functions as the primary observables. However, the detection of significant B-modes in the shear field makes the aperture measures the 'better' statistics to compare with predictions. They can be calculated from the shear correlation functions, as shown in (125). Calculating a likelihood function from the aperture mass dispersion proceeds in the same way as outlined in Sect. 6.4 for the correlation functions.

We have argued in Sect. 6.3 that $\langle M_{\rm ap}^2(\theta)\rangle$ provides very localized information about the power spectrum $P_\kappa(\ell)$ and is thus a very useful statistic. One therefore might expect that the aperture mass dispersion as calculated from the shear correlation functions contains essentially all the second-order statistical information of the survey. This is not true, however; one needs to recall that the shear correlation function ξ_+ is a low-pass filter of the power spectrum, and thus contains information of P_κ on angular scales larger than the survey size. This information is no longer contained in the aperture mass dispersion, owing to its localized associated filter. Therefore, in order to keep this long-range information in the comparison with theoretical predictions, it is useful to complement the estimates of $\langle M_{\rm ap}^2(\theta)\rangle$ with either the shear dispersion, or the correlation function ξ_+, at a scale which is not much smaller than the largest scale at which $\langle M_{\rm ap}^2(\theta)\rangle$ is measured. Note, however, that this step implicitly assumes that on these large angular scales, the shear signal is essentially free of B-mode contributions. If this assumption is not true, and cannot be justified from the survey data, then this additional constraint should probably be dropped.

The various constraints on parameters that have been derived from the cosmic shear surveys differ in the amount of prior information that has been used. As an example, we consider the analysis of van Waerbeke et al. (2002). These authors have considered a model with four free parameters: $\Omega_{\rm m}$, the normalization σ_8, the shape parameter $\Gamma_{\rm spect}$ and the characteristic redshift $z_{\rm s}$ (or, equivalently, mean redshift $\bar{z}_{\rm s}$) of their galaxy sample, assuming a flat Universe, i.e., $\Omega_\Lambda = 1 - \Omega_{\rm m}$. They have used a flat prior for $\Gamma_{\rm spect}$ and $\bar{z}_{\rm s}$ in a fairly wide interval over which they marginalized the likelihood function (see Fig. 46). Depending on the width of these intervals, the confidence regions are more or less wide. It should be noted that the confidence contours close if $\Gamma_{\rm spect}$ and $\bar{z}_{\rm s}$ are assumed to be known (see van Waerbeke et al. 2001), but when these two parameters are kept free, $\Omega_{\rm m}$ and σ_8 are degenerate.

The right panel of Fig. 46 shows the corresponding constraints as obtained from the RCS survey. Since this survey is shallower and only extends to magnitudes where spectroscopic surveys provide information on their redshift distribution, the range of $\bar{z}_{\rm s}$ over which the likelihood is marginalized is smaller than for the VIRMOS-DESCART survey. Correspondingly, the confidence region is slightly smaller in this case. Even smaller confidence regions are obtained if external information is used: Hoekstra et al. (2002a) considered Gaussian priors with $\Omega_{\rm m} + \Omega_\Lambda = 1.02 \pm 0.06$, as follows from pre-WMAP CMB results, $\Gamma_{\rm spect} = 0.21 \pm 0.03$, as follows from the 2dF galaxy redshift survey, and $\bar{z}_{\rm s} = 0.59 \pm 0.02$, for which the width of the valley of maximum likelihood narrows considerably. Jarvis et al. (2003) used for their estimate of cosmological parameters the aperture mass dispersion at three angular scales plus the shear dispersion at $\theta = 100'$, and they considered alternatively the E-mode signal, and the E-mode signal $\pm$ the B-mode signal, to arrive at constraints

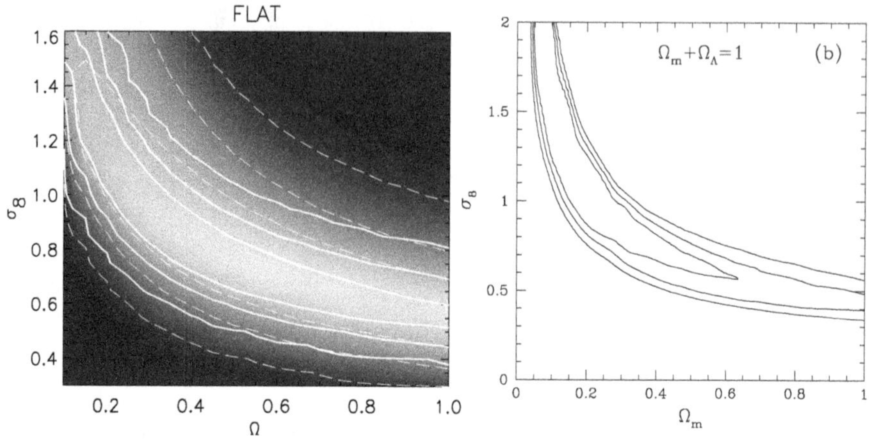

Fig. 46. Constraints on $\Omega_{\rm m}$ and σ_8 from two cosmic shear surveys. *Left*: The VIRMOS-DESCART survey (van Waerbeke et al. 2002). The grey-scale and dashed contours show the 68%, 95% and 99.9% confidence regions with a marginalization over the range $\Gamma_{\rm spect} \in [0.05, 0.7]$, and mean galaxy redshift in the range $\bar{z}_{\rm s} \in [0.50, 1.34]$, whereas the solid contours show the same confidence regions with the stronger priors $\Gamma_{\rm spect} \in [0.1, 0.4]$ and $\bar{z}_{\rm s} \in [0.8, 1.1]$. *Right*: The RCS survey (Hoekstra et al. 2002a,b,c,d), showing the 1, 2, and 3σ confidence regions for a prior $\Gamma_{\rm spect} \in [0.05, 0.5]$ and mean redshift $\bar{z}_{\rm s} \in [0.54, 0.66]$. In both cases, a flat Universe has been assumed

on the $\Omega_{\rm m}$–σ_8 parameter plane. Since the CTIO survey samples a larger angular scale than the other surveys (data at small angular scales are discarded owing to the large B-mode signal there), the results are much less sensitive to $\Gamma_{\rm spect}$; furthermore, for the same reason the Jarvis et al. results are much less sensitive to the fit of the non-linear power spectrum according to Peacock and Dodds (1996) which van Waerbeke et al. (2002) found to be not accurate enough for some cosmological models. In fact, if instead of the Peacock and Dodds fitting formula, the fit by Smith et al (2003) is used to describe the non-linear power spectrum, the resulting best estimate of σ_8 is decreased by 8% for the RCS survey (as quoted in Jarvis et al. 2003).

For the RCS and the CTIO surveys, the covariance matrix was obtained from field-to-field variations, i.e., ${\rm Cov}_{ij} = \langle (d_i - \mu_i)(d_j - \mu_j) \rangle$, where μ_i is the mean of the observable d_i (e.g., the aperture mass dispersion at a specific angular scale) over the independent patches of the survey, and angular brackets denote the average over all independent patches. The estimate of the covariance matrix for the VIRMOS-DESCART survey is slightly different, as it has only four independent patches.

To summarize the results from these surveys, each of them found that a combination of parameters of the form $\sigma_8 \Omega_{\rm m}^\alpha$ is determined best from the data, with $\alpha \sim 0.55$, where the exact value of α depends on the survey depth. If we consider the specific case of $\Omega_{\rm m} = 0.3$ which is close to the concordance value that was recently confirmed by WMAP, then the VIRMOS-DESCART survey yields $\sigma_8 = 0.94 \pm 0.12$, the RCS survey has $\sigma_8 = 0.91^{+0.05}_{-0.12}$, which improves to $\sigma_8 = 0.86^{+0.04}_{-0.05}$ if the stronger (Gaussian) priors mentioned above are used, and the CTIO survey yields $\sigma_8 = 0.71^{+0.12}_{-0.16}$, here as 2σ limits. Whereas these results are marginally in mutual agreement, the CTIO value for σ_8 is lower than the other two. The higher values are also supported by results from the WFPC2 survey by Refregier et al. (2002), who find $\sigma_8 = 0.94 \pm 0.17$, Bacon et al. (2003) with $\sigma_8 = 0.97 \pm 0.13$, and the earlier surveys discussed in Sect. 7.1. The only survey supporting the low value of the CTIO survey is COMBO17 (Brown et al. 2003; see also the reanalysis of this dataset by Heymans et al. 2004). Most likely, these remaining discrepancies will be clarified in the near future; see discussion below. It should also be noted that at least for some of the surveys, a large part of the uncertainty comes from the unknown redshift distribution of the galaxies; this situation will most likely improve, as efficient spectrographs with large multiplex capability become available at 10m-class telescopes, which will in the near future deliver large galaxy redshift surveys at very faint magnitudes. Those can be used to much better constrain the redshift distribution of the source galaxies in cosmic shear surveys.

7.6 3-D Lensing

As mentioned several times before, using individual source redshift information, as will become available in future multi-color wide-field surveys, can improve the cosmological constraints obtained from weak lensing. In this section we shall therefore summarize some of the work that has been published on this so-called 3-D lensing.

Three-dimensional Matter Distribution

Provided the redshifts of individual source galaxies are known (or estimated from their multiple colors), one can derive the 3-D matter distribution, not only its projection. The principle of this method can be most easily illustrated in the case of a flat Universe, for which the surface mass density $\kappa(\boldsymbol{\theta}, w)$ for sources at comoving distance w becomes – see (93)

$$\kappa(\boldsymbol{\theta}, w) = \frac{3H_0^2 \Omega_{\rm m}}{2c^2} \int_0^w \mathrm{d}w' \, \frac{w'(w - w')}{w} \frac{\delta(w'\boldsymbol{\theta}, w')}{a(w')} \; . \tag{129}$$

Multiplying this expression by w and differentiating twice yields

$$\frac{\mathrm{d}^2}{\mathrm{d}w^2} \left(w \, \kappa(\boldsymbol{\theta}, w) \right) = \frac{3H_0^2 \Omega_{\rm m}}{2c^2} \frac{w}{a(w)} \, \delta(w\boldsymbol{\theta}, w) \; ,$$

which therefore allows one to obtain the three-dimensional density contrast δ in terms of the surface mass densities κ at different source redshifts. As we have seen in Sect. 5, there are several methods how to obtain the surface mass density from the observed shear. To illustrate the 3-D method, we use the finite-field reconstruction in the form of (60), for which one finds

$$\delta(w\boldsymbol{\theta}, w) = \frac{2c^2}{3H_0^2\Omega_{\rm m}} \frac{a(w)}{w} \int \mathrm{d}^2\theta' \, \mathbf{H}(\boldsymbol{\theta}; \boldsymbol{\theta}') \cdot \frac{d^2}{\mathrm{d}w^2} \left[w \, \mathbf{u}_\gamma(\boldsymbol{\theta}', w) \right] \, . \tag{130}$$

Taylor (2001) derived the foregoing result, but concentrated on the 3-D gravitational potential instead of the mass distribution, and Bacon and Taylor (2003) and Hu and Keeton (2002) discussed practical implementations of this relation. First to note is the notorious mass-sheet degeneracy, which in the present context implies that one can add an arbitrary function of w to the reconstructed density contrast δ. This cannot be avoided, but if the data field is sufficiently large, so that averaged over it, the density contrast is expected to vanish, this becomes a lesser practical problem. For such large data fields, the above mass reconstruction can be substituted in favor of the simpler original Kaiser and Squires (1993) method. Still more freedom is present in the reconstruction of the gravitational potential. The second problem is one of smoothing: owing to the noisiness of the observed shear field, the w-differentiation (as well as the θ-differentiation present in the construction of the vector field $\mathbf{u}_\gamma$) needs to be carried out on the smoothed shear field. A discretization of the observed shear field, as also suggested by the finite accuracy of photometric redshifts, can be optimized with respect to this smoothing (Hu and Keeton 2002).

A first application of this methods was presented in Taylor et al. (2004) on one of the COMBO17 fields which contains the supercluster A 901/902. The clusters present clearly show up also in the 3-D mass map, as well as a massive structure behind the cluster A 902 at higher redshift. Already earlier, Wittman et al. (2001, 2003) estimated the redshifts of clusters found in their deep blank-field data by studying the dependence of the weak lensing signal on the estimated source redshifts, and subsequent spectroscopy showed that these estimates were fairly accurate.

Power Spectrum Estimates

A redshift-dependent shear field can also be used to improve on the cosmological constraints obtained from cosmic shear. Hu (1999) has pointed out that even crude information on the source redshifts can strongly reduce the uncertainties of cosmological parameters. In fact, the 3-D power spectrum can be constructed from redshift-dependent shear data (see, e.g., Heavens 2003, Hu 2002, and references therein). For illustration purposes, one can use the κ power spectrum for sources at fixed comoving distance w, which reads in a flat Universe – see (99)

$$P_\kappa(\ell, w) = \frac{9H_0^4\Omega_{\rm m}^2}{4c^4} \int_0^w \mathrm{d}w' \, \frac{(w-w')^2}{w^2 \, a^2(w')} \, P_\delta\left(\frac{\ell}{w'}, w'\right) \, . \tag{131}$$

Differentiating $w^2 P_\kappa$ three times w.r.t. w then yields (Bacon et al. 2004)

$$P_\delta(k,w) = \frac{2c^4}{9H_0^4 \Omega_\mathrm{m}^2}\, a^2(w)\, \frac{\mathrm{d}^3}{\mathrm{d}w^3}\left[w^2\, P_\kappa(wk,w)\right] \; . \tag{132}$$

In this way, one could obtain the three-dimensional power spectrum of the matter. However, this method is essentially useless, since it is both very noisy (due to the third-order derivatives) and throws away most of the information contained in the shear field, as it makes use only of shear correlations of galaxies having the same redshift, and not of all the pairs at different distances. A much better approach to construct the three-dimensional power spectrum is given, e.g., by Pen et al. (2003a).

In my view, the best use of three-dimensional data is to construct the shear correlators $\xi_\pm(\theta; z_1, z_2)$, as they contain all second-order statistical information in the data and at the same time allow the identification and removal of a signal from intrinsic shape correlations of galaxies (King and Schneider 2003). From these correlation functions, one can calculate a χ^2 function as in (116) and minimize it w.r.t. the wanted parameters. One problem of this approach is the large size of the covariance matrix, which now has six arguments (two angular separations and four redshifts). However, as shown in Simon et al. (2004), it can be calculated fairly efficiently, provided one assumes that the fourth-order correlations factorize into products of two-point correlators, i.e., Gaussian fields (if this assumption is dropped, the covariance must be calculated from cosmological N-body simulations).

Bacon et al. (2004) used the COMBO17 data to derive the shape of the power spectrum, using the redshift dependent shear correlations. They parameterize the power spectrum in the form $P(k,z) \propto A k^\alpha \mathrm{e}^{-sz}$, so that it is described by an amplitude A, a local slope α and a growth parameter s which describes how the amplitude of the power spectrum declines toward higher redshifts. In fact, the slope $\alpha = -1.2$ was fixed to the approximate value in ΛCDM models over the relevant range of spatial scales and redshifts probed by the COMBO17 data (since the data used cover only $1/2\ \mathrm{deg}^2$, reducing the number of free parameters by fixing α is useful). The evolution of the power spectrum is found with high significance in the data. Furthermore, the authors show that the use of redshift information improves the accuracy in the determination of σ_8 by a factor of two compared to the 2-D cosmic shear analysis of the same data (Brown et al. 2003).

The main application of future multi-waveband cosmic shear surveys will be to derive constraints on the equation of state of dark energy, as besides lensing there are only a few methods available to probe it, most noticeably the magnitude-redshift relation of SN Ia. Since dark energy starts to dominate the expansion of the Universe only at relatively low redshifts, little information about its properties is obtainable from the CMB anisotropies alone. For that reason, quite a number of workers have considered the constraints on the dark energy equation of state that can be derived from future cosmic shear surveys (e.g., Huterer 2002; Hu 2002; Munshi and Wang 2003; Hu and Jain

2004; Abazajian and Dodelson 2003; Benabed and van Waerbeke 2003; Song and Knox 2004). The results of these are very encouraging; the sensitivity on the dark energy properties is due to its influence on structure growth. With (photometric) redshift information on the source galaxies, the evolution of the dark matter distribution can be studied by weak lensing, as shown above. van Waerbeke and Mellier (2003) have compared the expected accuracy of the cosmic shear result from the ongoing CFHT Legacy Survey with the variation of various dark energy models and shown that the CFHTLS will be able to discriminate between some of these models, with even much better prospects from future space-based wide-field imaging surveys (e.g., Hu and Jain 2004).

7.7 Discussion

The previous sections have shown that cosmic shear research has matured; several groups have successfully presented their results, which is important in view of the fact that the effects one wants to observe are small, influenced by various effects, and therefore, independent results from different instruments, groups, and data analysis techniques are essential in this research. We have also seen that the results from the various groups tend to agree with each other, with a few very interesting discrepancies remaining whose resolution will most likely teach us even more about the accuracies of data analysis procedures.

Lessons for Cosmology

A natural question to ask is, what has cosmic shear taught us so far about cosmology ? The most important constraint coming from the available cosmic shear results is that on the normalization σ_8, for which only few other accurate methods are available. We have seen that cosmic shear prefers a value of $\sigma_8 \approx 0.8 - 0.9$, which is slightly larger than current estimates from the abundance of clusters, but very much in agreement with the measurement of WMAP. The estimate from the cluster abundance is, however, not without difficulties, since it involves several scaling relations which need to be accurately calibrated; hence, different authors arrive at different values for σ_8 (see, e.g., Pierpaoli, Scott and White 2001; Seljak 2002; Schuecker et al. 2003). The accuracy with which σ_8 is determined from CMB data alone is comparable to that of cosmic shear estimates; as shown in Spergel et al. (2003), more accurate values of σ_8 are obtained only if the CMB measurements are combined with measurements on smaller spatial scales, such as from galaxy redshift surveys and the Lyman alpha forest statistics. Thus, the σ_8-determination from cosmic shear is certainly competitive with other measurements. Arguably, cosmic shear sticks out in this set of smaller-scale constraints due to the fewer physical assumptions needed for its interpretation.

But more importantly, it provides a fully independent method to measure cosmological parameters. Hence, at present the largest role of the cosmic shear results is that it provides an independent approach to determining these parameters; agreement with those obtained from the CMB, galaxy redshift surveys and other methods are thus foremost of interest in that they provide additional evidence for the self-consistency of our cosmological model which, taken at face value, is a pretty implausible one: we should always keep in mind that we are claiming that our Universe consists of 4.5% normal (baryonic) matter, with the rest being shared with stuff that we have given names to ('dark matter', 'dark energy'), but are pretty ignorant about what that actually is. Insofar, cosmic shear plays an essential role in shaping our cosmological view, and has become one of the pillars on which our standard model rests.

Agreement, or Discrepancies ?

How to clarify the remaining discrepancies that were mentioned before – what are they due to ? One needs to step back for a second and be amazed that these results are in fact so well in agreement as they are, given all the technical problems a cosmic shear survey has to face (see Sect. 3). Nevertheless, more investigations concerning the accuracy of the results need to be carried out, e.g., to study the influence of the different schemes for PSF corrections on the final results. For this reason, it would be very valuable if the same data set is analyzed by two independent groups and to compare the results in detail. Such comparative studies may be a prerequisite for the future when much larger surveys will turn cosmic shear into a tool for precision cosmology.

Joint Constraints from CMB Anisotropies and Cosmic Shear

As mentioned before, the full power of the CMB anisotropy measurements is achieved when these results are combined with constraints on smaller spatial scales. The tightest constraints from WMAP are obtained when it is combined with results from galaxy redshift surveys and the statistics of the Lyα forest absorption lines (Spergel et al. 2003). Instead of the latter, one can use results from cosmic shear, as it provides a cleaner probe of the statistical properties of the matter distribution in the Universe. As was pointed out before (e.g., Hu and Tegmark 1999; see Fig. 34), the combination of CMB measurements with cosmic shear results is particularly powerful to break degeneracies that are left from using the former alone. Contaldi et al. (2003) used the CMB anisotropy results from WMAP (Bennett et al. 2003), supplemented by anisotropy measurements on smaller angular scales from ground-based experiments, and combined them with the cosmic shear aperture mass dispersion from the RCS survey (Hoekstra et al. 2002a). As is shown in Fig. 47, the constraints in the $\Omega_{\rm m}$-σ_8-parameter plane are nearly mutually orthogonal for the CMB and cosmic shear, so that the combined confidence region is substantially smaller than each of the individual regions.

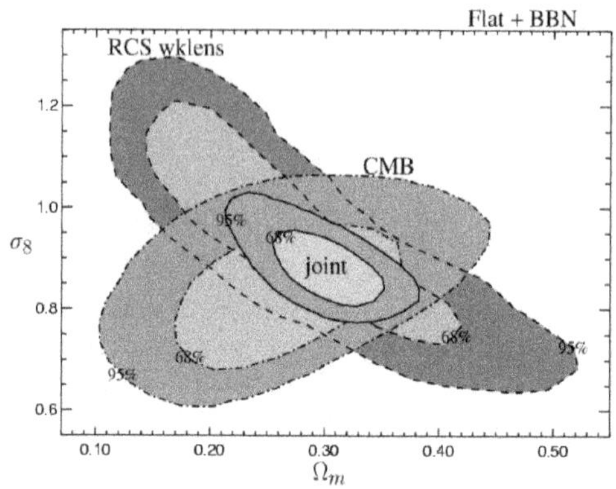

Fig. 47. The confidence region in the $\Omega_{\rm m}$-σ_8-plane obtained from the two-dimensional marginalized likelihood. Shown are the 68% and 95% confidence regions derived individually from the CMB and the RCS cosmic shear survey, as well as those obtained by combining both constraints (Contaldi et al. 2003)

Wide vs. Deep Surveys

In designing future cosmic shear surveys, the survey strategy needs to decide the effective exposure time. For a given total observing time (the most important practical constraint), one needs to find a compromise between depth and area. Several issues need to be considered in this respect:

- The lensing signal increases with redshift, and therefore with increasing depth of a survey; it should therefore be easier to detect a lensing signal in deep surveys. Furthermore, by splitting the galaxy sample into subsamples according to the magnitude (and/or colors), one can study the dependence of the lensing signal on the mean source redshift, which is an important probe of the evolution of the matter power spectrum, and thus of cosmology. If one wants to probe the (dark) matter distribution at appreciable redshifts ($z \sim 0.5$), one needs to carry out deep surveys.
- A wider survey is more likely to probe the linear part of the power spectrum which is more securely predicted from cosmological models than the non-linear part; on the other hand, measurement of the latter, when compared with precise models (e.g., from numerical simulations), can probe the non-linear gravitational clustering regime.
- Depending on the intrinsic galaxy alignment, one would prefer deeper surveys, since the relative importance of the intrinsic signal decreases with increasing survey depth. Very shallow surveys may in fact be strongly affected by the intrinsic signal (e.g., Heymans and Heavens 2003). On the other hand, for precision measurements, as will become available in the near future, one needs to account for the intrinsic signal in any case, using redshift information (at least in a statistical sense), and so shallow surveys

lose this potential disadvantage. In fact, the redshift estimates of shallower surveys are easier to obtain than for deeper ones.

- In this context, one needs to compromise between area and the number of fliters in which exposures should be taken. Smaller area means worse statistics, e.g., larger effects of cosmic variance, but this has to be balanced against the additional redshift information. Also, if a fixed observing time is used, one needs to account for the weather, seeing and sky brightness distribution. One should then devise a strategy that the best seeing periods are used to obtain images in the filter which is used for shape measurements, and bright time shall be spent on the longest wavelength bands.
- Fainter galaxies are smaller, and thus more strongly affected by the point-spread function. One therefore expects that PSF corrections are on average smaller for a shallow survey than for a deeper one. In addition, the separation between stars and galaxies is easier for brighter (hence, larger) objects.

The relative weight of these arguments is still to be decided. Whereas some of the issues could be clarified with theoretical investigations (i.e., in order to obtain the tightest constraints on cosmological parameters, what is the optimal choice of area and exposure time, with their product being fixed), others (like the importance of intrinsic alignments) still remain unclear. Since big imaging surveys will be conducted with a broad range of scientific applications in mind, this choice will also depend on those additional science goals.

Future Surveys

We are currently witnessing the installment of square-degree cameras at some of the best sites, among them MegaCam at the CFHT, and OmegaCAM at the newly built VLT Survey Telescope (the 2.6m VST) on Paranal (I present here European-biased prospects, as I am most familiar with these projects). Weak lensing, and in particular cosmic shear has been one of the science drivers for these instruments, and large surveys will be carried out with them. Already ongoing is the CFHT Legacy Survey, which will consist of three parts; the most interesting one in the current context is a $\sim$160 deg^2 survey with an exposure time of $\sim$1 h in each of five optical filters. This survey will therefore yield a more than ten-fold increase over the current VIRMOS-DESCART survey, with corresponding reductions of the statistical and cosmic variance errors on measurements. The multi-color nature of this survey implies that one can obtain photometric redshift estimates at least for a part of the galaxies which will enable the suppression of the potential contribution to the shear signal from intrinsic alignments of galaxies. A forecast of the expected accuracy of cosmological parameter estimates from the CFHTLS combined with the WMAP CMB measurements has been obtained by Tereno et al. (2005). It is expected that a substantial fraction of the VST observing time will be spend on multi-band wide-field surveys which, if properly designed, will be extremely useful for cosmic shear research. In order to complement results from the CFHTLS, accounting for the fact that the VST has smaller aperture than the CFHT (2.6m vs. 3.6m), a somewhat shallower but wider-field survey would be most reasonable. For both of these surveys, complementary near-IR data will become available after about 2007, with the WirCam instrument on CFHT, and the newly build VISTA 4m-telescope equipped with a wide-field near-IR camera on Paranal, which will yield much better photometric redshift estimates than the optical data alone. Furthermore, with the PanStarrs project, a novel method for wide-field imaging and a great leap forward in the data access rate will be achieved.

Toward the end of the decade, a new generation of cosmic shear surveys may be started; there are two projects currently under debate which would provide a giant leap forward in terms of survey area and/or depth. One is a satellite project, SNAP/JDEM, originally designed for finding and follow-up of high-redshift supernovae to study the expansion history of the Universe and in particular to learn about the equation of state of the dark energy. With its large CCD array and multi-band imaging, SNAP will also be a wonderful instrument for cosmic shear research, yielding photometric redshift estimates for the faint background galaxies, and it is expected that the observing time of this satellite mission will be split between these two probes of dark energy. The other project under discussion is the LSST, a 8m telescope equipped with a $\sim 9\,\mathrm{deg}^2$ camera; such an instrument, with an efficiency larger than a factor 40 over MegaCam@CFHT, would allow huge cosmic shear surveys, easily obtaining a multi-band survey over all extragalactic sky (modulo the constraints from the hemisphere). Since studying the equation of state of dark energy will be done most effectively with good photometric redshifts of source galaxies, the space experiment may appear more promising, given the fact that near-IR photometry is needed for a reliable redshift estimate, and sufficiently deep near-IR observations over a significant area of sky is not possible from the ground.

8 The Mass of, and Associated with Galaxies

8.1 Introduction

Whereas galaxies are not massive enough to show a weak lensing signal individually (see 19), the signal of many galaxies can be superposed statistically. Therefore, if one considers sets of foreground (lens) and background galaxies, then in the mean, in a foreground–background galaxy pair, the image ellipticity of the background galaxy will be preferentially oriented in the direction tangent to the line connecting foreground and background galaxy. The amplitude of this tangential alignment then yields a mean lensing strength that depends on the redshift distributions of foreground and background galaxies, and on the mass distribution of the former population. This effect is called galaxy–galaxy lensing and will be described in Sect. 8.2 below; it measures the mass properties of galaxies, provided the lensing signal is dominated by the galaxies themselves. This will not be the case for larger angular separations between foreground and background galaxies, since then the mass distribution in which the foreground galaxies are embedded (e.g., their host groups or clusters) starts to contribute significantly to the shear signal. The interpretation of this signal then becomes more difficult. On even larger scales, the foreground galaxies contribute negligibly to the lens signal; a spatial correlation between the lens strength and the foreground galaxy population then reveals the correlation between light (galaxies) and mass in the Universe. This correlated distribution of galaxies with respect to the underlying (dark) matter in the Universe – often called the bias of galaxies – can be studied with weak lensing, as we shall describe in Sect. 8.3 by using the shear signal, and in Sect. 8.4 employing the magnification effect. It should be pointed out here that our lack of knowledge about the relation between the spatial distribution of galaxies and that of the underlying (dark) matter is one of the major problems that hampers the quantitative interpretation of galaxy redshift surveys; hence, these lensing studies can provide highly valuable input into the conclusions drawn from these redshift surveys regarding the statistical properties of the mass distribution in the Universe.

8.2 Galaxy–Galaxy Lensing

The Average Mass Profile of Galaxies

Probing the mass distribution of galaxies usually proceeds with dynamical studies of luminous tracers. The best-known method is the determination of the rotation curves of spiral galaxies, measuring the rotational velocity of stars and gas as a function of distance from the galaxy's center (see Sofue and Rubin 2001 for a recent review). This then yields the mass profile of the galaxy, i.e. $M(\le r) \propto v_{\rm rot}^2(r)\, r$. For elliptical galaxies, the dynamics of stars (like velocity dispersions and higher-order moments of their velocity distribution, as a function of r) is analyzed to obtain their mass profiles; as the kinematics of stars in ellipticals is more complicated than in spirals, their mass profiles are more difficult to measure (e.g., Gerhard et al. 2001). In both cases, these dynamical methods provided unambiguous evidence for the presence of a dark matter halo in which the luminous galaxy is embedded; e.g., the rotation curves of spirals are flat out to the most distant point where they can be measured. The lack of stars or gas prevents the measurement of the mass profile to radii beyond the luminous extent of galaxies, that is beyond $\sim 10h^{-1}$ kpc. Other luminous tracers that have been employed to study galaxy masses at larger radii include globular clusters that are found at large galacto-centric radii (Coté et al. 2003), planetary nebulae, and satellite galaxies. Determining the relative radial velocity distribution of the latter with respect to their suspected host galaxy leads to estimates of the dark matter halo out to distances of $\sim 100h^{-1}$ kpc. These studies (e.g., Zaritsky et al. 1997) have shown that the dark matter halo extends out to at least these distances.

One of the open questions regarding the dark matter profile of galaxies is the spatial extent of the halos. The dynamical studies mentioned above are all compatible with the mass profile following approximately an isothermal law ($\rho \propto r^{-2}$), which has to be truncated at a finite radius to yield a finite total mass. Over the limited range in radii, the isothermal profile cannot easily be distinguished from an NFW mass profile (see IN, Sect. 6.2), for which measurements at larger distances are needed (the mass distribution in the central parts of galaxies is affected by the baryons and thus not expected to follow the NFW profile; see Sect. 7 of SL).

Weak gravitational lensing provides a possibility to study the mass profiles of galaxies at still larger radii. Light bundles from distant background galaxies provide the 'dynamical tracers' that cannot be found physically associated with the galaxies. Light bundles get distorted in such a way that on average, images of background sources are oriented tangent to the transverse direction connecting foreground (lens) and background (source) galaxy. The first attempt to detect such a galaxy–galaxy lensing signal was reported in Tyson et al. (1984), but the use of photographic plates and the relatively poor seeing prevented a detection. Brainerd et al. (1996) presented the first detection and analysis of galaxy–galaxy lensing. Since then, quite a number of surveys have measured this effect, some of them using millions of galaxies.

Strategy

Consider pairs of fore- and background galaxies, with separation in a given angular separation bin. The expected lensing signal is seen as a statistical tangential alignment of background galaxy images with respect to foreground galaxies. For example, if ϕ is the angle between the major axis of the background galaxy and the connecting line, values $\pi/4 \le \phi \le \pi/2$ should be slightly more frequent than $0 \le \phi \le \pi/4$ (see Fig. 48). Using the fact that the intrinsic orientations of background galaxies are distributed isotropically, one can show (Brainerd et al. 1996) that

$$p(\phi) = \frac{2}{\pi}\left[1 - \gamma_{\rm t}\left\langle \frac{1}{|\epsilon^{\rm s}|}\right\rangle \cos(2\phi)\right] , \qquad (133)$$

where $\phi \in [0, \pi/2]$ and $\gamma_{\rm t}$ is the mean tangential shear in the angular bin chosen. Thus, the amplitude of the cos-wave yields the (average) strength of the shear.

The mean tangential ellipticity $\langle \epsilon_{\rm t}(\theta) \rangle$ of background galaxies relative to the direction toward foreground galaxies measures the mean tangential shear at separation θ. Since the signal is averaged over many foreground–background pairs, it measures the average mass profiles of the foreground galaxies. For sufficiently large samples of galaxies, the lens sample can be split into several

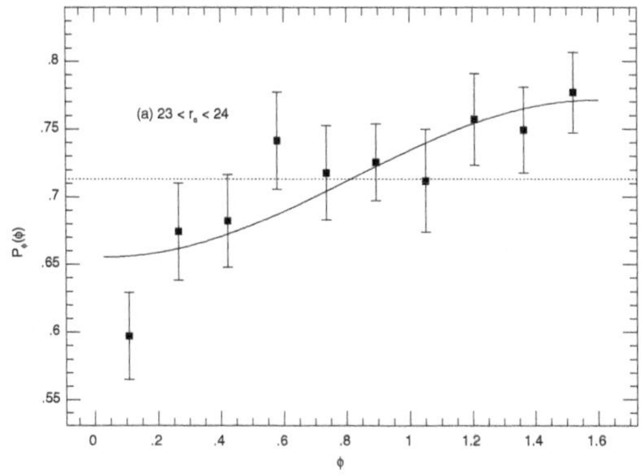

Fig. 48. The probability distribution $p(\phi)$ of the angle ϕ between the major axis of the background galaxy image and the connecting line to the foreground galaxy is plotted for the sample of Brainerd et al. (1996), together with the best fit according to (133). The galaxy pairs have separation $5'' \le \Delta\theta \le 34''$, and are foreground-background selected by their apparent magnitudes

subsamples, e.g., according to their color and/or morphology (early-type vs. late-type galaxies), or, if redshift estimates are available, they can be binned according to their luminosity. Then, the mass properties can be derived for each of the subsamples.

The distinction between foreground and background galaxies is ideally performed using redshift information. This is indeed the case for the galaxy-galaxy lensing studies based on the Sloan Digital Sky Survey, for which early results have been reported by McKay et al. (2001); all lens galaxies used there have spectroscopic redshifts, whereas the source galaxies are substantially fainter than the lens galaxies so that they can be considered as a background population. For other surveys, the lack of redshift information requires the separation of galaxies to be based solely on their apparent magnitudes: fainter galaxies are on average at larger distances than brighter ones. However, the resulting samples of 'foreground' and 'background' galaxies will have (often substantial) overlap in redshift, which needs to be accounted for statistically in the quantitative analysis of these surveys.

Quantitative Analysis

The measurement of the galaxy–galaxy lensing signal provides the tangential shear as a function of pair separation, $\gamma_t(\theta)$. Without information about the redshifts of individual galaxies, the separation of galaxies into a 'foreground' and 'background' population has to be based on apparent magnitudes only. In the ideal case of a huge number of foreground galaxies, one could investigate the mass properties of 'equal' galaxies, by finely binning them according to redshift, luminosity, color, morphology etc. However, in the real world such a fine binning has not yet been possible, and therefore, to convert the lensing signal into physical parameters of the lens, a parameterization of the lens population is needed. We shall outline here how such an analysis is performed.

The first ingredient is the redshift probability distribution $p(z|m)$ of galaxies with apparent magnitude m which is assumed to be known from redshift surveys (and/or their extrapolation to fainter magnitudes). This probability density depends on the apparent magnitude m, with a broader distribution and larger mean redshift expected for fainter m. Since the distribution of 'foreground' and 'background' galaxies in redshift is known for a given survey, the probabilities $p(z|m)$ can be employed to calculate the value of $D_{\rm ds}/D_{\rm s}$, averaged over all foreground–background pairs (with this ratio being set to zero if $z_{\rm s} \le z_{\rm d}$). For given physical parameters of the lenses, the shear signal is proportional to this mean distance ratio.

The mass profiles of galaxies are parameterized according to their luminosity. For example, a popular parameterization is that of a truncated isothermal sphere, where the parameters are the line-of-sight velocity dispersion σ (or the equivalent circular velocity $V_{\rm c} = \sqrt{2}\sigma$) and a truncation radius s at which the $\rho \propto r^{-2}$ isothermal density profile turns into a steeper $\rho \propto r^{-4}$ law. The velocity dispersion is certainly dependent on the luminosity, as follows from the Tully–Fisher and Faber–Jackson relations for late- and early-type galaxies, respectively. One therefore assumes the scaling $\sigma = \sigma_*(L/L_*)^{\beta/2}$, where L_* is a fiducial luminosity (and which conveniently can be chosen close to the characteristic luminosity of the Schechter luminosity function). Furthermore, the truncation scale s is assumed to follow the scaling $s = s_*(L/L_*)^{\eta}$. The total mass of a galaxy then is $M \propto \sigma^2 s$, or $M = M_*(L/L_*)^{\beta+\eta}$.

Suppose m and z were given; then, the luminosity of galaxy would be known, and for given values of the parameters σ_*, s_*, β and η, the mass properties of the lens galaxy would be determined. However, since z is not known, but only its probability distribution, only the probability distribution of the lens luminosities, and therefore the mass properties, are known. One could in principle determine the expected shear signal $\gamma_{\rm t}(\theta)$ for a given survey by calculating the shear signal for a given set of redshifts z_i for all lens and source galaxies, and then averaging this signal over the z_i using the redshift probability distribution $p(z_i|m_i)$. However, this very-high dimensional integration cannot be performed; instead, one uses a Monte-Carlo integration method (Schneider and Rix 1997): Given the positions $\boldsymbol{\theta}_i$ and magnitudes m_i of the galaxies, one can draw for each of them a redshift according to $p(z_i|m_i)$, and then calculate the shear at all positions $\boldsymbol{\theta}_i$ corresponding to a source galaxy, for each set of parameters σ_*, s_*, β and η. This procedure can be repeated several times, yielding the expected shear $\langle\gamma_i\rangle$ and its dispersion $\sigma_{\gamma,i}$ for each source galaxy's position. One can then calculate the likelihood function

$$\mathcal{L} = \prod_{i=1}^{N_{\rm s}} \frac{1}{\pi(\sigma_\epsilon^2 + \sigma_{\gamma,i}^2)} \exp\left(-\frac{|\epsilon_i - \langle\gamma_i\rangle|^2}{\sigma_\epsilon^2 + \sigma_{\gamma,i}^2}\right) , \tag{134}$$

where σ_ϵ is the intrinsic ellipticity dispersion of the galaxies. $\mathcal{L}$ depends on the parameters of the model, and can be maximized with respect to them, thereby yielding estimates of σ_*, s_*, β and η.

First Detection

The galaxy–galaxy lensing effect was first found by Brainerd et al. (1996), on a single $9'\!.6 \times 9'\!.6$ field. They considered 'foreground' galaxies in the magnitude range $m \in [20, 23]$, and 'background' galaxies with $m \in [23, 24]$; this yielded 439 foreground and 506 background galaxies, and 3202 pairs with $\Delta\theta \in [5'', 34'']$.[12] For these pairs, the distribution of the alignment angle ϕ is plotted in Fig. 48. This distribution clearly is incompatible with the absence of a lens signal (at the 99.9% confidence level), and thus provides a solid detection.

They analyzed the lens signal $\gamma_t(\theta)$ in a way similar to the method outlined above, except that their Monte-Carlo simulations also randomized the positions of galaxies. The resulting likelihood yields $\sigma_* \approx 160^{+50}_{-60}$ km/s (90% confidence interval), whereas for s_* only a lower limit of $25h^{-1}$ kpc (1σ) is obtained; the small field size, in combination with the relative insensitivity of the lensing signal to s_* once this value is larger than the mean transverse separation of lensing galaxies, prohibited the detection of an upper bound on the halo size.

Galaxy–Galaxy Lensing from the Red-Sequence Cluster Survey (RCS)

Several groups have published results of their galaxy–galaxy lensing surveys since its first detection. Here we shall describe the results of a recent wide-field imaging survey, the RCS; this survey was already described in the context of cosmic shear in Sect. 7.3. 45.5 square degrees of single-band imaging data were used (Hoekstra et al. 2004). Choosing lens galaxies with $19.5 \le R_C \le 21$, and source galaxies having $21.5 \le R_C \le 24$ yielded $\sim 1.2 \times 10^5$ lenses with median redshift of 0.35 and $\sim 1.5 \times 10^6$ sources with median redshift of ~ 0.53, yielding $\langle D_{ds}/D_s \rangle = 0.29 \pm 0.01$ for the full sample of lenses and sources. Figure 49 shows the shear signal for this survey.

The lens signal is affected by galaxies counted as lenses, but which in fact are in the foreground. As long as they are not physically associated with lens galaxies, this effect is accounted for in the analysis, i.e., in the value of $\langle D_{ds}/D_s \rangle$. However, if fainter galaxies cluster around lens galaxies,

[12] The lower angular scale has been chosen to avoid overlapping isophotes of foreground and background galaxies, whereas the upper limit was selected since it gave the largest signal-to-noise for the deviation of the angular distribution shown in Fig. 48 from a uniform one.

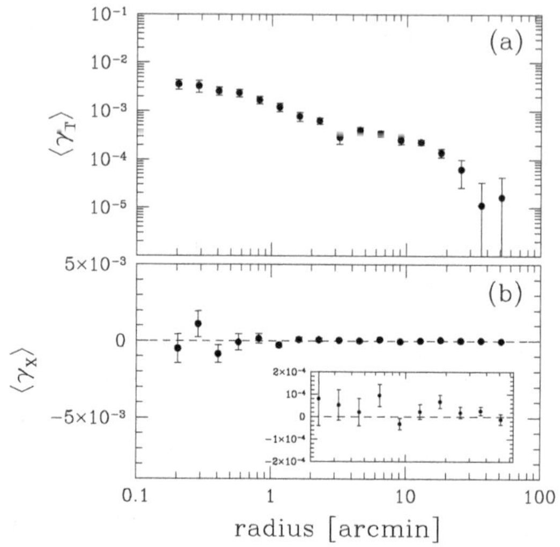

Fig. 49. (**a**) Tangential shear as a function of angular separation, obtained from the RCS survey; the shear signal is detected out to nearly one degree scale. (**b**) Cross shear signal, which is expected to vanish identically in the absence of systematic effects on the ellipticity measurements. As can be seen, the cross signal in indeed compatible with zero. The inset expands the scale, to better show the error bars (from Hoekstra et al. 2003)

this produces an additional effect. Provided the orientation of the associated faint galaxies are random with respect to the separation vector to their bright neighbor, these physical pairs just yield a dilution of the shear signal. The amplitude of this effect can be determined from the angular correlation function of bright and faint galaxies, and easily corrected for. Once this has been done, the corrected shear signal within $10'' \le \theta \le 2'$ has been fitted with an SIS model, yielding a mean velocity dispersion of the lens galaxies of $\sqrt{\langle\sigma^2\rangle} = 128 \pm 4\,\mathrm{km/s}$. If the scaling relations between galaxy luminosity and velocity dispersion as described above is employed, with $\beta = 0.6$, the result is $\sigma_* = 140 \pm 4\,\mathrm{km/s}$ for $L_* = 10^{10}h^{-2}L_\odot$ in the blue passband.

To interpret the shear results on larger angular scales, the SIS model no longer suffices, and different mass models need to be employed. Using a truncated isothermal model, the best-fitting values of the scaling parameters $\beta = 0.60 \pm 0.11$ and $\eta = 0.24^{+0.26}_{-0.22}$ are obtained, when marginalizing over all other parameters. Furthermore, $\sigma_* = 137 \pm 5\,\mathrm{km/s}$, in very close agreement with the results from small θ and the SIS model; this is expected, since most of the signal comes from these smaller separations. Most interesting, the analysis also yields an estimate of the truncation scale of $s_* = (185 \pm 30)h^{-1}\,\mathrm{kpc}$, providing one of only a few estimates of the scale of the dark matter halo. Hoekstra et al. also performed the analysis in the frame of an NFW mass model.

These results can then be used to calculate the mass-to-light ratio of an L_* galaxy and, using the scaling, of the galaxy population as a whole. Considering only galaxies with $M \geq 10^{10} h^{-1} M_\odot$, the mean mass-to-light ratio inside the virial radius of galaxy halos is about 100 in solar units.

The Shape of Dark Matter Halos

In the mass models considered before, the mass distribution of galaxies was assumed to be axi-symmetric. In fact, this assumption is not crucial, since the relation between shear and surface mass density, $\gamma_t(\vartheta) = \bar{\kappa}(\vartheta) - \kappa(\vartheta)$ is true for a general mass distribution, provided γ_t and $\kappa(\vartheta)$ are interpreted as the mean tangential shear and mean surface mass density on a circle of radius ϑ, and $\bar{\kappa}(\vartheta)$ as the mean surface mass density inside this circle (see 24). However, deviations from axial symmetry are imprinted on the shear signal and can in principle be measured. If the mass distribution is 'elliptical', the shear along the major axis (at given distance ϑ) is larger than that along the minor axis, and therefore, an investigation of the strength of the shear signal relative to the orientation of the galaxy can reveal a finite ellipticity of the mass distribution. For that, it is necessary that the orientation of the mass distribution is (at least approximately) known. Provided the orientation of the mass distribution follows approximately the orientation of the luminous part of galaxies, one can analyze the direction dependence of the shear relative to the major axis of the light distribution (Natarajan and Refregier 2000). Hoekstra et al. (2002b) have used the RCS to search for such a direction dependence; they parameterized the lenses with a truncated isothermal profile with ellipticity $\epsilon_{\rm mass} = f \epsilon_{\rm light}$, where f is a free parameter. The result $f = 0.77 \pm 0.2$ indicates first that the mass distribution of galaxies is not round (which would be the case for $f = 0$, which is incompatible with the data), and second, that the mass distribution is rounder than that of the light distribution, since $f < 1$. However, it must be kept in mind that the assumption of equal orientation between light and mass is crucial for the interpretation of f; misalignment causes a decrease of f. Note that numerical simulations of galaxy evolution predict such a misalignment between total mass and baryons, with an rms deviation of around 20° (van den Bosch et al. 2002). Given the above result on f, it is therefore not excluded that the flattening of halos is very similar to that of the light. Also note that this result yields a value averaged over all galaxies; since the lens efficiency of elliptical galaxies (at given luminosity) is larger than that of spirals, the value of f is dominated by the contributions from early-type galaxies.

Results From the Sloan Survey

The Sloan Digital Sky Survey (e.g., York et al. 2000) will map a quarter of the sky in five photometric bands, and obtain spectra of about one million galaxies. A large fraction of the data has already been taken by SDSS, and parts of this data have already been released (Abazajian et al. 2004). The huge amount of photometric data in principle is ideal for weak lensing studies, as it beats down statistical uncertainties to an unprecedented low level. However, the site of the telescope, the relatively large pixel size of $0\farcs4$, the relatively shallow exposures of about one minute and the drift-scan mode in which data are taken (yielding excellent flat-fielding, and thus photometric properties, somewhat at the expense of the shape of the PSF) render the data less useful for, e.g., cosmic shear studies: the small mean redshift of the galaxies yields a very small expectation value of the cosmic shear, which can easily be mimicked by residuals from PSF corrections. However, galaxy–galaxy lensing is much less sensitive to larger-scale PSF problems, since the component of the shear used in the analysis is not attached to pixel directions, but to neighboring galaxies, and thus varies rapidly with sky position. Another way of expressing this fact is that the galaxy–galaxy lensing signal would remain unchanged if a uniform shear would be added to the data; therefore, SDSS provides a great opportunity for studying the mass profile of galaxies.

Fischer et al. (2000) reported the first results from the SDSS, and a larger fraction of the SDSS data was subsequently used in a galaxy–galaxy lensing study by McKay et al. (2001), where also the spectroscopic redshifts of the lens galaxies were used. Their sample consists of $\sim$31000 lens galaxies with measured redshifts, and $\sim 3.6 \times 10^6$ source galaxies selected in the brightness range $18 \le r \le 22$. For this magnitude range, the redshift distribution of galaxies is fairly well known, leaving little calibration uncertainty in the interpretation of the shear signal. In particular, there is very little overlap in the redshift distribution of source and lens galaxies. The data set has been subjected to a large number of tests, to reveal systematics; e.g., null results are obtained when the source galaxies are rotated by 45° (or, equivalently, if $\gamma_\times$ is used instead of $\gamma_{\rm t}$), or if the lens galaxies are replaced by an equal number of randomly distributed points relative to which the tangential shear component is measured. Since the redshifts of the lens galaxies are known, the shear can be measured directly in physical units, so one can determine

$$\Delta\Sigma_+ = \bar\Sigma(\le R) - \Sigma(R) \tag{135}$$

in $M_\odot/{\rm pc}^2$ as a function of R in kpc.

Figure 50 shows the lensing result from McKay et al. (2001), where the lens sample has been split according to the type of galaxy (early vs. late type) and according to the local spatial number density of galaxies, which is known owing to the spectroscopic redshifts. The fact that most of the signal on small scales is due to ellipticals is expected, as they are more massive at given luminosity than spirals. The large spatial extent of the shear signal for ellipticals relative to that of spirals can be interpreted either by ellipticals having a larger halo than spirals, or that ellipticals are preferentially found in high-density environments, which contribute to the lens signal on large scales. This latter interpretation is supported by the lower panel in Fig. 50 which shows that the signal on large scales is entirely due to lens galaxies in dense environments. This then implies that the galaxy–galaxy lensing signal

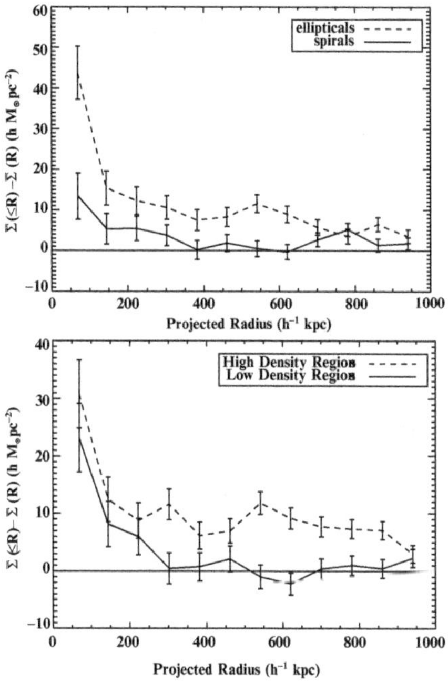

Fig. 50. The galaxy–galaxy lensing signal from the SDSS plotted against physical radius R. The lens sample has been subdivided into early- and late-type galaxies (*upper panel*), and in galaxies situated in dense environments vs. those with a smaller neighboring galaxy density (*lower panel*). The figure clearly shows that the lensing signal is dominated by elliptical galaxies, and by those located in dense environment. Owing to the morphology–density relation of galaxies, these two results are not mutually independent. Note that the lensing signal can be measured out to $1h^{-1}$ Mpc, considerably larger than the expected size of galaxy halos; therefore, the shear at these large separations is most likely caused by the larger-scale mass distribution in which the galaxies are embedded (from McKay et al. 2001)

on large scales no longer measures the density profile of individual galaxies, but gets more and more dominated by group and cluster halos in which these (predominantly early-type) galaxies are embedded.

A separation of these contributions from the data themselves is not possible at present, but can be achieved in the frame of a theoretical model. Guzik and Seljak (2001) employed the halo model for the distribution of matter in the universe (see Cooray and Sheth 2002) to perform this separation. There, the galaxy–galaxy lensing signal either comes from matter in the same halo in which the galaxy is embedded, or due to other halos which are physically associated (i.e., clustered) with the former. This latter contribution is negligible on the scales below $\sim 1h^{-1}$ Mpc on which the SDSS obtained a measurement. The former contribution can be split further into two terms: the first is from the dark matter around the galaxies themselves, whereas the second is due to the matter in groups and clusters to which the galaxies might belong. The relative amplitude of these two terms depends on the fraction of galaxies which are located in groups and clusters; the larger this fraction, the more important are larger-scale halos for the shear signal. Guzik and Seljak estimate from the radial dependence of the SDSS signal that about 20% of galaxies reside in groups and clusters; on scales larger than about $200h^{-1}$ kpc their contribution dominates. The virial mass of an early-type L_* galaxy is estimated to be $M_{200}(L_*) = (9.3 \pm 2.2) \times 10^{11} h^{-1} M_\odot$, and about a factor of three smaller for late-type galaxies (with luminosity measured in a red passband; the differences are substantially larger for bluer passbands, owing to the sensitivity of the luminosity to star formation activity in late types). From the mass-to-light ratio in red passbands, Guzik and Seljak estimate that an L_* galaxy converts about 10–15% of its virial mass into stars. Since this fraction is close to the baryon fraction in the universe, they conclude that most of the baryons of an L_* galaxy are transformed into stars. For more massive halos, the mass-to-light ratio increases ($M/L \propto L^{0.4 \pm 0.2}$), and therefore their conversion of baryons into stars is smaller – in agreement with what we argued about clusters, where most of the baryons are present in the form of a hot intracluster gas.

Yang et al. (2003) studied the cross-correlation between mass and galaxies using numerical simulations of structure formation and semi-analytic models of galaxy evolution. The observed dependence of the galaxy–galaxy lensing signal on galaxy luminosity, morphological type and galaxy environment, as obtained by McKay et al. (2001), is well reproduced in these simulations. The galaxy-mass correlation is affected by satellite galaxies, i.e. galaxies not situated at the center of their respective halo. Central galaxies can be selected by restricting the foreground galaxy sample to relatively isolated galaxies. The galaxy–galaxy lensing signal for such central galaxies can well be described by an NFW mass profile, whereas this no longer is true if all galaxies are considered. Combining the measurement with the simulation, they find that an L_*-galaxy typically resides in a halo with a virial mass of $\sim 2 \times 10^{12} h^{-1} M_\odot$.

With the SDSS progressing, larger datasets become available, allowing a more refined analysis of galaxy–galaxy lensing (Sheldon et al. 2004; Seljak et al. 2004). In the analysis of Seljak et al. (2004), more than 2.7×10^5 galaxies with spectroscopic redshifts have been used as foreground galaxies, and as background population those fainter galaxies for which photometric redshifts have been estimated. The resulting signal is shown in Fig. 51, for six different bins in (foreground) galaxy luminosity.

In a further test to constrain systematic effects in the data, Hirata et al. (2004) have used spectroscopic and photometric redshifts to study the question whether an alignment of satellite galaxies around the lens galaxies can affect the galaxy–galaxy lensing signal from the SDSS; they obtain an upper limit of a 15% contamination.

The SDSS already has yielded important information about the mass properties of galaxies; taken into account that only a part of the data of

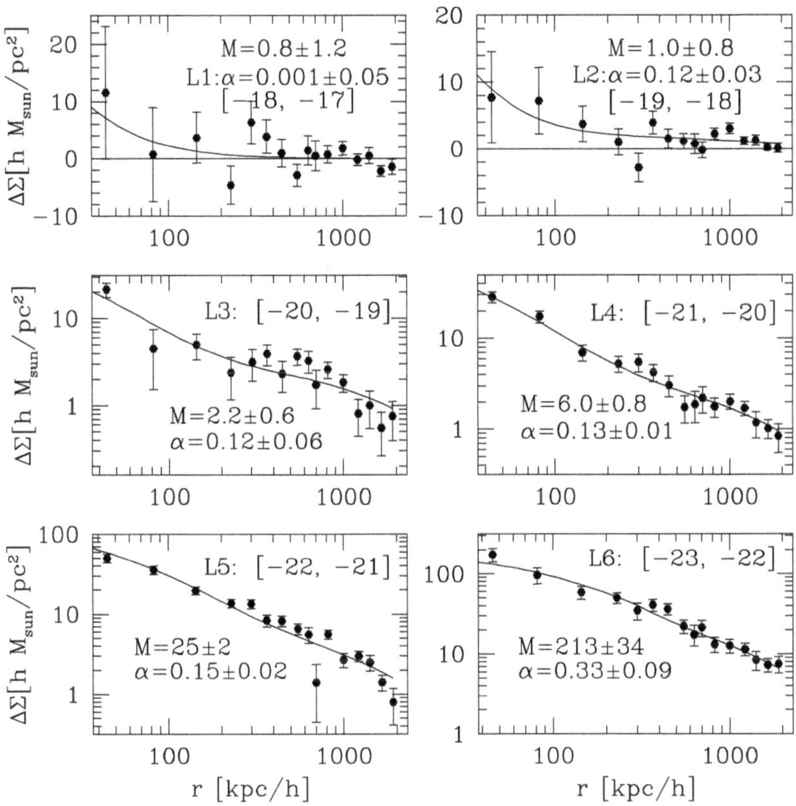

Fig. 51. The galaxy–galaxy lensing signal for six luminosity bins of foreground galaxies, as indicated by the absolute magnitude interval in each panel. The curves show a two-parameter model fitted to the data, based on the halo model, and the fit parameters are indicated: M is the virial mass of the halo (in units of $10^{11} h^{-1} M_\odot$) in which the galaxies reside, and α is the fraction of the galaxies which are not central inside the halo, but satellite galaxies (from Seljak et al. 2004)

the complete survey has been used in the studies mentioned above, an analysis of the final survey will yield rich harvest when applied to a galaxy–galaxy lensing analysis.

Lensing by Galaxies in Clusters

As an extension of the method presented hitherto, one might use galaxy–galaxy lensing also to specifically target the mass profile of galaxies in the inner part of clusters. One might expect that owing to tidal stripping, their dark matter halo has a considerably smaller spatial extent than that of the galaxy population as a whole. The study of this effect with lensing is more complicated than galaxy–galaxy lensing in the field, both observationally and from theory. Observationally, the data sets that can be used need to be taken in the inner part of massive clusters; since these are rare, a single wide-field image usually contains at most one such cluster. Furthermore, the number of massive galaxies projected near the center of a cluster is fairly small. Therefore, in order to obtain good statistics, the data of different clusters should be combined. Since the cores of clusters are optically bright, measuring the shape of faint background galaxies is more difficult than in a blank field. From the theoretical side, the lensing strength of the cluster is much stronger than that of the individual cluster galaxies, and so this large-scale shear contribution needs to be accounted for in the galaxy–galaxy lensing analysis.

Methods for performing this separation between cluster and galaxy shear were developed by Natarajan and Kneib (1997) and Geiger and Schneider (1998). Perhaps the simplest approach is provided by the aperture mass methods, applied to the individual cluster galaxies; there one measures the tangential shear inside an annulus around each cluster galaxy. This measure is insensitive to the shear contribution which is linear in the angular variable $\boldsymbol{\theta}$, which is a first local approximation to the larger-scale shear caused by the cluster. Alternatively, a mass model of the (smoothed) cluster can be obtained, either from strong or weak lensing constraints, or preferentially both, and subtracted from the shear signal around galaxies to see their signal. However, once the mass fraction in the galaxies becomes considerable, this method starts to become biased. Geiger and Schneider (1999) have suggested to simultaneously perform a weak lensing mass reconstruction of the cluster and a determination of the parameters of a conveniently parameterized mass model of cluster galaxies (e.g., the truncated isothermal sphere); since the maximum likelihood method for the mass reconstruction (see Sect. 5.3) was used, the solution results from maximizing the likelihood with respect to the mass profile parameters (the deflection potential on a grid) and the galaxy mass parameters.

Natarajan et al. (1998), by analyzing HST data of the cluster AC 114, concluded that the truncation radius of a fiducial L_* galaxy in this cluster is $\sim 15h^{-1}$ kpc; similarly, Geiger and Schneider (1999) showed that the best-fitting truncation radius for early-type galaxies in the cluster A 851 is $\sim 10h^{-1}$ kpc (see Fig. 52). Although the uncertainties are fairly large, these results indicate that indeed galaxies near cluster centers have a halo size considerably smaller than the average galaxy. The sample of clusters which can be investigated using this method will dramatically increase once the cluster sample observed with the new ACS camera onboard HST becomes available and gets properly analyzed.

8.3 Galaxy Biasing: Shear Method

On small scales, galaxy–galaxy lensing measures the mass profile of galaxies, whereas on intermediate scales the environment of galaxies starts to dominate the shear signal. On even larger scale (say, beyond $\sim 1h^{-1}$ Mpc), the host halo

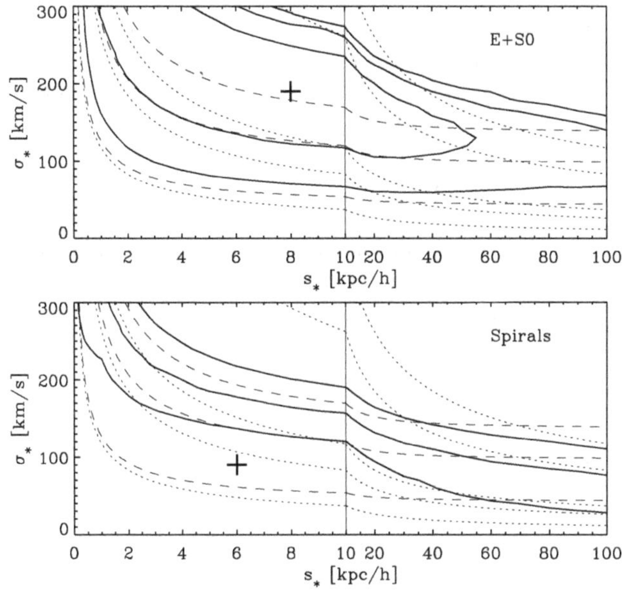

Fig. 52. Significance contours (*solid*) for galaxy properties obtained from galaxy–galaxy lensing of galaxies in the cluster Cl 0939+4713. The parameters are the velocity dispersion σ_* and the halo truncation radius s_* of an L_*-galaxy. Based on HST data (see Fig. 22), a simultaneous reconstruction of the cluster mass profile and the determination of the galaxy mass parameters was performed. No significant lensing signal is seen from the 55 late-type galaxies (*lower panel*), but a clear detection and upper bound to the halo size is detected for the 56 early-types. Dashed and dotted curves connect models with the same mass inside $8h^{-1}$ 1kpc and total mass of an L_*-galaxy, respectively (from Geiger and Schneider 1999)

contribution becomes negligible. Beyond that distance, any signal must come from the correlation of galaxy positions with the mass distribution in the Universe. This correlation, and the related issue of galaxy biasing (see Sect. 6.1 of IN), can ideally be studied with weak lensing. In this section we shall outline how these quantities can be determined from shear measurements, and describe some recent results. As we shall see, this issue is intimately related to galaxy–galaxy lensing. The next section deals with the magnification of distant sources caused by mass overdensities correlated with galaxies and thereby causing an apparent correlation between high-redshift sources and low-redshift galaxies; the amplitude of this signal is again proportional to the correlation between galaxies and the underlying dark matter.

An interesting illustration of the correlation between galaxies and mass has been derived by Wilson et al. (2001). They studied 6 fields with $30' \times 30'$ each, selected bright early-type galaxies from their $V - I$ colors and I magnitudes and measured the shear from faint galaxies. Assuming that mass is strongly correlated with early-type galaxies, these can be used to predict the shear field, with an overall normalization given by the mean mass-to-light ratio of the early-type galaxies. This correlation has indeed been found, at the 5.2-σ significance level, and a value of $M/L \approx 300h$ in solar units has been obtained, assuming a flat low-density Universe.

The Galaxy–Mass Correlation and the Bias Parameter

First, the concept of the correlation between galaxies and mass shall be described more quantitatively. The mass density inhomogeneities are described, as before, by the dimensionless density contrast $\delta(\mathbf{x}, w)$. In analogy to this quantity, one defines the number density contrast $\delta_{\rm g}(\mathbf{x}, w)$ of galaxies as

$$\delta_{\rm g}(\mathbf{x}, w) := \frac{n(\mathbf{x}, w) - \bar{n}(w)}{\bar{n}(w)} , \tag{136}$$

where $n(\mathbf{x}, w)$ is the number density of galaxies at comoving position $\mathbf{x}$ and comoving distance w (the latter providing a parameterization of cosmic time or redshift), and $\bar{n}(w)$ is the mean number density of galaxies at that epoch. Since the galaxy distribution is discrete, the true number density is simply a sum of delta-functions. What is meant by n is that the probability of finding a galaxy in the volume $\mathrm{d}V$ situated at position $\mathbf{x}$ is $n(\mathbf{x})\,\mathrm{d}V$.

The relation between δ and $\delta_{\rm g}$ describes the relative distribution of galaxies and matter in the Universe. The simplest case is that of an *unbiased* distribution, for which $\delta_g = \delta$; then, the probability of finding a galaxy at any location would be just proportional to the matter density. However, one might expect that the relation between luminous and dark matter is more complicated. For example, galaxies are expected to form preferentially in the high-density peaks in the early Universe, which would imply that there are proportionally more galaxies within mass overdensities. This led to the introduction of the concept of biasing (e.g., Bardeen et al. 1986; Kaiser 1984). The simplest form of biasing, called linear deterministic biasing, is provided by setting $\delta_{\rm g} = b\,\delta$, with b being the bias parameter. One might suspect that the relative bias is approximately constant on large scales, where the density field is still in its linear evolution (i.e., on scales $\gtrsim 10h^{-1}\,$Mpc today). On smaller scales, however, b most likely is no longer simply a constant. For example, the spatial distribution of galaxies in clusters seems to deviate from the radial mass profile, and the distributions of different galaxy types are different. Furthermore, by comparing the clustering properties of galaxies of different types, one can determine their relative bias, from which it is concluded that more luminous galaxies are more strongly biased than less luminous ones, and early-type galaxies are more strongly clustered than late-types (see Norberg et al. 2001, and Zehavi et al. 2002 for recent results from the 2dFGRS and the SDSS). This is also expected from theoretical models and numerical simulations which show that more massive halos cluster more strongly (e.g., Sheth et al. 2001; Jing 1998). In order to account for a possible scale dependence of the bias, one considers the Fourier transforms of δ and $\delta_{\rm g}$ and relates them according to

$$\hat{\delta}_{\rm g}(\mathbf{k},w) = b(|\mathbf{k}|,w)\,\hat{\delta}(\mathbf{k},w) \;, \tag{137}$$

thus accounting for a possible scale and redshift dependence of the bias.

Even this more general bias description is most likely too simple, as it is still deterministic. Owing to the complexity of galaxy formation and evolution, it is to be expected that the galaxy distribution is subject to stochasticity in excess to Poisson sampling (Tegmark and Peebles 1998; Dekel and Lahav 1999). To account for that, another parameter is introduced, the correlation parameter $r(|\mathbf{k}|,w)$, which in general will also depend on scale and cosmic epoch. To define it, we first consider the correlator

$$\left\langle \hat{\delta}(\mathbf{k},w)\,\hat{\delta}^*_{\rm g}(\mathbf{k}',w)\right\rangle = (2\pi)^3\,\delta_{\rm D}(\mathbf{k}-\mathbf{k}')\,P_{\delta{\rm g}}(|\mathbf{k}|,w) \;, \tag{138}$$

where the occurrence of the delta function is due to the statistical homogeneity of the density fields, and $P_{\delta{\rm g}}$ denotes the cross-power between galaxies and matter. The correlation parameter r is then defined as

$$r(|\mathbf{k}|,w) = \frac{P_{\delta{\rm g}}(|\mathbf{k}|,w)}{\sqrt{P_\delta(|\mathbf{k}|,w)\,P_{\rm g}(|\mathbf{k}|,w)}} \;. \tag{139}$$

In the case of *stochastic biasing*, the definition of the bias parameter is modified to

$$P_{\rm g}(|\mathbf{k}|,w) = b^2(|\mathbf{k}|,w)\,P_\delta(|\mathbf{k}|,w) \;, \tag{140}$$

which agrees with the definition (137) in the case of $r \equiv 1$, but is more general since (140) no longer relates the phase of (the Fourier transform of) $\delta_{\rm g}$ to that of δ. Combining the last two equations yields

$$P_{\delta{\rm g}}(|\mathbf{k}|,w) = b(|\mathbf{k}|,w)\,r(|\mathbf{k}|,w)\,P_\delta(|\mathbf{k}|,w) \;. \tag{141}$$

We point out again that galaxy redshift surveys are used to determine the two-point statistics of the galaxy distribution, and therefore $P_{\rm g}$; in order to relate these measurements to P_δ, assumptions on the properties of the bias have to be made. As we shall discuss next, weak lensing can determine both the bias parameter and the correlation parameter.

The Principle

In order to determine b and r, the three power spectra defined above (or functions thereof) need to be measured. Second-order cosmic shear measures, as discussed in Sect. 6, are proportional to the power spectrum P_δ. The correlation function of galaxies is linearly related to $P_{\rm g}$. In particular, the three-dimensional correlation function is just the Fourier transform of $P_{\rm g}$, whereas the angular correlation function contains a projection of $P_{\rm g}$ along the line-of-sight and thus follows from Limber's equation as discussed in Sect. 6.2. Finally, the cross-power $P_{\delta{\rm g}}$ describes the correlation between mass and light, and thus determines the relation between the lensing properties of the mass distribution in the Universe to the location of the galaxies. galaxy–galaxy lensing on large angular scales (where the mass profile of individual galaxies no longer yields a significant contribution) provides one of the measures for such a correlation. Hence, measurements of these three statistical distributions allow a determination of r and b.

As we shall consider projected densities, we relate the density field of galaxies on the sky to the spatial distribution. Hence, consider a population of ('foreground') galaxies with spatial number density $n(\mathbf{x}, w)$. The number density of these galaxies on the sky at $\boldsymbol{\theta}$ is then $N(\boldsymbol{\theta}) = \int \mathrm{d}w\, \nu(w)\, n(f_k(w)\boldsymbol{\theta}, w)$, where $\nu(w)$ is the redshift-dependent selection function, describing which fraction of the galaxies at comoving distance w are included in the sample. Foremost, this accounts for the fact that for large distances, only the more luminous galaxies will be in the observed galaxy sample, but ν can account also for more subtle effects, such as spectral features entering or leaving the photometric bands due to redshifting. The mean number density of galaxies on the sky is $\bar{N} = \int \mathrm{d}w\, \nu(w)\, \bar{n}(w)$; the redshift distribution, or more precisely, the distribution in comoving distance, of these galaxies therefore is $p_\mathrm{f}(w) = \nu(w)\, \bar{n}(w)/\bar{N}$, thus relating the selection function $\nu(w)$ to the redshift distribution. Using the definition (136), one then finds that

$$N(\boldsymbol{\theta}) = \bar{N}\left[1 + \int \mathrm{d}w\, p_\mathrm{f}(w)\, \delta_\mathrm{g}(f_K(w)\boldsymbol{\theta}, w)\right] . \tag{142}$$

We shall denote the fractional number density by $\kappa_\mathrm{g}(\boldsymbol{\theta}) := \left[N(\boldsymbol{\theta}) - \bar{N}\right]/\bar{N} = \int \mathrm{d}w\, p_\mathrm{f}(w)\, \delta_\mathrm{g}(f_K(w)\boldsymbol{\theta}, w)$.

Aperture Measures

We have seen in Sect. 6.3 that the aperture mass dispersion provides a very convenient measure of second-order cosmic shear statistics. Therefore, it is tempting to use aperture measures also for the determination of the bias and the mass–galaxy correlation. Define in analogy to the definition of the aperture mass M_ap in terms of the projected mass density the aperture counts (Schneider 1998),

$$\mathcal{N}(\theta) = \int \mathrm{d}^2\vartheta\, U(|\boldsymbol{\vartheta}|)\, \kappa_\mathrm{g}(\boldsymbol{\vartheta}) , \tag{143}$$

where the integral extends over the aperture of angular radius θ, and $\boldsymbol{\vartheta}$ measures the position relative to the center of the aperture. An unbiased estimate of the aperture counts is $\bar{N}^{-1} \sum_i U(|\boldsymbol{\theta}_i|)$, where the $\boldsymbol{\theta}_i$ are the positions of the galaxies. We now consider the dispersion of the aperture counts,

$$\left\langle \mathcal{N}^2(\theta) \right\rangle = \int \mathrm{d}^2\vartheta\, U(|\boldsymbol{\vartheta}|) \int \mathrm{d}^2\vartheta'\, U(|\boldsymbol{\vartheta}'|) \left\langle \kappa_\mathrm{g}(\boldsymbol{\vartheta})\, \kappa_\mathrm{g}(\boldsymbol{\vartheta}') \right\rangle . \tag{144}$$

The correlator in the last expression is the angular two-point correlation function $\omega(\Delta\vartheta)$ of the galaxies; its Fourier transform is the angular power spectrum

$P_\omega(\ell)$ of galaxies. Using the definition of $\kappa_{\rm g}$ together with the result (98) allows us to express P_ω in terms of the three-dimensional power spectrum of the galaxy distribution,

$$\begin{aligned} P_\omega(\ell) &= \int {\rm d}w \, \frac{p_{\rm f}^2(w)}{f_K^2(w)} \, b^2 \left(\frac{\ell}{f_K(w)}, w \right) P_\delta \left(\frac{\ell}{f_K(w)}, w \right) \\ &= \bar{b}^2 \int {\rm d}w \, \frac{p_{\rm f}^2(w)}{f_K^2(w)} \, P_\delta \left(\frac{\ell}{f_K(w)}, w \right) , \end{aligned} \tag{145}$$

where we made use of (140), and in the final step we defined the mean bias parameter $\bar{b}$ which is a weighted average of the bias parameter over the redshift distribution of the galaxies and which depends on the angular wave number ℓ. To simplify notation, we shall drop the bar on b and consider the bias factor as being conveniently averaged over redshift (and later, also over spatial scale). The aperture count dispersion then becomes

$$\left\langle \mathcal{N}^2(\theta) \right\rangle = \frac{1}{2\pi} \int {\rm d}\ell \, \ell \, P_\omega(\ell) W_{\rm ap}(\theta\ell) = 2\pi \, b^2 \, H_{\rm gg}(\theta) \, , \tag{146}$$

where $W_{\rm ap}$ is given in (109), and we have defined

$$H_{\rm gg}(\theta) = \int {\rm d}w \, \frac{p_{\rm f}^2(w)}{f_K^2(w)} \, \mathcal{P}(w, \theta) \, , \tag{147}$$

with

$$\mathcal{P}(w, \theta) = \frac{1}{(2\pi)^2} \int {\rm d}\ell \, \ell \, P_\delta \left(\frac{\ell}{f_K(w)}, w \right) W_{\rm ap}(\theta\ell) \, . \tag{148}$$

Using the same notation (following Hoekstra et al. 2002c), we can write the aperture mass dispersion as

$$\left\langle M_{\rm ap}^2(\theta) \right\rangle = \frac{9\pi}{2} \left(\frac{H_0}{c} \right)^4 \Omega_{\rm m}^2 \, H_\kappa(\theta) \, , \tag{149}$$

with

$$H_\kappa(\theta) = \int {\rm d}w \, \frac{g^2(w)}{a^2(w)} \, \mathcal{P}(w, \theta) \, , \tag{150}$$

where $g(w)$ (see 94) describes the source-redshift weighted efficiency factor of a lens at distance w. One therefore obtains an expression for the bias factor,

$$b^2 = \frac{9}{4} \left(\frac{H_0}{c} \right)^4 \frac{H_\kappa(\theta)}{H_{\rm gg}(\theta)} \, \Omega_{\rm m}^2 \, \frac{\left\langle \mathcal{N}^2(\theta) \right\rangle}{\left\langle M_{\rm ap}^2(\theta) \right\rangle} = f_{\rm b}(\theta) \Omega_{\rm m}^2 \, \frac{\left\langle \mathcal{N}^2(\theta) \right\rangle}{\left\langle M_{\rm ap}^2(\theta) \right\rangle} \, . \tag{151}$$

Note that $f_{\rm b}(\theta)$ depends, besides the aperture radius θ, on the cosmological parameters $\Omega_{\rm m}$ and Ω_Λ, but for a given cosmological model, it depends only weakly on the filter scale θ and on the adopted power spectrum P_δ (van Waerbeke 1998; Hoekstra et al. 2002c). This is due to the fact that both, $\langle \mathcal{N}^2(\theta)\rangle$ and $\langle M_{\rm ap}^2(\theta)\rangle$ are linear in the power spectrum, through the functions H, and in both cases they probe only a very narrow range of k-values, owing to the narrow width of the filter function $W_{\rm ap}$. Hence, the ratio $\langle \mathcal{N}^2(\theta)\rangle / \langle M_{\rm ap}^2(\theta)\rangle$ is expected to be very close to a constant if the bias factor b is scale independent.

Next we consider the correlation coefficient r between the dark matter distribution and the galaxy field. Correlating $M_{\rm ap}(\theta)$ with $\mathcal{N}(\theta)$ yields

$$\begin{aligned}\langle M_{\rm ap}(\theta)\mathcal{N}(\theta)\rangle &= \int {\rm d}^2\vartheta\; U(|\boldsymbol{\vartheta}|) \int {\rm d}^2\vartheta'\; U(|\boldsymbol{\vartheta}'|)\left\langle \kappa(\boldsymbol{\vartheta})\,\kappa_{\rm g}(\boldsymbol{\vartheta}')\right\rangle \\ &= 3\pi\left(\frac{H_0}{c}\right)^2 \Omega_{\rm m}\, b\, r\, H_{\kappa{\rm g}}(\theta)\;, \end{aligned} \tag{152}$$

with

$$H_{\kappa{\rm g}}(\theta) = \int {\rm d}w\; \frac{p_{\rm f}(w)\,g(w)}{a(w)\,f_K(w)} \mathcal{P}(w,\theta)\;. \tag{153}$$

It should be noted that $\langle M_{\rm ap}(\theta)\mathcal{N}(\theta)\rangle$ is a first-order statistics in the cosmic shear. It correlates the shear signal with the location of galaxies, which are assumed to trace the total matter distribution. As shown in Schneider (1998), the signal-to-noise of this correlator is higher than that of $\langle M_{\rm ap}^2\rangle$, and therefore was introduced as a convenient statistics for the detection of cosmic shear. In fact, in their original analysis of the RCS, based on $16\,{\rm deg}^2$, Hoekstra et al. (2001) obtained a significant signal for $\langle M_{\rm ap}(\theta)\mathcal{N}(\theta)\rangle$, but not for $\langle M_{\rm ap}^2(\theta)\rangle$. Combining (146) and (149) with (152), the correlation coefficient r can be expressed as

$$r = \frac{\sqrt{H_\kappa(\theta)\,H_{\rm gg}(\theta)}}{H_{\kappa{\rm g}}(\theta)}\;\frac{\langle M_{\rm ap}(\theta)\mathcal{N}(\theta)\rangle}{\sqrt{\langle M_{\rm ap}^2(\theta)\rangle\,\langle \mathcal{N}^2(\theta)\rangle}} = f_{\rm r}(\theta)\;\frac{\langle M_{\rm ap}(\theta)\mathcal{N}(\theta)\rangle}{\sqrt{\langle M_{\rm ap}^2(\theta)\rangle\,\langle \mathcal{N}^2(\theta)\rangle}}\;. \tag{154}$$

As was the case for $f_{\rm b}$, the function $f_{\rm r}$ depends only very weakly on the filter scale and on the adopted form of the power spectrum, so that a variation of the (observable) final ratio with angular scale would indicate the scale dependence of the correlation coefficient.

Whereas the two aperture measures $M_{\rm ap}$ and $\mathcal{N}$ can in principle be obtained from the data field by putting down circular apertures, and the corresponding second-order statistics can likewise be determined through unbiased estimators defined on these apertures, this is not the method of choice in practice, due to gaps and holes in the data field. Note that in our discussion of cosmic shear in Sect. 6.3, we have expressed $\langle M_{\rm ap}^2(\theta)\rangle$ in terms of the shear two-point correlation functions $\xi_\pm(\theta)$ – see (115) – just for this reason. In close analogy, $\mathcal{N}^2(\theta)$ can be expressed in terms of the angular correlation function

$\omega(\theta)$ of the projected galaxy positions, as seen by (144), or more explicitly, when replacing the power spectrum $P_\omega(\ell)$ in (146) by its Fourier transform, which is the angular correlation function, one finds

$$\left\langle \mathcal{N}^2(\theta) \right\rangle = \int_0^{2\theta} \frac{\mathrm{d}\vartheta\,\vartheta}{\theta^2}\,\omega(\vartheta)\,T_+\left(\frac{\vartheta}{\theta}\right) , \tag{155}$$

where the function T_+ is the same as that occurring in (115). Correspondingly, we introduce the power spectrum $P_{\mathrm{g}\kappa}(\ell)$, which is defined as

$$\left\langle \hat{\kappa}(\boldsymbol{\ell})\hat{\kappa}_{\mathrm{g}}^*(\boldsymbol{\ell}') \right\rangle = (2\pi)^2 \delta_{\mathrm{D}}(\boldsymbol{\ell}-\boldsymbol{\ell}')\,P_{\kappa\mathrm{g}}(|\boldsymbol{\ell}|) . \tag{156}$$

Applying (98), as well as the definitions of the bias and correlation functions, this projected cross-power spectrum is related to the 3-D density contrast by

$$P_{\kappa\mathrm{g}}(\ell) = \frac{3}{2}\left(\frac{H_0}{c}\right)^2 \Omega_{\mathrm{m}} br \int \mathrm{d}w \frac{g(w)p_{\mathrm{f}}(w)}{a(w)f_K(w)}\,P_\delta\left(\frac{\ell}{f_K(w)}, w\right) . \tag{157}$$

The angular correlation function $\left\langle \kappa(\boldsymbol{\vartheta})\kappa(\boldsymbol{\vartheta}') \right\rangle$ occurring in (152) can then be replaced by its Fourier transform $P_{\kappa\mathrm{g}}$. On the other hand, since the Fourier transform of the surface mass density κ is simply related to that of the shear, one can consider the correlation between the galaxy positions with the tangential shear component,

$$\begin{aligned}
\langle \gamma_{\mathrm{t}}(\theta) \rangle &:= \langle \kappa_{\mathrm{g}}(\mathbf{0})\gamma_{\mathrm{t}}(\boldsymbol{\theta}) \rangle \\
&= -\int \frac{\mathrm{d}^2\ell}{(2\pi)^2} \int \frac{\mathrm{d}^2\ell'}{(2\pi)^2} \mathrm{e}^{2\mathrm{i}(\beta'-\varphi)} \exp\left(-\mathrm{i}\boldsymbol{\theta}\cdot\boldsymbol{\ell}'\right) \left\langle \hat{\kappa}_{\mathrm{g}}(\boldsymbol{\ell})\hat{\kappa}(\boldsymbol{\ell}') \right\rangle \\
&= \frac{1}{2\pi}\int \mathrm{d}\ell\,\ell\,\mathrm{J}_2(\theta\ell)\,P_{\kappa\mathrm{g}}(\ell) \qquad (158) \\
\Rightarrow P_{\kappa\mathrm{g}}(\ell) &= 2\pi \int \mathrm{d}\theta\,\theta\,\langle \gamma_{\mathrm{t}}(\theta) \rangle\, J_2(\theta\ell) .
\end{aligned}$$

Note that $\langle \gamma_{\mathrm{t}}(\theta) \rangle$ is just the galaxy–galaxy lensing signal discussed in Sect. 8.2; this shows very clearly that galaxy–galaxy lensing measures the correlation of mass and light in the Universe. In terms of this mean tangential shear, the aperture mass and galaxy number counts can be written as

$$\langle M_{\mathrm{ap}}(\theta)\mathcal{N}(\theta) \rangle = \int_0^{2\theta} \frac{\mathrm{d}\vartheta\,\vartheta}{\theta^2}\,\langle \gamma_{\mathrm{t}}(\vartheta) \rangle\, T_2\left(\frac{\vartheta}{\theta}\right) , \tag{159}$$

where the function T_2 is defined in a way similar to $T_\pm$ and given explicitly as

$$T_2(x) = 576 \int_0^\infty \frac{\mathrm{d}t}{t^3}\,\mathrm{J}_2(xt)\,\left[\mathrm{J}_4(t)\right]^2 ; \tag{160}$$

this function vanishes for $x > 2$, so that the integral in (159) extends over a finite interval only. Hence, all three aperture correlators can be calculated from two-point correlation functions which can be determined from the data directly, independent of possible gaps in the field geometry.

Results from the RCS

Hoekstra et al. (2002c) have applied the foregoing equations to a combination of their RCS survey and the VIRMOS-DESCART survey. The former was used to determine $\langle \mathcal{N}^2 \rangle$ and $\langle M_{\rm ap}\mathcal{N} \rangle$, the latter for deriving $\langle M_{\rm ap}^2 \rangle$. As pointed out by these authors, this combination of surveys is very useful, in that the power spectrum at a redshift around $z \sim 0.35$ can be probed; indeed, they demonstrate that the effective redshift distribution over which the power spectrum, and thus b and r are probed, are well matched for all three statistics for their choice of surveys. 'Foreground' galaxies for the measurement of $\omega(\theta)$ and $\langle \gamma_{\rm t}(\theta) \rangle$ are chosen to have $19.5 \le R_C \le 21$, 'background' galaxies are those with $21.5 \le R_C \le 24$. In Fig. 53 the three aperture statistics are shown as a function of angular scale, as determined from their combined survey, whereas in the right panels, the ratios of these statistics as they appear in (151) and (154) are displayed. Also shown are predictions of these quantities from two cosmological models, assuming $b = 1$ and $r = 1$. The fact that these model predictions are fairly constant in the right-hand panels shows that the factors $f_{\rm b}$ and $f_{\rm r}$ are nearly independent of the radius θ of the aperture, as mentioned before.

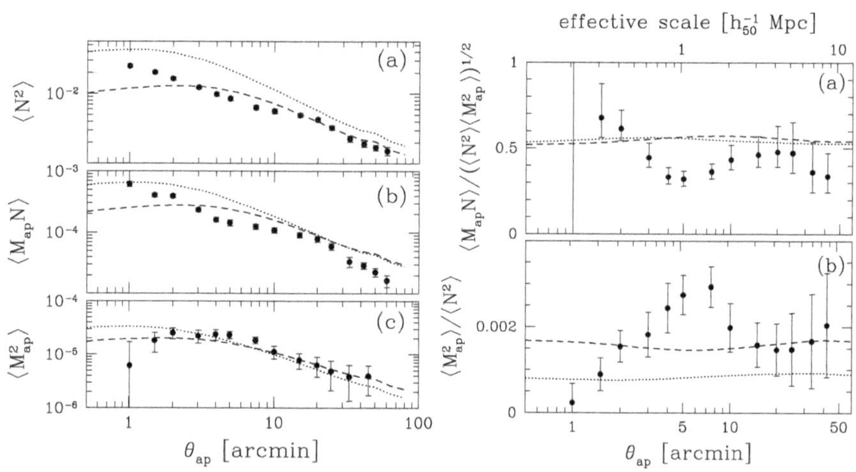

Fig. 53. The left figure displays the three aperture statistics as measured by combining the RCS and the VIRMOS-DESCART survey. Points show measured values, as determined from the correlation functions. The right panels display the ratios of the aperture statistics as they appear in (151) and (154). The dotted and dashed curves in all panels show the predictions for an OCDM and a ΛCDM model, respectively, both with $\Omega_{\rm m} = 0.3$, $\sigma_8 = 0.9$, and $\Gamma_{\rm spect} = 0.21$, for the fiducial values of $b = 1 = r$. The fact that the curves in the right panels are nearly constant show the near-independence of $f_{\rm b}$ and $f_{\rm r}$ on the filter scale. The upper axis in the right panels show the effective physical scale on which the values of b and r are measured (from Hoekstra et al. 2002c)

The results for the bias and correlation factor are shown in Fig. 54, as a function of angular scale and effective physical scale, corresponding to a median redshift of $z \sim 0.35$. The results indicate that the bias factor and the galaxy-mass correlation coefficient are compatible with a constant value on large scales, $\gtrsim 5h^{-1}$ Mpc, but on smaller scales both seem to change with scale. The transition between these two regimes occurs at about the scale where the density field at redshift $z \sim 0.35$ turns from linear to non-linear evolution. In fact, in the non-linear regime one does not expect a constant value of both coefficients, whereas in the linear regime, constant values for them appear natural. It is evident from the figure that the error bars are still too large to draw definite conclusions about the behavior of b and r as a function of scale, but the approach to investigate the relation between galaxies and mass is extremely promising and will certainly yield very useful insight when applied to the next generation of cosmic shear surveys. In particular, with larger surveys than currently available, different cuts in the definition of foreground and background galaxies can be used, and thus the redshift dependence of b and r can be investigated. This is of course optimized if (photometric) redshift estimates for the galaxy sample become available.

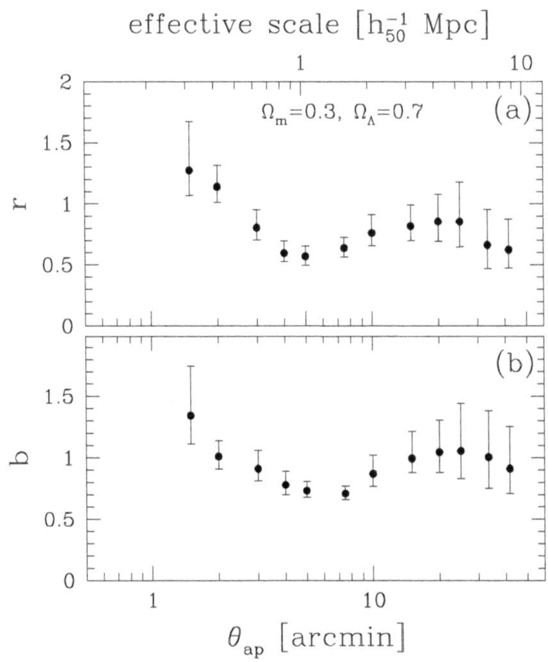

Fig. 54. The values of the bias and correlation coefficient, as determined from (151) and (154) and the results shown in Fig. 53; here, a ΛCDM model has been assumed for the cosmology dependence of the functions f_b and f_r. The upper axis indicates the effective scale on which b and r are measured (from Hoekstra et al. 2002c)

Results from the SDSS

The large sample of galaxies with spectroscopic redshifts already available now from the SDSS permits an accurate study of the biasing properties of these galaxies (see the end of Sect. 8.2). Two different approaches should be mentioned here: the first follows along the line discussed above and has been published in Sheldon et al. (2004). In short, the galaxy–galaxy signal can be translated into the galaxy-mass cross-correlation function $\xi_{\rm gm}$, due to the knowledge of galaxy redshifts. The ratio of $\xi_{\rm gm}$ and the galaxy two-point correlation function $\xi_{\rm gg}$ then depends on the ratio r/b. In Fig. 55 we show the galaxy-mass correlation as a function of linear scale, as well as the ratio b/r. Note that from the SDSS no cosmic shear measurement has been obtained yet, owing to the complex PSF properties, and therefore b and r cannot be measured separately from this data set.

The galaxy-mass correlation function follows a power law over more than two orders-of-magnitude in physical scale, and its slope is very similar to the

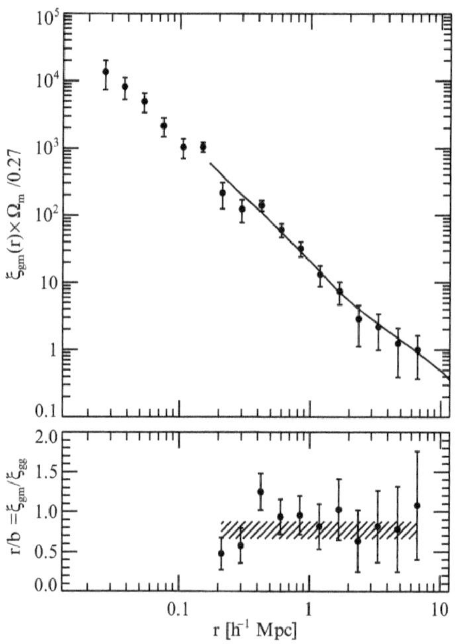

Fig. 55. The galaxy–mass cross-correlation function $\xi_{\rm gm}(r)$, as a function of linear scale (*dots with error bars*), scaled to a matter density parameter of $\Omega_{\rm m} = 0.27$, as well as the two-point galaxy correlation function obtained from the same set of (foreground) galaxies (*solid curve*). The ratio between these two is given in the lower panel, which plots b/r as a function of scale. Over the full range of scales, $\xi_{\rm gm}$ can be well approximated by a power law, $\xi_{\rm gm} = (r/r_0)^{-\gamma}$, with slope $\gamma = 1.79 \pm 0.06$ and correlation length $r_0 = (5.4 \pm 0.7)(\Omega_{\rm m}/0.27)^{-1/\gamma} h^{-1}$ Mpc. The ratio $r/b \approx (1.3 \pm 0.2)(\Omega_{\rm m}/0.27)$ is consistent with being scale-independent

slope of the galaxy two-point correlation function. Hence, the ratio between these two is nearly scale-independent. When splitting the sample into blue and red, and early- and late-type galaxies, the correlation length is larger for the red and the early-type ones. Furthermore, as expected, the lensing signal increases with the velocity dispersion in early-type galaxies.

An alternative approach was taken by Seljak et al. (2004). Their starting point is the fact that the biasing properties of dark matter halos is very well determined from cosmological simulations. This is of course not true for the biasing of galaxies. The bias parameter of galaxies with luminosity L is given as

$$b(L) = \int \mathrm{d}M \, p(M|L) \, b_\mathrm{h}(M) \; , \tag{161}$$

where b_h is the bias of halos of mass M relative to the large-scale matter distribution, and $p(M|L)$ is the probability that a galaxy with luminosity L resides in a halo of mass M. This latter probability distribution is then parameterized for any luminosity bin, by assuming that a fraction $1-\alpha$ of all galaxies in the luminosity bin considered are at the center of their parent halos, whereas the remaining fraction α are satellite galaxies. For the central galaxies, a unique mass $M(L)$ is assigned, whereas for the non-central ones, a mass distribution is assumed. The values of α and M for six luminosity bins are shown in the various panels of Fig. 51; they are obtained by fitting the galaxy–galaxy lensing signal with the model just described. The main reason why the mass spectrum can be probed is that the numerous low-mass galaxy halos contribute to the lensing signal only at relatively small scales, whereas at larger scales the higher-mass halos dominate the signal; hence, different halo masses appear at different separations in the galaxy–galaxy lensing signal. In this way, $b(L)$ can be determined, which depends on the non-linear mass scale M_* (see Sect. 6.2 of IN). The bias parameter is a relatively slowly varying function of galaxy luminosity for $L \lesssim L_*$, approaching a value $\sim$0.7 for very low-luminosity galaxies, but quickly rises for $L > L_*$.

Seljak et al. combined these measurements of the bias parameter with the clustering properties of the SDSS galaxies and the WMAP results on the CMB anisotropy, and derived new constraints on $\sigma_8 = 0.88 \pm 0.06$ and the bias parameter of an L_*-galaxy, $b_* = 0.99 \pm 0.07$; furthermore, the combination of these datasets is used to obtain new constraints on the standard cosmological parameters. This work has opened up a new way on how to employ the results from galaxy–galaxy lensing as a cosmological tool.

8.4 Galaxy Biasing: Magnification Method

High-redshift QSOs are observed to be correlated on the sky with lower-redshift galaxies and clusters. This topic has indeed an interesting history: The detection of very close associations of high-z QSOs with low-z galaxies (see Arp 1987, and references therein) has been claimed as evidence against the cosmological interpretation of the QSO redshifts, as the probabilities of observing such close pairs of objects which are physically unrelated were claimed to be vanishingly small. However, these probabilities were obtained a posteriori, and of course, any specific configuration has a vanishingly small probability. Since the cosmological interpretation of QSO redshifts is supported by overwhelming evidence, the vast majority of researchers consider these associations as a statistical fluke.

A physical possibility to generate the association of background sources with foreground objects is provided by the magnification bias caused by lensing: the number counts of background sources is changed in regions where a foreground lens yields magnifications different from unity – see Sect. 5 of IN. Thus, close to a galaxy where $\mu > 1$, the number counts of bright background QSOs can be enhanced since the slope of their counts is steeper than unity. There have been various attempts in the literature to 'explain' the observed QSO-galaxy associations by invoking the magnification bias, either with a smooth galaxy mass distribution or by including the effects of microlensing; see SEF for a detailed discussion of this effect. The bottom line, however, is that the magnification effect is by far not large enough to account for the small (a posteriori) probabilities of the observed individual close associations.

The topic has been revived, though in a different direction, by the finding that high-redshift AGNs are statistically associated with low-redshift galaxies. Fugmann (1990) provided evidence that radio-selected high-z AGNs from the 1-Jansky-catalog are correlated with relatively bright (and therefore low-z) galaxies taken from the Lick catalog, an analysis that later on was repeated by Bartelmann and Schneider (1993), using a slightly different statistics. Different samples of foreground and background populations have been employed in further studies, including the correlation between 1-Jansky AGN with bright IRAS galaxies (Bartelmann and Schneider 1994; Bartsch et al. 1997), high-z QSOs with clusters from the Zwicky catalog of clusters (Rodrigues-Williams and Hogan 1994; Seitz and Schneider 1995b), 1-Jansky AGNs with red galaxies from the APM catalog (Benítez and Martínez-González 1995; see also Norman and Impey 2001), to mention just a few. Radio-selected AGN are considered to be a more reliable probe since their radio flux is unaffected by extinction, an effect which could cause a bias (if the sky shows patchy extinction, both galaxies and QSOs would have correlated inhomogeneous distributions on the sky) or anti-bias (if extinction is related to the lensing matter) for flux-limited optical surveys of AGNs, and which therefore needs to be taken into account in the correlation analysis of optically-selected AGNs. However, most radio source catalogs are not fully optically identified and lack redshifts, and using incomplete radio surveys therefore can induce a selection bias (Benítez et al. 2001). These latter authors investigated the correlation between two completely identified radio catalogs with the COSMOS galaxy catalog, and found a very significant correlation signal.

The upshot of all these analyses is that there seems to be a positive correlation between the high-z sources and the low-z objects, on angular scales between $\sim 1'$ and about $1°$. The significances of these correlations are often not very large, they typically are at the 2–3σ level, essentially limited by the finite number of high-redshift radio sources with a large flux (the latter being needed for two reasons: first, only radio surveys with a high flux threshold, such as the 1-Jansky catalog, have been completely optically identified and redshifts determined, which is necessary to exclude low-redshift sources which could be physically associated with the 'foreground' galaxy population, and second, because the counts are steep only for high fluxes, needed to obtain a high magnification bias). If this effect is real, it cannot be explained by lensing caused by individual galaxies; the angular region on which galaxies produce an appreciable magnification is just a few arcseconds. However, if galaxies trace the underlying (dark) matter distribution, the latter can yield magnifications (in the same way as it yields a shear) on larger scales. Thus, an obvious qualitative interpretation of the observed correlation is therefore that it is due to magnification of the large-scale matter distribution in the Universe of which the galaxies are tracers. This view is supported by the finding (Ménard and Péroux 2003) that there is a significant correlation of bright QSOs with metal absorption systems in the sense that there are relatively more bright QSOs with an aborber than without; this effect shows the expected trend from magnification bias caused by matter distributions associated with the absorbing material.

We therefore consider a flux-limited sample of AGNs, with distance probability distribution $p_{\rm Q}(w)$, and a sample of galaxies with distance distribution $p_{\rm f}(w)$. It will be assumed that the AGN sample has been selected such that it includes only objects with redshift larger than some threshold $z_{\rm min}$, corresponding to a minimum comoving distance $w_{\rm min}$, which is larger than the distances of all galaxies in the sample. We define the AGN-galaxy correlation function as

$$w_{\rm Qg}(\theta) = \frac{\left\langle \left[N_{\rm g}(\boldsymbol{\phi}) - \bar{N}_{\rm g}\right] \left[N_{\rm Q}(\boldsymbol{\phi}+\boldsymbol{\theta}) - \bar{N}_{\rm Q}\right]\right\rangle}{\bar{N}_{\rm g}\,\bar{N}_{\rm Q}} \,, \tag{162}$$

where $N_{\rm g}(\boldsymbol{\phi})$ and $N_{\rm Q}(\boldsymbol{\phi})$ are the observed number densities of galaxies and AGNs, respectively. The former is given by (142). The observed number density of AGN is affected by the magnification bias. Provided the unlensed counts can be described (locally) as a power-law in flux, $N_{\rm Q,0}(>S) \propto S^{-\beta}$, then from (108) of IN we find that $N_{\rm Q}(\boldsymbol{\phi}) = N_{\rm Q,0}\,\mu^{\beta-1}(\boldsymbol{\phi})$, where $\mu(\boldsymbol{\phi})$ is the magnification in the direction $\boldsymbol{\phi}$. Then, if the magnifications that are relevant are small, we can approximate

$$\mu(\boldsymbol{\phi}) \approx 1 + 2\kappa(\boldsymbol{\phi}) = 1 + \delta\mu(\boldsymbol{\phi}) \,, \tag{163}$$

and the projected surface mass density κ is given by (93) with p_w in (94) replaced by $p_{\rm Q}$. Assuming that the magnifications do not affect the mean source counts $\bar{N}_{\rm Q}$, the cross-correlation becomes

$$w_{\rm Qg}(\theta) = 2(\beta-1)\bar{b}(\theta)\,\bar{r}(\theta)\,w_{\kappa{\rm g}}(\theta) \,, \tag{164}$$

where $\bar{b}$ and $\bar{r}$ are the effective bias factor of the galaxies and the mean galaxy-mass correlation function just as in Sect. 8.3, and $w_{\kappa g}$ is the correlation between the projected density field κ and the projected number density of galaxies κ_g, defined after (142), which is the Fourier transform of $P_{\kappa g}(\ell)$ defined in (141). Hence, a measurement of this correlation, together with a measurement of the correlation function of galaxies, can constrain the values of b and r (Dolag and Bartelmann 1997; Ménard and Bartelmann 2002).

The observed correlation between galaxies and background AGN appears to be significantly larger than can be accounted for by the models presented above. On scales of a few arcmin, Benítez et al. (2001) argued that the observed signal exceeds the theoretical expectations by a factor of a few. This discrepancy can be attributed to either observational effects, or shortcomings of the theoretical modeling. Obviously, selection effects can easily produce spurious correlations, such as patchy dust obscuration or a physical association of AGNs with the galaxies. Furthermore, the weak lensing approximation employed above can break down on small angular scales. Jain et al. (2003, see also Takada and Hamana 2003) argued that the simple biasing model most likely breaks down for the small scales where the discrepancy is seen, and employed the halo model for describing the large-scale distribution of matter and galaxies to predict the expected correlations. For example, the strength of the signal depends sensitively on the redshifts, magnitudes and galaxy type.

At present, the shear method to determine the bias factor and the galaxy-mass correlation has yielded more significant results than the magnification method, owing to the small complete and homogeneous samples of high-redshift AGNs. As pointed out by Ménard and Bartelmann (2002), the SDSS may well change this situation shortly, as this survey will obtain $\sim 10^5$ homogeneously selected spectroscopically verified AGNs. Provided the effects of extinction can be controlled sufficiently well, this data should provide a precision measurement of the QSO–galaxy correlation function.

9 Additional Issues in Cosmic Shear

9.1 Higher-Order Statistics

On the level of second-order statistics, 'only' the power spectrum is probed. If the density field was Gaussian, then the power spectrum would fully characterize it; however, in the course of non-linear structure evolution, non-Gaussian features of the density field are generated, which show up correspondingly in the cosmic shear field and which can be probed by higher-order shear statistics. The usefulness of these higher-order measures for cosmic shear has been pointed out in Bernardeau et al. (1997), Jain and Seljak (1997), Schneider et al. (1998a) and van Waerbeke et al. (1999); in particular, the near-degeneracy between σ_8 and Ω_m as found from using second-order statistics can be broken. However, there are serious problems with higher-order shear statistics, that shall be illustrated below in terms of the third-order statistics.

But first, we can give a simple argument why third-order statistics is able to break the degeneracy between $\Omega_{\rm m}$ and σ_8. Consider a density field on a scale where the inhomogeneities are just weakly non-linear. One can then employ second-order perturbation theory for the growth of the density contrast δ. Hence, we write $\delta = \delta^{(1)} + \delta^{(2)} + \ldots$, where $\delta^{(1)}$ is the density contrast obtained from linear perturbation theory, and $\delta^{(2)}$ is the next-order term. This second-order term is quadratic in the linear density field, $\delta^{(2)} \propto \left(\delta^{(1)}\right)^2$. The linear density field is proportional to σ_8, and the projected density $\kappa \propto \Omega_{\rm m}\sigma_8$. Hence, in the linear regime, $\left\langle\kappa^2\right\rangle \propto \Omega_{\rm m}^2\sigma_8^2$, where $\left\langle\kappa^2\right\rangle$ shall denote here any second-order shear estimator. The lowest order contribution to the third-order statistics is of the form

$$\left\langle\kappa^3\right\rangle \propto \left(\delta^{(1)}\right)^2 \delta^{(2)} \propto \Omega_{\rm m}^3\,\sigma_8^4\,,$$

since the term $\left(\delta^{(1)}\right)^3$ yields no contribution owing to the assumed Gaussianity of the linear density field. Hence, a skewness statistics of the form

$$\left\langle\kappa^3\right\rangle / \left\langle\kappa^2\right\rangle^2 \propto \Omega_{\rm m}^{-1}$$

will be independent of the normalization σ_8, at least in this simplified perturbation approach. In more accurate estimates, this is not exactly true; nevertheless, the functional dependencies of the second- and third-order shear statistics on σ_8 and $\Omega_{\rm m}$ are different, so that these parameters can be determined separately.

The Shear Three-Point Correlation Function

Most of the early studies on three-point statistics concentrated on the third-order moment of the surface mass density κ in a circular aperture, $\langle\kappa(\theta)\rangle$; however, this is not a directly measureable quantity, and therefore useful only for theoretical considerations. As for second-order statistics, one should consider the correlation functions, which are the quantities that can be obtained best directly from the data and which are independent of holes and gaps in the data field. The three-point correlation function (3PCF) of the shear has three independent variables (e.g. the sides of a triangle) and 8 components; as was shown in Schneider and Lombardi (2003), none of these eight components vanishes owing to parity invariance (as was suspected before – this confusion arises because little intuition is available on the properties of the 3PCF of a polar). This then implies that the covariance matrix has 6 arguments and 64 components ! Of course, this is too hard to handle efficiently, therefore one must ask which combinations of the components of the 3PCF are most useful for studying the dark matter distribution. Unfortunately, this is essentially unknown yet. An additional problem is that the predictions from theory are less well established than for the second-order statistics.

A further complication stems from a certain degree of arbitrariness on how to define the 8 components of the 3PCF. For the 2PCF, the vector between any pair of points defines a natural direction with respect to which tangential and cross components of the shear are defined; this is no longer true for three points. On the other hand, the three points of a triangle define a set of centers, such as the 'center of mass', or the center of the in- or circum-circle. After choosing one of these centers, one can define the two components of the shear which are then independent of the coordinate frame.

Nevertheless, progress has been achieved. From ray-tracing simulations through a cosmic matter distribution, the 3PCF of the shear can be determined (Takada and Jain 2003a; see also Zaldarriaga and Scoccimarro 2003; furthermore, the three-point cosmic shear statistics can also be determined in the frame of the halo model, see Cooray and Hu 2001; Takada and Jain 2003b), whereas Schneider and Lombardi (2003) have defined the 'natural components' of the shear 3PCF which are most easily related to the bispectrum of the underlying matter distribution. Let $\gamma^{\rm c}(\boldsymbol{\theta}_i) = \gamma_{\rm t} + {\rm i}\gamma_\times = -\gamma\,{\rm e}^{-2{\rm i}\zeta_i}$ be the complex shear measured in the frame which is rotated by the angle ζ_i relative to the Cartesian frame, so that the real and imaginary parts of $\gamma^{\rm c}$ are the tangential and cross components of the shear relative to the chosen center of the triangle (which has to be defined for each triplet of points separately). Then the natural components are defined as

$$\begin{aligned}\Gamma^{(0)} &= \left\langle \gamma^{\rm c}(\boldsymbol{\theta}_1)\,\gamma^{\rm c}(\boldsymbol{\theta}_2)\,\gamma^{\rm c}(\boldsymbol{\theta}_3)\right\rangle \;, \\ \Gamma^{(1)} &= \left\langle \gamma^{\rm c*}(\boldsymbol{\theta}_1)\,\gamma^{\rm c}(\boldsymbol{\theta}_2)\,\gamma^{\rm c}(\boldsymbol{\theta}_3)\right\rangle \;, \end{aligned} \tag{165}$$

and correspondingly for $\Gamma^{(2)}$ and $\Gamma^{(3)}$. Each of the natural components of the 3PCF constitutes a complex number, which depends just on the three separations between the points. Special care is required for labeling the points, and one should follow the rule that they are labeled in a counter-clock direction around the triangle. If such a unique prescription is not systematically applied, confusing and wrong conclusions will be obtained about the behavior of the shear 3PCF with respect to parity transformations (as the author has experienced painfully enough). In Schneider et al. (2005), explicit relations are derived for the natural components of the shear 3PCF in terms of the bispectrum (that is, the generalization of the power spectrum for the three-point statistics) of the underlying mass distribution κ.

Third-Order Aperture Statistics

Alternatively, aperture measures can be defined to measure the third-order statistics. Schneider et al. (1998a) calculated $\left\langle M_{\rm ap}^3\right\rangle(\theta)$ in the frame of the quasi-linear structure evolution model and showed it to be a strong function of $\Omega_{\rm m}$. van Waerbeke et al. (2001) calculated the third-order aperture mass, using a fitting formula of the non-linear evolution of the dark matter bispectrum obtained by Scoccimarro and Couchman (2001) and pointed out the strong sensitivity with respect to cosmological parameters. Indeed, as mentioned before, $\langle M_{\rm ap}^3 \rangle$ is sensitive only to the E-modes of the shear field. One might be tempted to use $\langle M_\perp^3 \rangle(\theta)$ as a measure for third-order B-mode statistics, but indeed, this quantity vanishes owing to parity invariance (Schneider 2003). However, $\langle M_\perp^2 \, M_{\rm ap} \rangle$ is a measure for the B-modes at the third-order statistical level. Jarvis et al. (2004) have calculated $\langle M_{\rm ap}^3(\theta) \rangle$ in terms of the shear 3PCF, for the weight function (110) in the definition of $M_{\rm ap}$. Schneider et al. (2005) have shown that this relation is most easily expressed in terms of the natural components of the shear 3PCF. On the other hand, Jarvis et al. (2004) have expressed $\langle M_{\rm ap}^3(\theta) \rangle$ in terms of the bispectrum of κ, and as was the case for the aperture dispersion in relation to the power spectrum of κ, the third-order aperture mass is a very localized measure of the bispectrum and is sensitive essentially only to modes with three wavevectors with equal magnitudes. For that reason, Schneider et al. (2005) have generalized the definition of the third-order aperture measures, correlating the aperture mass of three different sizes, $\langle M_{\rm ap}(\theta_1) M_{\rm ap}(\theta_2) M_{\rm ap}(\theta_3) \rangle$. This third-order statistics is again a very localized measure of the bispectrum, but this time with wave vectors of different magnitude $\ell_i \approx \pi/\theta_i$, and therefore, by considering the third-order aperture mass for all combinations of θ_i, one can probe the full bispectrum. Therefore, the third-order aperture mass correlator with three independent arguments (i.e., angular scales) should contain essentially the full third-order statistical information of the κ-field, since in contrast to the two-point statistics, the shear 3PCF does not contain information about long-wavelength modes.

Furthermore, the third-order aperture statistics can be expressed directly in terms of the shear 3PCF through a simple integration, very similar to the relations (125) for the two-point statistics. Finally, the other three third-order aperture statistics (e.g., $\langle M_\perp(\theta_1) M_{\rm ap}(\theta_2) M_{\rm ap}(\theta_3) \rangle$) can as well be obtained from the natural components of the shear 3PCF. These correlators are expected to vanish if the shear is solely due to lensing, but intrinsic alignments of galaxies can lead to finite correlators which include B-modes. However, as shown in Schneider (2003), $\langle M_{\rm ap}(\theta_1) M_{\rm ap}(\theta_2) M_\perp(\theta_3) \rangle$, as well as $\langle M_\perp(\theta_1) M_\perp(\theta_2) M_\perp(\theta_3) \rangle$, are expected to vanish even in the presence of B-modes, since these two correlators are not invariant with respect to a parity transformation. Therefore, non-zero results of these two correlators signify the violation of parity invariance and therefore provide a clean check on the systematics of the data and their analysis.

First Detections

Bernardeau et al. (2002) measured for the first time a significant third-order shear from the VIRMOS-DESCART survey, employing a suitably filtered integral over the measured 3PCF (as defined in Bernardeau et al. 2003). Pen et al. (2003a,b) used the aperture statistics to detect a skewness in the same data set. The accuracy of these measurements is not sufficient to derive strong constraints on cosmological parameters, owing to the limited sky area available. However, with the upcoming large cosmic shear surveys, the 3PCF will be measured with high accuracy. Determining the 3PCF from observed galaxy ellipticities cannot be done by straightforwardly considering any triple of galaxies – there are just too many. Jarvis et al. (2004) and Zhang and Pen (2005) have developed algorithms for calculating the 3PCF in an efficient way.

Based on the halo model for the description of the LSS, Takada and Jain (2003b) studied the dependence of the shear 3PCF on cosmological parameters. For relatively large triangles, the 3PCF provides a means to break the degeneracies of cosmological parameters that are left when using the second-order statistics only, as argued above. For small triangles, the 3PCF is dominated by the one-halo term, and therefore primarily probes the mass profiles of halos. Ho and White (2004) show that the 3PCF on small angular scales also contains information on the asphericity of dark matter halos. The full power of third-order statistics is achieved once redshift information on the source galaxies become available, in which case the combination of the 2PCF and 3PCF provides a sensitive probe on the equation-of-state of the dark energy (Takada and Jain 2004).

Beyond Third Order

One might be tempted to look into the properties of the fourth-order shear statistics (though I'm sure the reader can control her/himself in doing this – but see Takada and Jain 2002). OK, the four-point correlation function has 16 components and depends on 5 variables, not to mention the corresponding covariance or the redshift dependent fourth-order correlator. One can consider correlating the aperture mass of four different angular sizes, but in contrast to the third-order statistics, this is expected not to contain the full information on the trispectrum (which describes the fourth-order statistical properties of κ). Perhaps a combination of this fourth-order aperture mass with the average of the fourth power of the mean shear in circular apertures will carry most of the information. And how much information on cosmological parameters does the fourth-order shear statistics contain ? And even higher orders ?

Already the third-order shear statistic is not acccurately predictable from analytic descriptions of the non-linear evolution of the matter inhomogeneities, and the situation worsens with even higher order.[13] One therefore needs to refer to detailed ray-tracing simulations. Although they are quite time consuming, I do not see a real bottleneck in this aspect: Once a solid

[13] In the limits of small and large angular scales, analytic approximations can be obtained. For small scales, the highly non-linear regime is often described by the hierarchical ansatz and hyperextended perturbation theory (see Munchi and Jain 2001 and references therein), whereas on very large scales second-order perturbation theory can be used. Nevertheless, the range of validity of these perturbation approximations and their accuracy have to be checked with numerical simulations.

and accurate measurement of the three-point correlation function becomes available, certainly considerable effort will be taken to compare this with numerical simulations (in particular, since such a measurement is probably a few years ahead, in which the computer power will increase by significant factors). If we accept this point, then higher-order statistics can be obtained from these simulations, and several can be 'tried out' on the numerical data such that they best distinguish between different models. For example, one can consider the full probability distribution $p(M_{\rm ap};\theta)$ on a given data set (Kruse and Schneider 2000; Reblinsky et al. 1999; Bernardeau and Valageas 2000; Munshi et al. 2004). To obtain this from the observational data, one needs to place apertures on the data field which, as we have argued, is plagued with holes and gaps in the data. However, we can place the same gaps on the simulated data fields and therefore simulate this effect. Similarly, the numerical simulations should be used to find good strategies for combining second- and third-order shear statistics (and potentially higher-order ones) for an optimal distinction between comological model parameters, and, in particular, the equation-of-state of dark energy. Another issue one needs to consider for third- (and higher-)order cosmic shear measures is that intrinsic clustering of sources, and the correlation between galaxies and the dark matter distribution generating the shear field has an influence on the expected signal strength (Bernardeau (1998); Hamana (2001); Hamana et al. (2002)). Obviously, there are still a lot of important studies to be done.

Third-Order Galaxy–Mass Correlations

We have shown in Sect. 8 how galaxy–galaxy lensing can be used to probe the correlation between galaxies and the underlying matter distribution. With the detection of third-order shear statistics already in currently available data sets, one might expect that also higher-order galaxy-mass correlations can be measured from the same data. Such correlations would then probe, on large angular scales, the higher-order biasing parameters of galaxies, and thereby put additional constraints on the formation and evolution of galaxies. Ménard et al. (2003) considered the correlation between high-redshift QSOs and pairs of foreground galaxies, thus generalizing the methods of Sect. 8.4 to third-order statistics. The galaxy–galaxy-shear correlation, and the galaxy-shear-shear correlations have been considered by Schneider and Watts (2005). These correlation functions have been related to the underlying bispectrum of the dark matter and the third-order bias and correlation functions, and appropriate aperture statistics have been defined, that are related in a simple way to the bispectra and the correlation functions.

In fact, integrals of these higher-order correlations have probably been measured already. As shown in Fig. 50, galaxies in regions of high galaxy number densities show a stronger, and more extended galaxy–galaxy lensing signal than more isolated galaxies. Hence there is a correlation between the mean mass profile around galaxies and the local number density of galaxies, which is just an integrated galaxy–galaxy-shear correlation. In fact, such a correlation is only first order in the shear and should therefore be much easier to measure than the shear 3PCF. Furthermore, the galaxy-shear-shear correlation seems to be present in the cosmic shear analysis of the COMBO-17 fields by Brown et al. (2003), where they find a stronger-than-average cosmic shear signal in the A 901 field, and a weaker cosmic shear signal in the CDFS, which is a field selected because it is rather poor in brighter galaxies.

9.2 Influence of LSS Lensing on Lensing by Clusters and Galaxies

The lensing effect of the three-dimensional matter distribution will contaminate the lensing measurements of localized objects, such as galaxies and clusters. Some of the associated effects are mentioned in this section.

Influence of Cosmic Shear on Strong Lensing by Galaxies

The lensing effect of foreground and background matter in a strong lensing system will affect the image positions and flux ratios. As this 3-D lensing effects are not recognized as such in the lens modeling, a 'wrong' lens model will be fitted to the data, in the sense that the mass model for the lensing galaxy will try to include these additional lensing effects not associated with the galaxy itself. In particular, the corresponding predictions for the time delays can be affected through this effect.

Since the image separation of strong lens systems are less than a few arcseconds, the lensing effect of the LSS can be well approximated by a linear mapping across this angular scale. In this case, the effect of the 3-D matter distribution on the lens model can be studied analytically (e.g., Bar-Kana 1996). The lens equation resulting from the main lens (the galaxy) plus the linearized inhomogeneities of the LSS is strictly equivalent to the single-plane gravitational lens equation without these cosmological perturbations, and the mass distribution of the equivalent single-plane lens can be explicitly derived (Schneider 1997). For example, if the main lens is described by elliptical isopotential curves (i.e., elliptical contours of the deflection potential ψ) plus external shear, the equivalent single-plane lens will be of the same form. The orientation of the ellipticity of the lens, as seen by the observer, will be rotated by the foreground LSS by the same angle as the potential of the equivalent lens, so that no observable misalignment is induced. This equivalence then implies that the determination of the Hubble constant from time-delay measurements is affected by the same mass-sheet degeneracy transformation as for a single plane lens.

LSS Effects on the Mass Determination of Clusters

The determination of mass parameters of a cluster from weak lensing is affected by the inhomogeneous foreground and background matter distribution.

The effect of local mass associated with a cluster (e.g., filaments extending from the cluster along the line-of-sight) will bias the mass determination of clusters toward higher values, since clusters are likely to be located in overdense regions of the LSS, though this effect is considerably smaller than claimed by Metzler et al. (2001), as shown by Clowe et al. (2004a).

Hoekstra (2001, 2003) considered the effect of the LSS on the determination of mass parameters of clusters, using either SIS or NFW models. For the SIS model, the one parameter characterizing this mass profile (σ_v) can be obtained as a linear estimator of the shear. The dispersion of this parameter is then the sum of the dispersion caused by the intrinsic ellipticity of the source galaxies and the cosmic shear dispersion. For the NFW model, the relation between its two parameters (M_{200}, the mass inside the virial radius r_{200}, and the concentration c) and the shear is not linear, but the effect of the LSS can still be estimated from Monte-Carlo simulations in which the cosmic shear is assumed to follow Gaussian statistics with a power spectrum following the Peacock and Dodds (1996) prescription.

For the SIS model, the effect of the LSS on the determination of σ_v is small, provided the cluster is at intermediate redshift (so that most source galaxies are in the background). The noise caused by the finite ellipticity in this case is almost always larger than the effect by the LSS. There is an interesting effect, however, in that the relative contribution of the LSS and shape noise changes as larger aperture fits to the SIS model are considered: The larger the field over which the shear is fitted to an SIS model, the larger becomes the impact of cosmic shear, and this increase compensates for the reduced shape noise. In effect, cosmic shear and shape noise together put an upper limit on the accuracy of the determination of σ_v from shear data. The same is true for the determination of the mass parameters of the NFW model, as shown in Fig. 56. The uncertainties of the mass parameters of NFW profiles are about twice as large as if the effects from the LSS are ignored, whereas the effect is considerably smaller for the one-parameter model of the SIS. One should also note that a decrease of the shape noise, which can be obtained by using data with a fainter limiting magnitude, yields an increase of the noise from the LSS, since the fainter galaxies are expected to be at higher redshift and therefore carry a larger cosmic shear signal. For low-redshift clusters, these two effects nearly compensate.

The Efficiency and Completeness of Weak Lensing Cluster Searches

We take up the brief discussion at the end of Sect. 5.8 about the potential of deriving a shear-selected sample of galaxy clusters. The first studies of this question were based on analytical models (e.g., Kruse and Schneider 1999) or numerical models of isolated clusters (Reblinsky and Bartelmann 1999). Those studies can of course not account for the effects of lensing by the LSS. Ray-tracing simulations through N-body generated LSS were carried out by

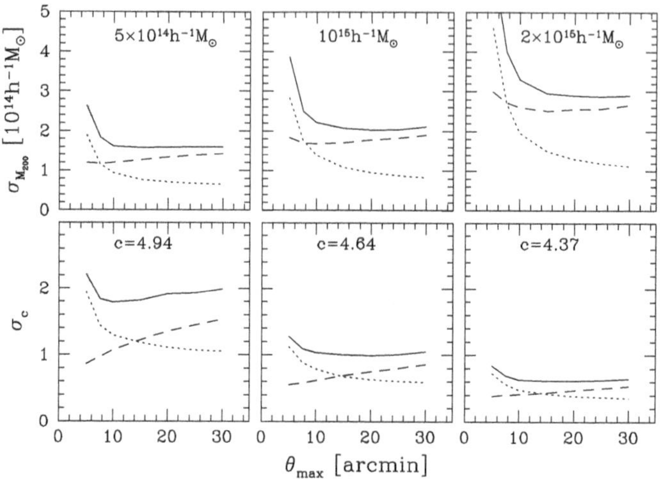

Fig. 56. The dispersion of the determination of the mass and concentration of three NFW halos at redshift $z_d = 0.3$. These parameters were derived by fitting an NFW shear profile to the shear simulated from an NFW halo with parameters indicated in the figure and adding shape noise and noise from cosmic shear. The outer angular scale over which the fit was performed is θ_{max}. Dotted curves show the effect from shape noise alone, dashed curves show the dispersion from cosmic shear, and the solid curves contain both effects. Surprisingly, the accuracy of the NFW parameters does not increase once $\theta_{max} \sim 15'$ is reached, as for larger radii, the cosmic shear noise more than compensates for the reduced ellipticity noise. Another way to express that is that the lensing signal at very large distance from the halo center is weaker than the rms cosmic shear and therefore does not increase the signal-to-noise any more (from Hoekstra 2003)

Reblinsky et al. (1999), White et al. (2002), Hamana et al. (2004), Vale and White (2003), Hennawi and Spergel (2005) and others. In these cosmological simulations, halos were identified based on their 3-D mass distribution. They were then compared to the properties of the lensing results obtained from ray tracing, either by considering the (smoothed) surface mass density κ (that could be obtained from a mass reconstruction from the shear field) or by studying the aperture mass M_{ap} which can be obtained directly from the shear. In both cases, noise due to the finite intrinsic source ellipticity can be added.

The two basic quantities that have been investigated in these studies are *completeness* and *efficiency*. Completeness is the fraction of dark matter halos above some mass threshold M_{min} that are detected in the weak lensing data, whereas efficiency is the fraction of significant lensing detections that correspond to a real halo. Both of these quantities depend on a number of parameters, like the mass threshold of a halo and the limiting significance ν of a lensing detection [in the case of the aperture mass, this would correspond to (80)], as well as on the choice of the filter function Q. Hennawi and Spergel

(2005) have pointed out that even without noise (from observations or intrinsic galaxy ellipticities), the efficiency is limited to about 85% – even under these idealized condition, the selected sample will be contaminated by at least 15% of spurious detections, generated by projection effects of the LSS.

To compare these predictions with observations, the six highest-redshift EMSS clusters were all detected at high significance with a weak lensing analysis (Clowe et al. 2000). Clowe et al. (2004b) have studied 20 high-redshift clusters with weak lensing techniques. These clusters were optically selected and are expected to be somewhat less massive (and potentially more affected by foreground galaxies) than the EMSS clusters. Only eight of these 20 clusters are detected with more than 3σ significance, but for none of them does the SIS fit produce a negative σ_v^2. Only for four of these clusters are the lensing results compatible with no shear signal.

10 Concluding Remarks

Weak lensing has become a standard tool in observational cosmology, as we have learned how to measure the shape of faint galaxy images and to correct them for distortions in the telescope and camera optics and for PSF effects. These technical issues are at the very center of any observational weak lensing research. It appears that at present, the accuracy with which shear can be measured is sufficient for the data available today, in the sense that statistical uncertainties are likely to be larger than potential inaccuracies in the measurement of unbiased shear estimates from faint images. This, however, will change quickly. The upcoming large cosmic shear surveys will greatly reduce statistical uncertainties, and then the accuracy of shear measurements from the data will be the essential limiting factor. Alternatives to KSB have been developed, but they need to undergo thorough testing before becoming a standard tool for observers. It should also be noted that the KSB method is applied differently by different groups, in particular with regards to the weighting of galaxies and other details. What is urgently needed is a study in which different groups apply their version of KSB to the same data set and compare the results. Furthermore, starting from raw data, the specific data reduction methods will lead to slightly different coadded images, and shear measurements on such differently reduced imaged should be compared. These technical issues will be a central challenge for weak lensing in the upcoming years.

The ongoing and planned wide-field imaging surveys mentioned at the end of Sect. 7.7 will allow us to investigate several central questions of cosmology. The two aspects that I consider most relevant are the investigation of the equation-of-state of the dark energy and the relation between galaxies and the underlying dark matter distribution. The former question about the nature of dark energy is arguably the central challenge of modern cosmology, and cosmic shear is one of the very few methods how it can be studied empirically.

The relation between dark matter and galaxies is central to our understanding of how galaxies form and evolve, and galaxy–galaxy lensing is the only way how this relation can be investigated without a priori assumptions.

Essentially all weak lensing studies today have used faint galaxies as sources, since they form the densest source population currently observable. The uniqueness of faint optical galaxies will not stay forever, with the currently planned future instruments. For example, there is a rich literature of weak lensing of the cosmic microwave background which provides a source of very accurately known redshift. Weak lensing by the large-scale structure enhances the power spectrum of the CMB at small angular scales, and the Planck satellite will be able to measure this effect. In particular, polarization information will be very useful, since lensing can introduce B-modes in the CMB polarization. The James Webb Space Telescope, with its large aperture of 6.5 meters and its low temperature and background will increase the number density of observable faint sources in the near-IR up to 5μm to several hundred per square arcminute, many of them at redshifts beyond 3, and will therefore permit much more detailed weak lensing studies, in particular of clusters (see Fig. 57; an observation of this huge number of arcs and multiple images will answer questions about the mass distribution of clusters that we have yet not even dared to ask). The envisioned next generation radio telescope Square Kilometer Array will populate the radio sky with very comparable source density as currently the deepest optical images. Since the beam (that is, the point-spread function) of this radio interferometer will be known very accurately, PSF corrections for this instrument will be more reliable than for optical telescopes. Furthermore, higher-order correlation of the shear field with sources in the field will tell us about non-Gaussian properties of galaxy–matter correlations and biasing, and therefore provide important input into models of galaxy formation and evolution.

Acknowledgments

I enjoyed the week of lecturing at Les Diablerets a lot. First, I thank Georges Meylan and his colleagues Philippe Jetzer and Pierre North for organizing this school so efficiently and smoothly; to them also my sincere apologies for not finishing these proceedings much earlier – but sure enough, lecturing is more fun than preparing lectures, and certainly very much more fun than writing them up into something to be published. Second, I thank my fellow lecturers Chris Kochanek and Joe Wambsganss for discussions, great company and good spirits. Third, and foremost, my compliments to the students who patiently sat through these lectures which, I'm sure, were not always easy to follow; nevertheless I hope that they grasped the essential points, and that these lecture notes help to fill in the details. These notes would not have been possible without the many colleagues and students with whom I had the privilege to collaborate over the years on various aspects of weak lensing, among them Matthias Bartelmann, Marusa Bradac, Doug Clowe,

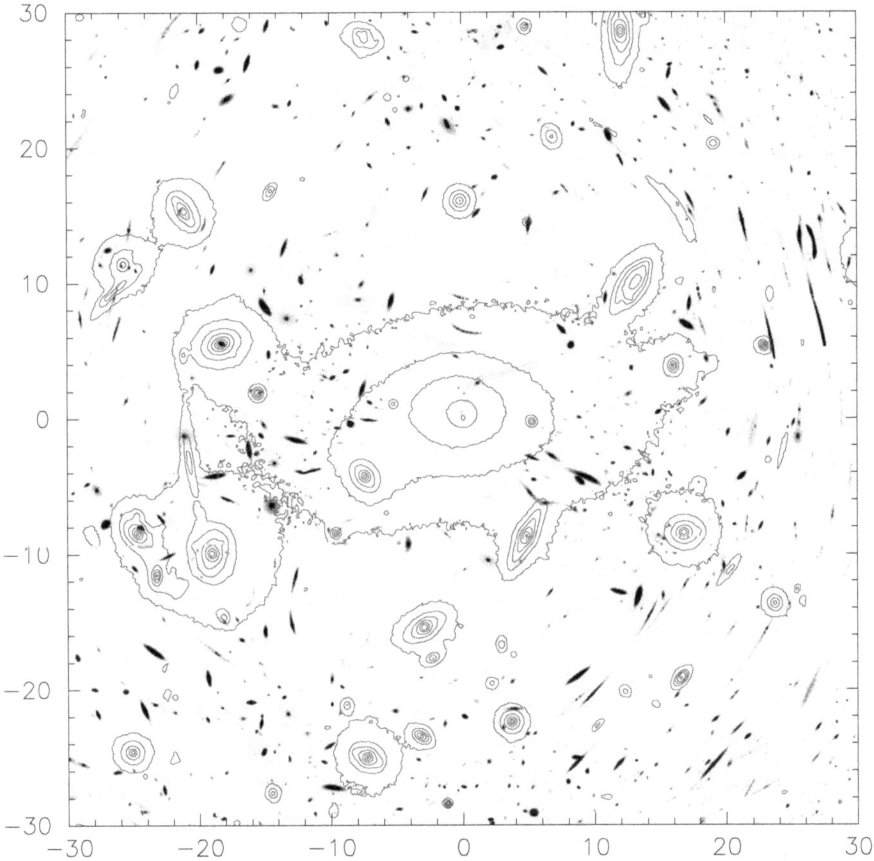

Fig. 57. Simulated image of lensed features in the very central part of the massive cluster A 2218, as observed with the future JWST. For these simulations, the mass profile of the cluster as constrained from HST observations and detailed modeling (Kneib et al. 1996) has been used. The number density of (unlensed) sources was assumed to be $4 \times 10^6 \mathrm{deg}^{-2}$ down to K=29. The redshift distribution assumed is broad and extends to redshift $z \sim 10$ with a median value $z_{\rm med} \sim 3$. The brighter objects (cluster galaxies and brightest arcs) seen by HST are displayed as contours, to make the faint galaxy images visible on this limited dynamic range reproduction. An enormous number of large arcs and arclets are seen; in particular, numerous radial arcs can be easily detected, which will allow us to determine the 'core size' of the cluster mass distribution. Due to the broad redshift distribution of the faint galaxies, arcs occur at quite a range of angular separations from the cluster center; this effect will become even stronger for higher-redshift clusters. It should be noted that this 1 arcminute field does not cover the second mass clump seen with HST; a JWST image will cover a much larger area, and more strong lensing features will be found which can then be combined with the weak lensing analysis of such a cluster. For this simulation, a pixel size of $0\farcs06$ was used; the JWST sampling will be better by a factor of 2 (from Schneider and Kneib 1998)

Jörg Dietrich, Thomas Erben, Bernhard Geiger, Hannelore Hämmerle, Bhuvnesh Jain, Martin Kilbinger, Lindsay King, Martina Kleinheinrich, Guido Kruse, Marco Lombardi, Yannick Mellier, Hans-Walter Rix, Mischa Schirmer, Tim Schrabback, Carolin Seitz, Stella Seitz, Patrick Simon, Ludovic van Waerbeke and Anja von der Linden. This work was supported by the German Ministry for Science and Education (BMBF) through the DLR under the project 50 OR 0106, by the German Ministry for Science and Education (BMBF) through DESY under the project 05AE2PDA/8, and by the Deutsche Forschungsgemeinschaft under the project SCHN 342/3–1.

References

Abazajian, K., Adelman-McCarthy, J.K., Agüeros, M.A. et al. 2004, AJ 128, 502

Abazajian, K.N. & Dodelson, S. 2003, PRL 91, 041301

Abell, G.O. 1958, ApJS 3, 211

Abell, G.O., Corwin, H.G. & Olowin, R.P. 1989, ApJS 70, 1

Allen, S.W. 1998, MNRAS 296, 392

Andersson, K.E. & Madejski, G.M. 2004, ApJ 607, 190

Arp, H. 1987, Quasars, redshifts, and controversies, Interstellar Media, Berkeley

Athreya, R.M., Mellier, Y., van Waerbeke, L., Pelló, R., Fort, B. & Dantel-Fort, M. 2002, A&A 384, 743

Bacon, D.J., Massey, R.J., Refregier, A.R. & Ellis, R.S. 2003, MNRAS 344, 673

Bacon, D.J., Refregier, A., Clowe, D. & Ellis, R.S. 2001, MNRAS 325, 1065

Bacon, D.J., Refregier, A.R. & Ellis, R.S. 2000, MNRAS 318, 625

Bacon, D.J. & Taylor, A.N. 2003, MNRAS 344, 1307

Bacon, D.J., Taylor, A.N., Brown, M.L. et al. 2004, astro-ph/0403384

Barber, A.J. 2002, MNRAS 335, 909

Barber, A.J., Thomas, P.A., Couchman, H.M.P. & Fluke, C.J. 2000, MNRAS 319, 267

Bardeen, J. M., Bond, J. R., Kaiser, N., Szalay, A. S. 1986, ApJ 304, 15

Bar-Kana, R. 1996, ApJ 468, 17

Bartelmann, M. 1995, A&A 303, 643

Bartelmann, M., Huss, A., Colberg, J., Jenkins, A. & Pearce, F. 1998, A&A 330, 1

Bartelmann, M., King, L.J. & Schneider, P. 2001, A&A 378, 361

Bartelmann, M. & Meneghetti, M. 2004, A&A 418, 413

Bartelmann, M., Meneghetti, M., Perrotta, F., Baccigalupi, C. & Moscardini, L. 2003, A&A 409, 449

Bartelmann, M. & Narayan, R. 1995, ApJ 451, 60

Bartelmann, M., Narayan, R., Seitz, S. & Schneider, P. 1996, ApJ 464, L115

Bartelmann, M., Perotta, F. & Baccigalupi, C. 2002, A&A 396, 21

Bartelmann, M. & Schneider, P. 1993, A&A 271, 421
Bartelmann, M. & Schneider, P. 1994, A&A 284, 1
Bartelmann, M. & Schneider, P. 1999, A&A 345, 17
Bartelmann, M. & Schneider, P. 2001, Physics Reports 340, 291 (BS01)
Bartelmann, M. & Steinmetz, M. 1996, MNRAS 283, 431
Bartelmann, M., Steinmetz, M. & Weiss, A. 1995, A&A 297, 1
Bartelmann, M. & Weiss, A. 1994, A&A 287, 1
Bartsch, A., Schneider, P. & Bartelmann, M. 1997, A&A 319, 375
Beckwith, S., Somerville, R. & Stiavelli, M. 2003, STScI Newsletter 20, Issue 04, p. 1
Benabed, K. & van Waerbeke, L. 2003, PhRvD 70, 123515, also astro-ph/0306033
Benítez, N. 2000, ApJ 536, 571
Benítez, N. & Martínez-González, E. 1995, ApJ 448, L89
Benítez, N., Sanz, J.L. & Martínez-González, E. 2001, MNRAS 320, 241
Bennett, C.L., Halpern, M., Hinshaw, G. et al. 2003, ApJS 148, 1
Bergmann, A.G. & Petrosian, V. 1993, ApJ 413, 18
Bernardeau, F. 1998, A&A 338, 375
Bernardeau, F., Mellier, Y. & van Waerbeke, L. 2002, A&A 389, L28
Bernardeau, F. & Valageas, P. 2000, A&A 364, 1
Bernardeau, F., van Waerbeke, L. & Mellier, Y. 1997, A&A 322, 1
Bernardeau, F., van Waerbeke, L., & Mellier, Y. 2003, A&A 397, 405
Bernstein, G.M. & Jarvis, M. 2002, AJ 123, 583
Bertin, E. & Arnouts, S. 1996, A&AS 117, 393
Blain, A.W., Kneib, J.-P., Ivison, R.J. & Smail, I. 1999, ApJ 512, L87
Blandford, R. & Narayan, R. 1986, ApJ 310, 568
Blandford, R.D., Saust, A.B., Brainerd, T.G. & Villumsen, J.V. 1991, MNRAS 251, 600
Bolzonella, M., Miralles, J.-M. & Pelló, R. 2000, A&A 363, 476
Bonnet, H. & Mellier, Y. 1995, A&A 303, 331
Bradac, M., Lombardi, M. & Schneider, P. 2004, A&A 424, 13
Brainerd, T.G., Blandford, R.D., & Smail, I. 1996, ApJ 466, 623
Bridle, S.L., Hobson, M.P., Lasenby, A.N. & Saunders, R. 1998, MNRAS 299, 895
Broadhurst, T.J., Taylor, A.N. & Peacock, J.A. 1995, ApJ 438, 49
Brown, M.L., Taylor, A.N., Bacon, D.J. et al. 2003, MNRAS 341, 100
Carlberg, R.G., Yee, H.K.C. & Ellingson, E. 1994, ApJ 437, 63
Clowe, D., De Lucia, G. & King, L. 2004a, MNRAS 350, 1038
Clowe, D., Luppino, G.A., Kaiser, N. & Gioia, I.M. 2000, ApJ 539, 540
Clowe, D. & Schneider, P. 2001, A&A 379, 384
Clowe, D. & Schneider, P. 2002, A&A 395, 385
Clowe, D., Schneider, P., Aragon-Salamanca, A. et al. 2004b, A&A, submitted
Connolly, A.J., Csabai, I., Szalay, A.S., Koo, D.C., Kron, R.G., & Munn, J.A. 1995, AJ 110, 2655
Contaldi, C.R., Hoekstra, H. & Lewis, A. 2003, PhRvL 90, 221303

Cooray, A. & Hu, W. 2001, ApJ 548, 7
Cooray, A. & Hu, W. 2002, ApJ 574, 19
Cooray, A. & Sheth, R. 2002. Phys. Rep. 372, 1
Coté, P., McLaughlin, D.E., Cohen, J.G. & Blakeslee, J.P. 2003, ApJ 591, 850
Crittenden, R.G., Natarajan, P., Pen, U.-L. & Theuns, T. 2001, ApJ 559, 552
Crittenden, R.G., Natarajan, P., Pen, U.-L. & Theuns, T. 2002, ApJ 568, 20
Croft, R.A.C. & Metzler, C.A. 2001, ApJ 545, 561
Czoske, O., Kneib, J.-P., Soucail, G., Bridges, T.J., Mellier, Y. & Cuillandre, J.-C. 2001, A&A 372, 391
Czoske, O., Moore, B., Kneib, J.-P. Soucail, G. 2002, A&A 386, 31
Dahle, H., Pedersen, K., Lilje, P.B., Maddox, S.J. & Kaiser, N. 2003, ApJ 591, 662
Dalal, N., Holder, G. & Hennawi, J.F. 2003, ApJ 609, 50, also astro-ph/0310306
de Blok, W.J.G., McGaugh, S.S. & Rubin, V.C. 2001, AJ 122, 2396
De Filippis, E., Schindler, S. & Castillo-Morales, A. 2003, A&A 404, 63
Dekel, A. & Lahav, O. 1999, ApJ 520, 24
Dietrich, J.P., Clowe, D.I. & Soucail, G. 2002, A&A 394, 395
Dietrich, J.P., Schneider, P., Clowe, D., Romano-Diaz, E. & Kerp, J. 2004, A&A submitted, also astro-ph/0406541
Dolag, K. & Bartelmann, M. 1997, MNRAS 291, 446
Donahue, M., Gaskin, J.A., Patel, S.K., Joy, M., Clowe, D. & Hughes, J.P. 2003, ApJ 598, 190
Doré, O., Bouchet, F.R., Mellier, Y. & Teyssier, R. 2001, A&A 375, 14
Dye, S. & Taylor, A. 1998, MNRAS 300, L23
Dye, S., Taylor, A.N., Greve, T.R. et al. 2002, A&A 386, 12
Dye, S., Taylor, A.N., Thommes, E.M. et al. 2001, MNRAS 321, 685
Ebbels, T., Ellis, R., Kneib, J.-P. et al. 1998, MNRAS 295, 75
Ellis, R.S. 1997, ARA&A 35, 389
Erben, T., Miralles, J.M., Clowe, D. et al. 2003, A&A 410, 45
Erben, T., van Waerbeke, L., Mellier, Y., Schneider, P., Cuillandre, J.C., Castander, F.J. & Dantel-Fort, M. 2000, A&A 355, 23
Erben, T., van Waerbeke, L., Bertin, E., Mellier, Y. & Schneider, P. 2001, A&A 366, 717
Ettori, S. & Lombardi, M. 2003, A&A 398, L5
Fabian, A.C. 1994, ARA&A 32, 277
Fahlman, G., Kaiser, N., Squires, G. & Woods, D. 1994, ApJ 437, 56
Fischer, P. 1999, AJ 117, 2024
Fischer, P., McKay, T.A., Sheldon, E. et al. 2000, AJ 120, 1198
Fluke, C.J., Webster, R.L. & Mortlock, D.J. 2002, MNRAS 331, 180
Fort, B. & Mellier, Y. 1994, A&AR 5, 239
Fort, B., Mellier, Y. & Dantel-Fort, M. 1997, A&A 321, 353
Fort, B., Mellier, Y., Dantel-Fort, M., Bonnet, H. & Kneib, J.-P. 1996, A&A 310, 705

Fort, B., Prieur, J.L., Mathez, G., Mellier, Y. & Soucail, G. 1988, A&A 200, L17
Fruchter, A.S. & Hook, R.N. 2002, PASP 114, 144
Fugmann, W. 1990, A&A 240, 11
Gautret, L., Fort, B., & Mellier, Y. 2000, A&A 353, 10
Gavazzi, R., Mellier, Y., Fort, B., Cuillandre, J.-C. & Dantel-Fort, M. 2004, A&A 422, 407
Geiger, B. & Schneider, P. 1998, MNRAS 295, 497
Geiger, B. & Schneider, P. 1999, MNRAS 302, 118
Gerhard, O., Kronawitter, A., Saglia, R.P. & Bender, R. 2001, AJ 121, 1936
Giavalisco, M. & Mobasher, B. 2004, STScI Newletter 21, Issue 01, p. 1
Gioia, I.M., Braito, V., Branchesi, M. et al. 2004, A&A 419, 517
Gladders, M.D., Hoekstra, H., Yee, H.K.C., Hall, P.B. & Barrientos, L.F. 2003, ApJ 593, 48
Gladders, M.D. & Yee, H.K.C. 2000, AJ 120, 2148
Golse, G., Kneib, J.-P. & Soucail, G. 2002, A&A 387, 788
Gray, M.E., Ellis, R.S., Refregier, A. et al. 2000, MNRAS 318, 573
Gray, M., Ellis, R.S., Lewis, J.R., McMahon, R.G. & Firth, A.E. 2001, MNRAS 325, 111
Gray, M., Taylor, A.N., Meisenheimer, K., Dye, S., Wolf, C. & Thommes, E. 2002, ApJ 568, 141
Gunn, J.E. 1967, ApJ 150, 737
Guzik, J. & Seljak, U. 2001, MNRAS 321, 439
Hamana, T. 2001, MNRAS 326, 326
Hamana, T., Colombi, S.T., Thion, A., Devriendt, J.E.G.T., Mellier, Y. & Bernardeau, F. 2002, MNRAS 330, 365
Hamana, T. & Mellier, Y. 2001, MNRAS 327, 169
Hamana, T., Miyazaki, S., Shimasaku, K. et al. 2003, ApJ 597, 98
Hamana, T., Takada, M. & Yoshida, N. 2004, MNRAS 350, 893
Hämmerle, H., Miralles, J.-M., Schneider, P. et al. 2002, A&A 385, 743
Heavens, A. 2003, MNRAS 343, 1327
Heavens, A.F., Refregier, A. & Heymans, C.E.C. 2000, MNRAS 319, 649
Hennawi, J.F. & Spergel, D.N. 2005, ApJ 624, 59, also astro-ph/0404349
Hetterscheidt, M. 2003, Diploma Thesis, Univ. Bonn.
Heymans, C., Brown, M., Heavens, A., Meisenheimer, K., Taylor, A. & Wolf, C. 2004, MNRAS 347, 895
Heymans, C. & Heavens, A. 2003, MNRAS 339, 711
Hirata, C.M., Mandelbaum, R., Seljak, U. et al. 2004, MNRAS 353, 529, also astro-ph/0403255
Hirata, C.M. & Seljak, U. 2004, PhRvD 70, 063526, also astro-ph/0406275
Ho, S. & White, M. 2004, ApJ 607, 40
Hoekstra, H. 2001, A&A 370, 743
Hoekstra, H. 2003, MNRAS 339, 1155
Hoekstra, H. 2004, MNRAS 347, 1337

Hoekstra, H., Franx, M. & Kuijken, K. 2000, ApJ 532, 88
Hoekstra, H., Franx, M., Kuijken, K., et al. 2001, ApJ 548, L5
Hoekstra, H., Franx, M., Kuijken, K., Carlberg, R.G. & Yee, H.K.C. 2003, MNRAS 340, 609
Hoekstra, H., Franx, M., Kuijken, K., & Squires, G. 1998, ApJ 504, 636
Hoekstra, H., Franx, M., Kuijken, K., & van Dokkum, P.G. 2002d, MNRAS 333, 911
Hoekstra, H., van Waerbeke, L., Gladders, M.D., Mellier, Y. & Yee, H.K.C. 2002c, ApJ 577, 604
Hoekstra, H., Yee, H.K.C. & Gladders, M.D. 2001, ApJ 558, L11
Hoekstra, H., Yee, H.K.C. & Gladders, M.D. 2002a, ApJ 577, 595
Hoekstra, H., Yee, H.K.C. & Gladders, M.D. 2004, ApJ 606, 67
Hoekstra, H., Yee, H.K.C., Gladders, M.D., Barrientos, L.F., Hall, P.B. & Infante, L. 2002b, ApJ 572, 55
Hu, W. 1999, ApJ 522, L21
Hu, W. 2002, PhRvD 66, 083515
Hu, W. & Keeton, C.R. 2002, PhRvD 66, 063506
Hu, W. & Jain, B. 2004, PhRvD 70, 043009, also astro-ph/0312395
Hu, W. & Tegmark, M. 1999, ApJ 514, L65
Hu, W. & White, M. 2001, ApJ 554, 67
Huterer, D. 2002, PhRvD 64, 063001
Jain, B., Scranton, R. & Sheth, R.K. 2003, MNRAS 345, 62
Jain, B. & Seljak, U. 1997, ApJ 484, 560
Jain, B., Seljak, U. & White, S.D.M. 2000, ApJ 530, 547
Jain, B. & Taylor, A. 2003, PRL 91, 141302
Jain, B. & van Waerbeke, L. 2000, ApJ 530, L1
Jarvis, M., Bernstein, G.M., Fischer, P. et al. 2003, AJ 125, 1014
Jarvis, M., Bernstein, G.M. & Jain, B. 2004, MNRAS 352, 338
Jing, Y.P. 1998, ApJ 503, L9
Jing, Y.P. 2002, MNRAS 335, L89
Kaiser, N. 1984, ApJ 284, L9
Kaiser, N. 1992, ApJ 388, 272
Kaiser, N. 1995, ApJ 439, L1
Kaiser, N. 1998, ApJ 498, 26
Kaiser, N. 2000, ApJ 537, 555
Kaiser, N. & Squires, G., 1993, ApJ 404, 441
Kaiser, N., Squires, G. & Broadhurst, T. 1995, ApJ 449, 460
Kaiser, N., Squires, G., Fahlman, G. & Woods, D. 1994, in *Clusters of galaxies, proc. XIVth Moriond astrophysics meeting, Méribel, France, 1994*, p. 269
Kaiser, N., Wilson, G. & Luppino, G. 2000, astro-ph/0003338
Kaiser, N., Wilson, G., Luppino, G. et al. 1998, astro-ph/9809268
Kilbinger, M. & Schneider, P. 2004, A&A 413, 465
King, L.J., Clowe, D.I., Lidman, C., Schneider, P., Erben, T., Kneib, J.-P. & Meylan, G. 2002a, A&A 385, L5
King, L.J., Clowe, D.I. & Schneider, P. 2002b, A&A 383, 118

King, L. & Schneider, P. 2001, A&A 369, 1
King, L. & Schneider, P. 2002, A&A 396, 411
King, L. & Schneider, P. 2003, A&A 398, 23
King, L., Schneider, P. & Springel, V. 2001, A&A 378, 748
Kleinheinrich, M. 2003, Ph.D. Thesis, University of Bonn
Kneib, J.-P., Ellis, R.S., Santos, M.R. & Richard, J. 2004, ApJ 607, 697
Kneib, J.-P., Ellis, R.S., Smail, I., Couch, W.J. & Sharples, R.M. 1996, ApJ 471, 643
Kneib, J.-P., Hudelot, P., Ellis, R.S. et al. 2003, ApJ 598, 804
Kochanek, C.S. 1990, MNRAS 247, 135
Kruse, G. & Schneider, P. 1999, MNRAS 302, 821
Kruse, G. & Schneider, P. 2000, MNRAS 318, 321
Kuijken, K. 1999, A&A 352, 355
Limber, D.N. 1953, ApJ 117, 134
Link, R. & Pierce, M.J. 1998, ApJ 502, 63
Lombardi, M. & Bertin, G. 1998, A&A 335, 1
Lombardi, M. & Bertin, G. 1999, A&A 342, 337
Lombardi, M. & Schneider, P. 2001, A&A 373, 359
Lombardi, M. & Schneider, P. 2002, A&A 392, 1153
Luppino, G.A., Gioia, I.M., Hammer, F., Le Fèvre, O. & Annis, J.A. 1999, A&AS 136, 117
Luppino, G.A. & Kaiser, N. 1997, ApJ 475, 20
Lynds, R. & Petrosian, V. 1989, ApJ 336, 1
Mackey, J., White, M. & Kamionkowski, M. 2002, MNRAS 332, 788
Maoli, R., van Waerbeke, L., Mellier, Y. et al. 2001, A&A 368, 766
Markevitch, M., Gonzalez, A.H., Clowe, D. et al. 2004, ApJ 606, 819
Marshall, P.J., Hobson, M.P., Gull, S.F. & Bridle, S.L. 2002, MNRAS 335, 1037
Marshall, P.J., Hobson, M.P. & Slosar, A. 2003, MNRAS 346, 489
McKay, T.A., Sheldon, E.S., Racusin, J. et al. 2001, astro-ph/0108013
Mellier, Y. 1999, ARA&A 37, 127
Mellier, Y., Fort, B., Soucail, G., Mathez, G. & Cailloux, M. 1991, ApJ 380, 334
Ménard, B. & Bartelmann, M. 2002, A&A 386, 784
Ménard, B., Bartelmann, M. & Mellier, Y. 2003, A&A 409, 411
Ménard, B. & Péroux, C. 2003, A&A 410, 33
Meneghetti, M., Bartelmann, M., Dolag, K. et al. 2005, NewAR 49, 111, also astro-ph/0405070
Meneghetti, M., Yoshida, N., Bartelmann, M. et al. 2001, MNRAS 325, 435
Metzler, C.A., White, M. & Loken, C. 2001, ApJ 547, 560
Miralda-Escudé, J. 1991, ApJ 380, 1
Miralles, J.-M., Erben, T., Hämmerle, H. et al. 2002, A&A 388, 68
Miralles, J.-M., Erben, T., Hämmerle, H. et al. 2005, A&A 432, 797, also astro-ph/0310716
Miyazaki, S., Hamana, T., Shimasaku, K. et al. 2002, ApJ 580, L97

Mould, J., Blandford, R., Villumsen, J., Brainerd, T., Smail, I., Small, T., & Kells, W. 1994, MNRAS 271, 31
Mulchaey, J.S., Dressler, A. & Oemler, A. 2004, eds, Clusters of Galaxies: Probes of Cosmological Structure and Galaxy Evolution, Carnegie Observatories Astrophysics Series, Vol. 3., (Cambridge University Press: Cambridge)
Munchi, D. & Jain, B. 2001, MNRAS 322, 107
Munshi, D., Valageas, P. & Barber, A.J. 2004, MNRAS 350, 77
Munshi, D. & Wang, Y. 2003, ApJ 583, 566
Narayan, R. & Nityananda, R. 1986, ARA&A 24, 127
Natarajan, P. & Kneib, J.-P. 1997, MNRAS 287, 833
Natarajan, P., Kneib J.-P., Smail, I. & Ellis, R.S. 1998, Ap 499, 600
Natarajan, P. & Refregier, A. 2000, ApJ 538, L113
Norberg, P., Baugh, C.M., Hawkins, E. et al. 2001, MNRAS 328, 64
Norman, D.J. & Impey, C.D. 2001, AJ 121, 2392
Oguri, M., Lee, J. & Suto, Y. 2003, ApJ 599, 7
Ota, N., Pointecouteau, E., Hattori, M. & Mitsuda, K. 2004, ApJ 601, 120
Peacock, J.A. & Dodds, S.J. 1996, MNRAS 280, L19
Pelló, R., Kneib, J.P.; Le Borgne, J.F. et al. 1999, A&A 346, 359
Pelló, R., Schaerer, D., Richard, J., Le Borgne, J.-F. & Kneib, J.-P. 2004, A&A 416, L35
Pen, U.-L., Lu, T., van Waerbeke, L. & Mellier, Y. 2003a, MNRAS 346, 994
Pen, U.-L., Zhang, T., van Waerbeke, L., Mellier, Y., Zhang, P. & Dubinski, J. 2003b, ApJ 592, 664
Pierpaoli, E., Scott, D. & White, M. 2001, MNRAS 325, 77
Premadi, P., Martel, H., Matzner, R. & Futamase, T. 2001, ApJS 135, 7
Press, W.H., Flannery, B.P., Teukolsky, S.A. & Vetterling, W.T. 1986, Numerical Recipes, (Cambridge: University Press)
Press, W.H., Teukolsky, S.A., Vetterling, W.T., Flannery, B.P. 1992, Numerical Recipes, (Cambridge: University Press)
Reblinsky, K. 2000, A&A 364, 377
Reblinsky, K. & Bartelmann, M. 1999, A&A 345, 1
Reblinsky, K., Kruse, G., Jain, B. & Schneider, P. 1999, A&A 351, 815
Refregier, A. 2003a, ARA&A 41, 645
Refregier, A. 2003b, MNRAS 338, 35
Refregier, A. & Bacon, D. 2003, MNRAS 338, 48
Refregier, A. Rhodes, J. & Groth, E.J. 2002, ApJ 572, L131
Refsdal, S. 1970, ApJ 159, 357
Rhodes, J., Refregier, A., Collins, N.R., Gardner, J.P., Groth, E.J. & Hill, R.S. 2004, ApJ 605, 29
Richards, E.A., Kellermann, K.I., Fomalont, E.B., Windhorst, R.A. & Partridge, R.B. 1998, AJ 116, 1039
Rix, H.-W., Barden, M., Beckwith, S.V.W. et al. 2004, ApJS 152, 163
Rögnvaldsson, Ö.E., Greve, T.R., Hjorth, J. et al. 2001, MNRAS 322, 131
Rodrigues-Williams, L.L. & Hogan, C.J. 1994, AJ 107, 451

Sand, D.J., Treu, T. & Ellis, R.S. 2002, ApJ 574, L129
Sand, D.J., Treu, T., Smith, G.P. & Ellis, R.S. 2004, ApJ 604, 88
Sarazin, C.L. 1986, RvMP 58, 1
Schirmer, M. 2004, PhD Dissertation, Univ. Bonn
Schneider, P. 1995, A&A 302, 639
Schneider, P. 1996, MNRAS 283, 83
Schneider, P. 1997, MNRAS 292, 673
Schneider, P. 1998, ApJ 498, 43
Schneider, P. 2003, A&A 408, 829
Schneider, P. & Bartelmann, M. 1997, MNRAS 286, 696
Schneider, P., Ehlers, J. & Falco, E.E. 1992, Gravitational Lenses (New York: Springer) (SEF)
Schneider, P., Kilbinger, M. & Lombardi, M. 2005, A&A 431, 9, also astro-ph/0308328
Schneider, P., King, L. & Erben, T. 2000, A&A 353, 41
Schneider, P. & Kneib, J.-P. 1998, in The Next Generation Space Telescope, Proceedings of the 34th Liège International Astrophysics Colloquium, ESA Sp-429, p. 89
Schneider, P. & Lombardi, M. 2003, A&A 397, 809
Schneider, P. & Rix, H.-W. 1997, ApJ 474, 25
Schneider, P. & Seitz, C. 1995, A&A 294, 411
Schneider, P., van Waerbeke L., Jain B., Kruse G., 1998a, MNRAS, 296, 873
Schneider, P., van Waerbeke, L. & Mellier, Y. 2002a, A&A 389, 729
Schneider, P., van Waerbeke, L., Mellier, Y., Jain, B., Seitz, S. & Fort, B. 1998b, A&A 333, 767
Schneider, P., van Waerbeke, L., Kilbinger, M. & Mellier, Y. 2002b, A&A 396, 1
Schneider, P. & Watts, P. 2005, A&A 432, 783
Schneider, P. & Weiss, A. 1988a, ApJ 327, 526
Schneider, P. & Weiss, A. 1988b, ApJ 330, 1
Schrabback, T. 2004, Diploma Thesis, Univ. Bonn
Schramm, T. & Kayser, R. 1995, A&A 299, 1
Schuecker, P., Böhringer, H., Collins, C.A. & Guzzo, L. 2003, A&A 398, 867
Scoccimarro, R. & Couchman, H. 2001, MNRAS 325, 1312
Seitz, C., Kneib, J.-P., Schneider, P. & Seitz, S. 1996, A&A 314, 707
Seitz, C. & Schneider, P. 1995a, A&A 297, 287
Seitz, C. & Schneider, P. 1997, A&A 318, 687
Seitz, S., Saglia, R., Bender, R., Hopp, U., Belloni, P. & Ziegler, B. 1998, MNRAS 298, 325
Seitz, S. & Schneider, P. 1995b, A&A 302, 9
Seitz, S. & Schneider, P. 1996, A&A 305, 383
Seitz, S. & Schneider, P. 2001, A&A 374, 740
Seitz, S., Schneider, P. & Bartelmann, M. 1998, A&A 337, 325
Seitz, S., Schneider, P. & Ehlers, J. 1994, Class. Quantum Grav. 11, 2345
Seljak, U. 1998, ApJ 506, 64

Seljak, U. 2002, MNRAS 337, 769
Seljak, U., Makarov, A., Mandelbaum, R. et al. 2004, PhRvD 71, 043511, also astro-ph/0406594
Sheldon, E.S., Johnston, D.E., Frioman, J.A. et al. 2004, AJ 127, 2544
Sheth, R.K., Mo, H.J. & Tormen, G. 2001, MNRAS 323, 1
Simon, P., King, L.J. & Schneider, P. 2004, A&A 417, 873
Smith, R.E., Peacock, J.A., Jenkins, A. et al. 2003, MNRAS 341, 1311
Sofue, Y. & Rubin, V. 2001, ARA&A 39, 137
Song, Y.-S. & Knox, L. 2004, PhRvD 70, 063510, also astro-ph/0312175
Soucail, G., Fort, B., Mellier, Y. & Picat, J.P. 1987, A&A 172, L14
Soucail, G., Kneib, J.-P. & Golse, G. 2004, A&A 417, L33
Soucail, G., Ota, N., Böhringer, H., Czoske, O., Hattori, M. & Mellier, Y. 2000, A&A 355, 433
Spergel, D.N. & Steinhardt, P.J. 2000, PRL 84, 3760
Spergel, D.N., Verde, L., Peiris, H.V. et al. 2003, ApJS 148, 175
Squires, G. & Kaiser, N. 1996, ApJ 473, 65
Takada, M. & Hamana, T. 2003, MNRAS 346, 949
Takada, M. & Jain, B. 2002, MNRAS 337, 875
Takada, M. & Jain, B. 2003a, MNRAS 340, 580
Takada, M. & Jain, B. 2003b, MNRAS 344, 857
Takada, M. & Jain, B. 2004, MNRAS 348, 897
Takada, M. & White, M. 2004, ApJ 601, L1
Taruya, A., Takada, M., Hamana, T., Kayo, I. & Futamase, T. 2002, ApJ 571, 638
Taylor, A.N. 2001, astro-ph/0111605
Taylor, A.N., Bacon, D.J., Gray, M.E. et al. 2004, MNRAS 353, 1176, also astro-ph/0402095
Taylor, A.N., Dye, S., Broadhurst, T.J., Benítez, N. & van Kampen, E. 1998, ApJ 501, 539
Tegmark, M. & Peebles, P.J.E. 1998, ApJ 500, L79
Tereno, I., Doré, O., van Waerbeke, L. & Mellier, Y. 2005, A&A 429, 383, also astro-ph/0404317
Torri, E., Meneghetti, M., Bartelmann, M., Moscardini, L., Rasia, E. & Tormen, G. 2004, MNRAS 349, 476
Trager, S.C., Faber, S.M., Dressler, A. & Oemler, A. 1997, ApJ 485, 92
Trentham, N. & Tully, R.B. 2002, MNRAS 335, 712
Tyson, J.A. 1988, AJ 96, 1
Tyson, J.A. & Fischer, P. 1995, ApJ 446, L55
Tyson, J.A., Valdes, F., Jarvis, J.F. & Mills Jr., A.P. 1984, ApJ 281, L59
Tyson, J.A., Valdes, F. & Wenk, R.A. 1990, ApJ 349, L1
Umetsu, K. & Futamase, T. 2000, ApJ 539, L5
Valdes, F., Jarvis, J.F. & Tyson, J.A. 1983, ApJ 271, 431
Vale, C. & White, M. 2003, ApJ 592, 699
van den Bosch, F.C., Abel, T., Croft, R.A.C., Hernquist, L. & White, S.D.M. 2002, ApJ 576, 21

van Kampen, E. 1998, MNRAS 301, 389
van Waerbeke, L. 1998, A&A 334, 1
van Waerbeke, L. 2000, MNRAS 313, 524
van Waerbeke, L., Bernardeau, F. & Mellier, Y. 1999, A&A 342, 15
van Waerbeke, L., Hamana, T., Scoccimarro, R., Colombi, S. & Bernardeau, F. 2001, MNRAS 322, 918
van Waerbeke, L. & Mellier, Y. 2003, astro-ph/0305089
van Waerbeke, L., Mellier, Y., Erben, T. et al. 2000, A&A 358, 30
van Waerbeke, L., Mellier, Y. & Hoekstra, H. 2005, A&A 429, 75, also astro-ph/0406468
van Waerbeke, L., Mellier, Y., Pello, R., et al. 2002, A&A 393, 369
van Waerbeke, L., Mellier, Y., Radovich, M. et al. 2001, A&A 374, 757
von der Linden, A. 2004, Diploma Thesis, Univ. Bonn
Wambsganss, J., Bode, P. & Ostriker, J.P. 2004, ApJ 606, L93
Wambsganss, J., Cen, R. & Ostriker, J.P. 1998, ApJ 494, 29
Weinberg, N.N. & Kamionkowski, M. 2002, MNRAS 337, 1269
Weinberg, N.N. & Kamionkowski, M. 2003, MNRAS 341, 251
White, M. & Hu, W. 2000, ApJ 537, 1
White, M., van Waerbeke, L. & Mackey, J. 2002, ApJ 575, 640
White, S.D.M., Navarro, J.F., Evrard, A.E. & Frenk, C.S. 1993, Nat 366, 429
Williams, R.E., Blacker, B., Dickinson, M. et al. 1996, AJ 112, 1335
Wilson, G., Kaiser, N. & Luppino, G. 2001, ApJ 556, 601
Wittman, D. 2002, in Gravitational Lensing: An Astrophysical Tool, F. Courbin & D. Minniti (eds.) (Springer-Verlag: Berlin), Lecture Notes in Physics, 608, 55
Wittman, D., Margoniner, V.E., Tyson, J.A., Cohen, J.G., Becker, A.C. & Dell'Antonio, I. 2003, ApJ 597, 218
Wittman, D.M., Tyson, J.A., Kirkman, D., Dell'Antonio, I. & Bernstein, G. 2000, Nat 405, 143
Wittman, D., Tyson, J.A., Margoniner, V.E., Cohen, J.G. & Dell'Antonio, I. 2001, ApJ 557, L89
Yang, X.H., Mo, H.J., Kauffmann, G. & Chu, Y.Q. 2003, MNRAS 339, 387
York, D.G., Adelman, J., Anderson, J.E. et al. 2000, AJ 120, 1579
Zaldarriaga, M. & Scoccimarro, R. 2003, ApJ 584, 559
Zaritsky, D. & Gonzalez, A.H. 2003, ApJ 584, 691
Zaritsky, D., Smith, R., Frenk, C. & White, S.D.M. 1997, ApJ 478, 39
Zaroubi, S., Squires, G., Hoffman, Y. & Silk, J. 1998, ApJ 500, L87
Zaroubi, S., Squires, G., de Gasperis, G., Evrard, A.E., Hoffman, Y. & Silk, J. 2001, ApJ 561, 600
Zehavi, I., Blanton, M.R., Frieman, J.A. et al. 2002, ApJ 571, 172
Zhang, L.L. & Pen, U.-L. 2005, New Astronomy 10, 569, also astro-ph/0305447
Zwicky, F. 1933, Helv. Phys. Acta 6, 110

Part 4: Gravitational Microlensing

J. Wambsganss

Gravitational Microlensing can be thought of as a version of strong gravitational lensing in which the image separation is too small to be resolved. Multiple images are formed, but their typical separation – $\Delta\theta \approx 2\,\theta_{\rm E}$ – is far below the limiting resolution determined by observational constraints. Given the dependence of the Einstein radius on lens mass and geometry, it is clear that microlensing will occur for sufficiently small masses and sufficiently distant lenses and sources. In very general terms, microlensing deals with the lensing effects of compact objects in the mass range $10^{-6} \leq m/M_\odot \leq 10^6$. This translates into Einstein radii/angular separations of a milli-arcsecond or smaller for the two main distance regimes: "galactic" – lens/source distances of order 10 kpc, and "extragalactic/cosmological" – lens/source distances of order Gpc. Both regimes will be discussed at length in the subsequent sections.

The mathematical possibility of microlensing – lensing effects of stellar mass objects on background stars – was already discussed many decades ago (Chwolson 1924; Einstein 1936). Right after the discovery of the first gravitational lens system in the late 1970s (Walsh et al. 1979), interest in lensing by stellar mass objects was revived: Chang and Refsdal (1979), Gott (1981), Paczyński (1986a,b), Kayser et al. (1986). Bohdan Paczynśki was the first to use the term "microlensing" for light deflection by stellar masses. The first observational detection of the microlensing effect came in 1989 (Vanderriest et al. 1989; Irwin et al. 1989) when individual stars in the lensing galaxies of gravitationally lensed quasars QSO 0957+561 and QSO 2237+0305 altered the magnification of one of the quasar images relative to other(s). The first Galactic microlensing events were reported in 1993: Alcock et al. (1993), Aubourg et al. (1993), and Udalski et al. (1993).

Most of the sections in this part of the proceedings deal with stellar or Galactic microlensing, the last one treats quasar or cosmological microlensing. There exist a number of good and more detailed reviews on microlensing, e.g. Paczyński (1996), Mao (2001), Mollerach and Roulet (2002), Courbin, Saha and Schechter (2002).

Wambsganss J (2006), Gravitational microlensing. In: Meylan G, Jetzer Ph and North P (eds) Gravitational lensing: Strong, weak, and micro. Saas-Fee Adv Courses vol 33, pp 453–540
DOI 10.1007/3-540-30309-X_4

1 Lensing of Single Stars by Single Stars

1.1 Brief History

Commonly it is assumed that light deflection is a modern phenomenon. However, more than 200 years ago scientists started to think about it. In the beginning of the 19th century, Johann Georg Soldner wrote an article entitled "Über den Einfluß der Schwerkraft auf die Ausbreitung des Lichtes"[1], in which he investigated the possibility that a light ray be attracted by the gravitation of a "heavenly body" (Soldner 1801). He even derived the deflection angle for a light ray passing close to the solar limb, arriving at half the correct value. In 1911, Einstein had thought about light deflection as well and published the same value (Einstein 1911). Only with the completion of the General Theory of Relativity, Einstein found the value that was later confirmed by the famous solar eclipse expedition. Chwolson (1924) mentioned a "fictitious double" star, an apparent illusion due to the light deflection of a foreground star by a background star, even considering a ring-like image for perfect alignment between lensing and lensed star. He was uncertain whether this might ever be observable. Years later, Einstein published again a letter on star–star lensing, initiated by a visit of the Czech engineer Mandl (Einstein 1936). He mentions the appearance of a luminous circle for perfect alignment between source and lens and derives the magnification properties. But he was skeptical regarding the observability: "of course, there is not much hope of observing this phenomenon directly". Renn et al. (1997) report that Einstein had dealt with the same question already as early as 1911/1912: in his notebooks he had derived the relations regarding double images, magnification, separation of images etc., but apparently had never bothered to publish it. Link (1937, 1967) had treated lensing of stars by stars as well, and produced tables for the magnification of finite sized background stars. With the seminal papers in the 1960s (Klimov 1963; Liebes 1964; Refsdal 1964a,b) lensing was put on firm theoretical footing and applicable to interesting astrophysical goals. Chang and Refsdal (1979, 1984) suggested that the lensing action of individual stars affect the apparent brightness of multiply imaged quasars. Gott (1981) suggested that lensing by stellar mass objects in halos of lensing galaxies can be used to detect compact dark matter, and finally Paczyński (1986a,b) introduced the term "microlensing", both for the action of stars in distant lensing galaxies on quasar images and for stars or dark matter objects in the Milky Way acting on background stars in the bulge or in the Large/Small Magellanic Clouds.

1.2 Theoretical Background

This section considers the mathematically simplest case: the lensing effect of a single foreground star on a single background star in the Milky Way or Local Group. It is typical and representative for most Galactic lensing systems. In

[1] "On the influence of gravity on the propagation of light"

this regime, lens and source distances are of the order kpc, the lenses have roughly stellar masses. This results in angular Einstein ring radii which are on the order of 10^{-3} arcsec, and D_s is small enough that we may reasonably set $D_{ds} = D_s - D_d$.

Point Source – Point Lens

Over a large range of current astrophysical interest, the stellar source subtends a considerably smaller angle than θ_E, so that it may be approximated as a point. Symmetry allows us the freedom to choose the origin as the position of the lens and a position along the positive θ_1 axis for the point source. The characteristic length scale is θ_E, by which we scale the one-dimensional lens equation

$$y = x - \frac{1}{x} \quad , \tag{1}$$

where $x \equiv \theta/\theta_E$ is a (normalized) image position corresponding to the (normalized) source position $y \equiv \beta/\theta_E$. The two solutions

$$x_\pm = \frac{1}{2}\left(y \pm \sqrt{y^2+4}\right) \tag{2}$$

to the quadratic lens equation correspond to positions which straddle the lens on the sky, with the positive parity image (+) on the source side of the lens, always magnified and further away from the lens than the negative parity image (−), which is the less magnified of the two.

Indeed, we can formulate the magnification of the two images as (cf. Introduction):

$$\mu_\pm = \frac{1}{\det \mathcal{A}_\pm} = \left(1 - \frac{1}{x_\pm^4}\right)^{-1} = \pm\frac{1}{4}\left[\frac{y}{\sqrt{y^2+4}} + \frac{\sqrt{y^2+4}}{y} \pm 2\right] . \tag{3}$$

Note that the image separation is $\Delta x \equiv |x_+ - x_-| = \sqrt{y^2+4}$. Relations for the total magnification μ, and the sum and ratio of the individual image magnifications can then be derived (however, only the total magnification is observable in (photometric) microlensing):

$$\mu \equiv \mu_+ + |\mu_-| = \mu_+ - \mu_- = \frac{1}{2}\left[\frac{y}{\sqrt{y^2+4}} + \frac{\sqrt{y^2+4}}{y}\right] = \frac{y^2+2}{y\sqrt{y^2+4}}, \tag{4}$$

$$\mu_+ + \mu_- = 1, \tag{5}$$

$$\left|\frac{\mu_-}{\mu_+}\right| = \left(\frac{y-\sqrt{y^2+4}}{y+\sqrt{y^2+4}}\right)^2 = \left(\frac{x_-}{x_+}\right)^2 . \tag{6}$$

For a point lens, the two-dimensional magnification distribution in the source plane – the magnification pattern – consists of circular contours of

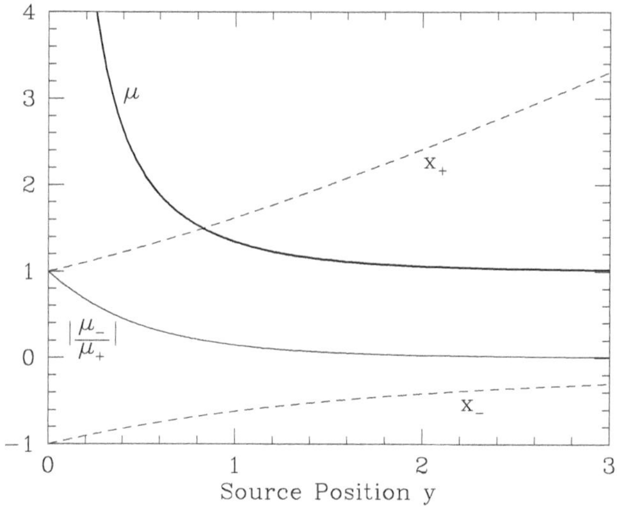

Fig. 1. Total magnification μ, normalized image positions $x_{\pm}$, and ratio of image magnifications for a point lens as a function of the normalized source position y (Figure courtesy Penny Sackett)

constant total magnification: the circles are centered on the lens position with a magnification value $\mu \rightarrow 1/y$ for small impact parameters y, and $\mu \rightarrow 1$ for large y. For source positions $y > 0$, the x_+ image is responsible for the lion's share of the total magnification μ for all source positions y (Fig. 1). At the fiducial source position $y = 1$ where $\mu = 1.34$, e.g., the positive parity image contributes 87% of the total magnification.

Microlensing "Events"

If the instantaneous magnification of a microlensing event were the only measurable quantity in a static stellar microlensing scenario, then the science of microlensing would be much less rich and its literature much less voluminous[2]. However, stars move around the Galactic center (and have an additional random velocity component with respect to one another). The relative velocities are such, that the time scale of the relative change of lens and source positions is of order of weeks or shorter. Hence this motion introduces a temporal component into the lensing geometry, causing the impact parameter and thus the normalized bending angle and the magnification to vary measurably as a function of time.

In general, the observer, the lens, and the source are all in motion, all with a certain three-dimensional velocity vector. The temporal behavior of

[2] In fact, without independent knowledge of the unlensed flux emanating from a source, it would not be possible at all to ascertain whether, or by how much, a background star is lensed.

the magnification – the "light curve" of the source – is dictated by the *relative* motion of the lens across the observer–source line-of-sight. The distances D_{d}, D_{s}, and D_{ds}, and thus the relative scalings of β, θ, and α, may also change with time, but in astrophysical situations these changes have a negligible effect on the lensing equation compared to changes in the impact parameter due to the projected relative motion.

The characteristic time scale for these changes is given by the Einstein time

$$t_{\mathrm{E}} \equiv D_{\mathrm{d}}\,\theta_{\mathrm{E}}/v_{\perp}, \tag{7}$$

where $v_{\perp}$ is the transverse speed of the lens relative to the source–observer line-of-sight. For $y < 1$, the total magnification of a point source can be expected to change appreciably over a time t_{E}. Stellar lenses in our own Galaxy are associated with typical t_{E} on the order of a month, and thus the change in the observed brightness of the source they induce are referred to as microlensing "events".

Since the magnification depends only on the source position in units of θ_{E} (cf. (4)), we need only describe how the source moves relative to the lens as a function of time to obtain a description of the light curve of the microlensed source. Assuming that for the duration of the observable event (say, several t_{E}) this motion is rectilinear, the relative motion of the source on the sky differs from that of the lens only by the sign. Taking the time t_0 to be that at which the source-lens separation y takes on its smallest value y_0, the trajectory of the source can be represented by

$$y(t) = \sqrt{y_0^2 + \left(\frac{t-t_0}{t_{\mathrm{E}}}\right)^2}\ , \tag{8}$$

since the line joining the lens to the source at time t_0 is perpendicular to the lens-source relative motion[3]. The corresponding light curve $F(t) = \mu(t)F_s$ of the source, examples of which are shown in Fig. 2, is then obtained by substituting $y(t)$ into (4) and multiplying by the unlensed source flux F_{s}.

Observables

A simple, point-source, point-lens microlensing light curve is thus described by four parameters: unlensed flux F_0, t_0, y_0 and t_{E}. Of these, F_0 can be measured in the absence of microlensing, t_0 sets an arbitrary time scale, and y_0 depends on the random placement of lens and source on the sky. Only $t_{\mathrm{E}} = D_{\mathrm{d}}\theta_{\mathrm{E}}/v_{\perp}$ contains physical information about the lensing system, albeit in a degenerate combination. Assuming that the source distance D_{s} can be determined from its properties (membership in a stellar system, spectral type and

[3] It may be worth noting that in the microlensing literature, the normalized source-lens separation on the sky, **y**, is often alternatively denoted as **u**. This convention is used below as well.

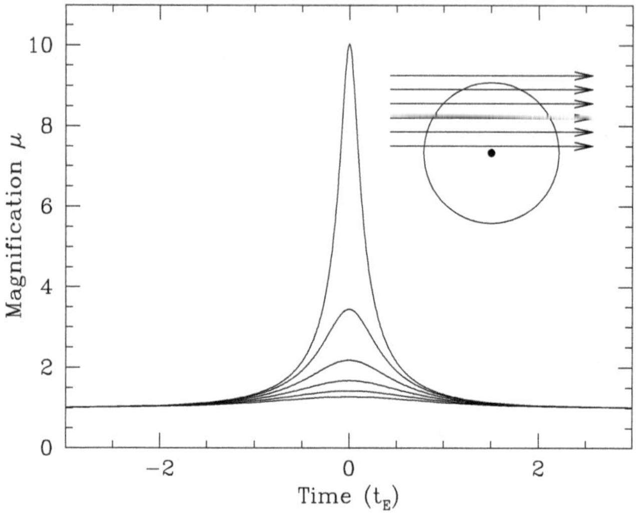

Fig. 2. Point-lens, point-source light curves for minimum impact parameters $y_0 = 0.1$ (*top*), 0.3,...,1.1 (*bottom*) and the corresponding trajectories across the Einstein ring (Figure courtesy Penny Sackett)

apparent magnitude), we are left with three physical parameters, lens mass M, lens distance D_d, and lens relative transverse velocity $v_\perp$, to determine from one observable. Other additional information is needed, if we want to learn more about the lens system. Unlike other forms of gravitational lensing, a microlens is not observed directly in general (however, cf. Subsection 'Direct Lens Detection' in Sect. 5).

As Fig. 1 illustrates, significant magnification occurs when the source lies within one angular Einstein radius of the lens. The microlensing event itself gives us very little possibility to measure or estimate the lens mass M, the lens distance D_d or the transverse velocity $v_\perp$ independently. To understand the severity of this degeneracy, consider a Galactic microlensing system in which the source is known to lie in the Galactic Bulge at, say, precisely 8 kpc and the Einstein time t_E has been precisely determined to be (a rather typical) 40 days. Assuming the lens to be bound to the Galaxy, it is likely that $0 < v_\perp < 600\ \mathrm{km}^{-1}$ with values nearer to the middle of the range statistically favored. Figure 3 shows the resulting degeneracy in the mass and distance of the lens. The distribution of lens masses ranges from those massive brown dwarfs ($\lesssim 0.1\mathrm{M}_\odot$) to that of a heavy stellar black hole ($\sim 10\mathrm{M}_\odot$).

1.3 How Good is the Point Lens – Point Source Approximation?

Information about individual image positions, magnifications and shape is entirely lost in a standard microlensing situation. This results in degeneracies in the lens–source combinations that can lead to the same observables. In order

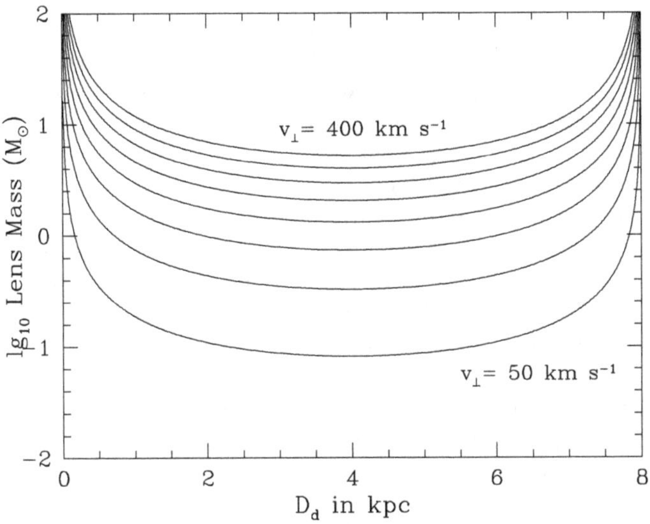

Fig. 3. Degeneracy between mass M and distance D_d for a Galactic microlens for different assumptions for its transverse speed ($50 \leq v_\perp \leq 400\ \mathrm{km\,s^{-1}}$, in steps of $50\ \mathrm{km\,s^{-1}}$) across the line-of-sight to the source (Figure courtesy Penny Sackett)

to understand what information remains, and how it is related to the physical parameters of the system, we begin with a simple model and gradually add complexity.

To a very good approximation, we can assume that the mass distribution of a star is spherically symmetric, hence the projected mass distribution is axisymmetric, independent of the direction. Since the mean free path of photons is quite short in all but the extreme outer layers of stellar atmospheres, we also assume that stellar lenses are not transparent. Taken together, this implies that we can model a single stellar lens as a point mass M as long as we consider only impact parameters larger than the stellar radius of the lens. The ratio of the angular size of the source compared to

$$\theta_\mathrm{E} = \sqrt{(4\,G\,M D_\mathrm{ds})/(c^2\,D_\mathrm{d}\,D_\mathrm{s})} \tag{9}$$

will determine whether it is appropriate to use the point source approximation.

Although a normal star acting as a lens emits light that could yield information about lens properties, this light is mixed with that of the source and any other luminous object lying in the same resolution element. However, most sources are bright giants, and most lenses are presumably faint (M-)dwarfs: Since bright stars are easier to detect at great distances, giants are much more likely sources. And since low mass stars are much more abundant than massive ones in the Milky Way disk (consequence of the the Initial Mass Function of stars), they are much more likely lenses. Combined with the luminosity-mass relation this means: in a microlensing event, the source light dominates lens light dramatically. Even if the lens contributes substantially to the total flux received during a microlensing event, the lens mass-distance degeneracy may not be reduced. The microlensing light curve is now a function of five parameters. F_s (actual source flux at baseline), t_0, y_0, t_E, and the blending flux F_b (constant flux contribution of some unrelated star) with the result that t_E is more difficult to determine.

The symmetry of the point-source, point-lens light curve aids in determining t_0 and the total baseline flux $F_0 = F_s + F_b$ can be measured well after the lensing event is over. The fraction of F_0 that is contributed by the source, $f_s \equiv F_s/F_0$, must be determined from subtleties in the light curve itself (e.g., color changes, or astrometric information), and thus can be strongly degenerate with y_0 and t_E.

Most of the measured microlensing events are indeed well fitted and described by the simple point lens–point source approximation and linear relative motion. Occasionally, some events were very well covered with hundreds of data points with small error bars. For some of them, the point lens–point source approximation did not produce satisfactory fits. In binary microlensing events with caustic crossings, the point source approximation breaks down. This can be used to determine source size and even source profile information. These cases and effects will be discussed in Sect. 5.

1.4 Statistical Ensembles

Gravitational microlensing offers the opportunity to measure the density and total mass of a population of objects – bright or dark – between a background population of sources and the observer on Earth. Paczyński (1986b) worked out this idea quantitatively, and applied it to objects potentially making up the dark matter halo of the Milky Way. If such objects had masses in the stellar range (very roughly from $10^{-6} \leq M/M_\odot \leq 10^2$), they would produce time-variable magnification of background stars in the Large or Small Magellanic Clouds. Quantitatively important in such a situation are: probability and duration of such events.

The optical depth to gravitational microlensing is equal to the ratio of surface mass density of microlensing objects to the critical mass density (cf. Introduction). For a variable mass density, the optical depth is an integral expression along the line-of-sight (Paczyński 1986b):

$$\tau = \int_0^{D_s} \frac{4\pi G D}{c^2} \rho(D_d) dD_d, \tag{10}$$

where $\rho(D_d)$ is the average microlensing matter density at distance D_d from the observer, and $D = (D_d D_{ds}/D_S)$. The resulting optical depth depends a bit on the exact direction and parameters of the isothermal halo. Assuming a simple isothermal sphere model ($M(R) = V_{\rm rot}^2 R/G$, $\rho(R) = V_{\rm rot}^2/4\pi G R_{\rm GC}^2$) for the dark halo of the Milky Way ($R_{\rm GC}$ is the distance to the Galactic Center),

Paczyński derived the numerical value of the optical depth turns out to be of order

$$\tau_0 = 5 \times 10^{-7}. \tag{11}$$

This means that roughly one out of a million stars in the nearby galaxies will be strongly lensed, i.e. the source is located within the Einstein radius of the lens and hence magnified by at least $\mu \geq 3/5^{0.5} \approx 1.34$. The concept of optical depth can easily be visualized in the following way: if all the lenses would be represented as dark disks with their respective Einstein radii, then the sum of the areas of all these disks would cover exactly the fraction τ_0 of the sky.

The event duration (defined as the time it takes to cross the Einstein radius) depends on the transverse velocity of the object and its mass (cf. (7)). A typical value for an object at a distance of $D_d = 10$ kpc and a tangential velocity of $v_\perp = 200\,\text{km/sec}$ is

$$t_0 \approx 6 \times 10^6 \text{ sec} \left(\frac{M}{M_\odot}\right)^{0.5} \approx 0.2 \text{ yr} \left(\frac{M}{M_\odot}\right)^{0.5}. \tag{12}$$

However, even assuming that all the lenses had the same mass and the same (randomly oriented) three-dimensional velocity, the event durations would cover a wide range. In particular, there should be a tail of relatively long events, because a fraction of lenses my have a three-dimensional velocity vector close to radial. Also, since the actual lens population most likely consists of a range of masses and velocities and distances, this would broaden the duration distribution even more.

If all events had the same time scale t_0, then the event rate N would be given as (cf. Paczyński 1996):

$$N = \frac{2}{\pi}\, n\, \tau\, \frac{\Delta t}{t_0}, \tag{13}$$

where n is the total number of sources monitored, τ is the optical depth, and Δt is the time interval of the monitoring campaign. In his review article, Paczyński (1996) derives the probability distribution of event durations; a more detailed analysis can be found in Mao and Paczyński (1996).

2 Binary Lenses

After treating the case of a single lens, the logical next step is the binary lens scenario. In the lens equation, the only change is that the deflection angle now consists of the sum of two point lenses:

$$\boldsymbol{\alpha}(\boldsymbol{x}) = \frac{4G}{c^2}\left(\frac{M_A(\boldsymbol{x}-\boldsymbol{x}_A)}{(\boldsymbol{x}-\boldsymbol{x}_A)^2} + \frac{M_B(\boldsymbol{x}-\boldsymbol{x}_B)}{(\boldsymbol{x}-\boldsymbol{x}_B)^2}\right), \tag{14}$$

where M_A, M_B are the masses of the two lenses and $\boldsymbol{x}_A$, $\boldsymbol{x}_B$ are their positions. Due to the non-linearity of the lens equation, the effect of replacing a single lens by two separate lenses does not at all have the effect of a simple sum or superposition of two single lens cases: the caustics and the two-dimensional magnification distributions in the source plane look *very* different compared to the single-lens case, and so do the lightcurves.

The major new phenomenon of a binary (or any asymmetric) lens, compared to the isolated point lens is the occurence of extended *caustics* in the source plane (cf. Introduction), due to the astigmatism of the lens. Caustics separate regions of different image multiplicities: when a source crosses a caustic, a new image pair is created or destroyed. Due to the very small image separation for stellar mass lenses (compared to the resolution of the telescope), these new images cannot be observed directly. However, since these new images are very highly magnified, the combined magnification of all images (which is an observable) is dominated by these bright new images: the lightcurve of a source undergoing a caustic crossing exhibits high peaks. Formally, a point source would even be infinitely magnified. Due to the realistic finite source size, the actual magnification remains finite but can get very high (events with magnifications of more than five magnitudes have been detected).

In a binary lens scenario, the second lens introduces three new parameters:

- the mass ratio $q = m_1/m_2$,
- the binary separation d (in units of the Einstein radius for the total mass $m = m_1 + m_2$),
- the angle ϕ between the source trajectory and the line connecting the two lenses.

This allows for a *very* large variety of binary lens lightcurves: "A double lens is vastly more complicated than a single one" (Paczyński 1996).

2.1 Theory and Basics of Binary Lensing

The properties of a system consisting of two point lenses have been explored in great detail in a seminal paper by Schneider and Weiss (1986). They derive analytically the critical curves and caustics for the binary lens with equal masses: $m_1 = m_2$, i.e. $q = 1$. They found three regimes for binary lensing: when the two lenses are widely separated, they act like two single lenses which are slightly perturbed: the degenerate point caustic of an isolated lens is slightly deformed into a small asymmetric asterisk with four cusps, and the circular critical line, the Einstein ring, is slightly deformed into an oval (see top left panel of Fig. 4 for a separation of $d = 1.2$). Once the separation of the two lenses approaches one Einstein radius, the two critical lines merge, forming the "infinity" sign, and the two separate caustics merge accordingly (cf. top right panel of Fig. 4). For further decreasing separation between the binary components, there is now one closed critical line and one closed six-cusp-caustic

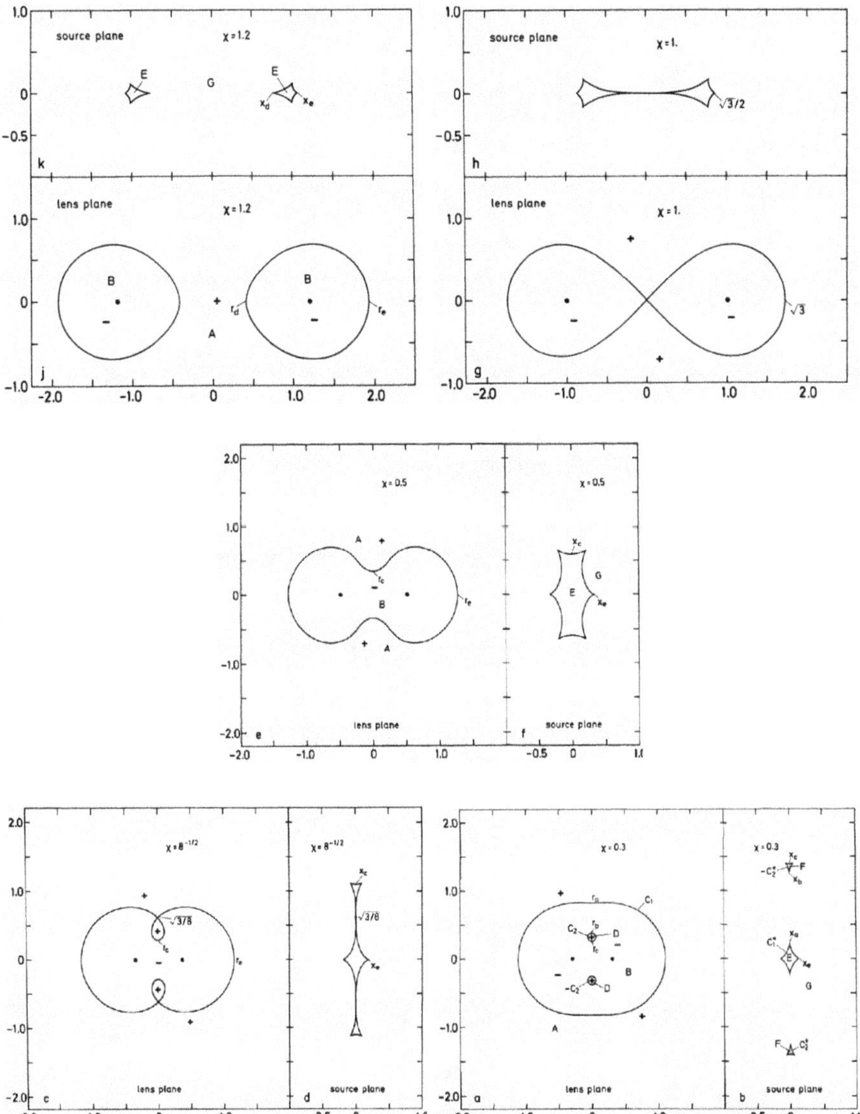

Fig. 4. The critical curves in the lens plane (*lower or left part* of each panel) and the caustics in the source plane (*upper or right part* of each panel) are depicted for a binary lens situation with equal masses ($M_A = M_B$) and decreasing separation: from 1.2 R_E (*top left*) to 0.3 R_E (*bottom right*). In particular the "transition cases" (separation 1.0 R_E and $8^{-1/2} R_E$) are of interest (from Schneider and Weiss (1986), Fig. 2)

(middle panel of Fig. 4). When the separation reaches $d = 8^{-0.5} \approx 0.35355$, another change of topology occurs: two regions inside the main critical line detach, and the caustic divides up into three parts as well: two triangular shaped caustics and one four-cusp asterisk (bottom left panels of Fig. 4). When the two lenses approach each other even further, the two triangular caustics move away from the main caustic very rapidly (bottom right panel of Fig. 4).

Figure 4 (a reproduction of Fig. 2 from Schneider and Weiss (1986)) also indicates the parity of the images in the lens plane with a plus or minus sign (the explanation of the additional labels can be found in the original paper). Accordingly, the image configuration for a binary lens can be very diverse, even for given mass ratio and separation. This was illustrated by Schneider and Weiss (1986) as well for an extended source and is reproduced here in Fig. 5 for equal masses ($q = 1$) and a separation of $d = 0.5$: A source inside the caustic (inset) has five images (top left panel in Fig. 5). When the source touches a caustic (other three panels), two or three images merge, respectively. In the panels, the size/area of an image is proportional to its magnification.

Schneider and Weiss (1986) studied also the effect of the source size on the magnification during a caustic crossing. In Fig. 6 (reproduced from their Fig. 9) the "lightcurve" of a variety of circular sources with different radii and

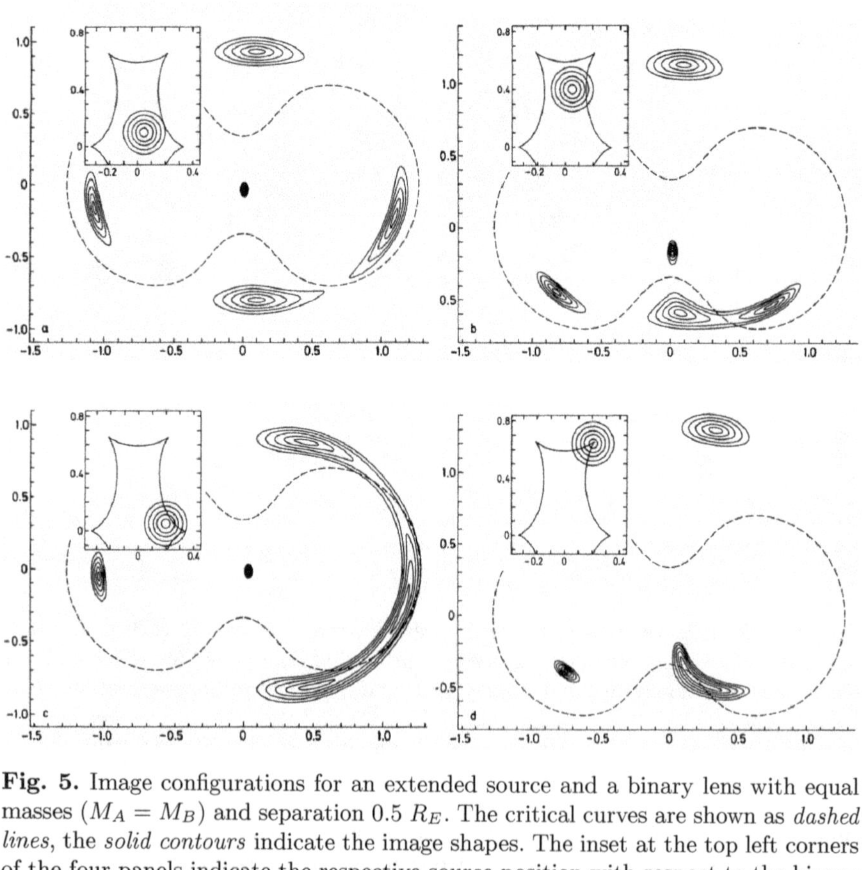

Fig. 5. Image configurations for an extended source and a binary lens with equal masses ($M_A = M_B$) and separation 0.5 R_E. The critical curves are shown as *dashed lines*, the *solid contours* indicate the image shapes. The inset at the top left corners of the four panels indicate the respective source position with respect to the binary lens caustic (from Schneider and Weiss (1986), Fig. 6)

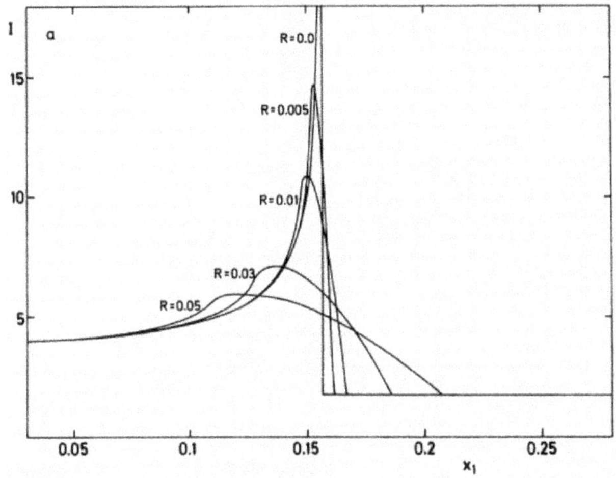

Fig. 6. Lightcurves of a caustic crossing (fold) for a point source ($R = 0.0$) and extended sources with radii varying from $R = 0.005R_E$ to $R = 0.05R_E$ (from Schneider and Weiss (1986), Fig. 9a)

constant surface brightness is shown: the maximum magnification as well as the exact shape of the lightcurve depend strongly on the source radius, here shown for a range from $\theta/\theta_E = 0.05, 0.03, 0.01, 0.005$ and for point-like source.

The study of Schneider and Weiss (1986) was generic in the sense that it was applicable to close pairs of galaxies as well as double stars. Mao and Paczyński (1991) concentrated entirely on microlensing of binary stars. Based on the observational fact that more than 50% of all stars are members of binaries or multiple star systems, they predicted that it is unavoidable that microlensing lightcurves of binary stars and star-plus-planet systems will be observed. However, due to the large range of separations, most of the physical binary stars will either act as two individual lenses (for large separation) or as a single lens (for very small separation). If the (projected) separation is of order the Einstein radius of the combined mass, it gets "interesting" for lensing (cf. also Figs. 4 and 5), i.e. deviations from single-lens lightcurves are to be expected. They concluded that about 10% of all observed stellar microlensing events should show signatures of the binarity of the lens.

Witt and Mao (1995) explored the binary lens further and found that the minimum total magnification for a source inside the caustic is three. They suggested that for an observed lightcurve in which this is apparently not the case, there are two possibilities: either there is light from another component ("blending"), which could be the lens itself or an unrelated background star, or the lens system consists of more than two stars (triple lens).

An illustration of the variety of lightcurves for a binary lens with separation $d = 0.5$ (and mass ratio $q = 1$) is shown in Fig. 7 for 5 parallel tracks and a

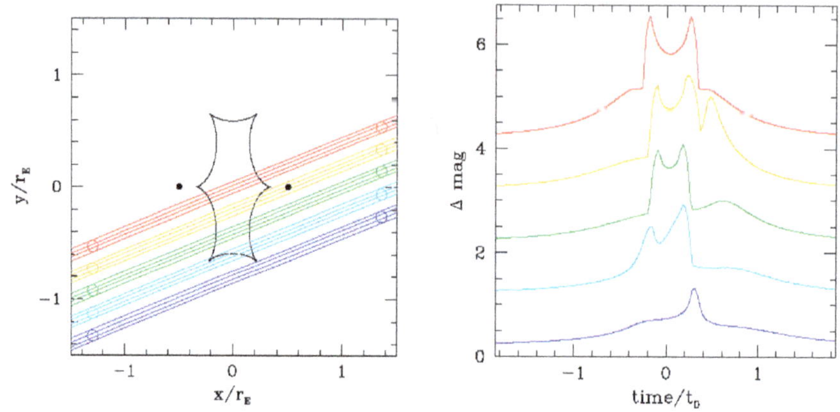

Fig. 7. The *left hand panel* indicates a binary lens caustic, the positions of the two (equal mass) lenses with separation $1.0R_E$, and the parallel tracks of an extended source (width of track corresponds to two source radii: $R = 0.05R_E$). On the *right hand panel*, the five corresponding lightcurves are displayed, offset by one magnitude for easier visibility (from Sackett [private communication], after Paczyński 1996)

finite source size. At `http://www.sns.ias.edu/~gaudi/movies.html`, Scott Gaudi's website, he provides a full suite of animations of binary microlensing scenarios with variable mass ratio and separation, indicating the critical lines, the caustics, the individual micro-images and the lightcurve for relative motion. A static example is shown in Fig. 8.

2.2 First Microlensing Lightcurve of a Binary Lens: OGLE-7

The first microlensing lightcurve of a binary lens to be detected was OGLE-7 (more on the OGLE-team in Sect. 3.3), for which two peaks were measured in the course of the 1993 season (see Fig. 9, top panel). This lightcurve was originally classified as "unusual", because the star had brightened by more than 2 magnitudes but deviated from the expected single-lens–single-source lightcurve: The flux as a function of time displayed a double-peak structure, following a completely flat and constant lightcurve at a low level in the previous season. This relatively bright phase lasted for about 60 days. The OGLE team (Udalski et al. (1994a,b)) found a simple binary lens solution (Fig. 9, bottom panel) with the following parameters: mass ratio $q = 1.02$, projected separation $a = 1.14R_E$, impact parameter $b = 0.050R_E$, angle $\theta = 48.3$ degrees, time scale $t_E = 80$ days, baseline magnitude $I_0 = 18.1$, and fraction of "blended" light $f = 56\%$.

The last value indicates that the measured apparent brightness must consist of two contributions, and so Udalski et al. (1994a,b) concluded that the lensed star should have a composite spectrum: the lens model required light from an additional (unresolved) star.

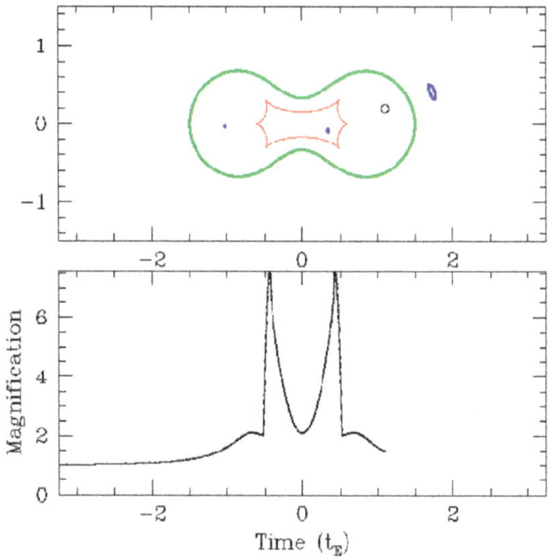

Fig. 8. Snapshot of a binary microlensing animation made available on the web by Scott Gaudi, under `http://www.sns.ias.edu/~gaudi/movies.html`. This particular situation illustrates the case of a binary lens with equal masses: in the top panel the critical line (*thick solid line*), the caustics (*thin line* with 6 cusps), the source size and position (*little circle*) and the three images (thick *circles/ellipses*) are shown for one particular instant of time; the source moves along a horizontal line from left to right. The bottom panel indicates the lightcurve (sum of all the micro-images) as it develops during the animation, with the current instant displayed at the top panel corresponding to the end of the black line

As mentioned above, Witt and Mao (1995) showed a few months later that the minimum magnification between the peaks of a double caustic crossing is three (for OGLE-7 it was factors of 2.2 and 2.4, respectively, in R and I filters). So in hindsight this fact showed unambigously that either there is blended light contributing to the lightcurve, or the event OGLE-7 is caused by lensing of a system with more than two components.

2.3 Binary Lens MACHO 1998-SMC-1

In the following years, the data reduction systems of the microlensing teams were dramatically improved in order to allow detection of microlensing events while they were still ongoing, with the goal of real-time detection. MACHO-98-SMC-1 was the first caustic crossing binary event toward the Magellanic Clouds which was 'caught in action' in this way (Alcock et al. (1999)). This allowed very good coverage of the lightcurve: Once an anomaly is recognized "on the run", the observing strategy can be changed immediately with a much more frequent monitoring of the active event. As a consequence, a prediction for the time of the second caustic crossing became possible.

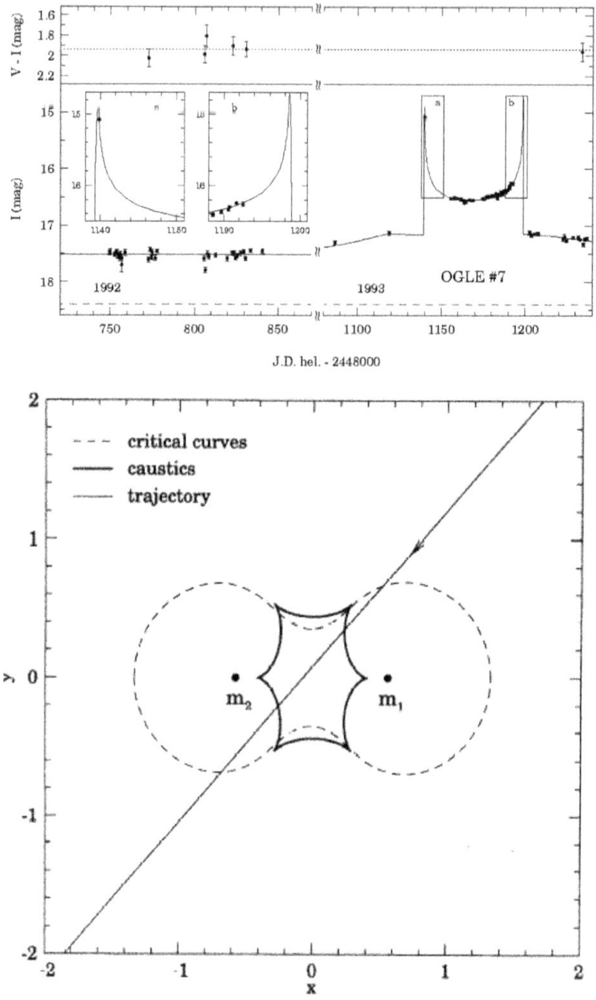

Fig. 9. Lightcurve obtained by the OGLE team of the first double-lens microlensing event OGLE-7 (*top*) and corresponding double lens configuration with caustics, critical lines and relative track (from Udalski et al. 1994a)

The event was originaly recognized on May 25.9, 1998 (UT), when it had brightened by 0.9 mag. At this time the first "alert" was activated (see Fig. 10, top panel). On June 6.5, 1998 (UT), a sudden brightening by another 1.5 mag was detected. This caused a level-2 alert, which meant that MACHO-98-SMC-1 was a likely caustic crossing event. An accurate prediction of the timing of the second caustic crossing was then a very important task (later it was shown by Jaroszyński and Mao (2001) that a reliable prediction of the exact timing of

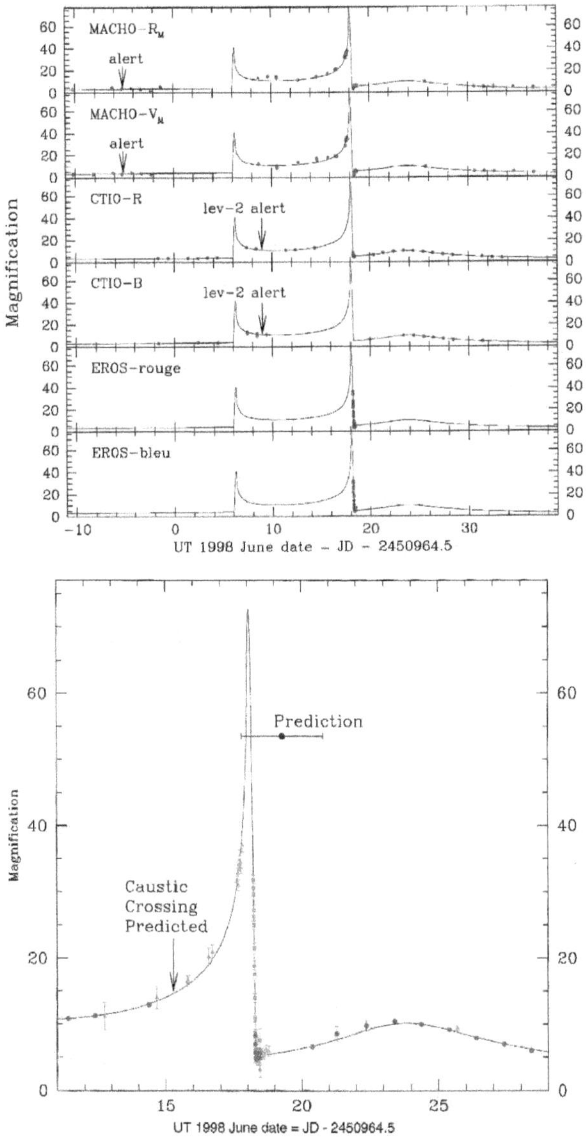

Fig. 10. Lightcurve obtained by the MACHO team of the first double-lens microlensing in the Small Magellanic Cloud, SMC-1 (*top*). The six panels represent data from three observing sites and two filters, respectively. The times of the first and second alerts are indicated by *arrows*. The *bottom panel* is a zoom around the second caustic crossing, indicated is the predicted epoch of the caustic crossing, and the time when it was announced (from Alcock et al. 1999)

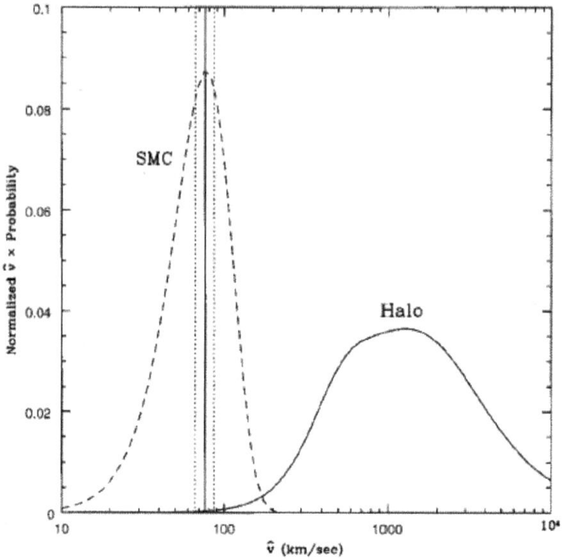

Fig. 11. Probability distribution for certain values of the transverse velocities of the lens in MACHO-SMC-1 (from Alcock et al. 1999), assuming that the lens resides in the LMC (*dashed*) or in the Milky Way halo (*solid*). The vertical solid (*dotted*) line shows the measured value (including the error bars)

the second caustic crossing is intrinsically difficult and possible only relatively late). The first prediction for the second caustic crossing was for UT June (19.3 ± 1.5), issued on June 15.3 (see Fig. 10, bottom panel). This value was revised on June 17 to UT June (18.2 ± 1.5), when it actually happened. The possibility to analysze the lightcurve while it is still ongoing made it hence possible to react quickly. This resulted in an amazing 1598 data points for this double peak microlensing event within about 50 days (Fig. 10)!

Measuring the caustic crossing time can help break the degeneracy between the lensing parameters. The method is very simple: assuming that locally the (fold) caustic is a straight line in the source plane, the duration of the caustic crossing is then just the time it takes the star to move its own stellar diameter, $2\ t_*$. As described in Alcock et al. (1999), they could make a very accurate measurement of the duration of the caustic crossing: $t_* = (0.116 \pm 0.010)$ days. With the knowledge of the physical size of the star, $R_* = (1.1 \pm 0.1) R_\odot$, from its spectral type ($T_{\rm eff} = 8000 K$), it became possible to determine the proper motion of the lens with respect to the source. Alcock et al. (1999) estimated the transverse velocity of the lens projected to the SMC distance to be $v = (76 \pm 10)$ km/s. This in turn allows to estimate the distance to the lens, one of the very interesting parameters which cannot be obtained in "normal" microlensing cases, due to the degeneracy of the parameters lens mass, distance and transverse motion (cf. Sect. 1.2).

In Fig. 11, the expected distribution of relative lens velocities projected to the source (SMC) plane are shown for two potential lens populations: if the lenses are in the Milky Way Halo, the typical (projected) velocities are in the range of about 1000 km/sec, whereas for a lens population in the SMC, it is rather around 60 km/sec.

The measured value of the projected transverse velocity hence clearly favors a lens position in the (foreground of the) SMC. Alcock et al. (1999) analyzed this quantitatively as well: The probability for a halo star to have such a low velocity is only 0.12%, whereas 38% of the SMC stars would have such a value or smaller. Hence Alcock et al. (1999) concluded that the lensing system responsible for MACHO-99-SMC-1 is much more likely to reside in the SMC rather than in the Galactic halo, hence it is a case of "self lensing" (Sahu 1994).

2.4 Binary Lens MACHO 1999-BLG-047

The well covered microlensing event MACHO 1999-BLG-047 displays a "nearly normal" lightcurve with a small but highly significant deviation close to the peak (Fig. 12). Since such small-amplitude deviations near the peak of a lightcurve can be produced by planetary lenses (cf. Griest and Safizadeh 1998), this event attracted a lot of interest. However, roughly equal mass binary lenses with either very small or very wide separation can introduce very similar features in the lightcurve. The analysis of Albrow et al. (2002) showed

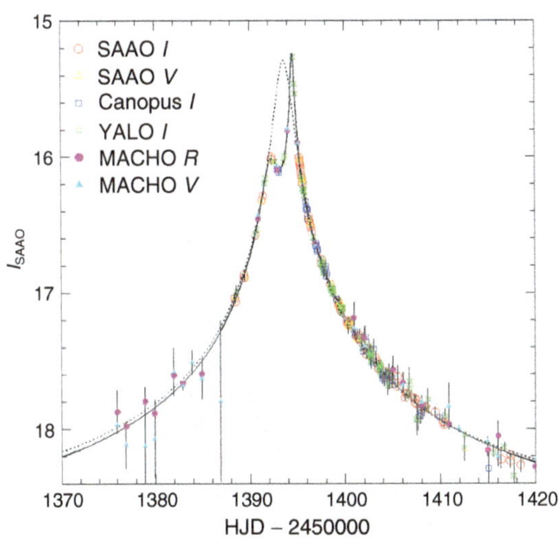

Fig. 12. Lightcurve of the binary microlensing event MACHO 99-BLG-47 obtained by the MACHO and PLANET teams (from Albrow et al. 2002). The symbols denote the data points from the various observatories. The *solid line* is the best fit, the *dotted line* is the best single lens fit, clearly not at all reproducing the high magnification data points

that these two cases can be distinguished with a high quality data set, and that in particular the event MACHO 1999-BLG-047 is produced by an extreme binary event. However, the analysis yielded two "islands" in the mass ratio versus separation diagram which both satisfied the observational data equally well (Fig. 13): it was not possible to find a unique solution. The two best fit models require the binary lens to be either a close binary with parameters $d/R_E = 0.134 \pm 0.009$ and $q = 0.340 \pm 0.041$ or $d/R_E = 11.31 \pm 0.96$ and $q = 0.751 \pm 0.193$ (more details see in Albrow et al. 2002).

2.5 Binary Lens EROS BLG-2000-005

The triple-peak microlensing event EROS BLG-2000-005 (Fig. 14, An et al. 2002) became one of the most spectacular examples of stellar microlensing (An et al. 2002). Originally detected by the EROS team, an alert was issued on May 5, 2000 for a possible microlensing event. On June 8, 2000, the MPS team (Microlensing Planet Search) sent an anomaly alert, stating that the star has changed its brightness by 0.5 mag compared to the previous night, and that it was still brightening at the remarkable rate of 0.1 mag per 40 minutes (!). The PLANET team increased the monitoring frequency of this event and kept it at a high level until about a year later.

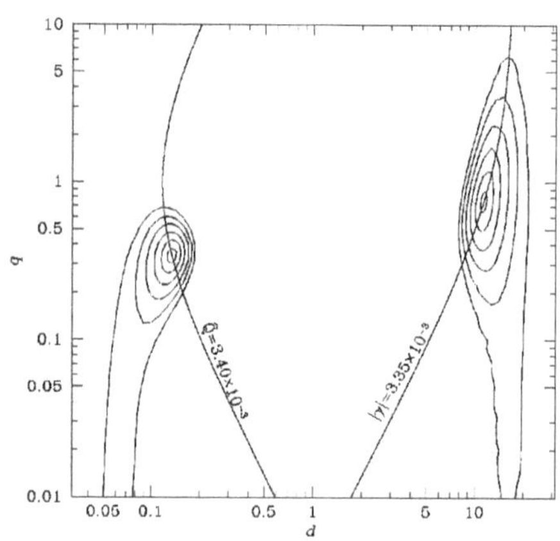

Fig. 13. Mass ratio versus separation for the binary lens parameters of the PLANET analysis of the binary microlensing event MACHO 99-BLG-47: shown are contour of good fits, based on PLANET and MACHO data. The binary separation d is in units of the Einstein radius of the combined mass, and the mass ratio q is the ratio of the farther component to the closer component to the source trajectory (i.e., q > 1 means that the source passes by the secondary). Contours are shown for $\Delta\chi^2 = 1, 4, 9, 16, 25, 36$ (with respect to the global minimum). It is obvious that there are two well separated minima. Also drawn are the curves of models with the same quadrupole moment $\hat{Q}$ as the best-fit close-binary model and the same shear γ as the best-fit wide-binary model (from Albrow et al. 2002)

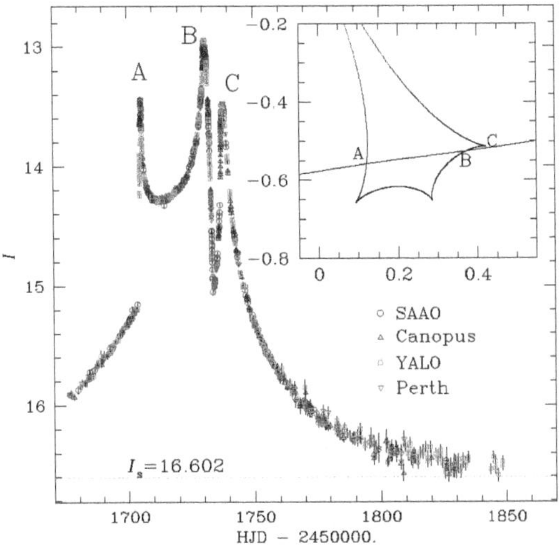

Fig. 14. Lightcurve of the binary microlensing event EROS BLG-2000-5 (from An et al. 2002). The I-band data points are displayed, the baseline magnitude I_s is given, the three peaks are labeled A, B and C. The inset at the *top right* indicates the geometry of the stellar track crossing the caustic, with the three caustic crossings labeled according to the maxima in the lightcurve

The I-band data of the PLANET team are presented in Fig. 14, containing 1286 data points (cf. An et al. 2002). The different symbols indicate the four PLANET telescopes in South Africa (SAAO), Tasmania (Canopus), Chile (YALO) and West Australia (Perth). The two main maxima in the lightcurve are more than 3.5 mag above the baseline and show very steep rising or dropping flanks, indicative of caustic crossings. The third peak is slightly less steep and shows the characteristics of a cusp passage. The inset in the top right part of Fig. 14 shows the track of the source star relative to the binary lens caustic. The two caustic crossings are labeled 'A' and 'B', whereas the cusp passage has the label 'C'.

The quality of the data is very good and the duration of the event long enough, in order to measure the parallax due to the motion of the Earth around the Sun. In Fig. 15, the geometry of the event as projected on the sky is shown (from An et al. (2002)). It shows very clearly the difference of the relative paths as seen from the Earth (solid track) and from the Sun (dashed track), i.e. the parallax effect.

In Fig. 16, a close-up of the previous figure is shown for the time of the cusp passage. The circle represents the source at the time of closest approach to the cusp (see also the inset panel).

The system could not be modeled satisfactorily without including the orbital motion of the binary (cf. Figs. 15 and 16). This made it possible to

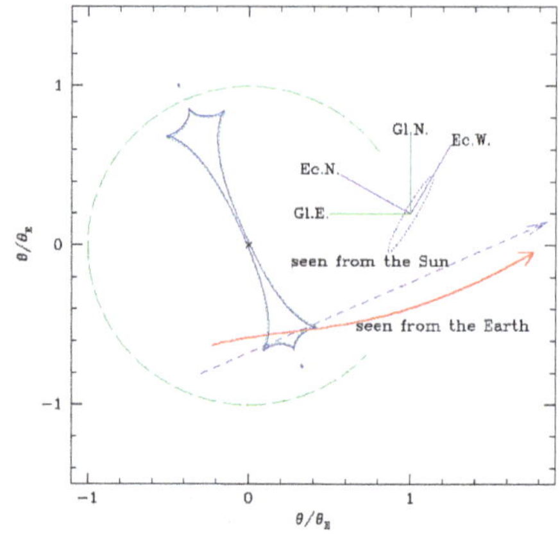

Fig. 15. Geometry of microlensing event EROS BLG-2000-5 as projected on the sky (from An et al. (2002)). The origin (marked with a *small cross*) is the center of mass of the binary lens. The path of the source relative to the lens as seen from the Earth is shown as the *solid curve*, whereas the relative proper motion as seen from the Sun is indicated as the *short-dashed line* (length of both trajectories corresponds to six months). The circle (*long-dashed line*) is the Einstein ring, and the lines within (*solid* and *dotted*) are the caustics of the binary system at two different epochs. The labels "Gl.E." and "Gl.N." indicate the directions East and North in galactic coordinates, "Ec.N." and "Ec.W." stand for North and West in ecliptic coordinates, respectively (from An et al. 2002)

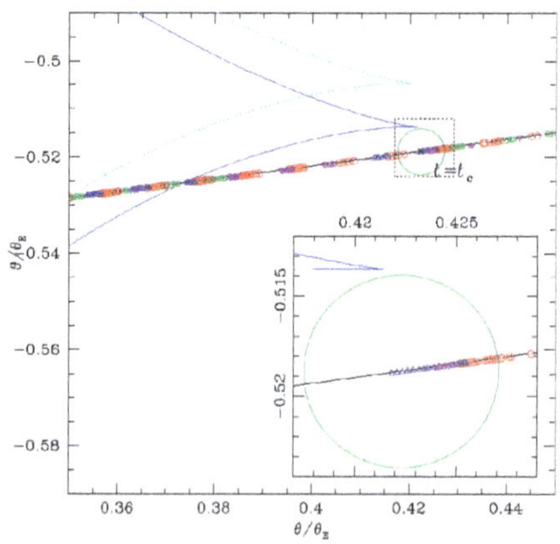

Fig. 16. Relative track of the source in the binary microlensing event EROS BLG-2000-5 during the time of closest approach to the cusp (from An et al. 2002). The symbols indicate the positions of the source center during data taking at the various observatories (cf. Fig. 14)

measure the projected Einstein radius $\tilde{r}_E = (3.61 \pm 0.11)$ AU. The angular Einstein radius, on the other hand, could be determined from the finite source effects on the lightcurve to $\theta_E = (1.38 \pm 0.12)$ mas with an estimate of the physical source size from its position in the color-magnitude-diagram, these two measurements result in a determination of the lens mass: $M_{\rm lens} = (0.612 \pm 0.057) M_\odot$. This is the first time that a microlens parallax was measured for a caustic crossing event, and also the first time that the lens mass degeneracy could be broken and that the mass of a microlens could be derived from photometric measurements alone.

3 Microlensing and Dark Matter: Ideas, Surveys and Results

3.1 Why We Need Dark Matter: Flat Rotation Curves (1970s)

Since the 1970s, measurements of the rotation curves of galaxies showed that the (rotational) velocity as a function of radius is roughly constant: galaxies have flat rotation curves (e.g., Bosma (1978); Rubin (1983)), see also Fig. 17, top left panel. This is a non-trival result: in the Solar System, as a contrast, the planets follow the Kepler law: velocity decreases with the square root of the radius (bottom left panel in Fig. 17). In general, for a stable circular orbit, gravitation is balanced by the centrifugal forces:

$$\frac{GM(r)}{r^2} = \frac{v(r)^2}{r}.$$

If $v(r) = \text{const}$ for a broad range of radii r, this implies: $M(r) \propto r$. In other words: flat rotation curves mean that the mass of the galaxy increases linearly with radius. In Fig. 17, three idealized rotation curves are shown for solid body rotation (top right), Keplerian rotation (bottom left) and a relation in which mass increases linearly with radius: $M(r) \propto r$.

Using the 21 cm Hydrogen line, the rotation velocity of spiral galaxies could even be measured far beyond the visible stellar part: The interesting – and very unexpected – result was: galactic rotation curves remain flat even outside the regions in which stars exist. These observations imply: more than 90% of the mass of a galaxy must be in an unknown and invisible form, soon to be called "dark matter". From its presumed roughly spherical distribution around the visible galaxies, the concept of "dark matter halos" was established[4].

For some time, an alternative explanation for the flat rotation curves was put forward: the concept that Newton's law of gravity (and also Einstein's

[4] To this day it is not really understood what dark matter is. The concept of "Dark Matter Halos", however, is so ubiquitous inside and outside physics, that it made it even into art/literature, see Reza (2000).

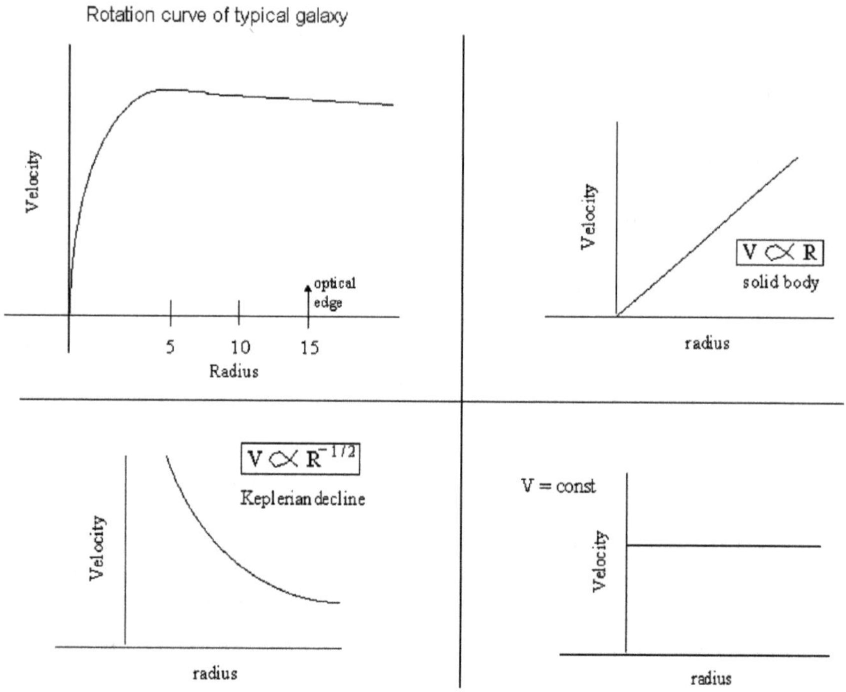

Fig. 17. Different types of rotation curves (schematically): observed flat rotation curve of spiral galaxies (*top left*); solid body rotation curve (*top right*): $v_{\rm rot}(r) \propto r$; Keplerian rotation curve (*bottom left*): $v_{\rm rot}(r) \propto r^{-0.5}$; rotation curve for constant velocity (*bottom right*): $v_{\rm rot}(r) = const$ (in other words: $\propto r^0$)

General Theory of Relativity) changes on large length scales. Two of the theories in the latter categories are the "MOdified Newtonian Dynamics", or MOND (Milgrom 2001), and the "conformal gravity" (Mannheim 1992). These theories change the relation between gravity and distance and try to avoid the concept of "dark matter" altogether. We cannot go into any more detail here, but rather refer the interested reader to a recent review on alternative theories of gravity (Sanders and McGaugh 2002). However, in the light of the latest results of the Wilkinson Microwave Anisotropy Probe (WMAP, Spergel et al. 2003), these alternative explanations seem not to be viable any more.

Though experimentalists have tried very hard for many decades (Rees 2003), no physical candidate for dark matter was detected. Two main dark matter candidate types were proposed: massive elementary particles and astrophysical compact objects. The list of elementary particle candidates for dark matter comprises many dozen candidates, among them many hypothetical ones: neutralino, Higgs particle, WIMPs (weakly interacting massive particles), axions. The suggested astrophysical candidates were stellar or sub-stellar mass black holes, neutron stars, white dwarfs, brown dwarfs, or planets.

3.2 How to Search for Compact Dark Matter (as of 1986)

In 1986, Bohdan Paczyński proposed a clean experiment for testing whether the latter type, compact astrophysical objects of roughly stellar mass, can make up the dark matter halo of the Milky Way (Paczyński 1986b). His idea was simple[5] and brilliant at the same time: if a class of compact objects in the mass range of very roughly $10^{-6} \leq m/M_\odot \leq 10^{6}$ exists in the Milky Way halo and makes up a fair fraction of the dark matter, then occasionally one of these objects must pass very close to the line-of-sight to a background star in the large magellanic cloud (LMC). As a consequence, the apparent brightness of this background star is magnified temporarily, in exactly the way that is explained in Sect. 1: single lens, single source.

Paczyński determined the fraction of background stars that would be within the Einstein radius of these MACHOs[6], the so called optical depth, to be in the range $p_{\mathrm{MACHO}} = 10^{-6}...10^{-7}$. This is a remarkably small number: it means that the apparent brightness of a few million stars has to be monitored very frequently, in order to find the handful of candidate lightcurves[7].

3.3 Just Do It: MACHO, EROS, OGLE et al. (as of 1989)

What sounded like science fiction at the time (Paczyński even refers to it this way in his original article), soon became reality, due to four developments:

1. Optical CCD chips got bigger, and it became possible to build cameras consisting of an array of such CCDs. This way one could determine the apparent brightness of many stars in "one shot".
2. Software could be developed for automatic data reduction pipelines, so that a large number of objects could be treated and analyzed with no or little human interaction.
3. Computer power kept increasing according to Moore's law, i.e. speed doubling roughly every 18 months, as well as data storage became available in sufficient amounts, so that by the mid 1990s literally tens of millions of stars could be monitored frequently enough with lightcurves being produced.
4. Scientists realized that the normal procedure of applying for a certain chunk of time at a certain telescope did not make much sense: they needed (and succeeded in getting!) dedicated telescopes.

[5] It was in fact so simple that the referee first rejected the paper; only after some discussion between author, referee and editor, the paper was published; and at the time of this writing, it has collected more than 500 references.

[6] MACHO - MAssive Compact Halo Object, an acronym coined for these dark matter candidates originally by Griest (1991)

[7] Paczyński's optimistic suggestion that this might be measurable was in stark contrast to Einstein's pessimistic view exactly 50 years earlier: he had derived the basic equation and estimated the probabilites and written that there is 'no hope of observing such a phenomenon directly' (Einstein 1936).

In the years following Paczyński's article, three teams formed and started to address the scientific question posed. Later on, a number of additional collaborations followed:

- the MACHO Team (USA/Australia), MAssive Compact Halo Objects: `http://wwwmacho.anu.edu.au/`
- the EROS Team (France), Expérience pour la Recherche d'Objets Sombres: `http://eros.in2p3.fr/`
- the OGLE Team (Poland/USA), The Optical Gravitational Lensing Experiment: `http://www.astrouw.edu.pl/~ogle/`
- the MOA Team (NZ/Japan), Microlensing Observations in Astrophysics: `http://www.physics.auckland.ac.nz/moa/`

They determined the apparent brightnesses of stars in the direction to the LMC/SMC and to the Galactic Bulge a few times per week, constructed lightcurves, identified the variable stars, and searched for the rare "needle-in-the-haystake" microlensing signal among the millions of stars.

3.4 "Pixel"-Lensing: Advantage Andromeda!

In 1992, Arlin Crotts had suggested to use the Andromeda Galaxy as a "unique laboratory for gravitational microlensing". M31 is roughly 15 times as distant as the LMC/SMC, hence individual stars cannot be resolved any more: only the combined flux of many stars can be measured in any resolution element of the CCD camera. This means that a possible microlensing event would be "buried" among an ensemble of constant or variable unrelated stars. The magnification consequently is diluted: "blending" dominates the lightcurve very heavily. Only very high magnification events would be detectable. However, Crotts (1992) pointed out that M31 does have a number of advantages compared to LMC/SMC searches: smaller angular size of source stars, (much) greater total mass, favorable geometry and foreground/background asymmetry, which should statistically allow to distinguish microlensing events due to Milky Way halo objects from those produced by M31-halo objects. This method was subsequently somewhat improperly called "pixel-lensing" and became popular under this name.

Subsequently, a number of teams jumped on the pixel-lensing train:

- AGAPE: Andromeda Galaxy Amplified Pixel Experiment (later POINT-AGAPE): `http://cdfinfo.in2p3.fr/Experiences/Agape/`
- MEGA: Microlensing Exploration of the Galaxy and Andromeda `http://www.astro.columbia.edu/~arlin/MEGA/`
- WeCAPP: Wendelstein Calar Alto Pixellensing Project, `http://www.usm.uni-muenchen.de/people/fliri/wecapp.html`.

For the "Theory of Pixel Lensing", see Gould (1996).

3.5 Current Interpretation of Microlensing Surveys with Respect to Halo Dark Matter (as of 2004)

In the more than ten years which have passed since the first publication of stellar microlensing events toward the LMC (Alcock et al. 1993; Aubourg et al. 1993), many more microlensing events have been discovered (and some of the first discovered events were retracted because they were later classified as misinterpretations of a rare kind of variable stars, so-called blue bumpers [Tisserand and Milsztajn, private communication 2004]). With a baseline of five years or longer, statistically quantitative results have been obtained. In the meantime, both MACHO and EROS have ended their campaigns. Final results and/or conference summaries have been published. The two robust results of these two experiments are:

- A certain relatively small number of lightcurves of LMC (and SMC) stars have been obtained which were definitely produced by microlensing of an intermediate single or binary star. "Intermediate" between source star and observer could mean in the Galactic Halo, in the disk of the Milky Way, or in the foreground of the LMC/SMC.
- The total number of these microlensing lightcurves (fewer than two dozen) is definitely far too small to explain ALL the dark matter in the Galactic halo by compact objects, even if all the lenses were objects in the Galactic halo. What fraction of the halo dark matter could still be explained by MACHOs is a matter of debate. The estimates range from about 20% to zero.

Here the results are summarized:

MACHO

The MACHO team ended their operation in 1999. Results are summarized in Alcock et al. (2000b): They were in operation for 5.7 years and had monitored 11.7 million stars in the LMC. The identification of microlensing events is not trivial: They applied two criteria and found 13 or 17 events, respectively. The time scales of these events range from $t_E = 34$ to 230 days. According to their modeling, they expected 2–4 events from known stellar populations in the Milky Way.

Their analysis results in an optical depth of

$$\tau_{\mathrm{LMC(MACHO)}} = 1.2^{+0.4}_{-0.3} \times 10^{-7},$$

plus an estimated systematic error of 20%. Their interpretation is that about 20% of the Milky Way halo could be made of dark matter objects in the mass range $0.15 \le m/M_\odot \le 0.9$, with a 90% confidence interval of 8% – 50%.

The lack of long duration events allows them to put limits on more massive objects, in particular potential black holes: they conclude that objects in the

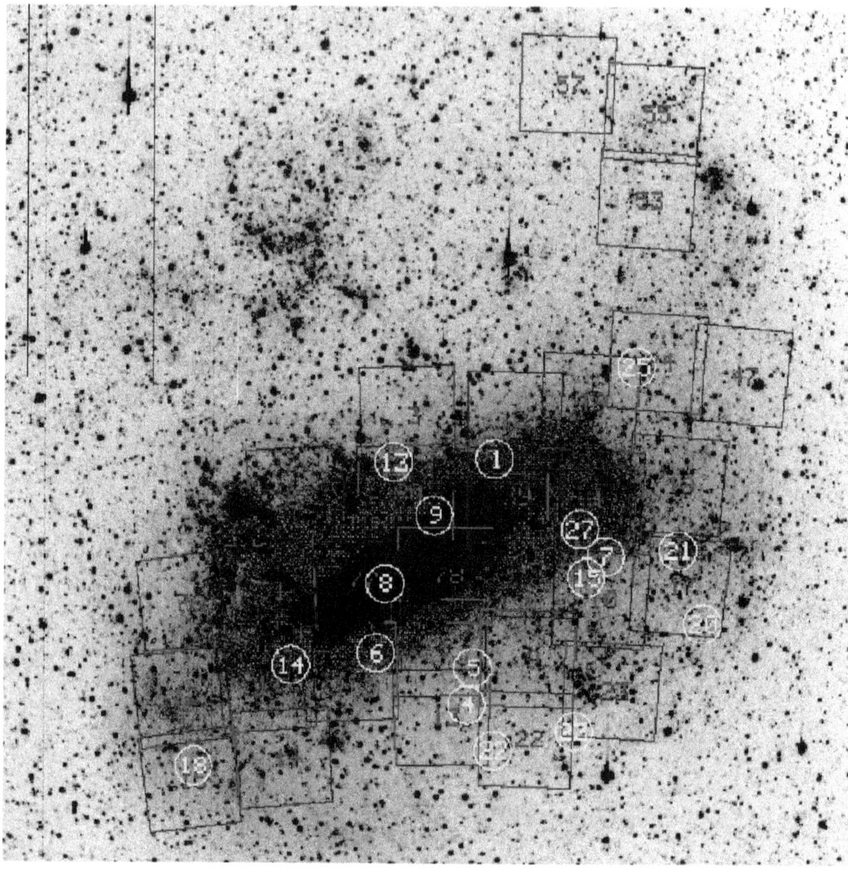

Fig. 18. MACHO fields with events indicated, from Alcock et al. (2000b): R-band LMC, 8.2 degrees at a side, 30 MACHO fields (squares), with 17 microlensing events (numbers in *circles*)

mass range $0.3 \leq m/M_{\odot} \leq 30$ cannot make up the entire dark matter halo (Alcock et al. 2000a,b).

In Fig. 18, the 30 central monitoring fields of the MACHO team are indicated on an R-band image, including the location of the 17 identified events[8]. The detection efficiencies of the MACHO team – defined as the fraction of events of a certain duration that would have been identified in the data, given the actual sampling and data quality – for microlensing events of certain duration are shown in Fig. 19 as a function of increasing coverage: 1 year, 2 years and 5 years (and two selection criteria for the latter). More information and detailed analyses can be found in Alcock et al. (2000a,b).

[8] The monitoring data of the MACHO team for 73 million stars in the LMC, the Small Magellanic Cloud and in the Galactic Bulge are available to the general public at `http://wwwmacho.mcmaster.ca` or `http://wwwmacho.anu.edu.au`.

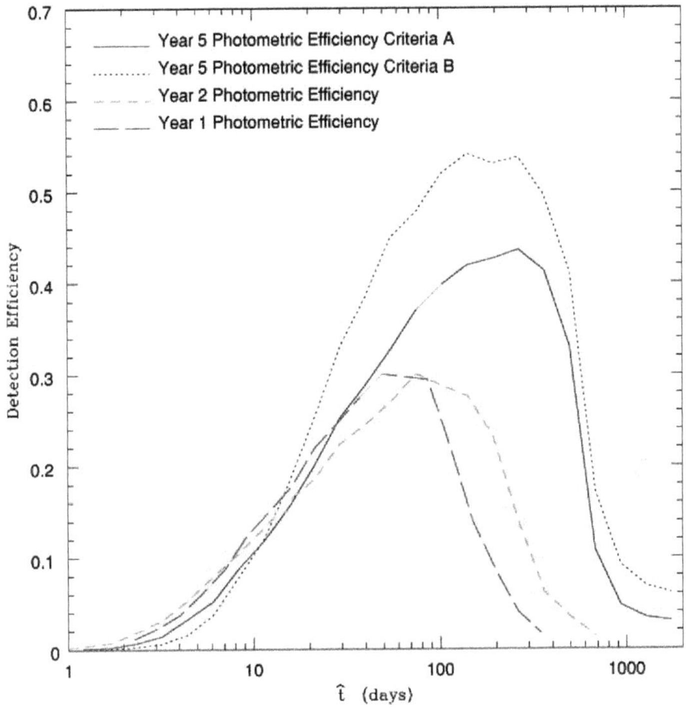

Fig. 19. Detection efficiencies of the MACHO experiment for one year, two years, and five years of operation, as a function of event duration (from Alcock et al. 2000b)

EROS

The EROS team ended their operation in February 2003. Preliminary results were already published in Lasserre et al. (2000): They had ruled out sub-solar mass dark matter objects as an important component of the Galactic Halo. In an analysis of 5 years of EROS data toward the Small Magellanic Cloud, Afonso et al. (2003a) concluded: Objects in the mass range from $2 \times 10^{-7} M_\odot$ to $1 M_\odot$ cannot contribute more than 25% of the total halo. They derived an upper limit on the optical depth:

$$\tau_{\mathrm{SMC(EROS)}} \leq 10^{-7}$$

(stating in addition that the long duration of all the EROS SMC candidates may point to the fact that they are more likely due to unidentified variable stars or self-lensing within the SMC, rather than due to halo objects). A preliminary analysis of the full 6.7 year EROS data set on the LMC strengthens this result (Glicenstein private communication 2003; Tisserand, private communication 2004): some of the formerly claimed EROS candidates turned out to be variable stars, and the derived optical depth toward the LMC is in the range

$$\tau_{\text{LMC(EROS)}} \approx 10^{-7}.$$

A graphic depiction of the EROS mass exclusion range for both the SMC and LMC directions can be found in Fig. 20 (from Afonso et al. 2003a). Similar results were obtained in an analysis of the microlensing experiments by Jetzer et al. (2004).

Some additional results of EROS are: they had identified 4 long duration events, which are most likely not produced by a halo population. The main EROS result concerning the dark matter fraction is: less than 25% of the standard Milky Way Halo can be in objects with masses between $2 \times 10^{-7} M_{\odot}$ and $1 M_{\odot}$.

Where are the Lenses?

Combining the EROS and MACHO data results in roughly 20 microlensing events in the directions toward the LMC and the SMC. Originally, the experiments were set up to find compact dark matter objects in the Galactic Halo. So the big question is: *are the lenses* that caused the microlensing events *dark matter objects*? This is difficult to answer, because this would involve to

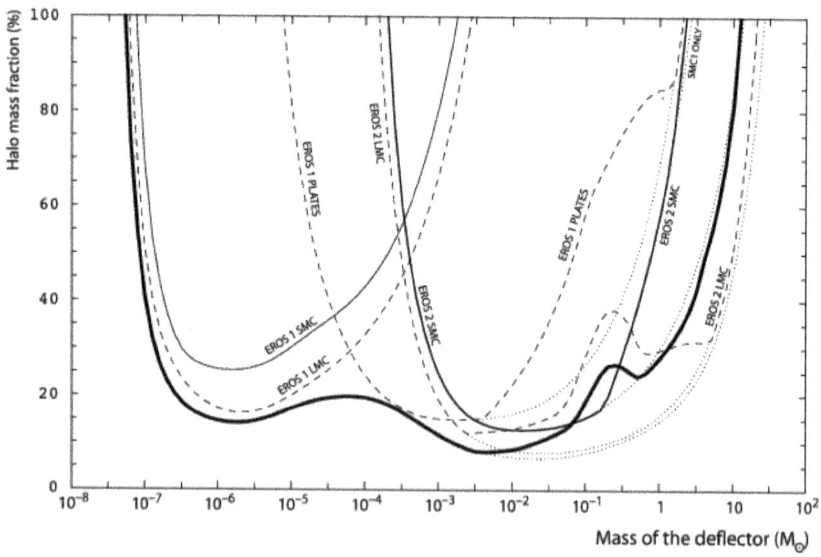

Fig. 20. Limits on Halo Mass fraction of EROS team: the 95% exclusion probability for the standard halo model are shown. *Dashed lines*: limits toward the LMC from EROS-1 and EROS-2; *thin line*: limit toward the SMC; *thick line*: combination of five EROS sub-experiments. The *dotted line* indicates the limits that would have been obtained without any detected events, it illustrates the overall sensitivity of the EROS experiment (from Afonso et al. 2003a)

uncover the nature of an invisible object, which is almost impossible to do. However, what may help answer this question is a related one: *where are the lenses?* And the answer to this latter question may be easier to obtain. From a purely observational point of view, the lenses could be at three distinct locations:

- in the Milky Way (thin/thick) disk: the lenses – in that case presumably normal stars – might become visible a few years after the event, once they have moved away from the bright background LMC/SMC star.
- in the Milky Way Halo: then they could be the searched for dark matter population. The density of events should be proportional to the density of stars in the LMC.
- inside the LMC/SMC: foreground stars could act as lenses on background stars: the density of events should be proportional to the star density squared.

Jetzer et al. (2004) conclude that the microlensing events are produced by various populations: a combination of self-lensing in the LMC, thick disk, spheroid, plus some "true machos" in the halo of the Milky Way and the LMC itself. Taking advantage of the apparent near-far asymmetry of the spatial distribution of the LMC events, Mancini et al. (2004) re-analyzed the possibility of self-lensing. Their main conclusion is that even considering this, self-lensing cannot account for all the observed microlensing events toward the LMC.

In an earlier conference proceedings contribution, Sahu (2003) discussed the issue: "Microlensing toward the Magellanic Clouds: Nature of the Lenses and Implications for Dark Matter" in some detail. In particular, Sahu investigated the question of the distance of the 17 MACHO events toward the Magellanic Clouds. As a first step he summarizes:

- for one of them, a binary-lens event, the distance could be determined securely via its caustic crossing time scale: it is within the SMC (cf. Fig. 10).
- for three more, the lens location could be estimated. This estimate is less certain than for the SMC event, but the evidence suggests that it is very likely that the lenses are located within the Magellanic Clouds as well.

As an independent second step, Sahu (2003) mentions that – assuming that most of the events are dark matter objects in the Galactic Halo – the time scales of the events toward the LMC would imply that masses are of the order of 0.5 $M_\odot$ (cf. Alcock et al. 2000a). However, with the same line of thought, the most likely masses for the events toward the SMC would be in the range 2 – 3 $M_\odot$. Could the mass distribution of objects with different masses be different from each other? No model of the Galaxy is consistent with such an inhomogeneous mass distribution. On the other hand, if one assumes that most of the events are caused by foreground objects in the LMC/SMC, then the expected masses would be of order 0.2 $M_\odot$ for both LMC and SMC.

A third line of argument uses the frequency of binary lenses. Two of the 17 events are caused by binary lenses. In both cases, the lenses are most likely objects located within the LMC/SMC. Assuming that roughly 50% of the potential lenses in the LMC/SMC are in binary systems (similar to the stars in the solar neighborhood), one would expect that 10% of all microlensing events would show binary characteristics (cf. Mao and Paczyński 1991 and Sect. 2.1). This implies that of order 20 events are expected to be caused by single stars within the LMC/SMC. So this would be perfectly consistent, if most of the observed microlensing events are caused by foreground stars within LMC/SMC.

A fourth argument of Sahu (2003): If the microlensing events are caused by 0.5 $M_\odot$ objects in the Galactic Halo (as claimed from the LMC observations), one would have expected to detect about 15 events in the direction toward the SMC, with time scales of about 40 days. Not a single event of this kind was detected: in fact, both SMC events are shown to be due to self-lensing.

Although each individual of these four arguments is not very strong, the combination of them provides relatively firm evidence against them being interpreted as mostly due to halo objects.

The conclusions in Sahu (2003) are: *"Close scrutinity of the microlensing results toward the Magellanic Clouds reveals that stars are major contributions as lenses, and the contribution of MACHOs to dark matter is 0% to 5%."*. This view might not be shared by everyone working in the field. However, it is certainly a viable one[9].

3.6 Microlensing toward the Galactic Bulge

As originally suggested by Paczyński (1986b), monitoring stars in the Galactic Bulge turned out to be a very fruitful enterprise. Originally meant as a safety measure[10], in the mean time the Bulge microlensing turned out to be a source of exciting astrophysical results in itself.

The angular distribution of the microlensing events lead to the re-discovery of the galactic bar (see Stanek et al. 1994; Paczyński et al. 1994). The microlensing optical depth in this direction turned out to be higher than expected: the original results by MACHO/OGLE were roughly

$$\tau_{\mathrm{Bulge(MACHO/OGLE)}} \approx 3 - 4 \times 10^{-6}$$

[9] In an earlier independent analysis, Graff (2001) had concluded: "Occam's razor suggests ··· that microlensing experiments have simply found a background of ordinary stars".

[10] Microlensing experiments toward the Bulge would produce microlensing events with certainty due to the known population of disk stars, which could be used as a test of the experimental setup; otherwise, the potential lack of microlensing events toward the LMC/SMC could always have two reasons: there are no MACHOs, or the experiment does not work properly.

(Udalski et al. 1994b; Alcock et al. 1997a,b). This caused some kind of problems of our understanding of the Galactic dynamics: the high optical depth for microlensing implies much more mass than people had thought there is in the inner part of the Galaxy. Quite a number of papers dealt with this issue and tried to solve the discrepancy.

Recently, EROS published their analysis for the optical depth toward the Galactic bulge, based on the identification of 16 microlensing events with clump giants from a region of 15 contiguous one-square-degree fields with a total of 1.42×10^6 clump giants. The distribution of the time scales of their microlensing events is displayed in Fig. 21: in a logarithmic presentation, most events were found with Einstein time scales of 10 – 30 days. EROS found a much lower value than what was previously favored:

$$\tau_{\mathrm{Bulge(EROS)}} = (0.94 \pm 0.29) \times 10^{-6}$$

(Afonso et al. 2003b), which lead to the remark "The issue of the optical depth to the bulge is solved" by one of the experts in the Galactic microlensing community (Andy Gould, private communication March 2003). Considering the error bars of the published values, the problem was never really severe: the new EROS result still agrees with most of the previous results at the 2σ

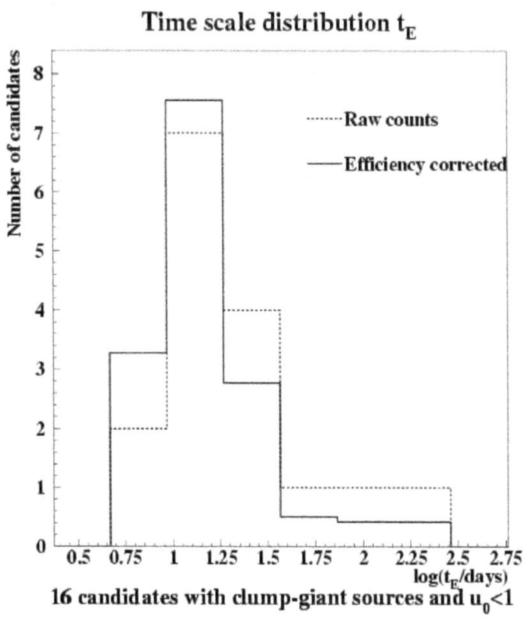

Fig. 21. Time scales of the 16 EROS bulge candidate events of clump giants: *dashed line* corresponds to raw data, the *solid line* shows the (rescaled) distribution, corrected for the detection efficiency (from Afonso et al. 2003b)

level. But it is closer to the many predicted values. For a detailed discussion of the differences between the various observational and theoretical analyses, see Afonso et al. (2003b).

4 Microlensing Surveys in Search of Extrasolar Planets

The very first time that microlensing by planets was mentioned in the literature was the paper by Shude Mao and Bohdan Paczyński from 1991: "Gravitational microlensing by double stars and planetary systems". This seminal paper with more than 150 citations by now (Dec. 2003) states the situation and explores the possibilities. Experiments for the detection of compact objects of stellar mass in the halo or the disk of the Milky Way via microlensing were planned and prepared at that time. Mao and Paczyński (1991) figured out that binary signatures of the lenses should be visible in some of the lightcurves. In addition, they stated that this microlensing technique will be able to detect planetary systems ultimately as well.

In this section, the current state of microlensing searches for extrasolar planets is summarized. The basics of the method are explained, the advantages and disadvantages are discussed and compared with other planet-search techniques. The teams active in the microlensing searches (OGLE, MOA, PLANET, MicroFUN) are presented. A number of recent observational and theoretical results on planet microlensing are mentioned. Good descriptions of the basics of planet microlensing can be found, e.g., in Paczyński (1996), Sackett (2001) and Gaudi (2003).

4.1 How Does the Microlensing Search for Extrasolar Planet Work? The Method

Only a few years after the original idea proposed by Paczyński (1986b) to use gravitational microlensing as potential test for stellar mass objects in the Galactic halo, Mao and Paczyński (1991) calculated that roughly 10% of all lensing events *had* to show the signature of a binary companion. So it only was a quantitative question: One had to monitor the apparent brightness of a very large number of stars in the Milky Way bulge, with the goal to detect the passage of a binary star or star-plus-planet system in the line-of-sight to one of these background stars, producing a very characteristic magnification lightcurve.

Compared to the situation of a single stellar lens, there are three additional parameters in a situation of a star-plus-planet lens (as shown in the binary lens case, Sect. 2.1): the mass ratio $q = M_{\mathrm{PL}}/M_*$, the projected separation between planet and star d, and the angle between the relative source track and the connecting line between star and planet. The binary-lens nature of the star-plus-planet system affects the observed lightcurves most strongly if the separation is in a certain range, the so-called lensing zone: $0.6 \leq d/R_E \leq 1.6$.

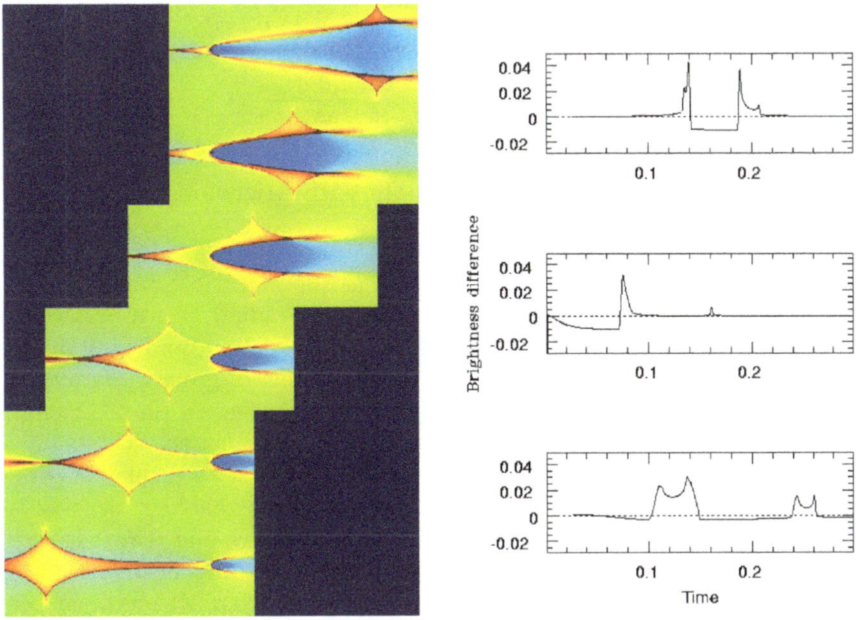

Fig. 22. *Left:* Caustics of a saturn-mass planet (mass ratio $q = 10^{-4}$) with a projected separation close to one Einstein radius: the six panels from top to bottom show parts of the magnification patterns for separations $d/R_E = 1.105, 1.051, 1.025, 0.975, 0.951, 0.905$; *right:* typical (difference) lightcurves obtained from the second, fourth and sixth panel; time scale is in units of the Einstein time t_E. The amplitude is given in magnitudes (after Wambsganss 1997)

In this case, the planet caustic(s) are within the Einstein radius of the host star. Due to a coincidence, this lensing zone corresponds to a projected distance range of order 1 AU. This means that the microlensing method is in principle capable of detecting planets at distances overlapping with the habitable zone[11].

In Fig. 22, six magnification patterns are shown for planet distances very close to the Einstein radius, so called "resonant lensing": $d/R_E = 1.11, \ldots, 0.91$ (Wambsganss 1997). The lightcurves on the right hand side (displayed is the "difference lightcurve" between the star-plus-planet lightcurve and the star-only lightcurve) show that the deviations are typically of small amplitude (few percent) and short duration (few percent of the Einstein time, i.e. order a day or shorter).

[11] 'Habitable zone' is defined as the distance range around a central star which would allow life to develop on a planet; because of lack of better criteria – and based on life as we know it – what is chosen in the simplest version is a temperature range between 0 and 100 degrees Celsius (centigrade), which allows water to be in the fluid phase.

Very nice animations of planet microlensing showing relative tracks, individual images and magnification as a function of time for various mass ratios q and separations d are provided by Scott Gaudi at:
http://cfa-www.harvard.edu/~sgaudi/movies.html.

4.2 Why Search for Extrasolar Planets with Microlensing? – Advantages and Disadvantages

Searching for extrasolar planets is a tough astrophysical enterprise. There are a number of different techniques being pursued: radial velocity variations or doppler wobble, transits, astrometric variations, pulsar timing, or direct detection. Each of those methods is used by a number of groups (more than 20 different teams, e.g., for transit searches alone, see review by Horne (2003)). So it is a fair question to ask: why bother applying yet another technique?

In this subsection, the microlensing method for planet searching is compared to the other indirect methods. It will be shown that microlensing is indeed a complementary method with different strengths, and that it is very worthwhile pursuing this search technique. As the starting point, here follows a list of commonly mentioned "disadvantages" of the microlensing planet searching technique (with a few comments added in parentheses):

1. The probability for an individual planet-lensing event is very small *(yes indeed, the chance for detecting a planet-microlensing event by monitoring an arbitrary background star in the galactic bulge is very roughly of order* 10^{-8} *or smaller).*
2. The duration of the planet-induced deviation in the microlensing lightcurve is very short *(yes, estimated typical durations for planet deviations are of order hours to days).*
3. The planets – once found – will be very distant *(true, most likely distance is a few kpc)*, and even worse: the exact distance determination will turn out to be very difficult or close to impossible *(true, unless we get additional information about the event).*
4. It is close to impossible to do subsequently more detailed investigations of the planet *(fair enough).*
5. The lightcurve shapes caused by extrasolar planets are diverse, occasionally there might be a parameter degeneracy when modeling the event, with no unique relation between lightcurve and planet parameters *(yes).*
6. Even when unambiguously detected, what can be determined is not the *mass* of the planet, but only the mass ratio between host star and planet *(true).*
7. No independent confirmation will be possible *after* the detection: it is a once-and-only event *(yes).*

These are fair points of critique toward using microlensing as a planet search technique. So, why bother anyway? Firstly I would like to emphasize and recall that almost all these arguments were put forward already more than a decade ago, then used against the "normal" stellar/dark matter microlensing which had been proposed by Paczyński (1986b) and produced the first results a few years later (Alcock et al. 1993; Aubourg et al. 1993; Udalski et al. 1993). Today no one has any doubts any more about the reality of the many stellar microlensing events, despite, say, their non-repeatability. Secondly, I now try to present one by one good reasons why the above arguments – though true to a large degree – are not really arguments *against* using the microlensing technique for planet searching:

1. *Small probability:* The probability for "normal" microlensing events in the galactic halo or disk (i.e., directions to the LMC/SMC or the galactic bulge) is already very small (of order $10^{-6}...10^{-7}$). Nevertheless, more than a dozen microlensing events have been found toward the LMC/SMC (Alcock et al. 2000a,b) and more than 1000 events (!) have been detected in the direction of the galactic bulge (see, e.g., on the OGLE web page `http://www.astrouw.edu.pl/~ogle/ogle3/ews/ews.html`). This shows: small probability in itself is certainly not a strong argument against using this technique. It is just a matter of statistics: even today it is possible to monitor of order 10^7 stars on a regular basis with sampling every few days on comparably small operational cost. Doubtlessly, this number will increase by an order of magnitude every few years.
2. *Short duration:* In the current "mode-of-operation", the planet-searching teams take advantage of the relatively coarse sampling in the time domain of the microlensing monitoring teams (in particular OGLE and MOA), they work "piggy-back": once a deviation indicative of a stellar microlensing event is detected by these monitoring teams, the planet-searching teams follow those alerted events with a very dense coverage in time. This can result in lightcurves with an average sampling of many data points per hour. A number of events with more than 1000 data points (An et al. 2002) with photometric accuracy of 1% or better have been observed. Due to a set-up of telescopes in Australia, South Africa and Chile, lightcurve coverage around the clock is possible, weather permitting (see 'The 24-Hour Night Shift', Sackett 2001). So even planetary deviations in the lightcurve lasting only a couple of hours can be covered very well with many data points.
3. *Large (and unknown) distance to the planet in general:* The distances to the microlensing planets will be larger by one or two orders of magnitude than those found with the conventional techniques. This is true, too, for the pulsar planets (Wolszczan 1994) and not a disadvantage in itself. The "not-well-determined" aspect can be treated in a statistical way for a sample of events. If there is additional information available (parallax, astrometric signatures), the distance can be determined for the individual events (cf. Alcock et al. 2001a; Gould 2001).
4. *More detailed investigation impossible:* Indeed, a more detailed study of the planet candidate will turn out to be very difficult. However, we may be able to get more information about the star which the planet is circling:

Alcock et al.,(2001a, b) show, that due to the relative proper motion, the projected positions of source star and lens star will move away from each other, so that we may be able to detect and measure the parent star and the relative proper motion of the star-plus-planet system, a few years after the event.

5. *Parameter degeneracy:* Lightcurves covering only the central caustic or only the outer caustic are likely to have two sets of solutions. However, there is a wide range of planetary lightcurves which will result in unique solutions/fits, if the data sampling and quality is good enough.
6. *Only mass ratios determinable:* Most stars in the disk of the Milky Way are low mass main sequence stars, M-dwarfs. Hence there is a relatively narrow range of absolute masses for most of the planets. Statistically, the planet mass distribution from microlensing can be determined to the same accuracy to which we know the mass function of the (host) stars. Furthermore, the most successful exoplanet search method to date – the radial velocity technique – also cannot determine the individual planet mass to better than a factor $1/\sin i$, due to the unknown inclination i of the orbital plane of the planetary system relative to the line-of-sight.
7. *Once-and-only event, no independent* confirmation: Most star-plus-planet microlensing events will not repeat, this is true. But whether the event is "believable" or not is just a question of signal-to-noise: once there are enough data points with small enough error bars, this is convincing. A lightcurve consisting of more than 1000 data points with accuracy of order of 1% or better (cf. PLANET team caustic crossing data of event EROS-BLG 2000-005, An et al. 2002) is beyond any reasonable doubt. In addition, lightcurves are often collected by two or more separate teams, which is a good independent confirmation. Furthermore, supernovae or gamma-ray bursts also do not repeat; no one takes this as an argument against them being real.

So all the arguments commonly used against microlensing as a useful planet search technique can be refuted or weakened. If the sampling and the photometric accuracy are good enough, planet microlensing deviations will be believed by the astronomical community. Occasionally there might still be model degeneracies. The most significant ones, though, just concern the projected separation between planet and host star: for each solution with separation d there is usually also one with separation $1/d$. We have to live with this, as well as we do with the unknown $\sin i$ of the radial velocity planets.

After having discussed in detail the potential or perceived disadvantages, let us now come to the positive aspects of planet searching with the microlensing technique, compared to the other methods:

- *No bias for nearby stars:* Almost all the conventional planet search techniques concentrate their efforts on nearby stars, mainly because the signals are stronger, the closer the host stars are. The solar neighborhood, however, might not be representative for the galactic planet population. Microlensing searches for planets are sensitive to stars anywhere along the line-of-sight to the source star in the galactic bulge at a distance of about 8.5 kpc, most sensitive for a lens position roughly half-way in between.

- *No bias for planets around solar-type stars/main sequence stars:* Almost all the conventional planet search techniques select and target the host stars. The very successful radial velocity technique cannot be applied to all stellar types, in particular not to active stars with broad and/or variable lines, so it has limited applications. Microlensing searches are "blind" for the characteristics of their host stars. Planet and host star will be found in proportion to their actual frequency in the Milky Way disk. The host stars of the microlensing planets will represent fair samples of the planet-carrying stars in the Milky Way. Planet microlensing is not constrained to any spectral type of host star, nor does it exclude any early type or active stars.
- *No strong bias for planets with large masses:* All conventional techniques are most sensitive to massive planets, with sensitivity strongly declining with decreasing planet mass. To first order, the microlensing signal – the amplitude of the lightcurve deviation – is independent of the planet mass. The duration and hence the probability for detection decreases, though, with decreasing planet mass. However, the size of the source star is important, and the lightcurve signal will be affected/smoothed by the finite source diameter, resulting in a lower amplitude signal (compared to a point source) and hence a lower detection probability.
- *Earth-bound method sensitive down to (almost) Earth-masses:* In principle, it is possible to detect even Earth-mass planets with ground based monitoring via microlensing. In practise, however, this would mean extremely high monitoring frequency and photometric accuracy. It is certainly true, though, that currently microlensing is able to reach down to lower planet masses than any other technique.
- *Most sensitive for planets in lensing zone, overlapping with habitable zone:* In the current mode-of-operation ("alerted" microlensing events being followed by dedicated planet-search groups), the most likely range of projected separations is the so-called lensing zone, roughly corresponding to a projected separation between 0.6 AU and 1.6 AU (Bennett and Rhie 1996). For low mass main sequence stars, this region overlaps with the habitable zone. This coincidence makes microlens-detected planets particularly interesting with regard to the question whether and how many planets exist in the habitable zone.
- *Multiple planet systems detectable:* There are two "channels", in which microlensing can even detect multiple planet systems: well sampled, very high magnification events have such small impact parameters that they pass the central caustic, which carries the signature of all the planets. Another channel would be the chance passage through two or more planet caustics, in case they happen to lie along the path of the background source star.

- *"Instantanous" detection of large semi-major axes:* The detection of long period planets is a long lasting process with the radial velocity or astrometry or transit techniques (years, decades?): ideally it takes at least one full period for confirmation, better two or three. Microlensing will find large-separation planets basically instantaneously. The measured (projected!) distance between planet and host-star is, though, only a lower limit to the real semi-major axis (statistically, the 3-dimensional distribution can be inferred under the assumption that there is no preferred direction of the planetary orbital planes in the Milky Way).
- *Detection of free-floating planets ("isolated bodies of planetary mass"):* The next generation of microlensing searches for planets most likely will not work in the two-step mode-of-operation described below, with one team sampling lightcurves coarsely and then follow-up teams sampling selected candidate frequently. Rather, they will do very massive photometry ground-based (cf. Sackett 1997), or potentially even continuously from space, as the satellite project "Microlensing Planet Finder" (MPF, formerly called GEST) promises to do (Bennett and Rhie 2002; Bennett et al. 2003). Once such an experiment is implemented, microlensing will also detect a potential population of free-floating planets, by the microlensing signature of single lenses with small mass, i.e. very short duration (Han and Kang 2003).
- *Ultimately best statistics of galactic population of planets:* Gravitational microlensing will ultimately provide the best statistics for planets in the Milky Way; it is not without biases, but the biases in the microlensing search technique are very different from those of all other methods and can easier be quantified.

So gravitational microlensing is a very powerful and promising method for the search for extrasolar planets. It is largely complementary to other planet search techniques and has relatively little sensitivity to the planet mass. It also has a number of not-so-favorable aspects, which, however, are more than balanced by the advantages listed above.

4.3 Who is Searching? The Teams: OGLE, MOA, PLANET, MicroFUN

The search for planets with the microlensing technique is currently done in a two-step process with shared tasks:

1. Stellar microlensing events have to be discovered while they are still in progress. This task is being done by two monitoring teams which measure the apparent brightness of a few million stars every few days:
    - MOA ("Microlensing Observations in Astronomy"; New Zealand/Japan, 60 cm telescope on Mt. John, NZ): covers about 20 square degrees few times per night; geared to high magnification events (Bond and Rattenbury et al. 2002): 10 events expected per season with $A_{\max} > 100$.

In total, 74 alerts in the whole 2003 bulge season. MOA alert page: `http://www.massey.ac.nz/`*sim*`iabond/alert/alert.html`.
From 2005 on, the MOA team will use a dedicated 1.8 m telescope for their microlensing searches, which will improve their efficiency dramatically.
- OGLE ("Optical Gravitational Lens Experiment", Poland/USA; 1.3 m telescope on Las Campanas, Chile): monitor 170 million stars regularly (Udalski 2003). In total 462 alerted events in the 2003 bulge season. OGLE alert page:
`http://www.astrouw.edu.pl/~ogle/ogle3/ews/ews.html`

These monitoring teams use the image subtraction technique (Alard and Lupton 1998) for accurate photometry and do basically online data reduction (Woźniaket al. 2001). Once they have discovered an ongoing microlensing event, these teams alert the community for follow-up observations, which involves the second step:

2. Two specialized teams concentrate only on follow-up monitoring of currently ongoing microlensing events:
   - PLANET ("Probing Lens Anomaly NETwork"; international team, various telescopes in Australia, South Africa and Chile): monitor selected on-going events around the clock. PLANET home page: `http://planet.iap.fr`
   - MicroFUN ("MICROlensing Follow-Up Network", US/SA/Israel/ Korea; 1.3 m telescope, Cerro Tololo): informal consortium of observers dedicated to photometric monitoring of interesting microlensing events in the Galactic Bulge. MicroFUN home page: `http://www.astronomy.ohio-state.edu/~microfun/`.

   Both follow-up teams monitor only *alerted events* with high frequency (ideally few times per hour) and high photometric accuracy. At any given time there are usually a few dozen interesting events being followed up.

4.4 What is the Status of Microlensing Planet Searches so far? The Results

At the time of the 33rd Saas Fee Advanced Course on Gravitational Lensing (April 2003), there were no definitive results on the detection of planets with the microlensing technique. A few candidates had been proposed, however, they remain controversial. Here a selected number of recent observational and theoretical results with respect to planet microlensing are presented:

PLANET Results

The PLANET team has put limits on Jupiters orbiting Galactic M-dwarfs (Gaudi et al. 2001, 2002): Analysis of 5 years of PLANET monitoring data toward the bulge with respect to short-duration events from single-lens light curves yielded a well defined sample of 43 intensely monitored events. The search for planet perturbations over a densely sampled region of parameter space (two decades in mass ratio and projected separation) resulted in no viable planetary lensing candidates. This analysis found that less than 25% of the primary lenses can have companions with mass ratio $q = 10^{-2}$ and separations in the "lensing zone": $0.6 \leq d/R_E \leq 1.6$. With a model for the mass, velocity, and spatial distribution of the stars/lenses in the bulge, astronomical limits could be obtained: less than 33% of the M dwarfs in the Galactic bulge can have companions with M_{Jupiter} between 1.5 AU and 4 AU; and, less than 45% of the M dwarfs in Galactic bulge can have companions with M_{Jupiter} between 1 AU and 7 AU.

Event OGLE-2002-BLG-055: Possibly Planetary?

The microlensing event OGLE-2002-BLG-055 was investigated by Jaroszyński and Paczyński (2002a,b). The lightcurve contains one data point which lies $\Delta m = 0.6$ mag above the "single-lens, single-source" fit (Fig. 23, left). There is only this one deviant point, but it is very reliable. The authors argue correctly, that there is no reason to ignore it. The simplest interpretation for the lightcurve is: a binary lens with parallax and mass ratio $q = 0.001 - 0.01$. The lower q-value would correspond to roughly a Jupiter-mass planet (depending on the exact mass of the primary). The authors caution, however: with a single deviant point, it is impossible to fit a unique model (cf. Fig. 23, right)!

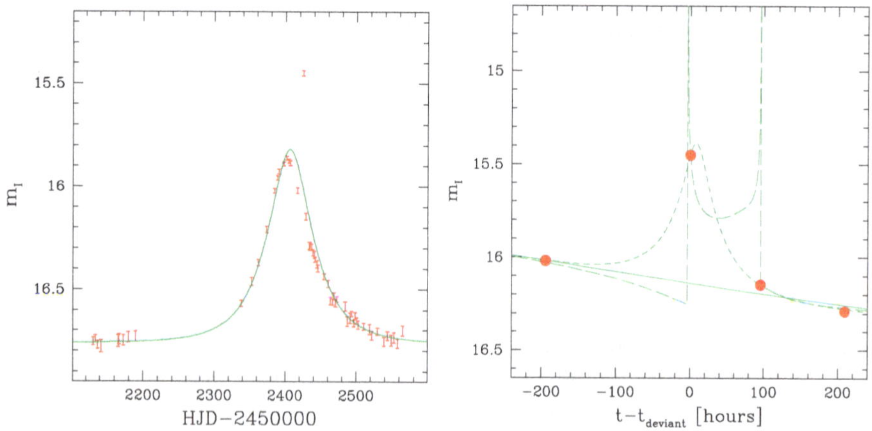

Fig. 23. *Left*: OGLE data of the microlensing event OGLE-2002-BLG-055 with the best fit single-lens-lightcurve including all data points. *Right*: zoom around the deviating data point with two well fitting binary lens models of mass ratios $q = 0.01$ (*short-dashed*) and 0.001 (*long-dashed*). The solid line indicates the best-fit single-lens lightcurve (from Jaroszyński and Paczyński 2002a,b))

Instead, they conclude: In order to make sure that for similar events in the future, more data points in the relevant epoch will be obtained, the OGLE observing strategy should be modified in the following sense:

1. instant verification of deviant points in a microlensing lightcurve,
2. in positive case (deviation confirmed): change observing strategy, follow this particular event by frequent time sampling to make unique model possible.

As a very fast consequence of this suggestion, Andrzej Udalski implemented a very fast check-and-verification system, the OGLE Early Early Warning System (EEWS). It uses the automatic data reduction and searches for deviations "on the fly", i.e. recognize and verify possible planetary disturbance in real time with instant follow up (Udalski 2003). Very impressively, OGLE is now able to verify or falsify a deviant data point within 5 minutes ! As shown by events OGLE-2003-BLG-170 and OGLE-2003-BLG-194 in June 2003 (cf. OGLE web page at `http://www.astrouw.edu.pl/~ogle`): it works ! This new developement is very promising indeed.

Limits on Number/Orbits of Exoplanets from 1998-2000 OGLE Data

Tsapras et al. (2003) analyzed the OGLE data base of the years 1998 to 2000. They put limits on the number and orbits of extrasolar planets. They focused on the frequency of "cool" Jupiters at a few astronomical units separation, based on 145 OGLE events. They used a maximum likelihood technique and found $n \leq 2$ realistic candidate events for a mass ratio of $q = 10^{-3}$. Their result: less than 21 × n% of all the lensing stars have Jupiter-mass planets within $1 < a/\mathrm{AU} < 4$. An additional result of their analysis is: it is more efficient to observe *many events less densely* in time, than intensively monitoring only a small number of events

New Theoretical Results

In a recent analysis, Gould, Gaudi, and Han (2003, 2004) looked into the question, how the different planet searching techniques fare in terms of sensitivity to Earth mass planets. Playing every method "to its strength", they found that only microlensing provides a realistic prospect with high signal-to-noise values for Earth-mass companions. In particular for orbital periods of order one year or larger, microlensing fares very well. Their conclusion is: microlensing has the best chances of all the methods studied for realistically detecting Earth-mass planets, with the above mentioned limitation that only the mass ratio is determined, which leaves the mass itself uncertain to within a factor of a few.

New Channels for Planet Detection

Already a few years ago, Di Stefano and Scalzo (1999a, b) had pointed out two other ways of finding planetary systems with microlensing: They showed that the microlensing signature of planets in wide orbits ($d > 1.5R_E$) could be seen as an isolated event of short duration. They figured that a distribution of events by stars with wide-orbit planets is necessarily accompanied by a distribution of shorter events. What is very important: very accurate photometry is necessary! Since the size of the star is comparable to the Einstein radius R_E of the planet, the amplitude Δm in the lightcurve will be low, the shape of the event is distorted and broader than the point source approximation (cf. Han and Kang 2003). In addition, for very wide planetary orbits, there could also be *repeating* single-lens events, in case the track of the source relative to the lens passed both within an Einstein radius of the stars *and* the planet. These events will be rare, but they must occur, and hence previous stellar microlensing events should be monitored with higher frequency in the following observing seasons!

Additional Future Ways for Planet Detection with Microlensing

Recently, one additional aspect of planet microlensing was discussed: Ashton and Lewis (2001) looked into the question whether planets accompanying the source stars can be detected. They found that during a caustic crossing, the (reflected) light of the planet can be *very* highly magnified due to the very small size of this secondary source, and hence potentially be detected. The deviation is proportional to $f \times \theta_P$, where f is the fraction of star light reflected by the planet, and θ_P is the angular radius of the planet in Einstein radii. They figured that even rings, satellites and atmospheric features on planets are detectable this way. Even in an optimistic scenario, though, it will take quite a number of years until such a measurement will be possible. But it is a very exciting possibility.

4.5 When will Planets be Detected with Microlensing? The Prospects

Considering that ...

- ... both OGLE and MOA have improved their alert efficiencies considerably, so that already now there are of order 1000 events per year measured,
- ... OGLE has implemented their early early warning system (EEWS),
- ... the PLANET team has improved their priority scheme for selecting between the events going on at the same time,
- ... PLANET and MicroFUN keep doing follow-up photometry with high sampling

- ... MOA follows very/extremely high magnification events,
- ... the new MOA 1.8 m dedicated telescope is under construction, (first light planned for 2004/05),
- ... microlensing IS sensitive down to Earth masses,

... the question is WHEN rather then WHETHER planets will be detected with the microlensing technique. My answer is: SOONER rather than LATER. I am very optimistic that within the next 2 to 3 years at the latest, the first convincing planet will have been detected with the microlensing technique.

4.6 Note Added in April 2004 (About One Year after the 33rd Saas Fee Advanced Course)

During the 33rd Sass Fee Advanced Course on Gravitational Lensing – which took place in April 2003 – the author (J.W.) had offered a bet (which was accepted by one of the student participants) that the first convincing detection of an extrasolar planet with the microlensing technique would take place within 12 months time.

Indeed, on a NASA press conference in April 2004, it was announced that MOA/OGLE/MICROfun had detected a microlensing event which can be explained only with a very low mass companion to the primary star: OGLE 2003-BLG-235 or MOA 2003-BLG-53. The result is published meanwhile as Bond et al. (2004), see also Fig. 24. In the original words of the authors:

> "A short-duration (∼7 days) low-amplitude deviation in the light curve due to a single-lens profile was observed in both the MOA and OGLE survey observations. We find that the observed features of the light curve can only be reproduced using a binary microlensing model with an extreme (planetary) mass ratio of 0.0039^{+11}_{-07} for the lensing system. If the lens system comprises a main-sequence primary, we infer that the secondary is a planet of about 1.5 Jupiter masses with an orbital radius of ∼3 AU."

The author considers this a very convincing planet microlensing event.

4.7 Summary

Microlensing as a planet search technique has stepped out of its infancy. It is a viable method which is complementary to other techniques. There is one unambiguous microlensing planet detection (Fig. 24; Bond et al. 2004) as of yet (April 2004): A star-plus-planet system with a mass ratio of $q = 0.004$. Furthermore, microlensing monitoring has put limits on the frequency of Jupiter-like planets at semi-major axes between 1 AU and 4 AU around M-dwarfs: PLANET results show that less than one third of M-dwarfs host them (Gaudi et al. 2001). Soon these limits will be pushed further down, maybe to the few percent level. Stellar/binary microlensing lightcurves with

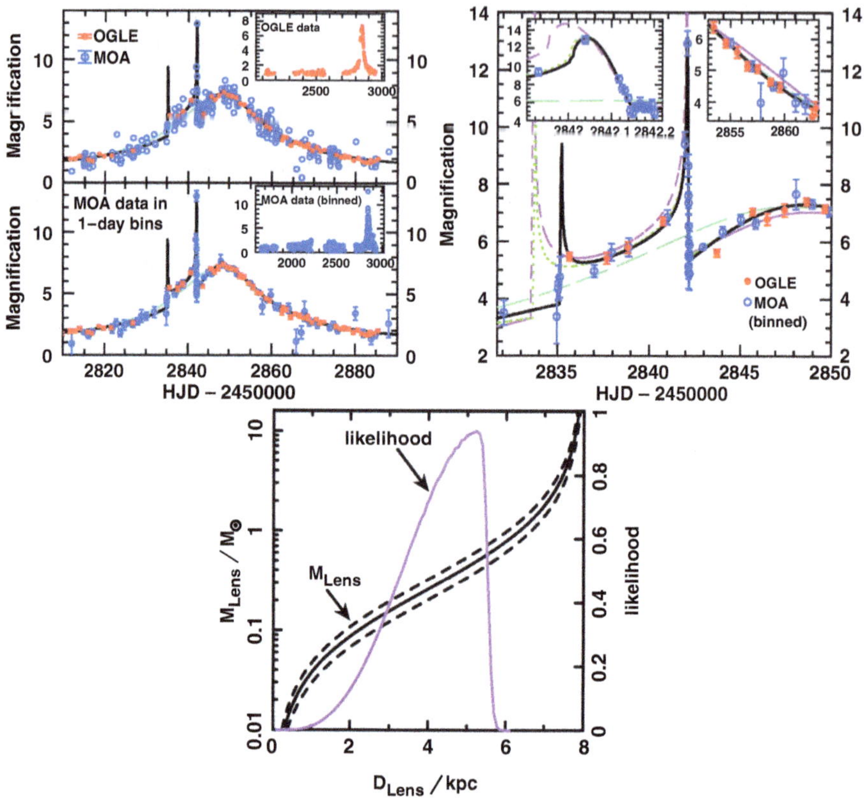

Fig. 24. *Top left*: Lightcurve of microlensing event OGLE 2003-BLG-235/MOA 2003-BLG-53: *open* (*filled*) symbols are MOA (OGLE) data points. Data points are shown individually in the top level, and binned in one-day intervals in the bottom panel. *Top right*: Data points and models covering about 18 days around the planetary deviation: *long-dashed line* – single lens case; *short-dashed line* – double lens with $q \geq 0.03$; *solid line* – best fit with $q = 0.004$. *Bottom*: Constraints and likelihood for the distance and mass of the lens: the *thick solid line* with the accompanying *dashed lines* as error limits shows the constraints on lens mass and distance from the measurement of the Einstein radius. The *thin line* (likelihood function) assumes the lens to be main sequence star (from Bond et al. 2004)

> 1000 data points have been obtained: binary/planetary signatures can be covered with very high signal-to-noise: non-repeatability is no problem. With improved detection software: OGLE/MOA produce routinely more than 1000 alerts per year of events caught "in action". Implementation of EEWS (OGLE) guarantees verification of any signification deviation "on-the-fly", within 5 minutes ! Microlensing remains the most promising method for the detection of Earth-mass companions, either ground- or space-based.

5 Higher Order Effects in Microlensing:

As originally worked out by Einstein (1936) and Paczyński (1986b), the magnification of a point source by a point lens is a very simple function of impact parameter or time (see also Sect. 1.2). The first observed microlensing lightcurves were well fit by this functional form. However, the point-lens–point-source ansatz with a linear relative motion between source, lens and observer is clearly a mathematical idealization.

In one sense, realistic situations are more complicated. In another sense, this helps us measure more/additional parameters and sometimes even break some of the degeneracies mentioned in previous sections. Some of the real world effects will be discussed here, for example:

- Blending – due to the dense star fields which are studied and the (very) large number of faint stars, often more than one star contributes to the light within the seeing disk. As a consequence, the measured microlensing lightcurve consists of two parts: a more-or-less constant background contribution (the "blending"), and the source star which is being microlensed.
- Parallax – For microlensing events with a duration of many months or longer (i.e. comparable to the orbital period of the Earth), the relative motion cannot be treated strictly as a straight line, but rather the changing observer position has to be considered. This leads to a modulation of the point source - point lens (PSPL)-lightcurve: it is not symmetric any more.
- Binary Lens – a binary lens clearly provides the most dramatic deviation from a PSPL-lightcurve: the caustic crossings are distincly different features; this situation was already treated in Sect. 2. For a binary lens in a short-period orbit, the caustic configuration may change both its shape and position in the course of the microlensing event.
- Finite Source Effects, Limb Darkening, Star Spots – for small impact parameter/high magnification microlensing events and for caustic crossing in binary events, the finite size of the source has to be considered. It can strongly affect the lightcurve. During caustic crossings, the one-dimensional surface brightness profile of a background source can be determined from a careful analysis/fitting of the well-sampled lightcurve. In principle, even star spots can be determined this way.
- Direct Detection of the Lens – most (micro-)lenses presumably are stellar objects; due to the relative motion between lens and source (which can be measured in units of angular Einstein radii per time), the lens will move away from the position of the source. By selection, the light contributing to the lightcurve is dominated by the background (giant) star. However, if the lens is a faint/low-mass star, at some angular distance from the source star, it may become visible. From the spectrum and the stellar type, the mass and maybe even the distance of the lens can be estimated, and hence the degeneracies can be broken.

- Binary Source – the source can also be a physical binary system. The measured lightcurve is then a superposition of two PSPL-curves, in general with different impact parameters, different colors and different times of closest approach.

Blending

The microlensing monitoring programs need to cover as many stars as possible on one CCD frame. Hence they select dense star fields, e.g. toward the Galactic bulge. The typical angular separation between stars in such fields is much smaller than the seeing disk. Hence it is unavoidable that flux of more than one star is contributing to the light measured for the light curve. The blending can be due to a physical companion of the source star, due to the lens itself, or due to a random superposition of a star along the line-of-sight (which is too far away in units of Einstein angles to affect the point-lens lightcurve, but still within the seeing disk).

The source stars for the lensing events are usually giants in the Galactic bulge, they dominate the light. However, the additional 'blending' light cannot be entirely neglected. Di Stefano and Esin (1995) investigated this question. They concluded that the optical depth for lensing of giants is greater than for the lensing of main-sequence stars, and that this effect can be quantified. The direct consequence of blending is that the measured lightcurve is not represented by the ideal point-lens-point-source model lightcurve. Di Stefano and Esin (1995) present methods to test whether the deviation from a PSPL-lightcurve can be attributed to blending. They also suggest that the effect of blending can be used to learn more about the lensing event than would be possible otherwise (e.g., it could be that without the blend contribution, this particular star may not have been above the brightness threshhold at baseline and hence not among the list of monitored stars). If blending is neglected, the lens mass distribution will be skewed toward lower masses than the actual underlying distribution of lenses.

Woźniak and Paczyński (1997) point out a strong degeneracy of the fitting procedure for single lensing events between blended and non-blended events. They conclude that it is practically impossible to identify blending by photometric means alone. Some blends might be detected astrometrically, but the majority has to be corrected for statistically.

Alard (1997) analyzed the situation in which the lensing event is not on the main star, but rather on an unresolved background star which represents only a 'blended' contribution to the light of the main (giant) star. He showed that such apparently short-duration events can be easily misinterpreted as brown-dwarf lensing events. Furthermore, Alard (1997) points out that there are ways to identify such events: usually, there is a color shift during the event. High resolution, dense, multi-band sampling helps identify such events and to estimate their contribution to the total lensing rates. He identifies OGLE-5 (Udalski et al. 1994a, b) as an obvious such event. Another method to identify and 'deblend' such events was suggested by Goldberg (1998): the shift of the center-of-light due to one of the stars in the seeing disk being magnified produces an astrometric signature which should be measurable in a fair fraction of such events.

Han et al. (1998) found that the contribution of the lens to the blending (suggested by Nemiroff 1997) has a small to moderate effect on the determination of the optical depth (decrease of 20% under the most extreme circumstances) and the Galactic mass distribution. Han and Kim (1999) derived analytical relations between the lensing parameters with and without the effect of blending and investigated the dependence of the derived lensing parameters on the amount of blended light and the impact parameter. In Han et al. (2000), it is shown that the difference image analysis method (Alard and Lupton 1998) is a very efficient way for the astrometric deblending of microlensing events, which was further developed by Gould and An (2002).

Parallax Effects

For microlensing events with a duration of many months or longer (i.e. comparable to the orbital period of the Earth) and a relative velocity between source, lens and observer comparable to (or smaller than) the orbital velocity of the Earth, the relative motion cannot be treated as a straight line any more. Rather, the changing observer's position in the course of the microlensing event influences the shape of the lightcurve: the PSPL-lightcurve is modified, it is not symmetric any more. Such events were predicted by Refsdal (1966) and Gould (1992), with the suggested applications to get more information and constraints on mass and transverse velocities of the lenses.

The first such event observed was reported by Alcock et al. (1995). It is the longest of their 45 microlensing events detected toward the Galactic bulge in their first year of observation (Alcock et al.1997a, b). In Fig. 25, the B-band and R-band lightcurves are shown, together with the best fit assuming only linear motion (dashed line), and the best fit including the motion of the Earth (solid line). Whereas the former clearly shows systematic deviations, the latter provides a very good fit. Since the event is achromatic, there is little doubt that this is a bona fide microlensing event, despite the deviation from the symmetric PSPL-shape.

Alcock et al. (1995) discuss the nature of the deviation from linear motion and emphasize that it is impossible to distinguish between the possibilities that motions of lens, source or observer lead to this modification. However, with the knowledge of the orbital parameters of the Earth, they tried to fit the lightcurve and argued that their reasonable fit with these assumptions is a strong argument in favor of Earth's motion really causing the deviations. This then allows them to compare the projected Einstein ring diameter crossing time with the size of the Earth orbit and hence obtaining a second constraint on the three unknown parameters of a typical microlens situation: lens mass M, lens distance D_d and relative transverse velocity v_t.

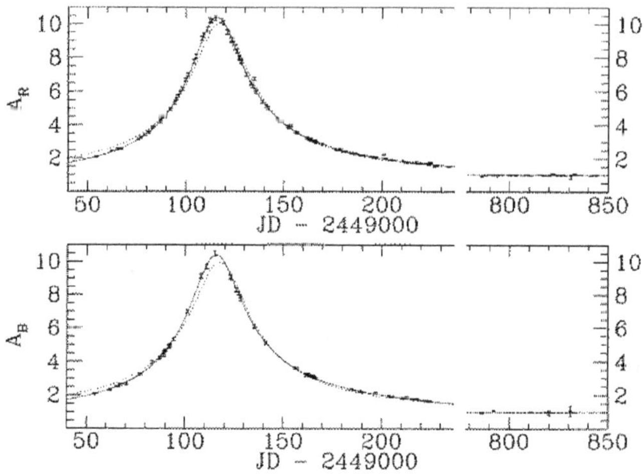

Fig. 25. Lightcurve of the first detected parallax lightcurve in R band (*top*) and B band (*bottom*); the (*linear*) magnification is shown as a function of time in days from JD 2,449,000. The *dashed curve* shows the best linear velocity point-lens–point-source fit, the *solid line* is the best fit allowing for the parallax effect, the motion of the Earth around the Sun, from Alcock et al. (1995)

In order to include the orbital motion of the Earth, the expression of the impact parameter as a function of time $u(t)$ gets more complicated than the standard form (Alcock et al. 1995):

$$\begin{aligned} u^2(t) = \quad & u_0^2 + \omega^2(t-t_0)^2 + \alpha^2 \sin^2[\Omega(t-t_c)] \\ & +2\alpha \sin[\Omega(t-t_c)][\omega(t-t_0)\sin\theta + u_0\cos\theta] \\ & +\alpha^2 \sin^2\beta \cos^2[\Omega(t-t_c)] \\ & +2\alpha \sin\beta \cos\theta[\Omega(t-t_c)][\omega(t-t_0)\cos\theta - u_0\sin\theta], \end{aligned} \tag{15}$$

where θ is the angle between the velocity vector v_t and the north ecliptic axis, the angular frequency $\omega = 2/\hat{t}$, and t_c is the time at which the Earth is closest to the line connecting Sun and source. The parameters α and Ω are defined as:

$$\alpha = \frac{\omega(1AU)}{\tilde{v}}\{1 - \epsilon\cos[\Omega_0(t-t_p)]\} \tag{16}$$

and

$$\Omega = \Omega_0 + \frac{2\epsilon\sin[\Omega_0(t-t_p)]}{(t-t_c)}. \tag{17}$$

Here t_p is the time of the perihelion, $\tilde{v} = v_t/(1-x)$ is the transverse speed of the lensing object projected to the solar position, $\Omega_0 = 2\pi/yr$ and $\epsilon = 0.017$ is the eccentricity of the Earth motion. This inclusion of the Earth's motion into their fitting procedure reduced the χ^2 per degree of freedom from roughly

10 to a value of order unity (for 206 degrees of freedom; more details see in Alcock et al. 1995).

With the additional parameter $v_t = \tilde{v}(1-x)$ replaced in the equation defining the Einstein crossing time, one obtains for the mass of the lens M(x) as a function of the lens distance:

$$M(x) = \frac{1-x}{x}\frac{\tilde{v}^2\hat{t}^2c^2}{16GL}, \tag{18}$$

which is displayed in Fig. 26. This curve shows that the lens could, e.g., be a low mass object (brown dwarf) in the Galactic bulge at large $D_{\rm lens}$, or an M dwarf in an intermediate distance range of 2 kpc to 6 kpc, or a solar type star (or even higher mass) nearby. However, from the limits on the apparent brightness of the lens (as a massive main sequence star only milli-arcseconds away from the background lensed star, it should contribute a significant amount of light within the seeing disk of the latter) one can constrain the mass at the upper end.

Alcock et al. (1995) tried to use even more constraints, namely on the velocities of lens and source and obtained two likelihood functions (dashed lines in Fig. 26) for the distance of the lens, based on some reasonable velocity limits. The most likely distance of the lens appears to be $D_L = 1.7^{+1.1}_{-0.7}$ kpc, corresponding to a mass range of $M = 1.3^{+1.3}_{-0.6}M_\odot$. With the assumption that the lens is a main sequence star, the constraints are slightly different: $D_{L,MS} = 2.8^{+1.1}_{-0.6}$ kpc and $M = 0.6^{+0.4}_{-0.2}M_\odot$.

For another case, Mao (1999) reported about an ongoing microlensing event toward the Carina spiral arm, discovered by the OGLE team (Udalski et al. 1998): OGLE-1999-CAR-1 (see Fig. 27). He showed that this long

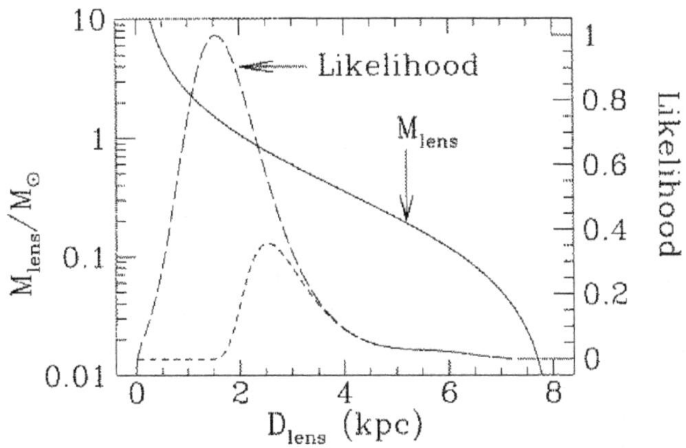

Fig. 26. Lens mass versus lens distance (*solid line, left scale*) and likelihood function for lens distance, using projected velocity and Galactic model (*long-dashed curve, right scale*) plus upper limit on brightness from a main-sequence lens (*short-dashed line, right scale*), from Alcock et al. (1995)

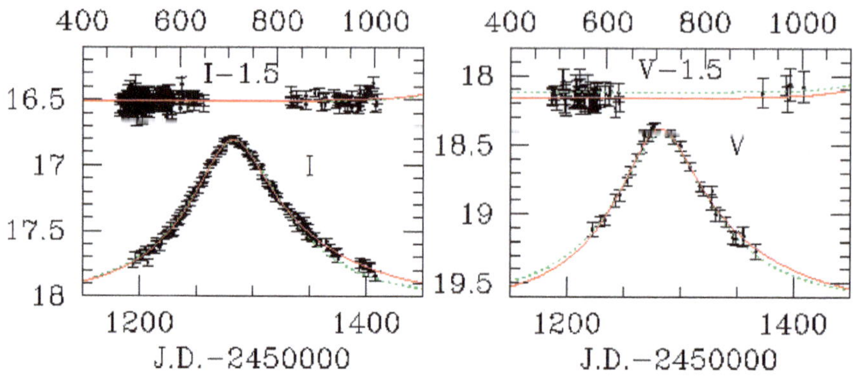

Fig. 27. Lightcurves of event OGLE-1999-CAR-1 in I band (*left*) and V-band (*right*): *dotted lines* are best fit PSPL-fits (linear motion), whereas the *solid lines* are best fits including both parallax and blending (from Mao (2001); more details there)

duration event exhibits strong parallax signatures, and determines the lens transverse velocity projected onto the Sun-source line to be about 145 km/s.

In a systematic search for parallax signatures among 512 OGLE-II microlensing from 1997-1999, Smith, Mao and Wozniak 2002a fitted both standard linear motion models and parallax models which included the motion of the earth around the Sun. Using additionally information on the duration of the events, they identified one convincing new candidate, sc33_4505, which is caused by a slow-moving and likely low-mass object, similar to other known parallax events (see Fig. 28). Smith et al. (2002a) emphasize that irregular sampling and gaps between observing seasons hamper the recovery of parallax events.

The first multi-peak parallax event (predicted by Gould 1994) was published by Smith et al. (2002b): the highly unusual microlensing lightcurve of OGLE-1999-BUL-19 (Fig. 29) exhibits multiple peaks which are *not* caustic crossings. The Einstein radius crossing time for this event is approximately 1 yr, which is unusually long. Smith et al. (2002b) show that a simple explanation for these additional peaks in the light curve is the parallax motion of the Earth. The fact that this effective transverse velocity between lens and source is significantly lower than the speed of the orbit of the Earth around the Sun ($v_{\rm Earth} \approx 30$ km/s) results in a periodic modulation of the impact parameter, superimposed on the linear motion: the motion of the Earth induces these multiple peaks. Smith et al. (2002b) also discuss binary-source signature but conclude that this is a less likely explanation.

In Fig. 29, the lightcurve of the multi-peak event OGLE-1999-BUL-19 is shown (bottom panel). The top panel shows the modulated apparent motion of the lens, projected in the observer plane relative to the observer–source line-of-sight, i.e. the location of the lens with respect to the Earth (denoted by the small cross).

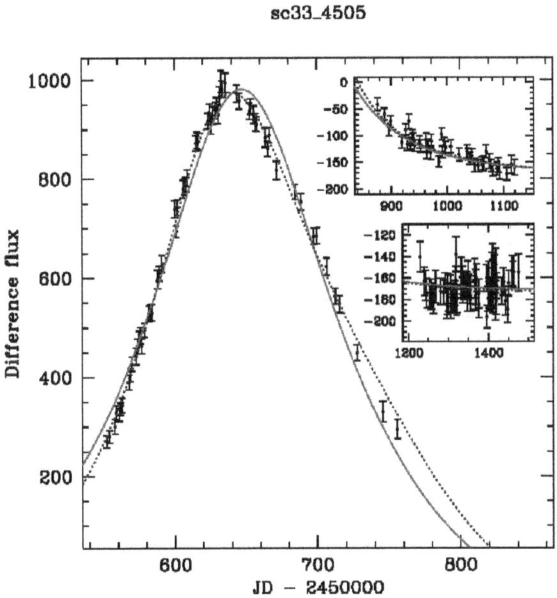

Fig. 28. Lightcurve of event OGLE-II event sc33_4505 toward the Carina spiral arm. The *solid line* is the best linear motion PSPL-fit, whereas the *dotted line* is the best parallax fit (from Smith et al. (2002a))

Short-Period Binaries

In Sect. 2, lensing by binary stars is considered. There, only the static situation is discussed: the lens configuration is assumed constant during the microlensing event, i.e. the binary period is much larger than the crossing time. However, this assumption will not always be true, there are small-separation binaries with periods of order years, months, or days.

Dominik (1998) investigated this case and discussed three scenarios: the rotating binary lens, rotating binary source, and observer on Earth orbiting the Sun (parallax, see above discussion). The most dramatic effects are expected in the case of a rotating binary lens, because the caustic structure changes with time. In the other two scenarios, the caustic configuration is static, the effect is only a modulation of the straight relative motion (parallax) and/or the superposition of two "static" lightcurves, which might cause some color changes as well (cf. Griest and Hu 1992). In this sub-section, only the rotating lens will be discussed further.

Dominik (1998) shows that the scenario of a rotating binary introduces five additional parameters, compared to a static binary: two rotation angles, the rotation period, the eccentricity and the phase. In Fig. 30, the effect of binary motion is illustrated on a lightcurve comparable to that of the event MACHO-LMC-1: the binary period varies between 365 days and 25 days. It is

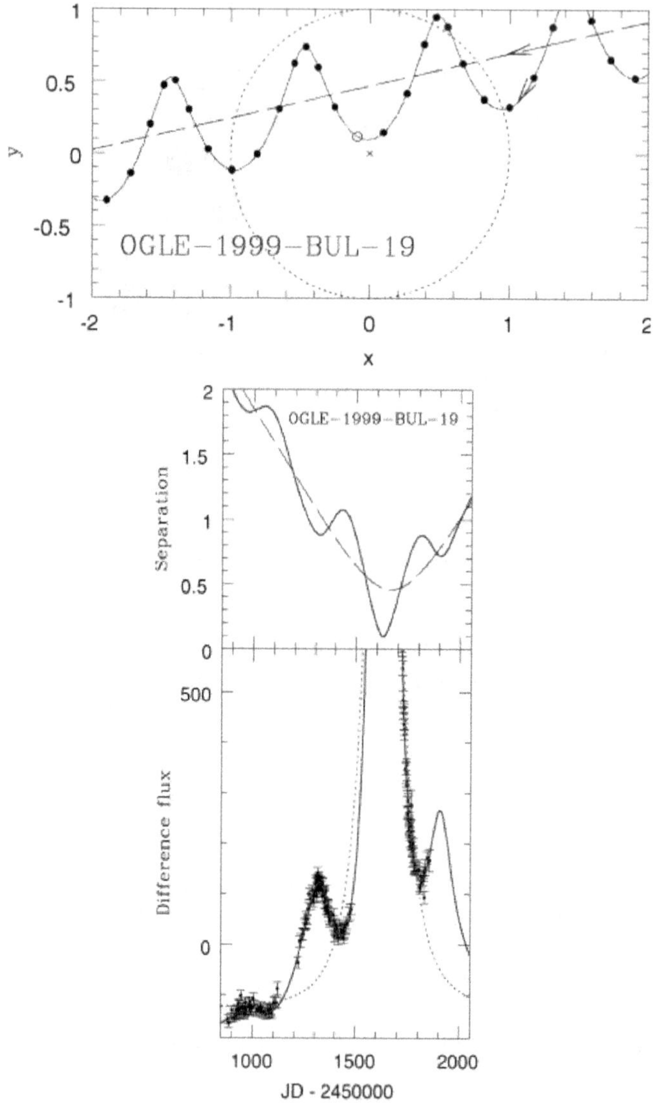

Fig. 29. Multi-peak lightcurve OGLE-1999-BUL-19: lens position relative to the observer as a function of time (*top panel*) and lightcurve plus corresponding impact parameter as a function of time (*bottom panel*) with best fit linear motion PSPL-curve (*dashed*) and best fit parallax model (*solid*), from Smith et al. (2002b)

obvious that the effect of binary rotation is most pronounced for short binary periods, compared to the event duration.

In 2000, the first detection of a rotating binary lens was published (Fig. 31): the lightcurve of MACHO 97-BLG-41 was the first event with a source crossing two physically distinct caustics (Albrow et al. 2000). Analyzing PLANET

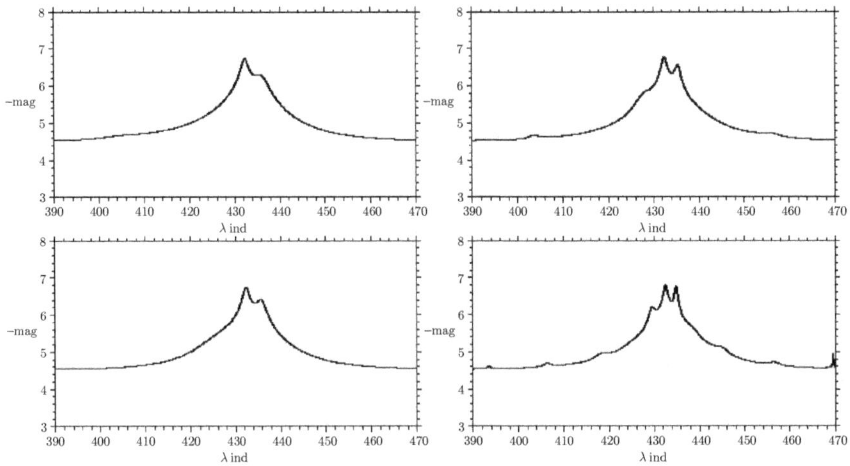

Fig. 30. Simulated lightcurves for a rotating binary lens, based on MACHO-LMC-1: binary rotation periods decrease from 365 days (*top left*) to 100 days (*top right*), 50 days (*bottom left*) and 25 days (*bottom right*), from Dominik (1998)

data for MACHO 97-BLG-41 (46 V-band and 325 I-band observations from five southern observatories), Albrow et al. (2000) showed that this data set is incompatible with a static binary lens. They do find a good model with a rotating binary lens of mass ratio q = 0.34 and angular separation $d = 0.5R_E$. The binary separation changes significantly in size during the 35.17 days between the separate caustic transits. Albrow et al. (2000) use this event to derive the first kinematic estimate of the mass, distance, and period of a binary microlens. The relative probability distributions for these parameters peak at a total lens mass of $M \approx 0.3M_\odot$, which would imply an M-dwarf binary system. The most likely lens distance is $D_L \approx 5.5$ kpc, and the binary period is $P \approx 1.5$ yr.

What made this model particularly convincing is the following: MACHO/GMAN data covering several sharp features in the light curve which are not probed by the PLANET observations and which did not enter the modeling, fall almost perfectly on the best fit lightcurve. This event MACHO 97-BLG-41 (see Fig. 31) had previously been modeled by a static binary lens plus a planetary companion. This much simpler and more robust rotating binary model makes a fit using a third lensing body less plausible.

Finite Source, Limb Darkening, Star Spots

When the source size is small compared to the Einstein radius of the lens *and* to the impact parameter, then the point source approximation is justified in a single lens scenario. However, if one of these conditions is violated, then the microlensing lightcurve is affected by the finite source.

Nemiroff and Wickramasinghe (1994) were the first to investigate this. They show that the central peak of the lightcurve is modified if the impact

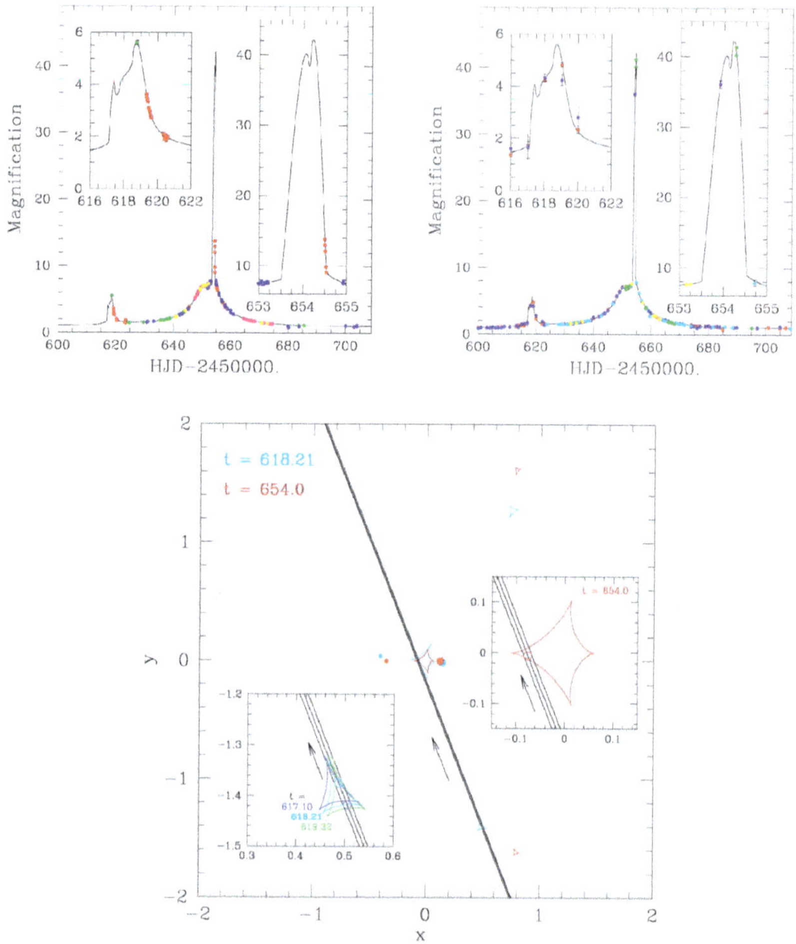

Fig. 31. Rotating binary event MACHO 97-BLG-41: (**a**) PLANET data (points) and best-fit rotating binary model (*solid line*); inset enlarges the two caustic crossing regions (*top left*); (**b**) same best-fit line as in (**a**), but here the data points of the MACHO/GMAN collaboration are added which did not enter the modeling procedure (*top right*); (**c**) caustic topology of best-fit rotating binary model, shown at time close to first and second caustic crossing. Straight line shows source trajectory. The positions of the two binary lens components are shown as large/small dots. Insets show regions close to the caustics, with the two additional lines indicating the finite source size (*bottom*) (from Albrow et al. 2000)

parameter u_0 is smaller than the source radius: Figure 32 (top) shows a circular source with uniform surface brightness crossing the point caustic behind a point lens centrally (top arrow and top lightcurve) and just barely within the source diameter (lower arrow and lightcurve).

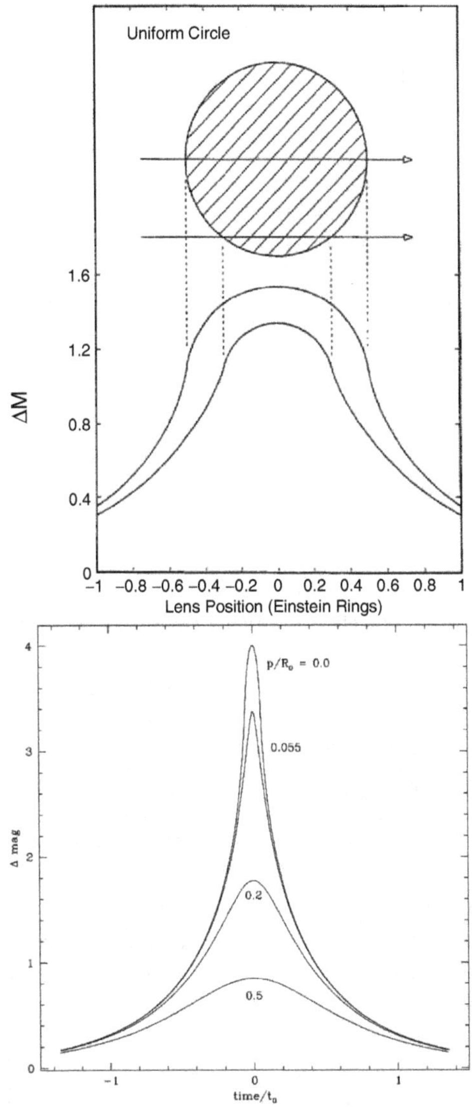

Fig. 32. Finite source effect: (**a**) central part for lightcurves with zero or very small impact parameter (*left*, from Nemiroff and Wickramasinghe 1994); (**b**) lightcurves for a limb darkened source with radius $r = 0.055R_E$, corresponding to star with $R = 10R_\odot$ at $D_S = 9$kpc and a $0.1M_\odot$ lens at $D_L = 8$kpc; four different impact parameters: $u_0 = 0.00, 0.055, 0.2$ and 0.5 (*right*, from Peng 1997)

The central peak of the microlensing lightcurve is clearly lower and broader than for a point source. Nemiroff and Wickramasinghe (1994) pointed out that an exact determination of the deviation from the PSPL-lightcurve can be used to determine the time it took the stellar disk to cross the central point caustic. With an independent determination of the source radius from knowledge of the stellar type and use of stellar evolution theory, this could be used to determine the transverse velocity. Witt (1995) estimated that at least 3% of all microlensing events in the Galactic bulge will be affected by finite source effects. He defined "being affected" by angular impact parameter being smaller than the angular source radius.

Peng (1997) looked into this question in more detail and presented the effects for a limb-darkened finite source with radius $r = 0.055 R_E$, corresponding to a star with $R = 10 R_\odot$ at $D_S = 9$kpc and a $0.1 M_\odot$-lens at $D_L = 8$kpc. Figure 32 (bottom) shows lightcurves for such a source with four different impact parameters: $u_0 = 0.00, 0.055, 0.2$ and 0.5. He finds that the source size can be fitted with reasonable accuracy only if the impact parameter u_0 of the event is smaller than the stellar radius.

The first observed finite source effect was reported by Alcock et al. (1997a,b) Their lightcurve of MACHO Alert 95-30 shows significant deviations from the point source lightcurve near the peak (cf. Fig. 33). They could determine the ratio between impact parameter and stellar radius to $u_0/R_* = 0.715 \pm 0.003$. With additional spectroscopic and photometric information they could identify the source as an M4 III star with a radius of $R = (61 \pm 12) R_\odot$ located at the far side of the Galactic bulge at about $D_S \approx 9$ kpc. The lens angular velocity could be determined relative to the source, to (21.5 ± 2.9) km/sec/kpc. With a likelihood analysis, the lens mass was determined to $m_L = 0.67^{+2.53}_{-0.46} M_\odot$.

Yoo et al. (2004) analyzed the short-duration event OGLE-2003-BLG-262, $t_E = (12.5 \pm 0.1)$ day. The lens is identified as a K giant in the Galactic bulge. The finite-source effects are used to measure the angular Einstein radius to be $\theta_E = (195 \pm 17)\mu$as. The lens mass could be constrained to the FWHM interval $0.08 < M/M_\odot < 0.54$, and the lens-source relative proper motion to $v_{\rm rel} = (27 \pm 2)$ km/s/kpc.

As described in the context of binary lenses, the lightcurve of a caustic crossing event offers the opportunity to determine the size of the source star, and even the surface brightness profile, i.e. to measure limb darkening of a star that is many kpc away ! This was successfully applied for the first time by the PLANET team on the event MACHO 97-BLG-28 (Albrow et al. 1999). The source star could be spectroscopically identified as a K giant. The observed lightcurve (Fig. 34) was modeled as being due to a cusp crossing of a binary lens caustic (cf. Fig. 35).

Modeling of the lens system resulted in a binary with mass ratio $q = 0.23$, and an instantaneous projected separation of d = 0.69 (for a lens in the Galactic bulge this corresponds to roughly 1 to 2 AU). The very good coverage of the lightcurve (696 data points in V and I from PLANET observatories in Chile, South Africa and Australia) made it possible to determine the radial surface brightness profile of the source star in the Galactic bulge. In particular the sharp central peak could be monitored with a time resolution of 3 to 30 minutes. The analysis resulted in a determination of the

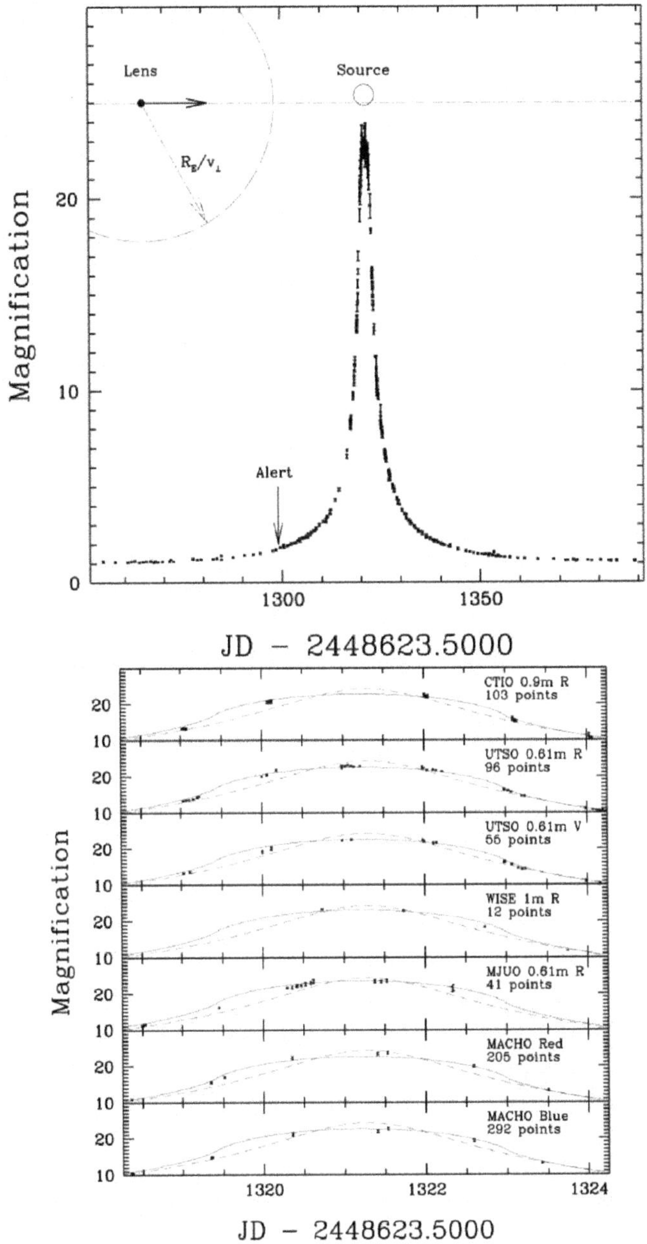

Fig. 33. Finite source effect: Full lightcurve of MACHO Alert 95-30 (*top*) and data close to the central peak (*bottom*) with point source and extended source fit; the arrow indicates when the alert was sent out (from Alcock et al. 1997a,b)

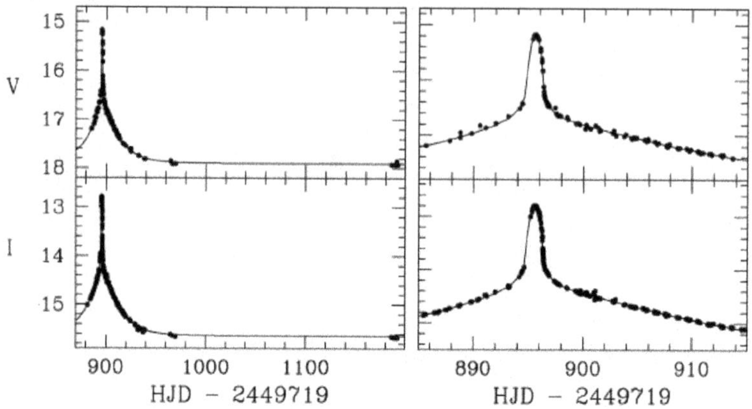

Fig. 34. PLANET lightcurve of event MACHO 97-BLG-28, covering a 300 day period (*left*) and a zoom of 30 days around the maximum (*right*). The V band lightcurve (*top*) consists of 155 data points, the I band lightcurve (*bottom*) comprises 431 data points. (from Albrow et al. 1999)

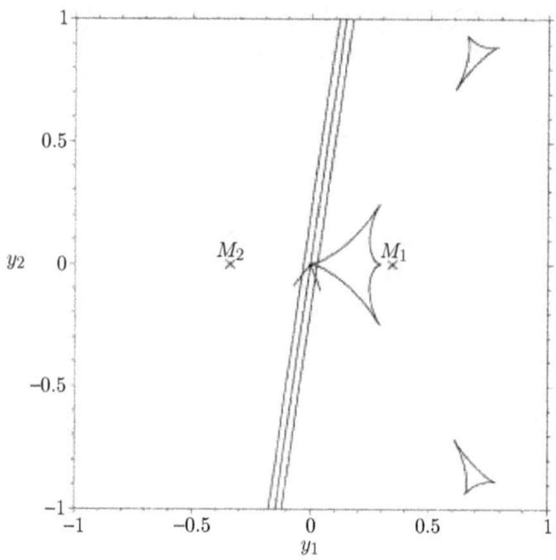

Fig. 35. Binary lens configuration (masses M_1 and M_2), caustic configuration and source track (width corresponds to diameter) of event MACHO 97-BLG-28 (from Albrow et al. 1999)

square-root limb darkening coefficient: for the assumed two parameter limb darkening law

$$I_\lambda(\theta) = I_\lambda(0)\ [1 - c_\lambda(1 - \cos\theta) - d_\lambda(1 - \sqrt{\cos\theta})], \tag{19}$$

where θ is the angle between the normal to the stellar surface and the line-of-sight, and I_λ is the intensity for wavelength λ, the parameters were determined for the two filters to $c_I = 0.40$, $d_I = 0.37$, and $c_V = 0.55$, $d_V = 0.44$. These values are in excellent agreement with the predictions for K giants from numerical modeling of stellar atmospheres. A source profile with a uniform surface brightness could be strongly ruled out.

Another example is event OGLE-1999-BUL-23 (Albrow et al. 2001), a binary lens system as well, for which modeling resulted in a mass ratio $q = 0.39$ and an instantaneous projected separation of $d = 2.42$. The source star is assumed to be a G/K subgiant in the Galactic bulge with an effective temperature of $T_{\rm eff} \approx 4800$K. The resulting limb darkening coefficients (a different limb darkening law was applied here) are consistent as well with theoretical predictions.

The possibility to detect star spots with microlensing has been explored by Hendry et al. (2002). It turns out that this will indeed be possible, though not in the immediate future. With sufficiently well-sampled lightcurves and high photometric accuracy, stellar microlensing will at least be able to put interesting constraints on the presence or absence of photospheric star spots.

Direct Lens Detection

In December 2001, for the first time the direct image of a lensing object in a microlensing event was published. Alcock et al. (2001a, b) reported the photographic image of a second object very close to the source star of the microlensing event LMC-5. The microlensing event had its maximum on February 5, 1993. Due to the relative motion of lensing object and source star, the angular separation was expected to increase with time. If the lens happens to be an ordinary main sequence star (rather than a "dark object"), there is a chance that after some time it will become visible next to the source star. The exact time cannot be predicted, because neither the transverse velocity nor the distance is known.

The MACHO team had successfully proposed for HST time to take very high resolution images of their previous microlensing events in the direction toward the LMC. On May 13, 1999 an image was taken of this particular source star, and 6.3 years after the peak in the microlensing lightcurve, this HST picture (see Fig. 36) revealed a "faint, red object displaced by 0.124 arcsec from the centre of an LMC main-sequence star that, on the basis of previous analysis, is the source star of this event" (Alcock et al. 2001a, b)[12].

The event LMC-5 had been a very high magnification and hence a very small impact parameter event. So it is safe to assume that at the time of the peak in the microlensing lightcurve, lens and source "coincided". So the relative proper motion of the lens can be easily determined to $\mu_{\rm rel} = (0.0214 \pm 0.0007)$ arcsec/yr.

[12] The authors emphasize that one normal disk star among the lenses of the 13 to 18 microlensing events toward the LMC is consistent with the number of expected "foreground" events (compared to the majority of the events which may be produced by MACHOs and which should correspond to no visible lens star in the future).

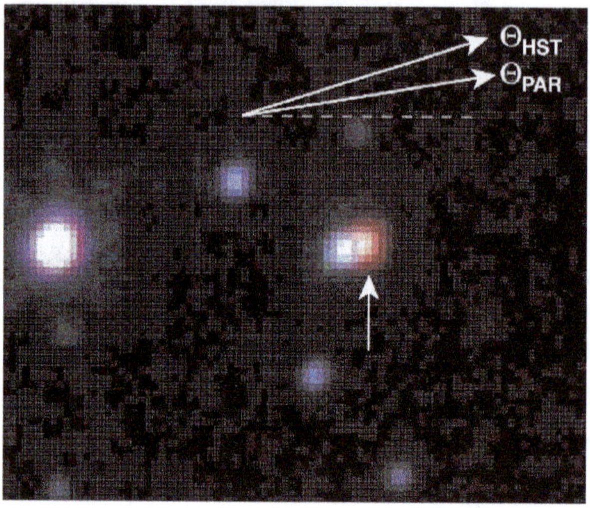

Fig. 36. Image of the region near the source star of microlensing event LMC-5 (peaked at February 5, 1993), taken with the HST WFPC2 camera on May 13, 1999. It reveals a faint red object (marked with an arrow) 0.124 arcsec to the top right from the center of the blue source star. The two arrows at the top ($\Theta_{\rm HST}, \Theta_{\rm PAR}$) indicate the direction of the lens motion determined by two methods (from Alcock et al. 2001a, b)

Alcock et al. (2001a, b) show that the HST and a parallax fit to the lightcurve data (which yields the transverse velocity projected on the observer's plane $\tilde{v} = -18.42^{+1.83}_{-1.91}$AU/yr) can be combined for a complete solution for this lens system. The lens mass is expressed by the observed parameters:

$$m_L = \frac{c^2}{16G}\tilde{v}\tilde{t}^2\mu_{\rm rel}. \tag{20}$$

This yields a value of $M_L = 0.039 M_\odot$, with a 3σ upper limit of $M_L \leq 0.097 M_\odot$. This is at or below the low end of the stellar mass scale.

In addition, the relation between the relative proper motion ($\mu_{\rm rel}$ in arcsec/yr) with the relative parallax ($\pi_{\rm rel}$, in arcsec) and the transverse velocity ($\tilde{v}$, in AU/yr) provides an estimate for the lens parallax:

$$\tilde{v}/\mu_{\rm rel} = 1/\pi_{\rm rel} \approx 1/\pi_{\rm L}. \tag{21}$$

The last approximate relation is valid because the lens as a disk star is much closer than the source star in the LMC. The lens parallax gives the distance to the lens: $d_L = \pi_L^{-1} \approx 200^{+40}_{-30}$pc. This leads to an absolute magnitude of $M_V = 16.2^{+0.6}_{-0.5}$. Finally a spectrum taken with the ESO VLT telescope revealed the star to be an M4-5 dwarf in the mass range $M_{\rm M4-5V} \approx (0.095–0.13) M_\odot$, inconsistent at the 3σ level with the parallax determined lens mass ! A photometric distance based on an empirical relation between color and absolute magnitude for M dwarfs leads to a distance determination of $D_{\mathrm{M4-5V}} \approx (650 \pm 190)$pc, inconsistent at the 2σ level with the parallax distance. It was obvious immediately that this apparent conflict should be resolvable with future HST ACS imaging.

The solution of this puzzle came about soon: Drake, Cook and Keller (2004) had used the new HST's Advanced Camera for Surveys (ACS) in order to get very high resolution images of lens and source stars of the MACHO-LMC-5 event. They measure the parallax and determine the distance of the lens star to $D_{\mathrm{L}} = 578^{+65}_{-53}$pc, and the proper motion to $\mu = (21.39 \pm 0.04)$mas/yr (Fig. 37). They conclude that the lens is an M dwarf which is more likely to be part of the thick disk than the thin disk population. In particular, they confirm Gould (2004) suggestion that the event MACHO-LMC-5 is a "jerk-parallax" event. Gould had found a second solution to the microlens parallax which was different from the one presented in Alcock et al. (2001a b).

Furthermore, Nguyen et al. (2004) obtained infrared images of the MACHO-LMC-5 region with the newly launched Spitzer Telescope. Their photometry established an infrared excess, hence confirming that the lens is a M5 dwarf star with a mass of about 0.2 $M_{\odot}$.

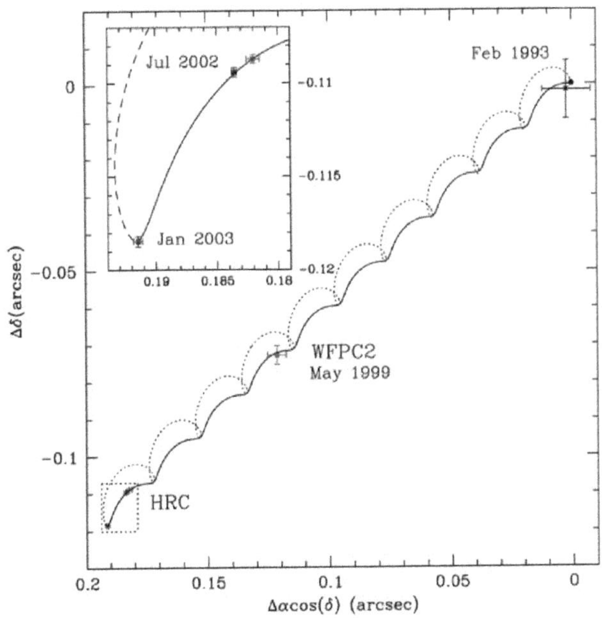

Fig. 37. Motion of the lensing star relative to the lensed star at the time of closest approach of event LMC-5. The inset is a zoom and shows the three measurements of the HST HRC camera in 2002. The source star is at location (0.00,0.00) (from Drake et al. 2004)

6 Astrometric Microlensing

In the point-lens–point-source scenario, there are always two images of a background star. Only when the impact parameter is of order a few Einstein radii, the secondary image very close to the point lens is magnified enough to become important (cf. (3)). Usually, only the combined magnification of the two images is measured, because the separation of the images is of order milli-arcseconds, unresolvable in most circumstances. Looking into the exact geometrical arrangement of image positions relative to the lens, one sees that the two images, the lens and the source always form a straight line in a point-lens–point-source scenario, as is illustrated in Fig. 38 (top panel). The line rotates by almost 180 degrees in the course of the lensing event. The corresponding center-of-light lies on this straight line as well (Fig. 38, bottom panel).

Paczyński (1998) has shown that the light centroid is displaced relative to the source position by a maximum of

$$\delta\phi_{\max} = 8^{-0.5}\phi_E \approx 0.354\phi_E \quad \text{for} \quad u_{\min} = \sqrt{2}. \tag{22}$$

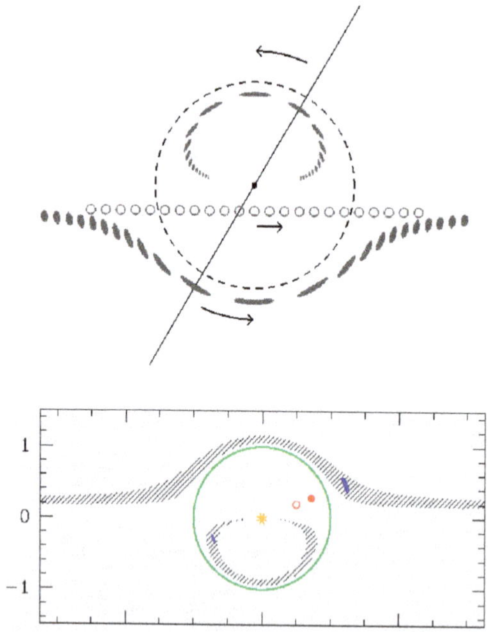

Fig. 38. *Top*: Alignment of the two images with the lens in a point-lens–point-source scenario: the source moves in a straight line in the background from left to right (*open circles*), the corresponding images (one inside, one outside the *dashed* Einstein circle), the source position and the lens lie on a straight line which apparently rotates around the lens position, here anticlockwise because the source position is *above* the lens (image courtesy Penny Sackett). *Bottom:* Similar situation, here the center-of-light is indicated (*filled circle*) for one particular arrangement (image courtesy Scott Gaudi)

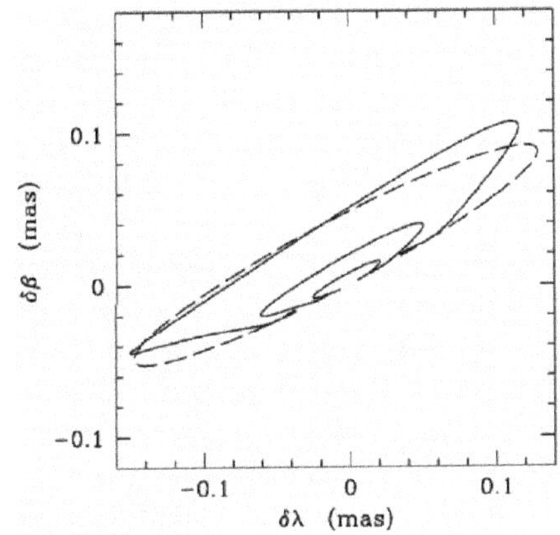

Fig. 39. Astrometric displacement (in ecliptic coordinates) caused by three microlensing events described in the text: The source is assumed to be a star in the LMC at a distance of 50 kpc, the lens has a mass of $M = 0.3M_\odot$, and the three *solid lines* correspond to three lens distances of $D_L = 10, 30$ and 45 kpc (the largest lens distance produces the innermost curve). The impact parameter is $u_{min} = 0.2$, corresponding to a maximum magnification of $\mu_{max} = 5.07$. The largest displacement of the curves is $8^{1/2}$ times smaller than the corresponding Einstein radius. The microlensing time scale is $t_0 = 50$ days in all three cases. The *dashed curve* corresponds to the $D_L = 10$ kpc case, with the effect of Earth's orbital motion suppressed (from Paczyński 1998)

In Fig. 39, three tracks are shown for $u_{\rm min} = 0.2$, $\mu_{\rm max} = 5.07$, $D_S = 50$ kpc, $D_L = 10$, 30 and 45 kpc, and $m = 0.3M_\odot$.

The fascinating aspect of this center-of-light variation is: if it is measured, the mass of the lens can be determined, the degeneracies can be broken. Combining the definition of the angular Einstein radius

$$\phi_E \approx 0.902\text{mas}\left(\frac{M}{M_\odot}\right)^{1/2}\left[10\text{kpc}\left(\frac{1}{D_d} - \frac{1}{D_s}\right)\right]^{1/2} \tag{23}$$

with the relative parallactic motion with the angular amplitude (in radians)

$$\pi_{ds} = 1\text{AU}\left(\frac{1}{D_d} - \frac{1}{D_s}\right), \tag{24}$$

(assuming linear relative motion between source and lens), one obtains

$$M = 0.123M_\odot\frac{\phi_E^2}{\pi_{ds}}, \tag{25}$$

where π_{ds}, is the measurable parallax of the lens-source !

Currently, the astrometric resolution of the best telescopes is of order 0.1 arcsec. Considering that light centroids can be determined to an accuracy of about 10% of that value, this still leaves us with a 10 milli-arcsec positional accuracy, at least two orders of magnitudes larger than the expected tens or hundreds of micro-arcsec from the astrometric microlensing.

However, this is not a lost case: there are instruments at the horizon which will make such a measurement possible: VLTI (Very Large Telescope Interferometry) and SIM (Space Interferometry Mission). The VLTI uses the combination of the four 8.2m-ESO telescopes in their interferometric mode, making use of the maximum 200m baseline between the unit telescopes. SIM, on the other hand, is a satellite project which will do astrometry with unprecedented accuracy: the specifications state that SIM will be able to do wide-angle astrometry with a nominal accuracy of 4 micro-arcsec, in the narrow-angle mode the accuracy should be as high as 1 micro-arcsec for a 20 mag star ! In addition, parallaxes will be measured with an accuracy of about 5 micro-arcsec, and proper motions down to 2 micro-arcsec/yr. Launch is planned for 2010, according to the web site `http://sim.jpl.nasa.gov`, where also much more information can be found.

Paczyński (1998) had derived the above relations and values for the SIM mission: SIM can measure the astrometric displacements of the light centroid of microlensing events which are discovered/detected photometrically from the ground. The amplitude of the center-of-light variation can reach a few tenths of the Einstein radius. Such a measurement will make it possible to determine the mass, the distance and the proper motion of almost any star or MACHO, capable of inducing a microlensing event toward the Galactic bulge or the LMC/SMC.

In addition, Paczyński (1998) pointed out another mode of operation: he suggested to select *lenses* rather than *sources*, in order to get interesting results, namely very high proper-motion stars. These stars are relatively nearby, therefore their angular Einstein radii are relatively large, which means they have a (very) large cross section for astrometric microlensing. As an example, Barnard's star with a parallax of $\pi_{\rm Barnard} = 0.522''$, a proper motion of 10.31"/yr, and an assumed mass of 0.2 $M_\odot$ has an angular Einstein radius of

$$\phi_{E,Barnard} = 30mas \frac{M_{\rm Barnard}}{0.2 M_\odot}.$$

Since the astrometric lensing effect only falls off with 1/r, a background star at an angular distance as large as 9 arcsec would still be displaced by 100 microarcseconds, a huge value for SIM. Another nice aspect of this suggestion by Paczyński: these events can be predicted ! There are more than 10,000 stars in the Hipparcos catalog with distances under 100pc and proper motion higher than 100 mas/yr !

Mao and Witt (1998a, b) treat finite source effects with respect to astrometric microlensing. They obtain analytically the centroid motion of a source with uniform surface brightness. They conclude that the finite source does affect the centroid shift significantly only when the angular impact parameter is comparable to the angular source size. In that case, the trajectories of the

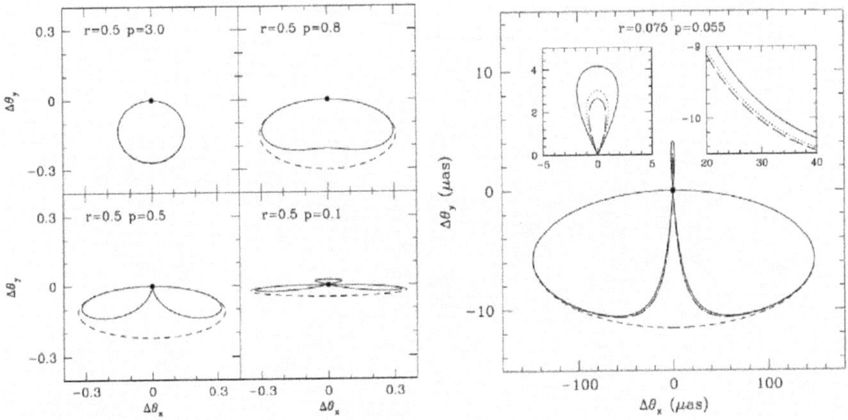

Fig. 40. *Left panel*: Four examples of astrometric trajectories showing finite source size effects: The source size is assumed to be $\theta_s = 0.5\,\theta_E$. In each example, the *black dot* marks the source position and the lens is moving from $-\infty$ to $+\infty$, parallel to the x-axis. The centroid motion starts at the origin and moves counter-clockwise. The impact parameter (here labeled p) of each trajectory is shown at the *top left* corner in each panel $p = 3.0, 0.8, 0.5, 0.1$. For each example, the *solid line* shows the trajectory that takes into account the finite source size effect while the *dashed lines* shows that for a point-source approximation. *Right panel*: Simulated astrometric trajectories for the first microlensing event (95-BLG-30) that shows photometric extended source effects (parameters taken from Alcock et al. (1997a,b)). The *dashed ellipse* is the centroid motion for a point source. The *solid line* shows the trajectory for a source with constant surface brightness, whereas the *dotted* and *long-dashed lines* show the predictions for the MACHO R and V bands, respectively. The two insets show magnified views of two regions to indicate the differences between various curves more clearly (both panels from from Mao and Witt 1998a,b)

light centroid become clover-leaf like. This offers the exciting possibility to detect stellar radii to very good accuracy. In Fig. 40, four examples of astrometric trajectories with a finite source size of $\theta_s = 0.5\theta_E$ are shown, with impact parameters $u_0 = 3.0, 0.8, 0.5, 0.1$. For comparison, the corresponding point source effect is shown as a dashed line as well. The second panel in this figure is a simulation of how the first microlensing event that showed extended source effects photometrically (MACHO 95-BLG-30) would have looked like astrometrically.

Han and Kim (1999) looked into another degeneracy: can astrometric microlensing help in uncovering blended microlensing events. They found that indeed, due to the high resolution of SIM, many blends will be directly identified: the imaging resolution of SIM is supposed to be 10 milli-arcseconds. But even for very close blends with a separation smaller than that (e.g. physical binary lenses), the blend signature can be identified in the blend contribution to the astrometric signature. In Fig. 41, the effect of the blended astrometric

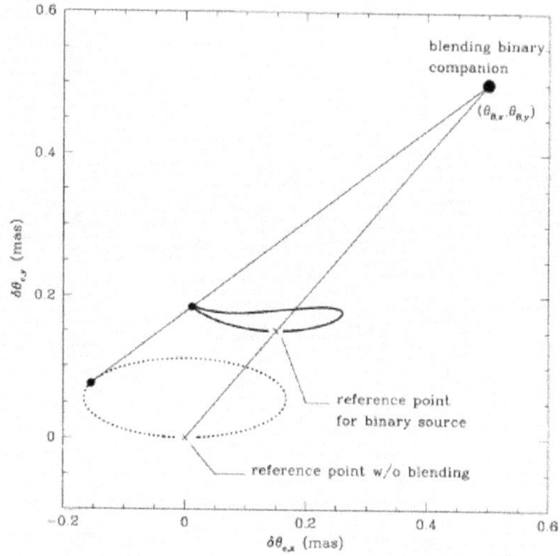

Fig. 41. Distortion of the astrometric trajectory of a binary companion: the *dotted line* is the astrometric shift without blending; with a binary companion a location $(\theta_{B,x}, \theta_{B,y})$, the light centroid will shift toward the companion. The *solid line* is the resulting trajectory. Parameters used here are time scale $t_E = 15d$, angular Einstein radius $\theta_E = 0.5$mas, contributing light fraction of the blending star is $f = 0.3$ (from Han and Kim 1999)

microlensing is illustrated qualitatively, in Fig. 42, various simulations are shown for increasing blend contribution (left hand column) and increasing binary separation (right hand column).

7 Quasar Microlensing

Quasars are affected by gravitational lensing in two ways: the "macrolensing" describes multiply imaged quasars, with angular separations of roughly one arcsecond. These cases are produced by typical galaxy lenses with masses of order 10^{12} $M_{\odot}$. About one out of 500 quasars is multiply imaged. Some 80 such cases are known to date (cf. `http://cfa-www.harvard.edu/glensdata`, the CASTLES web page), most of them consist of double or quadruple images. Once time delays between the images are determined and the mass distributions of the lenses are modeled properly, these quasar lenses can be used to measure the Hubble constant. Or one can turn this line of reasoning around: assuming one knows the Hubble constant, one can infer the mass profiles of the lensing galaxies (cf. Kochanek et al. 2003).

The second interesting regime is "microlensing": stellar mass lenses affect the apparent brightness of the quasar images. Microlens-induced variability can be used to study two cosmological issues of great interest, the size and brightness profile of quasars on the one hand, and the distribution of compact

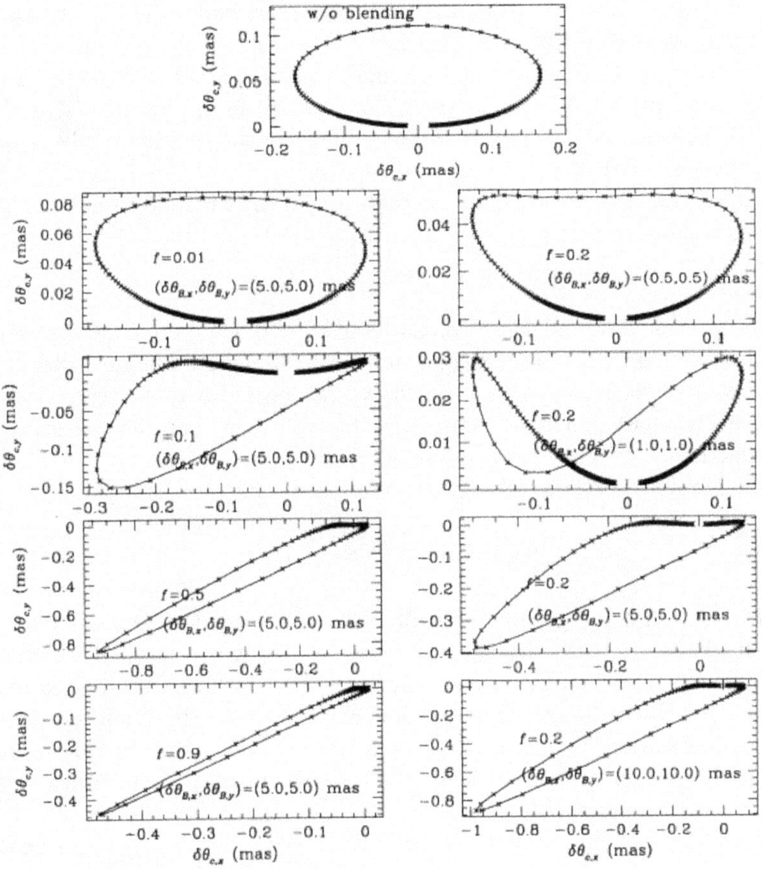

Fig. 42. Various forms of centroid shifts by binary-star blending: *top panel*: no blend contribution; *left column*: increasing blend contribution from top to bottom; *right column*: increasing binary separation from top to bottom (from Han and Kim 1999)

(dark) matter along the line-of-sight on the other hand. Here a summary of recent observational evidence for quasar microlensing is given, as well as a review of theoretical progress in the field. Particular emphasis is given to the questions which microlensing can address regarding the search for dark matter, both in the halos of lensing galaxies and in a cosmologically distributed form. A discussion of desired observations and required theoretical studies is presented at the end.

7.1 Microlensing Mass, Length and Time Scales

The lensing effects on quasars by compact objects in the mass range $10^{-6} \leq m/M_{\odot} \leq 10^{3}$ are usually called "quasar microlensing". The microlenses can be ordinary stars, brown dwarfs, planets, black holes, molecular clouds, globular clusters or other compact mass concentrations (as long as their physical size is smaller than the Einstein radius). In most practical cases, the microlenses are part of a galaxy which acts as the main (macro-)lens. However, microlenses could also be located in, say, clusters of galaxies or they could even be imagined "free floating" and filling intergalactic space.

The relevant length scale for microlensing is the Einstein radius of the lens in the quasar plane:

$$r_E \approx 4 \times 10^{16} \sqrt{M/M_\odot}\,\mathrm{cm},$$

where "typical" lens and source redshifts of $z_L \approx 0.5$ and $z_S \approx 2.0$ are assumed for the numerical value on the right hand side. Quasar microlensing turns out to be an interesting phenomenon, because (at least) the size of the continuum emitting region of quasars is comparable to or smaller than the Einstein radius of stellar mass objects.

The length scale translates into an angular Einstein scale of

$$\theta_E \approx 10^{-6} \sqrt{M/M_\odot} \quad \mathrm{arcsec}.$$

It is obvious that image splittings on such angular scales cannot be observed directly. What makes microlensing observable anyway is the fact that observer, lens(es) and source move relative to each other. Due to this relative motion, the micro-image configuration changes with time, and so does the total magnification, i.e. the sum of the magnifications of all the micro-images. This change in magnification can be measured over time: microlensing is a "dynamical" phenomenon.

There are two time scales involved: the standard **lensing time scale** t_E is the time it takes the source to cross the Einstein radius of the lens, i.e.

$$t_E = r_E / v_{\perp,\mathrm{eff}} \approx 15 \sqrt{M/M_\odot} \;\; v_{600}^{-1} \;\; \mathrm{years},$$

where the same assumptions are made as above, and the effective relative transverse velocity $v_{\perp,\mathrm{eff}}$ is parametrized in units of 600 km/sec: v_{600}. This time scale t_E results in discouragingly large values. However, in practice we can expect fluctations on much shorter time intervals. The reason is that the sharp caustic lines separate regions of low and high magnification. Hence, if a source crosses such a caustic line, we can observe a large change in magnification during the **crossing time** t_{cross} it takes the source to cross its own diameter R_{source}:

$$t_{cross} = R_{source} / v_{\perp,\mathrm{eff}} \approx 4 R_{15} \;\; v_{600}^{-1} \;\; \mathrm{months}.$$

Here the quasar size R_{15} is parametrized in units of 10^{15} cm.

In microlensing of multiple quasars, the normalized surface mass density is of order unity: $\kappa \approx 1$. This means that at any given time, a whole ensemble of microlenses is affecting the quasar. An illustration is shown in Fig. 43: in

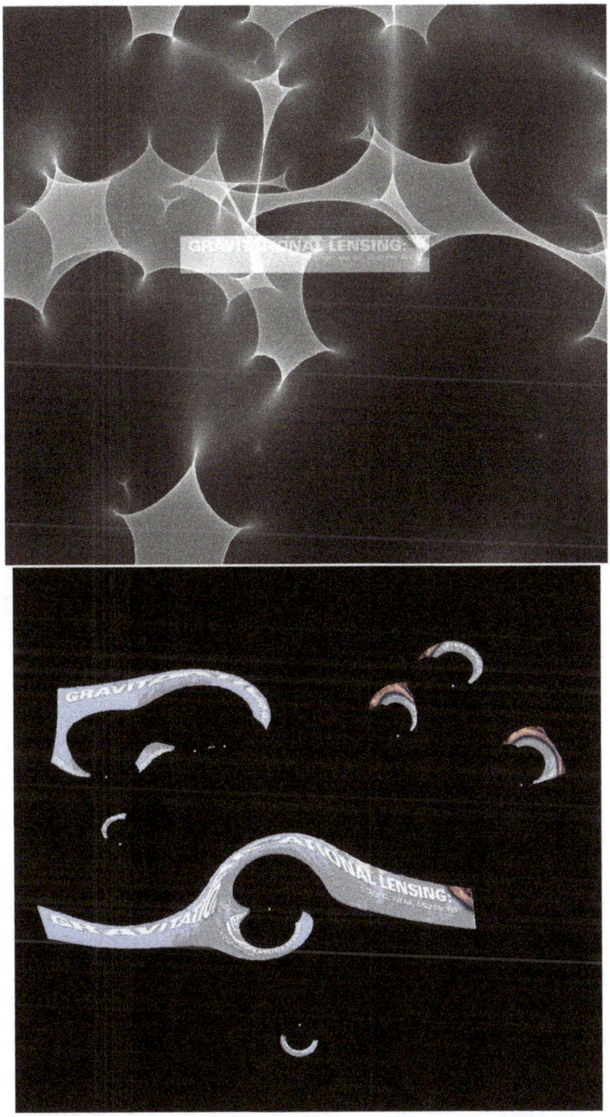

Fig. 43. Microlensing effect on an extended source: *top*: superposition of example source profile and microlensing magnication pattern produced by an ensemble of stellar lenses; *bottom*: corresponding image configuration

the top panel, an example source profile is superimposed on the magnification pattern of a randomly placed ensemble of microlenses for a surface mass density $\kappa = 0.5$. The bottom panel shows the corresponding micro-image configuration, which displays all effects that gravitational lensing can produce: offset of position, distortion, magnification and multiple images !

7.2 Early and Recent Theoretical Work on Quasar Microlensing

Right after the discovery of the first multiply imaged quasar, Chang and Refsdal (1979) suggested that the flux of the two quasar images can be affected by stars close to the line-of-sight. Gott (1981) proposed that a massive galaxy halo could be made of low mass stars and "should produce fluctuations of order unity in the intensities of the QSO images on time scales of 1–14 years." Young (1981) was the first to use numerical simulations in order to explore the effect of quasar microlensing.

Because the optical depth (or surface mass density) at the position of an image is of order unity, microlensing is expected to be going on basically "all the time", due to the relative motion of source, lens(es) and observer. In addition, this means that the lens action is due to a coherent effect of many microlenses, because the action of two or more point lenses whose projected positions is of order of their Einstein radii combines in a very non-linear way (cf. Wambsganss 1998). An illustration of this coherent action can be found in Figs. 44 and 45:

The magnification distribution produced by an ensemble of lenses is indicated in the quasar plane by different colors. The three dashed lines show the tracks of a quasar. In Fig. 45 the corresponding lightcurves are displayed, for two different source sizes. If the size of the quasar is small compared to the inter-caustic spacing, each caustic crossing is resolved individually, which results in relatively high maxima in the lightcurves (solid line). For a larger source (dashed line, factor 10 larger than solid line), the peaks are smoothed out, the character of the lightcurve is different.

The lens action of more than two point lenses cannot be easily treated analytically any more. Hence numerical techniques were developed in order to simulate the gravitational lens effect of many compact objects. Paczyński (1986a) had used a method to look for the extrema in the time delay surface. Kayser, Refsdal and Stabell 1986, Schneider and Weiss (1987) and Wambsganss (1990) had developed and applied an inverse ray-shooting technique that produced a two-dimensional magnification distribution in the source plane. An alternative technique was developed by Witt (1993) and Lewis et al. (1993); they solved the lens equation along a linear source track. All the recent theoretical work on microlensing is based on either of these techniques.

More recently, Fluke and Webster (1999) explored analytically caustic crossing events for a quasar. Lewis et al. (1998) showed that spectroscopic monitoring of multiple quasars can be used to probe the broad line regions (cf. also Lewis and Belle 1998). Wyithe et al. (2000a, b) investigated and found limits on the quasar size and on the mass function in Q2237+0305.

Agol and Krolik (1999) and Mineshige and Yonehara (1999) developed techniques to recover the one-dimensional brightness profile of a quasar, based on the earlier work by Grieger and Kayser (1988) and Grieger et al. (1991): Agol and Krolik showed that frequent monitoring of a caustic crossing event in many wave bands (they used of order 40 simulated data points in eleven filters

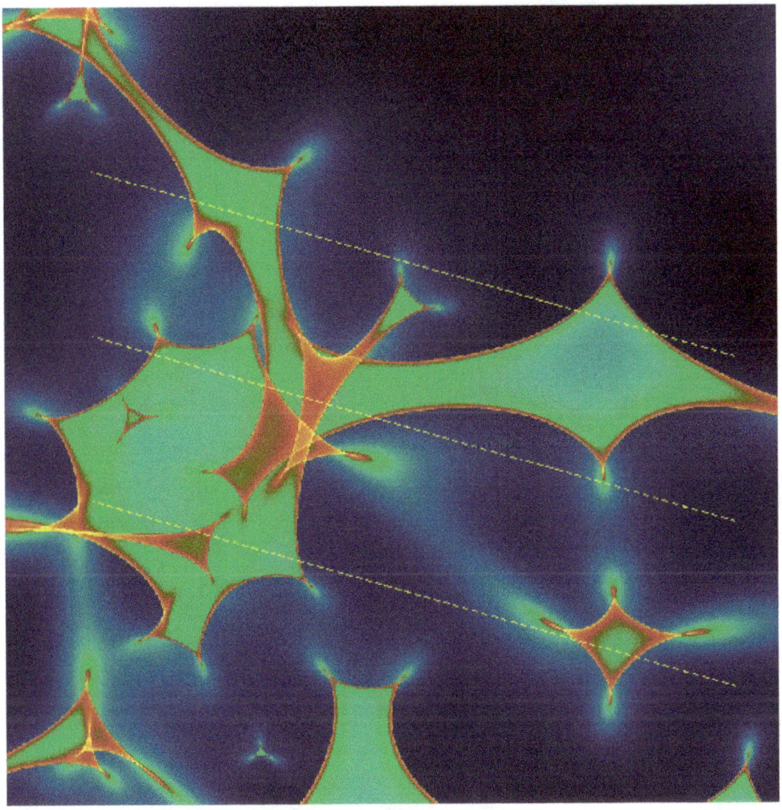

Fig. 44. Microlensing magnification pattern produced by stars in a lensing galaxy. The color steps represent different magnifications, with the sharp caustic lines corresponding to the highest magnification. The *dashed lines* indicates three tracks along which a background quasar moves. The corresponding lightcurves are displayed in Fig. 45

over the whole electromagnetic range), one can recover a map of the frequency-dependent brightness distribution of a quasar. Mineshige and Yonehara (1999) in a similar approach explored the effect of microlensing on two different accretion disk models. In another paper, Yonehara et al. (1998) showed that monitoring a microlensing event in the X-ray regime can reveal structure of the quasar accretion disk as small as AU-size.

Summarized, the theoretical papers exploring microlensing made basically four predictions concerning the potential scientific results. Microlensing should help us to determine:

1. the existence and effects of compact objects between the observer and the source,
2. the size of quasars,

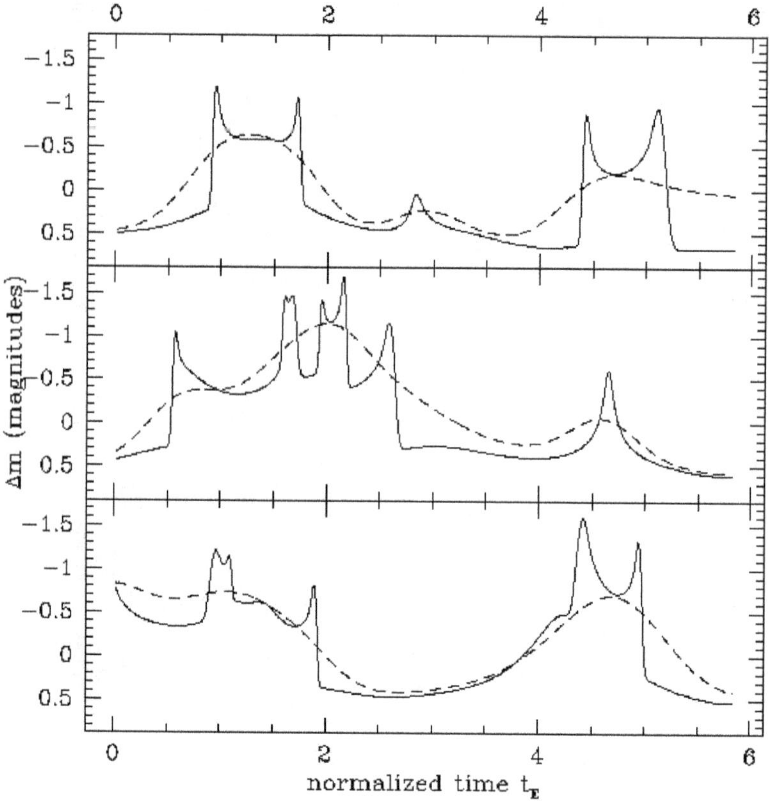

Fig. 45. Microlensing lightcurves for the three tracks shown in Fig. 44. The *solid line* correspond to a small source (Gaussian shape with width of about 3% of the Einstein radius), the *dashed line* represents a source that is a factor of 10 larger

3. the two-dimensional brightness profile of quasars,
4. the mass (and mass distribution) of lensing objects.

In the following sub-section the observational results to date will be discussed in some detail.

7.3 Observational Evidence for Quasar Microlensing

Fluctuations in the brightness of a quasar can have two causes: they can be intrinsic to the quasar, or they can be microlens-induced. For a single quasar (i.e., one that is not multiply imaged), the difference is hard to tell. However, once there are two or more gravitationally lensed (macro-)images of a quasar, we have a relatively good handle to distinguish the two possible causes of variability: any fluctuations caused by intrinsic variability of the quasar show up in all the quasar images, after a certain time delay[13]. So once a time delay is measured in a multiply-imaged quasar system, one can shift the lightcurves of the different quasar images relative to each other by the time delay, correct for the different (macro-)magnification, and subtract them from each other. All remaining incoherent fluctuations in the "difference lightcurve" can be attributed to microlensing. In a few quadruple lens systems we can detect microlensing even without measuring the time delay: in some cases the image arrangement is so symmetrical around the lens that any possible lens model predicts very short time delays (of order days or shorter), so that fluctuations in individual images that last longer than a day or so and are not followed by corresponding fluctuations in the other images, can be safely attributed to microlensing. This is in fact the case in the quadruple system Q2237 + 0305.

The Einstein Cross: Quadruple Quasar Q2237+0305

In 1989, evidence for cosmological microlensing was found by Irwin et al. (1989) in the quadruple quasar Q2237+0305: one of the components showed fluctuations, whereas the others stayed constant. In the mean time, Q2237 + 0305 has been monitored by many groups (Corrigan et al. 1991; Ostensen et al. 1996; Lewis et al. 1998). The most recent (and most exciting) results (Fig. 46, and Woźniak et al. 2000a, b) show that all four images vary dramatically, going up and down like a rollercoaster in the last three years: $\Delta m_A \approx 0.6$ mag, $\Delta m_B \approx 0.4$ mag, $\Delta m_C \approx 1.3$ mag, $\Delta m_D \approx 0.6$ mag. Comparison of these lightcurves with simulations (cf. Figs. 44 and 45) show that the continuum emitting region of the quasar is relatively small, of order 10^{14} cm (see, e.g., Wambsganss, Paczyński and Schneider 1990; Wyithe 2000b, Yonehara 2001).

The Double Quasar Q0957 + 561

The microlensing results for the double quasar Q0957 + 561 are not quite as exciting. Vanderriest et al. (1989) were the first to put attention on the observational evidence for potential microlensing in early lightcurve of the double quasar. In the first few years after its discovery, there is an almost linear change in the (time-shifted) brightness ratio between the two images (Schild 1996): $\Delta m_{AB} \approx 0.25$ mag over 5 years. But since about 1991, this ratio stayed more or less "constant" within about 0.05 mag, so not much microlensing has been going on in this system recently (Schild 1996; Pelt et al. 1998; Schmidt and Wambsganss 1998).

With numerical simulations and limits obtained from three years of Apache Point monitoring data of Q0957 + 561 (see Fig. 47), Wambsganss et al. (2000)

[13] This argument can even be turned around: the measured time delays in multiple quasars are the ultimate proof of the intrinsic variability of quasars.

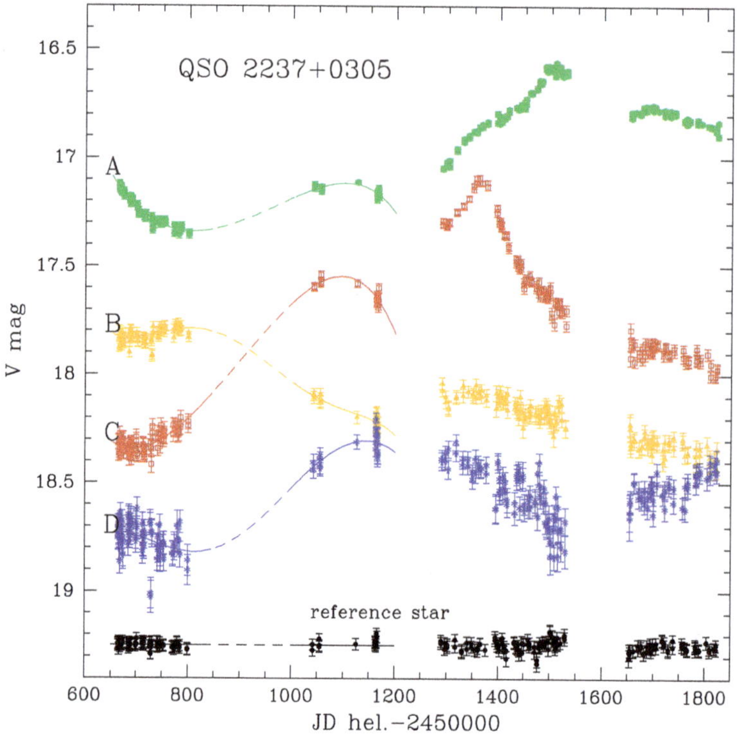

Fig. 46. Microlensing lightcurve of the quadruple quasar Q2237+0305, as measured by the OGLE team (Woźniak et al. 2000a, b; see also http://bulge.princeton.edu/~ogle/ogle2/huchra.html)

exclude a whole range of "MACHO" masses as possible dark matter candidates in the halo of the lensing galaxy in 0957+561. They extracted simulated lightcurves according to the timing of the observed ones and evaluated 100000 cases for seven different values for the lens mass (from $m/M_\odot = 10^{-7}$ to 10^0) and four different quasar sizes (10^{14} cm to 3×10^{15} cm): The small "difference" between the time-shifted and magnitude-corrected lightcurves of images A and B ($|\Delta m_{A-B,Q0957}| \leq 0.05$ mag) excludes a halo of the lensing galaxy made of compact objects with masses of $10^{-7} M_\odot - 10^{-2} M_\odot$ (cf. Figs. 48 and 49).

Refsdal et al. (2000) investigated the microlensing properties of the double quasar as well, using both the original linear change of 0.25 mag over a five year period and the subsequent 8 years of no or very little microlensing. They found constraints on the source size of $R \leq 6 \times 10^{15}$ cm, and the mass of the microlensing objects most likely to be in the range $10^{-6} \leq M/M_\odot \leq 5$.

Recently, the double quasar Q0957 + 561 was the target of a monitoring campaign particularly searching for short time scale variations. Colley et al. (2003a, b) report the observations and the result: making use of their very precise determination of the time delay in this system: $\Delta t = (417.09 \pm 0.07)$

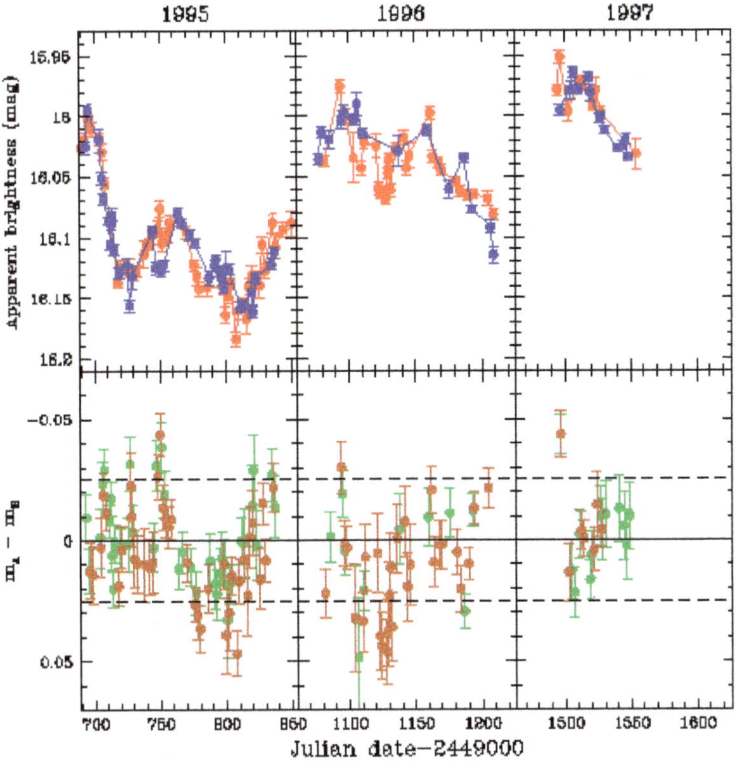

Fig. 47. Observed lightcurves of the double quasar Q0957 + 561; *top*: superposition of lightcurves of image A and (time shifted and magnitude shifted) image B; *bottom*: difference lightcurves (Wambsganss et al. 2000)

days, they found no microlensing fluctuations with amplitudes higher than 0.1 mag, and nearly rule out that objects in the mass range about $10^{-5} M_\odot$ make up a large fraction of the dark matter in the lens galaxy. In a further analysis of this data set, Colley and Schild (2003) report a microlensing signal at the 1% level with a time scale of 12 hours. If this result can be confirmed, it does provide a very interesting new window of exploration (and a challenge for theory).

Other Multiple Quasars/Radio Microlensing?

A number of other multiple quasar systems are being monitored more or less regularly. For some of them microlensing has been suggested (e.g. H1413+117, Ostensen et al. 1997; or B0218 + 357, Jackson et al. 2000). In particular the possiblity for "radio"-microlensing appears very interesting (B1600 + 434, Koopmans and de Bruyn 2000), because this is unexpected, due to the

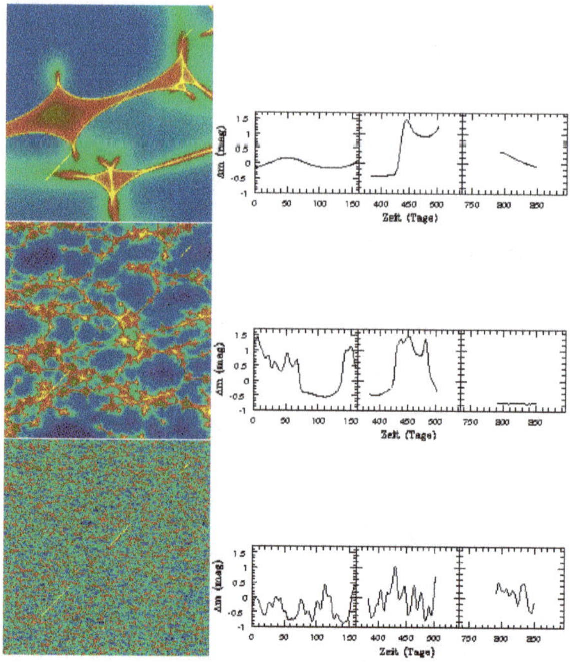

Fig. 48. Simulated microlensing lightcurves for the double quasar Q0957+561; *Left*: Magnification patterns for compact objects in three different mass ranges; the three-part straight line indicates the track of the background quasar. *Right*: corresponding three-part microlensing lightcurves (Wambsganss et al. 2000)

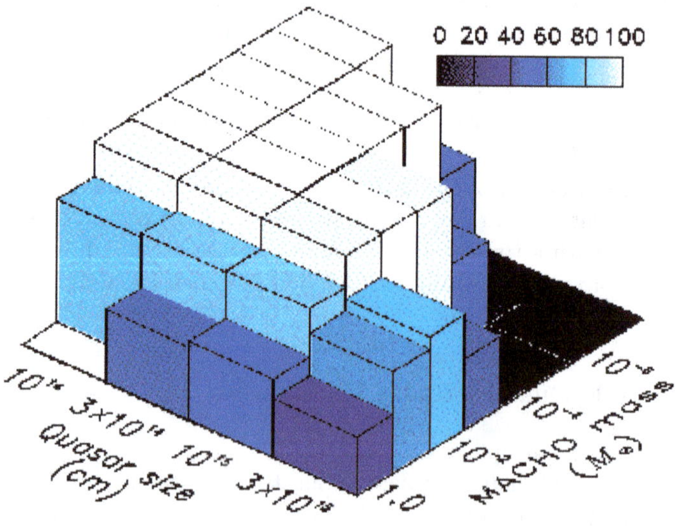

Fig. 49. Exclusion diagram: the highest (*white*) columns indicate values of quasar source size and "macho" mass which are excluded by more than 99.9% probability; the other columns show exlusion probabilities of between 40% and 85% (Wambsganss et al. 2000)

presumably larger source size of the radio emission region. The possibility of relativistic motion of radio jets may make up for this "disadvantage".

Unconventional Microlensing I: In Individual Quasars?

There were a number of papers interpreting the variability of individual quasars as microlensing (e.g., Hawkins and Taylor 1997, Hawkins 1998). Although this is an exciting possibility and it could help us detect a population of cosmologically distributed lenses, it is not entirely clear at this point whether the observed fluctuations can be really attributed to microlensing. After all, quasars are intrinsically variable, and the expected microlensing in single quasars must be smaller than in multiply imaged ones, due to the lower surface mass density. More studies are necessary to clarify this issue.

Unconventional Microlensing II: Centroid Shifts/Astrometric Microlensing

As in stellar microlensing (cf. Sect. 6), in quasar microlensing the astrometric signal of the lenses can be used and investigated as well. This was first put forward by Lewis and Ibata (1998), then further investigated by Treyer and Wambsganss (2004). At each caustic crossing, a new very bright image pair emerges or disappears, giving rise to sudden changes in the "center of light" positions (cf. Fig. 50).

The amplitude could be of order 100 micro-arcseconds or larger, which should be observable with the SIM satellite (Space Interferometry Mission), to be launched in 2010. This astrometric microlensing offers the exciting possibility to measure the mass of the lenses (in a statistical way) !

Unconventional Microlensing III: Million Solar Mass Objects or Sub-Structure: Milli-Lensing

A decade ago, the idea was popular that dark halos of galaxies could be made of black holes in the mass range of about a million solar masses. Wambsganss and Paczyński (1992) explored this effect on VLBI jets of multiply imaged quasars and suggested that this hypothesis could be tested: High signal-to-noise imaging of the two jet images of Q0957 + 561 should indicated clear lensing signatures, like kinks, holes, additional milli-images, if a significant fraction of the dark matter in the halo is made of such million solar mass objects (see Fig. 51). Garrett et al. (1994) presented such results and ruled out that the halo of the lensing galaxy in this double quasar consist of such objects.

Flux Ratio Anomaly: Microlensing or Substructure?

Microlensing may help solve another interesting issue: In a macrolensing scenario producing a quadruple quasar configuration with one close pair – corresponding to the source sitting inside but close to the (macro-)caustic – this

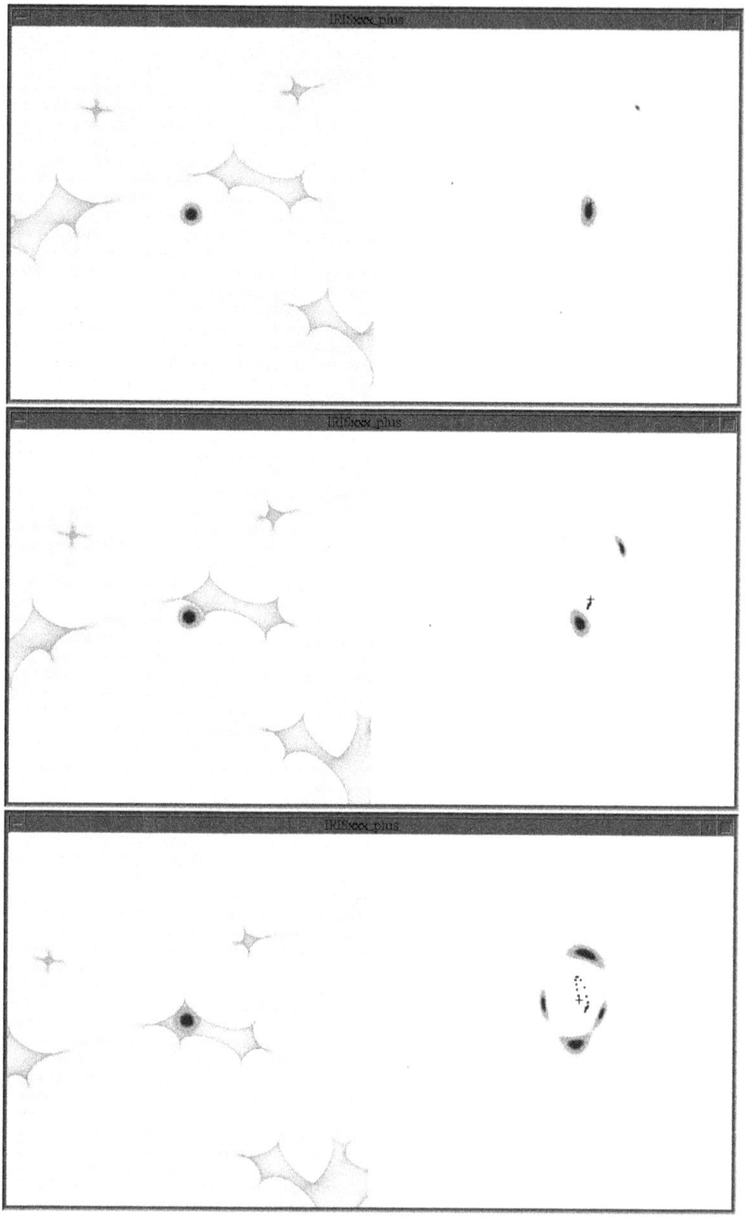

Fig. 50. Astrometric microlensing of quasars: due to the relative motion of quasar and foreground stellar lenses, the microimage configuration changes with time. This leads to a change of the light centroid which may be observable. The three panels show three epochs. On the *left* hand side, the caustics are shown superimposed with the quasar profile. On the *right* hand side, the micro-image configuration is shown. The *plus* sign indicates the "center-of-light"; the *points* next to the plus sign mark the light-centroid of previous epochs, i.e. the motion of the center-of-light for fixed quasar and moving microlenses (after Treyer and Wambsganss 2004)

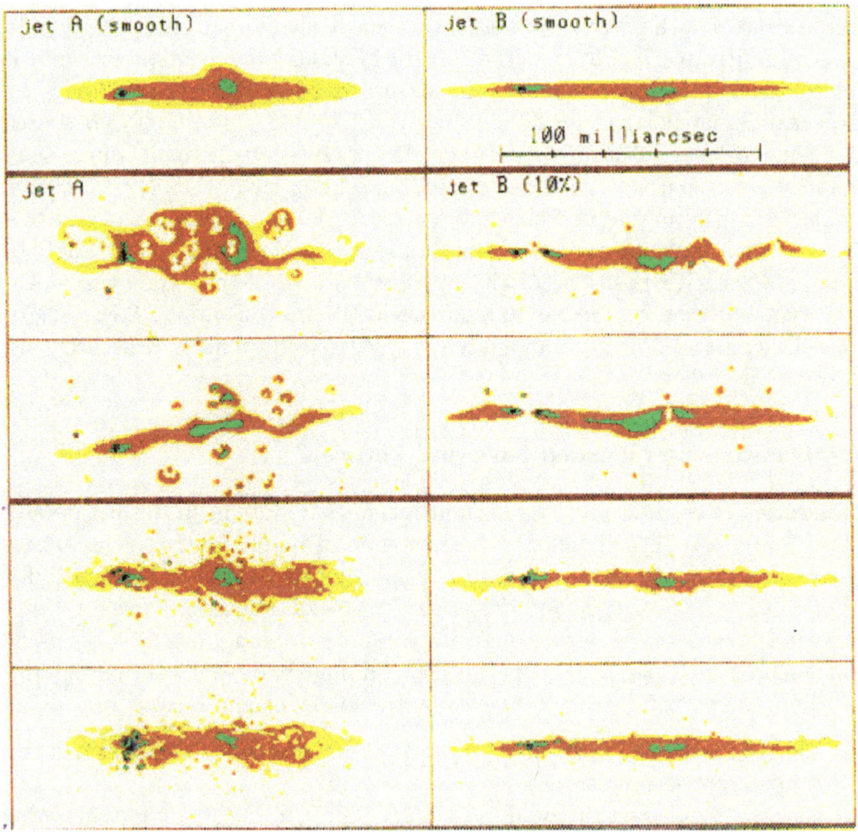

Fig. 51. Millilensing by million solar mass black holes affects the VLBI jets of multiply imaged quasars: the top row shows model VLBI jets for images A and B of the double quasar Q0957 + 561, as produced by the smooth lensing potential. If the halo of the lensing galaxy is made of million solar mass black holes (here: $m/M_{\odot} = 3 \times 10^5$), then the two jets should be affected by them *differently*, as shown in the four examples below: kinks, holes, additional milli-images should appear in both images uncoherently (Wambsganss and Paczyński 1991)

image pair should be highly magnified with very similar magnification of the two components. In most of the observed cases, however, this is not the case: close image pairs tend to have quite different magnifications. In almost all cases, the fainter (or demagnified) image seems to be the saddle point image. There are two competing explanations: Substructure in the macro-lens (galaxy) could introduce this flux ratio anomaly (Dalal and Kochanek 2002; Metcalf and Madau 2001; Metcalf and Zhao 2002). However, another possibility is microlensing by compact stellar mass objects plus smoothly distributed (dark) matter (Schechter and Wambsganss 2002). A nice thing about the two proposed mechanisms is that they make different predictions: If substructure is the cause for the flux ratio anomaly, then it should act the same way in basically all wave bands, and the flux ratio should be constant in time. If microlensing plus smooth matter is the origin of the discrepancy, then we expect different behavior in different wavebands, due to the fact, that source size changes as seen in different energy bands. In general, small source sizes (shorter wavelengths) should be affected more drastically then larger sources (which smooth out the microlensing caustics). A second consequence of this microlensing explanation is that the flux ratio should change with time, because the relative positions of the microlenses change over the course of a few years and hence produce fluctuations in the magnification. This issue of microlensing and flux ratio anomalies is discussed in detail in Schechter and Wambsganss (2002).

7.4 Quasar Microlensing: Now and Forever?

Monitoring observations of various multiple quasar systems in the last decade have clearly established that the phenomenon of quasar microlensing exists. There are uncorrelated variations in multiple quasar systems with amplitudes of more than a magnitude and time scales of weeks to months to years. However, in order to get close to a quantitative understanding, much better monitoring programs need to be performed. Summarized, and considering the "early promises" of quasar microlensing, the following can be stated:

1. the existence and effects of compact objects between the observer and the source: has been achieved;
2. the size of quasars: partly fulfilled, some limits on the size of quasars have been obtained;
3. the two-dimensional brightness profile of quasars: we are still (far) away from solving this promise;
4. the mass (and mass distribution) of lensing objects: it is fair to say that the observational results are consistent with certain (conservative) mass ranges.

Looking at the issue today, there are two important questions on the theoretical side: what do the lightcurves tell us about the lensing objects, and what can we learn about the size and structure of the quasar. As a response to the first question, the numerical simulations are able to give a qualitative understanding of the measured lightcurves (detections of microlensing in lightcurves of some multiply imaged quasars, and non-detections in others). The amplitudes and times scales of the events are in general consistent with "conservative" assumptions about the object masses and velocities. But due to the large number of parameters/unknowns (quasar size, masses of lensing objects, transverse velocity) and due to the large variety of lightcurve shapes, no satisfactory quantitative explanation or even prediction could be achieved. So far mostly "limits" on certain parameters have been obtained. The prospects of getting much better lightcurves of multiple quasars, as shown by the OGLE collaboration, should be motivating enough to explore this direction in much more quantitative detail.

The question of the structure of quasars deserves more attention. Here gravitational lensing is in the unique situation to be able to explore an astrophysical field that is unattainable by any other means. Hence more effort should be put into attacking this problem. This involves much more ambitious observing programs, with the goal to monitor caustic crossing events in many filters over the whole electromagnetic spectrum, and to further develop numerical techniques to obtain useful values for the quasar size and profile from unevenly sampled data in (not enough) different filters. Theoretically, Kochanek (2004) attacked this question in a brute force way: simulating microlensing for a large set of parameters and comparing with observed lightcurves for constraints on the input values. So far only a restricted data set of Q2237 + 0305 was used. This method should clearly be applied to more microlensing lightcurves. On the observational side, the way to go is building one (or more) dedicated telescope(s) for quasar monitoring. Moderate size is sufficient: 1 m to 2 m class. Excellent site is essential: median seeing better than one arcsecond. The use of robotic telescopes in a time-sharing mode is a much desired first step in this direction.

Acknowledgments

It is very great pleasure to thank a large number of people who contributed to the final version of this manuscript. First of all, I have to thank Penny Sackett, who unfortunately could not lecture at the 33rd Saas Fee Advanced Course on Gravitational Lensing, due to the fire at Mt. Stromlo. I have to thank her for providing me with text and figures and other support which were very useful for giving the lectures and for writing them up. I want to thank the organizers of the 33rd Saas Fee lectures, Georges Meylan, Philippe Jetzer and Pierre North, who did a marvellous job in planning, organizing and running this very nice and pleasant course ! It was a lot of fun to interact with my fellow lecturers Peter Schneider and Chris Kochanek, as well as with all the student and postdoc participants of the course. It is a pleasure to thank a number of colleagues who provided me with material for presentations and/or discussed these matters with me, including David Bennett, Ian Bond, Scott Gaudi, Andy Gould, Cheonho Han, Bohdan Paczyński, Nicholas Rattenbury, Andrzej Udalski, Phil Yock, and the PLANET team members. Finally, I want to thank the members of the gravitational lens group in Potsdam: Dijana Dominis, Pascal Hedelt, Janine Heinmüller, Andreas Helms, Susanne Hoffmann, Daniel Kubas, Robert Schmidt, Olaf Wucknitz for many discussions and in particular for reading the manuscript in various stages and providing valuable comments for corrections and improvements. Last but not least I want to thank our secretary in Potsdam, Andrea Brockhaus, who has always provided a very pleasant work atmosphere.

References

Afonso, C., Albert, J.N., Andersen, J., Ansari, R., Aubourg, E., Bareyre, P., Beaulieu, J.P., Blanc, G., et al. (The EROS Team) 2003a, A&A 400, 951

Afonso, C., Albert, J.N., Alard, C., Andersen, J., Ansari, R., Aubourg, E., Bareyre, P., Bauer, F., Beaulieu, J.P., Blanc, G., et al. (The EROS Team) 2003b, A&A 404, 145

Agol, E., Krolik, J. 1999, ApJ 524, 49

Alard, C., Lupton, R.H. 1998, ApJ 503, 325

Alard, C. 1997, A&A 321, 424

Albrow, M.D., Beaulieu, J.-P., Caldwell, J.A.R., Dominik, M., Greenhill, J. et al. 1999, ApJ 522, 1011

Albrow, M.D., Beaulieu, J.-P., Caldwell, J.A.R., Dominik, M., Gaudi, B.S. et al. (The PLANET Collaboration) 2000, ApJ 534, 894

Albrow, M.D., An, J., Beaulieu, J.-P., Caldwell, J.A.R., DePoy, D.L. et al. (The PLANET collaboration) 2001, ApJ 549, 759

Albrow, M.D., An, J.H., Beaulieu, J.-P., Caldwell, J.A.R., DePoy, D.L. 2002, ApJ 572, 521–539

Alcock, C., Akerloff, C.W., Allsman, R.A., Axelrod, T.S., et al. 1993, Nature 365, 621 (see also BAAS 182, 4004)

Alcock, C., Allsman, R.A., Alves, D., Axelrod, T.S., Bennett, D.P., Cook, K.H. et al. (The MACHO Collaboration) 1995, ApJ 454, L125

Alcock, C., Allsman, R.A., Alves, D., Axelrod, T.S., Bennett, D.P. et al. 1997a, ApJ 479, 119

Alcock, C., Allen, W.H., Allsman, R.A., Alves, D., Axelrod, T.S. et al. 1997b, ApJ 491, 436

Alcock, C., Allsman, R.A., Alves, D., Axelrod, T.S., Becker, A.C. et al. 1999, ApJ 518, 44

Alcock, C., Allsman, R.A., Alves, D.R., Axelrod, T.S., Becker, A.C. et al. 2000a, ApJ 541, 734

Alcock, C., Allsman, R.A., Alves, D.R., Axelrod, T.S., Becker, A.C. et al. 2000b, ApJ 542, 281

Alcock, C., Allsman, R.A., Alves, D.R., Axelrod, T.S., Becker, A.C., Bennett, D.P., Cook, K.H. 2001a, Nature 414, 617

Alcock, C., Allsman, R.A., Alves, D.R., Axelrod, T.S., Becker, A.C., Bennett, D.P., Cook, K.H. 2001b, ApJS 136, 439

An, J.H., Albrow, M.D., Beaulieu, J.-P., Caldwell, J.A.R., DePoy, D.L, Dominik, M. et al. 2002, ApJ 572, 521–539

Ashton, C.E., Lewis, G.F. 2001, MNRAS 325, 305

Aubourg, E. et al. 1993 Nature 365, 623

Bennett, D.P., Rhie, S.H. 1996, ApJ 472, 660

Bennett, D.P., Rhie, S.H. 2002, ApJ 574, 985

Bennett, D.P., Bally, J., Bond, I.A., Cheng, E., Cook, K. et al. 2003, SPIE 4854, 141

Bond, I.A., Rattenbury, N.J. et al. 2002, MNRAS 333, 71

Bond, I.A., Udalski, A., Jaroszynski, M., Rattenbury, N.J., Paczyński, B. et al. 2004, ApJ 606, L155
Bosma, A. 1978, PhD thesis, "The distribution and kinematics of neutral hydrogen in spiral galaxies of various morphological types", Groningen University
Chang, K., Refsdal, S. 1979, Nature 282, 561
Chang, K., Refsdal, S. 1984, A&A 132, 168
Chwolson, O. 1924, Astr. Nachrichten 221, 329
Colley, W.N., Schild, R.E. 2003, ApJ 594, 97
Colley, W.N. et al. 2003a, ApJ 587, 71
Colley, W.N. et al. 2003b, ApJ 588, 711
Corrigan, R.T. et al. 1991, AJ 102, 34
Courbin, F., Saha, P., Schechter, P.L. 2002, in Gravitational Lensing: An Astrophysical Tool, F. Courbin, D. Minniti (eds.), (Springer-Verlag: Berlin), Lecture Notes in Physics, Vol. 608, p. 1
Crotts, A.P.S. 1992, ApJ 399, L43
Dalal, N., Kochanek, C.S. 2002, ApJ 572, 25
Di Stefano, R., Esin, A.A. 1995, ApJ 448, L1
Di Stefano, R., Scalzo, A. 1999a, ApJ 512, 564
Di Stefano, R., Scalzo, A. 1999b, ApJ 512, 579
Dominik, M. 1998, A&A 329, 361
Drake, A.J., Cook, K.H., Keller, S.C. 2004, ApJ 607, L29
Einstein, A. 1911, Annalen der Physik 35, 898
Einstein, A. 1936, Science 84, 506
Fluke, C.J., Webster, R.L. 1999, MNRAS 306, 567
Garrett, M.A., Clader, R.J., Porcas, R.W., King, L., Walsh, D., Wilkinson, P.N. 1994, MNRAS 270, 457
Gaudi, B.S., Albrow, M.D., An, J.H., Beaulieu, J.-P., Caldwell, J.A.R., (The PLANET collaboration) 2001, in Microlensing 2000: A New Era of Microlensing Astrophysics, J.W. Menzies & Penny D. Sackett (eds.), (ASP: San Francisco), ASP Conf. Ser. # 239, p. 135
Gaudi, B.S., Albrow, M.D. et al. 2002, ApJ 566, 463
Gaudi, B.S. 2003, in Scientific Frontiers in Research on Extrasolar Planets, D. Deming & S. Seager (eds.), (ASP: San Francisco), ASP Conf. Ser, # 294, p. 79
Goldberg, D.M. 1998, ApJ 498, 156
Gott, J.R. III 1981, ApJ 243, 140
Gould, A. 1996, ApJ, 470, 201
Gould, A. 1992, ApJ 392, 442
Gould, A. 1994, ApJ 421, L75
Gould, A. 2001, Nature 414, 591
Gould, A. & An, J.H. 2002, ApJ 565, 1381
Gould, A., Gaudi, B., Han, C. 2003, ApJ 591, 53
Gould, A., Gaudi, B., Han, C. 2004, submitted to Astrophys. J., astro-ph/0405217
Gould, A. 2004, ApJ 606, 319

Graff, D.S. 2001, in Microlensing 2000: A New Era of Microlensing Astrophysics, J.W. Menzies and P.D. Sackett (eds.), (ASP: San Francisco), ASP Conf. Ser. # 239, p. 73
Grieger, B., Kayser, R. 1988, Sterne und Weltraum 27, 355
Grieger, B., Kayser, R., Schramm, T. 1991, ApJ 252, 508
Griest, K. 1991, ApJ 366, 412
Griest, K., Safizadeh, N. 1998, ApJ 500, 37
Griest, K., Hu, W. 1992, ApJ 397, 362 (Erratum 1993, ApJ 407, 440)
Han, C., Jeong, Y., Kim, Ho-Il 1998, ApJ 507, 102
Han, C. & Kim, T.-W. 1999, MNRAS 305, 795
Han, C., Park, S., Kim, H., Chang, K. 2000, MNRAS 316, 665
Han, C. & Kang, Y.W. 2003, ApJ 596, 1320
Hawkins, M.R.S., Taylor, A.N. 1997, ApJ 482, 5
Hawkins, M.R.S. 1998, A&A 340, L23
Hendry, M.A., Bryce, H.M., Valls-Gabaud, D. 2002, MNRAS 335, 539
Horne, K. 2003, in Scientific Frontiers in Research on Extrasolar Planets, D. Deming & S. Seager (eds.), (ASP: San Francisco), ASP Conf. Ser., # 294, p. 361
Irwin, M.J. et al. 1989, AJ 98, 1989
Jackson, N.J. et al. 2000, MNRAS 319, 7
Jaroszynski, M. & Paczyński, B. 2002a, Acta Astronomica 53, 361
Jaroszyński, M., Mao, S. 2001, MNRAS 325, 1546
Jaroszyński, M., Paczyński, B. 2002b, Acta Astronomica 52, 316
Jetzer, Ph., Milsztajn, A., Tisserand, P. 2004, astro-ph/0409496
Kayser, R., Refsdal, S. & Stabell, R. 1986, A&A 166, 36
Klimov, Y.G. 1963, Soviet Phys. Doklady 8, 119
Kochanek, C.S. 2004, ApJ 605, 58
Kochanek, C.S. et al. 2003, ApJ 595, 29
Koopmans, L.V.E., de Bruyn, A.G. 2000, A&A 358, 793
Lasserre, T., et al. 2000, A&A 355, L39
Lewis, G.F. et al. 1993, MNRAS 261, 647
Lewis, G.F., Belle, K.E. 1998, MNRAS 297, 69
Lewis, G.F., Irwin, M.J., Hewett, P.C., Foltz, C.B. 1998, MNRAS 295, 573
Lewis, G.F., Ibata, R.A.. 1998, ApJ 501, 478
Liebes, S. 1964, Phys. Rev. B 133, 835
Link, F. 1937, Bullettin Astronomique 10, 73
Link, F. 1967, Bull. Astron. Inst. Czechoslovakia 18, 215,
Mancini, L., Calchi Novati, S., Jetzer, Ph., Scarpetta, G. 2004, A&A 427, 61
Mannheim, P.D. 1992 ApJ 391, 429
Mao, S., Paczyński, B. 1991, ApJ 374, L37
Mao, S., Paczyński, B. 1996, ApJ 473, 57
Mao, S., Witt, H.J. 1998a, MNRAS 300, 1041
Mao, S. & Witt, H. 1998b, MNRAS 300, 1049
Mao, S. 2001, in Gravitational Lensing Recent Progress and Future Goals, T.G. Brainerd & C.S. Kochanek (eds.), (ASP: San Francisco), ASP Conf. Ser. # 237 p. 215

Mao, S. 1999, A&A 350, L19
Metcalf, R.B., Madau, P. 2001, ApJ 563, 9
Metcalf, R.B., Zhao, H.S. 2002, ApJ 567, 5
Milgrom, M. 2001, MNRAS 326, 126
Mineshige, S., Yonehara, A. 1999, ApJ 343, 41
Mollerach, S. & Roulet, E. 2002 Gravitational Lensing and Microlensing, World Scientific
Nemiroff, R.J. 1997, ApJ 486, 693
Nemiroff, R.J., Wickramasinghe, W.A.D.T. 1994, ApJ 424, L21
Nguyen, H.T., Kallivayalli, N., Werner, M.W., Alcock, C., Patten, B.M., Stern, D. 2004, ApJS 154, 266
Ostensen, R. et al. 1996, A&A 309, 59
Ostensen, R. et al. 1997, A&A 126, 393
Paczyński, B. 1986a, ApJ 301, 502
Paczyński, B. 1986b, ApJ 304, 1
Paczyński, B. 1996, ARA&A 34, 419
Paczyński, B. 1998, ApJ 494, L23
Paczyński, B., Stanek, K.Z., Udalski, A., Szymański, M, Kałużny, J., Kubiak, M., Mateo, M., Krzemiński, W. 1994, ApJ 435, L113
Pelt, J. et al. 1998, A&A 336, 829
Peng, E.W. 1997, ApJ 475, 43
Rees, M. 2003, Phil. Trans. Roy. Soc. Lond. 361, 2427-2434
Refsdal, S. 1964a, MNRAS 128, 295
Refsdal, S. 1964b, MNRAS 128, 307
Refsdal, S., 1966, MNRAS 134, 315
Refsdal, S., Stabell, R., Pelt, J., Schild, R. 2000, A&A 360, 10
Renn, J., Sauer, T., Stachel, J. 1997, Science 275, 184
Reza, Y. 2000, Trois Versions de la Vie, Albin Michel (Paris)
Rubin, V.C. 1983, Science 220, 1339.
Sackett, P.D. 1997, in Final Rep. ESO Working Group on Detection of Extrasolar Planets, F. Paresce & A. Renzini (eds.), (ESO: Garching)
Sackett, P.D. 2001, Gravitational Lensing: Recent Progress and Future Goals, Tereasa G. Brainerd & Christopher S. Kochanek (eds.), (ASP: San Francisco), ASP Conf. Ser. # 237, p. 227
Sahu, K.C. 1994, Nature 370, 275,
Sahu, K. 2003, in The Dark Universe: Matter Energy and Gravity, M. Livio (ed.), (Cambridge Univ. Press: Cambridge), p. 14, also astro-ph/0302325
Sanders, R.H., McGaugh, S.S. 2002, A&A 40, 263
Schechter, P.L., Wambsganss, J. 2002, ApJ 580, 685
Schild, R. 1996, ApJ 464, 125
Schmidt, R., Wambsganss, J. 1998, A&A 335, 379
Schneider, P., Weiss, A. 1986, A&A 164, 237
Schneider, P., Weiss, A. 1987, A&A 171, 49
Smith, M.C., Mao, S., Wozniak, P. 2002a, MNRAS, 332, 962
Smith, M.C., Mao, S., Wozniak, P., Udalski, A., Szymanski, M., Kubiak, M., Pietrzynski, G., Soszynski, I., Zebrun, K. 2002b, MNRAS 336, 670

Soldner, J. 1801, Berliner Astron. Jahrb. 1804, p. 161
Spergel, D.N., Verde, L., Peiris H.V., Komatsu E., Nolta M.R. et al. 2003, ApJS, 148, 175
Stanek, K.Z., Mateo, M., Udalski, A., Szymański, M, Kałużny, J., Kubiak, M. 1994, ApJ 429, L73
Treyer, M. & Wambsganss 2004, A&A 416, 19
Tsapras, Y., Horne, K., Kane, S. & Carson, R. 2003, MNRAS 343, 1131
Udalski A., Szymanski M., Kaluzny J., Kubiak M., Krzeminski W., Mateo, M., Preston, G.W. Paczyński, B. 1993, Acta Astronomica 43, 289
Udalski A., Szymanski M., Mao, S., Di Stefano, R., Kaluzny J., Kubiak M., Mateo M. & Krzeminski W. 1994a, Astrophys. J.L. 436, L103.
Udalski A., Szymanski M., Stanek, K.Z. et al. (The OGLE collaboration) 1994b, Act. Astr. 44, 165.
Udalski A., Szymanski M., Kubiak M. et al. 1998, Acta Astronomica 48, 147
Udalski A. 2003, Acta Astronomica 53, 291
Vanderriest, C., Schneider, J., Herpe, G., Chèvreto n, M., Moles, M., Wlérick, G. 1989, A&A 215, 1
Walsh, D. et al. 1979, Nature 279, 381
Wambsganss, J. 1990, PhD thesis (Munich University), also Max-Planck Report MPA-550
Wambsganss, J., Paczyński, B., Schneider, P. 1990, ApJ 358, 33
Wambsganss, J., Paczyński, B. 1991, AJ, 102, 864
Wambsganss, J., Paczyński, B. 1992, ApJ 397, 1
Wambsganss, J. 1997, MNRAS 284, 172
Wambsganss, J. 1998, Living Reviews in Relativity 1, No. 12, http://relativity.livingreviews.org/Articles/lrr-1998-12
Wambsganss, J., Schmidt, R.W., Colley, W.N., Kundić, T., Turner, E.L. 2000, A&A 362, L37
Witt, H.J. 1993, ApJ 403, 530,
Witt, H.J. 1995, ApJ 449, 42
Witt, H.J., Mao, S. 1995, ApJ 447, 105
Wyithe, J.S.B. et al. 2000a, MNRAS 315, 51
Wyithe, J.S.B. et al. 2000b, MNRAS 318
Wolszczan, A. 1994, Science 264, 538
Wozniak, P.R. et al. 2000a, ApJ 529, 88
Wozniak, P.R. et al. 2000b, ApJ 540, 65
Wozniak, P.R. et al. 2001, Acta Astronomica 51, 175
Wozniak, P. & Paczyński, B. 1997, ApJ 487, 55
Yonehara, A. et al. 1998, ApJ 501, 41
Yonehara, A. 2001, ApJ 548, 127
Yoo, J., DePoy, D.L., Gal-Yam, A., Gaudi, B.S., Gould, A. et al. 2004, ApJ 603, 139
Young, P. 1981, ApJ 244, 756

Index

www.ingramcontent.com/pod-product-compliance
Lightning Source LLC
Chambersburg PA
CBHW060744150926
51501CB00003B/84
* 9 7 8 3 5 4 0 3 0 3 0 9 1 *